Preparing STEM Teachers: The UTeach Replication Model

Preparing STEM Teachers: The UTeach Replication Model

Edited by

Joanne E. Goodell
Selma Koç

IAP

INFORMATION AGE PUBLISHING, INC.
Charlotte, NC • www.infoagepub.com

Library of Congress Cataloging-In-Publication Data

The CIP data for this book can be found on the Library of Congress website (loc.gov).

Paperback: 978-1-64802-166-4
Hardcover: 978-1-64802-167-1
E-Book: 978-1-64802-168-8

Printed and bound by CPI Group (UK) Ltd, Croydon, CR0 4YY

CONTENTS

PREFACE

Joanne Goodell and Selma Koç

This book chronicles the founding, growth and replication of the UTeach program since its inception at the University of Texas at Austin in 1997. Our collaboration on this book began, as many such projects do, over coffee at favorite local haunt one winter morning in early 2017 while discussing our plans for future joint publications and potential outlets for our efforts. We realized that while there are journals devoted to STEM education, including the International Journal of STEM Education, the Journal of Research in STEM Education, the Journal of STEM Outreach, Frontiers in Education, the Journal for STEM Education Research, and the International Journal of Science and Mathematics Education, none of them appeared to be a good fit for our planned work or for much of what had been presented at the annual UTeach conferences. One of the editors, Goodell, established the program at Cleveland State University in 2010 (see Chapter 3 of this volume for more details of this story), and has been actively involved in various aspects of the replication and program support efforts of the UTeach Institute since then, serving as a site visitor, a founding board member of the UTeach STEM Educators Association (USEA) (see Chapter 2 of this volume for more details), and a frequent annual conference presenter and attendee. The second editor, Koç, who had taught one of the courses in our UTeach program since its inception, suggested we work on a book, and this is the result.

Our goal for this book is to facilitate the dissemination of how different institutions have implemented and sustained the UTeach model in their local context

Preparing STEM Teachers: The UTeach Replication Model, pages ix–xii.

including innovations in the curriculum, the development of assessment instruments, and research on retention and program effectiveness.

The 29 chapters that follow are organized into three sections. The eleven chapters in Section I cover establishment, program, and replication stories from UTeach programs. In Section II, the eight chapters are focused on aspects of the instructional courses in the UTeach instructional program. The final ten chapters in Section III present current research focused on aspects of the UTeach program development and evaluation efforts.

In Chapter 1, Michael Marder chronicles the amazing story of how the UTeach program began at the University of Texas at Austin. In Chapter 2, Pamela Romero and Kimberly Hughes discuss the role the UTeach Institute has played in supporting the replication efforts. They also discuss the literature on fidelity of implementation, program replication, scaling innovations, and networked improvement which underpins the implementation strategy. In Chapter 3, Diane Bennett and Joanne Goodell compare and contrast their replication stories and distill key lessons learned for readers to consider. In Chapter 4, Shelly Rodriguez and Amy Moreland describe how the UTeach Internship program at the University of Texas at Austin collaborates with STEM-focused organizations, such as museums, after school clubs, makerspaces, and local schools to provide paid part-time work for STEM preservice teachers. In Chapter 5, Mary Urquhart and the UTeach Dallas Team describe how UTeach Dallas faculty and programs have been integrated into an ecosystem of support for STEM education stakeholders in which innovation thrives. In Chapter 6, Jennifer Dechaine and Ian Loverro discuss their experiences implementing the UTeach model in a large well-established STEM education program with a short implementation timeline and other contextual issues unique to their state education system. In Chapter 7, Kelli Allen discusses the many facets of induction support provided by the University of Texas at Austin to its program graduates during the first three years of their teaching career. In Chapter 8, Shelley Rodriguez, Jason Harron and Hannah Smith Brooks describe the development and implementation of the UTeach Maker program as well as the supports that have made it successful. Chapter 9 presents the story of UTeach Laboratory Schools as told by Michael Odell, Teresa Kennedy, Jo Ann Simmons and Jaclyn Pedersen from the University of Texas at Tyler. In Chapter 10, Anisha Campbell, Steven Karig, Kayla White and Julie Brenner describe a model of sharing master teachers' roles and responsibilities with their local school district partner. In Chapter 11, Kristen Apraiz, Gayle Nelson Evans, and Thomas Dana present the evolution of their introduction to teaching field-based course "Explorations in Teaching Secondary Mathematics and Science (*Explorations*)." Using a situated learning perspective, they explain how they worked towards developing a community of practice among program faculty and the pre-service teachers they teach.

Section II, Innovative Instructional Programming, starts with Chapter 12 by Kimberly Hughes, Ashley Welch and Amy Moreland, which provides information on the UTeach instructional program, elements and courses. The section

continues with Chapter 13 by Anita Sanyal Tudela, Sarah Henson-Darko, Steve Karig and Kayla White, which describes their efforts to integrate social justice perspectives into Step 2 coursework. Chapter 14 by Rachelle Haroldson, is about the journey of a UTeach program adapting the UTeach Step 1 and Step 2 course structure for a one-year post-baccalaureate program, with a specific focus on the approach taken to develop strategic partnerships and coordinate meaningful field experiences in a condensed summer format. Chapter 15 authored by Selma Koç describes the revisions the course instructor made on the clinical interview assignment in the Knowing and Learning in Mathematics and Science course. Chapter 16, by Patrick McGuire and Steve Ottmer, provides an overview of a co-teaching model involving a university faculty and in-service teacher in a pre-service STEM teacher preparation course, Classroom Interactions. Chapter 17, by Katherine Donaldson, Kate York, Robert Gagnon and Cory Gavitt, discusses best practices within the project-based instruction (PBI) courses, examines the road-blocks and successes of their parallel ten-year journeys, and brainstorms innovative next steps. In chapter 18, Dawn Woods and Candace Walkington investigate whether and how project-based learning impacts STEM learning outcomes and attitudes towards STEM. The literature review in this chapter examines the evidence behind project-based learning with traditionally underrepresented students in urban settings with the purpose of providing teachers, principals, districts, teacher educators and policy-makes with information to make convincing arguments for the inclusion of project-based learning in secondary STEM classrooms. Chapter 19, by Laurie Cavey, Jason Libberton, Tatia Totorica, Michele Carney and Patrick R. Lowenthal, includes the use of an iterative design-based approach in the development of four video-based learning modules that highlight a range of ways secondary students think about fundamental ideas related to functions and mathematical modeling. Section II ends with Chapter 20 by Gregory Gilson, which provides an overview of the curriculum, outcomes, assessments, and student projects of the UTeach content course titled Perspectives on Science and Mathematics (Perspectives) in order to explain why a liberal arts based course is an important and effective component of the UTeach STEM teacher preparation model.

Section III, Research and Future Directions, begins with Chapter 21 by Candace Walkington, Michael Marder and Elizabeth Adams. The purpose of this chapter is to provide practical information about using the UTeach Observation Protocol, a classroom observation instrument developed for STEM classrooms, and to present some real-world cases of how the UTOP has been used in practice. In Chapter 22, the authors Leslie Atkins Elliott, Michele Carney, Sara Hagenah and Tatia Totorica, describe their efforts to develop an instrument to measure teacher candidates' attentiveness to student thinking situated in key disciplinary-specific contexts. The instrument is designed to inform program improvement and evaluation. Chapter 23, by Doron Zinger, Therese Shanahan, Kris Houston, report the results of their qualitative study of program graduates using data collected using questionnaires and interviews. They analyzed aspects of preparation including

university coursework and field-based experiences during the program. Chapter 24, by Michael Horvath, Vasilios Kosteas and Joanne Goodell, discusses the potential effects of the UTeach model on teacher retention. Drawing on the literature in education, economics, and psychology, the authors propose a framework of factors that will affect a UTeach graduate's desire to remain in the profession. In Chapter 25, authors Melissa Dodson, Whitney Cade, Feng Liu, and Michael Vaden-Kiernan analyze data from seven UTeach programs to examine recruitment and retention in relationship to student demographics, enrollment, perceptions of financial incentives, the program, and teaching as a profession. Chapter 26, by Fernando Alegre, Juana Moreno, Jerry Weltman, David Kirshner, John Arthur Underwood, Frank Neubrander, James Madden, describes the development and piloting of two courses in a UTeach program, *Computational Thinking for Future Teachers* (CTFT) and *Computer Science Teaching Methods* (CS Methods). The CTFT course target undergraduates and will provide pedagogy and content knowledge to prepare them to teach introductory high school CS and CT courses. The CS Methods course targets Computer Science majors who are considering a career in teaching or are interested in CS pedagogy. Chapter 27, authored by David Sparks and Debbie Jackson, explores the perspectives and experiences of recent graduates of two urban UTeach teacher preparation programs while using project-based learning as a teaching pedagogy. Chapter 28, by Ellen Granger, Sherry Southerland and Robin Smith, reports their research about the practices of their UTeach program graduates' through the lens of the classroom visits that are a part of its induction program. Chapter 29, the final chapter, authored by Ann Cavallo, Gregory Hale and Erin Gonzales, examines data on UTeach Arlington graduates and their retention in high-need classrooms.

There are many elements to the UTeach story that are not represented in this book, however we are hopeful that this collection will be of interest to a variety of audiences, including teacher educators, teacher education researchers and education program policy makers. International teacher audiences and comparative education researchers will also be interested in this publication.

Finally, we wish to thank the authors for their wonderful contributions and the reviewers for their hard work and helpful suggestions. We also must sincerely thank our graduate students, Meng Yan (18–19) and Xiaona Jin (19–20) for their hard work and attention to detail needed to bring this project to fruition.

—Joanne Goodell and Selma Koç
August 23, 2020

SECTION I

THE UTEACH MODEL AND ELEMENTS OF SUCCESS

CHAPTER 1

THE DEVELOPMENT OF THE UTEACH MODEL

Michael Marder

UTeach is a secondary STEM teacher preparation program that began at the University of Texas at Austin in 1997. It was codeveloped by secondary STEM teachers, science and mathematics education research faculty, and science faculty. This chapter describes the process of development in words of some of the founders.

CONTEXT FOR FOUNDATION FOR UTEACH

UTeach originated at The University of Texas at Austin in 1997. It arose from a combination of specific circumstances, ideas, and people.

Texas Senate Bill 994 of 1987 (Texas, 1987) created some of the context. Authors of this bill were motivated by warnings about the mediocrity of education presented by *A Nation at Risk* (National Commission on Excellence in Education, 1983), by a report highly critical of teacher education from the Southern Regional Education Board (Simms & Miller, 1988) and also took guidance, at least in part, from a report of the Holmes Group, which had concluded that "the undergraduate education major must be abolished in our universities" (Lanier, 1986, p. 23).

Texas' SB 994 mandated that future teachers be required to take an "academic major or interdisciplinary academic major including reading, other than education…." This had the effect of abolishing education majors for secondary teachers, who would need to major in English, mathematics, history, or other subjects

Preparing STEM Teachers: The UTeach Replication Model, pages 3–16.

they wished to teach. A second provision of the bill was that no more than 18 semester credit hours of education were allowed, including student teaching.

Consequences at first unfolded so as to confirm the fears of the bill's opponents. The production of teachers in Texas began to drop. For example, in 1991–1992, UT Austin prepared 51 math and science teachers. By 1996–1997, the number had dropped to 31.

Heath (2005, p. 76) interviewed preservice science teachers preparing to become science teachers at UT Austin just before UTeach began. His description of their reflections is unsparing:

> [A] second brainstorming session took place about halfway through the student teaching experience. ... Now that they had acquired some "real world" (if somewhat limited) experience in teaching, the student teachers were asked to reflect on their preservice education and how it prepared them for the experience. The resulting nine affinities reveal how the student teachers feel about the utility of their education classes:
>
> **"A waste of time"**—This was by far the most thickly populated of the affinities that emerged from the second brainstorming session. All of the members of the small group chimed in some variation on this theme, in many instances describing their preservice education as "pathetic," a "joke," "stupid," "silly," and "worthless."

That the program inspired such negative reactions should not be too surprising since the legislature had stripped the College of Education faculty of much of their authority to prepare teachers, while at the same time leaving the responsibility in their hands. The number of secondary teacher preparation candidates was dropping, the ones in the program were dissatisfied, and the stage was set for something new.

THE IDEA FOR UTEACH, IN FOUNDERS' WORDS

The name UTeach originally belonged to a service-learning course where students received chemistry credit for teaching science lessons in elementary schools. David Laude, a faculty member in Chemistry at UT Austin, created this course in the mid-1990s.

David Laude [In 1997, Associate Dean for Student Affairs]:

> For me, UTeach began with my teaching an after-school science enrichment for my children's elementary school in the early to mid-90s. The enrichment was basically hands-on science lessons in a service-learning course. I plowed ahead without knowing the rules—with its peak in the late 90s, the outreach program enrolled 600 or 700 students in a single semester and was in pretty much every school in Austin without either UT Austin or AISD knowing about it. It was managed by a single TA and me, plus a lovely pyramid of student leaders.

> The name UTeach was offered to me by a young woman in the service-learning outreach program who thought that the name at the time "Super Science Sleuths" could be improved upon. I remember vividly where I was standing on the third floor of Welch right outside the Instrument Design and Repair lab when she made that comment. Let's see, "Super Science Sleuths" or "UTeach."
>
> Teaching that service-learning course was what introduced me to the idea that I could develop programs, which I found to be more fun than research. It also made me far more likely to reply to Mary Ann [Rankin]'s e-mail looking for a new student dean in the spring of 1997.

Mary Ann Rankin, Dean of the College of Natural Sciences at UT Austin in 1997 had become convinced after serving as a guest presenter in her daughter's classes that preparing teachers was an important part of the college mission, and that being able to teach well required substantial preparation. She shared her thoughts with an advisory group.

Jeff Kodosky [Co-founder, National Instruments]:

> I was a relatively new member of the Dean's Advisory Council (fall of '96) and I heard Mary Ann Rankin mention the idea of having a teacher training initiative within the College of Natural Sciences (CNS) for all the reasons we continue to espouse. In the spring '97 meeting I heard her mention the same thing again. A little while later I met with Mary Ann to ask about the progress that had been made on the initiative and was disappointed to learn that nothing had happened. We talked about possible ways to proceed. Somewhere along the way I asked if I made a $10K grant could something get started in the summer? She was enthusiastic so I made the grant but didn't pay much attention to what went on during the summer. I was delightfully surprised when Mary Ann established the first two classes starting in the fall of '97. Maybe stunned is a better word—I never expected the University to move so quickly, but it wasn't really the University, it was Mary Ann!
>
> I learned later on that Harry Lucas [Lucas Petroleum Group] had made a similar grant that first summer. I don't know if he had independently come to the same conclusion as I did or if Mary Ann reached out to him to add to my gift, but in any case, he was right there from the start.
>
> Seeing the early results of the program exceed expectations on all measures, including areas no one thought of (e.g., diversity), convinced my wife Gail and me to make the foundational endowment gift to UTeach. I still consider that to be my best investment ever (other than cofounding National Instruments).

THE INITITAL PLAN FOR UTEACH

Mary Ann Rankin hired Mary Long, the 1986 Texas Teacher of the Year, who identified additional master STEM teachers (David Journeay, Mary Walker, and Cathy Stacy) to work for six weeks in the summer of 1997 to design a new plan.

Mary Walker [In 1997 Administrative Supervisor for Science in the Austin Independent School District, later UTeach Master Teacher]:

> When I reflect on my work with UTeach, I am frankly struck by how successful it all turned out to be. When my long-time friend and co-worker, Mary Long, and I met at an Austin Independent School District (AISD) science administrators retreat in Spring 1997, we discussed how difficult it was to make any real changes in science teaching practices with 4 or 8 hour professional development (PD) days randomly scattered across the school year. Additionally, I was stunned to learn (from a presentation by a Texas A&M professor) that most elementary and middle school teachers had no formal science coursework in college and that their highest level science class in high school was Physical Science—taken in the 10th grade. No wonder the teachers in our PD sessions were so challenged to try to adopt Inquiry practices—it's hard to ask probing questions when you have no deep knowledge of the content. Worksheets provided by the textbook company that had answers provided in the teacher's manual were easier and much more comfortable for many.
>
> Talking with my colleagues in Mathematics admin in AISD, I found similar attitudes and frustrations with their attempts to change teaching practices through limited hours for PD. When I observed math classes on site visits, I watched teachers sitting next to an overhead projector in a darkened classroom, head bowed while solving problem after problem on the screen, rarely asking students any questions about what they did or did not understand about the process. On one occasion, accompanied by a math specialist, the teacher was asked why she did not use the algebra tiles that the district had provided so the students could work through these processes on their own. The teacher simply replied, "I tried that once and it didn't work."
>
> My observations of hundreds of math and science classes in AISD coupled with my experiences with PD strongly influenced my ideas about recruiting potential STEM teachers with classes like the Step course sequence—challenge their own educational experiences by forcing them to use innovative and non-traditional teaching practices. I had hoped we would see the courses selecting and sorting for those future teachers who would embrace Inquiry techniques and revolutionize the STEM classrooms where they taught. I am extremely pleased that for many of the teachers we produced, this has been the case and even more excited that the process has been successfully replicated across the nation. However, I'm aware that not all our graduates continue to use the UTeach model practices for a variety of valid reasons, including pressure to teach to high stakes tests.
>
> I still believe we made and are continuing to make significant changes in STEM teacher education and professional development. I will always consider my years building and improving on our work with UTeach as the high point of my career.

Mary Long [In 1997, first UTeach Master Teacher]:

> Dean Rankin issued only three directives to the working team: 1) Do not assume that anything about required courses, their content, sequencing, or other curriculum issues that now exist must necessarily continue to exist; 2) Do not worry about the

cost; 3) The entire preparation program must be doable within four years. The ad hoc working group formally presented a plan to the dean and the funders of the feasibility study, that included a philosophical rationale, descriptions of all required courses, and detailed degree plans.

The plan was for science teachers only. It was grounded in the personal experiences of the working team, and drew on sources such as the National Science Education Standards (National Research Council, 1996) and Tomorrow's Teachers (Lanier, 1986). It had many, but not all, of the elements that became part of UTeach. The course requirements for professional development were to be delivered by former secondary teachers, called Master Teachers, hired by the College of Natural Sciences. The sequence was

1. Professional Coursework
   - Teaching Methods "TAGs" (3 1-hr courses)
   - Laboratory Methods (3 hrs.)
   - Research Methods (3 hrs.)
   - Student Teaching (6 hrs.)
2. Supplemental Education Seminars
   - Teaching for Diverse Learners (3 hrs.)
   - Teaching Science from a Historical and Society Perspective (3 hrs.)
3. Composite Science Coursework

Students would go through the program in a cohort. Education courses would be paired with specific content courses—*TAG 1* was paired with *CHEM 301* for example—and since they were tagging along, they were the TAG courses. The plans for how much would be done in the courses were ambitious, and almost all were to be taught by Master Teachers. *TAG 1: Introduction to Discovery Learning* presented learning styles, gave students a chance to observe an elementary classroom, and then had them teach six elementary school science lessons, one of which students designed for themselves. *TAG 2: Planning for Instruction and Classroom Management* had students teach three middle school lessons, of which they designed one for themselves. *TAG 3: Constructivism and Assessment* did not contain a high school field experience, but was to have students design assessments that could be used in a high school physics class, working with students in introductory UT physics classes. *Laboratory Methods* provided students systematic exposure to the sorts of laboratory equipment available in high school laboratories, and then had them teach three middle school and three high school lab classes. *Research Methods* provided instruction in data collection and statistical analysis, and then required students to pose, investigate, and answer a question of their own, culminating in a paper and presentation on their results. *Teaching for Diverse Learners* was a seminar course, led by guest speakers such as school administrators and school counselors, to cover topics including dyslexia, English as a Second Language, Gifted and Talented students, and family literacy. Stu-

dents would conduct a research project including clinical interviews of students, teachers, parents, and administrators. In *Teaching Science from a Historical and Societal Perspective*, students would learn about eight milestones in the historical development of science, such as the Copernican Revolution and discovery of DNA. As students completed assignments in these courses, they would store artifacts in a portfolio that would constitute their record of progress towards becoming a professional teacher. Finally, students would take content courses to support Texas' composite science certification. The most common science certificate allowed teachers to teach Biology, Chemistry, Physics, and Earth Sciences. The Master Teacher working group felt that principals were under such pressure to hire teachers who could cover all these courses that it was better to try to provide preparation while students were still in college, rather than having teachers try to pick up new disciplines through professional development after leaving college.

INITIATION OF UTEACH

At the conclusion of the summer in 1997, the Master Teacher working group presented this plan, fully expecting it to sit on a shelf for several years. That is not what happened.

David Laude:

> I tend to plow ahead with stuff out of sheer ignorance. So not really knowing exactly what course and curriculum is, or what course inventory is, or what was even legal, I decided to push full steam ahead by sending an e-mail invitation to the 2000 or so incoming freshmen in CNS, asking it they would like to take part in a brand new teacher preparation program. We would pay the tuition for the first course, and we included an appropriately conditioned statement that the degree plan had not yet been approved.

Mary Long:

> To our surprise, Dean Rankin decided to initiate the program that very fall. Letters informed incoming CNS freshmen that an amazing opportunity was available to them immediately. They could sign up for a one-credit hour, tuition-free course called "TAG 1" that would, in a rudimentary way, prepare them to teach science or math lessons in a local elementary school. A "Master Teacher," an experienced person with an acknowledged excellent professional reputation, would coach the students. At this point, they would not need to have advanced content knowledge. There was very little risk. The students could try a bit of teaching under controlled conditions, but would they like it?
>
> Twenty-nine freshmen signed up for the first exploratory course, and all but one registered for the follow-up course the next semester. Regarding a teaching career, many were hooked! Having wondered if they were cut out to be teachers, they had

> discovered an aspect of who they truly were. Ultimately, seven students from the inaugural class completed the program and became teachers.

Having a quarter of the first class complete the program was not a bad start, considering it was being made up one semester at a time as they went through it.

Midway through the first semester of operation in the Fall of 1997, the College of Education learned of the existence of the teacher preparation program in an inconvenient way.

David Laude:

> The trouble began with a headline in the student newspaper, the Daily Texan, that CNS was creating a new teacher preparation program (yes, I was approached by a reporter who heard about TAG 1, and I was too young in the game to know you don't just blab to a student reporter, or any reporter, probably). I handed a copy of the headline to Mary Ann just outside her office, not thinking good or bad, and she screamed something like, "Oh No!!" and ran into her office, I assume to call [Dean of Education, Manuel] Justiz.

Justiz convened meetings of faculty and administrators in the two colleges to learn what was going on. Although the plan developed by the teachers had been quite detailed, certain things were left out. For example, the State of Texas allowed teachers to become certified only after they were formally recommended by a program, and for UT Austin, the ability to make this recommendation was housed in the College of Education. Thus, the College of Education was in a position to participate with or block the new approach. The first meetings were acrimonious, involving mutual suspicions between Education and Science faculty around some of the themes that a decade before had led to SB 994. Rankin and Laude decided to find a faculty member in the sciences to take over and try to work with the College of Education. They selected a professor in the physics department with no formal training in education.

Michael Marder [In 1997, Associate Professor of Physics, and Founding Co-Director of UTeach]:

> I got a call from Mary Ann Rankin in the fall of 1997 asking if I would apply to direct the Special Projects Office and run a new teacher preparation program. Although I had no formal preparation for this role, I had experiences as a child that gave me more background for it than I understood. I grew up in Champaign, Illinois and was bused to a magnet elementary school. Max Biberman, a mathematics educator at the University of Illinois and colleague of Seymour Papert, conducted educational experiments at my school. In my fifth-grade class, I was exposed to the world's first educational computer system (Plato V), and our class programmed turtles, plexiglas-domed robots that could be directed to move along computer-controlled paths (Papert, 1980). I was bused to the school as part of a desegregation order, but black and white students segregated themselves spontaneously within the classroom, and the special enrichment opportunities I received were not delivered to all students equally. I continued on to a magnet high school where the experiments

continued and became more extreme. For two years I was in a program called Form I where students had no classes at all (except foreign language) and instead designed projects. The goal was to see what would happen if one prevented teachers from standing between children and their learning. Most of the students played games most of the time, but a few of us plunged into learning as enthusiastically as educational theorists could have hoped. I learned electronics, taught myself calculus, and invented a language among other things. Therefore, by the time I left high school I had years of exposure to instructional technology and constructivist learning techniques, had experiences that let me reflect on equity and access, and had very positive memories of working, as a kid, with Education faculty. This is why I was offered the chance to direct the Special Projects Office.

EVOLUTION OF UTEACH IN ITS FIRST YEAR

During the spring of 1998, the teacher preparation program developed essentially into the form of today's UTeach program. This happened through weekly meetings between faculty in the Colleges of Natural Sciences and Education. It is interesting to compare the original plan developed by the Master Teachers with the program that emerged from the discussions.

In the original plan, Master Teachers were responsible for all the professional development courses. As a result of discussion between the two colleges, responsibility for half the instruction fell to the College of Education. Administrative decisions were negotiated between David Laude and an Associate Dean for Teacher Education, Marilyn Kameen. The key education faculty involved in curriculum decisions were Jere Confrey and Walter Stroup, both with expertise in mathematics education, and Tony Petrosino, with expertise in science education.

Jere Confrey [In 1997, Professor of Mathematics Education, UT Austin]:

> When I joined the faculty, the very beginning of a plan for UTeach was underway. Dean Rankin was committed to finding a way to attract more students into science and mathematics teaching. It should be noted that in a College of Education, the secondary teacher preparation numbers are far outweighed by elementary teacher education. And traditionally we prepared secondary teachers by requiring them to complete a mathematics or science major and then to take courses in a) psychology of education, b) sociology of education, c) methods and d) student teaching. Asked to represent and help define the College in the design of UTeach, I proposed an alternative structure. It would differ from the traditional approach in two ways. 1) all courses would be specific to math and science education rather than taking a general education orientation, and 2) the courses would be developed to conceptually model a sequence of how one might imagine the process of coming to be a fully prepared teacher. This orientation led me to propose a three-course sequence of a) Knowing and Learning in Mathematics and Science, 2) Classroom Interactions in Mathematics and Science and 3) Project-Based Instruction in Mathematics and Science, all followed by Student Teaching. The sequence assumed that first, our students needed to be much more aware of their own understanding of the content both epistemologically and psychologically in a reflective manner. This meant thinking about their

> own thinking, and when they were confident, and what they believed. The second course would begin to delve into interactions— understanding the dialectic between two or more students, and then students and instructors. That class would end with a model teaching experience, with math and science preservice students working together. In this experience, we did not expect perfect teaching but did expect for students to be able to examine their own practice critically and reflectively. The final course required the students to think about a larger curricular chunk, a unit. And it required them to think about how to embed science and mathematics learning in relevant and worthwhile contexts. It was a lovely and exciting opportunity to rethink teacher preparation in the company of faculty such as Tony Pertrosino and Jill Marshall. And it provided an opportunity for a deeper collaboration with scientists like Michael Marder and outstanding teachers. I remain indebted to Dean Rankin for the experience and I cherish my glass apple that reminds me of my involvement in co-leading the development of UTeach with Michael Marder.

Whereas the original plan concerned science teachers only, interaction with Confrey and Stroup made it obvious that both mathematics and science teachers needed to be involved. This meant convincing the mathematics department to join in—a significant step because there had long been a greater commitment to teacher preparation in mathematics than in the sciences, and the mathematics department housed a whole unit, the Dana Center, specializing in professional development and education policy. In the pedagogical sequence conceived by Confrey, Stroup, and Petrosino, pedagogical instruction in science and mathematics were integrated. This was before the acronym STEM became common, but it meant that the program design anticipated the desire to view STEM as a unified whole.

The program design evolved during the spring of 1998 to accommodate a balance between theory and practice. The courses designed by the College of Education faculty included a rich spectrum of readings from primary sources in education. Much of the theoretical framework came from *How People Learn* (National Research Council, 1999). For Knowing and Learning, characteristic readings included Duckworth (1972), Gardner, (1995), Schoenfeld, (1988), Skinner (1958), Vygotsky (1978) and many others. Classroom Interactions, with a focus on lesson development and equity featured readings such as Anyon (1980), Ma, (1999), and Rothstein (2004). Project-Based Instruction featured work such as Krajcik et al. (1998) and Barron et al. (1998).

At the same time, much of the original vision of the Master Teachers was retained. This was particularly true regarding field experiences. The cohort model was quickly abandoned, as the students taught everyone the impossibility in the actual university environment of sticking to a rigid plan about when they would take courses. The TAG concept was abandoned, and the first courses were renamed *Step 1: Inquiry Approaches to Teaching* and *Step 2: Inquiry-Based Lesson Design*, but they continued to feature elementary and middle school teaching experiences. The concept for UTeach, hammered out in months of discussions, was that students would learn about education in a constructivist fashion. They would begin by planning and delivering lessons in their first two one-hour courses, *Step*

1 and *Step 2*. In *Knowing and Learning in Mathematics and Science*, they would get an overview of educational thinking about how people learn science and mathematics. In both *Classroom Interactions* and *Project-Based Instruction*, theory and practice would be combined and both courses built in high school field experiences in which students taught lessons, assisted by Master Teachers. The preservice teachers would learn about lesson design while actually designing lessons they would deliver. They would learn about equity while actually going in and out of classroom settings where issues of equity naturally arose. Preservice teachers interacted both with STEM education faculty and with Master Teachers during *Classroom Interactions* and *Project-Based Instruction*. This was the way the program design resolved the tension between education practitioners and theorists.

RESEARCH METHODS, PERSPECTIVES, DOMAIN COURSES, AND APPRENTICE TEACHING

Two additional courses, *Research Methods* and *Perspectives on Science and Mathematics* were part of the teacher preparation program plan from the start. They had a different status from the other courses because the Master Teachers and STEM Education faculty did not have the same level of ownership, and it was not as completely evident to the students how they were connected to a future as teachers. They also were not subject to the state's 18-hour cap on education courses because of their content material.

David Laude:

> One contribution I did offer was to insist that a Research Methods course be part of the professional development sequence. I was insistent that the future of science education was going to be about experiential learning and that high school teachers needed to understand how research scientists thought, and that they should pass that perspective on to their students before they came to college. The first research methods course was taught, with some issues, by an astronomer. I taught the second research methods course and remember it as an environment in which I was attempting to get the UTeach students to perform independent inquiries (five of them in a semester), and the beginning of every lecture was 15 minutes of me telling them that everything was going to be okay.

The *Perspectives on Science and Mathematics* course was first developed by Bruce Hunt, a professor specializing in the History of Science in the Department of History. In basic outline it followed closely to the original vision of the Master Teachers, to challenge future teachers to understand the place of their discipline in the development of thought and society.

A concept that emerged from the months of discussion between Natural Sciences and Education was the 'domain course.' This was a course taught within a STEM discipline and concerned with STEM content, but also taught with attention to pedagogical content knowledge, and sensitive to the areas within the discipline where future teachers would need to be particularly well prepared. The

first courses developed within this framing were *Research Methods* and *Functions and Modeling* (first developed by Patrick Callahan, a mathematics instructor straddling mathematics and STEM education). *Functions and Modeling* provided mathematics majors the opportunity to revisit important topics in secondary mathematics through an inquiry approach. Another course that fits this description is *Physics by Inquiry*, taught by Physics Education Researcher Jill Marshall, which is in degree plans for students whose certification includes physics. The idea of the domain course was implemented inconsistently in Austin; consistent implementation would have meant bringing into existence courses such as Computer Science by Inquiry and securing instructors and places for them in degree plans. This was hard to do because, whereas in mathematics the thought of becoming a high school teacher was familiar both to students and faculty, in Computer Science the thought was completely unfamiliar, and the number of students taking this path was too small to support a course.

The final course in the UTeach sequence was Apprentice Teaching, a full-semester teaching experience in secondary schools. UTeach did not design this course until enough students came up through the previous courses to justify hiring someone to oversee it. This was a new Master Teacher, Pamela Powell. The most unconventional feature of the Apprentice Teaching course was that students were required to spend only a bit more than half a day in schools. This decision was originally made to increase the possibility that STEM majors could graduate in four years with their teaching certification, but instructors became convinced that this approach provided pedagogical advantages as well. It meant that Apprentice Teaching served as a bridge between college life and life as a full-time teacher. There was time for students to plan their lessons and respond to challenges in a way that would be very difficult at first with a full-time job. The attention to practical details of working in schools, key information coming in a weekly seminar, and intense support for every student made this course an embodiment of the Master Teacher contribution to the development of a UTeach teacher.

The question about whether UTeach would advance as a partnership between the Colleges of Natural Sciences and Education was taken up at the end of Spring 1998. The decision was made by Dean Justiz to continue to work together. This was ultimately due to the recognition that the collaboration between Master Teachers, STEM Education faculty, and Natural Sciences faculty, while frequently challenging, was producing a program that none of them could have developed on their own.

A FINAL WORD ON THE CREATION OF UTEACH

Walter Stroup [In 1997, an Assistant Professor of Mathematics Education]:

> In broad terms, there was a policy-related impetus for the creation of the UTeach program and that intersected with a then-ascendant framing of research in STEM

education, largely funded by the National Science Foundation, that carried with it a new conceptualization of what a STEM educator needed to be like.

At a policy level, the State Legislature had, some number of years prior to the creation of the UTeach program, done away with undergraduate education degrees and this could be seen to invite, if not outright require, moving the preparation of teachers, particularly in secondary education, beyond the institutional expectation that Colleges of Education would be the exclusive providers for programs culminating in certification to teach. Against this backdrop UTeach emerged as a new pathway to certification—existing tenuously at first and then more and more emphatically—between the poles of non-university based alternative certification programs—e.g., Teach for America—and certification programs run through traditional schools of education where programs are typically optimized for the preparation of elementary educators, and then only nominally adjusted to accommodate secondary educators.

Looking back, UTeach can be viewed as perhaps a remarkably successful instance of a new category of university-based alternative teacher preparation programs meant, on their best days, to programmatically integrate the multiple perspectives and contributions of research-active university faculty, experienced classroom educators, and, in the case of UTeach, early and on-going, schools-based field experiences. Three crosscutting themes and commitments were frequently referenced: equity, inquiry, and the domain-situated use of technology.

New, program-specific courses were created and offered in the College of Education and College of Natural Sciences. In the College of Education, three of the core courses required for UTeach students were developed and taught by tenured or tenure-track research faculty all of whom had, or were about to establish, a record of funding from the National Science Foundation and all of whom had previous experience as classroom teachers themselves. The three courses are Knowing and Learning in Mathematics and Science, Classroom Interactions, and Project Based Instruction.

In these early days we would frequently speak of "building the track just before the train arrived." This expression served both as a description of the practical reality of bringing a new program into existence and as a remarkably effective way of maintaining focus as we attempted to forge the UTeach program out of a plurality of convictions, commitments, and high-level competencies. What this looked like for a then very new assistant professor was that I was asked to develop the first of the thee-hour core courses, Knowing and Learning in Mathematics and Science out of whole cloth, on short notice, and with only a brief outline of the role the course was to play in framing an interdisciplinary perspective of how various significant accounts of knowing and learning can, and do, serve to support the classroom-based teaching and learning of STEM domains in classrooms.

As a brief summary of the challenge at the time, and a reminder of what I consider a still-developing conceptualization of what a STEM educator needed to be like, I quote from one of the course readings by an author who played a significant role in my own graduate work and who many of us involved in developing the core educa-

tion courses would view as making seminal contributions to the research traditions that have come to coalesce as the Learning Sciences. Especially as related to the integration of the three cross-cutting themes of UTeach mentioned earlier, Papert identifies the problem of "the supply of people" who can integrate, as examples, the kinds of knowledge "associated with academic psychology" with the "more emphatic" kinds of knowledge possessed by "creative artists and people who 'get along with children.'" He then goes on to frame what remains a vitally important summary of the problem central to the emergence of the UTeach program (Papert, 1980):

> The problem goes deeper than a mere short supply of such people. The fact that in the past there was no role for such people has been cast into social and institutional concrete; now there is a role but there is no place for them. In current professional definitions physicists think about how to do physics, educators think about how to teach it. There is no recognized place for people whose research is really physics, but physics oriented in directions that will be educationally meaningful. Such people are not particularly welcome in a physics department; their education goals serve to trivialize their work in the eyes of other physicists. Nor are they welcome in the education school—there, their highly technical language is not understood and their research criteria are out of step … [and] would be judged by whether it produced a 'measurable' improvement in a particular physics course. (pp. 187–188)

To the extent UTeach directly addressed, and continues to address, the problem identified by Papert and, in so doing, provides a welcoming place for future teachers to develop their abilities to integrate their perspectives and passions, including their major in a STEM domain, into becoming a highly effective classroom educators, THIS is, to my mind at least, absolutely central to how the UTeach model emerged. And to the extent that UTeach can continue to find ways around, and occasionally through, the "social and institutional concrete" that surround the preparation of teachers, the program will continue to advance a new, fully classroom-situated, conceptualization of what a STEM educator needed to be like.

REFERENCES

Anyon, J. (1980). Social class and the hidden curriculum of work. *Journal of Education*. https://doi.org/10.1007/BF02287080

Barron, B. J. S., Schwartz, D. L., Vye, N. J., Moore, A., Petrosino, A., Zech, L., & Bransford, J. D. (1998). Doing with understanding: lessons from research on problem- and project-based learning. *Journal of the Learning Sciences*. https://doi.org/10.1080/10508406.1998.9672056

Duckworth, E. (1972). The having of wonderful ideas. *Harvard Educational Review*. https://doi.org/10.17763/haer.42.2.g71724846u525up3

Gardner, H. (1995). Reflections on multiple intelligences: Myths and messages. *Phi Delta Kappan*. https://doi.org/10.1002/ss.327

Heath, J. E. (2005). *In their words, through their eyes: Novice teachers reflect on teaching and their preservice education*. Austin, TX: The University of Texas at Aus-

tin. Retrieved from: https://repositories.lib.utexas.edu/bitstream/handle/2152/2442/heathj87540.pdf?sequence=2

Krajcik, J., Blumenfeld, P. C., Marx, R. W., Bass, K. M., Fredricks, J., & Soloway, E. (1998). Inquiry in project-based science classrooms: initial attempts by middle school students. *Journal of the Learning Sciences*. https://doi.org/10.1080/10508406.1998.9672057

Lanier, J. E. (Ed.). (1986). *Tomorrow's teachers: A report of the Holmes Group*. Lansing MI: The Holmes Group.

Ma, L. (1999). Chapter 1: Subtraction with regrouping: Approaches to teaching a topic. *Teachers' Understanding of Fundamental Mathematics in China and the United States*. New York, NY & London, UK: Routledge.

National Commission on Excellence in Education. (1983). *A nation at risk*. Retrieved from: http://www2.ed.gov/pubs/NatAtRisk/index.html

National Research Council. (1996). *National science education standards*. Washington, DC: National Academies Press.

National Research Council. (1999). *How people learn: Brain, mind, experience, and school*. Washington DC: National Academy Press.

Papert, S. (1980). *Mindstorms: Children, computers, and powerful ideas*. New York, NY: Basic Books.

Rothstein, R. (2004). *Class and schools*. Washington DC: Economic Policy Institute.

Schoenfeld, A. H. (1988). When good teaching leads to bad results: The disasters of "well-taught" mathematics courses. *Educational Psychologist*. https://doi.org/10.1207/s15326985ep2302_5

Simms, R. L., & Miller, J. R. (1988). Assault on teacher education in Texas. *Journal of Teacher Education*, *39*(6), 17–20.

Skinner, B. F. (1958). Teaching machines. *Science*. https://doi.org/10.1126/science.128.3330.969

Texas. (1987). *Relating to teacher education.* Senate Bill 994, https://bit.ly/SB994_1987

Vygotsky, L. S. (1978). *Mind in society: The development of higher psychological processes*. London, UK: Cambridge University Press.

CHAPTER 2

IMPLEMENTING AND SUSTAINING UTEACH PROGRAMS

Lessons Learned from National Expansion

Pamela Romero and Kimberly K. Hughes

For more than a decade, the UTeach Institute has monitored and documented the implementation of the UTeach secondary STEM teacher preparation program at 43 universities in 21 states and the District of Columbia. What started as an effort to replicate a successful program created at The University of Texas at Austin has developed into a vibrant, national networked improvement community made up of more than 700 STEM and STEM education faculty all contributing their expertise to the ongoing development of what has become a national model for university-based, secondary STEM teacher preparation. We review how UTeach program implementation has progressed across diverse university settings and the extent to which UTeach programs across the nation are aligned to the original model. We highlight challenges, particularly regarding long-term sustainability and financial support, that programs continue to face and how, over time, programs have evolved, adapted, and innovated aspects of the program model.

Over the course of a decade, 43 U.S. universities across 21 states and the District of Columbia adopted the UTeach secondary STEM teacher preparation pro-

Preparing STEM Teachers: The UTeach Replication Model, pages 17–30.

gram model in order to strengthen their production of STEM teachers. During this time, the UTeach Institute carefully monitored and documented the implementation of UTeach across the country. In this chapter we argue that multiple years of data on program implementation; graduate production; and graduate entry into, and retention, in teaching indicate that UTeach programs are making important contributions in STEM education by increasing the number and the quality of teachers produced, leveraging the critical role of universities in the preparation of STEM teachers, making significant contributions to the field of STEM education, and, in doing so, improving and innovating the UTeach program model overall. Twelve years into national expansion, the majority of UTeach programs are well-established, are producing teachers, and remain closely aligned with the UTeach program model's core elements. Accordingly, we consider the national UTeach expansion effort a success. What started as an effort to replicate a successful teacher preparation program created at The University of Texas at Austin has developed into a vibrant, national networked improvement community made up of more than 700 STEM and STEM education faculty all contributing their expertise to the ongoing development of what has become a national UTeach model for secondary STEM teacher preparation.

In this chapter, we review how UTeach program implementation has played out across diverse university settings and the extent to which UTeach programs across the network are aligned to the original UTeach model. We highlight the challenges, particularly regarding long-term sustainability and financial support, that programs continue to face and also how, over time, programs have evolved, adapted, and innovated aspects of the program model.

THE UTEACH INSTITUTE'S APPROACH TO PROGRAM EXPANSION AND SUPPORT

In 2006, the UTeach Institute (Institute) was established and charged with taking the UTeach program to scale. Specifically, the Institute was to guide and support the successful implementation of UTeach at universities across the country. This ambitious charge, to replicate an entire academic program, is rarely observed in university settings. When the effort was begun, a great deal of uncertainty surrounded the degree to which the Institute could expect universities to implement the program with fidelity. Not only would UTeach program implementation be complex, requiring a number of activities and significant time to produce intended outcomes, but local conditions and state requirements would necessitate adaptations to fundamental program components, and the complexity inherent in higher education settings would present further challenges. Establishing a UTeach program involves the implementation of an entire academic program, made up of operational and instructional components on par with that of a small department, over a five-year period. It involves engaging key stakeholders, like deans and provosts, to make necessary institutional commitments to secure space, funding, faculty lines, etc. It involves engaging faculty members—experts in STEM and

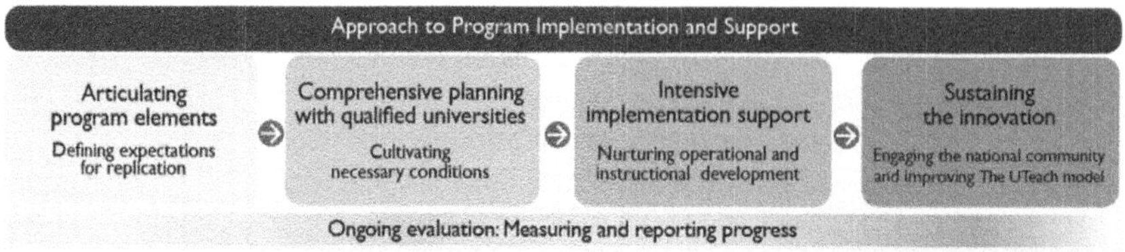

FIGURE 2.1. UTeach Institute's Approach to Program Implementation and Support. This figure depicts the UTeach institute's framework for supporting program implementation at partner universities.

STEM education themselves and peers of the developers of UTeach—in modifying or teaching new courses and becoming advocates for the new program (Beth, Romero, Lummus-Robinson, & Pérez, 2012).

Keenly aware of these challenges, the Institute developed an approach to implementing and scaling education programs (Beth, Hughes, Romero, Walker, & Dodson, 2011) that is aligned with recommendations from the research literature on fidelity of implementation, program replication, scaling innovations, and networked improvement (Bryk, Gomez, & Grunow, 2011; Century, Rudnick, & Freeman, 2010; Hall & Hord, 2010; Hill, Maucione, & Hood, 2007; Patton, 2011). To promote the dual goals of fidelity and long-term sustainability, the Institute's approach emphasizes (1) clear articulation of program elements and expectations for implementation, (2) comprehensive planning with qualified sites, (3) intensive implementation support, (4) ongoing evaluations of progress, and (5) long-term sustainability through community engagement and continuous, networked improvement.

ASSESSING ALIGNMENT TO THE UTEACH PROGRAM MODEL OVER TIME

UTeach Elements of Success

Typically, UTeach implementation begins with a comprehensive planning period, followed by four years of program implementation. Partner universities are expected to establish programs that incorporate all of the UTeach *Elements of Success*. These elements outline nine critical components of the UTeach model. As a whole, they articulate the operational structures and instructional philosophies that underlie the program (UTeach Institute, 2011).

The Elements of Success also form the basis for the Institute's evaluation activities. They are used to assess program implementation and to determine the degree of implementation fidelity and alignment to the UTeach model. Alignment to the UTeach model is assessed at three district time periods: (1) prior to implementing UTeach (baseline), (2) at the completion of the five-year implementation period (established), and (3) three years after programs are considered established.

Figure 2: The UTeach Program, Elements of Success

1. Distinctive Program Identity
2. Cross-College and School District Collaboration
3. Long-Term Institutional and Community Support
4. Compact and Flexible Degree Plans
5. Active Student Recruitment and Support
6. Dedicated Master Teachers
7. Rigorous, Research-Based Instruction
8. Early and Intensive Field Experiences
9. Continuous Program Improvement

FIGURE 2.2. The UTeach Program Elements of Success. This figure provides a list of critical components of the UTeach program model.

The nine *UTeach Elements of Success* are defined by more than 50 critical components that are mapped to more than 140 indicators that are the foundation for all data collection and evaluation activities. Every UTeach program was examined to determine (1) the degree to which each component criterion was implemented and (2) the degree to which program elements were adapted. While a high degree of fidelity to the model is desired during the early years of program implementation, local contextual adaptations and innovations are also expected in order to ensure long-term sustainability and institutionalization over time. As long as the theoretical basis of the UTeach program is not compromised, adaptations and innovations are highly desired. This balance, between fidelity and adaptation becomes critical as programs evolve over time. We believe that our implementation approach to first adopt, then adapt, ensures that a strong, shared foundational understanding of UTeach is developed by partner universities so that adaptations and innovations to the model are productive improvements and do not significantly diverge from the philosophical approach called for by the UTeach model.

Evidence of Alignment to the UTeach Model

Of the 48 universities who have begun UTeach programs since 2008, 44 are either established or currently in the start-up phase. Four programs were discontinued, in large part due to administrative reorganization and changes in leadership priorities. It is important to note that in all cases, these programs struggled to establish all program elements and attract sufficient numbers of students during the early years of implementation. Low enrollment made these programs especially vulnerable when changes in leadership and organizational restructuring occurred.

Alignment to the UTeach model is determined about each program at three district time periods: (1) prior to implementing UTeach coursework (planning period), (2) at the completion of the initial program implementation period (es-

tablished), and (3) three years after programs are considered established. Nine key program Elements of Success are mapped to more than 50 critical components and 130 individual indicators. Alignment to these indicators and critical components is determined though analysis of both regularly collected quantitative and qualitative data. Data are collected on the following: program, college, and university characteristics; program features; student characteristics; student recruitment, enrollment, and retention; graduate production; teaching certification; entry into, and retention, in teaching; program funding and expenditures; key course components and student artifacts; and student experience and satisfaction in the program. Each UTeach program was examined (1) to determine the degree to which each Element of Success was implemented and (2) identify any adaptations that were made to the UTeach model.

Upon completion of the initial implementation period (end of a grant-funded period), all UTeach programs were found to be in close alignment to the UTeach model, and implemented programs with a high degree of fidelity to the UTeach Elements of Success.

Figure 2.3 depicts the average degree to which each UTeach Element of Success has been implemented across all established programs. UTeach has been implemented in universities with very different characteristics and student populations. While larger universities tend to produce a greater number of UTeach graduates, the model has been successful in small and medium-sized universities as well. These universities represent several Carnegie Classification of Institutions of Higher Education categories, from *Doctoral Very High Research* to *Masters Large and Masters Medium* programs (Indiana University, 2018). They vary in size from low enrollment (5,000) to large (over 50,000), and represent various geographical settings, both urban and rural. These universities implemented pro-

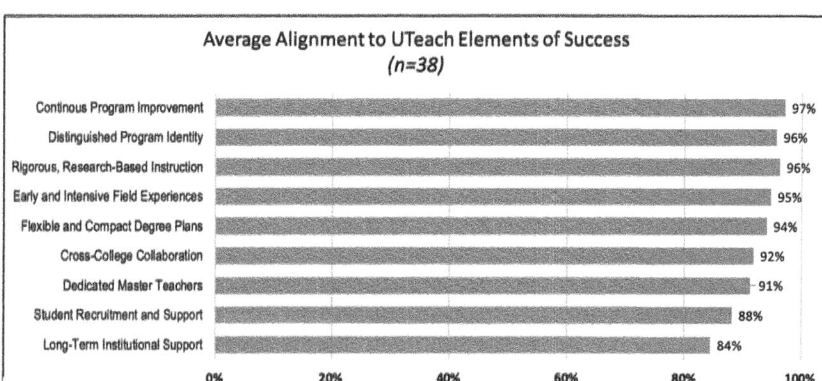

FIGURE 2.3. Alignment of UTeach to the UTeach Elements of Success. This figure illustrates that UTeach programs nationally are in overall close alignment to the UTeach model.

grams that bridge colleges of science and education, actively recruit and graduate STEM majors, promote early and intensive field experiences, incorporate relevant and authentic STEM content in their professional development courses, fundraise to ensure sustainability, and support students with a variety of benefits. These programs are implementing courses with similar content in roughly the same sequence, involving STEM, STEM education, and clinical faculty with similar areas and levels of expertise, collaborating on instructional support and program development, sharing best practices, and tackling shared challenges. While each program's local context is unique, all programs are clearly distinguishable as UTeach programs. Beyond the initial implementation phase, UTeach programs continue to be overall in close alignment with the foundational elements of the UTeach model, adapting as necessary, while also maintaining program effectiveness and meeting short-term program goals. In the next section, we discuss the implementation of UTeach program components and how these are articulated across the UTeach national network.

Evidence of Impact

The primary goal of national UTeach expansion is to increase the number of highly-qualified secondary STEM teachers. As of Spring 2019, 39 UTeach programs are producing graduates. Eight additional programs are on track to begin producing graduates within the next few years. As of 2019, more than 5,200 UTeach graduates have been produced and approximately 700 new graduates are produced annually. The Institute projects that more than 8,300 UTeach graduates will impact more than 6 million students by 2023. This significant growth in STEM teacher production has occurred despite overall national decreasing trends. According to U.S. Department of Education Title II teacher preparation data, the overall U.S. production of STEM teachers fell 20% between 2009 and 2014; during the same time, overall STEM teacher production at universities implementing UTeach grew 30%. (U.S. Department of Education, 2016).

Almost 84% of UTeach graduates enter the teaching profession. Of these, 92% are retained in teaching for at least three years. Approximately 20% of UTeach graduates are identified as underrepresented minorities and the majority of UTeach teachers (69%) are teaching in schools that are identified as serving economically-disadvantaged students. (UTeach Institute, 2019).

In terms of overall program impact, UTeach has been shown to be effective as demonstrated by two studies in Texas showing significant learning gains in mathematics and science by students of UTeach teachers when compared with students of non-UTeach teachers. Research by Backes, Goldhaber, Cade, Sullivan and Dodson (2018) found that secondary students taught by UTeach graduates outperformed all students in the state by an additional 4 months of learning in math and 5.7 months in science over the course of one academic year. In a second study, Marder and Hamrock (2016) found that secondary students taught by UTeach graduates gain an additional 9 months of schooling in both Algebra I and

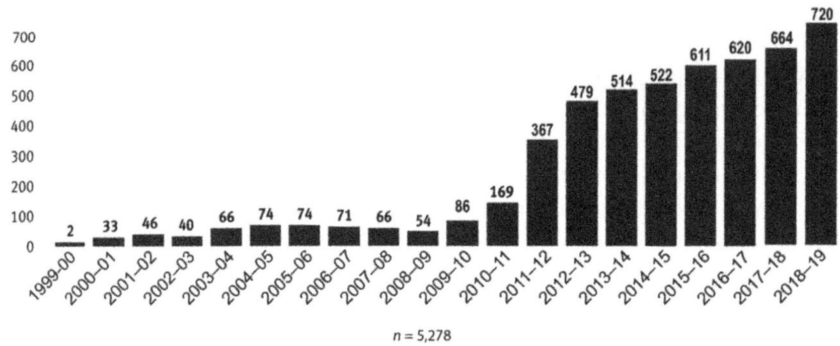

FIGURE 2.4. Annual National UTeach Graduate Production. This figure depicts the growth in UTeach graduate production from 1999 through 2019.

Biology for Gifted students and 5 months of learning in Biology for economically disadvantaged and Latino/a students. Combined, these studies demonstrate that UTeach teachers in Texas are more effective overall and with diverse, high-needs populations.

ESTABLISHING UTEACH PROGRAMS—LESSONS LEARNED

While all nine UTeach Elements of Success have overall been successfully adopted and implemented, there are notable differences and unique challenges presented by working across diverse university settings and in 22 states with unique teacher preparation requirements and regulations. Establishing a UTeach program involves putting in place a set of operational structures and instructional approaches over a four to five-year period. A discussion of how these operational and instructional components have been implemented across university settings follows.

Implementation of Operational Program Elements

A UTeach program operates on the scale of a small department with a distinctive program identity. It calls for a formally coordinated effort across colleges, in collaboration with local school districts, and supported by a broad range of stakeholders concerned with STEM education. Significant resources are provided to recruit STEM majors into the program and to maximize program and career retention.

Programs are typically structured as independent cost units and are housed in either the College of Education or Science. A few UTeach programs are housed in independent STEM centers. The location for the program appears to matter much less than does the fact that the program builds a distinctive identity on campus.

All programs have a unique, identifiable name and logo, and maintain a program website.

All UTeach programs are co-directed, typically by one faculty representative from the College of Education and another from one of the colleges preparing STEM majors. The UTeach model calls for a formal, cross-college steering committee that includes representatives from program faculty and staff and that meets regularly to develop program policies, monitor curriculum and instructional effectiveness, and manage student affairs and program operations. Early in the national expansion effort, we discovered that the structure and role of the steering committee was poorly understood and implemented. The most successful UTeach programs prioritize the role that the steering committee plays in fostering collaboration across the colleges and convene the committee with regularity to address both strategic program decisions as well as day to day program operations. They also regularly create opportunities to convene the entire group of faculty, staff, and program stakeholders to review data on program implementation and student success in an effort to foster continuous program improvement.

Student support is a critical component to recruiting and retaining students in UTeach programs. All UTeach programs work to ensure that all STEM majors are invited to participate in the program and are aware of its benefits. A common, aligned process is observed across the network. All established programs offer the one-credit hour recruitment courses to all STEM majors, offer some form of financial incentives, make presentations to introductory undergraduate math and science classrooms, promote the program online and through social media, and send targeted letters and post-cards to all in-coming STEM freshman. UTeach programs are most successful at recruiting students when they employ a variety of targeted communication strategies and recruitment activities in an ongoing fashion. Students reported that they hear about UTeach programs primarily through orientation sessions, posters and flyers, classroom presentations, and other program students (UTeach Institute, 2014–2018). The nature of the messaging used to recruit students is also very important. We have learned that students respond to the opportunity to expand career options through UTeach—"one degree, two careers" and invitations to "try out teaching," rather than being asked to make a firm commitment to teaching as a career through a formal application process in order to gain entry to the program.

These recruitment efforts have paid off. The model is unique in that it casts a wide net and invites STEM majors who may not necessarily have considered becoming a teacher as a potential teaching career option. Multiple sources, including national student survey data, alumni tracking data, and interview and focus group data indicate that UTeach programs do recruit students who were not necessarily thinking about pursuing a career in teaching. Of those students who complete the program, approximately 20% indicated that they were not committed to pursuing a career in teaching when they first enrolled in a UTeach course. (UTeach Institute, 2014–2018).

Student advising is another factor that has great bearing on student recruitment into and retention throughout the program. The most successful UTeach programs have dedicated advisors or assign formal advising duties to one or more of the program master teachers (clinical faculty). Because students are earning their degree through a STEM college, it is critical that UTeach programs work closely with advisors in the STEM disciplines so that they know about the program and how students can add the UTeach teaching option to their degree. While over 90% of students reported being satisfied with several aspects of their UTeach programs, 69% were satisfied with degree plan flexibility. Close to one-third of students were either neutral or dissatisfied with the options available to them for program completions. In addition, 19% of students that decided not to continue in the program, left because they were unable to complete the program coursework in a timely manner. (UTeach Institute, 2017).

The implementation of financial incentives to enroll in the first two exploratory, one-credit hour UTeach courses, while a component of the model, have proven to be unevenly implemented and difficult to sustain for established programs. Other, more significant opportunities for financial support through scholarships and paid internships, however, have shown to be very important to students' ability to remain in the program. Sixty-percent of program graduates reported that financial assistance was important in their ability to complete the program. The UTeach internship program provides students with paid opportunities to gain relevant work experience through flexible placements at nonprofit STEM or education-related organizations and local schools. Despite the fact that over 60% of students that complete the program reported that financial support is important to their decision to remain in the program, several UTeach programs have not developed internship programs to sufficiently meet the student demand for paid work. Many programs have not prioritized this program component or have struggled to raise additional funding in order to offer paid student internships. This is an area for future growth across UTeach programs and could lead to increased numbers of students enrolled in UTeach programs nationwide.

Once graduated, new UTeach teachers continue to receive support from the program during their initial years of teaching. This induction support is highly valued by new teachers. All (100%) teachers who responded to the alumni survey indicated that the support they received by master teachers was useful (19%) or very useful (81%). Examples of support include in-person meetings, classroom visits, communication via email, face-to-face professional development sessions, preparation for the first day of class, and access to materials and new teacher kits. Most (60% of new teachers) reported receiving approximately 5 to 10 hours of induction support annually. However, 12% reported receiving up to 40 hours of support annually. This indicates that for a small percentage of new program graduates, the induction support they required was intensive and possibly critical to their remaining in the profession. Induction support is another program component that typically requires annual gift or other external funding. Many UTeach

programs wish to strengthen their support to new graduates (UTeach Institute, 2014–2018).

Implementation of Instructional Program Elements

Across the network of UTeach programs, implementation of the UTeach instructional program is more variable than that of the operational elements. Some of this variation is necessitated by teacher preparation requirements unique to 22 states and the implications are discussed later in this chapter.

A fundamental principle of UTeach is to remove barriers to participation in the program. UTeach degree plans allow students to complete both a STEM degree and teaching certification in the same amount of time required by equivalent undergraduate STEM degrees, without the requirement and cost of additional semesters. All UTeach programs have developed four-year undergraduate STEM degree plans with the option to obtain teaching certification. Content course requirements are determined by STEM departments and vary quite a bit across UTeach programs. A key principle of the program design is flexible entry so that students entering as upperclassmen are able to double up on UTeach coursework and still complete the program in four years. A program's ability to offer UTeach courses every semester depends on sufficient student enrollment. Smaller UTeach programs have struggled to offer coursework with enough frequency to realistically accommodate students who enter the program after their sophomore year. Increasing pressure on undergraduates to graduate in four years makes this fast-track option even more important if program enrollments are going to be expanded to accommodate the greatest number of students. Recent national UTeach enrollment data indicate that, approximately, an equal percentage of underclassmen and upperclassmen enroll in the first UTeach course. Programs that are able to accommodate students that enroll in the program after their sophomore year are more successful at recruiting, retaining, and ultimately graduating students.

The streamlined, STEM-specific UTeach curriculum has been implemented across the network in very close alignment to the model. In accordance with the implementation approach, all courses are initially offered with high fidelity to the program model and then, over time, faculty adapt local courses to accommodate local needs and to incorporate individual faculty expertise. The overwhelming majority of programs have successfully implemented all of the UTeach coursework and continue to do so in close alignment with the UTeach instructional program's core intentions.[1]

[1] These core intentions include courses that: develop deep conceptual understanding and mastery of subject-area content; make explicit the underlying connections between mathematics and science; develop proficiency in core mathematics and scientific practices; develop research and information analysis skills; integrate content and pedagogy; build strong connections between educational theory and practice; build strong connections between educational theory and practice; integrate themes of assessment, equity, literacy, and technology; and cultivate reflective practice.

The active involvement of master teachers throughout the instructional program is a unique feature of UTeach. Master teachers—non-tenured clinical faculty with exemplary secondary teaching experience—are employed by all programs across the UTeach network and their roles are uniformly aligned with the model. Master teachers assume primary responsibility for student field teaching, provide ongoing coaching and support to students as they progress through the program, and are paired with STEM and STEM education faculty to co-teach UTeach courses. They provide students with their first introduction to teaching as instructors of the first two UTeach courses, and they are the last ones to work with students during their final teaching practicum and seminar. Approximately 85 percent of students across the network reported that master teachers provide useful feedback and are available to offer assistance. Students enrolled in Student Teaching reported higher percentages (92%) of satisfaction with the assistance provided my Master Teachers (UTeach Institute, 2014–2018). The degree to which STEM and STEM education faculty are involved in UTeach coursework varies more widely. The most successful UTeach programs involve a range of faculty across colleges of education, STEM, and liberal arts, and with varied STEM and STEM education expertise in the teaching of courses. Diverse faculty involvement is also key to continuous improvement and long-term sustainability of UTeach programs.

The nature of the student field experience in UTeach programs is fairly uniform across the network and well-aligned to the program model. Students experience paired and individual teaching in upper elementary, middle, and high school settings; prepare inquiry- and project-based STEM lessons; and receive intensive coaching, support, and feedback from master teachers, tenure-track faculty, and classroom mentor teachers. Clinical teaching is closely articulated with UTeach coursework and teaching opportunities are offered early and continue throughout the program. The actual number of field hours required by UTeach programs varies considerably across the network, due in large part to state requirements. The program model calls for a streamlined, intensive field teaching component that is highly supported and prioritizes teaching events over observation, but that also does not require an excessive number of hours so that students are able to effectively manage challenging STEM coursework, teaching requirements, and additional responsibilities within a four-year degree period. A typical recommended number of hours is approximately 50 leading up to a final student teaching experience that is half-day for a semester. Because of state requirements, the number of actual required total field experience hours vary across programs. Twenty-two percent of students reported leaving UTeach programs due to unmanageable field hour requirements. We continue to study the impact that changes in state teacher preparation requirements have on recruitment and retention of STEM majors into teaching.

For additional information about the UTeach curriculum, go to https://institute.uteach.utexas.edu/curriculum-overview.

Overall, students report being well-prepared by UTeach programs for their careers in teaching. In UTeach student surveys administered, all students (99%) reported that their felt prepared to enter the teaching profession overall. Likewise, 98% reported that they felt prepared to teach the subjects they were trained to teach (UTeach Institute, 2014–2018).

Sustaining UTeach Programs

Several key aspects of program sustainability, including clinical faculty and staff positions, are to various degrees, provided by recurring institutional funds across all programs. While program leaders continue to face several funding related challenges, all current UTeach programs are stable, sustained, and institutionalized. In general, program co-directors report that they have sufficient operating funds, but may not have sufficient funds to further expand their programs. All have had to make a variety of cuts to their programs following the initial implementation phase. Some budget challenges are not unique to UTeach—but are attributed to decreased state support for universities across the country. Recent budgetary pressures, combined with a challenging fundraising environment, have resulted in cuts primarily to student support elements, such as financial recruitment incentives, paid internships, and payments to cooperating teachers. As discussed earlier, these program elements are critical to recruiting and retaining students in UTeach programs. Many UTeach programs have the capacity to expand enrollment and teacher production if more funding was available.

One of the more significant challenges that UTeach programs face is changing institutional leadership and priorities. Program stakeholders must continuously educate new university and college-level leadership about the program and advocate for continued support. The most successful UTeach programs are able to weather leadership changes by demonstrating significant impact through a vibrant UTeach program characterized by a broad base of faculty and key stakeholder support, ongoing fundraising to provide student support program elements, and, most importantly, sufficient numbers of enrolled students and annual graduates.

SUSTAINING THE INNOVATION

There is substantive evidence that national expansion of UTeach has delivered on intended outcomes, and these have been sustained over time. We know that the UTeach model continues to be improved and strengthened under the continuous scrutiny of a national network of experts dedicated to STEM education. This growing community of more than 700 research and clinical faculty in colleges of science, mathematics, and education affiliated with UTeach programs presents a unique opportunity to bring collective action to bear in addressing even broader challenges related to improving STEM education. The UTeach Institute has been working deliberately to create a viable infrastructure to provide ongoing support for established UTeach programs and to collect UTeach graduate data in order to

provide information on the impact of UTeach programs and alumni nationwide. In 2014, we successfully established USEA, the UTeach STEM Educators Association. It is through USEA that we are able to sustain what is one of the best examples nationwide of a highly functioning networked improvement community collaborating to share best practices, addressing shared challenges, and continually improving a teacher preparation model. Several opportunities have emerged that engage network partners in collaborative projects. Examples include a collaborative partnership to strengthen the preparation of engineering and computer science certification pathways, a collaborative U.S Department of Education Supporting Effective Educator Development grant award to expand develop innovative and alternative pathways to make UTeach available for more audiences (career changers, returning veterans) and a collaborative project to expand Maker Education by developing maker-focused professional learning opportunities and integrating maker-centered pedagogies into the UTeach curriculum. USEA is also the mechanism through which we are able to continue monitoring program health and collecting program implementation data.

We are satisfied that institutional change through scaling programs that work is not only possible in a higher education setting, but that it is happening at the network of universities working to implement and improve the UTeach program. Further, we believe that the Institute's approach to program implementation and support is a viable model for expansion of other successful programs. The combination of key elements—defining expectations and cultivating conditions prior to replication, intensive and continuous operational and instructional support, ongoing evaluation, and fostering a national improvement community—appears to result in successful program implementation in higher education settings. Because the balance between fidelity and locally contextual adaptations is critical to long-term success, the Institute has learned to be much more explicit in communicating its expectations that the national community will adapt and improve, not simply adopt, the UTeach model. We recognize the concentration of expertise and commitment of our partners as a major advantage of program expansion in this setting.

In fact, engaging these experts in improvement activities and research surrounding STEM education may be the key to the long-term sustainability of the national community. We continue to work closely with our partner institutions to implement structures that facilitate the collection and sharing of meaningful and practical data that inform the ongoing improvement of the UTeach instructional program across a variety of institutional contexts.

REFERENCES

Backes, B., Goldhaber, D., Cade, W., Sullivan, K, & Dodson, M. (2018). Can UTeach? Assessing the relative effectiveness of STEM teachers. *Economics of Education Review, 64*, 184–198.

Beth, A. D., Hughes, K. K., Romero, P., Walker, M. H., & Dodson, M. M. (2011). *Replication as a strategy for expanding educational programs that work: The UTeach Insti-*

tute's approach to program replication. Paper presented at the annual meeting of the American Association of Colleges for Teacher Education, San Diego, CA. UTeach Institute. Retrieved from: http://uteach-institute.org/

Beth, A. D., Romero, P., Lummus-Robinson, M., & Perez, M. P. (2012, April). *Implementing a secondary STEM teacher preparation program at multiple universities: The UTeach Institute's evaluation approach and pilot summative data.* Paper presented at the 2012 American Educational Research Association Conference, Vancouver, B.C., Canada.

Bryk, A. S., Gomez, L. M., & Grunow, A. (2011). Getting ideas into action: Building networked improvement communities in education. In M. T. Hallinan (Ed.), *Frontiers in sociology of education.* Berlin: Springer.

Century, J., Rudnick, M., & Freeman, C. (2010). A framework for measuring fidelity of implementation: A foundation for shared language and accumulation of knowledge. *American Journal of Evaluation, 31*, 199–218.

Hall, G. E., & Hord, S. M. (2010). *Implementing change: Patterns, principles, and potholes.* Upper Saddle River, NJ: Prentice Hall.

Hill, L. G., Maucione, K., & Hood, B. K. (2007). A focused approach to assessing program fidelity. *Prevention Science, 8,* 25–34.

Indiana University. (2018). *Carnegie classification of institutions of higher education.* Menlo Park, CA: Indiana University School of Education Center for Postsecondary Research. Retrieved from: http://carnegieclassifications.iu.edu/

Marder, M., & Hamrock, C. (2016). *Math and science outcomes for students of teachers from standard and alternative pathways in Texas.* Austin: University of Texas at Austin. Retrieved from: https://pdfs.semanticscholar.org/0693/f28933b966f68a48813a75894353f954e639.pdf?_ga=2.103896832.1239140174.1510529724-172563159.1510529724

Patton, M. Q. (2011). *Developmental evaluation: Applying complexity concepts to enhance innovation and use.* New York, NY: Guilford Press.

U.S. Department of Education. (2016). *2016 Title II report: National teacher preparation data.* Retrieved from: https://title2.ed.gov/Public/SecReport.aspx

UTeach Institute. (2017). *UTeach vital signs: National data on UTeach program recruitment, retention, and teacher production November.* Unpublished Document.

UTeach Institute. (2019). *UTeach snapshot of UTeach expansion nationwide.* Retrieved from: https://institute.uteach.utexas.edu/

UTeach Institute. (2014–2018). *National UTeach student survey data. Fall 2014—Fall 2018.* Unpublished Data Sets and Documents.

UTeach Institute. (2011). *UTeach elements of success.* Retrieved at https://institute.uteach.utexas.edu/sites/institute.uteach.utexas.edu/files/uteach-elements-of-success-2011.

CHAPTER 3

DERIVATIVES OF THE STANDARD MODEL

Diane E. Bennett and Joanne E. Goodell

In this chapter we share stories of derivatives of the standard UTeach undergraduate program model. In both stories, you will see innovative and creative ways that we tailored the UTeach model for our university and community. It is our hope that you are inspired, enabled, and emboldened by these stories to create the most impactful teacher education programs possible for your particular institution.

PART I. THE CLEVELAND STATE UNIVERSITY STORY: HOW ONE EARLY REPLICATION SITE CHAMPIONED CHANGE

This is a story of how **Cleveland State University (CSU)** transformed its existing four-year undergraduate and two-year post-baccalaureate teaching programs to infuse the high-impact UTeach elements.

CSU is a comprehensive urban university located in the downtown area of Cleveland, Ohio. The campus of more than 85 acres and 40-plus buildings has the largest footprint in downtown Cleveland. Established in 1964, CSU enrolls 16,000 FTE undergraduate and graduate students in 200 major fields of study. The largely-commuter student body is comprised of 55 percent women and 45 percent men. Approximately 27 percent are minorities, of which two-thirds are African American. There are around 1,000 international students. The average age of the student body is 27, and about one-third are in graduate school or law school.

Preparing STEM Teachers: The UTeach Replication Model, pages 31–45.

Building Interest and Securing Funding for Our Replication of UTeach

In October 2004, I attended a National Research Council workshop for the National Science Foundation (NSF) Math/Science Partnerships (MSP) program awardees titled "Teacher Education for Effective Teaching and Learning." I attended a session about the UTeach program presented by Drs. Michael Marder and Mary Ann Rankin, and the vision of what a STEM teacher education program could be had a big impact on me. I knew then that I wanted to bring this exciting model to my university. The question was, how? In 2004, the Bush administration was nearing the end of its first term, the national report "Rising Above the Gathering Storm" (Committee on Prospering in the Global Economy of the 21st Century, 2007) was not yet begun, there was no UTeach Institute, and I was an untenured faculty member in the College of Education. It seemed like the impossible dream.

Then, early in 2007 when I was recently tenured and had just returned from a sabbatical, I received information about the request for proposals to replicate the UTeach program from the National Math and Science Initiative (NMSI). NMSI and the newly-formed UTeach Institute invited "states across the nation to participate in a dramatic initiative to increase the quality and quantity of mathematics, science, and computer science teachers (National Math and Science Initiative, 2007). I immediately remembered my experience at the MSP workshop and the excitement I had felt when I first learned about the UTeach program, and I began recruiting my colleagues across the university to get on board the UTeach train. Our three existing pathways to secondary teacher licensure for undergraduate, post-baccalaureate, and graduate students were based on STEM content degrees, and many of the UTeach classes aligned with classes in our existing pathway. How difficult could this be?

The RFP process was daunting, requiring state-level and university-level support. However, I felt we were well positioned to draw on the strong relationships I had been developing over the previous eight years during which I had led 20 successful grant projects focused on STEM teacher education. I had been very fortunate to have worked with a talented and dedicated grant writer on a number of these most recent successful proposals, and she was eager to jump in on this new project. I assembled the team, consisting of a core group of faculty members from the College of Science and the College of Education, and we ploughed ahead with the preliminary proposal. In that document, we listed 16 people who played a part in its development. From 50 applicants, we were one of 27 selected to submit a full proposal. A number of team members attended the applicant workshop in Austin, Texas in May. In July, we invited external partners we had worked with on other projects to campus to hear about our proposal, and we obtained letters of support from them. One of the key features of this proposal was that the state Board of Regents had to submit the proposal and the Provost had to be the Principal Investigator, demonstrating that we had the support of those offices. In all, 31 key people in three colleges, our partner school districts, and various community

organizations were named in the proposal, we had 17 letters of support, I had secured over $2.4 million in matching funds for the implementation, and it seemed like we were on track for success.

In September we found out that we were not successful. It was devastating to not be awarded the grant given the tremendous momentum that had developed in such a short space of time. However, we were given a lot of good feedback, and encouraged to not give up. Within a year, we were contacted by NMSI and invited to collaborate on two proposals, one to the NSF Robert Noyce scholarship program with a few other universities, and the other to the U.S. Department of Education Improving Teacher Quality (ITQ) program in a large consortium with 11 universities.

The potential partners on the Noyce proposal all dropped out, but we continued on our own and were successful in July 2009, securing funding for our first master teacher, student scholarships and other student supports. We were then invited to join the UTeach Institute, and I made my first of many trips to Austin, Texas in January to join "Cohort 2" of replication sites. The ITQ consortium, coordinated through NMSI, also dwindled, and we were the only university left after two resubmissions and very stringent matching funds requirements that we were able to meet, but others were not. The ITQ funding provided funding for five staff positions, and we provided matching funds from faculty salaries. Additionally, we had significant funding for scholarships from the Ohio Board of Regents through the Choose Ohio First program.

Implementation

Once we were awarded the Noyce funding in 2009, the planning for implementation began in earnest. We had set an aggressive target of awarding six Noyce scholarships per year, for four years. Since Noyce funding was only for the final two years, or for two years of a post-baccalaureate program, we had to implement our program in two years, not the usual four. That meant that we had to plan the new degree maps, provide sample syllabi for all nine UTeach courses, get both internal and state approval, recruit students for Noyce scholarships and the program in general, deal with transition issues for students in our existing programs and be ready to start at the beginning of Fall 2010 with a two-year pathway for transitioning students and post-baccalaureate students. We decided that we would not make any changes to the graduate program in the first few years as we had more than enough to deal with at that point.

As the 100-plus page curriculum proposal wound its way through the internal approval processes, I went to department, college, and university curriculum committee meetings. While we had built considerable support for the proposal, it was not universal. Cutting the existing program from almost 38 credits to 26 (science) or 29 (math) meant that three courses would be cut and field experience hours would decrease. There was strong resistance around one particular course. I dedicated myself to talking with key faculty members before the department

meeting at which they would vote on the proposal, and to my great relief, the proposal passed.

We were ready to begin the program in Fall 2010 after scrambling to get all of the courses and program approved through CSU and the Board of Regents in less than 6 months, lightning speed by higher education standards.

Innovations

In our case, the innovations or changes we made to the standard model were based not only on the need to align with state accreditation requirements but also to maintain some alignment with other programs in the college. The existing program had an 80-hour practicum in the course equivalent to Classroom Interactions, followed by a 15-week half day practicum taken concurrently with the content methods class in the senior year as a preparation for student teaching. A 15-week full time student teaching experience followed (the equivalent of Apprentice Teaching), usually but not always in the following spring. In comparison, the UTeach model had significantly lower field experience hours than our existing program. However, the existing program did not incorporate early field experiences in the introductory courses. We proposed a middle ground, with a reduction in overall hours in field experiences, but an increase in the number of placements from three to five with the addition of Step 1 and 2. We also proposed to make the Project-Based Instruction field experience a minimum of 80 hours, and for the candidates to remain in the same school for student teaching. We named these experiences Apprentice Teaching 1 (AT 1) and Apprentice Teaching 2 (AT 2), to emphasize the fact that the two experiences were linked. The year-long placement provided continuity and enabled candidates to transition from AT 1 to AT 2 and assume the majority of their mentor teacher's duties very early in the semester. In the previous model, the first three to four weeks of student teaching was spent observing and becoming oriented to the students and the culture of the school. With the new model, candidates were almost fully acculturated to the school as part of the AT 1 experience.

Sustaining the Initial Success

Initial recruiting for the first cohort was done by faculty during the summer of 2010. I personally met with around 50 current students, both undergraduate and post-baccalaureate who were in various stages of their program to inform them of the changes and plan for their transition to the CSUteach program.

In summer 2010 we hired our first master teacher, and we started the Fall 2010 semester with 56 students in Step 1 and 31 in Knowing and Learning. In 2011, we added three more staff, including a second master teacher, a dedicated program advisor and internship coordinator (one person split between these two roles), and a professional development and induction coordinator (one person split between these two roles). Students in both undergraduate and post-baccalaureate programs

took the same classes, and we quickly grew that to a steady state of around 100 active students for the next three years. Scholarships were available for both undergraduate and post-baccalaureate students through state and NSF Noyce grants. In 2011 we also received a Noyce grant for our master's degree program to infuse UTeach elements and provide full scholarships for up to 24 students. We have produced 130 graduates in 7 years from the undergraduate and post-baccalaureate program, which is almost double what we had produced in the previous 7 years.

Unfortunately, we have not been able to sustain this initial growth since the grant funding ended. We lost two of the four staff positions and all of the scholarship funding, including the state funding for scholarships, in the same year that our original Noyce grant concluded. Our original advisor no longer works with our students but did remain in the college, and now coordinates our teaching fellow's program that provides paid internships for students across all licensure programs. This was modeled on the internship program she had coordinated that was inspired by UTeach and funded from our ITQ grant.

Faculty in the non-STEM secondary licensure program also adopted some UTeach-inspired innovations, including Step 1 and 2 classes, a focus on inquiry teaching, and the year-long AT1 and AT 2 teaching experiences. All secondary licensure programs now follow a similar structure and are grouped together under the CSUteach banner as CSUteach-STEM and CSUteach-humanities.

Another innovation inspired by the adoption of the UTeach model was the development of a new clinically-based approach for the early childhood, middle childhood and special education programs at CSU (Banks, Jackson, & Harper, 2014). Faculty recognized the importance of early field experiences in the UTeach model and began the curriculum redesign process in 2013.

The number of students in our program has dropped precipitously since that time not only because our grant funding ended but also because of other local, state, and national trends. It is clear that we need to find ways to reverse this trend, and heed the advice of my co-author, given at the end of this chapter if we are to remain a viable program for the short term. Perhaps some of this attrition can be blamed on the cyclical nature of enrollments in teacher education, but we (the faculty) must re-dedicate ourselves to active recruiting of STEM majors, working with the college administration to provide more funding for external marketing and working with internal institutional funding programs to secure more scholarships for CSUteach students and thus ensure a viable future.

PART II. THE UW-RIVER FALLS STORY: CREATING THE FIRST UTEACH-INSPIRED, STAND-ALONE POST-BACCALAUREATE PROGRAM

The University of Wisconsin-River Falls is a regional comprehensive public university within the University of Wisconsin System. Teaching is a primary focus, and the student population consists of approximately 6,100 undergraduates and 400 master's degree students.

How We Came to be Interested in UTeach

In 2012 the new Dean of the College of Education and Professional Studies, Dr. Larry Solberg, shared with me his passion to bring a Robert Noyce Teacher Scholarship program to the University of Wisconsin-River Falls (UWRF) campus. The campus had applied for Noyce funding three years in a row and was denied each time. I was intrigued by this elusive grant and honored to be given the opportunity to lead an effort to bring teacher scholarships to a campus that was founded for the purpose of educating teachers. In looking for some fresh ideas and inspiration, I was reaching-out to Noyce-funded programs around the country. Of special interest were the Wisconsin and Minnesota Noyce programs. Through emails, phone calls, and examples of funded proposals I learned much about what it would take to write a successful Noyce proposal. (For all of this support, I am very grateful.)

The UWRF UTeach "replication" story started with a phone call to Dr. Mike Beeth in May of 2012. During the call, Professor Beeth, the director of an innovative teacher education program at UW-Oshkosh, asked "Have you heard about the UTeach model?" I had not. He graciously explained.

Learning about UTeach

Therefore, I continued with my plan of speaking with successful Noyce PIs by reaching out to Dr. Michael Marder, the Director of the (original) UTeach program. He was kind enough to share his Noyce Scholarship proposal. At that time, we were simply looking for "elements of success" which we could merge into our existing STEM education programs. We did not appreciate UTeach for what we now understand it to be.

The more we learned, the more we became convinced that we wanted to have a UTeach program at the UWRF. Our chair of teacher education and a senior physics faculty member agreed to go to the UTeach Open House to learn more. A very excited team returned. We formed a steering committee of campus stakeholders and spent nearly two years sharing about our ideas for replicating, while seeking input and buy-in. We found the lack of "stars" on the UTeach map in the upper Midwest to be a call for action. We were aware that there were some hurdles to overcome in order to replicate: historical campus programs, strict state licensure requirements, replication requirements, and money.

The strong undergraduate education programs that UWRF deploys are, in part, a result of the fact that the university started as a normal school in 1874 and has continued this commitment to the present day. While we initially saw this as a strength, in order to replicate, we would have needed to fundamentally change what we had built over the years. There was some resistance from the faculty to supporting an effort that would "fix something that wasn't broken."

There was another related, hidden hurdle. UWRF teachers primarily seek employment in Wisconsin and Minnesota, two states which hold some of the highest

teacher standards in the nation. As a result, there were examples of state licensure requirements that conflicted with the UTeach structure. One example was student teaching. The UTeach model called for half-day student teaching for ten weeks; the state of Wisconsin mandated full days for at least 18 weeks. Without the former structure, we could not create a four-year curriculum; without the latter structure, we could not be approved by the state. In fact, we wondered if the lack of UTeach replications in the Midwest was in part due to the existence of highly-regulated programs.

Examples of lower hurdles were questions of how to refund students for the cost of the Step 1 course since our students pay a flat fee for up to 18 credits, how to embed required environmental education coursework, and how to provide the state-mandated content area literacy course.

With our creative team and the invaluable support of Dean Solberg and the Wisconsin Department of Public Instruction (DPI), we were starting to find ways around each of these hurdles. The one hurdle that persisted was money.

We were seeking money to enable the cost of a five-year replication during a period of historic tightening of our state's education budget and layoffs at the university. We applied for money from the university, the state, regional foundations, and the UTeach Institute. These grant proposals provided opportunities for our team to do the important work of wrestling with what we wanted to do. We did receive some critical seed money, but not nearly enough for a replication effort. I started working on our next attempt at Noyce funding, admittedly without a clear project goal.

The UWRF Dilemma and the Paradigm Shift

Our journey started with seeking the Noyce Scholarship grant, but then our attempt to create the UTeach program that we had grown to deeply appreciate was starting to hinge on Noyce *Capacity Building* funding. With a hopeful proposal draft in hand, I asked our then director of grants and research to join me on a Noyce webinar. It was during that webinar that our plans changed dramatically. We learned that our undergraduate teacher licensing pathways did not qualify for the Noyce program because they were based on degrees in secondary education rather than in the content disciplines. (One reason for multiple Noyce proposal rejections in the university's history became suddenly clear.) Getting a Noyce grant meant either profoundly changing our long-standing undergraduate teacher education programs or creating a second pathway to STEM licensure. We chose the second path. That path meant a departure from the standard UTeach replication model.

After much discussion and with the "supported autonomy" provided by the UTeach Institute, it was decided that we would create a very different program than our undergraduate programs, for a very different group of future STEM teachers, with a different type of support from the UTeach Institute. The program, which eventually came to be called "STEM*teach*," was created for an impor-

tant and underserved "niche audience": STEM career changers and recent STEM graduates. This was a critical decision as it enabled us to create a program that did not directly compete with our existing education programs, but more importantly, created an attractive way for uniquely qualified individuals to become math, science, and computer science teachers. Not everyone believed that this would be possible.

The key to reaching the career changer audience was to design a program FOR this audience. This audience did not want to be treated like 18- year old undergraduate students. Having walked along a very non-linear career path myself, I was determined to create a program that not only accepted the uniqueness and past experiences of those interested in a new career in teaching but celebrated these things. In fact, our admissions process used a novel strategy that we developed in partnership with the Wisconsin DPI, called "equivalency standards." This allowed us to consider not only coursework and degrees as a means to qualify for licensures, but also work and volunteer experience in the content areas.

The fiscal climate limited us to a program that was financially self-sustaining, a type of programming we call "program revenue." That led to great autonomy (and responsibility) in all things related to the program, from marketing and recruiting to program costs to job placement and induction support. Budget, staffing, and administrative constraints also "forced" us to use a cohort model. That had wonderful broader impacts. Instead of our nontraditional students feeling out of place or disconnected, they would spend most of one year together in a small, supportive group. Career changers often have more factors to balance when considering a career in teaching than a typical undergraduate student. We have seen repeatedly that it's hard enough to boldly step out and make a career change, harder still if there is a lack of support from family, friends, or an educational institution. For STEM*teach*, our ability to provide quality induction support has much to do with our cohort structure, our master teachers, and our recruiting strategies.

It was indeed a constraint not to "hurt" our excellent undergraduate programming that led us to create something that was truly PARALLEL, meaning not intersecting with or interfering with, our historical programming. This was accomplished by, again, considering our audience: where they are and where they seek to be. With this in mind, STEM*teach* was designed with intentional and well-defined starting and ending points. The starting point is anyone with a STEM degree and the ending point is licensing in up to ten areas, with experience teaching in elementary through high school classrooms, informal STEM education experiences, and the option to complete a master's degree in education.

Bringing STEMteach to Life

The creation of STEM*teach* was made possible by a two-year Noyce Capacity Building Grant. The ambitious first year activities concluded with the approving of the new STEM*teach* initial licensure with optional master's degree program by both UWRF and the Wisconsin DPI, the hiring of the program's master teacher,

and the welcoming of the program's first cohort of teacher candidates. The eight candidates admitted during the development year were a testament to the need for such a program, the effective recruiting activities, the well-crafted marketing campaign, and the dedicated STEM*teach* steering committee.

During the course of designing the program and its nine new graduate courses the STEM*teach* steering committee met regularly with a group of UWRF representatives (the internal advisory group) as well as a team representing the STEM*teach* partner schools (the external advisory group). Both groups provided valuable input and support. The external advisory group made available school district contacts and ideas for creative field experiences, including working with the River Falls School Districts' elementary math and reading summer remediation program for the candidates' first field experiences.

A necessary element to the success of the project was a two-year resources and consulting contract with the UTeach Institute (funded by Dean Solberg). This provided the ability to use the extensive UTeach Institute online resources, attend course workshops, and have prompt one-on-one consultation with a UTeach Institute consultant (Dr. Amy Moreland). Dr. Moreland provided email and phone assistance in addition to in-person course development support before and/or after each of the course workshops that we attended.

One of our first major decisions was regarding the program structure. We realized that the most attractive and logical structure would either be a one-year or a two-year program. We considered the merits of a two-year program that students would complete part time and/or via night classes. Our strong desire for high-quality field experiences kept leading us back to the need to have a rigorous "quit your day job" program. Eventually we realized that there were inherent advantages to a one-year licensure program that began and ended in early summer. For example, teacher candidates would earn licensure with only one year of tuition costs and only one year of lost income. After completing the licensure program, they would have the summer to transition into a new teaching position. While we were able to create a quality one-year licensure program, we agreed that the master's degree would (and should) take longer to complete.

In year two of the capacity building grant, we turned our focus to refining the master's degree. Again, we learned a lot from key stakeholders, including cohort one, interested candidates, our external advisory team, and our professional network. We eventually created three *optional* master's degree completion pathways that are each flexible enough to be completed while working in the area or at a distance.

Once we had a strong foundation for our program, we worked on integration of computer science into the curriculum. (Cooper, Gover, Guzdial, & Simon, 2014; Scott & Bundy, 2015). This innovation was prompted by findings of a survey we sent to Wisconsin and Minnesota principals that showed a strong need for computer science teachers in our region. Also motivating was the President's call for "Computer Science for All," an initiative to "empower a generation of American

students with the computer science skills they need to thrive in a digital economy" (Smith, 2016). Among the highlights of our effort was the approval from the WI DPI to add CS licensure to the STEM*teach* licensure array.

As this section concludes, let me take a moment to emphasize the importance of marketing, recruiting, and "closing the deal." These three points may sound harsh, but I mean them in a very personal way. The marketing of STEM*teach* involves all the decisions that have been made to cause people to choose the program: product (cohort model, UTeach curriculum, face-to-face delivery, one year, master teacher, STEM-infused), place (UWRF), price (tuition rate, scholarships, referral to veterans programs like Troops to Teachers), and promotion (marketing communications). The marketing communications that the program has used to highlight these features are shown across the top of funnel in Figure 3.1. Recruiting can be defined as the time-consuming, personal, and strategic relationship building that we control firsthand. These activities encourage prospects to enroll in our program and are shown along the lower right-hand side. Loyalty and advocacy elements are shown at the lower left. These can lead to word of mouth referrals, and therefore serve as a powerful means of marketing communication, further filling the funnel.

In the program design work that I describe above, recruiting and marketing were key elements, for which we allocated about 10% of our budget. During this work, I was recruiting for the first and second cohorts at the same time as overseeing the building of the program. I was speaking to potential applicants via email, on the phone, and in person. As a result, we were able to build the program in a way that was responsive to our customer's needs!

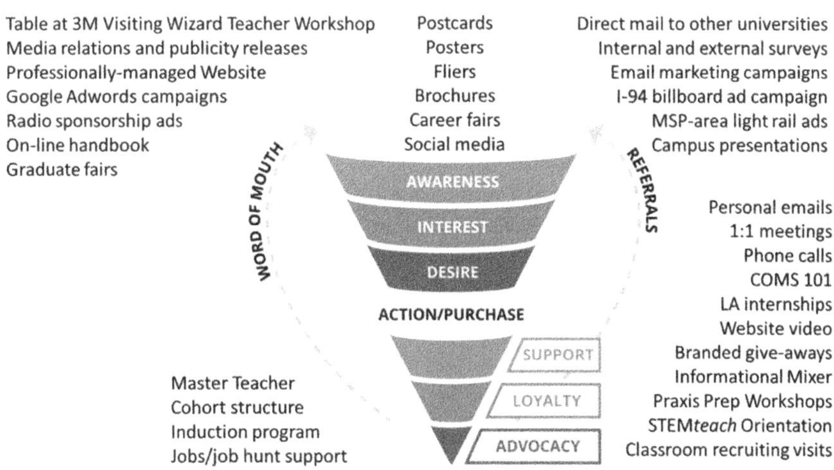

FIGURE 3.1 STEMteach marketing, recruiting, and advocacy activities that have proven to be successful.

STEM*teach* was built and continues to operate in much the same way as a small business. This means that letting people know about your "product" (the program) and helping them to discover whether your program will work for them is an on-going process. This is probably just as important as the program itself! The STEM*teach* audience is career changers and recent graduates. This audience does not NEED to choose a career in teaching, they do not NEED STEM*teach*. They choose STE*Mteach* because they want to make a difference in a new way, to give back, to mentor the next generation, to continue to learn. They WANT to become teachers. We need to help them to find our program. Without a dedication to high-quality recruiting, marketing, and nurturing of our audience, the program would soon cease to exist.

The Teacher Preparation Program

The STEM*teach* program at UWRF is an intensive program of 30 credit hours leading to initial teacher certification in the state of Wisconsin with the option of completing a Master of Science in Education degree with an additional six to ten credits of coursework. The compact one-year initial certification pathway is designed to be a rigorous and efficient route to preparing well-qualified science and math teachers using a UTeach-inspired curriculum that is inquiry-based and STEM-infused (National Research Council, 2010). The format of the program requires that candidates participate in challenging graduate coursework and extensive field preparation that promote an understanding and synthesis of theory and practice and a deep understanding of the challenges and rewards associated with teaching in diverse environments: urban/rural/suburban, high-needs/high performing schools, traditional, informal, and summer enrichment/summer remediation. Instruction occurs on campus as well as in partner schools.

Standard UTeach course content was modified in some of the courses, and a content literacy course was added, to ensure that the curriculum meets all of Wisconsin's requirements for licensure in mathematics and the sciences.[1, 2] You are encouraged to read Dr. Rachelle Haroldson's chapter in this volume about STEM*teach*'s "Teaching Steps 1/ 2 Combo." This course, developed with support from UTeach's master teachers, was one way that we were able to reduce time to licensure yet retain rigor.

In another example, we chose to deliver the Perspectives on Science and Mathematics course in a hybrid format during the January term, between the fall and spring terms. This online plus in-person format meant that our future teachers would be exposed to both learning and teaching in an online format. Moreover, we were able to have the students complete a course during the winter term, lessening their course load during the other terms, while reducing the driving required during a snowy time of year and giving them more scheduling freedom during

[1] http://docs.legis.wisconsin.gov/code/admin_code/pi/34

[2] http://docs.legis.wisconsin.gov/statutes/statutes/118/19

this time in which children are away from school and families are traveling. This course was another welcome outcome of the freedom that our derivative approach gave us to creatively design a program for our target audience.

Clinical Program

UWRF STEM*teach* is centered-around intentional and well-supported field experiences which utilize co-teaching. The one-year initial certification curriculum is shown in Table 3.1. Candidates take graduate-level courses with a small cohort (maximum 20).

Optional Master's Degree Pathways

There are three master's degree pathways which candidates can use to complete this optional degree: thesis, research paper, or capstone experience. Candidates are encouraged to follow the capstone or research pathway and select the one-year Professional Development Practicum course which was designed for STEM*teach*. This course provides induction support through weekly reflections, video discussions, and professional development planning. This course is intended to provide a way to complete the degree while being supported in the first year of teaching. (The 3-credit practicum course spans the fall and spring terms.)

New teachers are encouraged to tailor the master's degree to the needs of the district in which they are teaching or to their desired areas of professional enrichment. Some may be interested in exploring advanced topics in education or in their content area and others may desire to deepen their education or content area knowledge through research. Whatever path is chosen, the completion of the master's degree becomes a natural part of their professional development plan and eases and facilitates their transition into a leadership role in their school district. In summary, we believe that the optional master's degree enables our new STEM

TABLE 3.1, The STEMteach initial certification curriculum

STEMteach Course	Credits	Term
UTCH 701: STEMteach: Steps 1 and 2	3	Summer
UTCH 702: Knowing and Learning in Math and Science	3	Summer
UTCH 703: Classroom Interactions	3	Fall
UTCH 704: Project-based Instruction	3	Fall
UTCH 705: STEM Content Area Literacy	3	Fall
UTCH 706: Functions and Modeling	3	Fall
UTCH 707: Perspectives on Science and Mathematics	3	January
UTCH 708: Apprentice Teaching Seminar	3	Spring
UTCH 709: Apprentice Teaching for Secondary and Middle Grades	3	Spring

teachers to get into the classroom faster and leverage their real-world teaching experience to enable a more meaningful completion of the master's degree.

Part II Conclusion

The design of our UTeach-inspired, one-year post-baccalaureate program was based on the understanding that UTeach isn't about the replication *model* per se, but more about outcomes. Our unique version of the UTeach model combines the UTeach resources, support structures, and UTeach elements of success[3] with our local university needs, state regulations, and our local experiences as professional educators. UWRF has been in the business of preparing teachers for 140 years, but the vast majority of our teachers have been prepared as undergraduates. With the STEM*teach* program, UWRF instructors and mentor teachers are witnessing an exciting shift in the maturity, professionalism, and enthusiasm for teaching that our non-traditional students have. This shift is also evident in the extent and quality of the professional activity, such as presenting talks, serving on panels, submitting papers, and attending workshops and conferences, that the STEM*teach* candidates and STEM*teach* alumni have undertaken. By creating a UTeach-based program that specifically encourages and enables college graduates and career changers to become teachers, we have embraced the UTeach emphasis on excellent content preparation. STEM*teach* candidates must have a degree in a STEM discipline, but they typically also have life and professional experiences that positively impact their teaching. We look forward to continuing to contribute to the success of the UTeach model and its expansion into the talented pool of recent STEM graduates and career changers.

THE LESSONS OUR STORIES REVEALED

Taken together, our stories provide a number of key takeaways for readers, especially those involved in reforming teacher education or those who want to work towards significantly increasing the STEM teaching workforce. In Table 3.2 below, we summarize our shared experiences with a few cautionary notes.

It is our hope that these stories of UTeach derivatives inspire you to honestly and creatively consider how you can bring the most value to your university and community through excellence in STEM education as UWRF has done via STEM*teach* and CSU has done with a two-year post-baccalaureate program. Further inspiration can be found on the UTeach Institute website where you will find links to specific program websites. Also consider attending the annual conference held at the University of Texas at Austin in May every year. You will be inspired by the energy and achievements of the UTeach replication sites' faculty and staff.

[3] https://institute.uteach.utexas.edu/uteach-elements-success

TABLE 3.2. Lessons Learned and Key Takeaways

Opportunity	Keep in Mind
Post-baccalaureate programs can tailor programming to STEM professionals who are trying to re-enter the workforce and have an interest in teaching	• There may be limited access to scholarships, especially for participants in non-degree granting programs • Participation may not allow participants to work, thereby increasing the financial burden beyond tuition costs • Marketing and recruiting efforts can never rest
Grant funding is extremely helpful to enable a new program to get started	• Sustaining the initial growth after grant funding has ended may be more difficult than anticipated. The journey is never over, and continued effort is required to sustain the initial success. • Creating a new program requires a persistent commitment to reach the prize. • A champion who has the political capital necessary to smooth the way is critical during the building phase and continues to be important in an established program. • Program faculty leaders (co-directors) must also possess grit and be able to work around roadblocks as they occur.
There are different ways to structure the UTeach curriculum and field experiences to achieve university buy-in	• Examples of possible models include a one-year master's, a one-year post-bac, a two-year post-bac and a four-year undergraduate program
Modifications to the UTeach model can be replicated as well	• Start by articulating the elements that you wish to have in your "derivative" program and reach out to others via email, phone calls, or campus visits. Ask for copies of grant proposals and program documents. The UTeach community is here to help.

REFERENCES

Banks, T., Jackson, D., & Harper, B. (2014). Responding to the call to prepare highly effective teacher candidates in the United States: The curriculum redesign effort in advancing teacher education. *Higher Education Studies, 4*(2), 9–18.

Committee on Prospering in the Global Economy of the 21st Century. (2007). *Rising above the gathering storm: Energizing and employing America for a brighter economic future*. Washington, DC: National Academy of Sciences, National Academy of Engineering, Institute of Medicine.

Cooper, S., Gover, S., Guzdial, M., & Simon, B. (2014). A future for computing education research. *Communications of the ACM, 57*(11), 34–36.

National Math and Science Initiative. (2007). *UTeach request for proposals*. Dallas, TX: National Math and Science Initiative.

National Research Council. (2010). *Preparing teachers: Building evidence for sound policy* (The National Academies Press Ed.). Washington, DC: Committee on the Study of Teacher Preparation Programs in the United States, Center for Education. Division of Behavioral and Social Sciences and Education.

Scott, J., & Bundy, A. (2015). Creating a new generation of computational thinkers: Experiences with a successful school program in Scotland. *Communications of the ACM, 58*(12), 37–40.

Smith, M. (2016). Computer science for all. *The White House Blog*. Retrieved from: https://obamawhitehouse.archives.gov/blog/2016/01/30/computer-science-all

CHAPTER 4

THE UTEACH INTERNSHIP PROGRAM

Serving the Community and Promoting STEM Education

Shelly R. Rodriguez and Amy L. Moreland

As the percentage of science, technology, engineering, and mathematics (STEM) careers increases, the need for qualified teachers in these fields continues to grow. To help meet this demand, The UTeach internship program partners with STEM-focused organizations, such as museums, after school clubs, makerspaces, and local schools to provide a workforce of part-time STEM preservice teachers. UTeach interns participate in activities such as supporting individual or small groups of students in STEM fields, researching lesson ideas, working on curriculum projects, and assisting classroom teachers. This chapter offers an overview of the UTeach internship program and provides a compilation of personal vignettes from a variety of interns and community partners. The chapter also discusses emerging trends from program survey data as well as program challenges and next steps.

INTRODUCTION

As the percentage of science, technology, engineering, and mathematics (STEM) careers increases, the need for qualified teachers in these fields continues to grow (Department of Education, n.d.). Thousands are teaching out of field in STEM

Preparing STEM Teachers: The UTeach Replication Model, pages 47–60.

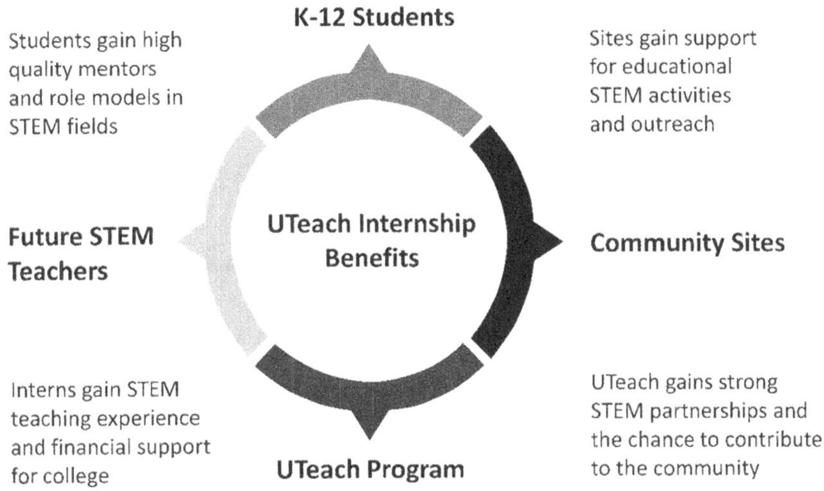

FIGURE 4.1. Multiple Benefits of the UTeach Internship Program

classrooms, and many do not have a standard teaching certificate (Texas Education Agency [TEA], 2018). These trends illustrate the continued need for high-quality STEM teacher preparation. Founded in 1997, UTeach is a nationally recognized secondary STEM teacher preparation program at The University of Texas that helps infuse the education system with highly qualified teachers who can educate the scientists and engineers of tomorrow.

The UTeach internship program partners with STEM-focused organizations, such as museums, after school clubs, makerspaces, and local schools to help fill the growing demand for high-quality educators by providing a workforce of part-time STEM preservice teachers. UTeach interns participate in activities such as supporting individual or small groups of students in STEM fields, researching lesson ideas, working on curriculum projects, and assisting classroom teachers. As Figure 4.1 indicates, the internship program not only supports UTeach interns, but also supplies mentors for K–12 students, provides no-cost support for the outreach efforts of internship sites, and fosters strong partnerships with the community.

This chapter offers an overview of the UTeach internship program and provides a compilation of personal vignettes from a variety of interns and community partners. The chapter also includes emerging trends from program survey data as well as a discussion of program challenges and next steps.

PROGRAM DEVELOPMENT

The UTeach internship program started in the fall of 1998 during the second year of UTeach program implementation. The rationale for the internship initiative

was a pragmatic one. The internships were seen as both a recruiting tool for the UTeach program as well as an educational enhancement. Internships were intended to supplement the income of UTeach interns while helping them develop additional instructional skills and providing high-quality STEM educational experiences for the community at large. Funds were obtained from a variety of private foundations that were interested in supporting both UTeach students and community organizations. UTeach used this private funding to pay interns and support the role of the internship coordinator.

Initial partner sites were obtained through cold calls to nonprofit organizations listed in an Austin guide to informal science opportunities. The organizations were contacted and asked if they would like future mathematics and science teachers to work as paid volunteers. The nonprofit organizations were pleased to participate and promised that the UTeach interns would only be asked to engage in activities relevant to their professional development, such as working with students, communicating with the public, or planning curriculum.

The notion of learning through real-world experience provides a rationale for the clinical field requirements in teacher preparation. This concept also underlies the design of the internship program. The reasoning behind this type of experiential preparation goes back to the days of Dewey. He made the case that teachers are learners and need experiences that support the construction of their learning (Dewey, 1938). It has also been argued that teachers need the opportunity to participate in authentic settings because educating others requires a complex set of skills that emerge in non-routine situations. Therefore, prospective teachers need the chance to encounter and work through these problems in authentic settings (Darling-Hammond, Gendler, & Wise, 1990). In this way, the interactions with STEM content, K–12 students, STEM professionals, and the public provided by the internship experience serve to better prepare preservice teachers for the complexities of the educational environment.

HOW IT WORKS

UTeach offers paid internships with local schools or nonprofit agencies which are associated with the STEM community. UTeach interns are paid employees who work in educational non-profit organizations to gain relevant experience. Internships are considered a part-time job and the work that interns do is not associated with any formal coursework or academic credit. UTeach internships are open to UTeach students at any point in the program. Thus, the program serves to both recruit and support future STEM teachers by 1) providing financial support while giving them additional experience working with K–12 students, and 2) helping them form connections to the STEM education community at large. Interns also gain valuable field experience while supporting STEM education for K–12 students, especially in underserved areas.

The UTeach internship program is coordinated by a master teacher who oversees program logistics and helps to foster a positive experience for both interns

and the sites they work with. The subsequent sections will share some of the experiences from sites, followed by the experiences of interns. All of the forms and resources discussed in this chapter can be accessed via the UTeach Internship website: https://austin.uteach.utexas.edu/internships

For Internship Sites

The internship coordinator develops and maintains relationships with non-profit, STEM-focused organizations that serve as internship sites. These connections are established through relationships with mentor teachers, school districts, community partners, and UTeach graduates. Figure 4.2 depicts the internship activities for partner sites. Pre-internship activities take place prior to the start of the semester, internship activities occur during the semester, and post-internship activities occur at the close of the semester.

To get started, interested organizations contact UTeach with a request to host interns. These sites are then provided with an overview of the program and an application. The evaluation of an application is based on alignment with the UTeach mission, which includes internship experiences that help students develop the content and pedagogical expertise needed to effectively teach STEM subjects to a diverse student population. Factors such as the geographic location of the site, the content area, the student population served, and the funding available are also considered. These factors help ensure that both the community and the students are equitably served. Once a site application is approved, each site is visited by the internship program coordinator and a memorandum of understanding (MOU) is drawn up.

Sites must assign a site supervisor who will have primary responsibility for creating an online job posting, providing on-site supervision/training, keeping a local record of intern hours, and verifying timesheets. Internship sites also have the option of attending the Internship Choice Fair. This fair is a three-hour event held on the university campus. Like a job fair, the purpose of the event is to allow sites to meet and greet possible interns, build rapport, and share information about their organization.

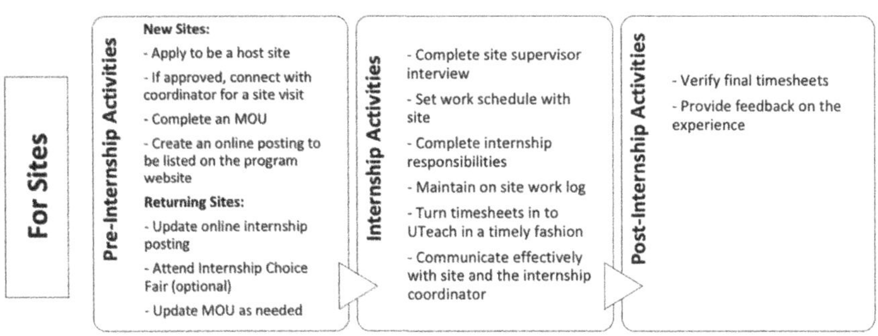

FIGURE 4.2. Internship Activities for Participating Sites

For UTeach Interns

For UTeach students, the internship coordinator has the role of recruiter, advisor, and mentor throughout their internship experience. Figure 4.3 shows the sequence of activities for students participating in the internship program.

At the beginning of each semester, the internship coordinator informs all students about the internship program and the list of possible sites, as found on the UTeach internship website. The website lists a description of the position, the desired intern candidate qualities, location, and hours for each posted position. Participating students select a site where they would like to work by completing an online request form.

Because students choose their internship location, there is no guarantee that a specific site will get an intern each semester. However, if a student is successful at an internship placement, then they are more likely to return. This can establish a pipeline of interns regularly flowing to that site. Once students and sites are matched, they are notified via email to set up a working schedule and complete a site supervisor interview form so that clear roles and responsibilities are established.

Internships are designed to provide UTeach students with the knowledge and skills needed for secondary classroom teaching in Math, Science, or other STEM areas and to give them experience working directly with students. Therefore, intern duties are kept in line with this mission. Common intern roles include working with an individual or small groups of students in STEM fields, researching lesson ideas for particular topics, assisting teachers during a lesson, and setting up lab materials for future lessons. Interns work with an organization for a semester and have the option to return in following semesters. At the end of each semester, interns complete a survey about their experience and are asked to complete a thank you card for their site supervisor and one for the program donors.

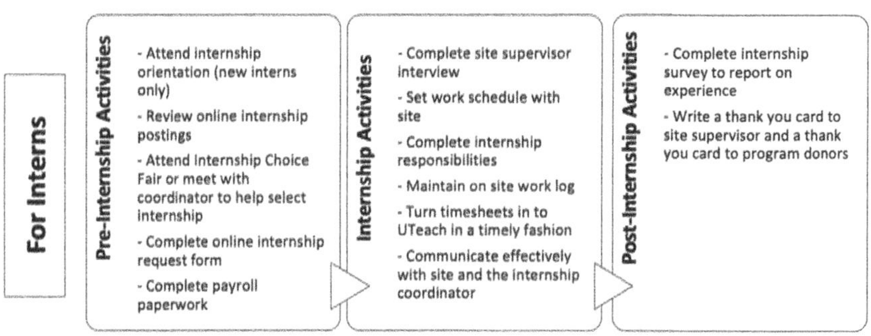

FIGURE 4.3. Internship Activities for Participating Interns

A WIN-WIN: STORIES FROM SITES AND INTERNS

Site Supervisor Stories

Internships have proven to be beneficial to both the interns and the sites where they work. This section provides stories from site supervisors about their experience with the internship program. The stories provided are composed fully of the words of the contributors.

Kathy—Girlstart. Senior Program Coordinator

> By interning through the UTeach program, students get to build skills and increase their confidence as a future teacher, while meeting their need for a job. These interns quickly become valuable team assets and leave lasting impacts where they work.

Girlstarts's mission is to increase girls' interest and engagement in STEM through innovative, nationally-recognized informal STEM education programs. Girlstart provides a year-round, intensive suite of STEM education programs for K–8 girls. Our core programs foster STEM skills development, an understanding of the importance of STEM as a way to solve the world's major problems, as well as an interest in STEM careers.

As the Senior Program Coordinator at Girlstart, I plan and coordinate partnerships with elementary campuses in central Texas to serve with Girlstart *After School*, organize logistics for Girlstart's Summer Camps across the country, hire and train STEM CREW (Creative, Resourceful, Empowered Women) to deliver programs, and create curriculum for the full suite of programs. Interns from UTeach are part of Girlstart's STEM CREW team that help deliver all of our programs in the community. With the classroom management foundation from UTeach, these interns are better prepared to work with diverse student populations, to manage their time while delivering lessons, and to engage students of all ages and their families at community STEM programs.

UTeach interns are a very positive addition to our team. For example, many students in Girlstart *After School* are English language learners. An intern leading the program at a local elementary school took note of a specific student struggling with a language barrier. She made the effort to translate key vocabulary words from each lesson, researched strategies to build the participant's language skills, and created an atmosphere of support for all of the girls in that program to help each other. This intern led the program at that campus for multiple semesters, providing consistency in that environment and served as an excellent role model for all of the elementary participants.

As a site supervisor, the biggest indicator of internship success is in the retention rate of interns and increased word-of-mouth recruitment through their testimonies. Having multiple UTeach interns work at Girlstart for many semesters of their college careers and also recruit their friends for internships continually validates the success of this partnership. As they graduate and become classroom

educators, our interns have gained valuable skills from the UTeach program and relevant experience from their internship which means they are better prepared to make a lasting impact on their students.

Paul—Magellan International School. Innovation Director and Founding Director of the i.lab for Design + Making.

> Educational innovation does not happen in a vacuum, and to successfully reach the whole child, every stakeholder plays an important role. The UTeach internship provides an invaluable bridge between the latest pedagogical approaches in education and best practices from current practitioners.

The Magellan International School provides the community with an inclusive, inquiry-based approach to learning in an immersive, multi-language environment. Creative thinking, passion, and purposeful action are the cornerstones of the student experience, which is founded on the principles of constructivism and culturally responsive education. Making is a constructivist practice that allows students to learn about content, tools, and materials through the creation of personally meaningful products. The UTeach internship program plays an integral part in the school's mission to provide maker-centered learning for its students as well as its ongoing effort to form lasting partnerships with the Austin area learning community. Since the school's makerspace serves every grade level, and maker-centered learning is taking place in classrooms as well, interns provide an additional layer of support for both students and staff.

As a site supervisor, one of the biggest challenges has been helping interns adjust to the high tempo of maker-centric projects that occur throughout the semester across a wide range of grade levels. While UTeach interns typically have prior experiences around making, the sheer volume and variety of projects, ranging from tinkering with linkages to digital fabrication on Computer Numerical Controlled (CNC) equipment, can be daunting. In addition to formal training, one of the ways that we've successfully addressed this is to provide interns with time, support, and materials to develop personally meaningful projects during their time at the i.lab. This affords the interns the opportunity to both build upon prior experiences and interests while encouraging them to try new ideas, tools, and approaches, which they can apply later when working with students.

Having interns of different genders, ethnicities, and backgrounds on campus has had a remarkable impact on how our students view of who can be a maker. For instance, at the beginning of the semester, one intern expressed difficulty in connecting making with her Japanese heritage. However, while brainstorming on ideas that could be developed into a personal project in the lab, she had a breakthrough based on her interest in animation and fond memories of the bento boxes (a single-portion meal in a box) her mother prepared when she was younger. Throughout the internship, she began to incorporate iterative design, rapid prototyping, and digital fabrication across several different projects. More importantly,

she shared this process with students, who were able to see the whole experience, including the successes and challenges that were part of her journey as a maker. By having a positive role model in the makerspace, Magellan students were able to further their understanding of what it means to express yourself through making.

Intern Stories

The previous section shared insights about the internship program from two site supervisors. The next section will share the perspective of a couple of interns working at educational nonprofits and local schools. As with the section above, the words provided are those of the interns.

Megan—River City Youth Foundation (RCYF)

Megan is a senior from Richardson, Texas She is majoring in Molecular Biology and hopes to pursue a career in forensic science or teaching. She began her work at RCYF as an intern in 2015.

> UTeach internships give interns valuable teaching experience while earning an income. They help participating organizations by providing additional staff and support services to their communities. I have seen so many lives changed at River City because of the work we are able to do through internships that would not have happened otherwise.

River City Youth Foundation, located in southeast Austin, is a community center that provides services to underprivileged families in the community. We offer tutoring throughout the week for students in all subjects during both semesters of the school year. We also offer programs on Fridays and Saturdays that teach students different life skills such as coding, fitness, cooking, and college preparatory sessions. Many of our students do not have computers or Wi-Fi at home, so they often come to RCYF to do homework or assignments. This year, many high school seniors did their college applications onsite. Almost all of the parents in the community work long hours. Their kids can come to RCYF after school or be dropped off on Saturdays, where they participate in educational programming and are provided food. [Internship Site] also helps parents apply for jobs and become more adept at using technology through its computer literacy programs. The UTeach interns provide program support for all of these efforts. Whether it be assisting in lessons, tutoring students, or cooking for students. Since the internships are funded by outside donors, we can contribute a lot to River City without being a drain on finances.

I have been working as an intern at River City for three years. This internship has taught me a lot about classroom management techniques and connecting with a wide variety of students. I have had to adapt to common teaching hurdles, such as last-minute lesson plan modifications, material shortages, and time constraints. Working at a non-profit in an underprivileged area comes with an unexpected set

of challenges. Over the years I have dealt with a lack of resources, exhausted management, and students facing multiple crises both inside and outside of school. However, it has been exciting to see the students' trust in me increase over the years and to watch the impact we have had on each other's lives. I have helped students do everything from setting up Pokémon Go accounts to navigating family situations and applying to college.

As an example, one of my regular students is a senior in high school this year, which means I met her as a freshman. During the years, she has had to deal with many personal and educational hardships. She has a hard time trusting people and asking for help, but because I have been around for so long, she and I have become close. This summer I helped her with personal challenges and checked in with her about her grades and studying for the SAT. She told me she was grateful to have met me as the 'older sister' she never had! This semester she got accepted into college and is on her way to pursuing her dream of becoming a speech pathologist, and that was really exciting to see!

Franco—Breakthrough

Franco is a third-year Biology student pursuing a composite science teaching certification. His dream is to teach in his hometown of Austin, Texas

> UTeach internships give students the chance to gain real-world experience and build teaching skills that cannot be learned in a college classroom. They also provide local educational organizations future educators who are eager to support their mission.

For the past three semesters, I have worked for Breakthrough, a nonprofit focused on helping low-income, first-generation students graduate from college. As a UTeach intern, my duties include supporting at various middle and high schools after-school programs throughout the week where I tutor students and help them with their homework and any other assignments they need to complete. I have provided support to Saturday programs by helping students complete a three-weekend project they work on throughout each semester. During the summer I was a teaching fellow who wrote and taught math and science curriculum. I have learned a significant amount from the Breakthrough staff. Working with students and their families remarkably shaped my teaching philosophy and focus on the classroom to be on individualized instruction.

At Breakthrough, I have the chance to serve as a role model while challenging students to develop their long-term goals and helping them to identify the resources needed to help them succeed. This past summer, I had the wonderful opportunity to be a Breakthrough teaching fellow at the UT Austin and teach math and science classes to middle school students. I shared with them my personal journey in reaching UT. Like many of the Breakthrough students, I grew up in a single-parent, low-income household. Being the first in my family to go to college came with several obstacles that I was able to overcome with the support of my teachers, peers, and family. After sharing, many of my students asked ques-

tions about how they might succeed in following their own dreams. As a first-generation college student, I am grateful for the opportunity to share my personal story because it helps students recognize that organizations like Breakthrough can provide the support they need to pursue their future dreams.

This internship has provided me the fantastic opportunity to grow as a teacher and person and to build my confidence in the classroom. Working with a passionate staff who are eager to support students succeed has motivated me to pursue a career as a teacher. Working with the students and their parents has also helped me to recognize the vital role that educators play in the community.

BY THE NUMBERS

The stories above paint a picture of the personal impact that the internship program has had on sites and interns. The upcoming section provides a snapshot of experiences reported by interns as they respond to our annual internship survey. Included in this section are emerging trends from the 2017–2018 internship program survey data report.

A 17-question online program survey was completed by 127 out of 175 active UTeach interns across 27 STEM-focused sites, resulting in a 73% response rate. The survey contained a combination of Likert-scale and open-ended items focusing on placements, nature of the work, teaching as a profession, and the importance of financial support. Based on survey responses, UTeach interns served over 10,150 K–12 students during the 2017–2018 academic year.

According to survey results, internships provide students with opportunities, facilitated and paid for by the program, to earn income and gain relevant work experience. They also act as a tool for program recruitment and retention. Out of 127 respondents, 31% were new to their internship program and 69% were returning interns. Of those returning, a full 72% chose to continue with their same placement. Students said they found value continuing their relationship with the mentors and seeing student growth in the succeeding year. Many enjoyed the schedule flexibility and familiarity offered by their recurring commitment to a site.

Regarding the financial support they received as an intern, 96% of survey respondents said this was very important. Because most students must work while in school, having the opportunity to work in positions that are relevant to STEM teaching is invaluable. An intern can "do the work that [they] enjoy and get paid for it" without having another job on the side. The financial resources provided by their internship work keeps students from having to engage in unrelated, minimum wage jobs or take out additional student loans. Intern survey responses show that getting paid for work while honing teaching skills within their content area is a substantial benefit for these college students.

It should be noted that a majority of interns reported that the internship experience, no matter the locale, benefited them as future teachers. Owing to the constant interaction with numerous students, UTeach interns reported feeling more comfortable and at ease around the classroom while building confidence in their

teaching skills. Other future-teacher affirmations included networking opportunities with other educators, developing classroom management skills, and practicing public speaking. This is supported by the sentiments of program graduates who have mentioned the importance of the internship experience in their work. For example, one recent graduate said, "During the internship, I had such an amazing time building my teamwork skills, classroom management, and experiences as a teacher." Another teacher said, "This internship helped me develop lessons that played a part in my love of curriculum design. I count the internship as one of my favorite undergraduate experiences, it has equipped me for years to come."

Lastly, to understand the nature of the daily work that the interns do, we asked them to briefly describe their jobs, especially focusing on the STEM activities incorporated into their internship. A review of this survey item revealed several overall themes, such as teaching STEM content at grade level, one-on-one tutoring, setting up labs and organizing laboratory materials, working within a makerspace environment, and conducting STEM activities in after-school, community-based, or weekend settings. By offering internships at a wide variety of STEM-focused sites, interns have the chance to select an experience relevant to the grade level and subject they want to teach, as well as an experience that provides support for the specific instructional skills they hope to develop.

DISCUSSION

This chapter shows the benefits of the internship program to UTeach interns, the sites they work with, the K–12 students they serve, and the UTeach program more broadly. The stories shared illustrate that work with the internship program has left a lasting impression on the UTeach interns and has the potential to impact their teaching practice for years to come. Our interns describe gaining valuable STEM knowledge and skills while learning how to develop positive relationships with students from all walks of life. Interns indicate that they have learned how to be responsible employees and how to translate the skills learned in their UTeach classes to a real-world setting. Results from the UTeach internship survey highlight the importance of financial support, schedule flexibility, and working in positions that are relevant for them as future teachers. As reported in the intern stories, many interns have continued with their same placement, allowing them to witness student growth year after year; a value-add difficult to replicate via isolated field experiences.

Challenges

The UTeach Internship program at The University of Texas has been successfully operating for over 20 years. The size of this initiative has grown in tandem with UTeach and we currently serve approximately 70 students per semester. However, there have also been challenges. As one might expect, the primary challenge revolves around securing sustainable funding. The program is fortunate

to have several donors and grants that fund the program annually, but we are constantly working with the development office to seek new sources of support. Funding limitations have led to another challenge which is providing enough interns to support community partner sites effectively. This is a particular challenge during the summer months when STEM outreach programs, such as camps, are at their zenith. UTeach prioritizes internship funding to support its students during the academic year, leaving little in reserve to fund summer internships. A final challenge relates to the transportation needs of our students. As a trend, the number of students who can drive and have access to a car has been in decline. Thus, we have had to prioritize internships at nearby locales or with convenient bus routes. This has negatively impacted some of our long-term partners. To help address this challenge, we are working with these sites and encouraging them to offer travel stipends or reimbursements for students using zip cars when possible.

Lessons Learned

In the many years that the UTeach Internship program has been in existence, there have also been some lessons learned. For example, the program has implemented several procedures to help ensure that UTeach students and the sites where they work have clear expectations of one another and also have tools for negotiating conflict. First of all, new interns must complete an in-person orientation before they can begin working. Veteran interns provide this service for new interns using a protocol provided by the internship program. Next, once an intern is placed, their first assignment is to complete a site supervisor interview. This form asks questions of the site and intern to help establish clear expectations between both parties and set up a successful experience for all. A copy of the completed form is sent to UTeach and retained as part of the intern's employment record. Furthermore, at times, either interns or sites fail to meet expectations. To this end, the program has developed an action plan to help supervisors and interns discuss issues or concerns that emerge and to help set specific targets for improvement. Though our interns are paid by UTeach, they ultimately work for the site and the site can choose to discontinue an internship if issues are not addressed.

The importance of face to face interactions in connecting our students with possible internships sites has also emerged as a valuable lesson. Thus, eight years ago, the program coordinator established the routine of hosting an Internship Choice Fair each semester. The internship coordinator runs this event and is present to advise and support current and prospective interns. This event has proven popular and has helped our lesser-known community partners recruit interns for their sites.

Other essential lessons learned center on the need for responsiveness and flexibility when working with interns. Internships are a first job for some interns and they need support in areas of professional conduct. This includes appropriate dress, addressing supervisors by their given title, responding to communication in a timely way, and more. Also, there are times when an internship placement ends

up as the wrong fit. In this case, the coordinator supports the intern as they communicate with their supervisor and helps them to find a new placement. Finally, at times interns are eager to gain experience at more than one location. To meet this demand, we have started offering split internships where interns can work at up to two sites each semester keeping with the 20-hour per week limit. This flexibility allows them to gain a wider variety of experience while serving more community organizations.

Next Steps

Moving into the future, there are plans to work more closely with research faculty in both the College of Natural Sciences and the College of Education to investigate ways that the internship program might intersect with their outreach efforts. This will help to leverage and maximize the community impact and benefits to the university. Additionally, there are targets to increase the fundraising effort for summer internships since this is a clear area of need. Finally, the program staff looks forward to deepening community partnerships and continuing to provide rich, real-world STEM experiences for UTeach students.

For those interested in starting their own internship program, we encourage you to use the information in this chapter as a guide. The graphic provided in Figure 4.1 can be useful when explaining the benefits of the program to development officers or potential donors. It illustrates how a single donation can have multiple impacts. In addition, the lessons learned section provides important insights to consider moving forward and the program website has links to copies of current program forms and protocols.

CLOSING

Providing paid internships for students comes with challenges; however, the many benefits make it a worthwhile effort. Through the internship program, UTeach students have the opportunity to learn by doing. They have the chance to exercise their skills in authentic settings in ways that are not financially burdensome to them or the sites where they work. Comments, stories, and surveys suggest the positive experiences of both interns and sites. Internships are always in high demand and the fact that sites partner with the internship program year after year reflects its success in promoting STEM education, serving the community, and meeting an unfilled need. Most importantly, the internship experience prepares UTeach students to respond flexibly in dynamic environments and provides them with a real-world skillset that will benefit them in any educational setting.

REFERENCES

Darling-Hammond, L., Gendler, T., & Wise, A. E. (1990). *The teaching internship.* Santa Monica, CA: Rand.

Department of Education. (n.d.) *Science, technology, engineering and math: Education for global leadership.* Retrieved October 14, 2018 from: https://www.ed.gov/stem

Dewey, J. (1938). *Experience and Education.* New York: Macmillan.

Texas Education Agency. (2018). *Out-of-Field data for all grade levels and subject areas SY 2017-18.* Retrieved October 13, 2018 from: https://tea.texas.gov/sites/default/files/Out-of-field%20data%20for%20all%20grade%20levels%20and%20subject%20areas%20SY%202017-18%20final%20accessible.pdf

CHAPTER 5

CROSS-POLLINATION IN AN UNUSUAL SETTING

The Success Story of UTeach Dallas

Mary L. Urquhart

The University of Texas at Dallas is a young and rapidly growing university in a major metropolitan area with more than 50 public school districts. With no School of Education, UTeach Dallas resides entirely within the School of Natural Sciences and Mathematics in a Science/Mathematics Education Department (SME). UTeach replication has had a dramatic impact on the nature of SME and its faculty, positively impacting its partnerships, resources, infrastructure, and academic programs. Much of the success of UTeach Dallas is due to the power of data, leveraging and cross-pollination between SME's portfolio of programs, and the wider roles Master Teachers have come to play as members of SME's faculty. The challenges of recruitment, sustainability, and building partnerships with local public schools remain ever-present. Yet UTeach Dallas continues to successfully meet these challenges, and has been interwoven into a foundation of an ecosystem to support STEM education in which innovation thrives.

The story of UTeach Dallas is both of one of success, and one of cross-pollination and leveraging in unusual replication circumstances. The success of UTeach Dallas can be measured by its expected ~35 new graduates/year, its 222 current graduates, its 90% five-year teaching retention rate, and the more than 150 STEM

Preparing STEM Teachers: The UTeach Replication Model, pages 61–76.

majors who give teaching a try each year. UTeach Dallas has balanced innovation with fidelity to the UTeach model. The program is well supported by the university administration with all fulltime faculty and staff funded on the university instructional budget. UTeach Dallas induction is ongoing, and has benefited from partnerships with other programs. Fundraising is an ever-present challenge that the UTeach Dallas leadership has been successful in navigating. Essential to the success of UTeach Dallas – and to the continued health of the small academic department that is its home – are shared faculty and resources within an ecosystem of efforts supporting STEM education.

The character and culture of a university in which a teacher preparation program resides influences both its challenges and its opportunities. UTeach Dallas at the University of Texas at Dallas (UT Dallas) is no exception. The UTeach replication process has dramatically transformed a small STEM education department, its faculty, and its infrastructure. The story of departmental transformation is also part of a story of replication success.

SME's family of programs, including UTeach Dallas, have been supported and shaped by the intricate interconnections between them in the service of STEM education, and in particular, STEM teachers. Leveraging programs and resources for in-service teachers to support UTeach Dallas, and vice versa, has led to greater stability, opportunity, and innovation. Much of the leveraging and cross-pollination in the UTeach Dallas success story was born from necessity. Nonetheless, the lessons learned are a valuable part of a conversation on how teacher preparation programs can thrive.

THE UTEACH DALLAS STORY

UT Dallas is a young, rapidly growing research university in a major metropolitan area of more than 7 million people (Hanna, 2018) and more than 50 independent school districts. Originally established as the Graduate Research Center of the Southwest in 1961, the university took its current name when it joined the University of Texas system in 1969. Junior and senior undergraduate students were first admitted in 1975 (The University of Texas at Dallas, 2017). The first freshman class enrolled at the university in 1990. Aside from its youth, the perhaps most unusual characteristic when compared with its UTeach peers is that UT Dallas has no School of Education. The Department of Science/Mathematics Education (SME), part of the School of Natural Sciences and Mathematics (NSM), had been home to teacher certification prior to the 1990s.

When UTeach replication began, SME's sole academic programs were the Master of Arts in Teaching (MAT) in Science Education and Mathematics Education. Both MAT programs were designed for in-service teachers with a heavy emphasis on content knowledge and pedagogical content knowledge. STEM content-focused, grant-supported outreach programs for in-service teachers, K–12 students, and the general public had also increasingly become an important component of the activities of the SME faculty. Teacher certification at UT Dallas

had moved to the School of General Studies (now the School of Interdisciplinary Studies), and the Teacher Development Center housed within. UTeach Dallas was founded as a new teacher preparation program within SME.

Changing Faculty Compositions

The UTeach replication process has dramatically transformed the composition of the faculty of SME. Today, SME's faculty include a mixture of science and mathematics Ph.D.s with expertise in—and a passion for – STEM education at all levels and Master Teachers with strong science or mathematics content backgrounds and extensive public-school leadership experience. Master Teachers, all of whom have been secondary science or mathematics teachers in local schools, bring relevance in their practitioner approach to teacher preparation. Master Teachers experience as leaders in the partner school districts in which field experiences take place further strengthens the credibility of the faculty with stakeholders.

The first steps towards UTeach replication at UT Dallas began prior to the founding of the UTeach Institute and announcement of national replication. In 2006, two SME Assistant Professors, Dr. Mary Urquhart and Dr. Homer Montgomery, were tasked by the university provost to bring back the UTeach model. UTeach Dallas was officially established after the winning of a successful Cohort 1 replication grant from the National Math and Science Initiative (NMSI) in the fall of 2007.

As shown in Table 5.1, the current faculty and staff composition of SME substantially differs from pre-replication. Support staffing has more than doubled. An administrative services officer and a program coordinator/materials manager now augment an administrative assistant and part-time secretary. The program coordinator herself holds a Ph.D. and participates in some faculty activities. Master Teachers, critical hires for the UTeach model, now make up the bulk of the SME faculty. One additional faculty member, an elementary science specialist ranked as a Senior Lecturer, shares many Master Teacher characteristics and duties.

The first Master Teacher for UTeach Dallas was hired in 2007 as the department prepared for UTeach replication. Without a School of Education, UT Dal-

TABLE 5.1. Faculty and Staff in SME Prior to UTeach Replication and in the Present Day

	Tenure/Tenure Track*			Non-Tenure Track		Staff		
	Assistant Professors	Associate Professors	Full Professors	Senior Lecturers	Master Teachers	Support Staff	Academic Advisor	Teaching Assistants
2006	3	1	1	1	0	1 ½	0	2
2019	0	2	0	3	7	3 ½	1	4

Note: *Loss in the number of tenure track faculty shown is not directly related to the UTeach replication process.

las Master Teachers have taken on duties more typical of regular faculty. Master Teachers serve as sole instructors of record not only for the introductory courses, but also for the majority of courses in the UTeach Dallas program. The instructional team for each course bears responsibility for finding mentor teachers, developing relationships with them, and scheduling pre-service teachers in their field experiences. Master Teachers play a leading role in the ongoing refinement, vertical alignment, and innovation in the UTeach Dallas curriculum.

Master Teachers in SME are full voting members of the faculty. Technically hired as staff, Master Teachers are considered to be equivalent in rank to Senior Lecturers by the NSM leadership and school bylaws. As with all fulltime faculty members, active participation in committee service and in the department's graduate and outreach programs is encouraged. The equality of Master Teachers to regular faculty members in rights and responsibilities has been a direct result of the advocacy of departmental and program leadership along with the support of the Dean of NSM.

As a consequence of no School of Education, UTeach Dallas has relied heavily on its parent UTeach program for its theoretical framework. UTeach Dallas, however, considers Master Teachers serving as the primary instructors for teacher preparation coursework to be a strength. Master Teachers have credibility from their time in the classroom and their modeling of exemplary teaching practices. However, due to the expanded role of the UTeach Dallas Master Teachers, leadership experience at the level of instructional specialist or higher is considered to be essential. All UTeach Dallas Master Teachers worked with local school district partners prior to joining the university. Most of these faculty members have served as instructional specialists in local partner school districts or at the local Texas Education Agency Region Service Center. Two Master Teachers have served as assistant principals of local high schools, and three (including one now retired) have served as curriculum directors in local partner school districts. To quote UTeach Dallas Master Teachers themselves, "district leadership experience provides a 30,000-foot view of education versus the teacher perspective" and "helps to prepare students for working with their future administrators." The UTeach Dallas Master Teachers have brought the benefits of their strong relationships with partner school districts to the program, and to the department as a whole.

Leadership in UTeach Dallas

Without a School of Education, leadership of UTeach Dallas differs from that of other UTeach sites. At the time of replication, the original Co-Directors were both assistant professors in SME holding Ph.D.s in science fields. Both Co-Directors were tenured soon after replication began. One Co-Director became SME Department Head in 2011, and upon the retirement of her colleague became the sole Director in 2018.

A lead Master Teacher has served in the role of Assistant Director since the inception of the program, and the current lead was promoted to Associate Director

in 2018. Master Teachers, as a whole, have taken on leadership roles well beyond a particular course and its curriculum. One Master Teacher is responsible for Apprentice Teaching and placements of graduates, another leads on teacher certification compliance, and yet another oversees induction. One Master Teacher leads the department's classroom technology innovations, whereas others have focused leadership in areas such as project-based learning (Donaldson, York, Gagnon, & Gavitt, 2020), working with English language learners, cooperative learning, and social and emotional learning. Sharing of expertise at conferences and with local schools, districts, and Region Service Centers is actively encouraged for all SME faculty, including Master Teachers.

Expanding Infrastructure

As with the makeup of the faculty, the infrastructure for SME has been transformed by UTeach replication. In 2006, the department had one small, dedicated classroom seating 14 students, one equipment storage room, and one storage closet. Currently, SME is home to the UTeach Dallas Center, built specifically for the program. The UTeach Dallas Center hosts the field-based courses in UTeach Dallas as well as SME mathematics courses and outreach programs. The center contains a 28-seat laboratory-style classroom suitable for group work and outfitted with technology similar to that found in the best local schools. Hands-on materials used in courses line one classroom wall. Interns assisting with materials management staff a workroom with configurable tables, student computers, printers, a large laminator, and copier. Two Master Teacher offices complete the space. In the adjacent Science Learning Center (SLC) building is a dedicated 42-seat laboratory classroom with its own storeroom and extensive laboratory equipment. The SLC space is used primarily for the department's science courses, including Research Methods, large courses, and outreach programs. Unlike the UTeach Dallas Center, the creation and outfitting of the SLC space was part of a new building constructed for the school of NSM. The original SME classroom remains part of dedicated departmental space, and is available for small classes, meetings, and breakout sessions. Additional storerooms have been provided to SME to house the extensive collection of UTeach Dallas materials kits, probeware, and other teaching resources.

UTeach replication substantially enhanced the number and quality of SME's library of teaching resources. Corporate in-kind contributions and external grant funds augmented the investment by the university in classroom technology. The materials kits purchased for UTeach Dallas field experiences fill multiple storerooms. The Program Coordinator hired for UTeach Dallas and UTeach Dallas interns assist with managing and preparing materials for not only field experiences, but for MAT courses, inductees checkout, and use by grant program participants.

MEETING THE CHALLENGES OF A UTEACH PROGRAM

Strategic collaborations between SME programs have been crucial for their success in meeting a number of UTeach-specific challenges including mentor teacher recruitment, induction support for graduates, and internship opportunities. These challenges are certainly not unique to UTeach Dallas. Likewise, needs for sustainability, constant fundraising, and building and maintaining relationships with schools and districts are commonalities among UTeach sites. These needs also apply to multiple other SME programs.

With 150–170 undergraduate students giving teaching a try each academic year, finding sufficient mentor teachers has to be an ongoing effort. UT Dallas is located at an intersection of cities in a major metropolitan area. As a result, UTeach Dallas suffers from an abundance of riches in regard to identifying potential schools and quality teachers with whom to place students. The approximately dozen districts with which UTeach Dallas works present challenges in relationship building and navigating the complexities of many different district leadership and human resources departments. However, establishing relationships with individual STEM teachers in local high-needs schools assists faculty in recruiting quality mentors. From the beginning, UTeach Dallas has actively recruited from among MAT students, graduates, and SME's grant-supported outreach programs for in-service teachers.

UTeach Dallas mentors, graduates, and students have, in turn, benefitted from SME's professional development and MAT course offerings. UTeach Dallas has strengthened the pre-replication relationships that existed between SME and local schools and school districts. Master Teachers, with their ongoing connections with schools and continued presence in observing students in field experiences, have been invaluable in growing support for the department.

Partnerships, Interactions, and Impacts

Interactions between programs in which SME faculty have overlapping responsibilities are a natural consequence of a small department. UTeach Dallas has benefited the other programs within SME, and in some cases, beyond the department. Likewise, other programs within SME have provided support to UTeach Dallas from its inception. Prior to UTeach replication, SME was home to a complex ecosystem of interacting academic and outreach efforts in the service of teachers, students, and the community. As UTeach Dallas grew it was interwoven into SME's STEM education portfolio. Faculty-driven cross-pollination developed into mutually beneficial leveraging between programs.

During the first few years of the replication process, the focus for UTeach Dallas and its faculty was on successfully implementing the UTeach model with high fidelity. As the program matured, more deliberate interaction between UTeach Dallas and SME's other activities began to take place. One example of the impact of UTeach replication on SME is changes in the MAT. In 2010, the faculty began a

redesign of the two MAT degree programs, bringing them together with a common vision. Replication heavily influenced the redesign effort in both philosophy and in the design of a new set of shared MAT core courses. Master Teachers have become increasingly involved in the MAT, including taking on teaching duties as the number of tenure track faculty within SME has declined. SME faculty members, with one exception, are active members of the UTeach Dallas team. UTeach Dallas/MAT cross-pollination has been essential in crafting UTeach Dallas pathways for STEM degree holders seeking certification. A grant from a generous UTeach Dallas funder includes MAT scholarship support for UTeach Dallas graduates, increasing MAT program stability. Increasingly, the twin MAT programs are serving graduates from UTeach Dallas and from other UTeach programs.

Coursework for both the MAT and UTeach Dallas continue to be influenced by one another, and by faculty participation in outreach programs. In 2019, a major redesign of a core MAT course drew together elements from UTeach Dallas courses in research methods and project-based learning. MAT course designs originally inspired by UTeach replication now, in turn, bring innovative practices back into UTeach Dallas. Sharing of practices and content creates new challenges, as does serving pre-service and in-service teachers side-by-side. In addition, graduate courses must be distinct from undergraduate courses in cases where either may be used to meet the same teaching certification preparation requirement. One approach utilized in the MAT is to expect greater independence and metacognition of graduate students. In addition to covering similar ideas with greater depth in the MAT, graduate students are expected to reflect more deeply on classroom practice in collaboration with in-service teachers in their courses. Instructors must work collaboratively and take care in course designs to ensure relevance and minimize repetition for students. Synergistic improvements of practice are not limited to interactions between academic programs. SME's curriculum at all levels has also benefited from the infusion of new ideas and resources derived from faculty involvement in non-academic programs. Individual and collaborative work on course improvement is actively encouraged, as is participation in professional development opportunities offered through grant partners and professional organizations. MAT and UTeach Dallas courses have integrated numerous activities and strategies acquired from the grant-supported outreach programs for teachers. Outreach programs have also served as fertile ground for in-house innovation. Faculty, including Master Teachers, are encouraged to share their creative professional work with broader communities outside the university.

Leveraging a Portfolio of Programs

A multitude of SME grant-supported activities have interacted with UTeach Dallas and the MAT in mutually beneficial ways. Some activities have been focused purely on outreach, such as for a NASA mission built at the university or science camps for secondary students that are organized and staffed by SME

faculty, students, and graduates. However, the most impactful interactions have occurred with programs designed to support in-service teachers.

Table 5.2 shows interactions between SME's longest-term activities focused on pre-service and in-service STEM teachers since the start of UTeach replication. Benefits to each SME program for pre-service and/or in-service teachers are shown by column. The Teacher Enhancement Academies in Math and Science (TEAMS) funded through the Texas Higher Education Coordinating Board were designed to support MAT and UTeach Dallas students. Other programs were designed specifically as outreach to support in-service teachers, including UTeach Dallas induction. UTeach Dallas induction follows the UTeach model. Graduates are provided up to three years of personalized support and invited to participate in Saturday workshops throughout the academic year.

Until the end of summer 2018, SME was also home to long-term, sustained, content-focused professional development programs through the Texas Regional Collaboratives (TRC) for Excellence in Science and Mathematics Teaching and WeTeach_CS. Each of these programs shared common leadership. From 2010–2018, the now-Director of UTeach Dallas also served as Project Director for the department's TRC and WeTeach_CS projects.

For 25 years, the TRC was a state-wide network supporting in-service science and mathematics teachers throughout with long-term, sustained, content-focused professional development (The University of Texas STEM Center, 2019). SME housed a science Collaborative continuously from 1996 until the end of the TRC in 2018, and a mathematics Collaborative from 2016–2018. In 2016, UT Dallas was selected to be part of the original small group of TRCs to expand into computer science. A goal of the WeTeach_CS Collaboratives was to prepare in-service science and mathematics teachers to certify in and teach computer science (The University of Texas at Austin Texas Advanced Computing Center, 2018). At UT Dallas, the WeTeach_CS Collaboratives were also designed to help teachers integrate computer science skills and computational thinking into their mathematics and science classrooms. Whether focused on science, mathematics, or computer science, each of these Collaboratives was required to provide professional development primarily to in-service teachers. In their later years, the UT Dallas Collaboratives were also intentionally designed at the local level to provide services directly to UTeach Dallas students and graduates.

Supporting New and Experienced Teachers Together

The needs of pre-service teachers, novice in-service teachers, and experienced in-service teachers often differ. UTeach Dallas induction workshops are based in part on surveys of recent graduates. As such these workshops focus on topics of particular interest to a novice teacher such as classroom management. Experienced teachers may be more interested in deepening their content in areas relevant to their current teaching assignments, or learning new classroom strategies, techniques, and technology. Yet there is also benefit to bringing novice and expe-

TABLE 5.2. Leveraging and Cross-Pollination Between SME Programs

	UTeach Dallas	Master of Arts in Teaching	UTeach Dallas Inductionl	UT Dallas Regional Collaboratives (TRC)	TEAMS
UTeach Dallas (2007– present)		Recruiting, curriculum, master teachers as faculty, resources, facilities	Support and retention of graduates, graduate tracking	Recruitment, internships, resources, professional development	Recruiting mentors, internships, scholarships, professional development
Master of Arts in Teaching in Science Education and in Mathematics Education (1975–present)	Recruiting mentors, recruiting degree holders, faculty, facilities and resources, STEM content coursework		Facilities, resources, data, access to experienced teachers	Contact hours, facilities, faculty, recruitment, resources	Program, faculty, resources
UTeach Dallas Induction (2011–present)	Retention, support and tracking of graduates, feedback on needs of teachers	Recruiting, resources		Content-focused professional development, materials, resources, community	
UT Dallas Regional Collaboratives (TRC) in Science (1996–2018) Math (2016–2018) & WeTeach_CS (2016-2018)	Recruiting mentors, graduate support, internships, professional development, interactions w/ experienced teachers	Recruiting, tuition assistance, materials, resources	Teacher and faculty professional development, materials and resources, access to experienced teachers		Recruitment
TEAMS (2009–2011, 2011–2013)	Recruiting mentor teachers, internships, scholarships, interactions with experienced teachers	Recruitment, scholarships, professional development			

Note: Example leveraging benefits to each program for pre-service and/or in-service teachers are indicated by column. For purposes of this table, multiple programs with overlapping characteristics are presented together.

rienced in-service and pre-service teachers together, and providing opportunities for quality peer interactions among them. As overlap between UTeach Dallas and the MAT increases, these populations are also mixing in the context of graduate coursework.

SME's outreach efforts serving experienced in-service teachers have positively impacted UTeach Dallas and the MAT in a number of ways. UTeach Dallas graduates and students have benefited from a wealth of opportunities made possible through grant-supported professional development. Infusions of teaching materials provided in conjunction with professional development allowed for immediate impacts on the classrooms of in-service teacher participants.

The longest term and most impactful interactions between grant-supported professional development programs and UTeach Dallas have been with the TRC. UTeach Dallas graduates participated in SME's TRC workshops as both as full members and inductee cadre members. UTeach Dallas graduates who joined the TRC benefited from support that went well beyond typical induction services. The TRC provided opportunities for sustained peer interactions with experienced teachers, materials to implement lessons in the classroom, and classroom technology. Peer mentoring was part of the TRC model, with participants actively supported in sharing expertise with another, and at local and state teacher conferences.

Participants in the TRC interacted regularly with SME faculty, as well as one another, in the context of research-based professional development with a state-identified content focus and through grant supported MAT courses. Additionally, UTeach Dallas students had opportunities to work alongside the in-service teachers in professional development workshops and in peer mentoring contexts. UTeach Dallas Noyce scholars, in particular, were invited to partake of SME's professional development offerings.

Despite the benefits of the TRC, full participation was often daunting to novice teachers. The TRC was funded in its later years primarily through a Department of Education Mathematics and Science Partnership (MSP) grant under No Child Left Behind and through the Texas Education Agency (TEA). MSP and TEA requirements for full members were sufficiently extensive to present a recruitment challenge to the programs, especially when participation was voluntary and did not include monetary compensation. TRC members committed to attending at least 100 contact hours of professional development, completion of demographic and teaching practices surveys, pre/post testing of content knowledge, and submission of state high-stakes test scores. Documented peer mentoring was required for most participants in the program overall and included a small stipend. The time commitment in particular had the unfortunate effect of dissuading many UTeach Dallas graduates from fully partaking of the benefits the TRC program could provide.

For those UTeach Dallas graduates who did join the TRC, SME's collaborative design provided in-service teachers with MAT tuition support to take content-

FIGURE 5.1. Bringing together experienced and novice teachers. UTeach Dallas inductees and students explore STEM concepts, classroom activities, and analyze sample student work side-by-side with experienced teachers –including UTeach Dallas mentor teachers – in joint workshops with a SME grant program for in-service teachers.

specific courses if they chose to do so. The WeTeach_CS grants, under the same management as those from the TRC, also provided MAT tuition support. MAT students, UTeach Dallas mentor teachers, and UTeach Dallas graduates were actively recruited as participants for the TRC and WeTeach_CS grants. As such, members were often in overlapping populations that contained MAT and UTeach Dallas students and graduates. The offering of graduate tuition support has provided UTeach Dallas graduates with opportunities to earn their MAT from UT Dallas at a much lower cost than would otherwise be possible. Graduate courses could also be taken in a non-degree seeking capacity specifically targeting the needs of an individual participant, including UTeach Dallas graduates. Such assistance can be particularly useful in a composite certification state, where a teacher with a biology degree may be assigned to teach high school physics or chemistry.

LEVERAGING PARTNERSHIPS TO GROW CAPACITY AND MAINTAIN STABILITY

As with UTeach replication itself, SME overall was strengthened through the department's multiple grant-programs supporting in-service and pre-service teachers. For both the TRC and WeTeach_CS programs, the professional development offered was designed and implemented locally within a framework of requirements from the parent organizations. The parent organizations of the TRC and WeTeach_CS provided extensive professional development opportunities to faculty, including Master Teachers, in their capacity as instructional team members. For example, faculty members serving on the instructional team for the science collaborative partook in a total of six different WestEd *Making Sense of SCIENCE*™ (WestEd, n.d.) facilitation academies funded by the TRC. *Making Sense of SCIENCE*™ workshops were subsequently offered to not only grant participants, but also to other UTeach Dallas graduates and UTeach Dallas students such as scholarship recipients in the National Science Foundation (NSF) Noyce Teacher Scholarship Program. A week-long *Making Sense of SCIENCE*™ *Force and Motion* course (Daehler, Shinohara, & Folsom, 2011) was offered twice to middle school teachers – including UTeach Dallas graduates and mentor teachers—in one UTeach Dallas partner school district as a summer institute (Romanillos, 2017).

Grant-funded resources and in-kind corporate contributions, in support of professional development, have enhanced the library of materials available for coursework and field experiences. A collection of resources and equipment available for loan to local teachers is largely based in materials acquired through the department's professional development programs. Faculty-designed professional development and teacher-ready lesson plans created for grant programs have been integrated into courses, just as courses have inspired professional development offerings.

Initiatives brought to the department via UTeach Dallas have likewise fostered beneficial interactions among SME's family of programs. For example, a joint foundation grant to the National Alliance for Partnerships in Equity (NAPE) and UTeach Dallas started with embedding the NAPE micromessaging equity training (National Alliance for Partnerships in Equity, 2019) into UTeach coursework. A later grant then enabled the infusion of micromessaging workshops into SME's TRC professional development offerings in the 2014–2015 grant year. SME faculty and NAPE subsequently partnered to share their experiences with the larger UTeach and TRC communities.

Efforts to serve future computer science teachers through UTeach Dallas had a natural synergy with the WeTeach_CS grant-supported projects to help prepare in-service mathematics and science teachers to bring computer science into their classroom. As with other SME projects, overlapping faculty in these efforts led to natural cross-pollination. Interactions with the WeTeach_CS and UTeach com-

munities beyond the university infused new ideas and resources into SME's work with both in-service and pre-service teachers.

Partnering to Support Graduates through Induction

Despite its successes, UTeach Dallas has at times depended on SME's other programs for sustainability. UTeach Dallas induction in particular has benefited from internal partnerships. From 2016–2018, induction professional development activities were largely supported through the TRC. The instructional teams for SME's science and math Collaboratives worked closely with the Master Teacher in charge of induction to build such support into their TRC programs. At the same time, the teams made an intentional effort to encourage interactions among science and math teacher participants through joint TRC workshops.

Non-UTeach Dallas teachers have likewise benefited from SME's TRC partnerships with UTeach Dallas induction and faculty. Overlap between the programs increased opportunities for experienced teachers, including mentor teachers, to interact with Master Teachers. Professional development sessions on project-based learning, educational technology, English language learners, and cooperative classrooms are a mere sampling of professional development Master Teachers have brought to a broader cross-section of teachers beyond inductees.

Partnering to Support Students Through Internships

Leveraging other programs to support UTeach Dallas students with internships, a student retention strategy in the UTeach model (UTeach Institute, 2011), has been ongoing since replication began. Although UTeach Dallas funding has been sufficient to provide stability in some internships directly supporting operations, additional opportunities available to students have varied greatly. Partnering has been especially important in offering internships beyond the freshman and sophomore years that are supported by the NSF Noyce Teacher Scholarship Program. Education and Public Outreach funds from the NASA CINDI mission (The University of Texas at Dallas, 2012) supported UTeach Dallas internships to staff the Women in Physics camps targeting middle and early high school girls (Women in Physics, 2019). In the second of the TEAMS grant programs funded by the Texas Higher Education Coordinating Board, a tiered mentorship model brought in-service MAT students together in regular professional development sessions with UTeach Dallas scholarship participants. UTeach Dallas students, in turn, mentored middle and early high school students in SME-supported outreach such camps and tutoring.

An NSF supplement to the UTeach Dallas Noyce grant to support innovations in a multivariable calculus course provided UTeach Dallas pre-service teachers with opportunities to serve as peer instructors. The WeTeach_CS grants allowed the first CS major in UTeach Dallas to not only assist faculty in an instructional capacity with in-service teachers, but also to help support instructional team

members for other WeTeach_CS projects throughout the state. UTeach Dallas TRC internships provided opportunities for students to work with faculty delivering professional development. UTeach Dallas interns have themselves provided invaluable support to programs. Beyond serving as instructional assistants, interns regularly work with faculty to prepare materials for in-service teacher professional development and prepare class sets of teaching materials for in-service teachers to take back to their classrooms.

UTeach Dallas and the Ever-Present Challenge of Funding

Several aspects of UTeach programs as implemented at UT Dallas require continuous infusions of external funding. Some examples include mentor teacher stipends, internships, scholarships, and induction. Acquiring external funding, in turn, places a fundraising burden on program leadership. Growing into a mature UTeach program presented its own funding challenges, both in raising external funds and in building financial support from university leadership.

When UTeach Dallas started, it was almost entirely externally funded. Development officers at both the school and university level were charged with assisting leadership in obtaining both support funding and growing the endowment to meet a Cohort 1 $1,000,000 challenge match by the NMSI. In 2012, at the end of replication, a significant investment from the dean of NSM brought most of the UTeach Dallas faculty and staff onto the recurring school budget. Additions of Master Teachers to the faculty were supported for at least three years through generous grants from a foundation, providing time for NSM to increase its investment. The UTeach Institute, through the replication process, required data collection on students and graduates that proved vital to the acquisition of the additional funding that has supported both program growth and sustainability. The value of this data is evident in the national replication of the parent UTeach program, and can be seen today in UTeach snapshots (UTeach Institute, 2019). UTeach Dallas has been, from its founding, a priority for university fundraising with a complementary culture of data collection and reporting.

Fundraising for UTeach Dallas has been vastly more successful than fundraising for SME's in-service teacher programs. Programmatic data collection and reporting, support from the UTeach Institute, and the priority of UTeach Dallas with the university Office of Development may all play a role in fundraising success. Yet, a struggle for SME leadership has been to maintain the identity for other programs within the department. While the faculty is proud that the administration has referred to UTeach Dallas as the crown jewel of SME, they have also struggled in communicating even within the university that SME is not simply the "UTeach department." Despite this challenge in departmental identity, the overall impact of UTeach replication has been overwhelmingly positive. UTeach replication has brought resources and attention to the STEM education efforts of SME that far exceed what existed prior to UTeach Dallas.

More than a decade after UTeach replication began, efforts by departmental leadership and increased collaborative interactions among faculty have made great strides towards equity in identity for SME's family of programs. The Office of Development now has placed the MAT as separate-yet-related funding priority alongside UTeach Dallas and UTeach Dallas Induction. Recently, with the support of the university, SME has been successful in actively drawing upon funder interest in STEM teacher preparation to obtain MAT scholarship funding for UTeach Dallas graduates.

BUILDING A THRIVING ECOSYSTEM TO SUPPORT STEM EDUCATION

UTeach Dallas is deeply interconnected with the broader efforts of SME and its faculty to support STEM education at all levels. Some aspects of the interweaving of a UTeach replication program within the existing STEM education ecosystem of this small department are unique to its unusual setting. Faculty, including Master Teachers, are not entirely focused on a single program. Rather, the individuals in SME work together on components of a broader effort to prepare teachers for, and assist teachers in, providing high quality STEM instruction. Faculty and students conduct and support outreach to K–12 students, and are active in efforts improve undergraduate STEM instruction.

The high degree of cross-pollination across the SME's programs may be largely due to the overlap in leadership and faculty. Sharing of resources across programs has often been driven by necessity. Continuous, positive interactions have also spurred innovation and broadened the support opportunities offered to both pre-service and in-service teachers. Innovations in coursework and professional development offerings have been drawn from a culture of continuous improvement, the intertwining of SME's STEM education efforts, faculty ownership of the curriculum, and leadership encouragement to grow and share expertise.

What can other teacher preparation programs take away from the success story of UTeach Dallas? Some messages are built into the process of replication itself. Collecting and sharing data with funders has been essential. Maintaining fidelity to the UTeach model while simultaneously encouraging faculty innovation is part of continuous program improvement. Empowering Master Teachers as faculty members has been part of the innovation success of UTeach Dallas. A continual infusion of new ideas and resources drives much of the success of SME's portfolio of programs. Leveraging has provided resiliency to help weather the inevitable ebb and flow of external funding. A potential weakness for SME is the dependence on the efforts of a few individuals to acquire the support necessary for sustainability. A strength of SME is that empowerment, ownership, and vision are broadly distributed within the small department. The story of UTeach Dallas continues to unfold. As the program looks to the future, it will continue its culture of cross-pollination and leveraging for the betterment of the broader STEM education community.

REFERENCES

Daehler, K. R., Shinohara, M., & Folsom, J. (2011). *Making sense of SCIENCE™: Force and motion.* San Francisco, CA: WestEd in Association with NSTA Press.

Donaldson, K., York, K., Gagnon, R., & Gavitt, C. (2020). A tale of two PBI programs. In A. J. Goodell & Selma Koç (Eds.), *Preparing STEM teachers: A replication model* (pp. 225–240). Charlotte, NC: IAP.

Hanna, B. (2018, March 22). DFW gains a whopping 146,000 residents in 2017 to lead the nation. *Fort Worth Star-Telegram*. Retrieved from: https://www.star-telegram.com/news/local/community/fort-worth/article206357404.html

National Alliance for Partnerships in Equity. (2019). *Micromessaging to Reach and Teach Every Student™*. Retrieved from: https://napequity.org/professional-development/professional-development-curricula/teacher-training/

Romanillos, R. (2017). *Improving science teachers' self-efficacy for science practices with diverse students.* (Unpublished doctoral dissertation). Baltimore, MD: John Hopkins University.

The University of Texas at Austin STEM Center. (2019). *Texas regional collaboratives in service of science and mathematics teachers*. Retrieved from: https://stemcenter.utexas.edu/trc

The University of Texas at Austin Texas Advanced Computing Center. (2018). *WeTeach_CS goals and outcomes.* Retrieved from: https://www.tacc.utexas.edu/epic/weteachcs/goals-and-outcomes

The University of Texas at Dallas. (2012). *CINDI Education.* Retrieved from: http://cssepo.utdallas.edu

The University of Texas at Dallas. (2017). *Printable timeline*. Retrieved from: https://www.utdallas.edu/create/print.html

UTeach Institute. (2011). *UTeach elements of success.* Retrieved from: https://institute.uteach.utexas.edu/sites/institute.uteach.utexas.edu/files/uteach-elements-of-success-2011.pdf

UTeach Institute. (2019). *UTeach impact.* Retrieved from: https://www.uteach-institute.org/uteach-impact

Women in Physics. (2019) *Women in Physics camp.* Retrieved from: https://wipphysicscamp.wixsite.com/wipcamp

WestEd. (n.d.). *Making Sense of SCIENCE*. Retrieved from: https://www.wested.org/service/making-sense-of-science/

CHAPTER 6

COLLABORATION, COMMUNICATION AND COMMUNITY

Transitioning from a Traditional Model

Jennifer M. Dechaine and Ian Loverro

Since it was founded as a State Normal School in 1891, Central Washington University (CWU) has trained Washington's teachers. CWU has some of the largest teacher education programs in the state, graduating up to 40 mathematics and science teachers annually in the last five years. Despite these successes, enrollments in STEM teacher education programs at CWU have declined, consistent with state- and nation-wide trends. These patterns and a related funding opportunity prompted us to pursue UTeach implementation over 2016–2019. Implementing the UTeach model in a new state in the context of a large, established STEM teacher education program created a unique array of opportunities and challenges. This chapter details our implementation process from receiving grant funding through when our first cohort of students graduated. We then report enrollment trends from pre- and post-UTeach student cohorts. Finally, we present key lessons learned that are likely to inform universities with similar institutional contexts, such as large or established STEM education programs, short program implementation timelines, and high field hour and assessment requirements.

Preparing STEM Teachers: The UTeach Replication Model, pages 77–90.

INTRODUCTION

Established in 1891 as the Washington State Normal School, Central Washington University (CWU) has a long history of offering well-regarded teacher education programs. CWU has the largest, first credential teacher education program in Washington State, certifying 381 to 584 teachers annually in the last four years (Professional Educator Standards Board [PESB], 2019). CWU is also one of the largest producers of mathematics and science teachers in Washington State and graduated 25 to 40 middle level and secondary science and mathematics teachers per year in the five years before UTeach implementation (2013–2018). Despite these comparably high numbers of graduates, CWU's STEM teaching enrollment had peaked and then decreased in recent years, consistent with nationwide trends. In Washington State, STEM teaching endorsements have remained stagnant for the last 20 years and lag behind the number of teachers leaving the field (Goldhaber, Krieg, Theobald, & Brown, 2014). Mathematics has been listed as a Teacher Shortage Area in Washington State every year since 1996, as have science fields since 1998 (U.S. Department of Education, 2019).

Coinciding with this STEM teacher shortage, is a drive in Washington State to increase the number of residents pursuing STEM degrees. The 2018 Washington State STEM Education Report Card ranks Washington as the #2 state in the nation for STEM job concentration but #46 in the proportion of high school graduates going directly to college (Washington State STEM Education Innovation Alliance, 2018). Washington's focus on STEM job growth led the state to create the Washington State Opportunity Scholarship Opportunity Expansion Fund (OEF) through legislative action. In 2016, the OEF invited all public institutions of higher education in Washington to apply for funding to increase graduates from undergraduate STEM programs, including STEM education. CWU assembled a faculty and administration team representing education, science education, and mathematics and wrote an OEF grant proposal to be the first institution in Washington State to implement the UTeach model. Key reasons that we chose the UTeach model were its demonstrated success at increasing the number of STEM graduates at adopting institutions, its over 80% five year retention rate (compared to 50% retention in Washington State at the time), and its ease of adoption due to the UTeach Institute and the diversity of institutions implementing the model.

The overarching goal we outlined in the grant was to increase the number of qualified STEM teachers in Washington State in the next 5–10 years. Our measurable outcomes toward this goal were to: double the number of enrolled STEM teaching students at CWU (3–4 years, funding period) and double the number of STEM teaching graduates at CWU (5–7 years). In June 2016, we received a three-year $2.19 million OEF grant to support the implementation of CWU Teach STEM, a UTeach model for Washington State. Due to the grant timeline and our over 50% population of transfer students, we implemented UTeach in an accelerated three years. In spring 2019, we graduated our first cohort of CWU Teach STEM students. This chapter outlines our UTeach implementation process,

progress toward our goal, and lessons learned. Our process and analysis should inform universities with similar contexts who are looking to implement the UTeach model or other institutional changes within STEM teacher preparation.

IMPLEMENTATION

Institutional Context

CWU follows a "decentralized" model in which secondary teacher education majors and teaching endorsement-specific faculty are housed primarily outside of the college or department of education. Instead, the endorsement-specific pedagogy and content area classes are divided across colleges. For secondary education endorsements, the Department of Curriculum, Supervision, and Educational Leadership (CSEL), housed within the College of Education and Professional Studies, offers a 50 credit Professional Education Program (PEP), which, prior to CWU Teach STEM, was taken by all teacher education candidates and includes foundational courses, such as Introduction to Education, Instructional Methods, and Student Teaching. In addition to the PEP, most candidates complete a teaching major in a different department, often in a different college. Prior to implementing UTeach, the majority of middle level and secondary science and mathematics candidates completed teaching majors (e.g., Biology Teaching, BS) within their home departments (e.g., Department of Biological Sciences), that differed from the majors of non-teaching peers in terms of breadth and depth of study. Field experiences before UTeach were largely concentrated in the last years of an undergraduate's career. These included a variable number of content-specific teaching opportunities within junior-senior year science or mathematics education courses, as well as a pre-fall field experience—with the good intention of being completed in sophomore or junior year—but too often left until senior year, and a final 10-week quarter of student teaching in the PEP.

Although most at CWU would agree that STEM education students have always graduated well-prepared to teach, we hoped that the UTeach model would help us improve in several areas. These included: increasing degree flexibility and reducing time to graduation; more evenly distributing field experiences and increasing their integration with content; and, perhaps most importantly, improving overall program cohesiveness by breaking down established silos that segregated science, mathematics, and foundational education faculty and courses. We tackled our challenge areas by organizing the implementation of the nine UTeach Elements of Success (UTeach Institute, 2011) around four categories (Table 6.1), discussed next.

Curriculum and Courses

One of the first decisions we made was to retire the STEM teaching majors and have all students pursue the standard biology, chemistry, geology, mathematics, and physics majors (e.g., Biology, BA or BS). This provided more flexibility

TABLE 6.1. CWU Teach STEM Implementation Timeline Aligned to UTeach Elements of Success (EoS).

	Category	Curriculum & Courses	Institutional Support & Sustainability	Field Experiences & Master Teachers	Recruiting & Student Support
	EoS*	2, 4, 7	3, 9	6, 8	1, 5
Year 1 (2016–17)	Summer	Visioning meeting Working groups outlined major/minor structure and STEM Teaching minor	Received OEF grant funding	Working groups outlined field experience structure	
	Fall-Winter	Submitted final curriculum for University approval	Moved to new space	Identified faculty member to oversee field experience development	Drafted website, program mark, and recruiting materials Hired Advisor/Recruiter
	Spring-Summer	Co-developed program syllabus template, assessment structure, and 17–18 course syllabi	Signed MOUs between colleges of science and education	Met with local school districts to plan STEP 1 & 2	Recruited in STEM intro courses, CWU orientations, and STEM outreach events
Year 2 (2017–18)	Fall	STEP 1 & 2 combination	Assembled Steering Committee	Hired 1st Master Teacher	First cohort starts courses Produced promotional video
	Winter	Knowing and Learning, STEP 1	Developed endowment goal and fundraising materials	Met with regional junior high to plan spring field experience	Hired student interns focused on community support
	Spring	Classroom Interactions I (co-taught), Functions and Modeling for STEM Teaching (co-taught), STEP 2		Met with regional school districts to plan year-long placements	Started Teach STEM club
	Summer	Co-developed 18–19 course syllabi			

Year 3 (2018–19)	Fall	Classroom Interactions II (co-taught), Perspectives, Functions and Modeling for Secondary Mathematics		Hired 2nd Master Teacher Mentor teacher training and year-long placements begin	Developed intern program ambassadors Hired additional interns
	Winter	Research Methods, Project-Based Instruction	Received first major endowment gift		
	Spring	Apprentice Teaching (full time)	Final UTeach evaluation visit	First cohort apprentice teaches	Celebrate first cohort graduation

*EoS: 1. Distinctive program identity, 2. Cross-college and school district collaboration, 3. Long-term institutional and community support, 4. Compact and flexible degree plans, 5. Active student recruitment and support, 6. Dedicated master teachers, 7. Rigorous, research-based instruction, 8. Early and intensive field experiences, 9. Continuous program improvement. Bolding identifies CWU adaptions or additions to the original UTeach model. Some adaptations are unique to CWU and a number are shared by other UTeach adopting institutions.

for students to switch into and out of teacher education. The second curricular decision was not as simple. How do we adapt the PEP to the UTeach Elements of Success? After much negotiation between departments and college Deans, we decided to co-develop a STEM Teaching Minor degree. STEM education students would then complete their major and the STEM Teaching Minor, replacing the PEP and science and mathematics pedagogy courses for those students. To be a minor, the program could be at most 44 quarter (~30 semester) credits. Together, the new majors and minor would decrease time to graduation by reducing the total credits by 8–15, depending on the endorsement area.

Curriculum planning for the STEM Teaching Minor was the most important aspect of the implementation process to many faculty. Almost all involved faculty were tenure-track or tenured with multiple years of experience working in teacher preparation. This experience created the challenge of faculty holding varied and sometimes conflicting values but also provided an opportunity to draw on their deep and broad expertise. We used several structures to break down silos and facilitate faculty collaboration. First, shortly after receiving the OEF grant funding, we held a large visioning meeting that included faculty and administrators from all affected departments, as well as school administrators, teachers, and STEM coordinators from our region. This meeting asked: "If you could design the ideal undergraduate STEM teacher education program from the ground up, what would it look like?" Looking back, the most important outcomes of this meeting and much of the curriculum design process discussed in the following paragraphs were to surface stakeholders' core values, recognize that those values were generally in line with each other and UTeach, and thus generate buy-in for the UTeach program from faculty and school districts who were not involved in developing the grant.

After the visioning meeting, fifteen faculty and administrators formed two working groups: curriculum and field experiences. The working groups largely focused on addressing several, intersecting challenges resulting from our institutional context and Washington State requirements. These included: honoring faculty and stakeholder values, adapting the UTeach courses to six quarters (to accommodate our 50% transfer students), aligning course curriculum to InTASC standards (Interstate New Teacher Assessment and Support Consortium [InTASC], 2013), incorporating Next Generation Science Standards (NGSS, 2013), and integrating practice edTPA assignments (American Association of Colleges for Teacher Education [AACTE], 2019). To address faculty values, the curriculum working group outlined three guiding themes, each with three subthemes (Figure 6.1), and aligned these to InTASC and the UTeach curriculum. At the same time, the field experience working group planned how 100 clinical practice hours could be distributed across the five quarters before Apprentice Teaching. Both working groups then considered how the learning cycle (planning—instruction—assessment), NGSS, and the edTPA integrated with these themes and field experiences. Finally, the working groups used all of this information to write theme-aligned

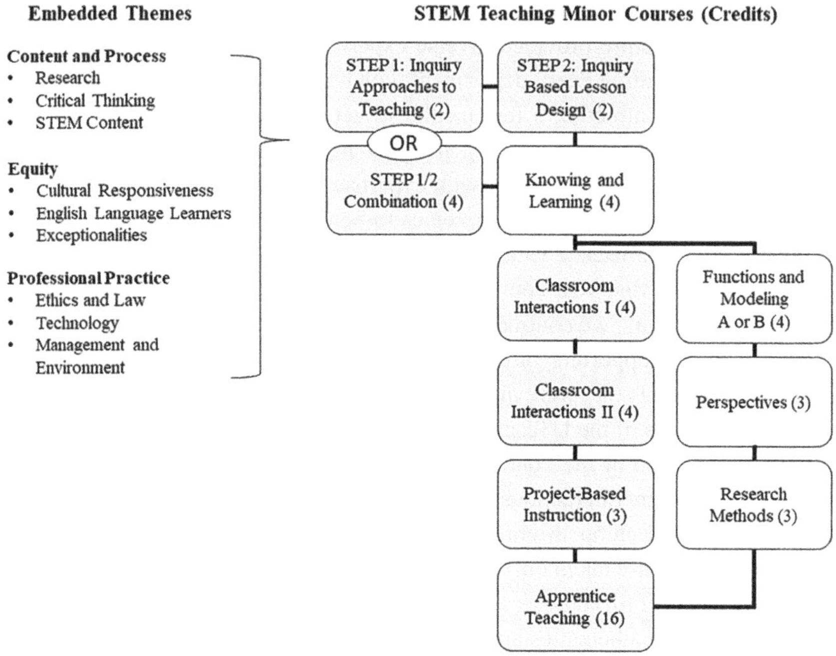

FIGURE 6.1. STEM Teaching Minor degree with curricular themes. The three curricular themes with three subthemes each are embedded throughout the courses. The right side of the figure shows the STEM Teaching Minor course sequence with course names and number of quarter credits in parentheses. Grey fill indicates courses with field experiences. STEP 1 through Apprentice Teaching must be taken in sequence. Functions and Modeling, Perspectives, and Research Methods can be taken any time after Knowing and Learning. Course descriptions and outcomes are publically available in CWU's online Undergraduate Catalog (CWU, 2019).

course outcomes and plan major activities and assessments across all courses in the STEM Teaching Minor, which was submitted to the University electronic approval process by the new program deadline in December 2016.

With the curriculum submitted, it was time to turn course outcomes into syllabi. This was accomplished by a new set of working groups, working over summer 2017 through the following academic year. A program working group created a syllabus template, lesson plan template, program reading library, and Canvas (University learning management system) shells to be used for all program courses. Then, working groups for each course we were offering in 2017–2018 developed the full syllabus, course schedule with major activities, and all major assessments with rubrics. A field experience working group worked in tandem to develop a detailed plan for how to initiate and manage field experiences. It was very important that all working groups included at least one faculty member

from each core contributing department: CSEL, Science Education and Mathematics. This structure provided diverse expertise to all development areas and built trust by providing every department with representation. Finally, we further built buy-in by contracting a few faculty to act as theme consultants, meaning they worked with all course working groups to help embed each curricular theme. For example, one faculty member with expertise working with English language learners provided resources and strategies to be used in several STEM Teaching Minor courses. We used a similar structure over summer 2018 through spring 2019 to revise the first-year courses and develop the remaining program courses. Over these two years, we contracted 30 faculty to contribute to program development, generating supporters, often champions, representing all ten (education and STEM) departments that were directly affected by the STEM Teaching Minor.

The adaptations of the UTeach model most unique to the STEM Teaching Minor are the addition of the Functions and Modeling for STEM Teaching course, splitting Classroom Interactions into two courses, co-development of program themes, and co-teaching in core courses (Table 6.1). UTeach's Functions and Modeling is typically taken only by secondary mathematics students. We created a second version of this course, Functions and Modeling for STEM Teaching, for the middle level mathematics and all science students to learn basic mathematical modeling, NGSS-aligned engineering design, and experience project-based instruction by completing a locally-relevant environmental engineering project. We split Classroom Interactions into two courses, because there was simply too much content and field hours to fit it all into a single quarter-long course, and faculty were passionate about providing candidates a comprehensive overview of equitable classroom management principles within the context of diverse STEM field experiences.

Creating two Classroom Interactions courses, co-developing themes, and co-teaching contributed to breaking down silos, building trust, and increasing program cohesiveness. One indicator of program buy-in is that 18 faculty (6 CSEL, 9 Science Education, 3 Mathematics) have now attended the annual UTeach meeting, and faculty from all three of these departments presented there in 2019. Co-teaching, in particular, has been instrumental in developing trust among faculty. Knowing and Learning and Classroom Interactions I and II are co-taught by two faculty: one with STEM education expertise from the College of the Sciences, and one with foundational education expertise from CSEL. Co-teaching across colleges at our institution is ground-breaking and continually challenged by barriers, such as the University budgeting. Overall, involving and compensating as many faculty and stakeholders as possible in the visioning and curriculum design process may have been our most important decision toward creating institutional support and thus sustainability.

Institutional Support and Sustainability

Administrative champions were also instrumental in creating institutional support for CWU Teach STEM implementation. The University President, Provost, and science and education college Deans provided a strong letter of support for our OEF grant. The College of the Sciences turned over a large, brand new physical space to the CWU Teach STEM program. Included within the space was a classroom, conference room, student study and collaboration spaces, offices for Master Teachers and program staff, as well as an equipment room holding a large array of STEM teaching materials and instructional technology. The College of the Sciences also agreed to pay salaries for three new positions, an Advisor/Recruiter and two Master Teachers, after the conclusion of the OEF grant (starting in 2019–2020). In addition to salaries, we have begun working toward long term financial support for CWU Teach STEM by collaborating with the University Foundation to create fundraising materials and set an over $3 million endowment goal. Although still far from that goal, we are ecstatic that an endowment has been established by a large legacy gift from Howard E. Anderson (Burke, 2019).

Deans and department heads drafted a memorandum of understanding (MOU) to support the co-teaching model discussed above, identify Co-Director positions from the colleges of science and education, and establish the CWU Teach STEM Steering Committee, headed by the Co-Directors and populated with equal numbers of faculty or administrators from each college. The Steering Committee's original role was to review and approve all course syllabi and has now matured into reviewing major program decisions to support long-term sustainability. The core elements for a sustainable program are now in place. In addition to the salaries, space and equipment discussed above, we have started to create a strong network of committed mentor teachers, and our STEP courses now meet General Education requirements–recruiting is strong without student stipends. In the next few years, we plan for our sustainability efforts to focus on four areas: continued strong student recruitment and support services, streamlined systems (e.g., field placements and mentor teacher training), additional fundraising, and ongoing faculty engagement. For example, we are starting a faculty-led brown bag series in fall of 2019.

Field Experience and Master Teachers

Washington State requires 450 clinical practice hours for all teacher education programs, over 100 more than the 320 hours embedded in the original UTeach model (UTeach Institute, 2014). We met 350 hours by retaining our full time, 10-week Apprentice Teaching quarter. In addition to the typical 10 hours for STEP 1 and 2, we increased field hours in Classroom Interactions I to 15 (from 8) and in Classroom Interactions II and Project-Based Instruction to 80 hours total (from 23). For Classroom Interactions II, Project-Based Instruction, and Apprentice Teaching, we designed a field model in which one local mentor teacher has two

students: one for fall/winter and one who stays for the entire year through spring Apprentice Teaching. Students choose their placement length (partial or year-long) before they are assigned to a mentor teacher. We chose this model so that students could get enough field hours, some students and local mentor teachers could get the benefit of collaborating for an entire year, and students who want to go closer to home for Apprentice Teaching would still have that option. Students may be placed up to a one-hour drive away from CWU, because our location in Ellensburg, WA, is a small rural town with only one middle and one high school.

In line with the UTeach model, field experiences before Apprentice Teaching are largely supervised by our Master Teachers. We hired our first Master Teacher (science/engineering focus) in fall 2017 and the second (mathematics focus) the following fall. Both are Full-Time Non-Tenure Track Clinical Faculty. The Clinical Faculty designation was created in our faculty Collective Bargaining Agreement in 2017, largely because of the Master Teacher position. To ease the on-boarding of our Master Teachers, our grant also funded some time for a faculty member to oversee field experience development during implementation. Each of our Master Teachers has over 20 years of experience teaching STEM in the middle and high schools, and that expertise has been invaluable to the program.

Recruiting and Student Support

Instead of adding to the responsibilities of the Master Teacher position, as is common in UTeach programs, we hired a dedicated Advisor/Recruiter for CWU Teach STEM. She was the first position we hired after receiving grant funding, and she worked closely with CWU Public Affairs to develop several distinctive marks, recruiting materials, a video, and our website (Central Washington University [CWU], 2017). In the first year of implementation, she focused on recruiting in introductory courses for all the STEM majors, attending orientations and STEM recruiting events and working with students to help them transition into CWU Teach STEM. She also helped develop and launch our internship program and the Teach STEM student club, both of which she continues to oversee. Our internships have now supported twelve students who work on community STEM outreach and serve as student ambassadors. Interns oversee several robotics clubs and events, a Kid Wind club, community Math Circles, and numerous outreach events. They also work with our Advisor/Recruiter as CWU Teach STEM Ambassadors to recruit and work as tutors and mentors to other CWU Teach STEM students.

PROGRESS TOWARD OUR OUTCOMES

In June 2019, our first CWU Teach STEM cohort of 22 students graduated. We expect 10 students to graduate next fall or winter. Retention for this first student cohort from STEP 1 to their most recent CWU Teach STEM course (PBI or AT, depending on graduation quarter) was 71%. From fall 2017 to spring 2019, 170 students have taken STEP 1. Based on current enrollments and adjusted for our

71% retention rate, we project a 68% or more increase in the number of graduates on average for the three years post-UTeach implementation versus the three years prior (Table 6.2), which is good when compared to CWU's small drop in first credentials awarded from 2016–17 to 2017–18 (PESB, 2019). Our projected number of graduates for 2020–2021 is also double the pre-UTeach three-year mean (52 vs. 26). These numbers indicate that we are on track to meet our outcome of doubling the annual number of STEM teaching graduates from CWU.

We also observe interesting trends in endorsement areas. The data indicate increases for low enrollment science endorsement areas: Chemistry, Earth Science, Physics, and Middle Level Science (Table 6.3). The percentage of students seeking Middle Level Mathematics endorsements may have decreased, but this trend is likely partially attributable to comparing graduates (pre-UTeach) to enrolled students (post-UTeach) in Table 6.3. The percentage of students seeking each mathematics endorsement is likely to become more similar as they approach graduation because several students change from Secondary to Middle Level Mathematics each year. We are also excited that three Computer Science students took the STEP courses and one student has continued with the Computer Science teaching endorsement in the first year that program has been available, although we have learned from these low numbers that we need to create other pathways to this endorsement. Several students, including the majority in low enrollment areas (e.g., computer science), have reported that they would not have pursued teaching certification under the old teaching majors model and that trying teaching in STEP 1 was essential in their choice to continue with the program. The UTeach model has been instrumental in moving us toward our enrollment goals and increasing the number of STEM teaching graduates at CWU. We hope our lessons learned continue this trend by lessening the burden of adopting UTeach or elements of the model at more Washington State institutions of higher education.

TABLE 6.2. Graduates Pre- and Post-UTeach Implementation

Pre UTeach		Post UTeach	
Year	**Grad #**	**Year**	**Grad #**
2015–2016	22	2018–2019	22
2016–2017	18	2019–2020	39 (Projected)
2017–2018	37	2020–2021	52 (Projected)
Mean (SD)	26 (8)	Mean	38 (12)

The left two columns show the number of graduates in each of the three years before UTeach implementation, as well as mean graduates over the same dates with standard deviation (SD). The right two columns show the same in the three years post-UTeach implementation. Projected graduates assume 71% retention from STEP 1 to graduation, reflecting what we observed for our first student cohort.

TABLE 6.3. Percentage of Students in Each Endorsement Area Pre- and Post-UTeach Implementation

Endorsement Area	Pre UTeach %	Post UTeach %
Biology	16	16
Chemistry	3	6
Earth Science	0	2
Physics	1	2
Middle or Secondary Science	1	5
Middle Math and Science	9	4
Middle Math	33	20
Secondary Math	36	44
Computer Science	NA	1

Pre-UTeach reflects the three years prior to UTeach implementation. Post-UTeach shows all active students in 2018–2019.

LESSONS LEARNED

Including many voices in the planning and teaching process is a wise investment for breaking down barriers. We used collaboration and communication across as many stakeholders as possible to build a trusting community. Stakeholders who had rarely interacted previously developed a shared vision, collaborated on working groups and co-taught courses. This approach generated buy-in and trust. It incorporated diverse values and expertise and improved relationships across departments and colleges. Although shared vision is always a recommended first step in institutional change (Elrod & Kezar, 2017), much more than that was needed to break down the cross-college silos that had been established over many years.

Co-teaching, in particular, has contributed to building trust. Faculty better appreciate each other's expertise. Students see how teaching professionals can collaborate and address a variety of instructional issues from more than one perspective. Faculty demonstrate that pedagogy and content have equal value. However, co-teaching is not without challenges. Faculty values are not always congruent, and co-teaching takes a lot of extra time to plan and coordinate. University structures make it difficult to compensate faculty equally, and even assessing two instructors for a single course can be challenging. Co-teaching has greatly contributed to our program's success, but we also recommend that institutions that wish to try it invest heavily in professional development and faculty compensation to smoothly launch a co-teaching system.

Institutional change requires administrative champions. Although the Co-Directors had little prior experience in institutional change, much of what worked is consistent with institutional change literature (Elrod & Kezar, 2017). University

structures create barriers to working across colleges, which could not have been broken down without the collaboration of CWU's administration. We built early administrative support by emphasizing how the program aligned to University and college focus areas, showing how it would support Washington State's interest in STEM workforce development, presenting data on how combining two smaller programs (science and mathematics education) would reduce expenses due to increased course enrollments, and obtaining external funding (OER grant) from a group that reports to our state legislature and major STEM companies. We also invited administrators to the first planning stages. For example, our new Provost at the time attended the first visioning meeting a couple of weeks before she had officially started work. Our Provost was instrumental in empowering the Deans and supporting our work with Public Affairs and the CWU Foundation. The college Deans negotiated a MOU, agreed to cross-college co-teaching, and continue to both support and promote CWU Teach STEM. The administration made CWU Teach STEM a University funding priority, and without that, we doubt we would currently have an endowment.

Establishing a strong and clear management and communication structure is essential. Our working group structure during implementation provided a chain of command and communication plan. We quickly learned in our second summer to provide even more detailed expectations and frequent communication to the working groups. These have now morphed into a Steering Committee that monitors course syllabi and approves changes. This structure prevents individual courses from drifting away from the UTeach principles, yet allows for needed change. Since the Steering Committee is headed by the Co-Directors and populated by faculty and administration from both colleges, all decisions are collaborative. Despite the differences in norms across colleges, regular meetings and frequent conversation allow us to address any issues that arise, which contributes to the long-term sustainability of CWU Teach STEM.

Two Master Teachers are better than one. At first, we thought it might be possible to save money and only hire one Master Teacher, and while we managed, it was a false economy. Now that we have a second Master Teacher, we are not relying on a single individual to be expert in both mathematics and science. We also now recognize how essential it is for the Master Teachers to have a peer team to support each other since their position is so unique to the University. Finally, we believe that for a program as large as ours, two Master Teachers would not be able to handle the typical UTeach workload if we did not also have the Advisor/Recruiter.

It pays to have money. So many bumps in the road to implementation were smoothed out due to having a large external grant. From paying faculty to consult on the working groups, feeding in-service teachers when we brought them in to discuss what our Teach STEM program would expect of them, paying for interns to conduct STEM nights, and even purchasing swag to promote the program; none of these would have been possible under the traditional financial constraints

of our pre-UTeach model. Minimally, we would recommend that an institution considering this large of an institutional change provide implementation funding to support: refreshments, and ideally stipends, for several collaboration meetings over two years; recruiting materials and swag; stipends for all collaborating mentor teachers; and at least three student interns. At our institution, lack of launch funding is a barrier to other programs interested in radically overhauling their teacher education programs.

REFERENCES

American Association of Colleges for Teacher Education. (2019). *edTPA*. Retrieved from: http://edtpa.aacte.org/. AACTE, Washington, DC.

Burke, R. (2019). *Harold E. Anderson leaves $800,000 to CWU Teach STEM program.* Retrieved from: https://www.cwu.edu/foundation. CWU Foundation, Ellensburg, WA.

Central Washington University. (2017). *CWU teach STEM.* Retrieved from: https://www.cwu.edu/teachstem/. CWU, Ellensburg, WA.

Central Washington University. (2019). *2019–2020 undergraduate catalog.* Retrieved from: http://catalog.acalog.cwu.edu/. CWU, Ellensburg, WA.

Elrod, S., & Kezar, A. (2017). Increasing student success in STEM: Summary of a guide to systemic institutional change. *Change: The Magazine of Higher Learning, 49*(4), 26–34.

Goldhaber, D., Krieg, J., Theobald, R., & Brown, N. (2014). *The STEM and special education teacher pipelines: Why don't we see better alignment between supply and demand?* (CEDR WP #2014-3). Seattle, WA: Center for Education Data & Research. University of Washington.

Interstate New Teacher Assessment and Support Consortium. (2013). *Model core teaching standards and learning progressions for teachers 1.0: A resource for ongoing teacher development.* Washington, DC: Council of Chief State School Officers.

NGSS Lead States. (2013). *Next generation science standards: For states, by states*. Washington, DC: The National Academies Press.

Professional Educator Standards Board. (2019). *Goal 4 data: Qualified educators*. Retrieved from: https://www.pesb.wa.gov/resources-and-reports/monitoring-strategic-goals/goal-4/. State of Washington, PESB, Olympia, WA.

U.S. Department of Education. (2019). *Teacher shortage areas nationwide listing 1990–1991 through 2017–2018.* Washington, DC: Office of Postsecondary Education.

UTeach Institute. (2011). *UTeach Elements of Success.* Austin, TX: The University of Texas at Austin.

UTeach Institute. (2014). Chapter 5: The UTeach Instructional Program. In *The UTeach operations manual.* Austin, TX: The University of Texas at Austin.

Washington STEM Education Innovation Alliance. (2018). *2018 STEM education report card.* Olympia, WA: Washington State.

CHAPTER 7

THE MANY FACETS OF INDUCTION SUPPORT

Kelli Allen

The overarching purpose of the UTeach induction program is to support STEM teachers in the first critical years of teaching. An integral part of the UTeach induction program is its emphasis on providing one-on-one, intensive support to recent graduates since the first three years are critical for new teacher retention. Teacher recruitment and retention are ever-growing issues, so to address this difficult problem and increase retention rates, the UTeach induction program provides extensive, "on-demand" induction support to graduates of the program.

UTeach induction support makes the following assumptions: Teachers new to the profession need between one to three years of support to develop expertise in the teaching field. Induction support should be provided using multiple methods of delivery. Sufficient resources are needed to deliver appropriate induction support. Support should leverage 21st century technologies. This chapter will address these assumptions and discuss resources needed to support these ideas.

UTeach induction support is a natural extension of the UTeach program. After preparing pre-service teachers for the classroom, this program supports them in the first crucial years of their professional growth. This induction model serves and assists new UTeach graduates in STEM areas by providing personalized support and resources throughout the first few years of teaching. An integral part of the UTeach induction program is its emphasis on providing one-on-on, intensive

Preparing STEM Teachers: The UTeach Replication Model, pages 91–101.

support to the program's recent graduates since the first three years in the classroom are critical for new teacher retention. In Texas and across the nation teacher recruitment and retention are ever-growing issues. According to Texas Education Agency reports, from 2013–2018, "the retention rate of teachers coming from all types of certification routes is 68.7% after five years of teaching." On a national scale, according to the National Education Association, approximately 50 percent of new teachers will leave the profession within their first five years of teaching. To address this difficult problem and increase retention rates, the UTeach induction program provides extensive, "on-demand" induction support to graduates of the program.

UTEACH ELEMENTS OF SUCCESS

The UTeach induction support program demonstrates the UTeach elements of success in a variety of ways. One of the key UTeach elements of success is a distinctive program identity. Induction support serves as a distinctive piece of the program's identity and has been in place since the first UTeach Austin graduates obtained employment in the Austin area. Master teachers who have established relationships throughout the UTeach program have always served the role of Induction Support Specialist. The qualifications for UTeach master teachers include a content-heavy master's degree at least three years of classroom teaching experience with a demonstrated ability to use best practices as evidenced by local, state and/or national awards. In addition to having deep content knowledge in a STEM related field and practical experience, a master teacher should be a recognized leader in the local school districts, ensuring knowledge of STEM curriculum, personnel, district policies and procedures. Ensuring the same master teacher supports both apprentice teachers and the induction program is a key element in building trustful relationships among both entities.

Collaborating with school districts is another key element of the UTeach program. Many of the UTeach Austin graduates acquire jobs in schools where they have conducted previous field experiences and have established early and ongoing relationships with these districts. Induction support provides another layer of assistance during the critical first few years of teaching. Even though new teachers are assigned an on-campus mentor, the added support provided by UTeach includes varied resources to foster success, as well as an unbiased sounding board for new teachers.

UTeach master teachers—non-tenured clinical faculty with exemplary secondary teaching experience—are exclusively dedicated to teaching, student support, and induction support program success. The UTeach program at UT Austin prides itself in maintaining the same master teacher presence in both student teaching and induction support for the past twelve years. This consistency ensures the preservation of effective support practices, the implementation of new strategies to ensure program progression, and attributes to the fact that UTeach induction support is also part of Continuous Program Improvement. UTeach systematically collects

and analyzes both student- and program-level data to make informed decisions about the induction program development and continual improvement.

PROGRAM GOALS

The UTeach induction support program includes the following goals: **1) retain new mathematics, science, and computer science teachers for at least five years.** According to a recent study by Ronfeldt and McQueen (2017), "receiving induction supports in the first year predicts less teacher migration and attrition, suggesting that using induction to reduce new teacher turnover is a promising policy trend." **2) help transform novice teachers into expert teachers through continuous professional development, enabling them to reach their full potential as well-prepared, highly-qualified, and highly-effective teachers.** Data collected by the UTeach Institute boasts a 91% three-year retention rate for UTeach Austin graduates and a 92% three-year retention rate for combined UTeach programs across the nation. **3) improve student achievement scores in mathematics and science by assuring that UTeach graduates have the training and resources needed to teach effectively**. A study conducted by the National Center for Analysis of Longitudinal Data in Education Research in 2018 includes "We find that students taught by UTeach teachers perform significantly better on end-of-grade tests in math and end-of-course tests in math and science by 8% to 14% of a standard deviation on the test, depending on grade and subject." **4) encourage the development of leadership among UTeach graduates by providing personalized support to help meet their individual goals.** Of the 1047 UTeach Austin graduates, 104 hold some type of leadership position. This represents 10% of all UTeach Austin graduates.

UTeach Induction Support makes the following assumptions: Teachers new to the profession need between one to three years of support to develop their expertise in the teaching field. Induction support should be provided using multiple methods of delivery. Sufficient resources, including personnel and money, are needed to deliver appropriate induction support. Induction support should leverage 21st century technologies. The following elements discussed are ways that the UTeach induction support model addresses these assumptions.

PROGRAM STAFFING AND FUNDING

Program staffing and funding are important elements to consider when developing an Induction support program. In the UTeach Austin induction program, one master teacher is entirely in charge of support services for new UTeach graduates teaching math, science, and computer science in Central Texas. The same master teacher has been involved with induction support for the past twelve years. This master teacher has other responsibilities, including serving an instructor for Apprentice Teaching. At UTeach Austin, the master teacher directly supports inductees in the Austin area and supports those further away through virtual induction

meetings, phone calls, email, and online and face-to-face professional development offerings. Master teachers are reimbursed mileage for travel from school to school. The number of miles will vary from month-to-month, so master teachers must keep a detailed record of the schools traveled to and the mileage between schools. Reimbursement for mileage occurs monthly.

The UTeach Austin program spends approximately $2,000 per inductee each school year. This covers at least part of the master teachers' salary plus New Teacher Starter Kits, monthly workshops, and other resources. This value varies greatly depending on the available funds. Funding availability will determine the number of monthly workshops that are conducted. UTeach graduates have access to the UTeach resources, as they are able to check out the resources from the workroom, research methods lab or Step classrooms. The master teacher in charge of induction support may provide delivery services to and from the induction teacher's classroom. This service supports the use of innovative resources already available throughout UTeach Programs.

The program also employs a database supervisor who provides supervision of data entry and data processing to ensure the smooth and efficient handling of information about UTeach graduates. The database supervisor verifies a variety of data in appropriate formats. They track and verify the quality of data entry from all sources and resolves problems regarding the data. The ability to generate useful reports about the success of UTeach graduates who go on to teaching jobs is a primary duty of the database supervisor. They must also be able to maintain a complete and accurate database of information about the activities of the new teachers. Since information is subject to frequent changes, the supervisor must make diligent effort to stay in touch with all the graduates to obtain updates in their contact information and job status. The data supervisor works to establish a friendly and personal relationship with each of the graduates in order to encourage them to stay in touch with the program.

PROGRAM RECRUITMENT

Program recruitment is necessary to inform teachers of the services available through UTeach induction support. The induction program is not required—new teachers can choose not to participate. However, since induction support is a valuable service, the master teachers work to recruit new teachers to the program. Induction support for new teachers actually begins with Apprentice Teaching. At the end of the semester, students are asked to provide permanent contact information so that they can be reached by the program once they leave the university.

Each summer, master teachers invite all UTeach graduates who have completed their Apprentice Teaching in the previous fall or spring to attend a Blastoff meeting before the beginning of the school year. The purpose of the Blastoff is to outline further the services provided by the induction program and highlight the benefits of participating. A typical agenda includes the following activities; 1) Provide a welcome letter listing the benefits of the induction program, so new

teachers will have the information in writing, 2) distribute teacher New Starter Kits, 3) have the new teachers fill out an information sheet including school, classroom number, teaching schedule, etc. 4) schedule the first master teacher classroom visit with each new teacher. Once they know the number of inductees they will be working with and their individual needs, the master teachers can start scheduling school visits and professional development activities. Traditionally, approximately 80% of the most recent graduates who are working in the area participate in the UTeach Blastoff event held at the start of the new school year.

The busiest time of year for master teachers supporting inductees is August through September when most of the new teachers are just beginning their work and need lots of advice and resources. Many of the December graduates begin teaching in January at the start of the second semester, although no formal "Blast-off" is held for these graduates. The induction support specialist gathers information from induction teachers for new teacher starter kit delivery, to start scheduling school visits and construct desired professional development activities. The ratio of induction support specialist to new teachers is typically 1:20. Higher ratios will degrade the quality of personalized support.

SUPPORT SERVICES

The two major support services of the UTeach Austin Induction Program are intensive one--on--one support and formal professional development. The one--on--one support component is crucial and should be the core of any UTeach induction program. However, both one--on--one support and professional development can be approached differently depending on the needs of a university's local inductees.

ONE-ON-ONE SUPPORT

Because many participants in the UTeach Austin induction program work within a reasonable driving distance from the university, master teachers are able to work frequently with them in person. Some UTeach induction programs cover areas as large as an entire state, in which case in--person support is not feasible. Nonetheless, a lot of additional support takes place by virtual meetings, phone calls or email. UTeach Austin also has some reciprocation with other Texas programs; programs support each other's graduates if, for example, a graduate of UTeach Dallas works in Austin or an Austin graduate works in Houston (home of teach-HOUSTON).

One university has addressed this issue by developing a graduate induction course that is composed of a mix of online and face--to--face components. Others provide support almost entirely by phone, email, Skype or Zoom, websites, and other electronic media.

Types of Support

One-on-one support includes a variety of approaches. Master teachers may visit a new teacher before or after school for an informal chat; make scheduled or unscheduled observations; help with curriculum and instruction; deliver resources; and provide emotional (sometimes emergency) support. The one--on--one support is on—demand—based on the immediate needs of the new teacher—and therefore needs to be flexible. Master teachers try to meet inductees at their schools, at times that are most convenient (before school or after school) or most needed (during class for observations). They tend to visit each inductee in the classroom at least twice a semester, though if an inductee needs more support, a master teacher might visit more often (two or more times a month) or call or email frequently.

The master teachers develop ways to assess each inductee's needs from the personal conversations conducted with each new teacher. They also track their contact with inductees in various ways, from making notes about the type of support each inductee has received (to help further assess need), maintaining a communication log, and noting their gas mileage to and from schools (for reimbursement).

Schools tend to welcome the presence of UTeach master teachers. The resources offered by UTeach induction support only add to what the school system has in place. Having multiple UTeach graduates positioned at the same school is beneficial for many reasons. First and foremost, UTeach grads collaborate to create inquiry-based 5E lessons or project-based lessons. These types of activities have a trickle-down effect for non-UTeach teachers who discover the value in these types of learning activities. The induction support specialist is able to serve groups of UTeach grads more efficiently, and special projects can be carried out at schools with where gaggles of UTeach grads are working.

Additionally, it is important to develop a relationship with school administrators. At times, master teachers may need to advocate for changes in an inductee's assignment. It is also possible that a master teacher has different advice about what is needed in an inductee's classroom than the administrators; in these cases, it is important for the master teacher to be able to adjust his or her expectations and work productively with the administrators to find a solution agreeable to everyone. Administrators also often seek out additional UTeach graduates to fill vacant positions. Open communication between administrators and induction support personnel is helpful in establishing cohorts of UTeach graduates teaching in schools.

Chats are a popular form of one-on-one support that master teachers use to check in with inductees. Chats are short, informal talks that usually take place virtually, over the phone, or in person. Personal meetings often occur if a master teacher is visiting a school where several inductees are working and wants to check in on each of them. The information shared throughout these chats vary depending on the teacher's personal setting and confidence in their performance

as a new teacher. For various reasons, some inductees are reluctant to meet with master teachers at first to ask for help. If an inductee is initially reluctant, master teachers might schedule a few chats to establish a comfortable and trusting relationship. These ongoing, casual conversations help the master teacher keep tabs on how each inductee is progressing and help predict what types of support might be needed immediately or in the near future.

Although informal chats are beneficial, class observations are essential to meeting the individual needs of teachers new to the profession. No two teachers are the same. Classroom management issues will differ from classroom to classroom. First-hand observations allow a master teacher to determine areas of need and offer the appropriate support for the new teacher. Class observations can be scheduled for specific lessons an inductee would like feedback on or when an inductee calls the master teacher for some emergency help. Observations also often occur when the master teacher has dropped by for a quick chat and the inductee asks the Master Teacher to stay and observe a particular subset of students or to provide classroom management advice.

Observations can last two hours or more if the master teacher observes a class, then stays afterward to give immediate feedback and suggestions. If the master teacher notes some areas for improvement, he or she may follow up with lesson plans or model a lesson, depending on the support needed. Later, the master teacher will return to the inductee's class and observe again, to make sure the support was effective and offer any other suggestions and help.

New teachers should be assured that classroom visits are completely confidential and will not be shared with the school administrators. The formal observational feedback provided goes directly to the teacher and no one else. The observations are used only to help the new teachers and do not play a role in any evaluation.

Curriculum and Instruction Assistance

Master teachers also provide assistance with planning for and developing curriculum and instruction. Following the curriculum guidelines of the school or district, inductees and master teachers often work together to develop curricular plans embedded with innovative teaching ideas, differentiated instruction, technology, and strong content incorporation. Most new teachers need support in this area, but there are some situations in which a teacher may need very concentrated support. The following are examples of when concentrated support is needed: 1) A teacher may be the only person at his or her campus teaching a particular content or grade level, so there are few opportunities to get support from or collaborate with colleagues. 2) A teacher may be assigned to teach multiple subjects, which significantly increases the content he or she needs to know and the preparation time. 3) Teachers at private schools or charter schools likely do not have the same kind of support that teachers in districts have, with support teams or coaches for content areas and grade levels.

Regardless of their training in the UTeach program and the level of district or school support, most new teachers need help integrating student--centered inquiry instruction with the day--to--day pressures of navigating multiple challenges. They also need advice from experienced teachers about ways to maintain order and good humor in the classroom and create a spirit of shared endeavor and discovery, as well as ways to cut down on time spent grading papers, doing lesson plans, gathering resources, and analyzing student data.

Resource Assistance

Almost all new teachers request help with resources. As states and districts face shrinking budgets, many schools lack basic resources like office supplies as well as lab and technological resources. The UTeach program requires students to use technology in their inquiry--based instruction. However, new teachers often find their schools, especially high--needs schools, do not have the necessary resources to implement these lessons.

The UTeach Austin induction program offers resource assistance in these ways; 1) A Starter Kit is provided to all new teachers. Kits can be distributed at the annual Blastoff meeting for new teachers held in August or personally delivered by the UTeach Induction Support Specialist. Personal deliveries serve as a great opportunity to visit induction teacher classrooms. The kits contain general teaching supplies like paper, markers, rulers, pens, staplers, etc. All induction teachers receive a new teacher starter kit regardless of location.2) Materials and equipment can be checked out from the UTeach supply/resource center. Master teachers check out the materials and deliver them to the inductees. Some examples of equipment needed by teachers in the field are laptop computers, iPads, probes, electronic scales, calculators, chemicals, GEMS kits, microscopes, and prepared slides.

Emotional support is also an integral part of an induction program. Master teachers are impartial, provide a positive point of view, and work in complete confidentiality. There are meetings that occur with new teachers when the entire meeting is spent listening to the teacher talk about concerns regarding classroom management issues, pressure from administrators or problems with parents or other teachers. Master teachers can suggest positive actions that the new teachers can use to improve their situations. They can also help navigate school politics and communications with department chairs, administrators, and parents, often helping script conversations.

Master teachers also serve as advocates for inductees. If a placement is extremely problematic, they will work with the new teacher to finish out the school year at the present placement and provide guidance for seeking a new position in another district, charter school, or private setting depending on openings and teacher preference. If an inductee is really struggling, a master teacher spends extra time with that teacher, sometimes on a daily basis, to provide emotional and instructional support. The master teachers' response is to meet with the inductee

and provide any support necessary through observations, help with curriculum and instruction, resources, advice on addressing any identified areas for improvement, or just being a sympathetic listener.

Assessing new teacher needs is an ongoing process. Master teachers in the UTeach Austin induction program typically find that inductees need the most help with developing and implementing instructional strategies, developing classroom management skills and routines, and obtaining resources. However, the master teachers are constantly assessing and reassessing the individual needs of each inductee.

Though most UTeach graduates are very well prepared and need little support beyond observations and occasional advice or resources, a few may need intensive support. The UT Austin master teachers estimate that about 80% of their inductees are low--needs teachers, about 15% are medium--needs, and about 5% are high--needs. Though there are not many high--needs inductees, they do require a significant amount of a master teacher's time.

PROFESSIONAL DEVELOPMENT

Professional development is the other major support system provided by the UTeach Induction Support Program. UTeach advocates lifelong learning for teachers as well as encouraging teachers to seek leadership roles at their campus or within their district. The types of professional development services provided have changed throughout the course of the program, but professional development has always been a key feature in supporting the goals of retaining new teachers, transforming novice teachers into expert teachers, assisting with the improvement of student achievement and encouraging leadership among graduates.

UTeach Austin provides free formal face-to-face professional development to inductees over the course of the school year that are held monthly and/or on Saturdays. These sessions are typically held on campus, but some events have occurred at different campuses where induction teachers are employed. New teachers are busy and overwhelmed, and though they know they can benefit from this professional development, they often need to be encouraged to attend. The master teachers at UTeach Austin encourage inductees to attend the professional development by providing additional perks, including classroom resources, door prizes, continuing professional education (CPE) credits and food.

One of the most popular professional development events is the UTeach Induction Panel that occurs each semester. This panel involves induction teachers coming to the apprentice teaching seminar to provide advice and support for student teachers who are gearing up for full-time employment. This is an opportunity for the UTeach alumni to support induction efforts in a professional setting. Each induction teacher receives a prompt to address and share personal expertise regarding topics such as, "When communicating with parents, I always make sure…., "I routinely differentiate instruction by….," "I consistently integrate inquiry-based instruction by…., etc. The Induction teachers enjoy sharing insight with their

younger peers and also take advantage of collaborating with fellow graduates. The Induction Panel is always a fun night of camaraderie that is enjoyed by all who attend.

Monthly virtual meeting opportunities using the Zoom meeting room tool are also available through the UTeach Induction Program. New teachers are invited to meet virtually to discuss topics such as communicating with parents, classroom management, supporting English language learners, etc. The virtual meetings serve teachers who are working away from the Austin area, but local teachers are also invited to attend. Virtual meetings accommodate teacher workday schedules and are typically held after school.

UTeach Austin provides online professional development opportunities for graduates and other in-service K–12 teachers. The online courses were developed based on communicated needs of teachers in the field. The online courses cover many topics such as blended learning, flipped classrooms, differentiated instruction, supporting ELL students, classroom strategies for inquiry-based learning, etc. These courses are self-paced and asynchronous to help meet the professional learning needs of busy teachers.

There is also the presence of an online community, the UTeach Nation Virtual Network, which serves as a vibrant learning community where sharing of ideas and resources does take place. This resource is available to all graduates to connect with fellow alumni from the 45 UTeach programs across the country. The National UTeach Alumni Network allows teachers to collaborate with fellow grads from partner universities across the country and learn about news from other alumni, upcoming events, and teaching tips.

UNIQUENESS OF UTEACH AUSTIN INDUCTION PROGRAM

The UTeach Austin induction program differs somewhat from other induction programs at non-UTeach institutions. Elements of the UTeach Austin induction program that make it unique are 1) A Master Teacher is in charge of Induction Support who has developed positive relationships through previous interactions with students during the UTeach Program, especially during the student teaching semester. This is an integral aspect of Induction Support to have these pre-established relationships established for a smooth transition into Induction Support. 2) UTeach induction teachers have access to a wide variety of resources already made available to them throughout the UTeach inventory of materials. Recall one of the assumptions made by the UTeach induction program is that induction support should leverage 21st-century technologies. The UTeach inventory contains the most recent 21st-century technologies to support teachers in developing and integrating innovative learning activities that incorporate technology for enhanced engagement and implementation. All UTeach graduates have access to these resources. 3) UTeach at UT Austin is one of the few sites that has a Maker Endorsement option. An extension of this has been integrated into UTeach Maker Induction. This endeavor supports UTeach graduates who want to bring project-

based, constructionist, Maker education into their own classrooms. UTeach Maker Induction supports UTeach graduates who want to bring project-based, constructionist, Maker education into their own classrooms. Through this program, teachers can learn new skills and earn an endorsement as a UTeach Maker Teaching Fellow. UTeach grads who are interested in becoming a UTeach Maker Teaching Fellow apply to be accepted into the Maker Program and upon acceptance their first task is to complete on online professional development course entitled Makerspace in the Classroom. Once the online course is completed, teachers then participate in the Maker events that are held throughout the year. These teachers are also responsible for completing the requirements for the UTeach Maker Teaching Fellow endorsement. Monthly Maker meetings are held to support individual efforts to enhance and foster the acquisition of an endorsement as a UTeach Maker Teaching Fellow.

CONCLUSION

The UTeach induction program is constantly evolving to meet the needs of our graduates. Decisions on the types of induction support services to offer are made based on data collected by student and alumni surveys as well as informal discussions with mentors and administrators who have hired UTeach graduates. UTeach induction support is a vital component of the program that helps to ensure that our graduates stay in teaching longer, improve student performance in STEM courses, while also influencing students to enter STEM fields.

REFERENCES

Backes, B., Goldhaber, D., Cade, W., Sullivan, K., & Dodson, M. (2018). *Can UTeach? Assessing the Relative Effectiveness of STEM Teachers.* CALDER Working Paper No. 173. Washington, DC: American Institutes for Research.

Educator Data. (2019). *Texas Education Agency*. Texas Education Agency. Retrieved from: tea.texas.gov/Reports_and_Data/Educator_Data/

Ronfeldt, M., & McQueen, K. (2017). Does new teacher induction really improve retention? *Journal of Teacher Education*, *68*(4), 394–410. Retrieved from: https://doi.org/10.1177/0022487117702583

CHAPTER 8

UTEACH MAKER

Preparing Future STEM Educators for Classroom-Based Making

Shelly R. Rodriguez, Jason Harron, and Hannah Smith Brooks

Making is an iterative process of creation that allows for self-expression through the construction of meaningful artifacts. UTeach Maker is an embedded micro-credentialing program within the UTeach licensure pathway that supports preservice teachers who express interest in integrating making into STEM classrooms or K–12 makerspace environments. Participants in the program develop a Maker Showcase, as an open portfolio to highlight their growth as a maker and educator. The sections of the Maker Showcase include the development of a maker philosophy, the creation of a maker project, the implementation of a maker-centered lesson, and documentation of connection to the broader maker community. Showcases are reviewed by national leaders and publicly presented to peers and mentors. Once the Showcase is complete, participants earn the title of UTeach Maker Teaching Fellows. This chapter describes the development and implementation of the UTeach Maker program as well as the supports that make it successful.

Preparing STEM Teachers: The UTeach Replication Model, pages 103–117.

INTRODUCTION

> Making, as an educational tool, can be used to engage students' minds and imaginations. This hands-on approach lets students push the boundaries of their knowledge. Making breaks through the barrier between the classroom and the real world.
>
> —*Grant, UTeach Maker*

Grant is a senior physics major in his second year in UTeach Maker. His quote expresses how "making," as an instructional approach, can connect the STEM classroom to the authentic interests and experiences of students. Drawing from a "do-it-yourself" mindset, *making* is an iterative process that allows students to bring their original ideas to life. Through making, students can engage in collaborative practices to design, build, and share something meaningful (Rodriguez, Harron, Fletcher, & Spock, 2018). Thus, maker-centered instruction occurs when the tools, skills, and processes of making are harnessed to promote specific learning outcomes.

Making and UTeach

The goal of inquiry-based teaching is to increase student engagement in learning through the development of critical thinking skills. An inquiry approach to teaching values the knowledge and experience of students, prioritizes active problem solving, and highlights the shared construction of new ideas. This approach is core to the UTeach program and is reflected in the program's commitment to inquiry-rich instructional models such as project-based instruction and the 5E learning cycle. As with these models, making and maker-centered instruction can help students learn content as they engage in designing solutions and building products. These products can address real-world challenges or be items of personal interest that students are inspired to create. Thus, maker-centered instruction is the latest iteration of the UTeach program's commitment to student inquiry and exploration.

Supported by an NSF Noyce grant (Award 1557155), the UTeach STEM teacher preparation program at The University of Texas is working with the Austin maker community to engage students in meaningful, authentic, and relevant projects and activities to explore how "making" can improve education. To this end, UTeach students have the opportunity to participate in UTeach Maker, an innovative micro-credentialing program that helps bring cutting-edge, project-based, maker practices into STEM learning environments. Each program participant has a personal mentor, engages in ongoing maker professional development, and creates a Maker Showcase to serve as a portfolio of their work. In UTeach Maker, participants can select from a suite of activities to create their own maker-centered experience. The menu of activities includes workshops, internships, participation in national and local maker events, and connection with a mentor active

in maker education. These program elements help students explore making in terms of tools, resources, community, authorship, and the promotion of constructivist instruction.

LITERATURE REVIEW

The section below reviews literature related to school-based making, the connection between making and inquiry-based learning, and the role of making in teacher preparation.

The Call for School-Based Making

Over the past decade, a growing number of classrooms, libraries, and community centers have been transformed into makerspaces. These makerspaces provide room for educators to implement maker-centered activities that encourage learners to engage in creative, higher order problem-solving through the iterative process of design and construction (Freeman, Adams Becker, Cummins, Davis, & Hall-Giesenger, 2017). In addition, maker-centered instruction emphasizes the development of student empowerment through creative, hands-on learning experiences that are both exciting and motivating (Clapp, Ross, Ryan, & Tishman, 2016). Thus, there has been an increase in national support for integrating the practices and tools associated with making into science, technology, engineering, and mathematics (STEM). For example, in 2017, The National Science Foundation (NSF) recognized that making has the potential to foster innovation, increase student retention, and broaden participation in STEM education (National Science Foundation, 2017).

Making as Inquiry

As in project-based learning (PBL) and other inquiry-based teaching practices, maker-centered instruction promotes learning by doing and can help instructors realize three-dimensional instruction that integrates scientific practices, disciplinary core ideas, and crosscutting concepts (National Research Council, n.d.). As an inquiry-based teaching strategy, maker-centered instruction has many elements in common with *High Quality Project Based Learning* (HQPBL, 2018). Both maker-centered instruction and HQPBL recommend that projects include intellectual challenge and accomplishment, authenticity, collaboration, project management, the creation of a public product, and reflection. While these elements overlap significantly with features of maker-centered instruction (Cohen, Jones, Smith, & Calandra, 2017; Rodriguez et al., 2018) there are also important differences. Maker-centered instruction is rooted in constructionism (Papert, 1991), where learners gain knowledge as they actively conceptualize and create tangible artifacts. Furthermore, maker-centered instruction places emphasis on personal relevance, the use of tools and materials in construction, and the process of iterative design. Maker-centered learning draws from design thinking protocols like

those of IDEO and the Stanford d.school, which promote skills such as empathy, defining problems, ideation, and proficiency with technological tools needed for prototyping and testing (Institute for Design at Stanford, 2010); all of which are needed for success in modern society. Maker-centered instruction also aims to promote valuable habits of mind such as playfulness, resilience, collaboration, and reflection, known as a "maker mindset" (Martin, 2015). So, unlike PBL, maker-centered instruction gives primacy to notions of personal relevance, design and fabrication, and the maker-mindset.

Studies in the secondary science classroom have shown that maker-centered instruction can help students explore science concepts and phenomena. There are several examples of making being used to promote student inquiry (Bevan, 2017; Peppler, Halverson, & Kafai, 2016). In one instance, copper tape, LEDs, and coin cell batteries have provided science teachers a way to use the creation of interactive pop-up books and user-friendly paper templates to support student explorations with electronics (Qi & Buechley, 2010, 2014). In another instance, investigations with sewable circuits have been shown to increase student interest in science (Tofel-Grehl et al., 2017). These sewable circuits can be used in combination with instruments, such as the Arduino-based Lilypad, to introduce students to coding through the creation of e-textiles (Fields, Kafai, Nakajima, Goode, & Margolis, 2018). In addition to digital tools, making includes traditional crafts such as welding, knitting, wood working, and other skills which can be used to promote inquiry into the natural world through the acts of invention and creation.

Making and Teacher Preparation

While making and maker-center instruction show promise for promoting inquiry, agency, and creativity in the STEM classroom, more must be done to prepare future STEM educators to implement these practices. For instance, a national survey reported that only half of undergraduate teacher preparation programs in the United States provided an opportunity to learn about maker-education and the associated technologies, and only 17% had a makerspace available for preservice teachers (Cohen, 2017). This suggests that future STEM educators are not being exposed to professional learning experiences related to making. Research has shown that STEM teachers often uptake and implement the inquiry-based practices with which they have personal experience (Windschitl, 2003) and that those who are new to making require thoughtful support in order to use the processes of making in their own teaching (Blikstein & Worsley, 2016). Thus, a lack of exposure to maker-centered instructional practices during teacher preparation means that future educators may remain unaware of the benefits of making for their students and may be wary of adding making to their curriculum. UTeach Maker hopes to fill this gap by providing long term, high quality maker professional development and learning experiences for its participants. In addition, by moving making from the hobbyist space into the classroom, UTeach Maker hopes to

provide access to innovative, constructionist practices to students from all backgrounds and walks of life.

THE UTEACH MAKER PROGRAM

UTeach Maker is a micro-credentialing program embedded within the UTeach licensure pathway. It supports preservice teachers who express interest in integrating making into STEM classrooms and/or promoting learning within K–12 makerspace environments. Since micro-credentials for teachers are competency-based forms of certification, the UTeach Maker program helps its participants to develop knowledge and skills around making. This extra skillset, not only makes them more appealing to potential employers, but ultimately enhances their professional practice in the classroom. It should be noted that all program materials, rubrics, and examples referred to in this paper can be accessed on the UTeach Maker website (https://maker.uteach.utexas.edu).

Recruitment and Application

UTeach students interested in joining the UTeach Maker program start by completing an online application. The application asks students questions about their previous experience with making, why they are interested in the program, and what personal strengths and resources they bring. Applications are reviewed by program leadership and new participants are selected each semester. UTeach Maker requires a minimum two semester commitment. UTeach students enrolled in *Step 2* through *Project-based Instruction* (PBI) are eligible. Students further along in the UTeach course sequence are given priority with others being placed on a waiting list. Based on resources, UTeach Maker can support a maximum of 24 participants at any given time.

The Program

Making is a complex endeavor where unforeseen challenges, rewards, and vulnerabilities emerge. Thus, a primary goal of UTeach Maker is to build a trusting community where participants feel safe in asking questions, pushing their boundaries, and taking risks. The program fosters community through monthly cohort meetings where participants share meals, resources, and ideas. In addition, UTeach Maker offers internships, weekend workshops, and special maker events to help participants develop proficiency with the tools and skills of making and maker-centered instruction.

During the program, each participant creates a Maker Showcase to serve as an open portfolio of their work. This is a publicly accessible website that allows participants to capture, curate, and share their accomplishments in ways that highlight their unique voice and abilities. As participants build their showcase, they reflect on their work, make critical choices about what to share, and create artifacts that allow both themselves and others to recognize the skills and ideas they

have to offer. To support the creation of the Maker Showcase, each participant has a veteran maker educator serving as a personal mentor. Additional details about the structure of the UTeach Maker program can be found in the 2019 NSF STEM For All Video Showcase archives (https://stemforall2019.videohall.com/presentations/1406).

Micro-Credentialing

Once complete, the Maker Showcase is submitted to a national leader in maker education for review and feedback using a rubric. This feedback supports the participant in reflecting on and strengthening their product. Once feedback is received, participants are asked to reflect and make any needed revisions or additions. Each participant also presents their Maker Showcase to peers in the UTeach Maker cohort, community members, friends, family, and with a broader audience that may include potential employers. Once the presentation is successfully completed and rubric criteria are met, then the micro-credential is issued. Program graduates earn the title of UTeach Maker Teaching Fellow, indicating that they have completed the process and developed key skills enhancing their resumes. Those who have earned the title are qualified to serve as future mentors or advisers.

PROGRAM SUCCESS FACTORS

Several components have contributed to the success of UTeach Maker. These factors include the structure provided by the Maker Showcase, the community support available for participants, and an orientation toward external oversight and public access to resources.

The UTeach Maker Showcase

Participation in the UTeach Maker micro-credential requires that students complete an open portfolio over a minimum of two semesters, while drawing on a wealth of support from a variety of individuals, institutions, and resources. The portfolio is comprised of four sections describing the maker's individual journey into making and maker-centered learning. To complete their Maker Showcase each participant must provide documentation in the areas of maker philosophy, a maker project, maker community, and maker education.

Maker Philosophy

The philosophy section promotes an understanding of the ideological and educational roots that underlie maker-centered education. It encourages participants to explore the philosophical and pedagogical origins of making in the context of other progressive models like constructivism and constructionism. In this section, participants also consider how making is situated both historically and culturally. Finally, participants are asked to articulate a personal philosophy and provide a

rationale for their instruction. As an example, in this section, one maker says of his instructional stance,

> My Maker philosophy doesn't only revolve around the idea of choice in education, but also the idea of accessibility and agency… Finding problems in the community and working together as future scientists and engineers allows students to work to make their world a better place through innovation and creation. (Kevin, Fall 2017)

Maker Project

The project section captures the personal experiences of participants engaged in making a product of their own. Here participants have the chance to show elements of their "maker mindset" and the dispositions involved in their approach to making. As they move through the construction of their project, they document major milestones in a narrative format with photos, reflections, and videos. Participation in this project allows participants to authentically engage in the growth-oriented mindset and playful nature of making. An excerpt from a Spring 2018 showcase provides an example from Nikki who created a mathematical quilt. Of her project, Nikki says,

> This project displays four proofs of the Pythagorean Theorem in the form of a quilt. My heart for this project was two-fold in that diversity would be celebrated through the inclusion of mathematicians from varied backgrounds, and that all of my future students would realize that they too could create a new, unique proof of this famous theorem. (Nickie, Spring 2018)

Examples of projects from other participants include a stained-glass periodic table with individually addressable LEDs for a chemistry classroom, a powered automatic water system for plants called the "Garduino," and custom board games or cosplay items that highlight the personal passions of participants.

Maker Community

For the community section of the Maker Showcase, participants are asked to document connection with the maker community. This may include things like attending monthly cohort meetings, virtual or face-to-face maker meet-ups, or Maker Faires. Participants might use pictures, reflections, social media posts, or other artifacts to provide evidence of their engagement. For example, May's Showcase provides pictures and a reflection of her time working at the 2019 Bay Area Maker Faire. Of her time there she says,

> Guests decorated fractals and created necklaces by decorating a compass rose and the spiral of Theodorus. The booth was so crowded at times that the volunteers had to step out to make room for the guests. I had the opportunity to work with Cara, who is a UTeach student from UC Berkeley. She told me that she did not know what Making was before volunteering at Maker Faire, but by the last day of the event, she seemed genuinely interested in it. (May, Spring 2019)

Since making occurs within diverse communities connecting to the broader network exposes participants to a wide range of expertise and perspectives. By connecting to like-minded others, participants forge relationships that will sustain their interest in making after their time with the UTeach Maker program ends.

Maker Education

Situated within a preservice teacher education program, UTeach Maker places heavy emphasis on the role that making and a maker mindset play in the classroom context. The Maker Education section of the showcase asks participants to make connections between personally meaningful making and K–12 education. This section encourages participants to consider educational issues that arise and frame making within those systems. Examples include issues of access and equity, strategies for meeting content specific standards, and a consideration of management tools that support maker-centered instruction. The showcase excerpt below is a reflection from one participant on the success of her maker lesson. In this lesson she asked students to create pictures using conics sections. The pictures were first graphed by hand then, using Desmos, the pictures were printed and turned into personally meaningful 3D creations using craft materials. In her reflection Helen says,

> Something that I actually don't think went well was introducing making to my students. I think the video was a lot as it shows over-the-top projects. I think instead showcasing my own small projects and asking students what they have made would be more accessible. We are all makers, we just might not know it. This is also something I struggled with before I joined UTeach Maker, so I understand students thinking it's something they can't do. (Helen, Spring 2019)

Each of the facets of the Maker Showcase inform both personal and professional identity and promote an understanding of the intersection of making with deeper educational principles. In addition to framing the open portfolio, the four domains for the Maker Showcase also serve as an organizing framework for the UTeach Maker program. The domains and their corresponding criteria provide a focus for group and individual maker pursuits. Therefore, all program activities are designed to further participant work with regards to one or more of these areas. In this way, the Maker Showcase forms the backbone of the program.

Community Support

Given the comprehensive nature of the Maker Showcase and the central role that community plays for makers, providing adequate support to participants was a key consideration in the development of the model. These support structures include mentorship from a maker in the community, monthly cohort meetings, use of makerspaces in the community, participating in maker-centered internships, and meeting other makers through local venues and events. These support struc-

tures are designed to promote agency, while still providing guidance and access to skills and knowledge.

Maker Mentors

Each UTeach Maker participant is paired with a maker mentor. These individuals are maker educators who are active in local and/or global maker communities. Many have access to makerspaces where students can work and practice skills like laser cutting or woodworking. Each mentor is grouped with four preservice teachers based on area of interest and availability. Mentors play an important role in guiding the participants along a personal journey of exploration.

Once paired, the mentor regularly checks-in with their mentees, ensuring that the participant is making progress on the Showcase. They work together to set timelines and define goals in order to complete the Showcase by graduation. As the participants complete each component of the Showcase, mentors review and provide valuable feedback. Following the external Showcase review, mentors work with their mentees to distill and respond to feedback.

UTeach Maker mentors possess a wealth of knowledge about classroom interactions and connections that exist between K–12 settings and the maker community. They help their mentees consider classroom management strategies, implementing lesson plans, access to tools and technology, and the integration of STEM content through making. Furthermore, because they are often connected to local maker groups, they have insight on professional development opportunities and events that serve to better inform participants as they move into educational spaces.

Cohort Meetings

The UTeach Maker community meets once a month for a cohort meeting. These meetings are held on campus or at local makerspaces, and, in addition to participants and mentors, often include stakeholders and guest presenters from the local maker community.

Each cohort meeting follows a similar agenda, starting with meals and mentor-mentee check-ins. These valuable conversations allow time for the mentors to evaluate the progress of mentees on maker projects and the larger Showcase product. Following dinner, participants engage in maker-centered professional development, including topics such as project-based learning, elements of design, engineering practices, and other maker techniques and tools. Makers are given time to reflect and think about application of the cohort theme before the meeting is adjourned, allowing each student to tie the learning to their own educational journey.

Local Makerspaces, Internships, and Community Events

Despite hosting the UTeach Maker program, UTeach Austin does not have its own makerspace. Participants have access to the university makerspaces located

in the Fine Arts and Engineering Departments. We have also partnered with local K–12 public and private schools, encouraging participants to explore school -based makerspaces while building relationships with educators and working with pupils. In addition, UTeach Maker has worked to identify community makerspaces to provide access for those in our program. By encouraging participants to engage with resources outside of the university, we hope to provide access to making that is not bounded by their status as a student.

Many participants opt to partake in maker-centered internships. Offered as part of the larger UTeach Internship program, these internships are part-time, paid positions where students are placed with schools or non-profit organizations that support making. During their internship, participants are encouraged to practice using new equipment and work on their skills for facilitating making with groups of pupils. These field experiences help participants to develop necessary instructional skills while at the same time providing support and role models for underserved spaces in the community. For example, existing internships have provided an opportunity to work with a local non-profit to engage in museum learning activities related to mathematics and to the development of maker-based math tools such as golden ratio calipers, laser-cut unit circles, and templates for hands-on mathematical activities. Maker interns have also worked at several local middle and high schools and helped to set up and run often under-staffed makerspaces.

UTeach Maker participants also have the opportunity to participate in community events like maker meetups and conferences. For example, as a group, UTeach Maker regularly attends socials and events as part of SXSW EDU and volunteers and presents at the Austin Maker Faire. By attending and engaging in these opportunities, participants increase the visibility of making in the educational community and foster meaningful connections between teachers and makers.

External Review and Public Presentation of Work

The open source nature of the maker community highlights the pride that many makers feel in sharing and showcasing their work. Thus, UTeach Maker was intentional in building the program with a careful eye towards public-facing products.

The Maker Showcase was designed as an openly networked product that would allow for members of the public to view each makers journey in ways that provide inspiration and insight. In addition to being accessible online, during their final semester, each participant's Maker Showcase is sent out for external review from a national expert in the field of maker education. During this time, participants personally present their work to a group of peers, family, friends, and mentors. The public presentations occur during the last cohort meeting of the semester. At these presentations there is time for the participants to highlight their maker accomplishments as well as time for questions and feedback from the audience. This structure promotes two-way conversations and audience engagement. The presentations also help UTeach Maker participants, just starting the program, to

better understand the domains of the Maker Showcase and begin to formulate ideas about their own maker trajectory. Both the external review and the public showcase presentation encourage feedback and reflection, requiring participants to consider next steps and/or the implications of their work.

UTeach Maker has also been intentional about ensuring the public and free availability of the materials developed as part of the program. To this end, the program website now serves as a repository for materials with open access to all. For example, UTeach Maker hosts a Maker Lesson Bank. This bank serves as a reservoir of lessons, instructional ideas, and planning tools. It is openly available and can be used by any educator ready to explore making.

LESSONS LEARNED

UTeach Maker is entering its fourth year. In that time, many lessons have been learned. The section below outlines the insights gained in developing a maker-centered framework for the program and navigating logistical considerations. The section also highlights the benefits of adding inservice teachers to the program and describes the emergence of our participants as teacher leaders.

Developing The Elements of Making Framework

A major challenge faced by UTeach Maker was to define what we, as a program, meant by the term 'making.' In some ways, making by its nature defies a singular definition but for program cohesiveness it was important to establish a shared meaning. This task was taken on in the Spring of 2016 and approached in several ways. In one approach, Dr. Jill Marshall, a co-PI for the Noyce grant, engaged in a literature review around the essential elements of making. She included articles from peer-reviewed journals, practitioner journals, and books on making and maker education. She also conducted a survey of well-known maker educators and leaders in the field and used this information to create a framework to evaluate making in STEM classrooms (Marshall & Harron, 2018). At the same time, UTeach Maker's newly formed advisory group of local maker educators from university, informal STEM, and K–12 settings approached the question in another way. The group was charged to develop a consensus statement about making. This process started by each advisory group member submitting their own conceptual statement. The submissions were then combined into a central statement which was sent to the group for comment. After several rounds of debate and revision, the UTeach Maker advisory group settled on a statement in the Fall of 2016: "Making is an iterative process of tinkering and problem solving that draws on a DIY mindset. Making is collaborative and allows for self-expression through the creation of a personally meaningful artifact that is shared with a larger community." (UTeach Maker, 2016).

The advisory group statement, in line with literature on maker education and Dr. Marshall's findings, became the guidepost for UTeach Maker. However, over

time, it became clear that the participants needed even more detail about what it meant to 'make' in STEM classrooms. Thus, using the UTeach Maker advisory group statement along with our experience supporting novice maker educators, a framework for classroom-based making was created. This framework, called *The Elements of Making*, was published in the February 2018 issue of The Science Teacher (Rodriguez et al., 2018). It suggests that learners as makers: 1) create original, personally meaningful products, 2) engage in iterative design and fabrication, 3) demonstrate characteristics of a maker mindset, 4) collaborate and connect with community, 5) present their work publicly, and 6) utilize science and engineering practices. The framework articulates each of these elements with varying degrees of scaffolding and provides a range of options to help support students as they move toward more independent, sophisticated making. These elements now form the backbone of the UTeach Maker stance on classroom-based making.

The development of the Elements of Making framework was a turning point for the program. It gave our makers, their mentors, and the classroom teachers they work with a common language for discussing making, planning lessons, and evaluating their work. It helped explain to schools and administrators what was meant when they were asked to try maker-centered lessons in classrooms. It also gave our program graduates a vocabulary to use when explaining making to others and highlighting various aspects of their Maker Showcase to employers. Finally, the Elements of Making framework has become the foundation of the maker lesson planning tools and the published maker lessons produced by the program.

Lessons on Logistics

While developing a framework for making was a major conceptual challenge, running a micro-credentialing program like UTeach Maker also comes with logistical challenges. As of the spring of 2019, there were 24 makers, six maker mentors, and 16 maker fellows. Each of these stakeholders is at a different point in their career and in the UTeach Maker program. It takes significant coordination and effort to keep everyone together. To that end, the UTeach Maker website has been invaluable. This site holds everything from promotional videos, to meeting summaries, to calendars, to forms and applications. In addition, having a UTeach master teacher as the program director has also been useful in meeting this challenge. The relationships formed with students throughout their time in the UTeach program is an essential part in the effectiveness of monitoring and support efforts. The director consistently strives to stay in personal contact with each program participant.

Benefits of Inservice Teachers

It is important to note that not all participants are preservice teachers. Shortly after UTeach Maker was established, several UTeach graduates asked to apply. Though inservice teachers were not included in the original program conception,

we were eager to add them because of the potential to maintain strong relationships with our graduates, foster a continuum of learning, and promote maker practices in the classrooms of practicing teachers. We have learned that the inclusion of inservice teachers provides a critical link between the discussion of making on a theoretical level and the integration of making within classrooms. Having inservice teachers as participants offers opportunities for collaboration between novice and more seasoned teachers. It promotes cross-pollination between UTeach and the local schools. Their familiarity with maker tools is a resource that UTeach Maker participants can bring to the schools who often have access to maker technologies but lack the expertise in their use. Thus, the inservice teachers in the program provide an immediate value-add to local schools, help pave the way to bring making to these venues, and expose more students to creative STEM instruction.

Emergence of Teacher Leaders

The opportunity for participants to obtain a micro-credential boosts their own understanding and increases attractiveness to potential employers. However, the UTeach Maker experience offers other ways for participants to have an impact. We have found that, with encouragement and support from the program, many participants have become teacher leaders in the area of maker education. For example, several participants have started Maker Faires at their schools and delivered maker professional development to teachers on their campuses. Program participants have also presented at regional and national conferences and have facilitated national conversations around making in education. In addition, several makers have submitted the maker-centered lessons that they have created to peer-reviewed journals aimed at K–12 STEM educators. To date, program participants have authored or co-authored five maker-centered articles for NSTA journals, The Science Teacher and Science Scope, with two more articles currently under review. While these experiences are valuable in their own right, they also provide makers with valuable opportunities to see themselves as educational front-runners and to develop skills for communication and collaboration that can foster future innovation and leadership.

DISCUSSION

In summary, UTeach Maker is a selective micro-credentialing program that supports passionate UTeach students interested in bringing innovative, maker practices into their STEM classroom. Each participant has a maker mentor throughout their experience, receives an opportunity to develop key maker skills, develops a portfolio of their work, and enhances their teaching careers through the hands-on experience gained in this inventive program. Participants in the program become part of a community of like-minded others interested in reimagining traditional education. They become part of the larger maker community and establish a network of maker friends and colleagues to support them.

Moving forward, UTeach Maker continues to look for ways to support participants as UTeach students, as classroom teachers, and a as part of a network of educational leaders. As previously mentioned, participants in UTeach Maker are tasked with completing and submitting a Maker Showcase. While participants work on this artifact throughout their time in the program, there is often significant effort needed for the final push. Thus, a new Maker Fellow Friends initiative is being piloted. In this effort, recent Maker graduates support participants who are in their final semester and help them as they document their accomplishments. Participants are matched with Fellow Friends based on content area specialty and interests. Since the Fellow Friends have recently been through the Showcase process, they are well positioned to serve as peer mentors. UTeach Maker is also partnering with other maker-centered organizations working in education to provide ongoing support for participants teaching outside of the local area.

Finally, in the spring of 2019, 22 UTeach programs signed a Make For All commitment to help expand maker-centered learning for all students. These programs join a group of over 50 other organizations spanning 149 communities nationwide. The programs in this initiative are positioned to reach more than 768,000 students in the next three years. As part of this effort, several UTeach programs have plans to implement their own UTeach Maker initiatives. These universities intend to integrate making into UTeach coursework, join with existing maker-based university initiatives, connect with community and K–12 partners, and/or establish their own maker micro-credentialing program. By bringing making into the classroom, participants of UTeach Maker programs across the nation will pave the way for a new vision of STEM education and help ensure that students from all backgrounds have an opportunity to engage in making as an authentic and empowering way of learning.

REFERENCES

Bevan, B. (2017). The promise and the promises of making in science education. *Studies in Science Education, 53*(1), 75–103. doi:10.1080/03057267.2016.1275380.

Blikstein, P., & Worsley, M. (2016). Children are not hackers: Building a culture of powerful ideas, deep learning, and equity in the Maker Movement. In K. Peppler, E. R. Halverson, & Y. B. Kafai (Eds.), *Makeology: Makerspaces as learning environments* (Vol. 1, pp. 64–79). New York, NY: Routledge.

Clapp, E. P., Ross, J., Ryan, J. O., & Tishman, S. (2016). *Maker-centered learning: Empowering young people to shape their worlds.* San Francisco, CA: Jossey-Bass.

Cohen, J. (2017) Maker principles and technologies in teacher education: A national survey. *Journal of Technology in Teacher Education, 25*(1), 5–30. Retrieved from: https://www.learntechlib.org/primary/p/172304/

Cohen, J., Jones, W. M., Smith, S., & Calandra, B. (2017). Makification: Towards a framework for leveraging the maker movement in formal education. *Journal of Educational Multimedia and Hypermedia, 26*(3), 217–229.

Fields, D. A., Kafai, Y., Nakajima, T., Goode, J., & Margolis, J. (2018). Putting making into high school computer science classrooms: Promoting equity in teaching and learn-

ing with electronic textiles in exploring computer science. *Equity & Excellence in Education, 51*(1), 21–35. doi:10.1080/10665684.2018.1436998.

Freeman, A., Adams Becker, S., Cummins, M., Davis, A., & Hall Giesinger, C. (2017). *NMC/CoSN horizon report: 2017 K–12 edition*. Austin, TX: The New Media Consortium. Retrieved from: https://cdn.nmc.org/media/2017-nmc-cosn-horizon-report-k12-EN.pdf

High Quality Project Based Learning (HQPBL). (2018). *A framework for high quality project based learning*. Retrieved from: https://hqpbl.org/wp-content/uploads/2018/03/FrameworkforHQPBL.pdf

Institute for Design at Stanford. (2010). *An introduction to design thinking process guide*. Retrieved from: https://dschool-old.stanford.edu/sandbox/groups/designresources/wiki/36873/attachments/74b3d/ModeGuideBOOTCAMP2010L.pdf

Martin, L. (2015). The promise of the maker movement for education. *Journal of Pre-College Engineering Education Research, 5*(1), 30–39. doi:10.7771/2157-9288.1099.

Marshall, J., & Harron, J. R. (2018). Making learners: A framework for evaluating making in STEM education. *Interdisciplinary Journal of Problem-based Learning 12*(2). doi:10.7771/1541-5015.1749

National Research Council (NRC). (n.d.). *Three dimensional learning*. Retrieved from: https://www.nextgenscience.org/three-dimensions

National Science Foundation (NSF). (2017). *The National Science Foundation and making*. Retrieved from: https://www.nsf.gov/news/news_summ.jsp?cntn_id=131770

Papert, S. (1991). Situating constructionism. In I. Harel & S. Papert (Eds.), *Constructionism* (pp. 1–11). Norwood, NJ: Ablex Publishing Corporation.

Peppler, K., Halverson, E., & Kafai, Y. B. (2016). Introduction to this volume. In K. Peppler, E. Halverson, & Y. B. Kafai (Eds.), *Makeology: Makerspaces as learning environments* (Vol. 1, pp. 1–11). New York, NY: Routledge.

Qi, J., & Buechley, L. (2010, January). Electronic popables: Exploring paper-based computing through an interactive pop-up book. In *Proceedings of the fourth international conference on Tangible, embedded, and embodied interaction* (pp. 121–128). Cambridge, MA: ACM.

Qi, J., & Buechley, L. (2014, April). Sketching in circuits: Designing and building electronics on paper. In *Proceedings of the SIGCHI Conference on Human Factors in Computing Systems* (pp. 1713–1722). Toronto, ON, Canada: ACM.

Rodriguez, S., Harron, J., Fletcher, S., & Spock, H. (2018). Elements of making: A framework to support making in the science classroom. *The Science Teacher, 85*(2). doi:10.2505/4/tst18_085_02_24

Tofel-Grehl, C., Fields, D., Searle, K., Maahs-Fladung, C., Feldon, D., Gu, G., & Sun, C. (2017). Electrifying engagement in middle school science class: Improving student interest through e-textiles. *Journal of Science Education and Technology, 26*(4), 406–417.

UTeach Maker. (2016). *What is making?*. Retrieved from: https://maker.uteach.utexas.edu/why-make/what-is-making

Windschitl, M. (2003). Inquiry projects in science teacher education: What can investigative experiences reveal about teacher thinking and eventual classroom practice?. *Science Education, 87*(1), 112–143. doi:10.1002/sce.10044.

CHAPTER 9

LABORATORY SCHOOLS TO SUPPORT THE PREPARATION OF UTEACH PRESERVICE TEACHERS

Michael Odell, Teresa Kennedy,
Jo Ann Simmons, and Jaclyn Pedersen

Replicating the UTeach STEM model can create challenges for smaller universities. The most significant challenge is funding to ensure the sustainability of the program once initial grant funds are exhausted. Traditional university funding formulas can be problematic for smaller specialized teacher preparation programs due to essential program elements that require investment. These elements include the hiring of Master Teachers to support preservice teachers in their coursework and clinical experiences, as well as securing adequate numbers of Mentor Teachers and Cooperating Teachers for clinical placement sites implementing inquiry-based instructional strategies in the teaching of mathematics and science. To address these challenges, the University of Texas at Tyler created a STEM focused K–12 Laboratory School. The laboratory school serves as self-supporting high-fidelity placement site for UTeach clinical and apprentice teaching experiences improving learning outcomes for preservice teachers.

Keywords: STEM Academy, Laboratory School, Program Sustainability, Inquiry-based learning, Project-based Learning, Problem-based Learning, Clinical.

Preparing STEM Teachers: The UTeach Replication Model, pages 119–133.

Since the launch of Sputnik in 1957, there have been concerns about the lack of STEM competitiveness in the United States, along with the fear of our potential inability to meet the demands of the rapidly changing STEM-based economy. As a result, calls for educational reform surfaced, such as through the landmark report entitled, *A Nation at Risk* (United States National Commission on Excellence in Education. Department of Education, 1983). A more recent report, *Rising Above the Gathering Storm,* reiterated the same concerns. According to the report, if the U.S. is to remain competitive in the 21st century economy, there must be a serious effort to "enlarge the pipeline of students who are prepared to enter college and graduate with a degree in STEM" (National of Academy of Sciences, National Academy of Engineering, & Institute of Medicine, 2007, p. 6).

Two recommendations in *Rising Above the Gathering Storm* are specifically relevant to K–12 STEM education reform and the UTeach model replication. The first recommendation called for the "annual recruitment of 10,000 science and mathematics teachers" (National of Academy of Sciences, National Academy of Engineering, & Institute of Medicine, 2007, p. 6). In addition to growing the teaching workforce, another recommendation in the report was to "enlarge the pipeline of students who are prepared to enter college and graduate with a degree" in STEM disciplines by increasing the number of students who pass AP, IB, and dual credit coursework. The implication is that new STEM teachers will need to be better prepared to provide high school students with advanced STEM coursework, with the goal that they will be STEM College Ready upon graduation. The report recommendations specifically call for the development of specialty STEM high schools:

> Specialty secondary education can foster leaders in science, technology, and mathematics. Specialty schools immerse students in high-quality science, technology, and mathematics education; serve as a mechanism to test teaching materials; provide a training ground for K–12 teachers; and provide the resources and staff for summer programs that introduce students to science and mathematics (National of Academy of Sciences, National Academy of Engineering, & Institute of Medicine, 2007, p. 6).

In 1997, the University of Texas at Austin created an innovative STEM teacher preparation program in response to the need for more and better prepared STEM teachers, establishing the UTeach Program, a rapid response initiative established and successfully functioning ten years prior to the publication *Rising Above the Gathering Storm*. "UTeach was created expressly to attract significant numbers of diverse science, technology, engineering, and mathematics (STEM) majors and prepare them to become exemplary secondary science, mathematics, and computer science teachers" (The University of Texas at Austin, 2011). UTeach prepares teachers with deep content knowledge and inquiry-based pedagogical strategies. Overall, research has shown that Texas UTeach graduates stay in teaching longer, improve student performance in math and science, and influence students to

enter STEM fields (Backes, Goldhaber, Cade, Sullivan, & Dodson, 2018; Beth, Hughes, Romero, Walker, & Dodson, 2011; University of Texas at Austin, 2019).

UTeach Preservice teachers complete four pedagogical courses prior to enrolling in apprentice teaching with their assigned cooperating teacher. One of the important features of the UTeach model is placing potential STEM preservice teachers in classrooms to observe and teach. This can occur as early as the freshman year of college. Typically, STEM majors in the freshman and sophomore years of their degree program enroll in two introductory courses developed to provide an opportunity for students to "try out teaching." The first course, Step 1: Inquiry Approaches to Teaching, provides an opportunity for UTeach Preservice Teachers to learn about the teaching profession by completing courses covering theory and practice behind inquiry-based science and mathematics instruction, as well as teach lessons in an elementary classroom and experience curriculum planning and implementation. During Step 2: Inquiry-Based Lesson Design, preservice teachers are provided experiences with developing three middle school inquiry-based mathematics and/or science lessons that are taught in sixth, seventh and eighth grade classrooms. Preservice teachers that elect to continue in the UTeach Program after completion of Steps 1 and 2 are required to complete two additional courses: Classroom Interactions and Project-Based Instruction (PBI) prior to Apprentice Teaching. Field experiences are integrated into both courses to provide real-world teaching experiences prior to engaging in apprentice teaching. During Apprentice Teaching, the culminating student teaching experience, UTeach preservice teachers are provided with opportunities to explore leadership opportunities beyond the classroom. They are encouraged to join professional organizations, attend conferences, and collaborate with others through the UTeach STEM Educators Association to share their experiences at professional conferences and professional development opportunities.

The UTeach model has been replicated at over U.S. 45 universities with an enrollment of 6,938 preservice teachers in academic year 2016–17 (University of Texas at Austin, 2019). Most UTeach replication sites are located at large universities where economies of scale may allow for program sustainability on typical student credit hour production formulas. To support replication institutions, the UTeach Institute developed a guiding document, the UTeach Elements of Success (University of Texas at Austin, 2011). The UTeach Elements of Success provide criteria for measuring implementation progress at each site. More information on the UTeach Elements of Success can be found in Chapter X (Hughes and Romero Eddington).

The University of Texas at Tyler UTeach Replication Model

The University of Texas at Tyler established the UTeach Tyler replication site in August 2010 with faculty from the School of Education and STEM departments in the College of Arts and Science. Initially, the replication at UT Tyler was funded with grants from the Texas Education Agency, the University of Texas

System, and private foundations. Once grant funding ended, the expectation of UT Tyler Administration was that the program would be budget neutral. Therefore, the newly established UTeach Program would have to abide by the same budget formulas as the traditional teacher certification programs offered by the university or find other sources of funding to make up the difference in anticipated needed resources. This expectation created challenges, both financially and programmatically. While funding is a specific challenge for UT Tyler, the strategies that follow may provide insights for other programs that may be struggling to replicate with fidelity since budget allocation formulas vary from university to university.

In the case of UTeach Tyler, the most costly element of UTeach replication is the addition of full-time Master Teachers, veteran teachers that teach pedagogy courses and supervise clinical experiences. Master Teachers are a necessary and vital component of the UTeach Program. A related factor is the requirement for increased supervised observations throughout the UTeach Program as compared to the traditional preservice teacher education program. For example, there are five observations of apprentice teachers in the UT Tyler traditional preservice education program compared to eleven observations for UTeach Tyler apprentice teachers. The large number of observations was a compromise with the administration in the School of Education to provide UTeach preservice teachers with six observations using the UTeach focused observation protocols and five observations using a modified version of the Texas-Teacher Evaluation and Support System (T-TESS) utilized in school districts to evaluate teachers. In addition, there is the general expectation that Mentor Teachers and Cooperating Teachers hosting preservice teachers for clinical experiences and apprentice teaching receive compensation for their time, expertise and effort. More observations increases the cost of placing students in classrooms.

Costs are not the only challenge. UT Tyler is located in a rural location in East Texas. Finding enough placement sites for clinical teaching observations and apprentice teaching is often problematic. The UTeach model requires clinical experiences early in the program. Providing teacher education candidates with early field experiences leads to a better understanding of the significance of learning contexts and classroom diversity, as well as the demands of the teaching profession (Yarmus & Begum, 2013). It is common for freshmen and sophomores to enroll in the STEP 1: Inquiry Approaches to Teaching course and the follow-on course STEP 2: Inquiry-based Lesson Design. The early field experience courses create a need for additional placement sites when sites are already limited. Making the case to local schools and the traditional teacher preparation programs to place freshman and sophomores in early clinical experiences to introduce STEM majors to teaching as a recruitment strategy, specifically inquiry-based teaching and inquiry-based lesson design, are a lower priority for local schools and the School of Education placement office when compared to finding placements for juniors and seniors that have been admitted to the teacher certification program and are committed to graduating with a teaching credential.

Even when placements are available, finding STEM classrooms that engage in inquiry are at a premium. Historically, there has been the lack of connection between coursework and clinical experiences. Although most teacher preparation programs include multiple and diverse clinical experiences, there is often a disconnect between what teacher candidates are taught in their pedagogy courses and their opportunities to learn to enact these practices in their clinical placements (Bullough et al., 2002; Bullough, Kauchak, Crow, Hobbs, & Stokes, 1997; Francis, 2017; Kleinsasser, 2000; Zeichner, 2010; Zeichner & Bier, 2012). Finding STEM placement sites where preservice teachers can implement inquiry-based pedagogies for the entire UTeach course sequence is essential if the model is to be replicated with fidelity.

Direct instruction tends to be the common teaching methodology in area schools. One of the contributing factors that has led to a predominance of direct instruction is the impact of state testing. School accountability measures in Texas are based in large part on high stakes exams. High stakes testing often results in the unintended consequence that instructional strategies used in the classroom become more teacher-centered versus student-centered. Studies have shown that teachers showed dramatic increases in the amount of teacher-centered, fact-driven instruction in subjects included in state mandated tests (Au, 2007; Gerwin & Visone, 2006). These practices in the clinical setting can create a disconnect between what preservice teachers learn in their pedagogy courses and what they experience in their clinical placements. This is especially problematic during the spring semesters since many schools focus on test preparation in the spring to prepare their students for the State of Texas Assessments of Academic Readiness (Texas Education Agency, 2019). As a result, the Texas testing season typically involves a significant amount of time diverted from instruction to benchmark testing and student interventions. Students in traditional schools often complete packets of test-prep worksheets, with the hope that the more exposure to test material, the better they will perform. The instructional strategies tend to become focused on the test prep items.

UNIVERSITY LABORATORY SCHOOLS

In 2011, UT Tyler university faculty and administrators met to discuss STEM teacher preparation, sustainability of the UTeach Tyler Program, and education reform in general. After considerable deliberation, the university submitted an application to the Texas State Board of Education and the Commissioner of Education to open three University Charter Schools. Discussion initially focused on the costs of sustaining the UTeach program once grant funded expired. That said, there were a number of factors outside of UTeach concerns that opened the door for considering opening laboratory schools. One of the primary factors included increasing campus facility utilization numbers. This was especially attractive at that time as the university had excess space. This included the main campus and satellite campuses. A second factor was recruiting. Opening a STEM focused lab

school provided a new recruitment strategy and allowed the university to think or recruitment in terms of P–16 versus solely high school. Finally, opening the laboratory schools provided a pool of master teachers whose job description included working with preservice UTeach students in addition to teaching K–12 students. It should be noted that there was pushback from local school districts. The administration moved forward in spite of these political challenges and the long-term impacts have been minimal.

These charter schools, deemed as Laboratory STEM Schools, would implement the Texas STEM Academy Blueprint (Avery, Chambliss, Pruiett, & Stotts, 2010) and incorporate the tenants of the UTeach Program. Creating public university Laboratory Schools would also provide additional placement sites for preservice teachers to focus on STEM and inquiry-based instructional strategies and experience the methodologies taught in the UTeach course sequence with higher levels of fidelity when compared to placement sites in local public schools. The Laboratory Schools initially opened in fall 2011 under the name the University of Texas at Tyler Innovation Academies (UTTIA). In 2018, the academies were renamed the University of Texas at Tyler University Academies (UTTUA).

As public schools, all students enrolled in the laboratory schools must take the STAAR assessment and each school is rated on the same accountability criteria required of all Texas public schools. Although the state standards and assessments are the same for the academies and traditional public schools, teacher practices and student experiences can differ significantly from school to school. Research has shown that "teaching in a "high-stakes" assessment environment impacts the implementation, fidelity, and sustainability of new teaching methods" (Boardman & Woodruff, 2004, p. 1). However, when teachers understand the intermediate conditions that must be met to incorporate test content and format into established teaching methods, successful implementation can occur (Chapman & Snyder, 2000; Flessner, 2012). Because the university oversees the academies, the teaching methodology implemented can be consistently implemented across all academy locations. The academies focus less on test preparation and focus more on inquiry-based teaching that incorporates test content during instruction. In this student-centered learning environment, the interests of the students take center stage and the teacher facilitates student voice and choice. Students have a more active role in their learning. This environment helps students develop the skills for independent problem solving and lifelong learning.

Strategically, because the laboratory schools are public schools, they are eligible for state funding. This solves many of the cost barriers of implementing UTeach at a small university. UTeach Tyler and the University Academies leverage personnel and expertise to support STEM teacher preparation. Furthermore, having school personnel working closely with university faculty in a laboratory school setting provides opportunities for practical and theoretical issues to be addressed in the classroom resulting in consistent messaging to students enrolled in UTeach. This minimizes disconnects between coursework and practice during

clinical observations and experiences. An added benefit to opening the university charter schools is the creation of a pipeline of future college students interested in and prepared for STEM majors at the university, including STEM teaching.

University laboratory schools are not new. The first laboratory school in the U.S. was opened at the University of Chicago in 1894. The purpose of the first lab school was to develop theories of child development and education (Wilcox-Herzog & McLaren, 2012). The UT Tyler University Academies have a similar purpose. The schools serve as a research platform with a focus on inquiry-based STEM teaching and learning, primarily Problem-based and Project-based learning pedagogies (PBL). The design of the schools follows the Texas STEM Academy Design blueprint (Avery et al., 2010; Educate Texas, 2015).

All three UT Tyler University Academies are designed to serve as high fidelity placements for UTeach students for clinical observations and teaching. UTeach preservice teachers placed in the schools conduct observations and are immersed in classrooms that utilize inquiry-based instructional strategies taught in the UTeach courses. Most importantly, UTeach preservice teachers are placed with high quality Mentor and Master Teachers who work at these schools and support the UTeach Tyler STEM teacher preparation program.

The state of Texas allows for the creation of university charter schools. The process is not competitive and interested universities can apply for a charter if they can demonstrate adequate need for opening a school. The commissioner of education may grant open-enrollment charters to public universities as long as they meet certain criteria as outlined in the Texas Education Code. These schools may operate on a campus of the university (See Texas Education Code (TEC), Chapter 12 (Outside Source), Subchapter E §12.152.). Universities may be granted a charter if the following conditions are met. The specific criteria that university educational programs must follow includes the following:

- Be structured in a manner that incorporates innovative teaching methodologies;
- Be implemented under the direct supervision of a member of the teaching or research faculty; and
- Be designed to meet specific goals described in the charter, including improving student performance, and each aspect of the program must be directed toward the attainment of the goals as laid out in the charter application.

UT Tyler submitted a charter application to open schools located on the main campus in Tyler and its two satellite campuses in Longview and Palestine. To meet the conditions set forth by the commissioner, the laboratory schools would implement inquiry-based instruction as the primary "innovative" instructional methodology for all core subject areas. Inquiry-based instruction is a teaching strategy where teachers create scenarios where students solve problems. Lessons

are designed to provide students an opportunity to make connections to previous knowledge and interests.

University charter schools are open enrollment public schools. There is no cost for attendance and any student may apply to attend. If there is an empty seat, students may enroll. If classes are full, there is a lottery for any available seats that open annually. Each academy is allowed to balance classes by gender to increase the number of girls in STEM education pathways.

The T-STEM Academies Design Blueprint (Educate Texas, 2015) serves as a road map for benchmarks, program requirements and indicators to facilitate academy growth. The T-STEM Academy model was selected because it aligns with the university strategic plan for growing STEM programs in Engineering and Biomedical sciences. And, as a result, the academies have become feeder programs to STEM degree programs at the university.

The University Academies serve as model schools that utilize inquiry-based instructional strategies for STEM focused schools. As such, they are designed to support clinical experiences of UTeach students. The UTeach Preservice Teachers interact with K–12 students that are enrolled in vertically aligned STEM pathways. The academies only offer STEM tracks. As called for in the report *Rising Above the Gathering Storm*, the University Academies are STEM-focused specialty schools that prepare K–12 students for STEM majors at the university. Students attending can complete up to 42 credits of university coursework prior to high school graduation at no cost to the student. Unlike Early College High Schools that focus on students earning Associates degrees, the University Academies provide dual credit courses that meet the requirements of Engineering, Chemistry, Biology, and Biomedical Science degrees at the university. Consequently, UTeach Preservice Teachers have the opportunity to complete clinical experiences in STEM-focused schools and practice designing and delivering inquiry-based lessons. This opportunity sets them apart from other traditional teacher preparation programs, making them more prepared and experienced to maintain rigorous inquiry-based STEM instruction.

UTeach Tyler is committed to providing preservice teachers with diverse placements for clinical observation and teaching. Therefore, UTeach preservice teachers are placed at the University Academy Lab schools and in traditional public schools. Multiple placements assure that UTeach students experience teacher-centric traditional instruction and classroom management strategies that are the primary methodologies in local public schools. They also experience the University Academies that are student-centric and utilize inquiry-based instruction such as Project-based Learning and Problem-based Learning (PBL), as well as the 5Emodel of instruction. Classroom management strategies in the academies are focused on collaborative student-centered learning. Student-centered classroom management requires more responsibility on the part of the K–12 student. Developing relationships with students that help K–12 students become thoughtful about their contributions or detraction from the class community can be ob-

served and practiced versus traditional classroom discipline techniques. The lab school setting also assures that UTeach Preservice Teachers experience STEM instructional technology to enhance learning that may not be available in traditional schools. The academies provide a 1:1 computer to student ratio and blended learning is implemented in most classrooms.

Each Laboratory School campus serves unique demographics. The Tyler Campus is representative of a small urban school population. The Palestine Campus population is representative of a small rural school. The Longview Campus is representative of a suburban school. UTeach students leave the program having experienced distinct populations and school cultures when considering the university academy and local school placements.

The University Academies are K–12 schools and serve a total population of 740 students in 2018–19. Each school serves one class section per grade. UTeach students experience classrooms at the elementary, middle, and high school levels. Mentor teachers at the academies teach across grade levels vertically. A typical mathematics or science teacher will teach at multiple grade levels instead of the same class at the same grade level repeated all day as commonly occurs when compared to teaching loads at large schools. This allows Uteach Tyler Preservice teacher to be mentored by a STEM teacher that teaches multiple grade levels versus a single grade or subject.

Many of the Mentor Teachers at the academies are UTeach graduates. Having graduated from the program provides clinical experience consistency. Teachers at the academy understand the UTeach model and the types of experiences UTeach students are expected to complete. As university employees and laboratory school teachers, working with clinical teachers is part of their assigned load which helps reduce the overall expense of the program. By opening the University Academies, UT Tyler is able to address *all* the UTeach Elements of Success, including those that would prove challenging as the program moved toward sustainability. This includes the following challenging Essential Elements:

- Long-term Institutional and Community Support;
- Cross College and School District Collaboration;
- Dedicated Master Teachers;
- Rigorous Research-based Instruction; and
- Early and Intensive Field Experiences

Long-Term Institutional Support

As described earlier, once initial grant support ended, maintaining the UTeach Program to fidelity would be difficult based on typical budget outlay formulas at the university. Operating Laboratory Schools allows UT Tyler to cost-effectively implement the UTeach Program with higher fidelity. Because the Laboratory Schools are self-supporting with funding from the state, supporting Master Teachers, Mentor Teachers, and Cooperating Teachers employed by the university at the

academies leverages the cost of the UTeach model. The UTeach Tyler Program is not large enough to support hiring full-time Master Teachers. Rather than support full-time Master teachers, the university has the flexibility to buy-out a portion of a teacher's contract to dedicate time to supporting the UTeach Program. The university also recruits teachers at the laboratory school who are graduates of the UTeach Program to serve as Mentor and Cooperating teachers, building a pipeline of teachers that experienced the UTeach preparation model and have implemented inquiry-based teaching strategies in their classrooms as the primary mode of instruction for secondary STEM students. As university employees, teachers in the school can work with preservice teachers as part of their employment contract. In addition, instead of hiring full-time Master Teachers, the university has leveraged laboratory school personnel for teaching UTeach pedagogy courses. Teachers at the school who have completed their Masters or Doctorate degrees are eligible for adjunct faculty status within the School of Education.

RIGOROUS RESEARCH-BASED INSTRUCTION AND INTENSIVE FIELD EXPERIENCES

Since the School of Education has oversight of the curriculum, instructional strategies, and input on hiring of personnel, the fidelity of practice in clinical experiences can be consistent and align to inquiry-based instruction taught in the UTeach courses. In some cases, the UTeach courses are delivered in the laboratory schools providing an authentic setting for the courses.

To accommodate diverse clinical settings that allow UTeach students to experience multiple school environments, preservice teachers are assigned to clinical experiences based on the content of the UTeach courses, the desired certification level and subject, and the clinical experience expectations. UTeach Tyler students can earn certification in Middle School Math 4–8, High School Math 7–12, Middle School Science 4–8, Biology 7–12, or Chemistry 7–12. By first placing UTeach preservice teachers in the academies for early experiences (UTeach Steps 1 & 2) followed by enrollment in the PBI course, the program is able to demonstrate to the preservice teachers the importance of inquiry-based pedagogies. The experiences are authentic and provide experiential credibility. In addition, since the teachers at the University Academy supervise as part of their contract, the budget impact is neutral. The academy teachers also benefit from a steady supply of prepared STEM clinical interns that enhance the collaborative school learning environment.

According to the UTeach Elements of Success, "in order to promote confidence and accelerate professional development, UTeach students begin a carefully scaffolded sequence of intensive teaching opportunities in their first semester of the program and continue these field experiences throughout" (University of Texas at Austin, 2011, p. 2). There are a number of logistical and pedagogical challenges of placing preservice teachers in traditional clinical teaching placements, especially when there are limited number of placement sites. Teacher preparation programs

have little or no control over the curriculum and teaching strategies that preservice teachers will experience. Replicating UTeach requires that clinical experiences be "domain specific," and "tightly articulated with the UTeach curriculum" (University of Texas at Austin, 2019). There is no guarantee that a school's curriculum and teaching strategies will align. Most local schools emphasize teacher-centered instructional strategies. As a result, preservice teachers come to see those as validated strategies versus those learned in the university pedagogy courses. Clinical experiences should be closely supervised by UTeach "course instructors—both tenure-track faculty and master teachers—to promote full integration of critical knowledge and skills" (University of Texas at Austin, 2019). This is difficult to achieve when placing students in traditional settings.

In rural settings, traveling to schools can be time intensive and inefficient. Creating laboratory schools eliminated these challenges. UTeach Faculty, Master Teachers, Mentor Teachers, and Cooperating teachers are all employed by the university. The academies and the UTeach Program are both housed primarily in the college. In practice, the laboratory schools are treated as a college department. This organizational arrangement promotes articulation between the UTeach Program and the academies.

UTeach requires that preservice teachers "develop their own lesson plans, using research-based instructional materials and strategies, with intensive coaching and feedback from both master teachers and tenure-track faculty who are experts in STEM content and pedagogy, in order to ensure UTeach expectations for accuracy and inquiry-based practice are met" (University of Texas at Austin, 2019). University faculty and Master Teachers work together and teach the Project-based Instruction course on location at the academies. Master teachers at the academy also serve as instructional coaches for the campus and UTeach preservice teachers.

UTeach preservice teachers are placed in the laboratory schools during their STEP 1 and 2 early field experiences. They develop their own lessons under the guidance of Master Teachers and University faculty in collaboration with the classroom mentor teachers. All parties are aligned in purpose to allow the preservice teacher to develop and deliver their inquiry-based lesson. All parties are trained in the UTeach observation forms to assure consistent feedback and reinforce inquiry-based instruction as a valid and preferred strategy for STEM teaching and learning. The laboratory school environment facilitates preservice teachers and ensures their "time in classrooms is carefully structured, from focused observations of authentic teaching to clinical interviews of students regarding problem solving strategies to their own experiences teaching, receiving formative feedback, and revising and re teaching lessons" (University of Texas at Austin, 2019). The laboratory school setting provides a "safe place" to practice inquiry-based teaching. Preservice teachers are also able to receive feedback from University Academy K–12 students about their lessons. This is especially helpful during the UTeach Project-based instruction course that is taught on-site.

UTeach replication promotes compensating mentor teachers that host UTeach preservice teachers for their "collaboration and creating supportive classroom environments, reviewing lesson plans, and provide oral and written feedback to UTeach students after observing them teach" (University of Texas at Austin, 2019). Being a teacher at the University Academy comes with the expectation that teachers will work with preservice teachers. Compensation is built into the teaching contract at the Laboratory Schools.

The Laboratory School setting allows UTeach preservice teachers to experience inquiry-based instruction to a high degree of fidelity compared to traditional classrooms. However, we recognize that UTeach preservice teachers must "experience multiple STEM teaching opportunities in high-need and diverse elementary, middle, and high school settings to gain an understanding of current K–12

TABLE 9.1. Clinical Experience by Course and Setting

Course	Experience	Classroom Setting	Clinical Setting	Lessons and Teaching Evaluated by:
STEP 1	Teach 3 Lessons 4 Hours Observations	University	University Academy Elementary Classroom	Mentor Teacher Master Teacher or University Faculty
STEP 2	Teach 3 Lessons 4 Hours Observations	University	University Academy Middle School or High School Classroom	Mentor Teacher and Master Teacher or University Faculty
Classroom Interactions	Teach 3 Lessons 30 Hours Observing and Assisting	University	High Need Title I Traditional School Site Local Middle School or High School	Mentor Teacher and Master Teacher
Project-based Instruction	Teach 3 Lessons 30 Hours Observing and Assisting	University Academy	University Academy	Mentor Teacher and Master Teacher
Apprentice Teaching	One Semester Student Teaching Week 1: Observe Week 2: Teach 1 Class Period Week 3: Teacher 2 Class Periods Week 4: Teach 3 Class Periods Week 5: Teach 4 Class Periods until semester ends.	Traditional School Site	Requested School Site if Available. Most Public Schools in East Texas are High Need.	Cooperating Teacher and University Supervisor

public school environments and student populations" (University of Texas at Austin, 2019). Since the vast majority of UTeach Tyler preservice teachers will be employed in traditional school districts, it is essential that they also have clinical experiences in traditional high needs public schools.

The UTeach Tyler Program has also identified a few traditional high needs schools for clinical settings that have employed UTeach Tyler graduates. Even though the schools may focus on teacher-centered direct instruction, having former UTeach graduates as Mentor and Cooperating Teachers improves the likelihood that during the UTeach Classroom Interactions class, preservice teachers will be able to practice inquiry-based instruction in a traditional setting. The laboratory schools make it easier to locate a few highly aligned placements in traditional schools that are used as clinical sites each semester. The following Table 9.1 illustrates the clinical assignments by course.

CONCLUSIONS

The creation of laboratory schools has allowed UTeach Tyler to implement the UTeach replication model with high levels of fidelity. Since universities may apply for a public school charter, this new funding source can be leveraged to prepare new teachers trained in inquiry-based instructional strategies. By completing clinical experiences in schools that actually utilize inquiry-based instructional models, UTeach Preservice Teachers are able to practice these strategies and receive feedback focused on delivering high-quality STEM lessons.

We recognize that not all states have the same laws concerning charter schools and that many universities that replicate the UTeach may not wish to create and manage schools. That said, the creation of the University Academies has been integral to the sustainability of the UTeach Tyler Program. Universities in states that allow higher education institutions to open and operate charter schools should investigate opening a laboratory charter that can support UTeach preservice teachers' preparation. Compared to traditional laboratory school models that existed prior to charter school funding, laboratory schools supported by state funds through a charter are by law open enrollment public schools. Open enrollment schools are not magnet or specialty schools and therefore can attract a diverse student body of varying abilities and socio-economic status. Charter schools are not private schools that can limit what students enroll, except under very limited circumstances as allowed by state law. As public schools, funding is based on enrollment and average daily attendance. Charter school funding covers all costs associated with managing and operating the schools. As a result, the operation is budget neutral and in some cases can be revenue generating if the university rents space to the laboratory school.

The University Academies also provide a research platform for university faculty to develop and investigate new instructional strategies, curricula, as well as professional development strategies. Conducting research in traditional public schools is often difficult since there are often barriers to gaining access to schools

for research purposes. In addition, traditional schools often have high rates of turnover in administration and teachers making longitudinal research difficult. Conducting research is a core activity of laboratory schools. Having a laboratory school on campus can also help in bridging the gap between theory and practice. Lab schools are another tool to reform STEM education.

REFERENCES

Au, W. (2007). High-stakes testing and curricular control: A qualitative metasynthesis. *Educational Researcher, 36*(5), 258–267. Retrieved from: https://pdfs.semanticscholar.org/f5a6/7da0ed2e47c4b4a07f67fe5a70fd75d0d093.pdf.

Avery, S., Chambliss, D., Pruiett, R., & Stotts, J. L. (2010). *T-STEM academy design blueprint, rubric, and glossary.* Retrieved from: http://www.edtx.org/uploads/general/pdf-downloads/misc-PDFs/2011_TSTEMDesignBlueprint.pdf

Backes, B., Goldhaber, D., Cade, W., Sullivan, K., & Dodson, M. (2018). Can UTeach? Assessing the relative effectiveness of STEM teachers. *Economics of Education Review, 64*, 184–198. Retrieved from: https://reader.elsevier.com/reader/sd/pii/S0272775717306969?token=6C07E608B8A5B0BB216D8D72C72593E62E326AC1906F1BB52CE7826BB4FD29DDEC15772CE9D5A030E6945777A24810CF.

Beth, A. D., Hughes, K. K., Romero, P., Walker, M. H., & Dodson, M. M. (2011, February). *Replication as a strategy for expanding educational programs that work: The UTeach Institute's approach to program replication.* Paper presented at the annual meeting of the American Association of Colleges for Teacher Education, San Diego, CA. Retrieved from: http://uteach-institute.org/publications.

Boardman, A. G., & Woodruff, A. L. (2004). Teacher change and high stakes assessment: What happens to professional development? *Teaching and Teacher Education, 20*, 545–557. Retrieved from: https://www.sciencedirect.com/science/article/pii/S0742051X04000605?via%3Dihub.

Bullough, R. B., Kauchak, D., Crow, N. A., Hobbs, S., & Stokes, D. (1997). Professional development schools: Catalysts for teacher and school change. *Teaching and Teacher Education, 13*(2), 153–169. Retrieved from: http://citeseerx.ist.psu.edu/viewdoc/download?doi=10.1.1.474.8264&rep=rep1&type=pdf

Bullough, R. V., Young, J., Erickson, L., Birrell, J .R., Clark, D.C., Egan, M. W., Berrie, C. F., Hales, V., & Smith, G. (2002). Rethinking field experience: Partnership teaching versus single-placement teaching. *Journal of Teacher Education, 53*(1), 68–80. Retrieved from: http://citeseerx.ist.psu.edu/viewdoc/download?doi=10.1.1.841.5152&rep=rep1&type=pdf

Chapman, D. W., & Snyder, C. W. (2000). Can high stakes national testing improve instruction: Reexamining conventional wisdom. *International Journal of Educational Development, 20*(6), 457–474. Retrieved from: https://doi.org/10.1016/S0738-0593(00)00020-1

Educate Texas. (2015). *T-STEM academy design blueprint.* Retrieved from: http://www.ndisd.net/hs/2015-blueprint.pdf

Flessner, R. (2012). *Scholarship and Professional Work—Education. 42. Addressing the Research/Practice Divide in Teacher Education.* Retrieved from: http://digitalcommons.butler.edu/coe_papers/42

Francis, A. T. (2017). Reforming only half: A study of practice-based teacher education in traditional field placements. *Mid-Western Educational Researcher, 29*(3), 235–260. Retrieved from: https://www.mwera.org/MWER/volumes/v29/issue3/V29n3-Francis-FEATURE-ARTICLE.pdf

Gerwin, D., & Visone, F. (2006). The freedom to teach: Contrasting history teaching in elective and state-tested course. *Social Education, 34*(2), 259–282. Retrieved from: https://www.researchgate.net/publication/254336614_The_Freedom_to_Teach_Contrasting_History_Teaching_in_Elective_and_State-Tested_Courses

Kleinsasser, R. C. (2000). A professional development school's evolution potential: Teachers' initial perceptions of their teaching and learning community. *Learning Environments Research, 2*, 265–289. Retrieved from: https://link.springer.com/content/pdf/10.1023%2FA%3A1009958924071.pdf

National of Academy of Sciences, National Academy of Engineering, & Institute of Medicine. (2007). *Rising above the gathering storm: Energizing and employing America for a brighter economic future.* Washington, DC: The National Academic Press. Retrieved from: https://www.nap.edu/catalog/11463/rising-above-the-gathering-storm-energizing-and-employing-america-for.

Texas Education Agency. (2019). *STAAR Resources.* Retrieved from: https://tea.texas.gov/student.assessment/staar/.

Texas Education Code 12.152 (c-12), (2009).

United States. National Commission on Excellence in Education. Department of Education. (1983). *A nation at risk: the imperative for educational reform: A report to the Nation and the Secretary of Education.* United States Department of Education. Washington, D.C.: The Commission: [Supt. of Docs., U.S. G.P.O. distributor]. Retrieved from: https://www.edreform.com/wp-content/uploads/2013/02/A_Nation_At_Risk_1983.pdf.

University of Texas at Austin. (2011). *UTeach Elements of Success.* Retrieved from: https://institute.uteach.utexas.edu/sites/institute.uteach.utexas.edu/files/uteach-elements-of-success-2011.pdf.

University of Texas at Austin. (2019). *The UTeach Program.* Retrieved from: https://uteach.utexas.edu/.

Wilcox-Herzog, A. S., & McLaren, M. S. (2012). Lessons learned: Building a better laboratory school. *NALS Journal, 4*(1), 1–8. Retrieved from: https://digitalcommons.ric.edu/cgi/viewcontent.cgi?article=1021&context=nals

Yarmus, R. L. & Begum, N. N. (2013). Field experiences and teacher education programs: What is and what if? *Journal of the European Education Network, 8*, 52–61. Retrieved from: http://jeten-online.org/index.php/jeten/article/download/26/pdf.

Zeichner, K. (2010). Rethinking the connections between campus courses and field experiences in college- and university-based teacher education. *Journal of Teacher Education, 61*(1–2), 89–99. Retrieved from: https://journals.sagepub.com/doi/pdf/10.1177/0022487109347671.

Zeichner, K., & Bier, M. (2012). The turn toward practice and clinical experience in U.S. Teacher Education. *Beiträge Zur Lehrerbilgung, 30*(2), 153–170. Retrieved from: https://www.academia.edu/25555326/The_Turn_Toward_Practice_and_Clinical_Experience_in_U.S._Teacher_Education.

CHAPTER 10

INNOVATIVE SHARED MASTER TEACHER MODEL

The University of Maryland and a Local School District

Anisha Campbell, Steven Karig, Kayla White, and Julie Brenner

In this chapter, we describe an innovative shared master teacher (SMT) initiative developed by Terrapin Teachers at the University of Maryland (UMD). The initiative involved hiring additional master teachers, but rather than the master teachers being full-time employees of UMD, the individuals were teachers in a local school district who were assigned to work 60% of their time with Terrapin Teachers. The SMTs served as liaisons between Terrapin Teachers and their respective high schools and contributed to the development of the Professional Development School partnership. The high schools benefited by the SMTs sharing practices that were developed or tuned by their collaborative work with our full-time master teachers and university faculty. Although the SMT initiative was created contextually for UMD and focused on STEM, the model could be an option for other universities in the UTeach network and include other disciplines, *e.g.*, English, social studies, world languages.

Preparing STEM Teachers: The UTeach Replication Model, pages 135–145.

INTRODUCTION AND RATIONALE

In this chapter, we describe an innovation developed by Terrapin Teachers at the University of Maryland (UMD). The Shared Master Teacher (SMT) initiative brought high school teachers into the teacher preparation process in a way that permitted their roles in both schools and teacher preparation communities to be carried out simultaneously. While the Master Teacher model in UTeach implementation sites places significant value on classroom experience, the SMT model pushed further toward the integration of school experiences and teacher learning by maintaining legitimate participation and voice in both communities.

Shared Master Teachers are teacher leaders in their schools, which are the Professional Development Schools (PDS) in which the teacher preparation program collaborates. As we discuss in this chapter, SMT effectively bridged the gap between UMD and PDS through their current, intimate knowledge of school processes and cultures as well as their participation in decision making at the schools and the university. The teachers contribute to deepening and equalizing the relationships between the two communities. Additionally, it helps in practical terms for meeting concrete demands such as staffing multiple sections of Step 1.

Zeichner (2010) synthesized the ongoing challenges university-based teacher preparation programs faced as they navigated the terrains of the school and the university. He argued that "colleges and universities continued to maintain hegemony over the construction and dissemination of knowledge, and schools remained in the position of "practice fields" where teachers are to try out the practices provided by the university." He further argued that border crossing and hybridity was crucial for moving beyond traditional structures to lean toward the integration of knowledge. The SMT model fulfills these demands, and in their dual and simultaneous roles, strengthens the connection between Terrapin Teachers and the schools.

In the following sections, we describe the conception of the SMT model, the benefits to both communities, and the roles and duties performed by the shared master teachers. We close with testimonies from our SMTs to allow an intimate look into their daily schedules.

Implementation and Structure of the Shared Master Teacher Initiative

In 2014, Terrapin Teachers established an advisory council composed of renowned local leaders from education, academia, and policy, all of whom had a deep interest in enhancing Science, Technology, Engineering, Mathematics (STEM) education and the Terrapin Teachers initiative. The members of the advisory group whose support was critical to the implementation of the SMT model included the Provost of UMD, and the Superintendent of Montgomery County Public Schools (MCPS). The idea of the SMT partnership between UMD and the school district was inspired by the superintendent's desire to expand the PDS net-

work within one of our collaborating school districts, and to influence instruction in science and mathematics at the school level.

Montgomery County Public Schools is the most extensive school system in Maryland and the 14th largest school system in the United States, with 162,680 students (MCPS, 2017–2018a). The superintendent strongly recommended that the SMT initiative be implemented in two schools: Northwood High School and Paint Branch High School. The superintendent recognized the growth potential through the SMT model in the science and mathematics departments for both schools.

Northwood and Paint Branch High School (HS) have similar demographics and student data for the 2016–2017 school year (MCPS, 2017–2018b). Northwood HS had 1,731 students with 74.9% classified as receiving free and reduced meals (FARMS) and 15.7% meeting University System of Maryland (USM) entrance requirements. Paint Branch HS had 2,000 students with 63.6% as FARMS, and 32.6% percent of students meeting the USM entrance requirements.

The SMT model was approved by the Provost, the Terrapin Teachers Steering Committee, which consisted of the Co-Directors, full-time Master Teachers, and Assistant Deans from the College of Education and College of Computer, Mathematics, and Natural Sciences. The Co-Directors then met with the Superintendent, Director of Curriculum and Instruction, and Chief Financial Officer of MCPS to discuss the initial rollout of the shared master teacher model. A Memorandum of Understanding (MOU) was created between the school district and UMD that described the rollout of the SMT initiative. The initiative involved hiring additional Master Teachers, but rather than the Master Teachers being full-time employees of UMD, the teachers were "shared" with the MCPS and assigned by their school district to work part-time with Terrapin Teachers.

One critical aspect of the model was that the SMT retained their benefits and seniority during a three- to- five-year rotation where they were in-service teachers in the schools as well as instructors with Terrapin Teachers. The University was responsible for reimbursing the school district for 60% of the salary and benefits. After one year and the initial hiring of our first SMT, addenda were added in 2017 to the original MOU to represent a revised negotiation between UMD and the school district. The district paid 40% and 70% (a larger portion for the district) of the salary and benefits for the second and third SMT, respectively. The addenda were demonstrative of the superintendent's commitment and support of the SMT model.

The Terrapin Teachers Co-Directors met with both principals from Northwood and Paint Branch HS. The goal was to establish deadlines to ensure that the hiring process was completed before the school district's voluntary transfer season. The voluntary transfer is a period (usually one-two weeks) of time when public school teachers can interview for teaching positions with other schools within and outside of their school districts. The importance of hiring before the transfer process

began was to ensure that the most highly qualified candidates would be available during the SMT search.

The job description for the SMT was advertised on the websites for MCPS and Terrapin Teachers. The search/interview committee was composed of representatives from the school district (principals and resource teachers) and Terrapin Teachers (full-time Master Teachers and Associate Director).

Cross-Institutional Benefits

The Shared Master Teachers were advantageous for the interest in expanding the district-university partnerships through the addition of PDS schools. As a part of the agreement, the SMTs' schools joined the PDS network. The College of Education cultivated the PDS partnership between the university and the local school district, and the mutually beneficial relationship provided the foundation to expand the effort to include the SMT initiative.

Our model incorporated research (Darling-Hammond, 1992, 2005) that supported professional growth activities that consisted of team-teaching, collegial work settings, and school-university collaboration. The SMTs became a part of an exciting teacher preparation pathway while experiencing college teaching. Childs-Bowen, Moeller, and Scrivener (2000) found that "teachers are leaders when they function in professional learning communities to affect student learning; contribute to school improvement; inspire excellence in practice, and empower stakeholders to participate in educational improvement."

Similarly, Carpenter and Sherretz (2012) indicated that PDS partnerships activities encouraged teachers to assume leadership functions to improve their schools' teaching and instructional strategies for student growth. The SMTs have brought learning from the university into the school setting by modeling and implementing new and progressive practices. Our SMTs have been Professional Learning Community (PLC) leaders, mentors, and curriculum developers in their schools. They have brought practices to their schools that were developed or tuned by their collaborative work with our full-time master teachers and university faculty.

One advantage of the SMT initiative for Terrapin Teachers was that the College of Education faculty, Terrapin Teachers' full-time master teachers, and undergraduates(pre-service teachers) gained experienced members of their team who were intimately involved in the secondary schools. In our Step 1 and Step 2 courses, undergraduates had access to schoolteachers who not only had years of secondary classroom experience but who taught high school that very same morning. SMT brought authenticity to the lessons delivered to Step 1 students, along with relatable commentary on how the techniques played out in their classrooms. Undergraduates see SMTs as resources to get a real sense of a teacher's daily life so they can make more informed decisions about their teaching practices and futures in education.

On a program level, SMT bring a new perspective to the university. In general, SMTs provide updates about the daily life and logistics of secondary schools

as well as knowledge of school structures and cultures, classroom technology, standardized testing, local curriculum, and current changes in district policies or priorities. As experienced teachers and leaders in highly collaborative PLCs at their respective high schools, the SMT's team- approach to planning and reflection enhanced the existing collaborative culture of the Terrapin Teacher Master Teachers. For example, they are instrumental to an intensive course revision of the Step 1 course including rethinking course assignments, in-class discussions, and fieldwork structures, as well as the design of new Step 1 material.

SMT contributed to providing a coherent learning experience between the university and the PDS. They collaborated with the full-time master teachers on the Terrapin Teachers curriculum to add, among other improvements, an emphasis on district initiatives that involved the Next Generation Science Standards science and engineering practices and the Common Core Standards of mathematical practice. SMT also mentor university students at PDS,. helping them to understand the HS culture and navigate school politics as well as modeling for them responsive STEM teaching practices in their own classrooms.

Shared Master Teacher Responsibilities

The Shared Master Teachers were expected to serve as liaisons between Terrapin Teachers and their respective high schools and contribute to the development of the PDS partnership. As part of their 60% workload, SMT were responsible for teaching early field experiences courses (Step 1 and Step 2) with the possibility of co-teaching upper-level Terrapin Teachers curriculum courses depending upon their discipline and the needs of the program. The shared master teachers were on-boarded for Step 1 and Step 2 using a co-teaching model. The full-time master teachers were the "lead" instructor, and the SMT assumed an "observer" role in the classroom and the field experiences. Additional duties included undergraduate advising/recruitment, attending meetings, and serving on committees for the College of Education and Terrapin Teachers.

Responsibilities at the high schools include teaching two classes (during the morning) as a 0.4 (40%) full-time employee. As actively engaged faculty members within their departments, the SMT designated hours during each week for tutoring, participating in departmental meetings, and attending extracurricular activities at their high schools.

SMT served as mentors to our undergraduate and graduate interns during apprenticeship teaching. SMT hosted graduate interns as part of the 60% shared time with Terrapin Teachers. The graduate students were hired by MCPS to fill part-time (0.4 or 0.6 FTE) vacancies at the schools of our shared master teachers. In this part-time paid internship model, the interns were the teachers of record and held the title of long-term substitutes. Unlike the "traditional" teacher preparation university-based internship where a mentor teacher is in the classroom during the entire school day, the SMT were not able to fulfill the same mentor role because they are teaching their courses during the mornings and then teach Step 1 or Step

2 at the university during the afternoons. The mentor role was modified to accommodate the shared master teachers' schedules. The SMT mentor role incorporated reduced numbers of observations and evaluations of the part-time paid interns while still fulfilling the Master's program requirements, *e.g.*, performance-based assessments, edTPA. When SMT return to their high schools full-time following their collaborative work with UMD, they are trusted placement options for undergraduate student teachers.

The SMT initiative provides opportunities for professional instructional growth that contribute to SMT's content and pedagogical knowledge. Terrapin Teachers sought individuals who could identify innovation in teaching with the potential to contribute to their secondary school's improvement plan. During their time with UMD, SMT pilot an innovative idea in their high school classrooms and share what they learned with colleagues in their departments, school buildings, and at local and national conferences. Additional professional growth was extended to the SMT by offering the opportunity for them to enroll in graduate courses at UMD.

Perspectives from Shared Master Teachers

Our first SMT, Teacher A, who began fall 2016, was a National Board-Certified Teacher (NBCT) mathematics teacher with eighteen years of public-school experience. She participated as a SMT for three years, after which time she returned full-time to Northwood HS. In her first year returning, Linda served as a mentor for one of the Terrapin Teachers undergraduate students who were enrolled in apprenticeship teaching. In 2017, using the same hiring process, we added two more shared master teachers; Teacher B, who was a NBCT mathematics teacher at Paint Branch HS with ten years of teaching experience, and Teacher C who was a NBCT science teacher with twelve years of experience including traditional public and public charter secondary schools.

Teacher A: I have the time and opportunity to attend conferences and develop rich lessons to test in my classroom. I share lessons and strategies that are successful with my high school colleagues. Additionally, I can share materials from the university to use in our high school classrooms.

Teacher B: Having the opportunity to collaborate, plan, and observe one of our Master's-level student interns allowed me to take on a new leadership and mentor role within my department. I shared my best practices and lesson resources with a future teacher and also tried out new lesson ideas and activities from a current university student.

Teacher C: With the title of a shared master teacher, I am seen as a resource at my school and therefore am invited to collaborate on and lead professional development. In my department, I designed and

> led half-day and whole-day professional development sessions related to designing rigorous assessments. At the whole school level, I created and offered sessions to interested faculty on the topic of problem-based learning, which is a component of the Terrapin Teachers coursework. The role of SMT provides me time and resources, like access to conferences and webinars, to create and carry out these leadership opportunities. As a known resource at my school, observers, including teachers- and administrators-in-training as well as outside guests, often visit my classroom which provides another opportunity for me to share my practice. I also use my network at UMD to facilitate exciting opportunities for my colleagues' students, including connecting our AP Biology teacher with guest lecturers from the College of Computer, Mathematical, and Natural Sciences.

Shared Master Teachers were seen as resources and leaders within their buildings, with time and space to reinforce the ideals of Terrapin Teachers, including equitable access to rigorous inquiry instruction; and SMT, who were veterans of their respective high schools, and have the opportunity to reflect on and improve their instruction. As they highlighted the virtues of practices such as productive discourse in Step 1 and Step 2, those strategies were reinforced in their secondary classrooms. Through collaboration with university faculty, SMT reconnected with science and mathematics education research.

The SMT initiative could be a model used to retain seasoned teachers who want to remain in the classroom. At this stage in their secondary teaching career, the SMT position was a valuable chance to grow professionally while remaining in the classroom. The SMTs partnered with a variety of dedicated UMD faculty who had different experiences than what they encountered at their high schools. At the university, SMTs reconnected with science and mathematics education research, including recent innovations in teaching and teacher training because they collaborated with full-time university colleagues who had their pulse on the latest findings. The SMT worked with dedicated experts on equity, diversity, and social justice and were invited to collaborate on meaningful projects and presentations; they were fine-tuning their critical lenses and growing as educators.

As instructors to future STEM teachers, sharing a vision for high-quality pedagogy reinforced and improved their STEM teaching in secondary classrooms. As veteran teachers, the SMT may rely on practices and lessons they've used for several years, many of which were effective, though some of which were left unexamined. By revisiting the fundamentals of teaching in Step 1 and Step 2, they identified areas of their teaching that could be refreshed. The SMTs also had opportunities and support to attend and present at local and national education conferences.

Teacher A: I have presented at national conferences as well as co-lead regional Professional Development courses. This work helps me to rethink what I do in the classroom. As a result, I have implemented computer science and STEM-based lessons into my Algebra 2 curriculum. Students can make connections and feel that the work is more relevant to their lives.

Teacher B: Terrapin Teachers makes it an early priority for undergraduates to examine how equity is at play in the classrooms they are teaching and visiting. This experience allowed me to take a critical perspective of my teaching and further reinforce the need for equity to be at the forefront of my classroom teaching.

Teacher C: Step 1 focuses heavily on discourse, and I often give students feedback that they take control of classroom discourse while teaching their lessons. Upon reflection, I noticed in my practice that I can be heavy-handed in my approach to facilitation due to the age-old teacher concern: 'There's not enough time to cover everything!' As a shared master teacher, I have a wonderful opportunity to ensure that if I talk the talk, that I also walk the walk, and to make deliberate moves to give students ownership of discourse.

Challenges for the Shared Master Teacher

The transition from a full-time high school teacher at one location to a SMT position working at both a high school and a university presented some challenges.

SMT had to adjust to scheduling differences which included balancing their time among their various roles and teaching courses at their high school and the university. Navigating schedules between two organizations, both of which featured non-negotiable elements like classes as well as other moving parts, *e.g.*, tutoring high schoolers, serving on committees, proved challenging. TSMT had to practice self-advocacy and communicate with both institutions the time constraints and their availability. Assemblies, inclement weather, and meetings with staff, students, and parents are common occurrences in public schools that impacted their time at the university. At Terrapin Teachers, ad-hoc meetings occurred throughout the week with colleagues, and outside collaborators, and SMT traveled to several elementary schools to observe undergraduates during Step 1 and Step 2 fieldwork placements, all of which competed with high school commitments. No one week looked the same. The maintenance of an extensive calendar and detailed to-do list was imperative and required constant attention.

Community was the foundation for working and learning, and while the SMT enjoyed belonging to two exceptional education communities, they were only part-time members of each. Their physical presence was reduced at both the high school and the university; consequently, SMT were not available as much as they

desired to be at either location. Their abilities to hold consistent office hours for their high schoolers and undergraduates and to meet with colleagues in-person was limited:

> The role of the SMT was constantly evolving. Managing the benefits and challenges of being a team member between the two organizations was a puzzle, but it was one that we felt fortunate to be involved with.
>
> — *Teacher A, Teacher B, and Teacher C*

Time management challenges connected to a larger dilemma of determining what "part-time" looked like at each location. There were rough guidelines in the SMT job description, but to feel truly successful at both places, they wanted to be involved as much as possible at each location. As time progressed, the SMT adjusted to their schedules and determined their levels of involvement at the high schools and the university. This required negotiating with/between Terrapin Teachers' co-directors and secondary school principals.

Expanding the Shared Master Teacher Model

The SMT model supported recommendations by the National Council for Accreditation of Teacher Education (NCATE, 2010) -Recommendation #10:

> Strategic partnerships are imperative for powerful clinical preparation: School districts, preparation programs, teacher unions, and state policymakers must form strategic partnerships based on the recognition that none can fully do the job alone. Each partner's needs can be met better by defining clinically based teacher preparation as common work for which they share responsibility, authority, and accountability covering all aspects of program development and implementation.

In our shared master teacher model, university faculty and high school teachers provided different attributes and worked as co-learners and colleagues. Higher education faculty contributed additional content knowledge to complement the instructional expertise of K–12 teachers (Hamos et al., 2009).

Although there was a mixture of evidence linking teachers' subject matter (content) knowledge to their classroom practice and their students' achievement (Ball, 1988; Wilson, Floden, & Ferrini-Mundy, 2001), there was certainly research that supported the notion that teachers' level of understanding and knowledge of subject matter was linked to their ability to teach the content (Nadelson, Louis, Seifer, Moll, & Coats, 2012) effectively. With the population of students (more than 90% underrepresented groups) at the SMT's PDS schools, one goal was to establish a STEM pipeline which addressed the future of American global competitiveness (Ralston, Hieb, & Rivoli, 2013).

Terrapin Teachers expanded the SMT initiative to include other STEM subjects. For fall 2019, we hired two computer science SMTs. In this model, the SMTs will not only work with Terrapin Teachers, but also teach introductory,

non-major and major courses in the computer science department at UMD. This model was more complicated with the SMT's time split between three stakeholders (school district/TT/computer science department). We'll use best practices learned from our form SMT model, which included clear, concise communication and expectations for the school district, the computer science department, and the SMT, to ensure a smooth transition.

It is our goal not only to provide an exemplary SMT model for UMD, but one that can be exported to other universities in the UTeach network. Although the SMT initiative was created contextually for UMD, the model could be an option for other universities in the UTeach network and include other disciplines, *e.g.*, English, social studies, world languages.

REFERENCES

American Association of Colleges for Teacher Education. (2004). *Professional development schools at a glance.* Retrieved from: http://aacte.org/index.php?/Programs/Professional-Development-Schools/professional-development-schools-at-a-glance.html

Ball, D. L. (1988). *Research on teaching mathematics: Making subject matter knowledge part of the equation* (Research Report No. 88-2). East Lansing, MI: National Center for Research on Teacher Learning, Michigan State University.

Byrd, D., & McIntyre, D. (1999). *Research on professional development schools*. Teacher Education Yearbook VII. Thousand Oaks, CA: Corwin Press.

Carpenter, B. D., & Sherretz, C. E. (2012). Professional development schools partnerships: An instrument for teacher leadership. *School-University Partnerships*. *5*(1), 89–101.

Childs-Bowen, D., Moller, G., & Scrivner, J. (2000). Principals: Leaders of leaders. *NASSP Bulletin, 84*(616), 27–34.

Conzemius, A., & O'Neill, J. (2001). *Building shared responsibility for student learning*. Alexandria, VA: ASCD.

Darling-Hammond, L. (1992). *Professional development and restructuring*. Pleasantville, NY: DeWitt Wallace/Reader's Digest Fund.

Darling-Hammond, L. (2005). *Professional development schools: Schools for developing a profession*. New York, NY: Teachers College Press.

Hamos, J. E., Bergin, K. B., Maki, D. P., Perez, L. C., Prival, J. T., Rainey, D. Y., Rowell, G. H., & VanderPutten, E. (2009). Opening the classroom door: Professional learning communities in the math and science partnership program. *Science Educator, 18,* 14–24.

Lewis, A. C. (2002). School reform and professional development. *Phi Delta Kappan, 83*(7), 488.

Marder, M., Brown, C. R., & Plisch, M. (2017). Recruiting teachers in high needs STEM fields: A Survey of current majors and recent STEM graduates. *American Physical Society Panel on Public Affairs*, 1–37. Retrieved from: http://aps.org/reports

Montgomery County Public Schools. (2017–2018a). *Schools at a glance*. Retrieved from: https://www.montgomeryschoolsmd.org/departments/sharedaccountability/glance/index.aspx

Montgomery County Public Schools. (2017–2018b). *Schools at a glance*. Retrieved from: https://www.montgomeryschoolsmd.org/departments/regulatoryaccountability/glance/currentyear/SAAG2018.pdf

Nadelson, L. S., Seifer, A., Moll, A. J., & Coats, B. (2012). I-STEM summer institute: An integrated approach to teacher professional development in STEM. Boise State University Scholarly Works. *Journal of STEM Education, 13*(2), 69–83.

National Council for Accreditation of Teacher Education. (2010). *Transforming teacher education through clinical practice: A national strategy to prepare effective teachers. Report of the Blue-Ribbon Panel on Clinical Preparation and Partnerships for Improved Student Learning*. Washington, DC: ERIC Clearinghouse.

Ralston, P. A. S., Hieb, J. L., & Rivoli, G. (2013). Partnerships and experience in building STEM pipelines. *Journal of Professional Issues in Engineering Education and Practice. 139*(2), 156–162.

Wilson, S., Floden, R., & Ferrini-Mundy, J. (2001). *Teacher preparation research: Current knowledge, gaps, and recommendations: A research report prepared for the US Department of Education and the Office for Educational Research and Improvement, February 2001*. Seattle, WA: University of Washington, Center for the Study of Teaching and Policy.

Zeichner, K. (2010). Rethinking the connections between campus courses and field experiences in college-and university-based teacher education. *Journal of Teacher Education, 61*(1–2), 89–99.

CHAPTER 11

ALIGNING FIELD AND CLASSROOM EXPERIENCES FOR SECONDARY STEM TEACHER PREPARATION

Kristen Apraiz, Gayle Nelson Evans, and Thomas Dana

One of the continued struggles shared by teacher preparation programs across the nation is recruiting and retaining preservice teachers. Within our UFTeach secondary mathematics and science education minor, we have engaged in curricular revision for the field based courses, starting with our combined Step 1 and Step 2 course, entitled Explorations in Teaching Secondary Mathematics and Science (*Explorations).* Since 2016, program faculty have implemented this redesigned introductory course to better meet the needs of the program, faculty, and students. The situated learning perspective is employed to explain how we worked towards developing a community of practice among program faculty and the PSTs we teach. The changes we have made to the initial *Explorations* course have led to guiding principles that we use throughout our UFTeach program as we turn our sights to revising subsequent courses to better align field and classroom experiences.

UFTeach at the University of Florida was founded in 2008 as a replication site of the University of Texas' UTeach model of teacher preparation for secondary mathematics and science, emphasizing the connection between field experiences

Preparing STEM Teachers: The UTeach Replication Model, pages 147–159.

and university based pedagogical learning (UTeach Institute, 2011). In enacting this model first, as a UTeach replication, and now as a program incorporated into our college's School of Teaching and Learning, UFTeach faculty have engaged in cycles of curricular revision for the field-based courses: Step 1, Step 2, Classroom Interactions, Project Based Instruction and Apprentice Teaching to better meet the needs of our local context. In this chapter, we will focus on the field placement for our introductory course, *Explorations in Teaching Secondary Mathematics and Science* (*Explorations*) as an exemplar of the revisions we have embarked upon for all of our field-based courses as we address programmatic needs of recruitment and retention while remaining true to our commitment to prepare highly qualified secondary mathematics and science teachers.

UFTeach is an undergraduate minor for students majoring in STEM fields of study (mathematics, science, engineering, computer science, statistics, etc.) leading to temporary teacher certification for secondary (grades 6–12) mathematics or science. Our initial cohort was a class of 34 students enrolled in the first course—Step 1. Ten of preservice teachers (PSTs) completed the required course sequence earning a UFTeach minor; the highest cohort number of PSTs completing the program over the course of ten years. As we watched class sizes shrink over the years, we couldn't help but wonder, how is it possible in a university of more than 40,000 undergraduates, that we were struggling to graduate ten or fewer students from our program each year?

One of the first trends we noticed was a steep initial decline in retention, with more than half of students enrolled in Step 1 choosing not to continue into Step 2 and beyond. Due to a decrease in enrollment and changes in available faculty, we found it necessary to create a course combining the curriculum of Step 1 and Step 2 into a single, 3 credit course titled, *Explorations in Teaching Secondary Mathematics and Science* (*Explorations*), in spring 2011. One expectation for this change was to remove a barrier to completion of the UFTeach minor by reducing the required number of courses from ten to nine, however, enrollment declines remained consistent with the trend we were seeing prior to the move to *Explorations*.

Given the challenges we face as an established program struggling to maintain sufficient enrollment to justify our continuing existence, program coordinators and faculty have been engaged in reflection and evaluation using multiple sources of data to gather information about what aspects of our program are working well, and where improvements can be made. We began in 2011 by administering informal end of semester student surveys in *Explorations* asking students about what they did and did not like about the course and whether they planned on continuing in the minor and why. In reading through these surveys, faculty also engaged in reflection based on their interactions with students and the in-service teachers who mentored them in their classrooms. The area that most stood out as problematic was the field experience. In moving from Step 1 to *Explorations* as our initial course, we changed the first field experience from working in an elementary class-

room to a middle school classroom. Reports from our students, faculty, and the in-service mentors indicated that an initial teaching experience in a middle school classroom was overwhelming to pre-service teachers (PSTs). We were also challenged to find sufficient numbers of mentor teachers who were willing and able to host our initial field placement, being a relatively small county with only five middle schools easily accessible within a ten-mile radius from campus. In 2015, we conducted a study with surveys sent to all middle and high school mathematics and science teachers in our school district (71 of 196 teachers participated), and interviews were conducted with 5 high school and 10 middle school administrators within the district (Evans, McTureous, Weber, & Crippen, 2017). From this study, we learned the majority of teachers who chose not to mentor cited lack of time (45%) and pressures due to standardized testing (36%), however, a number of teachers who chose not to mentor (27%) were making that choice based on a prior negative experience with mentoring.

These data led us to focus on the initial field experience as the area with potential for the greatest positive impact on our program enrollment. For the originally conceived *Explorations* field experience, PSTs were placed in pairs in a middle school classroom to observe and to teach a minimum of two 5E lessons (Bybee, 2014). Each lesson was observed by a university-based master teacher and the classroom mentor teacher with a short conference held with PST pairs to provide feedback afterwards. End of semester surveys indicated that this field placement was perceived by several of our PSTs as intimidating and problematic. PSTs reported feeling unprepared to face the challenges of the classroom environment in their first semester of coursework. They struggled to adequately plan and implement inquiry-based lessons incorporating appropriate grade level content standards while also trying to handle classroom management of middle school students, management of time and materials, and negotiating a relationship with a mentor teacher who may have a more traditional educational philosophy than espoused in UFTeach courses. Mentor teachers indicated frustration with the need to predict exactly what would need to be taught on a given day a week or more in advance to allow PSTs adequate time to plan and refine their lessons.

Our guiding question as we considered restructuring the initial field experience was, "How can we give PSTs an experience where they will have opportunities to engage in productive struggle, planning and teaching lessons, managing student behaviors, materials and time, all while feeling adequately prepared and supported by their instructors and supervisors?" In considering this question, we knew selecting an appropriate context for the teaching experience was going to be important. Reflecting upon the original course sequence of Step 1 and Step 2, our faculty felt there was real value in an initial experience with elementary students. We have found elementary students to be enthusiastic and more willing to accept a PSTs authority with less tendency to question their legitimacy as a student teacher as compared with older students. Inspired by the mathematics and science curriculum topic study process (Keeley, 2005; Keeley & Rose, 2006), and the 2020

NSTA Standards for Teacher Preparation (National Science Teacher Association [NSTA], 2020) we also see the value in giving our PSTs opportunities to develop an understanding of how mathematics and science content is taught in the elementary grades through planning and implementing standards based elementary STEM lessons. This builds a foundation from which they will be able to trace the progression of curriculum topics from elementary grades through middle to high school coursework, to see how content understanding unfolds across the grade levels. In preparation for the fall of 2016, we planned for a complete restructuring of the field component for the introductory course, *Explorations*, to offer a more supportive experience for PSTs potentially interested in secondary mathematics or science teaching.

THEORETICAL FRAMEWORK

As we worked to conceptualize an improved initial field experience for UFTeach, we were guided by the notion of building our teacher preparation program as a community of practice for STEM teaching and learning. Communities of practice are formed by people coming together through shared proximity, and common goals or interests (Wenger, 1998). Usually held together by informal associations, members of these communities grow together and interact in ways that contribute to the success of one another by offering opportunities for learning and identity formation that would not be possible in isolation (Lave, 1991; Wenger & Lave, 2001); "developing an identity as a member of a community and becoming knowledgeably skillful are part of the same process" (Lave, 1991, p.65). Through participation in a community of practice, individuals shape their identity formation within the cultural context of the community. Beginning as legitimate peripheral participants, learners enter into communities of practice through observation, and participation in basic tasks associated with their developing expertise, and gradually, through forms of cognitive and cultural apprenticeship (Collins, Brown, & Holum, 1991; Lave, 1991), take on more sophisticated roles as they move from "newcomers" eventually growing into the replacements for the "old timers" (Lave, 1991). Operating within a community of practice allows learners to develop skills, knowledge and perspectives that are relevant to real world applications. Situating learning in a social and cultural context allows learners to be "exposed to the use of a domain's conceptual tools in authentic activity—to teachers acting as practitioners and using these tools and wrestling with problems of the world" (Brown, Collins, & Duguid, 1989, p. 43).

ROLE OF FIELD EXPERIENCE IN TEACHER PREPARATION

In our program, we define field experience as "situating learning [about] teaching within the context of classrooms and communities for particular students" (Hollins, 2015, p. 93) through offering an organized opportunity for PSTs to engage in work with K–12 students in a sustained manner within an instructional envi-

ronment. According to the National Council for Accreditation of Teacher Education [NCATE] (2010) report, the clinical experience, or in our case the field experience, should be the central focus of PST education. The clinically based preparation model for teacher preparation "fully integrates content pedagogy, and professional coursework around a core of clinical experiences" (NCATE, 2010, p. 6). The field experience gives PSTs opportunities to work with students and encounter problems of practice. To provide PSTs with a meaningful field experience, teacher educators need to ensure the instruction they provide is aligned with reform-based educational practices (Korthagen, Loughran, & Russell, 2006; Shymansky, 1992).

Frequently, field placements in K–12 settings do not support the efforts of the university programs in implementing reform-based teaching methods, which potentially leads frustrated PSTs to revert back to the more transmissive practices adhered to by those teachers who taught them as students (Putnam & Borko, 2000). Too often in teacher education programs, coursework and field experiences are presented as two separate entities with few overt connections (Zeichner, 2010). Field experiences for preservice teachers should be situated in an environment that incorporates reflection upon the knowledge presented in college courses, inquiry into students' mathematical and scientific thinking, and the opportunity to see this content taught in ways other than prescriptive manners (Association of Mathematics Teacher Educators [AMTE], 2017; Windschitl, Thompson, Braaten, & Stroupe, 2012).

In the NCATE (2010) report, field experiences are included among the three "aspects of teacher preparation that are likely to have the highest potential for effects on outcomes for students" (p. 1). Field experiences are most impactful when occurring concurrently with pedagogically focused university coursework (Korthagen & Kessels, 1999). The field experience can provide PSTs valuable opportunities to learn and apply newly-acquired knowledge from the coursework through first-hand experiences in the field placement, see mathematics and science teaching in action, and examine how students learn and think about new concepts. Creating opportunities for PSTs to be "innovators and problem solvers, working with colleagues constantly seeking new and different ways of teaching students who are struggling" will lead to continual growth in the teaching profession (NCATE, 2010, p. 5).

Learning to teach is not a passive activity. Teacher candidates benefit from a clinically based preparation model which provides opportunities for them to participate in communities of teaching practice, develop their pedagogical knowledge and skills, and focus on how students learn (NCATE, 2010). Furthermore, teacher preparation programs and school districts have the opportunity to build partnerships through facilitation of field experiences; and these partnerships are a key aspect in preparing teacher candidates to meet the unique educational needs within their communities (Carlsen, 1999; Council for the Accreditation of Educator Preparation [CAEP], 2013; Horn & Campbell, 2015).

DETAILS OF THE REVISED FIELD EXPERIENCE

The field experience is emphasized as the central organizing feature of our teacher preparation model and realized through concurrent learning of pedagogy and teaching practice (Korthagen, 2010; Korthagen & Kessels, 1999); closely situating field-based courses within local contexts in public schools and informal educational settings. In the Fall 2016 semester, a new field experience was launched with the 25 PSTs enrolled in this iteration of *Explorations*. Working in partnership with the extended day enrichment program (EDEP) at a nearby elementary school, PSTs engaged in weekly cycles of collaborative lesson planning and implementation. For two hours each week, in the company of the course instructor, PSTs taught small groups of elementary students and engaged in reflection upon the teaching episode. In an additional weekly class meeting at the university, PSTs deconstructed their teaching experiences framed by discussion of research and practitioner articles related to the specific teaching challenges they faced, and applied this knowledge and understanding to planning their next lesson to be taught the following week.

Each week, in teaching groups of three to four, PSTs started with a mathematics, science or engineering concept provided by the course instructor, identified relevant content standards in the grade band of students they would teach, and using a 5E planning template, composed a lesson including concepts and applications of STEM to teach the following week. Kindergarten through 5th grade students in the EDEP were given an option to participate in these lessons as one of their activities each week. We grouped elementary students into four groups based on grade levels (K–1, 1–2, 3–4, 4–5) and maintained an approximate ratio of two students per PST.

GUIDING PRINCIPLES FOR PROGRAM DEVELOPMENT

Informed by Design Based Implementation Research (DBIR; Fishman, Penuel, Allen, Cheng, & Sabelli, 2013; Penuel & Fishman, 2012), our work reflects a desire to organize research and program development to understand and respond to challenges in sustaining programs that work in one setting but possibly not in another. Our work as STEM teacher educators has been iterative and, although systematic, also reflects the reality of just-in-time adaptation to continuously improve what we do and how we do it. Each semester, program faculty engage in reflection and study of the semester as one DBIR cycle. In *Explorations*, we employ a pre-post exercise in which PSTs "create a concept map that represents how you think about teaching mathematics and science to children" on the first day of class, then they repeat this exercise on the final day of class for the semester. After PSTs have completed their final map, we return the initial map and ask them to reflect upon the two together and write a brief paragraph comparing and contrasting the two concept maps. This leads to the final reflection assignment where PSTs revisit written weekly reflections from each teaching episode and compose

a narrative of their educational journey of learning as it relates to the field experience, assigned readings, and class discussions. Through analysis of these artifacts each semester, we continuously adjust each new iteration of the course, such as rethinking how we assign PSTs into teaching groups, how to ensure each PST has experience teaching a variety of students and grade levels, and refining the scaffolding we provide for standards-based lesson planning.

Through our own cycles of research and reflection on UFTeach courses and associated field experiences, UFTeach faculty have developed four guiding principles informing program development and our work to align field experiences with overarching program goals.

> **Guiding Principle One:** Even with adaptation to local context, our underlying philosophy of teacher education remains consistent with the UTeach model, in employing the 5E lesson plan, emphasis on formative assessment, and building a habit of reflective teaching from PSTs first field experience through Apprentice Teaching.

Aligned with the UTeach Model, PSTs use the 5E cycle (Bybee, 2014) as the basis of their lesson planning throughout their coursework in our program. Even in their first semester, taking *Explorations*, PSTs began to see the value of the 5E lesson plan through their initial teaching experiences.

> ...at the beginning of the course we learned about the 5E model, and I thought "this is definitely not for me, I can't do that" but as time went through, I quickly learned that the 5E model was very efficient. I started being conscious of how I could engage my students, how I could help them explore the lesson, how I could explain the lesson, and the hardest one how I could extend the lesson. (NY62/ 2016)

We agree with Korthagen and colleagues' (2006) statement that "universities generally, and university-based teacher educators particularly, have no right to recommend to teachers any teaching practices that they have not themselves used successfully at the university" (p. 1036). UFTeach faculty also use the 5E model in planning our own lessons for courses, providing PSTs opportunities to experience the inquiry cycle both as students and as teachers.

Another critical component of learning how to teach is using formative assessment (FA) to drive instructional decisions. FA is introduced to PSTs in *Explorations* through modeling examples of FA in the university class by using techniques such as "Fist to Five," and "Predict-Observe-Explain (PEO) Probes" (Keeley, 2015; Keeley & Tobey, 2011); assessment for learning through questioning and discourse (Harrison, 2015; Keeley & Tobey, 2016; Windschitl, et al., 2012) and through emphasis of the evaluation aspect the 5E Model as an ongoing process present in every phase of an inquiry-based lesson (Bybee, 2014). During the field experience, PSTs use FA to evaluate both their own teaching and their students' learning. PSTs in *Explorations* read and reflect upon an article about the benefits of FA (Keeley, 2017) in week eight, and in the weeks following, PSTs are then

challenged to include one or more FA strategies in each subsequent lesson plan. One PST explained,

> Not only was I teaching students and interacting the whole time, but I now also know how to gather data and measure the success of my lesson. I am now aware how to prepare, how to meet common core state standards, and how to measure success. (YS37/ 2016)

During the debrief following lesson implementations, faculty also ask PSTs to consider the information they learned from students based on their FA strategy, and how that leads to possible next steps in designing future lesson plans for their students.

As far back as John Dewey, the value of reflection has been emphasized in learning and teaching, especially as it relates to teacher education. (Calderhead & Gates, 2003; Dewey, 1933; Loughran, 2002; Procee, 2006). Reflective teaching is an essential element in our teacher education philosophy. Our goal is to instill the importance of reflection in the PSTs as they continue to refine their craft of teaching; beginning in *Explorations* as weekly reflections following each teaching event and reinforced in later courses. One PST commented about the benefit of reflection:

> I also really liked the weekly reflections, because it helped us think about what we did right, and what we could improve upon in our teaching methods. I was able to grow from those reflections. (JU57/ 2016)

PSTs engage in sharing reflections in small groups during class meetings and whole class conversations. This sharing allows PSTs to learn from their peers as well as their own teaching experiences. In essence, we have created professional learning communities within each course to encourage peer support systems (Wenger, 1998).

> **Guiding Principle Two:** Problems of practice enhance PSTs understanding of and interest in reform-based pedagogy. This is addressed through explicit incorporation of newly learned pedagogies into teaching episodes, and emphasis on building relationships with students as PSTs develop their pedagogical knowledge through frequent opportunities to practice teaching.

Beginning in *Explorations*, we scaffold the dual importance of practice and pedagogy (Korthagen & Kessels, 1999; Zeichner, 2010). Each week PSTs are assigned to read an article emphasizing some pedagogical consideration such as questioning, building relationships with students, collaborative group roles, addressing the needs of diverse learners, use of precise mathematical language, and how to manage engineering design lessons. Each article provides the PST with an opportunity to read about ways practicing teachers conduct research in their own classrooms and then consider how they can apply this process to their own developing teaching practice. Before class meets, PSTs respond to a discussion post

in which they are asked to reflect upon the article and how they see themselves incorporating these new ideas into their next teaching experience. One PST in *Explorations* course shared her perspective:

> The Strengthening Discussions article should be discussed in depth because it helped me change my mindset, such as the fact that "correct ideas" are not the only ideas that should receive regular attention. When original ideas that are incorrect are discussed, then the math that is explored expands. (YP66/2016)

Following each week's lesson, in their written teaching reflection, PSTs revisit the concepts in the article for that week and discuss the extent to which they were successful in incorporating what they learned from the reading into their lesson.

Future secondary mathematics and science teachers are drawn to teaching because they enjoy sharing their content knowledge with others and see themselves as being confident in this knowledge (Ball, 1990; Manuel & Hughes, 2006). A love of the content areas sometimes overshadows the necessary teacher/student relationships that need to be in place for effective teaching to occur. From the first course in UFTeach, we place an emphasis on developing student relationships by positioning PSTs to build relationships with the elementary students they teach during the ten 5E lessons (Bybee, 2014). At the beginning of every lesson, we recommend an engagement including an opening activity or icebreaker where students share something about themselves allowing PSTs to make connections to students' interests within their lessons. One PST shared:

> One particularly memorable moment for me this semester was when I tried to explain the concept of infinity to a girl. Although the incident seems comical, as the girl was quickly disinterested, it taught me two important lessons. Number 1: when teaching, you must be patient. Although the student may not get it first, you must slowly and thoroughly try to fix their misunderstandings. Number 2: you must somehow grab their attention and get them interested in the subject. Had I somehow piqued the girl's curiosity, maybe she would have been interested in the concept of infinity, regardless of how obscure it may be. (VJ97/2016)

Merely learning mathematical and /or science content will not prepare preservice teachers for success in the classroom. Preservice teachers need opportunities to develop and practice the appropriate pedagogical knowledge which will allow them to "create, select, or modify tasks" for students and learn from the implications of their instructional decisions (Conference Board of the Mathematical Sciences [CBMS], 2012, p. 1). The Standards for Mathematical Practice (National Governors Association, 2010) and science practices (National Research Council [NRC], 2012), are introduced early in the *Explorations* course and provide scaffolding for the learning activities PSTs use for their lesson plans.

> **Guiding Principle Three:** Shared teaching experiences build a strong community among beginning preservice teachers.

Field experiences provide a common area of inquiry for PSTs and teacher educators allowing us to build a community, sharing problems of practice, seeking advice from one another, and strengthening reform-minded teaching practices (Dana & Yendol-Hoppey, 2020; Lave, 1991; Wenger, 1998). One PST explained:

> Overall, I have grown because I listen to and learn, not just from my teachers but also from peers and fellow classmates around me. (YP66/2016)

We have seen retention as high as 60% of PSTs from the initial course continuing in the program since we increased efforts to build peer support, offer common field experiences and emphasize collaboration with peers. For example, PSTs are given a T-shirt featuring our program logo to wear each week to the *Explorations* field experience. This investment in shirts has led PSTs to identify fellow UFTeach members across our college campus and within courses; cultivating a stronger, more connected community within our program. Social events, such as a welcome back lunch and an annual banquet to acknowledge our graduates provide opportunities for program PSTs to meet, share stories and experiences, and strengthen ties of friendship and support.

> **Guiding Principle Four:** Leveraging community partnerships ensures a diversity of field placements.

The diverse range of field experiences we offer would not be possible without building strong community partnerships. Establishing and maintaining community partnerships takes extra effort from university based teacher educators, community based schools, nonprofit organizations, museums, and other partners. We strive to ensure that partnerships mutually benefit all parties in fostering PST development as well as students' learning. For example, the partnership for the *Explorations* field experience began from a mutual desire on the part of our program and an EDEP to provide elementary students with enhanced opportunities to engage in learning mathematics and science. The director of the EDEP fully supports the work the PSTs do with the elementary students by providing space at the school, resources to support student learning and enthusiastic help in encouraging students to participate in our PSTs lessons each week. Our PSTs benefit from this arrangement by having a nearby, diverse and enthusiastic population of students to teach each week in a group setting. In a similar manner, the success we have seen with our *Explorations* partnership has led us to build unique partnerships serving the particular needs of each field placement in our course progression.

CONCLUSION

Within this chapter, we have shared our journey with re-envisioning and enacting an initial field experience for STEM preservice teacher preparation. Our goal is to cultivate well-prepared beginning mathematics and science teachers who are ready to confront the challenges in secondary education. This work in re-envi-

sioning and refining the field experiences for secondary mathematics and science teacher preparation is not done. Each semester, we continue to meet with community partners, mentor teachers, and program faculty to re-evaluate our progress. Based upon our enrollment and completion data, we feel the changes we describe here have positively impacted PSTs decision to complete the UFTeach minor and enter the teaching profession prepared to implement reform-based teaching strategies, and we expect to see even greater gains in the future. The changes and shifts made in our program are a good start as we continue to evaluate and improve upon the remaining field-based courses in our secondary mathematics and science preparation program.

REFERENCES

Association of Mathematics Teacher Educators [AMTE]. (2017). *Standards for preparing teachers of mathematics*. Retrieved from: amte.net/standards

Ball, D. L. (1990). The mathematical understandings that prospective teachers bring to teacher education. *The Elementary School Journal*, *90*, 449–466.

Brown, J. S., Collins, A., & Duguid, P. (1989). Situated cognition and the culture of learning. *Educational researcher*, *18*(1), 32–42.

Bybee, R. W. (2014). The BSCS 5E instructional model: Personal reflections and contemporary implications. *Science and Children*, *51*(8), 10–13.

Calderhead, J., & Gates, P. (Eds.). (2003). *Conceptualizing reflection in teacher development*. Bristol, PA: Taylor & Francis Inc.

Carlsen, W. (1999). Domains of teacher knowledge. In J. Gess-Newsome & N. Lederman (Eds.), *Examining pedagogical content knowledge* (pp. 133–144). Norwell, MA: Kluwer Academic Publishers.

Collins, A., Brown, J. S., & Holum, A. (1991). Cognitive apprenticeship: Making thinking visible. *American Educator*, *15*(3), 6–11.

Conference Board of the Mathematical Sciences [CBMS]. (2012). *The mathematical education of teachers II* (Vol. 17). Providence, RI: American Mathematical Society.

Council for the Accreditation of Educator Preparation [CAEP]. (2013). *2013 CAEP standards*. Retrieved February 20, 2019 from: http://www.ncate.org/standards/standard-2

Dana, N. F., & Yendol-Hoppey, D. (2020). *The reflective educator's guide to classroom research: Learning to teach and teaching to learn through practitioner inquiry* (4th ed.). Thousand Oaks, CA: Corwin Press.

Dewey, J. (1933). Philosophy and civilization. *Philosophy, 8*(31), 360–361.

Evans, G., McTureous, R., Weber, G., & Crippen, K. (2017). Assessing the needs of site-based mentors as a means of improving partnerships for STEM teacher development, *2017 international conference proceedings*, Des Moines, Iowa. Retrieved from: https://theaste.org/publications/proceedings/2017-proceedings/

Fishman, B. J., Penuel, W. R., Allen, A. R., Cheng, B. H., & Sabelli, N. O. R. A. (2013). Design-based implementation research: An emerging model for transforming the relationship of research and practice. *National Society for the Study of Education*, *112*(2), 136–156.

Harrison, C. (2015). Assessment for learning in science classrooms. *Journal of Research in STEM Education*, *1*(2), 78–86.

Hollins, E. R. (2015). *Rethinking field experiences in preservice teacher preparation: Meeting new challenges for accountability*. New York, NY :Routledge.

Horn, I. S., & Campbell, S. S. (2015). Developing pedagogical judgement in novice teachers: mediated field experience as a pedagogy for teacher education. *Pedagogies: An International Journal, 10(20)*, 149–176.

Keeley, P. (2005). *Science curriculum topic study: Bridging the gap between standards and practice*. Thousand Oaks, CA: Corwin Press.

Keeley, P. (2015). *Science formative assessment, volume 1: 75 practical strategies for linking assessment, instruction, and learning, Second Edition*. Thousand Oaks, CA: Corwin Press.

Keeley, P. (2017). Embedding formative assessment into the 5E instructional model. *Science and Children*, *55*(4), 28–31.

Keeley, P., & Rose, C. M. (2006). *Mathematics curriculum topic study: Bridging the gap between standards and practice*. Thousand Oaks, CA: Corwin Press.

Keeley, P., & Tobey, C. R. (2011). *Mathematics formative assessment, volume 1: 75 Practical strategies for linking assessment, instruction, and learning*. Thousand Oaks, CA: Corwin Press.

Keeley, P., & Tobey, C. R. (2016). *Mathematics formative assessment, Volume 2: 50 More practical strategies for linking assessment, instruction, and learning*. Thousand Oaks, CA: Corwin Press.

Korthagen, F. A. (2010). Situated learning theory and the pedagogy of teacher education: Towards an integrative view of teacher behavior and teacher learning. *Teaching and Teacher Education*, *26*(1), 98–106.

Korthagen, F. A. J., & Kessels, J. P. A. M. (1999). Linking theory and practice: Changing the pedagogy of teacher education. *Educational Researcher*, *28*(4), 4–17.

Korthagen, F., Loughran, J., & Russell, T. (2006). Developing fundamental principles for teacher education programs and practices. *Teaching and Teacher Education, 22*(8), 1020–1041.

Lave, J. (1991). Situating learning in communities of practice. In L. B. Resnick, J. M. Levine, & S. D. Teasley (Eds.), *Perspectives on socially shared cognition* (pp. 63–82). Washington, DC: American Psychological Association. http://dx.doi.org/10.1037/10096-003

Loughran, J. J. (2002). *Developing reflective practice: Learning about teaching and learning through modelling*. Bristol, PA: Routledge.

Manuel, J., & Hughes, J. (2006). 'It has always been my dream': Exploring pre-service teachers' motivations for choosing to teach. *Teacher Development*, *10*(1), 5–24.

National Council for Accreditation of Teachers Education [NCATE]. (2010). *Transforming teacher education through clinical practice: A national strategy to prepare effective teachers*. Washington, D.C.: National Council for Accreditation of Teachers Education (ED512807).

National Governors Association. (2010). *Common core state standards*. Retrieved from http://www.corestandards.org/

National Research Council [NRC]. (2012). *A framework for K–12 science education: Practices, crosscutting concepts, and core ideas*. Washington, DC :National Academies Press.

National Science Teachers Association [NSTA]. (2020). *2020 NSTA standards for science teacher preparation*. Retrieved October 9, 2019, from http://static.nsta.org/pdfs/2020NSTAStandards1-6.pdf

Penuel, W. R., & Fishman, B. J. (2012). Large-scale science education intervention research we can use. *Journal of Research in Science Teaching*, *49*(3), 281–304. doi.org/10.1002/tea.21001

Procee, H. (2006). Reflection in education: A Kantian epistemology. *Educational Theory*, *56*(3), 237–253.

Putnam, R., & Borko, H. (2000). What do new views of knowledge and thinking have to say about research on teacher learning? *Educational Researcher*, 29(1), 4–15.

Shymansky, J. A. (1992). Using constructivist ideas to teach science teachers about constructivist ideas, or teachers are students too!. *Journal of Science Teacher Education*, *3*(2), 53–57.

UTeach Institute. (2011). *UTeach elements of success*. Retrieved from: https://institute.uteach.utexas.edu/sites/institute.uteach.utexas.edu/files/uteach-elements-of-success-2011.pdf

Wenger, E. (1998). Communities of practice: Learning as a social system. *Systems Thinker*, *9*(5), 2–3.

Wenger, E., & Lave, J. (2001). Legitimate peripheral participation in communities of practice. In *Supporting lifelong learning* (pp. 121–136). Routledge.

Windschitl, M., Thompson, J., Braaten, M., & Stroupe, D. (2012). Proposing a core set of instructional practices and tools for teachers of science. *Science Education*, *96*(5), 878–903. https://doi.org/10.1002/sce.21027

Zeichner, K. (2010). Rethinking the connections between campus courses and field experiences in college-and university-based teacher education. *Journal of Teacher Education*, *61*(1–2), 89–99.

SECTION II

INNOVATIVE INSTRUCTIONAL PROGRAMMING

CHAPTER 12

THE UTEACH INSTRUCTIONAL PROGRAM, ELEMENTS, AND COURSES

Kimberly K. Hughes, Ashley R. Welch, and Amy L. Moreland

The UTeach instructional program offers an innovative approach to university-based STEM teacher preparation, and the program design is key to attracting undergraduate STEM majors to teaching. UTeach provides students with rigorous, clinically intensive preparation through a streamlined course sequence that is specialized for secondary STEM teaching and that prioritizes classroom teaching from the very first course. UTeach preparation is also well-aligned with current education and teacher preparation standards. The UTeach secondary STEM teacher preparation program has been preparing math and science teachers to design and implement inquiry- and project-based instruction that emphasizes application of mathematics and science concepts and practices since 1997, and has refined and updated these practices to stay current with research literature over the last 22 years. Current STEM education Sand teacher preparation standards call for the essential approaches to teaching and learning on which the UTeach instructional program was originally founded, and continues to exemplify.

The population of undergraduate STEM majors attending U.S. colleges and universities represents a significant, untapped pool from which to recruit and prepare STEM teachers. More than 355,000 students graduated with a STEM degree

Preparing STEM Teachers: The UTeach Replication Model, pages 163–174.

in 2016 (National Center for Education Statistics, 2016). And a 2017 national survey demonstrated that many of them would consider teaching if provided the right opportunity (Marder, Brown, & Plisch, 2017). The UTeach secondary STEM teacher preparation program provides just such an opportunity.

The UTeach instructional program offers an innovative approach to university-based STEM teacher preparation, and the program design is key to attracting undergraduate STEM majors to teaching. Over and over again, students have reported—through annual UTeach surveys—that the opportunity to expand their career options—the "one degree, two careers" aspect of the program—and still graduate within four years heavily influences their decision to enroll in UTeach programs and explore teaching as a career. In recent years, the cost of college has made graduating in four years even more important for even more students. And widespread acknowledgement of the need for a more diverse teaching population further necessitates efficient, but effective, teacher preparation program design so that pursuing secondary teaching becomes a viable option for all undergraduates, not just those who can afford more coursework. Removing barriers to participation in UTeach programs has always been fundamental to the program design, and can be largely credited with increasing the number of STEM teachers produced by universities that adopt the UTeach model. A recent study demonstrated that UTeach preparation also results in more effective STEM teachers (Backes, Goldhaber, Cade, Sullivan, & Dodson, 2018).

UTeach is designed to make it possible for undergraduate STEM majors to add a teaching credential without adding time or cost to their degrees. It provides students with rigorous, clinically intensive preparation through a streamlined course sequence that is specialized for secondary STEM teaching and that prioritizes classroom teaching from the very first course. UTeach preparation is also well-aligned with current education and teacher preparation standards. Current U.S. STEM education content standards emphasize developing depth of content knowledge and conceptual understanding; building skills in fundamental science, mathematical, and engineering practices; and applying science and mathematics to solve authentic problems. This in turn necessitates the preparation of teachers with deep content knowledge who have experience designing, carrying out, analyzing, and presenting independent scientific inquiries; who can apply mathematical tools and practices to solve complex problems and model scientific phenomena; and who are able to design problem- and project-based learning environments for all learners.

The Common Core State Standards for Mathematics (CCSS-M) demand increased rigor and emphasis on problem-solving and mathematical thinking. There is also a focus on the application of conceptual understanding to authentic contexts (National Governors Association Center for Best Practices, Council of Chief State School Officers, 2010). Likewise, the Next Generation Science Standards (NGSS) call for science instruction that more closely reflects the real work of scientists and engineers. They emphasize depth of conceptual understanding of core

ideas over breadth of content coverage and the integration of science knowledge and ability with the practices needed to engage in scientific inquiry and engineering design (National Research Council, 2012).

The Council for the Accreditation of Educator Preparation (CAEP), in their 2015 recommendations, emphasizes the link between instruction and content, calling for teacher preparation programs to develop candidates' content knowledge and pedagogical knowledge so that "candidates develop a deep understanding of the critical concepts and principles of their discipline and, by completion, are able to use discipline-specific practices flexibly to advance the learning of all students toward attainment of college- and career-readiness standards" (Council for the Accreditation of Educator Preparation, 2015, p. 2).

This "deeper learning" called for by the CCSS-M and CAEP is exemplified by:

- An understanding of the meaning and relevance of ideas to concrete problems;
- An ability to apply core concepts and modes of inquiry to complex real-world tasks;
- A capacity to transfer knowledge and skills to new situations, to build on and use them;
- Abilities to communicate ideas and to collaborate in problem solving; and
- An ongoing ability to learn to learn (p. 5)

Likewise, the InTASC Model Core Teaching Standards were updated in 2013 in part by "the new imperative that every student can and must achieve high academic standards" (Council of Chief State School Officers, 2013, p. 3). They call for a stronger focus on application of knowledge and skills so that teachers can have a "deep and flexible understanding of their content areas and be able to draw upon content knowledge as they work with learners to access information, apply knowledge in real world settings, and address meaningful issues to assure learner mastery of content" (Council of Chief State School Officers, 2013, p. 8). Content Standards 4 and 5 are of particular importance:

- Standard #4: Content Knowledge. The teacher understands the central concepts, tools of inquiry, and structures of the discipline(s) he or she teaches and creates learning experiences that make the discipline accessible and meaningful for learners to assure mastery of the content. (p. 8)
- Standard #5: Application of Content. The teacher understands how to connect concepts and use differing perspectives to engage learners in critical thinking, creativity, and collaborative problem solving related to authentic local and global issues. (p. 8)

The UTeach secondary STEM teacher preparation program has been preparing math and science teachers to design and implement inquiry- and project-based instruction that emphasizes application of mathematics and science concepts and

practices since 1997, and has refined and updated these practices to stay current with research literature over the last 22 years. Current STEM education and teacher preparation standards call for the essential approaches to teaching and learning on which the UTeach instructional program was originally founded.

UTEACH PROGRAM DESIGN

The UTeach pre-service instructional program is characterized by a rigorous content major paired with a streamlined, field-intensive UTeach curriculum that is firmly situated within the STEM domains. The UTeach curriculum, pictured in Figure 12.1, consists of eight (nine in the case of math majors) courses and 24 (27 for math majors) credit hours and includes the following groups of courses:

- UTeach STEM teaching exploratory courses (*Step 1: Inquiry Approaches to Teaching* and *Step 2: Inquiry-Based Lesson Design*) are one-hour courses taught by UTeach Master Teachers (experienced, successful K–12 STEM teachers who are hired on full-time at the university as clinical instructors), and designed to encourage STEM majors to "try out teaching" through a series of teaching experiences in local elementary and middle school classrooms;
- UTeach STEM education courses (*Knowing and Learning in Mathematics and Science, Classroom Interactions,* and *Project-Based Instruction*) are based on current research in teaching and learning specifically within the STEM domains and are taught by science and mathematics education faculty. UTeach Master Teachers assist with field teaching components associated with these courses;
- UTeach STEM content courses (*Perspectives on Science and Mathematics, Research Methods,* and *Functions and Modeling*) are generally taken concurrently with STEM education courses and are designed to emphasize content and develop scientific and mathematical practices of particular importance to secondary math and science teachers;
- *Apprentice Teaching* consists of a semester-long clinical teaching experience and seminar that provides a culminating opportunity for students to demonstrate proficiencies required for certification; and
- The UTeach portfolio, which is not a course but embedded within the coursework, requires students to provide evidence that they are proficient across a number of criteria ranging from subject-matter knowledge to effective instructional design and classroom management. Along with proficiencies required during the final *Apprentice Teaching* experience, this collection of evidence must satisfy minimum criteria in order for students to be recommended for certification.

The UTeach instructional program is designed with several core intentions:

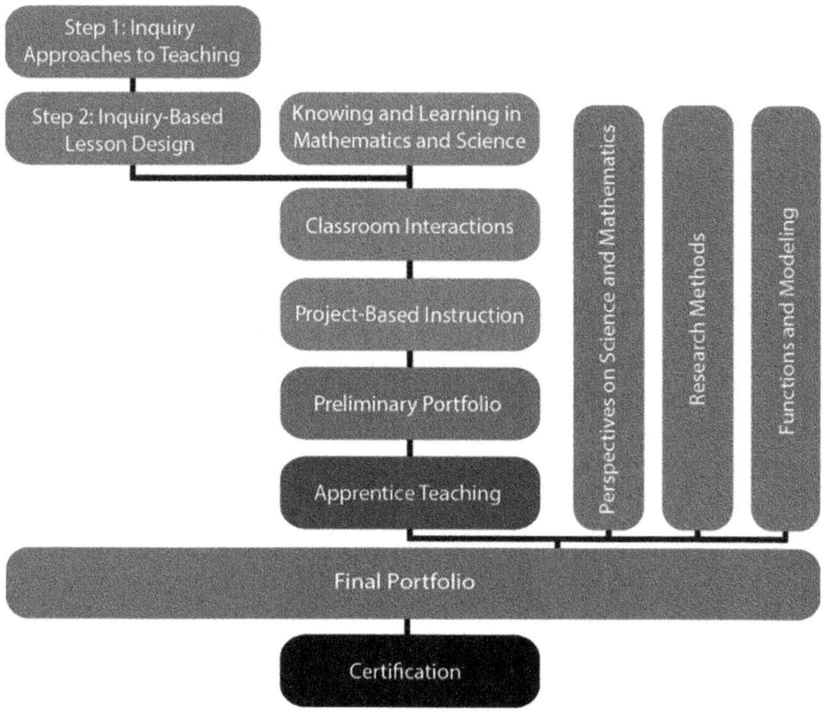

FIGURE 12.1. The UTeach Curricular Components. The UTeach curriculum includes a tightly articulated set of courses focused specifically on the needs of future secondary STEM teachers. Recruitment courses are shown in orange, STEM education courses in blue, STEM content courses (in addition to major requirements) in green; the capstone Apprentice Teaching course in purple; and the student portfolio that is developed concurrently in and across courses in gray.

1. Provide Early STEM Teaching Experiences in an Exploratory Context

All STEM majors are invited to explore teaching as early as their freshman year. Two, one-credit courses, *Step 1: Inquiry Approaches to Teaching* and *Step 2: Inquiry-Based Lesson Design*, provide students with the opportunity to "try out teaching" in a low-stakes setting, one that requires minimal time and no commitment to pursue a teaching credential. During the first course, students learn about inquiry using the Biological Sciences Curriculum Study (BSCS) 5E lesson structure, which is an inquiry-based instructional model that enables learners to construct their own understandings of a concept (Bybee et al., 2006). They create inquiry-based lessons, working with research-based Full Option Science System (FOSS) kits developed by the Lawrence Hall of Science, and deliver them in upper-elementary classrooms.

Under the guidance of Master Teachers, they work in pairs to develop and teach three lessons. Students are observed and provided written feedback each time they teach. They also formally reflect on their teaching experiences.

During the second course, student pairs work closely with middle school classroom teachers to develop inquiry-based lessons. Again, they teach three lessons, are observed and provided written feedback by both the classroom teacher and UTeach Master Teachers, and formally reflect on their teaching. They also learn about adolescent development, focus more on classroom management strategies, integrate technology into one lesson, and revise a lesson based on student assessment data they collect. By the end of the second course, students determine for themselves whether or not teaching is something they enjoy and wish to pursue.

By design, the majority of students who enroll in these early courses do not continue in the program (UTeach Institute, 2017). They decide that the teaching profession is not for them. Having had this unique experience, however, prepares these students to be informed advocates for education, whatever their chosen vocation. The UTeach approach of inviting all STEM majors to explore teaching through these one-credit courses does result in significant numbers of students certifying as STEM teachers who would not otherwise have considered teaching. Nationwide, a full third of students who complete the program reported they were uncertain they would teach when they enrolled in the first UTeach course (UTeach Institute, 2017). UTeach successfully recruits STEM majors into teaching who have not come to the university to teach, but who have the potential to be great teachers. When they are provided the right opportunities to explore teaching, many of them discover that they have an interest.

2. Develop Deep Conceptual Understanding and Mastery of Subject-Area Content

UTeach courses provide rigorous instruction based on current research and emphasize the development of deep-level understanding of both subject material and pedagogy. UTeach students take all the coursework required to obtain a rigorous STEM major along with their teaching certification. Content knowledge is assessed by enforcing a cumulative minimum GPA across STEM major and UTeach pedagogy courses. Content knowledge is also a component of UTeach classroom observation protocols and must be demonstrated through evidence submitted through a required portfolio.

3. Make Explicit the Underlying Connections Between Mathematics and Science

UTeach courses help students explore the relationships within and between the disciplines of mathematics and science and how that content is learned by novices and experts. Mathematics and science students take most of their UTeach courses together, creating ongoing opportunities for collaboration and discussions of the commonalities and differences in how each discipline is learned and taught.

All students design and carry out a series of independent scientific inquiries that necessitate the application of mathematics for data analysis and modeling. All students receive instruction on the historical and philosophical development of mathematics and science disciplines.

4. Develop Research and Information Analysis Skills

Students are required to enroll in UTeach courses that include an explicit focus on data and information analysis and synthesis. The *Perspectives on Science and Mathematics* course is a history course designed to develop sophisticated research and information analysis skills through the analysis of primary sources of historical research and application to problems related to STEM teaching, building student understanding of the historical development of the kinds of questions scientists and mathematicians have tried to answer and why. In the *Research Methods* course, developing the habits of mind and practices of scientists and mathematicians, students conduct statistical analysis of investigational data, apply scientific arguments in matters of social importance, write scientific papers, and give oral presentations of scientific work.

5. Integrate Content and Pedagogy

There are no generic education courses in the UTeach course sequence. Rather, UTeach education courses are specifically designed to meet the needs of future secondary teachers of science, mathematics, computer science, and engineering, and are discipline specific. Discipline-focused content courses such as *Research Methods*, *Functions and Modeling*, and *Perspectives on Science and Mathematics* provide content knowledge of particular importance for teachers and also count toward a STEM degree major. For example, students in *Research Methods* not only carry out their own independent investigations, but review and critique the work of their peers to gain experience providing feedback and guidance that can be applied in their future secondary classrooms.

6. Build Strong Connections Between Educational Theory and Practice

Following the first two exploratory courses, students who choose to continue with the program receive their first in-depth exposure to educational theory in the *Knowing and Learning in Mathematics and Science* course. This course focuses on knowing and learning in secondary STEM subjects, resting on the premise that formal research in these disciplines constitutes a robust line of inquiry and design of its own. This line of inquiry has tended to be situated relative to classroom practice and draws on significant insights from many fields, including psychology, anthropology, critical literacy, sociology, biology, linguistics, neuroscience, philosophy, developmental theory, artificial intelligence, and the domains of mathematics, science, and computer science. In this course, students analyze

domain-specific problem-solving activities and approaches in an applied fashion, such as through the clinical interview process.

Knowing and Learning in Mathematics and Science is followed by the *Classroom Interactions* course. This course continues the process of preparing students to teach mathematics and science in secondary settings by providing opportunities to see how theories explored in the *Knowing and Learning* course play out in instructional settings. Students design and implement instructional activities informed by their own understandings of what it means to know and learn mathematics and science, and then evaluate the outcomes of those activities on the basis of student artifacts (i.e., what students say, do, or create). The course introduces ways in which curricula and technologies are used in classroom settings to build relationships among teachers and students. In essence, *Classroom Interactions* is centered on a close examination of the interplay between teachers, students, and content, and how such interactions enable students to develop deep conceptual understanding. Students learn how content and pedagogy combine to make effective teaching. *Classroom Interactions* is followed by *Project-Based Instruction*. It is discussed in detail later. *Apprentice Teaching* concludes the pedagogical course sequence. This final teaching practicum reinforces and expands teaching strategies that students have developed through their coursework and field experiences and provides a final, intensive opportunity to reflect on the application of educational theory to classroom practice. The UTeach program also attempts to fill any gaps in students' professional development during this final semester. In particular, *Apprentice Teaching* focuses on classroom management and time management strategies, parent-teacher communication strategies, school culture, effective middle school and high school dynamics, legal and logistical issues in teaching, the final student assessment portfolio, and state certification examinations. Student performance is evaluated using the UTeach Observation Protocol (Walkington & Marder, 2018).

The design of the UTeach program field components provides students with early and ongoing opportunities to implement and reflect on the instructional strategies applied in grades 4–12 classrooms, about which they are learning in their university pedagogy courses. All practice teaching experiences are tightly coupled with the content and concepts presented in the five required field-based courses—*Step 1: Inquiry Approaches to Teaching, Step 2: Inquiry-Based Lesson Design, Classroom Interactions, Project-Based Instruction*, and *Apprentice Teaching*. Course assignments and projects involve detailed analysis of student field experiences, and significant class time is devoted to preparing for, practicing, debriefing, and reflecting on those teaching experiences.

UTeach Master Teachers assume primary responsibility for supervising field experiences and provide ongoing support and coaching to students. UTeach field experiences are designed so that time in the classroom is purposeful. Significant time is devoted to preparing for, debriefing, and reflecting on students' teaching experiences to help them integrate their learning, develop confidence, and de-

velop clearly identified proficiencies. Master Teachers, STEM education faculty, and classroom-based mentor teachers provide feedback during lesson development. Further feedback is provided when lessons are practiced in the university classroom prior to being taught in the field. During field teaching, students are observed by course instructors, Master Teachers, and their classroom mentor teacher, all of whom provide written feedback. Throughout the course of field teaching, students are formally observed a minimum of 20 times. While limited, purposeful classroom observations are a component of the field experience, the majority of student time in the field is devoted to teaching lessons. Additionally, in all field-based courses, there is significant emphasis on peer review and individual student reflection throughout all phases of lesson planning and implementation.

7. Emphasize Inquiry-, Problem- and Project-Based Instructional Approaches

Students begin the program designing, implementing, and reflecting on a series of inquiry-based lessons utilizing the BSCS 5E Instructional Model, focusing initially on the development of effective questioning strategies. This emphasis continues throughout the program as a variety of research-based instructional strategies are introduced, explored, and analyzed for effectiveness in secondary school settings. As a culminating experience, UTeach includes an entire field-based course devoted to project-based instruction (PBI). During PBI, students design a complete project-based unit while they design and implement multi-day, problem-based lessons in the field. Student proficiency in this area is a component of UTeach classroom observation protocols and must be demonstrated through video evidence submitted in the required exit portfolio.

8. Integrate Themes of Assessment, Equity, Literacy, and Technology

Rather than have stand-alone courses on each of these topics, the UTeach curriculum integrates these important themes into all courses, developing students' competency as they progress through the program. UTeach student competencies are assessed developmentally in each of these areas through the use of an observation protocol during all clinical teaching experiences, and students are required to reflect on and analyze their own growth by repeatedly providing evidence of effective use of assessment strategies, integration of technology, and lesson design that promotes learning for all students. Student proficiency in this area is a component of UTeach classroom observation protocols and must be demonstrated through evidence submitted in the required exit portfolio.

9. Cultivate Reflective Practice

UTeach students are observed and provided feedback by clinical faculty (Master Teachers), course instructors, and classroom mentor teachers a minimum of 10 times prior to the *Apprentice Teaching* semester and another 10 times during

the final *Apprentice Teaching* practicum. Students prepare formal, written reflections each time they teach in the five field-based courses. Peer feedback is another explicit component of all UTeach courses and ranges from collaborative lesson plan development to feedback on practice teaching, writings, and oral presentations of scientific or mathematical concepts. Students are required to modify lessons based on evidence of student learning and provide written rationales for the changes made, based on assessment of student data. In-depth student video and written analyses of their own teaching are major components of the *Classroom Interactions* and *Project-Based Instruction* courses. Evidence of reflective practice is a component of the required exit portfolio as well.

10. Develop Proficiency in Mathematical, Scientific, and Engineering Practices

One of the greatest strengths of the UTeach instructional model is the emphasis placed on developing teachers' abilities to 1) apply core scientific and mathematical practices to authentic, complex problems and 2) design instructional learning environments to engage K–12 learners in these same practices. The UTeach curriculum explicitly develops candidates' proficiency in the eight Mathematical Practices called for by the CCSS-M and the eight Scientific and Engineering Practices called for by the NGSS. An underlying UTeach design principle, the integration of math and science, creates a rich learning environment in all UTeach courses, where mathematics students are continually exposed to scientific applications and science students develop facility in the use of mathematical tools. Furthermore, beginning in 2012, a formal effort was begun to integrate engineering design activities into UTeach course curriculum (Marshall & Berland, 2012; Marshall & Edelmon, 2012). An acknowledgement that all STEM teachers should also be equipped with foundational knowledge in computational thinking has led to a recent effort to integrate computer science concepts into the UTeach curriculum. There are a few UTeach courses in particular that stand out with regard to the emphasis on teaching practices highlighted in the CCSS-M and the NGSS. They are discussed below.

Research Methods. All students, including mathematics majors, take the *Research Methods* course where students learn how to design and carry out independent investigations to answer questions they pose themselves. Students design experiments to answer scientific questions and to reduce systematic and random errors. They incorporate statistics to interpret experimental results and deal with sampling errors, and do mathematical modeling of scientific phenomena. They also present their scientific work orally. Thus, the course content is organized into five units that correspond to the development of their inquiries and a presentation on a scientific topic of choice: Curiosity and Scientific Inquiry, Experimental Design and Analysis, Statistics, Scientific Information, and Modeling. During this course, students experience inquiry-based learning for themselves. In the subse-

quent *Project-Based Instruction* course, they learn to design these same learning environments for K–12 students.

Project-Based Instruction. The UTeach instructional sequence includes an entire course devoted to project-based instruction (PBI). A major focus of this course is on developing an approach to designing, implementing, and evaluating problem- and project-based curricula and processes by employing approaches that have emerged from collaborations between teachers and researchers. Skills in integrating the STEM disciplines and STEM-specific pedagogical content knowledge are developed as UTeach students work toward the design of project-based units. Competency is continually built as students read about and discuss the principles of PBI; reflect on observations of project-based learning environments in high school settings; and incorporate what they are learning into the design of problem-based lessons, and ultimately, an entire project-based unit. An intensive field component includes observation of well-implemented project-based instruction in local schools as well as implementation of problem-based lessons with area high school students in an out-of-classroom field setting.

Functions and Modeling. In addition to their major course requirements, mathematics majors take the UTeach course *Functions and Modeling*, designed to engage future math teachers in challenging, non-routine problem solving, inquiry-based learning, and applications of mathematics. In this mathematics course, students collaborate to collect data and explore a variety of situations that can be modeled using linear, exponential, polynomial, and trigonometric functions. Topics involving function properties and patterns, complex numbers, parametric equations, polar equations, vectors, and exponential growth and decay are investigated. Explorations involve the use of multiple representations, transformations, data analysis techniques (such as curve fitting), and interconnections among topics in algebra, analytic geometry, statistics, trigonometry, and calculus. The lab investigations use various technologies, including computers, calculators, and computer graphing software.

CONTINUOUS IMPROVEMENT OF THE UTEACH MODEL

The professional learning community of more than 700 faculty and Master Teachers across UTeach programs nationwide offers a significant advantage; namely that a collaborative infrastructure has been established that fosters sharing of best practices and community problem-solving with the explicit goal of continuous improvement of the UTeach secondary STEM teacher preparation model. In this way, all UTeach programs nationwide benefit from the insights gained by individual programs as they encounter new challenges related to preparing strong STEM teachers. The UTeach Institute serves as the hub for this network, and maintains UTeach instructional program content, including complete course curriculum, through a secure website. UTeach courses and instructional support resources are regularly updated based on contributions from UTeach partner programs. Annual Master Teacher retreats and the UTeach conference also provide opportunities for

UTeach programs to share adaptations and innovations to the instructional model. Ongoing contributions by the UTeach network of experts in STEM and STEM education ensures that the UTeach instructional program will continue to evolve to reflect the latest research on STEM teaching and learning.

REFERENCES

Backes, B., Goldhaber, D., Cade, W., Sullivan, K., & Dodson, M. (2018). Can UTeach? Assessing the relative effectiveness of STEM teachers. *Economics of Education Review, 64*, 184–198.

Bybee, R. W., Taylor, J. A., Gardner, A., Van Scotter, P., Powell, J. C., Westbrook, A., & Landes, N. (2006). *The BSCS instructional model: Origins and effectiveness: A report prepared for the National Institutes of Health, Office of Science Education.* Colorado Springs, CO: BSCS.

Council for the Accreditation of Educator Preparation. (2015). *CAEP Accreditation Standards, as approved and amended by the CAEP Board of Directors.* Washington DC: Council for the Accreditation of Educator Preparation.

Council of Chief State School Officers. (2013, April). *Interstate teacher assessment and support consortium InTASC model core teaching standards and learning progressions for teachers 1.0: A Resource for ongoing teacher development.* Washington, DC: CCSSO.

Marder, M. P., Brown, R. C., & Plisch, M. (2017). *Recruiting teachers in high-needs STEM Fields: A survey of current majors and recent STEM graduates*. College Park, MD: American Physical Society Panel on Public Affairs.

Marshall, J. A., & Berland, L. (2012). Developing a vision of engineering education. *Journal of Pre-College Engineering Education, 2*(2), 36–50.

Marshall, J. A., & Edelmon, G. (2012). *An optics design challenge tying preservice teachers to physics classrooms*. Paper presented at the American Association of Physics Teachers Winter Meeting, Ontario, CA.

National Center for Education Statistics, U.S. Department of Education. (2016). *Status and trends in the education of racial and ethnic groups: Indicator 26— STEM Degrees.* Retrieved from: https://nces.ed.gov/programs/raceindicators/indicator_reg.asp#

National Governors Association Center for Best Practices, Council of Chief State School Officers. (2010). *Common core state standards for mathematics.* Washington DC: National Governors Association Center for Best Practices, Council of Chief State School Officers.

National Research Council. (2012). *A framework for K–12 science education: Practices, crosscutting concepts, and core ideas*. Washington, DC: The National Academies Press.

UTeach Institute. (2017). *UTeach vital signs: National data on UTeach program recruitment, retention, and teacher production.* Austin, TX: UTeach Institute.

Walkington, C., & Marder, M. (2018) Using the UTeach observational protocol (UTOP) to understand the quality of mathematics instruction. *ZDM Mathematics Education, 50*(3), 507–519. Retrieved from: https://doi.org/10.1007/s11858-018-0923-7

CHAPTER 13

STEM TEACHING FOR SOCIAL JUSTICE

Experiences in Early Fieldwork Courses

Anita Sanyal Tudela, Sarah Henson-Darko,
Steve Karig, and Kayla White

In this chapter, we describe our effort to integrate social justice perspectives into our Step 2 course. The story we tell outlines how we came to decide to modify Step 2 and the specific modifications we made to the course in our efforts to center social justice. We pay special attention to describing why we believe it was timely and necessary to design instruction to address questions related to social justice and why we are doing this in Step 2. We also describe the specific ideas about why we believe this work should start at the beginning of teacher preparation. We then outline our approach to carrying out the course revision and present a few examples of modifications we made in our instruction. We conclude with some personal reflections on the process we took in revising the course.

INTRODUCTION

This chapter reports on our effort to integrate equity and social justice into our Step 2 course at the University of Maryland, College Park. We are convinced for many reasons that it is worthwhile to spend time in an introductory teacher recruitment course on questions related to diversity, equity, and social justice. The

Preparing STEM Teachers: The UTeach Replication Model, pages 175–188.

story we tell in this chapter outlines how we came to decide to modify Step 2 and the specific modifications we made to the course in our efforts to center social justice. We pay special attention to describing why we believe it was timely and necessary to design instruction to address questions related to social justice and why we are doing this in Step 2. We also describe the specific ideas about why we believe this work should start at the beginning of teacher preparation. We then outline our approach to carrying out the course revision and present a few examples of modifications we made in our instruction. We conclude with some personal reflections on the process we took in revising the course. Throughout our chapter, we will refer to our students as "undergraduates" and K12 students as "students."

OUR CONTEXT

Our university's implementation of the UTeach model (we began the rollout of UTeach courses in Fall 2014) responds to our specific institutional context in a few ways that are relevant to mention. First, we have multiple pathways that STEM majors can opt for to get certified to teach. These include an undergraduate and graduate pathway for secondary STEM certification, but also pathways for middle grades STEM certification, and the option for elementary certification. Our Step courses function as recruitment courses for science and math majors who are interested in exploring the field of teaching. However, we also see them as foundational courses since nearly all future secondary STEM teachers that the university will graduate begin with the Step courses, regardless of which certification pathway they pursue. The Step courses effectively form the starting point for learning about teaching that is then refined and built upon over the course of the program.

Beyond being a standard topic in many teacher preparation curricula, the need to address equity and diversity is directly related to the complexity of our context. Our local K12 schools enroll a majority of students of color (Maryland State Department of Education, 2018), while our university is predominantly white (University Office of Institutional Research, Planning, and Assessment, 2018). Our fieldwork placement schools have high percentages of students receiving Free and Reduced Meals (Maryland State Department of Education, 2018), even as the county is home to some of the most affluent Black communities in the U.S. (Brown, 2015). Many of our undergraduates navigate their first teaching experiences in classrooms that look drastically different from their own experiences, while, for others, these settings are more familiar. Our own undergraduate classrooms are comprised of people who have both benefited and been marginalized by the systems and structures that we are trying to pay attention to.

This requires us to pay close attention to what undergraduates are thinking as they make sense of teaching through their fieldwork experiences. Addressing common and deep-rooted cultural assumptions can begin from these first experiences students have in classrooms (Sleeter, 2008). We see the early field experi-

ences offered in the Step courses as opening new possibilities for learning opportunities and trajectories, including an opportunity to address deep rooted social assumptions about children and teaching right from the start. We recognize that deep rooted assumptions are difficult to change and require sustained attention and revisiting. In our courses, we enact this idea by encouraging content instruction that views students' ideas and language as resources for building conceptual understanding and seeking to uncover societal and institutional structures that contribute to the maintenance of deficit perspectives in the context of classroom interactions.

OUR GROUNDING

On a broad level, the STEM fields have been widely critiqued for cultural practices that work to exclude women and other marginalized communities (Brickhouse, 2001; de Sousa Santos, 2007; Harding, 2016). Such critiques point to education and the formal experiences students have learning in STEM areas as fundamental for encouraging greater representation and better experiences of marginalized communities (Barton, 2001). We see social change in the classroom as contributing toward re-shaping representation and practice in the STEM fields. In teacher education there are widespread calls to integrate questions of diversity and social justice across courses (Cochran-Smith, 2010; McDonald & Zeichner, 2009; Sleeter, 2008) as well as within the content areas (Barton, 2001; Kumashiro, 2001). Prospective STEM teachers should have opportunities in their teacher preparation to grapple with complex, difficult or otherwise sensitive political questions that relate to the learning experiences children have (and teachers create) in classrooms.

We choose "social justice" to talk about what we are doing, since we are interested in more than "celebrating," "acknowledging" difference or "achieving" equity (Cochran-Smith, 2010; Kumashiro, 2001; McDonald & Zeichner, 2009). We are interested in pushing our students (and ourselves) to begin to understand the specific ways structural or institutional structures shape individual and group student experiences and academic success, and to grapple with teachers' role in the promotion or interruption of these (McDonald & Zeichner, 2009). We believe "meeting students' needs" and giving everyone a fair chance at success (equity) requires identifying, understanding, and countering oppressive mechanisms in schools and classrooms. For this reason, we see ourselves as promoting the grounding for the development of anti-oppressive pedagogies (Kumashiro, 2001). Since understanding the ways structural elements impact individual experience has to do with uncovering and unpacking common sense or normalized ideas and practices in teaching and learning, we assume it to require sustained attention and opportunities to think about explicitly across the teacher preparation trajectory.

We also are attentive to the stakes involved in assuming this work at such an early point in teacher preparation. We understand that putting explicit attention to questions of social justice forward could open uncomfortable conversations or

cause conflicts/questions for undergraduates' personal (and future professional) identities. At times, this feels in tension with the recruitment agenda of the Step courses, which brings us to a more explicit and purposeful positioning. We are intentional about how and when we enter into these "courageous conversations" (Singleton, 2015).

WHY STEP 2?

In our version of the Step 2 course, some notable adaptations have been made to the UTeach model. Step 2 is a two-credit course, which offers time and space to situate their fieldwork experiences in a larger context. Additionally, the fieldwork experiences associated with Step 2 have been developed in order to allow undergraduates to engage deeply with the same middle school curricular content material for the entire semester. These fieldwork experiences include a one-on-one conversation with a middle school student, a classroom observation, the planning and enactment of a lesson, and the planning and facilitation of a whole-class discussion to extend or revisit components of the original lesson. Additionally, the principal assignments include an analysis of the conversation with a student, observation and written teaching reflections, an analysis of student work collected in the context of the lesson, and a personal teaching philosophy.

As the Step courses evolved over a few years of implementation, we began to feel that there was a need to address the persistent assumptions our undergraduates were making about teaching and students, which we noticed in our class discussions and in their writing. We will describe examples in more detail in the next section, but at the heart of our intent was wanting to help them understand that their thinking about students and teaching shapes teaching practices that can work to either maintain or deepen advantages and disadvantages in the STEM classroom.

In Spring 2018, we were presented with a unique opportunity for us to explore how we might address these concerns. We offered two sections of Step 2, where each section was co-taught by a "Shared Master Teacher," (part-time high school teacher and part-time university-based Master Teacher; see Chapter 10 of this book), and a "university-based Master Teacher" (the traditional position of Master Teacher in the UTeach model). The two university-based Master Teachers had taught Step 2 in previous semesters and had engaged in some exploration of equity and justice within the course. We delved into topics like gender in the field experiences, the labels of "special education" and "ESL," and how notions of "control" shape assumptions about students, teachers, and classrooms. This work helped put into action our commitment to making space for addressing such topics as they came up in both planned and unplanned moments (Henson-Darko & Sanyal Tudela, 2017). We also were affirmed by very positive feedback from our undergraduates about these courageous conversations and how they shaped their experiences in Step 2. In Spring of 2018, all four of us felt personally committed to social justice and were interested in pushing our own teaching and learn-

ing toward making this an explicit part of our course. And we decided to take a systematic approach toward educating ourselves, designing new instructional interventions, and piloting new topics of discussion.

OUR APPROACH

We adopted a mode of work that permitted us to read and reflect as we planned our week-to-week work. Our work was carried through over the semester in two-hour weekly team meetings of the four course instructors. In this meeting, we discussed academic articles, reflected on our own understandings and teaching practice, we designed learning experiences to carry out in class, modified core assignments, and planned our next steps. The format for meetings was standard and followed the sequence of beginning with a seven-minute silent reflective writing period where we set prompts related to attending to our own thinking about readings we had done or the ways our new activities and class discussions were playing out in class sessions. Following this reflection and a brief round of sharing, we moved into more direct instructional planning, which included both revising the core assignments of the course, as well as designing new learning experiences that permitted us to help connect field experiences with critical analysis (e.g. reading, online discussion posts, and in-class discussions). Throughout all this work, we allowed the centering of social justice to be a guiding force in our decision making.

We started with close consideration of where our undergraduates are in their pre-service teacher learning trajectory. We rely on Britzman's (1986) idea that undergraduates come into teacher preparation with strongly held ideas about what it means to learn and to teach that are based in their extensive personal experience in schools. Essentially, they construct images of teaching based on the aspects of teaching that are visible to them as students in their own school experiences. These notions embed tacitly accepted ideas about teaching, the relationship between teaching and learning, and the roles of teachers and students. We see our undergraduates as possessing an educational "common sense" that guides the assumptions of what teaching is and what teachers and students do (Kumashiro, 2015; Sanyal Tudela, 2014). We are interested in our undergraduates unpacking these intuitive notions with which they enter into the program, what Britzman (1986) calls "myths" about teaching. "Myths" include, for example, common assumptions such as students that are "paying attention" are learning, or even some commonly used and accepted ideas in the profession (e.g. "these are my *tough kids*" or "they should *get it*...we covered that last month"). Such myths reinforce stereotypical images that undergraduates have about students in ways that correspond to assumed school-related capacities and status hierarchies. The maintenance and citation of such myths serves to uphold structures of power (Butler, 1997; Kumashiro, 2000). Myths serve as premises on which practices are built and are especially reinforced if experiences in the field rely upon them. Since intuitive notions are difficult to change, we believe that disrupting such narratives

as early as possible in the teacher preparation program can encourage candidates to begin to construct a vision and practice that is positioned to address oppressive mechanisms in the classroom.

We will use two specific examples to share the impact this work has had in Step 2. What becomes visible in our description is the way the concepts became seeds for developing more ideas, and how later in the semester, were transformed into conceptual resources for thinking about and describing their own teaching.

The Myth of the Perfect Classroom

In our first session of class, we introduced a reading titled "Rejecting the Myth of the Superteacher" (Chafee, 2017) that outlines the author's experience being observed as a teacher and her reflections on the need for making the "messy" in teaching more visible and valuable. We used this article to help our candidates challenge the dominant narrative of a "perfect" classroom. This seed allowed us to make visible the kinds of generalizations and assumptions that get enacted when we keep in place the standard of "perfect" for a classroom, teacher, or student, as well as who is excluded from or marginalized within this narrative. This piece also seemed like a strong choice for helping us begin the installation of a vision of teaching that relies on interaction, that responds to students and their ideas, and that is grounded in beliefs that see the inherent worth and possibility of each student. The online discussion post asked students to distill out the enduring message they take from the article and to make connections between this reading and a discussion started in a prior class. The prompts for this online discussion were: *Share at least one of your takeaways from this article, and describe at least one connection between this article and the practices/beliefs posted on the board during Session 1.* An excerpt of an undergraduate's response to this prompt helps us see the kinds of connections that can be made as undergraduates think about Chafee's (2017) points:

> The idea that really resonated with me was the author's main message that 'we must embrace the messiness inherent in teaching and learning.'..I want things to be perfect, or at least almost perfect, or else I feel like I've failed my students...it's okay to make mistakes, specifically in regard to students making mistakes whilst learning materials...there is no perfect student. The author says it well in her last line, stating that 'sometimes, the best learning happens when things don't go perfectly.

In the class following the discussion post assignment, we revisited students' enduring takeaways and we also spent time describing the image of the classroom and students that the notion of perfection instills. From this conversation, we were able to build images that fit generalized descriptions of the perfect classroom (e.g. "students all paying attention," "there are no disruptions," "students all understand in the same way and at the same pace," "the teacher doesn't have to repeat herself") and the perfect student (who "sits still," "pays attention," and "never disagrees with the teacher"). These conversations permitted us to ask our under-

graduates who does not fit into the images we built, for which they responded with descriptions of students that do not fit an assumed ideal (e.g. "students who need more time to learn something," "students who are on a different grade level"). Ultimately, the act of naming what the perfect classroom and perfect student looks like and then critically viewing our lists to see who is excluded in these notions felt like a powerful starting point for a few reasons. First, the reading was accessible for our undergraduates—they could read and understand it, and they had a notion of perfect that was easily recallable in the context of the class discussion. In painting these pictures, undergraduates were also invited to personally confront some of their own implicit biases by considering further details about who was being included and excluded from this vision, like race, gender, socioeconomic status, language, etc. Finally, these conversations helped us open the possibility of deeper critical analysis as we moved forward. It helped solidify the enduring question of who is benefited and marginalized by the perpetuation of the myth as a fundamental consideration in viewing and planning in classrooms, and subsequently made possible much more sensitive and difficult discussions that addressed specific structures of exclusion in more detail (e.g. race).

From here we decided to take a look at the observation assignment, one of the core assignments in Step 2, to focus on the ways we could play up the ideas and critical perspectives that we named during class in an effort to adjust our undergraduates' lenses to view a different set of things that occur in the classrooms. In this assignment, our goal was to move students from making sweeping conclusions upon first glance in the classroom. As a way to disrupt the expectations of perfect classrooms, we tried to cast light on the fact that paying attention to any one student will likely show different ways in which this student engages with or experiences the class. In addition, bringing into view a few different observations of one student also casts light on the student as a multidimensional subject. That is, that the student themself is not static. Our revised prompt read as follows:

> At the very beginning of the class you are observing, select one student that will be easy for you to closely observe. Describe three different, specific observations you can make about this student's **participation** in this class, keeping in mind that there are **many** different ways students can **participate** in a class. What does this student's participation look and/or sound like? What did you learn about this student based on your three observations?

This example of an undergraduate response captures the outcome from this prompt and demonstrates a multidimensional look at a student and her/his participation, and mediated the potential deficit perspectives that might have come through upon a first glance:

- Observation 1: Ben works better on the iPad than with pen and paper. The iPad offers a "Prodigy" game that is simultaneously a social and educational platform. The game lets you interact with others and earn awards,

but in order to do so, you must solve multiple math problems. The possible rewards of the game increased Ben's interest in the class. When he was at a station that required pencil and paper, he did not seem too enthusiastic about the work.

- Observation 2: Ben works very well with his peers. When he is in a group with only students, he cooperates and values the idea of teamwork. Whether his input is right or wrong, he is always willing to contribute to his group's effort.
- Observation 3: Although Ben works well with his peers, he freezes up when in contact with a teacher. When Ben was working on simplifying an equation, he could not figure out that he had to multiply in order to isolate a variable. He knew that the opposite of division is multiplication, but he did not understand that in order to get a variable by itself, you have to perform inverse procedures. When I tried to guide him to an answer, he froze up, not saying a word. I let him continue and [he] left his table, but as soon as I left, he began to ask his peers questions. Ms. Deal said that he does that a lot.
- Summary: Ben is an enthusiastic student, but shies away from student-teacher interaction. He also prefers to do work on the iPad than through pen and paper. It seems as though Ben does not have a problem doing work and making possible mistakes, but when he is with someone of higher authority, he feels like a mistake will get him in some sort of trouble. I tried to tell him that it is okay to get things wrong as long as he tries and it worked to a very small extent. Hopefully, we can incorporate group discussions and the iPads in our lesson in order to get the maximum engagement out of him. I do plan on calling on him even if his hand is not raised just to see how he responds.

As the semester continued, a critique of the "perfect classroom" was revisited independently by some undergraduates upon reflecting on their teaching experiences and within their personal teaching philosophies at the end of the semester. Here, an undergraduate reflects on "perfect" in her personal teaching philosophy:

> The biggest idea of teaching, to me, that has evolved over the semester is that it is okay to make mistakes and not have a "perfect" lesson. At the beginning of the semester my sole focus was on how I can deliver a lesson that will completely fulfill all of my goals. I found myself worrying so much about the lesson running perfectly when, in reality, it is the imperfect lessons that teach you the most. I have realized that it is okay to make mistakes and it is okay for not everything to go as planned. I learned so much from teaching my lesson and was able to make huge improvements moving forward into my follow-up discussion.

Seeing Privilege in Our Teaching

We introduced another reading in the tenth week of class titled "Privilege 101: A Quick and Dirty Guide" (Ferguson, 2014). This reading explains and gives examples of privilege, oppression, and intersectionality in a short and accessible manner. It offers undergraduates a way to understand the systemic dynamics that shape individual experiences and opportunities. Since the reading emphasizes that privilege/oppression happens as an effect of accepted everyday practices and knowledge (and not individual intent), we see its possibility for helping challenge some of the assumptions with which students enter the programs, and for opening an opportunity for undergraduates to see themselvs as actors within a system of privilege/oppression. By this time in the semester, students had read and unpacked notions of perfection in the classroom, deficit perspectives, and bias. We felt ready to push students toward more engagement with structures that maintain privilege and marginalization in the classrooms. The online discussion post included the following prompts:

> Describe privilege and oppression in your own words. And reflect on your own life experiences and the systems of privilege and oppression as described in the article. Discuss one way you benefit from privilege. Discuss other aspects of your identity that intersect with this privilege.

The second prompt from the online discussion post was specifically oriented to try to push our undergraduates into seeing themselves within frames of privilege and oppression. This was a chance for our undergraduates to highlight how their own intersectionality produces multiple narratives of privilege and oppression:

> The fact that I am a male plays a big role in how I relate with others. Because I am a male, people tend to appreciate my strengths and look at my weaknesses as things that I can easily work on and improve. For a female, it is quite the opposite; people try to find reasons to belittle a female's strengths and highlight their weaknesses. On another note, I am a minority and do face oppression because of my race, but I do receive privilege because I am of Indian descent. I have noticed that I benefit [in some ways] more than my black counterparts do. When companies try to diversify the work environment, they would tend to look for someone of my race and background opposed to someone who may be African American. Personally, such examples puts everything into perspective, showing that although it may seem hard for myself, like I have to work harder than my white counterpart to get what they have, there is someone who has to work twice as hard as me to get what I have.

These online posts lent themselves to a productive discussion in the classroom where structures of privilege and oppression in terms of race, gender, sexuality, ethnicity/language, and ability were able to be named. One effect was that undergraduates were able to build some shared understandings of each others experiences and perspectives. Another effect was that these terms and ideas became operationalized and were more likely to be referred to in subsequent discussions as a result.

Our next move was the revision of the teaching reflection to include specific questions related to privilege and the context of their own teaching experiences. The initial concept and discussions primed students to be able to reflect back on their own teaching experiences with an eye toward naming the potential ways they may have been complicit in the maintenance of privilege as they taught. The new prompt read: *Consider the entirety of your lesson (may include the content, task, presentation, interactions with and between students, etc.) and our recent "Privilege 101" reading. Share a reflection of how the idea of privilege was present during your lesson.* The following response from an undergraduate demonstrates a complex reflection on how privilege and oppression may have been at play within both the undergraduates and the students in the classroom:

> Many of the people who were volunteering and more willing to share their answers within class today were the males as compared to maybe one or two of the females. This is especially significant, seeing as though most of the STEM field is still significantly male dominated. Although this may be a stretch, I feel like part of this may be of my own fault. Being the only girl in my teaching group, I have a feeling had I spoken up more throughout the lesson, there may have been a slightly different ambiance in the room. That aside, I do not know the normal environment within [this] classroom except for what I witnessed during the observation, so I can only speak on ideas and what I saw versus actual trends.

Privilege also became a resource for some students for thinking about their own teaching and came up independently in the final teaching reflection.

> [In my field placement], I learned that teachers have to be aware of unconscious bias when working directly with the students. In my lesson, people with American names were, at first, more likely to be called on than those with more "obscure" names. Although I was not trying to do this intentionally, I noticed that students like Jamie and Carson were more likely to be called on than Djimon. This limited answers, though, from students who may have the same or more knowledge than those being called on. I improved as a teacher, however, by admitting my faults and taking responsibility to make sure everyone was equally as likely to be called on...One of the first steps to being a better teacher is holding ourselves accountable for our unconscious bias, and actively trying to allow students an equal opportunity.

Overall, the evidence we see in the assignments suggest a few things. Our belief in the possibility for engaging students in these difficult or sensitive topics as possible and reasonable for undergraduates in their first teaching courses was confirmed. The accessibility of the readings made way for specific thinking in preparation for class discussion, which was subsequently enriched and progressed into more challenging terrain. There was variation in the ways that students engaged with topics and we presented here a few of the responses that felt most powerful. Some students engaged in more superficial or less personal ways, and others found that the concepts and perspectives we presented resonated closely with them. Finally, we believe the readings, class discussions and writing provid-

ed valuable opportunities to engage with the ways systemic features shape school experiences and for developing conceptual tools and criteria (McDonald, 2008) for viewing more specific STEM related classroom dynamics.

REFLECTIONS

As we proceeded through the semester, we also kept track of our own thinking and processes. The documentation of our thinking occurred in our reflective writing moments during meetings and audio-recorded meetings. We were actively sense-making ourselves about what it means to center a social justice perspective in this course. We each grappled in different ways.

We have differing experiences with teaching in a social justice frame. For two of us, this semester systematized a commitment we had made over the previous few semesters in Step 2. Our courageous conversations work helped us explore difficult themes and uncomfortable topics, which then helped us imagine what taking a more explicit focus might look like. Two of us are classroom teachers, as well as university-based Master Teachers. What became clear was the potential for our own learning through centering social justice in our teaching to extend beyond the university classroom and into our high school contexts.

As a group, we had a few enduring takeaways that, in general, caused us to have far more questions than answers and pushed us to embrace discomfort and uncertainty in our own teaching (Singleton, 2015). We have recognized from the beginning that this work is difficult and complex. And we experienced some resistance to the critical perspectives we were advancing, and especially when conversations touched on questions of race. For example, we experienced students holding tight to the idea of the achievement gap, despite reading and hearing critiques for how it is defined, measured, and the assumptions it promotes about ability and race. We anticipated this and we understand these resistances as both inevitable and as opportunities for all of us to develop a critical perspective. In this sense, we constantly questioned ourselves about our priorities and about what enduring message we wanted to leave with our students. An ongoing tension existed between feeling compelled to disrupt harmful narratives in class discussions and creating safe spaces for undergraduates to feel comfortable in taking risks about sharing their personal thinking. This dilemma also brought us to question how and when we were meeting our students "where they were," even if our goal was to push beyond our own and their comfort zones.

Our roles in later parts of the teacher preparation program (for example, in teaching PBI or mentoring in the apprenticeship experiences) also permitted another set of questions to emerge. We understood that the development of a professional teaching identity happens over the course of the program and beyond, as teachers enter the labor force. We also know that fully engaging in understanding one's privilege or how one is complicit in the propagation of oppression requires sustained involvement with interrogating oneself related to questions of personal identities (including race, gender, sexuality, and ability). In Step 2, we scratched

the surface in this area. We do wonder what possibilities can be integrated into the later learning experiences that support this sort of inquiry.

Finally, we are encouraged about how this work was received by our students. We received positive feedback from many students about what such perspectives brought to their learning experiences.

Extending Beyond the University

Since two of us are practicing teachers, this work opened an opportunity for us to attend to the way our thinking was pushed in the context of our roles as high school teachers and members of a school community. We, individually, narrate a few examples.

> For me, as a practicing high school teacher who also works at the university, engaging with our social justice work impacted me in a variety of ways. As I worked to help undergraduates in Step 2 develop a critical lens, I too improved my own lens, with particular growth in my ability to notice deficit perspectives and dominant narratives. For example, at a professional development session I attended, I felt particular elements were problematic with regards to race and gender. The focus of the PD was a particular labeled "subgroup" with several students from that subgroup in attendance to participate. The session and ensuing conversation with students played out in a way that felt like these students were a "problem" to be solved and that teachers might be searching for "solutions" based on assumed inherent differences of these learners, who despite meeting a few criteria, are not defined by one label. Reflecting individually and with trusted colleagues at the university, I felt empowered to open up dialogue with the facilitators of the session. I shared my comments not as an expert on issues of equity and diversity, but as someone who is working to develop a critical lens. Without my work at the university, I do not feel I would have viewed the session in the same way.
>
> — *Steve Karig, Shared Master Teacher*

> For me, the high point of being a part of the social justice work in Step 2 was that it did not remain confined to Step 2. In being a part of this work, social justice and equity came to the forefront of my mind. Once this work became the center of Step 2, I found myself explicitly paying attention to it in my work teaching math at the high school level. Specifically, in moments where I am around other teachers who are "venting" in the office. While in the past, I would have picked up on the deficit language some teachers use, now not only do I notice it, but I feel empowered to not sit in silence. I feel like I have the knowledge and voice to help other teachers pay attention to the language we use to describe our students and the impact that language is having on the messages students receive. As a result of my participation in this work, I have become extremely conscious of not only others' actions, but also my own. When planning and reflecting on my own teaching, I take explicit time to focus on my own practices in the classroom, the alignment with the beliefs I have, and the language I use with and about students.
>
> — *Kayla White, Shared Master Teacher*

FINAL THOUGHTS

Our work to engage in deep questions about social justice in teaching has helped to prioritize the topic for our team. This work continues to grow and morph as it becomes something we repeatedly come back to in larger program meetings, as well as smaller meetings related to individual courses. On the horizon for our program is to begin to imagine trajectories of work that relate to equity and social justice spanning across multiple courses, as well as to continue to have conversations among ourselves about our program position in this respect.

REFERENCES

Barton, A. C. (2001). Science education in urban settings: Seeking new ways of praxis through critical ethnography. *Journal of Research in Science Teaching: The Official Journal of the National Association for Research in Science Teaching, 38*(8), 899–917.

Brickhouse, N. W. (2001). Embodying science: A feminist perspective on learning. *Journal of Research in Science Teaching, 38*(3), 282–295.

Britzman, D. (1986). Cultural myths in the making of a teacher: Biography and social structure in teacher education. *Harvard Educational Review, 56*(4), 442–457. doi:10.17763/haer.56.4.mv28227614l44u66

Brown, D. L. (2015, January 23). Prince George's neighborhoods make 'Top 10 List of Riches Black Communities in America.' *The Washington Post*. Retrieved from: https://www.washingtonpost.com/

Butler, J. (1997). *Excitable speech: A politics of the performative*. New York, NY: Routledge.

Chafee, S. (2017, May 2). Rejecting the myth of the 'superteacher.' *Cognoscenti*. Retrieved from: http://www.wbur.org/cognoscenti/2017/05/02/no-perfect-teacher-sydney-chaffee

Cochran-Smith, M. (2010). Toward a theory of teacher education for social justice. In *The Second International Handbook of Educational Change* (pp. 445–467). Dordrecht, The Netherlands: Springer.

de Sousa Santos, B. (Ed.). (2007). *Cognitive justice in a global world: Prudent knowledges for a decent life*. Plymouth, UK: Lexington Books.s

Ferguson, S. (2014, September 29). Privilege 101: A quick and dirty guide. *Everyday Feminism*. Retrieved from: https://everydayfeminism.com/2014/09/what-is-privilege/

Harding, S. (2016). *Whose science? Whose knowledge?: Thinking from women's lives*. Ithaca, NY: Cornell University Press.

Henson-Darko, S., & Sanyal Tudela, A. (2017, May). *Courageous conversations: Attending to issues of classroom equity in early fieldwork courses,* Paper presented at the 2017 UTeach National Conference, May 23–25, Austin, TX.

Kumashiro, K. (2000). Toward a theory of anti-oppressive education. *Review of Educational Research, 70*(1). doi:10.2307/1170593

Kumashiro, K. K. (2001). "Posts" perspectives on anti-oppressive education in social studies, English, mathematics, and science classrooms. *Educational Researcher, 30*(3), 3–12. https://doi.org/10.3102/0013189X030003003

Kumashiro, K. K. (2015). Against common sense: Teaching and learning toward social justice. New York, NY: Routledge.

Maryland State Department of Education. (2018). *Maryland State Report Card*. Retrieved from: http://reportcard.msde.maryland.gov/Entity.aspx?K=16AAAA

McDonald, M. A. (2008). The pedagogy of assignments in social justice teacher education. *Equity & Excellence in Education, 41*(2), 151–167.

McDonald, M., & Zeichner, K. 2009. Social justice teacher education. In W. Ayers, T. Quinn, & K. Stovall, K. (Eds), *Handbook of social justice in education*. Mahwah, NJ: Lawrence Erlbaum.

Sanyal Tudela, A. (2014). Discourses of experience: The disciplining of identities and practices in student teaching. *Australian Journal of Teacher Education, 39*(3), n3.

Singleton, G. (2015). Courageous conversations about race: *A field guide for achieving equity in schools* (2nd ed.). Thousand Oaks, California: Corwin, A SAGE Company.

Sleeter, C. (2008). Equity, democracy, and neoliberal assaults on teacher education. *Teaching and Teacher Education, 24*(8), 1947–1957. doi:10.1016/j.tate.2008.04.003

University of Maryland Office of Institutional Research, Planning & Assessment. (2018, October). *UMD undergraduate student profile*. Retrieved from: https://irpa.umd.edu/CampusCounts/Enrollments/stuprofile_allug.pdf

CHAPTER 14

STEPS 1/2 COMBO

Summer Field Experiences on Campus and in the Community

Rachelle Haroldson

The courses Step 1: Inquiry Approaches to Teaching and Step 2: Inquiry-Based Lesson Design build the foundation for students in the UTeach model. In these beginning courses students have field experiences where they explore teaching practice. These courses become a pivotal point in students' decision to pursue education as a career. This chapter describes the journey of one program adapting the UTeach Step 1 and Step 2 course structure for a one-year post-baccalaureate program, with specific focus on the approach taken to develop strategic partnerships and coordinate meaningful field experiences in a condensed summer format.

The courses Step 1: Inquiry Approaches to Teaching and Step 2: Inquiry-Based Lesson Design build the foundation for students in the UTeach model. In these beginning courses students have field experiences where they explore teaching practice. These courses become a pivotal point in students' decision to pursue education as a career. This chapter describes the journey of one program adapting the UTeach Step 1 and Step 2 course structure for a one-year post-baccalaureate program, with specific focus on the approach taken to develop strategic partnerships and coordinate meaningful field experiences in a condensed summer format.

Preparing STEM Teachers: The UTeach Replication Model, pages 189–202.

The chapter starts with an overview of the STEM*teach* program at the University of Wisconsin-River Falls, rationale for combining Step 1 and Step 2, and differences between this program and the traditional UTeach model. Following this background information are descriptions of the field experiences and suggestions for developing relationships within and around campus to foster rigorous practica during the summer, a time when finding teaching opportunities can be a challenge. The chapter concludes with a description of the field experiences' impact on the teacher candidates' development and with key takeaways for teacher preparation programs looking to develop field-based courses in the summer.

OVERVIEW OF STEM*TEACH*

Background of STEMteach

The alternative delivery of the new STEM*teach* initial licensure program at the University of Wisconsin-River Falls was borne of the desire to create a more compact curriculum for the post-baccalaureate audience. For this reason, STEM*teach* is a one-year UTeach-inspired program. Candidates enter the program with a degree in a STEM field, planning to change careers and obtain licensure to teach at the secondary level. By the end of the year-long program, they have 24 credits towards their master's degree as well their teaching license(s). In order to implement the UTeach model in this format, we had to make some adjustments.

Rationale for Combining Step 1 and Step 2 Courses

As STEM*teach* is a graduate level program, there is no need for Step 1 and Step 2 in the traditional recruitment role and neither course is offered. We decided to merge the key concepts of the beginning courses, Step 1 and Step 2, into one intensive three-week session of Steps 1/2 Combo (Steps Combo). This combined version of the course provides students with the educational foundation they need while meeting the one-year timeline of the program. Therefore, Steps Combo is offered during the summer, as the first course in the program.

After consulting with Master Teachers from UTeach about how they taught Steps Combo, STEM*teach* designed our Steps Combo course to follow the format of teaching the Step 1 content followed by teaching the Step 2 content. This allows proper scaffolding of the material, since the concepts in Step 1 (inquiry-based lessons, standard-objective alignment, instructional equity, safe classroom practices, probing questions, reflection) support the concepts in Step 2 (effective teaching strategies, pre- and post-assessment alignment, peer feedback).

The curriculum and field experiences for Steps Combo focus on the same objectives as the UTeach Step 1 and Step 2 model, with modification to the instructional model and the STEM content focus. See Table 14.1 for a detailed comparison of the UTeach Step 1 and Step 2 courses and STEM*teach*'s Steps Combo course. The most notable differences include 1) STEM*teach* does not have an emphasis on technology integration in the Steps Combo lessons and instead

TABLE 14.1. Comparison of UTeach Step 1 and Step 2 Courses and STEMteach Steps Combo Course

	UTeach Step 1 and Step 2 Courses	STEMteach Steps Combo Course
Purpose	Recruitment Introduction to teaching	Foundational course and preparation for fall courses
Semesters Taught	Summer, Fall, Spring	Summer Only (no separate Step 1 or Step 2 options)
Objectives/ Topics	Inquiry-based lessons (using BSCS 5E Instructional Model) Diversity in the classroom and instructional equity Writing performance objectives aligned with STEM standards Reflection Questioning strategies Safe classroom practices Pre-and post-assessments Technology integration	Inquiry-based lessons (using a modified BSCS 5E Instructional Model) Diversity in the classroom and instructional equity Writing performance objectives aligned with STEM standards Reflection Questioning strategies Safe classroom practices Pre-and post-assessments Computer science education
Formal Field Experiences	Step 1 2 observations 3 inquiry-based lessons taught Step 2 1 observation 3 inquiry-based lessons taught	Steps Combo 2 observations 2 inquiry-based lessons taught
Informal Field Experiences	Not part of the field experience component	Summer camps Summer college preparatory program Educational outreach event with local business

includes an emphasis on computer science education, and 2) Steps Combo has fewer formal inquiry-based lessons taught in the field experience and has more informal fieldwork opportunities. Additionally, STEM*teach* candidates write their own inquiry-based lessons and are not given pre-written exemplars to adapt. They go through both rough and final draft phases, attending to detailed feedback from the Master Teacher to improve their lessons. This is appropriate for a rigorous, condensed program as students will have to write their own lesson plans in the fall and gaining that experience now in Steps Combo is vital preparation for their future success.

In both the traditional UTeach model and Steps Combo, candidates focus on writing inquiry-based lessons using the BSCS 5E Instructional Model (Bybee et al., 2006; Bybee, 2014; Eisenkraft, 2003) guided by the Next Generation Science Standards and Common Core State Standards in Mathematics (National Governors Association Center for Best Practices, Council of Chief State School Officers, 2010; NGSS Lead States, 2013). In the lesson plan template, candidates

TABLE 14.2. Progression of Candidate Skill and Knowledge Development in Field Experiences

Skill/Knowledge	Application in Summer Field Experiences	Application in Fall Field Experiences
Instructional Model	BSCS 5E Instructional Model (Bybee at al., 2006; Bybee, 2014; Eisenkraft, 2003).	Launch-Explore-Summarize Teaching Model (Annenberg Learner, n.d.) The Learning Cycle (Lawson, 2002)
Assessment	Pre-assessment and post assessment (as part of elicit and evaluate, respectively, based on the BSCS 5E Instructional Model)	Formative assessment and summative assessment (Garrison & Ehringhaus, n.d.; Stiggins & Chappuis, 2006)
Academic Language	Vocabulary (SCALE, 2018)	Syntax, discourse, language function (SCALE, 2018)
Groupwork	Using procedural roles and task cards in groups (Cohen & Lotan, 2014)	Understanding and mitigating status issues in groups (Boaler & Staples, 2008; Cohen & Lotan, 2014) Cooperative learning (Johnson, Johnson, & Holubec, 1994)
Equity	Current reform efforts in STEM (AMTE, 2015; Januszyk, Miller, & Lee, 2016; Lottero-Perdue, Lovelidge, & Bowling, 2010; NCTM, 2014; Strachan, 2017)	Accommodations and modifications for special education students and English language learners (Dong, 2004; Dong, 2009; Gonzalez & Moll, 2002; Tomlinson, 2014; Wiest, 2008)
Questioning	Purpose and structure of questioning (Johnson, 1982; Manouchehri & Lapp, 2003)	Coordination of effective small and large group discussions with an emphasis on assessment (Cartier, Smith, Stein, & Ross, 2013; Smith & Stein, 2018)
Classroom Management	Focus on procedures, interventions, and relationships (Marzano, Marzano, & Pickering, 2003; Wong & Wong, 1998; Wong, Wong, Jondahl, & Ferguson, 2014)	Focus on middle school transition, relationships, learning environment, and social forces (Cushman & Rogers, 2008)

are required to collect pre- and post- assessment data to be analyzed in their final project. For this reason, the structure they use is a hybrid of the 5E model (engage, explore, explain, elaborate, evaluate) from BSCS (Bybee, 2014) and the 7E model (elicit, engage, explore, explain, elaborate, evaluate, extend) proposed by Eisenkraft (2003). In the 7E model, Eisenkraft unpacks *engage* into elicit and engage. Adding *elicit* allows for a specific place in the lesson plan where candidates collect evidence about what students come into the lesson knowing. We call this the 5E+Elicit model. Candidates use engage, explore, explain, elaborate, and evaluate, as well as *elicit* before the engage to ensure they are actively gathering pre-assessment data. The evaluate stage serves as the part in the lesson when candidates collect post assessment data. This instructional approach makes explicit

the gathering of data prior to the lesson and at the end of the lesson in order for candidates to later reflect on student learning and the effectiveness of their instruction.

Scaffolding of Material in Steps Combo

Because Steps Combo is not used as a recruitment class, the goal is to establish the core foundation for the candidates to be successful in the fall and spring semesters of the program, rather than preparing them in Step 1 for Step 2 as happens with the traditional UTeach model. See Table 14.2 for a detailed outline of the concepts covered and applied in the summer field experiences and how they are built upon in the fall coursework and field experiences.

Rationale for Field Experiences in Steps Combo

Following the UTeach philosophy of having field experiences early and often, STEM*teach* candidates enter the classroom within their second week of class in Steps Combo. Rather than waiting until after they have had a series of foundational courses, candidates have field experiences while developing the core skills of planning, instruction, and assessment. Candidates' decision to enter the teaching profession is quickly confirmed; STEM*teach* has a 98% candidate retention rate from the summer to fall semester.

SUMMER FIELD EXPERIENCES IN STEPS COMBO

As Steps Combo occurs during the summer, finding meaningful field experiences was challenging and rewarding. What began as a search for a single school partner resulted in opportunities in districts, on campus, and with a local business to offer a rich array of formal and informal field experiences. In this section we provide an overview of our journey to offer insight to teacher preparation programs considering offering Steps Combo in the summer.

In the early planning stages of STEM*teach*, it was hoped that the program could start in the summer term. The immediate question then became the availability of summer field experience opportunities. Prior to the launch of STEM*teach*, districts and schools were invited to be our partners with administrators and teachers serving on our external advisory group. When we started brainstorming ideas for summer field experiences, we reached out to these partners first. Partners provided placements and made suggestions for local resources.

Summer Field Experiences in the Schools

The first field experience relationship developed with our local school district partner. Upon reaching out to the Director of Academic Services, we discussed ways to have candidates teach in the summer school math and reading remediation program.

Having candidates begin by delivering lesson plans with elementary school aged children follows the UTeach model. By working with elementary students, candidates see the learning and developmental progression from elementary to middle school to high school. (Candidates have an informal field experience with older students later in the summer, followed by a formal field experience in the middle school and then in the high school in the coming school year). Elementary students tend to create a low risk environment and a safe starting point for the candidates to dip their toes into teaching; the classes are smaller, and the children are welcoming.

During the spring, the STEM*teach* Master Teacher coordinated with the Director of Academic Services to find teachers working in the summer school program to mentor candidates. Mentor teachers were all experienced teachers, capable of modeling best practices and positive student-teacher relationships. Each mentor was assigned a pair of candidates. The Master Teacher recruited teachers at a planning meeting in May and finalized placements by June.

During the first week of class, the candidates attended a meeting to pitch their lesson ideas and get to know their mentor. The mentors provided feedback on the standards the candidates would be addressing and the activities they planned to implement. The Master Teacher also addressed expectations of the mentors, calendar of observations and lessons, mentor agreement form, and background of the program.

Pairs of candidates were expected to teach two different lesson plans aligned with the Next Generation Science Standards (NGSS) and/or Common Core State Standards in Mathematics. For each formal observation, the pairs completed a rough and final draft of their lesson plans that were pre-approved by the Master Teacher. Each lesson contained a learning sequence of 75 to 90 minutes. Because many of these students were behind grade level academically, candidates chose standards for the grade the students had just finished. Their lessons reviewed and/or reinforced concepts for the students. Prior to teaching each of their lesson plans, candidates completed an observation in the classroom. They were expected to help and support in any way the mentor teacher suggested.

Because of being outside of the academic year, the mentor teachers had fewer administrative responsibilities and were able to spend more time providing suggestions and observations during the lesson. Mentor teachers used a form, provided by the Master Teacher, that structured the feedback based on the different parts of the lesson (elicit, engage, explore, explain, elaborate, evaluate). Following the lesson, the mentor met with the teaching pair to provide verbal feedback. During this meeting, many mentors elaborate on the written feedback they made when the candidates were leading the lesson.

Summer school students offered a unique and rewarding population for the candidates' first lessons. Most of the students attending summer school were working on reaching grade-level skills and comprehension. During the school year, many of these students were pulled from their science classes for extra reading or math support. For this reason, the STEM lessons that the candidates taught

were viewed as special. Students looked forward to these hands-on activities focused on engineering, science, and math. In many cases the mentor teachers asked permission to implement the candidates' lessons into their classes in the coming school year. Overall, the mentors enjoyed having more support in the classroom to assist with their students.

Summer Field Experiences on Campus

As we continued to pursue options for field experiences, we found two other opportunities on our campus. These field experiences came to us from professors on the internal advisory group who worked with these programs as well as from colleagues in the Office of Continuing Education where the programs are housed. These programs included Camp Badger, an engineering camp for pre-8th graders, and Upward Bound Trio, a college preparatory program.

Camp Badger

Serving as a satellite campus for the University of Wisconsin-Madison, the University of Wisconsin-River Falls taught a weeklong engineering camp, Camp Badger, for students entering 8th grade. Students stayed on campus, attended engineering field trips, and engaged in engineering activities. For this summer program, middle school teachers often recommend and encourage students to attend.

Because the programming for the camp was previously established, STEM-*teach* found areas within the existing structure to fit into the program. Candidates' roles varied depending on the topics and where in the course we collaborated with Camp Badger. Candidates went on field trips as chaperones to power plants, sewage plants, wastewater treatment plants, bridge tours, or cheese factories. They were assigned a group of five students. During the trip candidates were encouraged to get to know the students and practice asking probing questions connected to the engineering concepts. This was also an opportunity to practice small group management, re-directing students when they were off task or meandering from the larger group. Candidates saw engineering in action, the ebb and flow of field trips, and the coordinating of middle school students. One year when Camp Badger overlapped with a later week of Steps Combo, candidates prepped and taught a mini 5E lesson (using just elicit, engage, and explore) to build foundational knowledge before a field trip.

The goal of working with Camp Badger was for candidates to gain more exposure to students in informal STEM learning environments, reflect on what they observed in students' behavior and interactions, and determine how those observations aligned with their readings about the adolescent brain (Hall & Brier, 2007). Because the middle school students are at a different place developmentally, in class discussion the candidates were asked to compare the cognitive development and STEM comprehension of the middle school students with that of the elementary students they were working with in their formal teaching experience.

Upward Bound

Upward Bound is a college preparatory program aimed to support high school students who are first generation and/or low income. The students Upward Bound serves come primarily from a high school in the Twin Cities, which is also one of our partner schools. The Upward Bound Director was excited to find ways for the candidates to encourage her students to pursue STEM classes and careers. This partnership blossomed in two key ways; the development of a STEM Day in partnership with 3M (discussed more in the next section) and supporting a computer science class.

Computer Science

With the significance of computer science in education on the rise (Smith, 2016), and the option for qualified incoming candidates to pursue a computer science license in STEM*teach*, we integrated specific professional development in computer science. This started by encouraging all candidates to take the Bootstrap workshop. Bootstrap is a programming language that uses algebra to teach mathematical concepts and computational thinking (Bootstrap, n.d.). As a result of this experience, two candidates who took this workshop were asked to teach Bootstrap the following summer to the Upward Bound students in a computer science camp. To incorporate computer science into Steps Combo, we turned this into an informal field observation experience for all candidates.

In Steps Combo, candidates prepared for this field experience through computer science modules around computational thinking, learning the seven core computer science practices from the K–12 Computer Science Framework (2016), and introductory lessons from the Bootstrap curriculum. Two days a week they went into the class with the Upward Bound students. During this time, they made observations about the classroom management techniques, instructional strategies, and assessment practices of the instructors. When the students had work time, the candidates moved around the room asking questions about the video games they were working on to gauge what the students were learning. As the students worked in pairs to model the practice of collaborating around computing through pair programming (K–12 Computer Science Framework, 2016), candidates were also able to observe group work in action, look for inequities, and discuss ways to promote productive group work. Even though most of the candidates were not pursuing computer science licensure, they were tasked with looking for evidence of the computer science practices at work in the classroom and thinking of ways to integrate computational thinking into their non-computer science classes.

This field experience supported STEM*teach* alumni with a paid teaching opportunity and offered current candidates exposure to students engaged in computational thinking. Given the dearth of computer science field experiences, providing creative options such as this one guaranteed computer science candidates at least one field experience in this area before apprentice teaching in the spring.

Summer Field Experiences with the Community

Another significant field experience partnership was with the corporation 3M. 3M has an educational outreach program, the 3M Visiting Wizards. This program encourages young people to learn about science and technology through hands-on activities and demonstrations. Current and retired 3M employees (referred to as "Wizards") go into K–12 classrooms to present various STEM kits on topics such as: optics, electricity and circuits, polymers, air and vacuum, wind energy, groundwater, scents and perfumes, vibrating robot cups, and cryogenics.

STEM*teach* and the Visiting Wizards then collaborated to host a STEM Day for the Upward Bound students. The candidates would learn about one of the kits and co-teach sessions with their Wizard. Thus, another field experience was added to the course. The goal of this field experience was for candidates to get experience working with new groups of students for short amounts of time. Essentially, they learned how to quickly form relationships and hook students into the task, repeating this process with a new group every 45 minutes.

Prior to the STEM Day, the Visiting Wizards hosted a 3-hour workshop with the STEM*teach* candidates. All the Wizards had a passion for education and exposing students to STEM, and some of them had classroom teaching experience. For this reason, they had a good synergy with the candidates, who were also eager to share their passion with students. Each Wizard worked with a pair of candidates. Candidates spent time with their Wizards learning the details about how to teach their kits. By the end of the workshop, the candidates and Wizards had planned how they would co-teach the kit, safety procedures, and materials set-up.

The STEM Day happened at the end of Steps Combo, so up this point candidates had only taught single lessons. Candidates now gained experience teaching the same lesson multiple times, learning how to refine after each iteration. While no formal structure like the 5E model was used, STEM day activities used an interactive demonstration and/or hands-on activity format. Some of the kits even had a "make and take" quality with the students coming away with vibrating cup robots or homemade perfumes. The students rotated every 45 minutes, experiencing three to four kits throughout the morning.

Upward Bound recently shifted to a new activity focused more on STEM Careers. STEM*teach* adjusted to collaborate with a local partner middle school to host the STEM day with middle school students in the summer school program. By collaborating with the Director of Academic Services, the middle school summer school teachers, and the 3M Visiting Wizards, we were able to provide a half day of activities for the middle school students as a new method of maintaining the Visiting Wizards' mission of getting science in the hands of young people.

IMPACT OF SUMMER FIELD EXPERIENCES ON CANDIDATES' DEVELOPMENT

The summer field experiences supported the Steps Combo model by gradually scaffolding candidates' skills. The candidates began the summer with a formal field experience where they taught two lesson plans they designed. They learned about the core aspects of teaching: planning, instruction, and assessment. They developed an understanding of unpacking dense content standards and how to translate that into performance objectives. One candidate reported that having a field experience early in his teacher preparation allowed him to clearly identify the need for equity strategies in classroom discussions. He felt confident in pursuing a career in education because he could see this need, was already learning strategies to address it, and had practice and quality feedback on implementing these strategies. Candidates welcomed the opportunity for real-life teaching situations with actual students and the chance to begin to develop their own teaching style. Teaching the lessons they wrote and being fully immersed in a classroom allowed them to put theory into practice and see for themselves how the various instructional strategies and assessments impacted student learning.

The candidates taught these lessons with a teaching partner and gained experience with the co-teaching model. Most candidates used a combination of strategies like team teaching (both candidates actively involved in the lesson), station teaching (each candidate leading a different learning station), and "one teach, one assist" (one candidate takes the lead during instructional time and the other candidate supports and monitors students) (St. Cloud State University Teacher Quality Enhancement Center, 2011).

Following both formal teaching experiences, the candidates completed a reflection. Being a reflective practitioner is a key skill for teachers and a core part of UTeach, both in Step 1 and Step 2 and throughout the course sequence. By starting this process early, candidates began to develop this practice of evaluating their effectiveness. The reflection process started with a discussion of feedback with their mentor teacher, a written reflection including assessments from students, and then a discussion with the Master Teacher to review feedback and identify strengths and areas of improvement from the lesson. One candidate reported that she strengthened her implementation of ethical learning strategies through reflection on early teaching opportunities. She learned the importance of structured questions, intentional lesson plans using the 5E format, and the need for equitable practices to avoid situations where students were marginalized from the learning experience (as she witnessed in teaching her first lesson, then remedied in designing and teaching her second lesson).

In the informal field experience where the candidates collaborated with the Visiting Wizards, they built on the skills they developed with the formal field experience (assessing student learning, engaging students, co-teaching, classroom management, questioning) and developed different skills. Candidates learned how to take an existing lesson and adapt it to their teaching style. The STEM day

additionally allowed candidates to teach a lesson multiple times, offering the opportunity for reflection and revision to improve the lesson through the day. The candidates now interacted with older students, either middle schoolers or high schoolers. The STEM day follows their formal experience with the elementary students, thus allowing them to make comparisons between the cognitive development and growth of the different age groups.

The placement of the Upward Bound and Camp Badger informal field experiences within Steps Combo changed depending on the year. For this reason, the impact on the teacher candidates varied. For experiences earlier in Steps Combo, they offered a chance to interact with more students. Candidates practiced developing relationships and working on understanding the adolescent mind. They focused in on the ways the instructors for these programs engaged and coordinated the students to optimize the learning environment. For example, in the Bootstrap workshop candidates observed an experienced teacher engaging students in designing their own video games using algebra concepts in programming. For experiences later in the course, the candidates applied what they had learned. Their line of questioning involved getting to know the students as well as assessing their knowledge about the STEM activity. For example, one year the candidates designed mini-lessons with just the elicit, engage, and explore stages of the 5E model before going on a field trip to the wastewater treatment plant and water filtration plant. Throughout the day, candidates asked the students questions connecting what was happening at the different sites to their earlier lessons about water. All these experiences set the foundation for the upcoming schoolyear fieldwork experiences they would have in the fall and spring.

KEY TAKEAWAYS

In conclusion, we offer these key takeaways when designing field experiences for the Steps course, whether that be the individual Step 1 and Step 2 courses or Steps Combo. While these field experiences were pursued with summer in mind, these thoughts can be applied to almost any timeframe with a flexible mindset.

- Leverage partners and local schools for the programs they offer. With the more flexible nature of summer school, mentor teachers are receptive to candidates developing their practice and exploring with inquiry-based methods.
- Campus is rich with resources and hosts to various programs serving K–12 youth. Work across departments and offices to find ways for candidates to enrich existing camps and services. Both groups benefit. Students get more exposure to STEM mentors and candidates get more exposure to working with students of all ages and backgrounds.
- Businesses often have an outreach or education branch. Seek those partners out, especially in the local community. Discuss ways to collaborate and bring people together who share the desire to invest in young people.

- Maintain relationships with directors, coordinators, and instructors, even when funding structures and needs change. Opportunities for new and different collaborations may emerge in the future.
- Teacher preparation programs need to be on the forefront of promoting computer science education. Computer science coursework and field experiences for all STEM candidates need to be integrated into the program. Focus on how computer science overlaps with other course objectives rather than treating it as an addition. For example, computer science emphasizes fostering an inclusive classroom. Adjust the discussion around diversity and instructional equity through the lens of computer science.
- A challenge for Master Teachers is going to be the balance of course content and field experiences. Master Teachers benefit from looking at the foundational core components that run through the program and aligning the Steps' field experiences with those components.
- Master Teachers need to be creative and engage in "out of the box" thinking about field experiences to meet the needs of all of the candidates, especially those licensure areas, like computer science, that do not have plentiful options. Master Teachers want to seek out and pursue advocates of STEM education. These people might be in the community, local businesses, school districts, or other faculty and staff on campus.

REFERENCES

Annenberg Learner. (n.d.). *The missing link teacher planning tools: Planning a math unit: launch-explore-summarize teaching model.* Retrieved from: http://www.learner.org/workshops/missinglink/support

Association of Mathematics Teacher Educators (AMTE). (2015). *Position: Equity in mathematics teacher education (A position paper of the Association of Mathematics Teacher Educators).* Raleigh, NC: AMTE.

Boaler, J., & Staples, M. (2008). Creating mathematical futures through an equitable teaching approach: The case of Railside School. *Teachers College Record, 110*(3), 608–645.

Bootstrap. (n.d.). *Bootstrap.* Retrieved from: https://www.bootstrapworld.org

Bybee, R., Taylor, J. A., Gardner, A., Van Scotter, P., Carlson, J., Westbrook, A., & Landes, N. (2006). *The BSCS 5E instructional model: Origins and effectiveness.* Colorado Springs, CO: BSCS.

Bybee, R. W. (2014). The BSCS 5E Instructional model: Personal reflections and contemporary implications [Guest Editorial]. *Science and Children, 51*(8), 10–13.

Cartier, J., Smith, M. S., Stein, M. K., & Ross, D. (2013). *5 Practices for orchestrating task-based discussions in science.* Reston, VA: NCTM.

Cohen, E., & Lotan, R. (2014). *Designing groupwork: Strategies for the heterogeneous classroom* (3rd ed.). NY, New York, NY: Teachers College Press.

Cushman, K., & Rogers, L. (2008). *Fires in the middle school bathroom: Advice for teachers from middle schoolers.* New York, NY: The New Press.

Dong, Y. (2004). Getting at the content. *Educational Leadership, 62*(4), 14–19.

Dong, Y. (2009). Linking to prior learning. *Educational Leadership, 66*(7), 26–31.

Eisenkraft, A. (2003). Expanding the 5E model. *The Science Teacher, 70*(6), 56–59.

Garrison, C., & Ehringhaus, M. (n.d.). Formative and summative assessment in the classroom [Association for Middle Level Education Article]. Retrieved from: https://www.amle.org/BrowsebyTopic/WhatsNew/WNDet/TabId/270/ArtMID/888/ArticleID/286/Formative-and-Summative-Assessments-in-the-Classroom.aspx

Gonzalez, N., & Moll, L.C. (2002). Cruzando el puente: Building bridges to funds of knowledge. *Educational Policy, 16*(4), 623–641.

Hall, M., & Brier, G. (2007). From frustrating forgetfulness to fabulous forethought. *The Science Teacher, 74*(1), 24–27.

Januszyk, R., Miller, E. C., & Lee, O. (2016). Addressing student equity and diversity. *Science Scope, 39*(8), 16–19.

Johnson, D. R. (1982). *Every minute counts: Making your math class work.* Milwaukee, WI: Dale Seymour Publications.

Johnson, D., Johnson, R., & Holubec, E. (1994). *The new circles of learning: Cooperation in the classroom and school.* Alexandria, VA: ASCD.

K–12 Computer Science Framework. (2016). *K–12 Computer Science Framework.* Retrieved from: http://www.k12cs.org

Lawson, A. E. (2002). The Learning Cycle. In R. G. Fuller (Ed.), A *love of discovery: Science education, the second career of Robert Karplus.* (pp. 51–66). New York, NY: Kluwer Academic/Plenum Publishers.

Lottero-Perdue, P., Lovelidge, S., & Bowling, E. (2010). Engineering for all: Strategies for helping all students succeed in the design process. *Science and Children, 47*(7), 24–27.

Manouchehri, A., & Lapp, D.A. (2003). Unveiling student understanding: The role of questioning in instruction. *Mathematics Teacher, 96*(8), 562–566.

Marzano, R, J., Marzano, J. S., & Pickering, D. (2003). *Classroom management that works: Research-based strategies for every teacher.* Alexandria, VA: Association for Supervision and Curriculum Development

National Council of Teachers of Mathematics (NCTM). (2014). *Access and equity in mathematics education: A position of the National Council of Teachers of Mathematics.* Reston, VA: NCTM.

National Governors Association Center for Best Practices, Council of Chief State School Officers. (2010). *Common core state standards mathematics.* Washington, DC: National Governors Association Center for Best Practices, Council of Chief State School Officers.

NGSS Lead States. (2013). *Next generation science standards: For states, by states*. Washington, DC: The National Academies Press.

Smith, M. (2016). *Computer science for all* [web log]. Retrieved from: https://obamawhitehouse.archives.gov/blog/2016/01/30/computer-science-all

Smith, M. S., & Stein, M. K. (2018). *5 Practices for orchestrating productive mathematics discussions* (2nd ed.). Reston, VA: NCTM.

St. Cloud State University Teacher Quality Enhancement Center. (2011). *Co-teaching definition and examples*. Retrieved from: https://www.stcloudstate.edu/oce/_files/documents/coteaching/Co-Teachingdefinitionsandexamples.pdf

Stanford Center for Assessment, Learning and Equity (SCALE). (2018). *Understanding academic language in edTPA: Supporting learning language development.* Re-

trieved from: https://concordia.csp.edu/teachered/wp-content/uploads/sites/3/K–12-PE-Academic-Language.pdf

Stiggins, R., & Chappuis, J. (2006). What a difference a word makes. *Journal of Staff Development, 27*(1), 10–14.

Strachan, S. L. (2017). The science standards and students of color. *The Science Teacher, 84*(4), 47–50.

Tomlinson, C. A. (2014). *The differentiated classroom: Responding to the needs of ALL learners* (2nd ed.). Alexandria, VA: ASCD.

Wiest, L. (2008). Problem-solving support for English language learners. *Teaching Children Mathematics, 14*(8), 479–484.

Wong, H. K., & Wong, R. T. (1998). *The first days of school: How to be an effective teacher*. Mountain View, CA: Harry K. Wong Publications, Inc.

Wong, H. K., Wong, R., Jondahl, S., & Ferguson, O. (2014). *The classroom management book*. Mountain View, CA: Harry K. Wong Publications, Inc.

CHAPTER 15

INCORPORATING THE CLINICAL INTERVIEW METHOD INTO THE KNOWING AND LEARNING IN MATHEMATICS AND SCIENCE COURSE

Selma Koç

In the Knowing and Learning in Mathematics and Science course, preservice teachers explore "what it means to know" in which they engage in activities that directly relate to understanding what the students know and how they came to know. The Clinical Interview method is an essential component of the course to help preservice teachers examine students' problem-solving and thought processes in 7–12 grade mathematics and science. Through the clinical interview process, preservice teachers "analyze domain-specific problem-solving activities and approaches in an applied fashion" (UTeach Institute, 2014, p. 18). This paper describes the revisions the course instructor made on the clinical interview assignment to improve preservice teachers' understanding of their prospective students and the course concepts as they apply theory into teaching practice.

INTRODUCTION

In the Knowing and Learning in Mathematics and Science course, preservice teachers explore the foundations of how people learn, how to understand student

Preparing STEM Teachers: The UTeach Replication Model, pages 203–210.

thinking processes, and how theories of learning inform instructional decisions with a specific emphasis on mathematics and science. The course "focuses on issues of what it means to know and learn science and mathematics, and students develop a powerful toolkit of relevant approaches." (UTeach Institute, 2014, p. 18). The Knowing and Learning in Mathematics and Science course description includes: a) How knowing and learning are structured, b) The tensions between general, cross-disciplinary characterizations of knowing (e.g., intelligence) and the specifics of coming to understand powerful ideas in mathematics and science, c) Connections between kinds of assessments and theories of knowing, and d) Various uses of technology associated with specific approaches to learning. The course emphasizes a novice-to-expert paradigm and the importance of conceptual learning with a focus on improving preservice teachers' knowledge and skills in making meaningful connections between theory and practice.

The Knowing and Learning in Mathematics and Science course prepares preservice teachers for the subsequent Classroom Interactions course by "providing opportunities to see how theories explored in the *Knowing and Learning* course play out in instructional settings." (UTeach Institute, 2014, p. 22, para. 1). The knowledge and skills gained in the Knowing and Learning course prepares preservice teachers "design and implement instructional activities informed by their own understandings of what it means to know and learn mathematics and science, and then evaluate the outcomes of those activities on the basis of student artifacts" such as what students say, do or create. (UTeach Institute, 2014, p. 22, para. 2). The Knowing and Learning course employs the clinical interview method for preservice teachers to observe and analyze students' reasoning in science or mathematics, what the students know and how they came to know something. The clinical interview study is important in preparing preservice teachers to incorporate effective instructional strategies in their UTeach program sequence as well as future teaching.

Science and mathematics teachers face numerous challenges as they enter the classroom including understanding learners and instruction (Davis, Petish, & Smithey, 2006). The clinical interview method, originally developed by Piaget, is "a powerful method for entering the child's mind" (Ginsburg, 1997, p. xii) and understanding children's thought processes. This method allows "researchers to actively investigate students' reasoning and problem-solving strategies, clinical interviews have become indispensable tools for mathematics education researchers" (Groth, Bergner, & Burgess, 2016, p. 48). This paper presents how the clinical interview method was incorporated into the Knowing and Learning in Mathematics and Science course and what revisions were made to improve preservice teachers' analysis of their experience and students' knowing and learning in mathematics and science.

THE CLINICAL INTERVIEW

In its first-year implementation, the course was developed and taught by two instructors who have expertise in educational psychology, educational technology, and science education. They primarily used the UTeach course repository for the course materials and activities. The students who were enrolled in the course consisted of preservice teachers specializing in 7–12 grade mathematics and science.

As a writing across the curriculum course, the Knowing and Learning course included three case study analyses, a lesson plan and peer-reviews, theory comparison and analysis, and a final project. The final project required the students to write a comprehensive paper analyzing and reflecting on classroom teaching incorporating course content such as physical, cognitive, psychosocial development, teaching strategies, and motivation theories. In addition to these, the clinical interview assignment, developed by Stroup (2007), was used for preservice teachers to examine essential questions such as "What does it mean to know something?," "What does it mean to learn something?" ,"How can we understand what students are thinking?." The purpose of the clinical interview assignment was to increase preservice teachers' understanding of K–12 learners' problem-solving and thought processes through the implementation of one-on-one clinical interviews with students on given mathematics or science problems. The clinical interview study required preservice teachers write a reflective analysis of students' thought processes while making connections between theory and instructional practice. Through this experience, it was expected the preservice teachers challenge their own assumptions about how students learn mathematics and science. The following was provided to the class for the clinical interview assignment:

> The purpose of this project is to engage you in the inquiry of understanding a student's thought processes, conceptual understanding of the subject matter, and the factors that may influence his/her thinking when posed a mathematics problem. During the interview, keep in mind the guidelines for clinical interviews and make mental or written notes of what concepts, terms, perspectives, etc. you can incorporate when you analyze and interpret the process. You will also need to include a reflection of your experience as a future teacher, how you might use this experience in your teaching, how you might approach teaching a subject similar.

In addition, excerpts from Stroup's (2007) clinical interview study were included in the provided to the class:

> A significant theme in current math and science education reform is that "good teaching" has as its foundation learning to listen to, engage and extend learners' thinking and sense-making. We'd like you to see if you can explore with someone … his/her understandings of some mathematics usually labeled "basic..".. One of the things we will be looking at in this course is what it means to know and/or understand mathematics and science (para. 1).

> ... Try and keep the interview going long enough for you to get a sense of how the person is thinking (and push to get behind quick answers) but don't force the process. For every one of the examples ... the interviewee is likely to have a simple procedure (e.g. "invert and multiply") they follow to "get the answer." Your task is to try and get behind this rote procedure and see how the interviewee has made sense (or not) of this procedure. When you ask something like, "Okay, how do you make sense of moving the decimal?" or "Could you draw me a picture or tell a story related to the 1/5 ÷ 2/3?" be ready for a range of responses. (Stroup, 2007, para. 3)

In Stroup's (2007) clinical interview study, there were four math questions such as a long division problem, calculating the tip at a restaurant and a decimal multiplication. The questions included what if, how and why type of problems which aimed at understanding students' thinking processes when solving mathematics problems. In the Knowing and Learning class, to help the preservice teachers develop interview techniques and skills, the class read and discussed an article on clinical interview guidelines by Ginsburg (1997). The class collectively developed a PowerPoint presentation on the key points that they needed to pay attention to when conducting their interviews. The instructor emphasized the importance of using an interview protocol and building trust with the students and how to ask open-ended questions while encouraging student effort in solving a given problem. Prior to the class activities related to clinical interview guidelines, the class watched clinical interview videos on child development and learning which modeled conducting the clinical interview method. The preservice teachers then were put into pairs for mock clinical interviews on the four mathematics questions from the UTeach course repository. Primarily the science majors volunteered to be interviewed since their background did not focus on teaching mathematics.

After the class was coached on conducting their clinical interviews and writing their projects, they were asked to interview a student in grades 7–12. The course repository provided by UTeach did not include an assessment rubric or criteria. The instructors developed assessment criteria to guide the preservice teachers to write their clinical interview assignments. They were required to submit a draft of their assignment so that the instructor could provide feedback and scaffold the preservice teachers' application and analysis of the course concepts and theories in their study. Although there were several exceptional analyses of the clinical interviews, it was found that a large number of the preservice teachers included analyses of few concepts in their draft submissions. They had a tendency to use similar concepts or terms in their papers that seemed easier for them to explain in their analyses particularly in making connections between theory and practice.

The instructor gave detailed feedback on the preservice teachers' draft submissions pointing out concepts, terms, or theories they could have used in their papers. In this Midwestern state, teacher candidates are required to take an educator licensure exam which include pedagogical knowledge assessment multiple-choice questions and written essay responses. In this context, it was essential for the Knowing and Learning course to address the knowledge and skills needed for

preparing preservice teacher candidates for this licensure test as well as preparing them to become effective teachers.

One of the instructors continued to teach the course in its second year of the UTeach program. Upon reflecting on her first-year teaching experience, the instructor decided to revise the clinical interview guidelines and how she should coach her students. Rather than using the four mathematics problems the preservice teachers previously used, she asked them to write their own mathematics and science problems. The preservice teachers were prompted to construct mathematics or science problems that allowed for opportunities in exploring students' thought processes. The instructor told the class to be sure to use questions that assess multiple proficiencies. She suggested them to include questions that are ambiguous. Ambiguous questions allow the opportunity to observe how a student arrives at that particular answer and the student's reasoning and strategies he/she uses for solving the problem.

Once the questions were developed, the preservice teachers discussed and revised the questions and ranked them in order of suitability for their interviews. The majority of the problems proposed were real-life problems such as calculating percentages, interest rates, and life science. The preservice teachers selected 11 problems out of the 18 they have constructed. Even though these preservice teachers had high level of mathematics and science knowledge and skills, they reported difficulty coming up with problems that were suitable for their prospective interviewees. They used the Web for questions that they can include in their clinical interviews. This challenge seems to benefit the preservice teachers to reflect on their readiness to teach secondary mathematics and science as well as preparing them for writing and selecting problems that are effective in assessing student learning.

The instructor observed that this process allowed the preservice teachers to practice constructing mathematics and science problems appropriate for the age and knowledge level of their future students. It also allowed them to reflect on and talk about students' problem-solving skills, concepts, and processes that may be difficult for them and any misconceptions they may have. The activity helped preservice teachers' ownership of the questions posed as they practiced constructing and writing mathematics and science problems. In addition, it was interesting to observe how the preservice teachers went through the process of eliminating questions that were not well-written or were dependent on rote learning. One additional benefit was that as the class wrote their questions, discussed, and reflected on them, they also thought about the theories, concepts from the Knowing and Learning course which helped them prepare for their clinical interview study assignment.

The preservice teachers were instructed to use four questions from the 11 problems the class agreed on or from Stroup's (2007) mathematics problems from the UTeach course repository. In addition to those, the instructor suggested that they can ask their interviewees to pick one more question from their list of questions.

They were instructed to ask their interviewee why he/she chose that particular question in order for the preservice teachers to explore students' motivational reasoning.

Writing and selecting their mathematics and science problems allowed the preservice teachers to focus on topics in their specialization areas. In addition, they were able to use problems that they were most curious about exploring with respect to their students' problem-solving and reasoning processes. From an epistemological point of view, constructing their own mathematics and science problems provided the preservice teachers an opportunity to find out how their students attain knowledge.

When the course ended, the instructor concluded that the assessment guidelines and criteria should be improved to provide her students with an in-depth and clearer description of the clinical interview assignment. She revised the assessment criteria, added more explanation of the assignment expectations, and assigned more point values for the analysis and reflection components of the clinical interview assignment (Table 15.1). During the class, she frequently emphasized the importance of making meaningful connections between theory and practice in the analysis of their clinical interview experience.

In addition, during her first time teaching the Knowing and Learning course, the course instructor observed that providing the guiding concepts and terms for the preservice teachers to incorporate in their clinical interview study analyses

TABLE 15.1. Revised Clinical Interview Assessment Criteria

Grading Criteria	Points
A description of the participant (using a pseudonym) characteristics and their experiences with mathematics and science (age, education level, use of math at the workplace -if applicable-, etc.). (Possible questions: Do you like math/science? Why/why not? Do you think they're good at math/science? Do you think you understand math/science or do you follow directions? Do you use math/science on a regular basis?)	6
Analysis of the participant's thought process (thought process: i.e. deductive reasoning, reflection, pausing for thought, speaking the problem out loud, recall of what's learned, recall of what was told as a rule) about all problems (include examples, notes, diagrams, questions the interviewee or you've posed—can be submitted as an appendix). Your interpretation of the participant thinking and problem-solving method specific to each problem. What you think the learner knows and how he/she proceeds with that knowledge, her/his responses. Incorporate terms, concepts, theoretical perspectives whenever applicable.	50
Your thoughts and reflection on how you might teach someone within the context of the interview questions and the thinking/reasoning processes involved; how you would approach teaching the participant next time. Incorporate terms, concepts, theoretical perspectives whenever applicable.	40
Your writing is free of spelling, grammar, and/or citation errors. Use APA style formatting.	4

TABLE 15.2. Possible Concepts and Terms Expected to be Incorporated into the Clinical Interview Study

Concepts/terms you may use but not limited to: Schemata, assimilation, accommodation, disequilibrium (cognitive dissonance), constructivism, formal/concrete operations, Bloom's taxonomy/cognitive domains, misconceptions, expert vs novice thinking, metacognition, direct instruction, rote learning, scaffolding, cognitive tool, manipulatives, zone of proximal development, coach/guide/facilitate, performance-goal approach, mastery learning/goal, self-efficacy, self-talk, self-regulation, intrinsic/extrinsic motivation, probing question, self-concept, deductive reasoning, elaboration, reconstruction error, retrieval, storage, lower/higher order thinking, state anxiety, debilitating or facilitating anxiety, attribution, reinforcement, reinforcers, hyperactivity, prior knowledge, cause and effect relationship, negative transfer, conceptual understanding, conditioned response, comprehension monitoring, declarative knowledge, discovery learning, help-seeking, automaticity, spatial intelligence, stereotype threat, gender differences, self-handicapping, learned helplessness, encoding/decoding, implicit knowledge, procedural knowledge, knowledge construction, meaningful learning, meaningful learning set, wait time, mnemonics...

helped them to make deeper connections between theory and practice and allowed them to incorporate more concepts from the course (e.g. scaffolding, cognitive tool, zone of proximal development, situated motivation, performance-avoidance, deductive reasoning, reconstruction error, debilitating anxiety, self-efficacy, formal operations stage). She added the concepts and terms she expected the preservice teachers to include in their analysis paper (Table 15.2) as well as adding excerpts from previous successful preservice teachers' analysis as examples.

The clinical interview method served its purpose in engaging the preservice teachers on the importance of understanding students' "thought processes" and "conceptual learning" in teaching mathematics and science. Much growth was observed in the preservice students' reflective analysis of their clinical interviews; particularly, in incorporating more concepts that were prompted by their instructor. In their analysis of their clinical interviews, they made well-constructed connections integrating theory into practice. Including both mathematics and science problems and having the students develop their own questions may have contributed to meeting the instructor's expectations. Allowing the preservice teachers to choose their mathematics and science questions made it easier for them to have a better analysis of their clinical interview experience. Based on her teaching experience of this UTeach course, the instructor started using the clinical interview method and several readings and activities that were suggested by UTeach in her teaching of other educational psychology courses.

CONCLUSION

Helping preservice teachers advance from novice teachers to expert teachers is crucial for them to teach their students effectively. In the Knowing and Learning Mathematics and Science course, the instructor's efforts in revising the assignment, guiding and scaffolding her students in conducting clinical interviews, and

write a reflective analysis of their observations improved the preservice teachers' use of the concepts, terms, and theories from the course. In doing so, she felt that she was able to help her students make progress in their process of becoming good teachers.

REFERENCES

Davis, E. A., Petish, D., & Smithey, J. (2006). Challenges new science teachers face. *Review of Educational Research, 76*(4), 607–651.

Ginsburg, H. (1997). *Entering the child's mind: The clinical interview in psychological research and practice.* Cambridge, MA: Cambridge University Press. doi:10.1017/CBO9780511527777.005.

Groth, R. E., Bergner, J., & Burgess, C. R. (2016). Exploration of prospective teachers' learning of clinical interview techniques. *Mathematics Teacher Education and Development,18*(2), 48–71.

Stroup, W. (2007). Clinical interview: Understanding the basics. *UTeach Course Preparations: Knowing and Learning in Mathematics and Science.* The University of Texas at Austin. Austin, TX.

UTeach Institute. (2014). Chapter 5: The UTeach instructional program. *Uteach Institute.* Retrieved from: https://institute.uteach.utexas.edu/sites/institute.uteach.utexas.edu/files/uteach-operations-ch05-instructional-program-2014.pdf

CHAPTER 16

CLASSROOM INTERACTIONS

Course Overview and Description of an Innovative Co-Teaching Model

Patrick McGuire and Steve Ottmer

This chapter provides an overview of a co-teaching model involving a university faculty and in-service teacher in a pre-service STEM teacher preparation course, *Classroom Interactions*. Fundamentally, the chapter focuses on our unique co-teaching framework implemented in the course and highlights the value added features of including a full-time high school teacher as a co-instructor. We share quantitative data and qualitative student comments collected over the course of 15 semesters, which suggest this co-teaching model is an effective way to administer the course. General implications related to our co-teaching model, including the positive benefits as well as challenges and barriers, (e.g., cost, scheduling, etc.) are highlighted. The chapter concludes with recommendations for other pre-service teacher preparation programs interested in implementing co-teaching models involving in-service K–12 teachers and university-based faculty members.

Teacher preparation programs, particularly those based in university settings, have historically prioritized theory over practice, often missing opportunities to discuss common challenges and realities that confront new teachers (Le Maistre & Paré, 2010). An unintended consequence of this theoretical prioritization is the disconnect between what pre-service teacher candidates learn and what they actually need to know and do to be a successful in-service teacher (Cochran-Smith,

Preparing STEM Teachers: The UTeach Replication Model, pages 211–224.

2013). This prioritization also results in most university-based courses being taught exclusively by faculty members with theoretical expertise, often without consultation or collaboration with full-time teachers who possess authentic experiences and knowledge.

Despite the generally accepted notion that in-service teachers can add significant value to pre-service teacher preparation (Darling-Hammond, 2006; Teitel, 1999) examples of in-service teachers working in tandem with university-based faculty to co-design and co-teach courses for pre-service teacher candidates are rare (Mason, 2013). While most pre-service teacher preparation programs develop strong partnerships with in-service teachers, unfortunately, these partnerships are often limited to on-site supervision and mentoring of teacher candidates. As Grundoff and colleagues suggest, in-service teachers are unfortunately viewed more as "providers of classrooms for students to teach in" (Grundoff, Haigh, & Mackisack, 2017, p. 231) versus true collaborative partners with a voice in the program courses, activities, and assessments (Hagger & McIntyre, 2006; Mason, 2013; Sewell, Cody, Weir, & Hansen, 2018). In effect, the traditional role reserved for in-service teachers minimizes their ability to have a broader impact on teacher preparation programs as a whole and prevents "boundary crossing" between K–12 classrooms and university-based activities (Akkerman & Bruining, 2016).

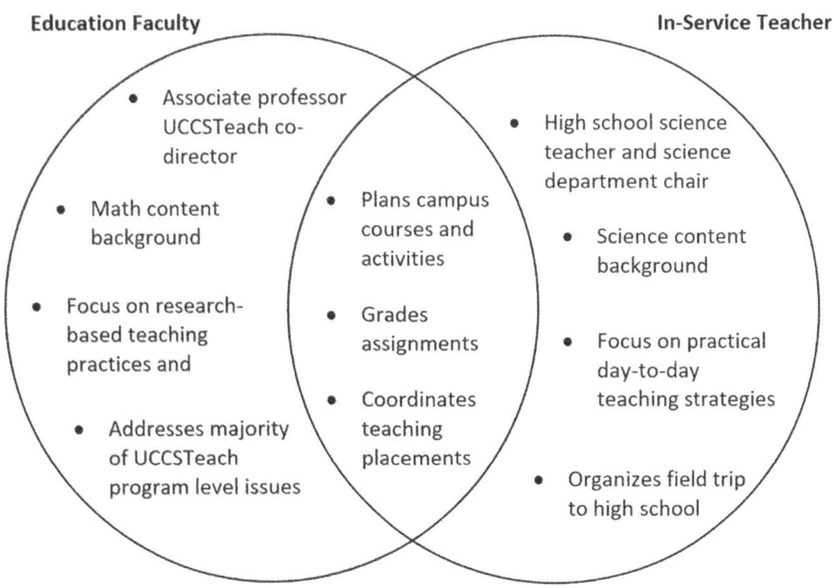

FIGURE 16.1. Co-teaching model for the UCCSTeach Classroom Interactions Course

In this chapter, we outline a unique co-teaching model being implemented in a pre-service teacher preparation course, *Classroom Interactions,* offered within the UCCSTeach program at the University of Colorado Colorado Springs (UCCS). In Figure 16.1, we provide an overview of our innovative co-teaching model and describe the complementary roles of each co-instructor.

For the past four years, our *Classroom Interactions* course has been co-taught by a full-time high school science teacher and tenured STEM education faculty member. As a result of this symbiotic teaching relationship, we describe the course innovations that have emerged and how this partnership has helped to bridge the gap between university-based teacher preparation and the high school classroom.

CLASSROOM INTERACTIONS COURSE OVERVIEW

Classroom Interactions is a three credit hour, field-based course required within the larger UCCSTeach STEM teacher preparation program. UCCSTeach is a 2010 replication site of the national UTeach model described in detail in other chapters within this book. At the most fundamental level, the *Classroom Interactions* course explores interactions that occur in high school classrooms between teachers, students, and STEM content. The UCCSTeach *Classroom Interactions* course is built upon three major themes: (1) Building relationships and establishing classroom culture; (2) Effectively managing classroom interactions, and (3) Integrating high impact teaching strategies to support diverse learners. The *Classroom Interactions* course meets 16 times per semester on Tuesday evenings from 4:45–7:20 pm. In Table 16.1 we provide a basic overview of major topics covered throughout the semester.

Classroom Assessment Scoring System Secondary (CLASS-S) as an Anchor

The *Classroom Interactions* course leverages the *Classroom Assessment Scoring System Secondary (CLASS-S)* model as an anchor (Pianta, Hamre, & Mintz, 2012).The *CLASS-S* model is based on many years of developmental theory (Bronfenbrenner & Morris, 1998) and empirical research studies suggesting that the quality of teacher-student interactions in classroom strongly influence student learning outcomes (Allen, Gregory, Mikami, Lun, Hamre, & Pianta, 2013; Allen, Pianta, Gregory, Mikami, & Lun, 2011; Kane, Staiger, McCaffrey, Cantrell, Archer, Buhayar, & Parker, 2012). UCCSTeach pre-service teacher candidates are required to purchase the *CLASS-S* dimensions guide which effectively serves as their textbook for the course. The *CLASS-S* model includes three domains, Emotional Support, Classroom Organization, and Instructional Support. Within these domains are twelve dimensions (e.g., Positive Climate, Productivity, and Quality of Feedback). The *CLASS-S* dimensions guide provides UCCSTeach teacher candidates with a valid and reliable observational framework, but also a common language for analysis and discussion of interactions.

TABLE 16.1. Classroom Interactions Course Content Coverage

Course	Topic Coverage
1	Course Overview & CLASS-S Emotional Support Domain
2	CLASS-S Classroom Organization Domain
3	CLASS-S Instructional Support Domain Part I
4	CLASS-S Instructional Support Domain Part II
5	Motivation & Mindset
6	Cultivating Equity and Diversity
7	Supporting English Language Learners
8	Special Education, 504 plans, IEPs
9	Supporting Classroom Interactions with Technology (Appy Hour)
10	Differentiated Instruction
11	Guest Teacher Panel
12	Doherty High School Field Trip (Course meets at Doherty High School)
13	Assessment & Feedback
14	Portfolio Night and Professional Dispositions
15	Snow make up day or "teacher workday"
16	Video analysis presentations and exit interviews

Graded Course Activities

The UCCSTeach *Classroom Interactions* course includes 500 total possible points (see Table 16.2). Graded course activities are intentionally designed to be authentic in nature, but also to capture teacher candidate's emergent understanding of the most important aspects of the teaching profession (e.g., professionalism, lesson planning, engaging in reflective teaching, etc.).

Field Experiences and Practice Teaching

In conjunction with the campus-based course requirements, teacher candidates enrolled in *Classroom Interactions* complete observation hours in a local high school math or science classroom. Teacher candidates are matched with an in-service high school teacher (mentor teacher) in the class they will practice teach in, watching for specific teaching behaviors as well as aspects of teacher-student interactions that align with the *CLASS-S* model and language. Each teacher candidate is required to observe a minimum of eight observation hours over the first month of the semester, answer questions about their experiences, and collect teacher and student artifacts related to the interactions they observe. The on-site observations conducted by teacher candidates serve as an excellent way to synthesize their understanding of the *CLASS-S* model by allowing them to see the dimensions and specific classroom interactions "in action." The field-based ob-

TABLE 16.2. Classroom Interactions Graded Course Assignments

Assignment	Description	Points
Participation and Professionalism	Ten points per week x 15 course meetings. This score is based on an individual teacher candidate's attendance, preparedness, and professionalism	150
Ten Pointers	There are five unique "ten pointer" assignments associated with the course. UCCSTeach teacher candidates are required to complete at least three. Ten pointers cover topics such as building relationships with students, classroom management, equity and diversity, growth mindset, and engaging in reflective teaching.	30
Field Observation Log	Teacher candidates are required to complete eight observation hours in conjunction with their assigned mentor teacher from a local school. Candidates observe and analyze teacher-student interactions through the CLASS-S lens, answer a set of predetermined questions, and collect artifacts from both teachers and students to highlight their field-based experiences.	70
Lesson Plans and Practice Teaching (x3)	Teacher candidates work together, in pairs, to develop three inquiry-based lesson plans and teach at a local high school with their assigned mentor teacher. Lessons plans are based on the 5E model (Bybee et al., 2006). Teaching pairs are observed by a UCCSTeach supervisor using the CLASS-S rubric and given substantive feedback. Lesson plans and practice teaching scores increase as the semester progresses (e.g., lesson #1 = 25 points; lesson #2 = 50 points; and lesson #3 = 75 points)	150
Video Analysis Project	The culminating project for the course engages teacher candidates in a video analysis of their final teaching demonstration. Teacher candidates videotape their interactions and are responsible for coding their video using the CLASS-S rubric. Teacher candidates write a substantive reflection paper (15+pages) describing the strengths and weaknesses observed in their interactions and highlight dimensions of their teaching practice they would like to improve upon.	100
	Total Semester Points Possible	500

servation assignment also challenges teacher candidates to reflect on *CLASS-S* dimensions that they have identified as areas for personal growth.

All UCCSTeach teacher candidates are enrolled in *Classroom Interactions* are responsible for designing and teaching three separate inquiry-based lesson plans in a high school math or science classroom throughout the semester. These lesson plans are based on the 5E model (Bybee et al., 2006) and require teacher candidates to further synthesize their understanding of the *CLASS-S* dimensions and how these are operationalized in a high school classroom. The co-instructors for *Classroom Interactions* supervise UCCSTeach teacher candidates as they teach these lessons in the field and provide feedback after each teaching demonstration. The lesson plans and practice teaching demonstrations are fundamental to the course and constitute 30% of the grade.

CLASSROOM INTERACTIONS CO-TEACHING MODEL

Since the spring of 2012, *Classroom Interactions* has been offered for a total of 15 semesters at UCCS (note: this course is only offered in the fall and spring semesters due to the field-based requirements). The in-service teacher/university-based faculty co-teaching model involving the two authors of this chapter, Pat McGuire and Steve Ottmer, began in the spring of 2015 and has been implemented a total of seven semesters (see Table 16.3). It should be noted that the in-service teacher, Steve Ottmer, has co-taught the UCCSTeach *Classroom Interactions* course every semester since spring of 2015. In the spring of 2017 and 2018 the university faculty member, Pat McGuire, was unable to teach the course due to scheduling constraints. For these two semesters, Steve Ottmer co-taught the course with a UCCSTeach Master Teacher.

Co-teacher Backgrounds and Experiences

University Faculty Member

Pat McGuire is an associate professor at the University of Colorado Colorado Springs (UCCS) in the Department of Teaching and Learning. Since 2012, he has served as co-director of UCCSTeach. McGuire was the UCCS faculty member who was originally responsible for developing the UCCSTeach *Classroom Interactions* course in the spring of 2012 and has taught the course a total of 13 semesters.

In-Service Teacher

Steve Ottmer is a full-time science teacher at Doherty High School in Colorado Springs, Colorado. Over his sixteen years at Doherty High School, Steve has taught a variety of science courses, served as the science department chair, and assisted as an instructional coach. Steve began serving as a mentor for UCCSTeach Apprentice Teacher candidates and later became a co-instructor for *Classroom In-*

TABLE 16.3. Classroom Interactions Co-Teaching History

Semesters Co-Taught with an In-Service Teacher and University Faculty Member (n = 7)	Semesters not Co-Taught with an In-Service Teacher and a University Faculty Member (n = 8)
• Spring 2015	• Spring 2012
• Fall 2015	• Fall 2012
• Spring 2016	• Spring 2013
• Fall 2016	• Fall 2013
• Fall 2017	• Spring 2014
• Fall 2018	• Fall 2014
• Spring 2019	• Spring 2017
	• Spring 2018

teractions with McGuire in the spring of 2015. Subsequently, Steve has co-taught the course which he has co-taught for a total of nine semesters, seven with Pat.

Selected Course Innovations

In this section we describe course innovations and related outcomes that have emerged as a result of the in-service teacher/university-based faculty co-teaching model. We argue these selected innovations infuse a sense of authenticity into the course by having two complementary professionals working together to design and deliver the course.

"TALES FROM THE FIELD" SCENARIO DISCUSSIONS

Tales from the field scenarios provide teacher candidates with an opportunity to discussion situations that typically arise at the high school level. These scenarios are based on interactions from actual classrooms (often from the in-service teacher's classroom) and provide teacher candidates with opportunities to widen their thinking and engage in authentic learning. Tales from the field scenarios cover topics such as school violence, bullying, cheating, cell phone challenges, and other contentious issues that arise in the classroom. Tales from the field scenarios such as those described in Table 16.4 below are presented throughout the semester. After the tale is introduced, a 15–30 minute structured Socratic Seminar discussion follows, led by the co-instructors. This discussion allows teacher candidates to synthesize their understanding of the scenarios, engage in personal reflection, and discuss potential ways to react to these situations should they arise in their own classroom one day.

TABLE 16.4. Tales from the field Scenario Examples

Example 1: You are teaching a high school biology class and have begun a unit on evolution. Students in your class have differing viewpoints on the subject, and a few have refused to participate in the lessons, citing their religious beliefs. How do you handle the situation?
Example 2: You are a high school math or science teacher. It is the fourth week of the school year and up to this point, with a few minor exceptions, your students have been relatively well behaved. However, you have noticed an increase in your students texting and using their cell phones during class, despite the clear cut school policy that mobile devices are not permitted during instructional time. These students are clearly off track and not as engaged in your class as they could be. Discuss how you would respond to this situation.
Example 3: In your last period class, one third of your students play sports, which means that at least once a week, one third of your class is missing. Since they play different sports, the students are not missing the same class, so attendance is an issue during sports seasons. What might you do to ensure the absent students are able to keep pace with the course material without holding back the student who are in attendance?

Doherty High School Field Trip Night

Another significant course innovation, and perhaps the most well-received by our UCCSTeach teacher candidates based on anecdotal evidence and end-of-semester surveys, is the Doherty High School "Field Trip" Night. For one night each semester of our *Classroom Interactions* course, normally about three quarters of the way through the semester, we hold class off campus at the local high school where the in-service teacher is employed. Because our course meets from 4:45–7:20 pm and Doherty High School is located just 4 miles from the UCCS campus, the logistics of this trip are relatively straightforward. Plus, there is free parking!

The objective of the field trip is to provide teacher candidates with a learning experience very different than our normal campus-based interactions and orient them to the inner-workings of a school. In Figure 16.2 below we include our playlist (i.e., agenda) for the Doherty High School Field trip night designed to provide teacher candidates with experiences that are complementary to the traditional observation hours they complete as part of the course.

The *Classroom Interactions* field trip begins with a welcome to the school, discussion of who to know in the front office, sign-in policies and security, etc. We then make our way to Mr. Ottmer's science classroom (see Figure 16.3 below) to engage in the first half of class.

The focus of this first hour is on classroom setup including the organization of the physical space. This provides an excellent opportunity for pre-service teachers to observe several of the *CLASS-S* dimensions "in action." For example, they are able to observe how the physical design of the space is intentionally set up to increase productivity and facilitate instructional dialogue among the high school students (e.g., the desk arrangement). It also brings to life our campus-based discussions about how to build a positive learning and safe environment (e.g., posting pictures of students on a cork board, a class turtle, and other random decorations around the room classroom intended to promote a positive and collaborative culture.

Classroom Interactions Playlist

1. Welcome to Doherty!
2. Classroom setup
3. Emergency procedures jigsaw
4. Steve's Schoology example

--------------------------break---------------------------

5. Walking tour/ observing different classrooms
6. Classroom redesign
7. Housekeeping

FIGURE 16.2. Classroom Interactions Field Trip "Playlist"

FIGURE 16.3. Mr. Ottmer's Science Classroom—Doherty High School

During the first half of the field trip teacher candidates engage in an emergency procedures jigsaw activity, learning about the mandated district policies for various emergencies such as an active shooter, bomb threat, lockdowns, weather alert, etc. Unfortunately, in this day and age, it is increasingly important that pre-service teachers are introduced to these situations and understand how to react in order to keep students and themselves safe. Discussing emergency procedures in an actual classroom provides an added sense of relevance and authenticity, one that is not as easily simulated in a normal campus-based discussion. The final part of the field trip time in Mr. Ottmer's classroom is spent reviewing and modeling the technology he leverages on a daily basis, including the Schoology platform now being implemented by many teachers in high school.

In-service Teacher as Lead Facilitator for Field Placements

The final innovation, one that occurs largely off campus, involves the field placement experiences for UCCSTeach pre-service teacher candidates. As previously mentioned, *Classroom Interactions* is a field-based course and requires pre-service teacher candidates work in a local high school classroom with an assigned mentor teacher. UCCSTeach teacher candidates complete a minimum of eight observation hours and also teach three inquiry-based STEM lessons with their assigned mentor.

For the past several years Steve Ottmer, the in-service high school teacher, has taken the lead on placing UCCSTeach candidates for their field work in *Classroom Interactions*. Because Steve currently serves as the acting science department chair of Doherty High School, a large school with a science department consisting of 13 total teachers, he is well-positioned to facilitate the field placements

of UCCSTeach *Classroom Interactions* teacher candidates within his department. Over the past few years Steve has continued to build relationships with his colleagues in other STEM departments (e.g., mathematics and engineering) and has recently begun facilitating these placements as well. Having an in-service teacher facilitate the placements within his own school minimizes many of the logistical and communication issues that often arise when a university-based faculty member or staff is solely responsible.

In addition to facilitating the placements and coordinating the training of other in-service mentors, Steve normally serves as a mentor for one or two UCCSTeach pre-service teaching pairs (normally teacher candidates pursuing science degrees). The fact that Steve simultaneously serves as a mentor teacher and co-instructor for the campus based course has several advantages. First, UCCSTeach pre-service teacher candidates have the opportunity to learn new skills at the university and then observe how those same skills are applied in an actual high school science classroom with an in-service teacher serving as a guide. This allows teacher candidates see the connection between campus-based theories and practical applications, learning about both the *why* and the *how* at the same time. Second, an in-service teacher working simultaneously as a mentor teacher and co-instructor to support a subset of UCCSTeach pre-service candidates minimizes a significant amount of the communication and scheduling issues that periodically arise with mentor teachers who are not as directly connected to the UCCSTeach program. Because Steve is able to meet with his assigned teaching pairs each week during the campus class, he is able to communicate in person, ask questions, and provide quality feedback to UCCSTeach pre-service teacher candidates regarding their field-based experiences at Doherty High School. This personal relationship and opportunity to engage with pre-service teacher candidates in multiple roles at the same time, cannot be understated.

COURSE SUMMARY DATA: QUALITATIVE AND QUANTITATIVE RESULTS

To provide evidence of the effectiveness of the co-teaching model implemented in our UCCSTeach *Classroom Interactions* we have provided snapshot data collected over the past 14 semesters (Spring 2012 through Fall 2018). Representative data include both qualitative comments (see Table 16.5) and quantitative data (see Table 16.6) compiled from the university-mandated end-of-semester Faculty Course Questionnaires (FCQs) that are anonymously completed by students. Data collected on FCQs range from 1 (low) to 6 (high).

Quantitative Faculty Course Questionnaire (FCQ) Data

In addition to the qualitative student comments and anecdotal praise our *Classroom Interactions* course has received over the years, we were curious to see if there were any marked differences in FCQ data when comparing the results

TABLE 16.5. Qualitative Data (Student Comments from FCQs)

FCQ Student Comment 1: "Steve has experience that is beyond valuable to this course. Being a current in-service science teacher allowed us as students to truly pick his brain, learn about the day-to-day things he experiences, and get to know a different high school in the city. Steve and Pat also work so well together and make sure we as students get the greatest benefit possible from this course."

FCQ Student Comment 2: "I really enjoy the interactions that Pat and Steve add to the course. Having two teachers really keeps me more engaged and interested in all of the subjects. I also like that Steve is a current high school teacher. This adds a different perspective to the course that is very helpful."

FCQ Student Comment 3: "The class gave opportunities for us to gather valuable experience in the field teaching, the ability to talk about our experiences as a class, and the information learned each week was used in the field and will be used in the future. Everything about this course was helpful and it has been my favorite of all the UTEACH courses."

FCQ Student Comment 4: "I really enjoyed this class and it was often the highlight of my week. The variety of activities were great and especially helpful in maintaining interest so late in the day. I enjoyed our discussions and felt that I learned a lot from the topics presented. It was really helpful to have a forum in which to ask for practical advice. The questions presented for the observations were really helpful."

TABLE 16.6. Quantitative Data (FCQ Comparison Teaching Models)

	# of Students	Student Perception of Amount Learned	Course Overall	Instructors Overall
Classroom Interactions sections co-taught with high school teacher and university faculty member	n = 52	M = 5.54 (SD = .22)	M = 5.55 (SD = .05)	M = 5.88 (SD = .07)
Other Classroom Interactions sections	n = 71	M = 5.08 (SD = .43)	M = 5.20 (SD = .39)	M = 5.58 (SD = .37)

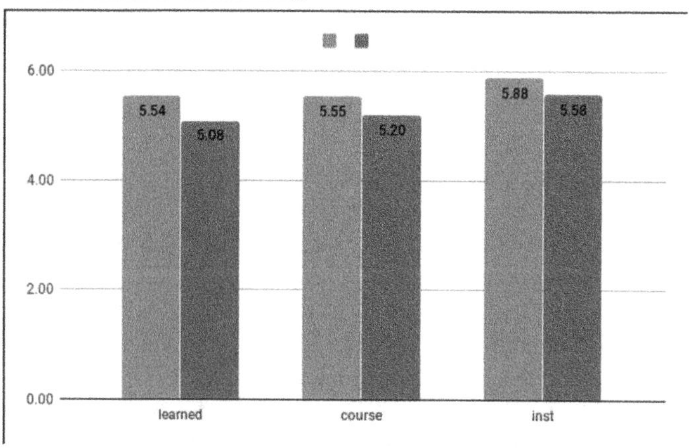

from semesters where the co-teaching model involving an in-service teacher and university faculty member was implemented vs. other semester. In Table 16.6 and associated graph, we provide a brief overview of the aggregate mean course data and standard deviations from three FCQ categories including the *student perception of amount learned, course rating overall, and instructor rating overall.* Means and standard deviations are provided to compare the innovative co-teaching model to other semesters in which the *Classroom Interactions* course has been offered. Note, the FCQ data presented in the table were calculated using the standardized campus-scoring rubric that uses a range of 1 to 6 (low to high). FCQ data are included for the semesters of spring 2012 through fall 2018 and include 123 total student responses.

DISCUSSION

It is clear from the quantitative FCQ data, qualitative student comments, and anecdotal evidence that the co-teaching model implemented in our *Classroom Interactions* course has resulted in significant benefits for our UCCSTeach program. However, there are limitations and barriers worth discussing. First, and perhaps the most obvious limitation, is that co-teaching is expensive. The UCCSTeach program is fortunate to have received base funding from the university and anticipates the ability to sustain this co-teaching model into the foreseeable future. Additionally, the UTeach model calls for most courses within the program to be co-taught, so this co-teaching model is not only supported at the UCCSTeach program level with our internal course data, but by the larger UTeach Institute and the fundamental "elements of success." Second, co-teaching a university-based course with an in-service teacher presents scheduling challenges in terms of when the course can be offered. In our case, the *Classroom Interactions* course is offered on Tuesday evenings from 4:45–7:20 pm to accommodate the in-service teacher's schedule. Some of our UCCSTeach students, in particular those with part-time jobs or children of their own, comment that the time the course is offered is one of their least favorite elements of the experience.

Recommendations

In this section we provide recommendations for other pre-service teacher programs interested in exploring co-teaching models involving in-service teachers and university faculty. First, we hypothesize the advantages of this particular co-teaching model would be particularly beneficial in a field-based course. Including in-service teachers in these courses present a unique opportunity for teachers to share their wisdom and make practical, real-time connections that serve as a vehicle to bridge the gap between the university and K–12 environments (Allsopp, DeMarie, Alvarez-McHatton, & Doone, 2006). Second, we recommend the co-instructors, to the best extent possible, possess complementary backgrounds. In this case the in-service teacher is a science expert with vast practical experi-

ences and the university faculty member is a mathematics education expert with strong theoretical knowledge. These complementary backgrounds not only support the "symbiosis" described in this chapter, but also allow the co-instructors to meet the needs of the UCCSTeach candidates in terms of content, pedagogy, and theoretical research. Finally, we recommend others implementing this co-teaching model commit to a minimum of at least two semesters. Co-teaching is not a trivial endeavor, and through our experiences we have realized it takes a significant planning, tweaking, and practice before the model is properly calibrated! Despite these challenges, we are confident that our unique co-teaching model will continue to provide significant benefits to our UCCSTeach pre-service candidates and facilitate boundary crossing between in-service teachers and university teacher preparation.

REFERENCES

Akkerman. & Bruining, T. (2016). Multilevel boundary crossing in professional development school partnership. *Journal of the Learning Sciences, 25*(2), 240–284.

Allen, J. P., Pianta, R. C., Gregory, A., Mikami, A. Y., & Lun, J. (2011). An interaction-based approach to enhancing secondary school instruction and student achievement. *Science, 333*, 1034–1037.

Allen, J. P., Gregory, A., Mikami, A. Y, Lun, J., Hamre, B. K., & Pianta, R. C. (2013). Observations of effective teacher-student interactions in secondary school classrooms: Predicting student achievement with the classroom assessment scoring system. *School Psychology Review, 42*, 76–98.

Allsopp, D. H., DeMarie, D., Alvarez-McHatton, P., & Doone, E. (2006). Bridging the gap between theory and practice: Connecting courses with field experiences. *Teacher Education Quarterly*, *33*, 19–35.

Bronfenbrenner, U., & Morris, P. A. (1998). The ecology of developmental processes. In W. Damon & R. M. Lerner (Eds.), *Handbook of child psychology, Vol. 1: Theoretical models of human development* (5th ed., pp. 993–1023). New York, NY: John Wiley and Sons, Inc.

Bybee, R. W., Taylor, J. A., Gardner, A., Van Scotter, P., Powell, J. C., Westbrook, A., & Landes, N. (2006). The BSCS 5E instructional model: Origins and effectiveness. *BSCS, 5*, 88–98.

Cochran-Smith, M. (2013). Trends and challenges in teacher education: National and international perspectives. In F. Waldron, J. Smith, T. Dooley, & M. Fitzpatrick (Eds.), *Reimagining initial teacher education: Perspectives on transformation* (pp. 29–53). Dublin: Liffey Press.

Darling-Hammond, L. (2006). Constructing 21st-century teacher preparation. *Journal of Teacher Education*, *57*(3), 1–15.

Grundoff, L., Haigh, M., & Mackisack, V. (2017). Re-envisaging and reinvigorating school-university practicum partnerships. *Asia-Pacific Journal of Teacher Education*, *45*(2), 180–193.

Hagger, H., & McIntyre, D. (2006). *Learning teaching from teachers: Realising the potential of school-based teacher education.* UK: McGraw-Hill Education.

Kane, T., Staiger, D., McCaffrey, D., Cantrell, S., Archer, J., Buhayar, S., & Parker, D. (2012). *Gathering feedback for teaching: Combining high-quality observations with student surveys and achievement gains* (Technical report). Seattle, WA: Bill & Melinda Gates Foundation, Measures of Effective Teaching Project.

Le Maistre, C., & Paré, A. (2010). Whatever it takes: How beginning teachers learn to survive. *Teaching and Teacher Education, 26*(3), 559–564.

Mason, K. O. (2013). Teacher involvement in pre-service teacher education. *Teachers and Teaching, 19*(5), 559–574.

Pianta, R. C., Hamre, B. K., & Mintz, S. (2012). *Classroom assessment scoring system manual middle/secondary version*. Charlottesville, VA: University of Virginia.

Sewell, A., Cody, T. L., Weir, K., & Hansen, S. (2018). Innovations at the boundary: An exploratory case study of a New Zealand school-university partnership in initial teacher education. *Asia-Pacific Journal of Teacher Education, 46*(4), 321–339.

Teitel, L. (1999). Looking toward the future by understanding the past: The historical context of professional development schools. *Peabody Journal of Education*, *74*(3–4), 6–20.

CHAPTER 17

A TALE OF TWO PBI CLASSES

Katie Donaldson, Kate York,
Robert Gagnon, and Cory Gavitt

UTeach Dallas at The University of Texas at Dallas and UCCSTeach at The University of Colorado-Colorado Springs discuss best practices within their project-based instruction (PBI) courses, examine the road-blocks and successes of their parallel ten-year journeys, and brainstorm innovative next steps. Despite differences in the typical makeup of the student body, the challenges have been similar for both programs. This chapter also examines the lessons learned from site visit experiences shared between both programs as a result of a collaboration grant awarded by the 100kin10 organization. Takeaways include the necessity of incorporating an authentic audience when developing a unit design in project-based learning (PBL) format and the importance of providing preservice teachers with a true PBL field experience. Both of those components are enhanced by technology integration and global/informal education partnerships.

True inquiry-based instructional design, a hallmark of the UTeach model, is a work of art when implemented correctly. One of its most exciting and innovative forms is Project Based Learning (PBL). PBL challenges both the traditional transmission–reception model of classroom instruction and the roles typically assigned to *teacher* and *student.* In PBL classrooms,

> students drive their own learning through inquiry, as well as work collaboratively to research and create projects that reflect their knowledge. From gleaning new, viable technology skills, to becoming proficient communicators and advanced problem

Preparing STEM Teachers: The UTeach Replication Model, pages 225–239.

> solvers.... PBL is a key strategy for creating independent thinkers and learners." (Bell, 2010, p. 39)

Because of this, the inclusion of Project Based Instruction (PBI) is a crucial component of the UTeach model. It requires a fair balance of interpersonal skills, pedagogical sophistication, creativity, and just down right persistence. However, instructors who have watched their students engage with and learn content by solving real-world problems know it is a powerful tool. Students in PBI classrooms gain proficiency with their content standards at the same level or better than peers in traditional education settings (Holm, 2011; Larmer, Mergendoller, & Boss, 2015), while also gaining soft skills that are valuable to success in today's global marketplace (Bell, 2010; Kaldi, Filippatou, & Govaris, 2011; Krajcik & Czerniak, 2018; Moylan, 2008; Pearlman, 2010; Rotherham & Willingham, 2009).

This chapter tells the stories of two PBI courses in two UTeach replication programs. The programs of UTeach Dallas at the University of Texas at Dallas (UT Dallas) and UCCSTeach at the University of Colorado, Colorado Springs, were members of the first and second UTeach replication cohorts, respectively. After navigating the recruitment and funding process, supporting the initial early field courses, and creating a distinct program identity, the next step was establishing a successful PBI course. The PBI course within the UTeach model introduces preservice teachers to a pedagogical approach grounded in providing relevancy to students' lives, engaging students in inquiry-based investigations, facilitating collaborative processes, and using emergent technologies. These tenets are not only reflected in the organization and instructional delivery within the class, but are put into practice as preservice teachers participate in active teaching fieldwork in a K–12 classroom. In addition to strong STEM pedagogical content acquisition, this PBI course equally supports the development of critical soft skills in preservice teachers. The following details each program's course, along with the challenges and successes of implementing PBL.

THE UTEACH DALLAS STORY

Our history is unique. Once upon a time (not so very long ago in 1961), there was an empty grain field just north of Dallas. The three founders of T\exas Instruments, with headquarters in the area, envisioned a regional "intellectual atmosphere" where they could develop the next generation of STEM talent to provide the expanding company with engineers and scientists (The University of Texas at Dallas, 2019, para. 3). With an ideal building site location, coupled with a clear vision and need, this was the birth of what would eventually become The University of Texas at Dallas.

Officially chartered exclusively as a graduate institution in 1969, the University's first undergraduates (junior and senior level only) were admitted in 1975.

The first freshman class was admitted in 1990. The university currently enrolls over 28,000 students, 70% of which are undergraduates.

The University is unique in that is lacks a college of education. The UTeach Dallas program was established in 2008 as part of the Department of Science/Mathematics Education (SME), housed in the School of Natural Sciences and Mathematics (NSM). The department itself has a decades-long history of producing quality Master of Arts in Teaching graduates in both science and mathematics and has been a state leader in providing STEM teacher professional development.

UTeach Dallas recruits are generally undergraduate, freshman STEM majors, many of whom indicate pursing a teaching certification as a secondary option to other career preferences; in essence, a teaching certification is often regarded as a back-up plan. Few indicate an intention to teach prior to hearing about the program during freshman orientation.

Most universities implementing the UTeach model rely on education faculty to provide the course instruction, while master teachers organize and facilitate the practical field experiences. In the absence of a college of education at UT Dallas, UTeach Dallas master teachers serve as both course instructors and fieldwork coordinators and supervisors. The PBI course, which is the fourth and final field-based course prior to apprentice (clinical) teaching, averages 15–20 students each semester. PBI is a three credit-hour course where students complete a minimum of 11 hours of fieldwork, of which at least three hours are active teaching. Currently, it is taught by master teachers with extensive PBL training and experience who provide a strong theoretical foundation, but with a practitioner focus.

Class Structure: How We Scaffold PBI

The UTeach Dallas PBI course consists of three discrete project experiences used to scaffold learning. To begin, preservice teachers participate in a full-immersion mini-PBL experience. During this experience, preservice teachers play dual roles—that of student and teacher—to learn and reflect on the components and the process. This mini-PBL introduces preservice teachers to facilitating STEM content and soft skills acquisition using full PBL lesson design elements (see Figure 17.1). For example, one introductory mini-PBL used in the class requires preservice teachers to engage with and master chemistry and algebra standards pertaining to gas laws and their application to ballooning. They use this learning to create educational resources for a regional hot air balloon festival held each in year in the area. At the conclusion of this project, instructors and preservice teachers spend time reflecting on the experience and deconstructing the steps of PBL.

The second major project consists of preservice teachers designing a mini-PBL lesson to teach in a K–12 classroom within a local school district. This is their field experience for the course. This project lasts six weeks (five weeks for lesson design and up to one week for lesson delivery in the K–12 classroom). We (instructors) deliver the content and instruction for developing this mini-PBL lesson using PBL design, providing preservice teachers the opportunity to see the process

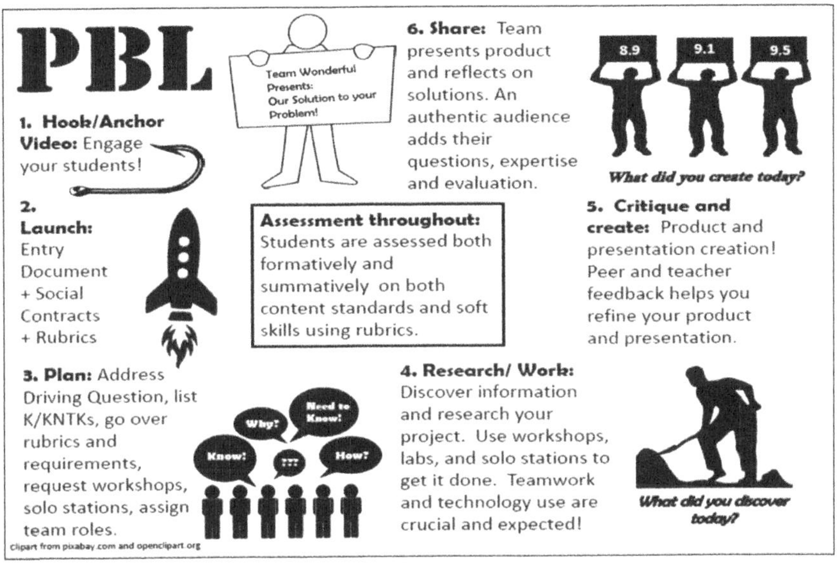

FIGURE 17.1. The PBL Process. This Figure Provides an Overview of the PBL process.

modeled again as they develop their own lesson. This project accomplishes two goals. First, it allows preservice teachers to take a deep dive into the PBL design elements (ex. aligning content standards, crafting driving questions, incorporating an authentic audience, eliciting *need to knows*, designing content instruction, and assessing student learning). Second, it has preservice teachers examine effective PBL implementation processes (ex. project facilitation, classroom management, collaborative/cooperative teaming structures, supports for diverse learners, etc.).

The PBL model comes to life for preservice teachers as they begin to develop their personal *need to knows*, enlisting master teachers for design support, advice, and resources. We hold workshops and/or provide materials to assist preservice teachers in finding what they need to develop successful lessons that display best practices in PBL. Their mentor teacher in the school system also serves in a key role, providing them with the content standards for which they will be responsible, support and feedback on lesson design, and information about their practice classroom and pupils. This project culminates in lesson delivery to the students in their mentor's classroom. They learn by doing as they give their students in the field the same type of learning experiences that they had in their first project in the course. After the lesson teach, preservice teachers give written and oral reflections on their PBL design and teaching experiences. They also receive frequent feedback and teaching evaluations from the mentor teacher and instructors.

The last seven weeks of class revolve around the third and final project, where preservice teachers become curriculum developers, creating a full, extended

STEM PBL unit. Since one best practice of PBL involves using a real-world audience, we attempt to provide this each time in the final project. Several examples include writing STEM lessons for teachers at an orphanage in Zambia, designing integrated secondary science and mathematics units for regional informal education partners (more on this in a following section), and creating STEM PBLs delivered at professional development events for in-service teachers. This project finishes as does any PBL lesson, with a public presentation to an expert audience (their authentic audience). Our preservice teachers complete the course with a full PBL unit that could possibly be used during their apprentice teaching semester, in a future classroom of their own, or by other classroom teachers across the state, nation, or world.

It is important to note that technology is firmly and seamlessly embedded in the course. Most of our preservice teachers are comfortable with basic educational technology they experienced in earlier UTeach Dallas courses. Multiple digital platforms and programs are used for reflection, collaboration, research, exploration, and presentation. Since many preservice teachers in our program conduct their field experiences in 1:1 technology schools that use interactive platform, our PBL preservice teachers are expected to use one to run their created PBL lessons.

Additionally, preservice teachers read and investigate many of the topics set out by the UTeach model for PBI outlined in The UTeach Secondary STEM Teacher Preparation Model and Current Standards Reform Initiatives (Hughes, 2013). Readings for the class consist of text selections from various research journals, practitioner journals/manuals, and textbooks. Topics include constructivist theory, alternative assessment, classroom management in PBL, collaborative learning, equity, integrating relevant technology, and motivation. Preservice teachers are responsible for using a research-based educational strategy to assist in leading a 30-minute discussion session on the topic of the week during class.

UTeach Dallas Challenges

A major key to the UTeach model's success is the early and often field experience throughout the program. This is particularly true for the PBI course. The nature of PBL lesson design requires numerous observation experiences and sustained teaching practice. Fieldwork is where preservice teachers put theory into practice, and it serves to reinforce learning.

However, one of the biggest challenges for any PBI course with a field experience component is finding school programs that are far enough along in their own PBL journeys to provide mentor teachers for their preservice teachers. UTeach Dallas is fortunate to be within 30 minutes driving distance of two PBL immersion schools. In addition, several other local ISD's have campuses with classrooms whose instructional practices range from occasional PBL projects to full PBL implementation.

Even with these advantages, it can be difficult to place all teaching teams in a classroom where they will observe quality PBL. We typically have six to ten

teams distributed across three or four school districts each semester. This distribution, coupled with the different implementation practices and levels of our partner districts, campuses, and individual mentor teachers, means that our individual preservice teachers may have vastly different fieldwork experiences in terms of the consistency, frequency, and/or quality of PBL offered in the mentor teacher's classroom. To counteract this possibility, we have two safeguards in place. First, each semester we schedule a two-hour field trip to a school we *know* is engaged in a rigorous, research-based PBL program. We are unable to schedule all of our fieldwork at this campus, but we do provide our preservice teachers with a meaningful experience by allowing them to tour the school and talk with students, faculty, and administrators who are immersed in a PBL model. Second, we structure in-class debriefings of their fieldwork experiences, discussing the level of PBL they are seeing and why buy-in and implementation may differ between individual classrooms, campuses, and districts. These are rich conversations and our preservice teachers are open about their observations and how they influence their comfort level with PBL.

Interestingly, however, we have witnessed a phenomenon we have termed *inquiry inoculation*. When only limited or partial PBL is occurring in a mentor's classroom, the excitement, quality lesson design, and implementation success often achieved by our preservice teachers influences these teachers to further their own PBL practice; preservice teachers become teacher-leaders before they graduate from their preparation programs.

From our experience, we suggest recruiting strong classroom teachers at least willing to allow preservice teachers to experiment with this type of instructional method, even if they are PBL novices themselves. Consider also organizing professional development in PBL practices for these teachers during mentor orientation; providing instructional training to mentor teachers has for us proven beneficial in supporting our preservice teachers.

UTeach Dallas PBI Course Changes: High Value Add-ons

While UTeach Dallas has remained faithful to the original UTeach model, there have been several addendums along the way, particularly to the PBI course. The two most significant changes have been the addition of global and informal education elements.

Global Education

Providing preservice teachers with coursework and experiences that foster global awareness, appreciation, and collaboration opportunities in this day and age is a necessary, worthy goal (Zong, 2009). As part of the our current PBI course, preservice teachers are able to examine global content and pedagogy, engage with global partners, and address global problems, knowing this is of growing importance to both their personal and their students' lives.

Many preservice teacher programs incorporate additional perspectives classes or provide opportunities for study abroad and/or international student teaching experiences (Aydarova & Marquardt, 2016; Walters, Garii, & Walters, 2009; Zong, 2009). However, given the nature of our program, it can be difficult for our preservice teachers to engage in these types of experiences. Rigorous STEM degree requirements, coupled with state certification requirements, make it challenging to include any additional course work dedicated to global perspectives or expanding views on diversity. Rather, we use creative means to bring the benefits of global education into the PBI classroom through classroom-based (virtual) global collaboration experiences and find that PBL provides an ideal framework for our preservice teachers to engage with and work collaboratively on projects with other in-service and preservice teachers and K–12 students from around the globe. We have used various technology platforms to engage in PBL projects that include international partners from countries such as Belarus, Taiwan, Brazil, Zambia, and the Bahamas. This has been an exciting adventure in our PBL coursework that our students, along with instructors, say they have found beneficial (York, Hite, & Donaldson, 2019) and serves as a small step to ensuring that future teachers enter the classroom are not only globally competent, but capable of designing and providing global collaboration experiences to their own students.

Informal Education

Additionally, the initial iteration of the UTeach model PBI course incorporated a preservice teacher–led student field trip component as an immersion experience in informal education. Informal education involves learning in a non-school, learner-centered environment, such as a nature center or museum. Students can explore their interests along with the interactive space for learning, therefore obtaining the sense that science is personally relevant and rewarding (Bell, Lewenstein, Shouse, & Feder, 2009).

Many of the first and second replication cohorts have had success with this imbedded experience. From the beginning, though, this was just not feasible for UTeach Dallas. Working with multiple districts in several field site locations made the financial and logistical challenges difficult to overcome. However, UTeach Dallas wanted to ensure we did not deny our preservice teachers exposure to the wonderful possibilities that can happen when informal education providers partner with local teachers, especially through a PBL project. Our solution was to take our preservice teachers on an informal education field trip as part of our course and have those organizations discuss the available resources and potential benefits of formal–informal education partnerships. We have taken private tours of the Perot Museum in Dallas, fished and dissected trout at the Texas Freshwater Fisheries Center (TFFC), and banded birds at the Lake Lewisville Environmental Education Center (LLELA).

During most semesters, as part of the final innovative PBL unit design project, our preservice teachers incorporate a global education or informal science com-

ponent into their PBL unit. Many of these final projects have been written directly for these global education or informal science partners. Having our preservice teachers engage as budding curriculum writers for these partners has resulted in a project focus and product quality that we have rarely seen. As a result, numerous PBL lessons written about or for these partners are now in the process of being further developed for publication and classroom use.

THE UCCSTEACH STORY

Like the University of Texas at Dallas, the University of Colorado, Colorado Springs (UCCS) started from humble beginnings. Using land purchased for $1, UCCS was established in 1965. With only a few buildings on location, classes started as part of a commuter campus. In fact, it was not until 1996 when UCCS built its first dormitory that UCCS was able to establish on-campus housing. Currently, only approximately 13% of students live on campus; the university is still very much a campus of commuters.

The student body of UCCS is relatively non-traditional. UCCS hosts a large military population, with over 19% of undergraduates having a military affiliation due to the five military bases located within Colorado Springs. Nearly 30% undergraduates are first-generation college students. The average age of undergraduate students at UCCS is 22.8 years. Class sizes at UCCS are considered small compared to other Colorado universities, as well, averaging 17 students. When UCCSTeach began in 2010, our student demographics mirrored those of the university. Put simply, our students are unique and have different needs than a traditional four-year university.

While the initial goal of UCCSTeach was to mirror the UTeach replication model closely, the needs of the student population often dictated the changes within the program. Today, much of the program looks the same, but the requirements of field courses differ. These different elements were created in order to meet the unique needs of our preservice teacher population within UCCSTeach.

Class Structure: How We Began PBI

PBI was first taught in the spring of 2013 at UCCS. Taking into consideration that the entire concept of PBL was just gaining a foothold in education, there were not many resources. Since this was the first university course in the state dedicated solely to PBL, there was not an opportunity to learn from any similar courses at other institutions of higher education in Colorado. UCCSTeach, therefore, closely adhered to the fundamentals of the UTeach model. As previously described by UTeach Dallas, the UTeach model for PBI focuses heavily on the theory behind PBL. In that framework, the instructors seek out a solid field experience for their preservice teachers to develop and teach a PBL unit. After the field experience, the preservice teachers reflect upon their time in local PBL classrooms and schools in order to fashion individual meaning from what they had just created and taught.

This is a solid process, but one we found difficult to easily replicate in our community.

Initially, it was thought the best way to develop rapport and buy-in with the school districts was to show that the PBL process had been done before in multiple schools across the country with solid results. Unfortunately, it was quickly determined that our local school districts were neither willing, nor able, to support a PBL class within their schools. The concept, while not new or unique, is not one that Colorado Springs area schools had adopted. In fact, the reception to the idea was a cool one, at best. There were no schools at the time labeled as project-based, and it was extremely difficult to connect our ideals to create a field-based component. Indeed, there was so much pushback from local school districts that UCCSTeach considered using community-based organizations, such as the local zoo, in lieu of going into traditional public schools for their field experience.

However, after careful deliberation and some concessions made, it was determined best to create a course that *did* have a field-based component within the public school setting. The focus changed slightly from a robust, field-based PBL experience to a strong PBL process within the UCCSTeach PBI course itself and providing a weeklong quasi-PBL experience in the field. UCCSTeach education courses prior to PBI only included individual lessons taught in a field-based setting. UCCSTeach believed that creating an opportunity for students to teach a full project unit was of value. This provided a "tomorrow" for our students; they could learn from the events from one lesson, reflect on these details, and prepare for the next day. In addition, the time requirement was such that preservice teachers were to be at their partner school as much as schedules permitted. We requested they be on location for at least 25 hours. Simply put, if they were not attending a UCCS class, they were expected to be at their site school for that week. Post-teach reflections revealed that preservice teachers found this arrangement extremely valuable. They were able to see what a typical week looked like in a school before they entered their apprentice teaching experience.

The first iteration of our course utilized a variety of resources provided by the UTeach Institute. This included the course texts, the suggested assignments, and even the final exam. We quickly learned, however, that much of their suggested ideas focused around a solid field experience model that did not work in our context. This resulted in designing a modified field experience. Using a *Know*, *Learn*, and *Do* decision-making process (detailed further), we adapted other resources to more appropriately address the needs of our preservice teachers.

We started with the *Know*. Within the first six weeks of class, preservice teachers explored the text and engaged in classroom experiences, covering topics such as the rationale for PBL, theoretical frameworks, and knowing your students in order to serve them well. This gave our preservice teachers the opportunity to study situations where they were expected to demonstrate their understanding of the PBL process.

PBL is effective at increasing student achievement, engagement, and motivation (Holm, 2011), so the first six weeks of the course largely focused on unlearning comfortable, but less effective methods of classroom instruction so ingrained in common practice. This was particularly relevant in light of several decades of research regarding secondary student engagement, which has shown that students are often bored, unchallenged, and disconnected to the more passive teaching methods traditionally used in classrooms, a trend which persists even in the most recent data (National Association of Independent Schools, 2017; Yazzie-Mintz, 2010). Instead, we presented preservice teachers with opportunities for students to learn by doing. PBL is the process of active learning, not just the completion of the project itself. The product and showcase at the end are certainly part of the process, but the entire learning process is what counts.

After our sales pitch of PBL, we moved to the *Learn.* The goal of this portion of the course was to provide a chance for preservice teachers to create and teach as close to a PBL experience for their students in the field as was possible given the constraints of the mentor teachers and school settings. Recall the pushback from schools our preservice teachers would often hear, "That sounds great, but we can't do it that way," "We need to show different evidence of student learning," or "Our school requires…." It became difficult to create a complete PBL experience for students; thus, the major focus shifted to creating a full unit of study with pieces of PBL embedded. Our preservice teachers created a weeklong, full day experience and implemented it in real time at their assigned campus. The results were mixed. The field experience was found to be helpful. Our preservice teachers were pleased to have the opportunity to teach in a classroom setting for an extended period of time. There were also some true PBL experiences created, such as a preservice teacher team that gave their class the opportunity to design and build the baseball dugouts for their local school. They even brought in a member of the community, the facilities manager of the local Triple A baseball team, to discuss logistics! Overall, however, the PBL process became what we feared- an extended project, not a PBL experience.

The final portion of the course was the *Do.* In keeping with the tenets of the UTeach model, our preservice teacher teams were tasked with creating a month-long PBL unit of study, which served as a theoretical extension of their PBI field-based experience. We asked them to look at the content standards, determine how long it would take to extend their original PBL field-based lesson, and from that develop a comprehensive PBL unit. If they could not teach it in schools, at least they could write the curriculum as if they would. This gave UCCSTeach students a chance build upon their experience and answer the question, "If I had the opportunity to create an authentic PBL unit, what would it look like?"

Their final PBL units focused on the process of learning and served as their final grade in the course. We also asked them to include a grant proposal along with their unit of study to mirror actions an in-service teacher might take to make PBL possible in their school. UCCSTeach also subscribes to providing authentic

learning experiences, so we invited stakeholders from local school districts to listen to our preservice teachers' final PBL unit and grant proposal presentations.

UCCSTeach PBI Course Changes: PBI 2.0

Portions of our initial PBI implementation worked. Preservice teachers gained valuable classroom experience before apprentice teaching, an opportunity the UTeach model is truly fantastic at creating. But, our preservice teachers were still missing designing and teaching a true PBL experience. After six semesters of facilitating the course in its original iteration, faculty reevaluated the objectives UCCSTeach had for its PBI course.

The summer of 2016 provided the time needed to redefine the purpose and significantly overhaul our PBI course. While there were many positive aspects to the original PBI sequence, the course had inadvertently strayed from its mission. Using the MoSCoW method (*Must Have, Should Have, Could Have, Won't Have*), we identified three main detractors from the course's intent: 1) excess teacher preparation content; 2) time constraints; and 3) field site cultures. Specifics pertaining to each are discussed below.

Excess Teacher Preparation Content

As our program developed and grew during replication, it was necessary to find ways to merge accreditation, graduation, licensing, and other requirements within the UTeach model. PBI became a catchall for some of these, as its placement near the end of the program meant it was a last chance opportunity to address these topics. This created a situation where we not only lost time to provide PBL experiences, but also led to the class being taught in a more traditional instructional style which weighed heavily on the instructors. Likewise, the irony was not lost on the preservice teachers .

The redesign examined teacher preparation elements that were valuable for preservice teachers, but were detracting from the PBI course objectives. Some of these topics were incorporated into new aspects of the course or relocated to more appropriate courses. Others were retired as *Could Haves* or *Won't Haves*. Removing these elements freed up both preservice teacher and instructor time, which was allocated to more important elements.

Time Constraints

The original field component provided an excellent opportunity for preservice teachers to experience the rigors of a full week of planning, assessing, and teaching, but the lack of an extended teaching experience created more traditional projects, activities, and labs than actual PBL experiences for them, the mentors, and their students. To address this, we switched from having our preservice teachers instruct full days to instead having them focus on one class block or period during the school day. This allowed for greater flexibility with scheduling, which is important for all stakeholders. For example, it provided our students with more

control over their university and work schedules, who as noted earlier, are mostly non-traditional commuters. Likewise, it allowed our faculty and mentors more opportunities to actively facilitate the field experiences. By reducing impacts to our partner school's typical day, we were able to generate greater buy-in and establish more productive school partnerships.

Field Site Cultures

A major hurdle to overcome in the Colorado Springs area is a lack of schools implementing PBL. Additionally, a majority of instructional and curricular decisions are made by and delegated from district-level personnel with the expectation of compliance from their campuses. This allows little flexibility in implementing innovative instructional design beyond the scope of district mandates. Making inroads and gaining trust with PBL pedagogy has been and continues to be one of the greatest barriers to implementing PBL in the local public and charter schools.

When we rolled out our new PBI 2.0, we were fortunate to find a school that was piloting a teacher-choice class period. The mentors at this school were enthusiastic to collaborate with our preservice teachers and design units for this new course. It was fascinating to watch as the mentors and our preservice teachers worked together to define and adapt our PBL skeleton to suit each of their mentors' unique needs. The teams developed PBLs that ranged from three to nine weeks, covering topics such as science, art, engineering, and even standardized test review. Students in the partner school designed lessons for their peers, simulated natural disasters, and built and tested scale building models. The vast breadth of ideas and implementation strategies was inspiring. This process was followed up with exit interviews in which students reflected on personal struggles, teamwork skills, and areas of improvement in team interactions. Unfortunately, this school had a dip in standardized test scores after the state implemented new content standards and a new test. Despite the success of these courses, the school converted the teacher-choice period into a test prep period. We were forced to look for other campus partners.

We reached out to Paul Tweed and Liz Schubert, authors of *An Improbable School* (2015) (a required reading for this course) and teachers at the Wildlands School, a PBL secondary campus in Fall River, Wisconsin. They agreed to provide us with a virtual field experience that we were not able to provide locally to our preservice teachers. Using online synchronous sessions, our preservice teachers were provided access to the authors and other instructors and were allowed real time feedback from Wildlands students. This was an eye-opening experience which provided preservice teachers with validation that these techniques are not just theoretical, but can be implemented successfully by practitioners.

We have been fortunate to find hard working teachers and administrators who have cracked open doors for us and have had many successes in small doses. However, we still lack a place to provide the support and experience from a fully functioning PBL school in the Colorado Springs area. Yet, we are encouraged by

several new partnerships we have formed. One is a relatively new charter school still developing their identity. They have allowed UCCSTeach to provide instructional support while providing our preservice teachers unprecedented access to their classrooms. This has been enlightening for our preservice teachers, as they get to observe the process of creating a school culture, curriculum, and instructional focus from the ground up.

In addition, several other schools we work with have recently developed individual courses in a PBL fashion. One such school has created an applied physics pathway with a PBL twist. Although we have not yet met our goal of creating widespread partnerships with dedicated PBL schools in our area, it appears the PBL concept is spreading. The future looks bright, even if it will take some time to get there.

COLLECTIVE TAKEAWAYS FROM A SHARED PBL EXPERIENCE

One of the wonderful side benefits of the UTeach network is the ability to connect and collaborate with others teaching the same courses and doing similar work. In 2018, we (the authors) had a unique opportunity to form a collaborative partnership. The collaboration between UTeach Dallas and UCCSTeach began at a session we co-presented at the annual UTeach Conference. A very productive chat about the deepest needs of our course led to a decision that our chances to learn from one another should continue in a more strategic, intentional manner. After applying for and receiving a 100Kin10 collaboration grant (https://100kin10.org), we visited each other's campuses with our goal being to better our courses by experiencing firsthand best practices in each program.

UCCSTeach first traveled to UTeach Dallas. Our (UCCSTeach) biggest takeaway was how UTeach Dallas organized their final innovative unit design project. As part of that semester's project, we went with UTeach Dallas on their field trip to the local nature center and watched our counterparts establish an authentic audience for their preservice teachers by having them design a series of PBL lessons for their informal science partners to use with classroom teachers. While UCCSTeach had previously considered a similar design, we had never moved forward. However, as a result of this collaboration, we connected with a local organization, Challenger Learning Center of Colorado Springs (CLCCS). Our classes now review some of their pre-lessons given to teachers, modify them with a PBL feel, and present their findings to the CLCCS team.

We were also able to visit a PBL immersion high school in the Dallas area. Since Colorado Springs does not have an actual PBL school, this was a great opportunity to engage with administrators, faculty, and students who experience PBL daily and bring that learning back to our preservice teachers.

UTeach Dallas then traveled to Colorado Springs, curious about the training and partnership that UCCSTeach experienced with a local charter school, the extended field experience that we were able to see action, and the influence that the UCCSTeach preservice teachers and faculty had on establishing and supporting

PBL in non-PBL environments. Specifically, the latter spurred us (UTeach Dallas) to develop an affordable, high quality two-day PBL professional development for local in-service teachers who could then serve as mentors to our preservice teachers. It also helped in building rapport with campus and district leaders in our most accessible district.

This collaboration allowed us to explore what was (and was not) working, validated some of our own practices, and provided a more formal means of sharing creative ideas. Although we had the additional fortune of a grant to support our work, these types of UTeach replication program collaborations can occur with minimal or no financial assistance. Attending and networking at conferences (primarily the UTeach Institute, but others, as well) is an ideal way to connect with colleagues involved in similar work. Establishing virtual means to share ideas and practices is made easy with collaborative technology. It would also be worth exploring small teaching and learning grants from the respective universities or outside organizations that might support more robust collaborations.

Regardless of the level of the collaboration, it is incredibly important to share ideas, innovations, and insights with other PBL trailblazers. In addition to collaborating with colleagues, for faculty not well-versed in PBL practices, consider attending professional development that provides research-based PBL training. Schedule visits to schools and districts using PBL methods (this can also be a good way to establish district and campus partners). And, for preservice teachers, provide frequent opportunities for them to write PBL lessons and practice lesson implementation in a local classroom, even if the schools are just dipping their toes into the PBL process. Teaching is an art; artists learn from the works of other artists. Find collaborators who believe in the PBL process and have seen it in action. Your greatest PBI course masterpiece is just waiting for you to begin.

REFERENCES

Aydarova, E., & Marquardt, S. K. (2016). The global imperative for teacher education: Opportunities for comparative and international education. *FIRE: Forum for International Research in Education, 3*(1), 23–40.

Bell, P., Lewenstein, B., Shouse, A. W., & Feder, M. A. (2009). *Learning science in informal environments: People, place, and pursuits*. Washington, DC: The National Academies Press.

Bell, S. (2010). Project-based learning for the 21st century: Skills for the future. *The Clearing House, 83*(2), 39–43.

Holm, M. (2011). Project-based instruction: A review of the literature on effectiveness in prekindergarten through 12th grade classrooms. *Rivier Academic Journal, 7*(2), 1–13.

Hughes, K. K. (2013). *The UTeach secondary STEM teacher preparation model and current standards reform initiatives*. Austin, TX: The UTeach Institute.

Kaldi, S., Filippatou, D., & Govaris, C. (2011). Project-based learning in primary schools: Effect on pupils' learning and attitudes. *Education 3–13, 38*(1), 35–47.

Krajcik, J. S., & Czerniak, C. M. (2018). *Teaching science in elementary and middle school: A project-based approach*. New York, NY: Routledge.

Larmer, J., Mergendoller, J., & Boss, S. (2015). *Setting the standard for project based learning*. Alexandria, VA: ASCD.

Moylan, W. A. (2008). Learning by project: Developing essential 21st century skills using student team projects. *The International Journal of Learning, 15*(9), 287–292.

National Association of Independent Schools. (2017). *NAIS report on the 2016 high school survey of student engagement.* Retrieved from: https://www.fcis.org/uploaded/Data_Reports/2016-HSSSE_Final_1.pdf

Pearlman, B. (2010). Designing new learning environments to support 21st century skills. In J. Bellanca & R. Brandt (Eds.), *21st century skills: Rethinking how students learn* (pp. 117–147). Bloomington, IN: Solution Tree Press.

Rotherham, A. J., & Willingham, D. (2009). 21st century skills: The challenges ahead. *Teaching for the 21st Century, 67*(1), 16–21.

The University of Texas at Dallas. (2019). *A pioneering past and a promising future*. Retrieved from: https://www.utdallas.edu/create/narrative.html

Tweed, P., & Seubert, L. (2015). *An improbable school: Transforming how teachers teach and students learn.* n.p.: CreateSpace.

Walters, L. M., Garii, B., & Walters, T. (2009). Learning globally, teaching locally: Incorporating international exchange and intercultural learning into pre-service teacher training. *Intercultural Education, 20*(sup1), S151–S158.

Yazzie-Mintz, E. (2010, March). Leading for change. *Principal Leadership*, *10*(7), 54–58.

York, M. K., Hite, R., & Donaldson, K. (2019). Lessons learned from going global: Infusing classroom-based global collaboration (CBGC) into STEM pre-service teacher preparation. *Innovations in Science Teacher Education*, *4*(4). Retrieved from: https://innovations.theaste.org/lessons-learned-from-going-global-infusing-classroom-based-global-collaboration-cbgc-into-stem-preservice-teacher-preparation/

Zong, G. (2009). Global perspectives in teacher education research and practice. In T. Kirkwood-Tucker (Ed.), *Visions in global education: The globalization of curriculum and pedagogy in teacher education and schools* (pp. 71–89). New York, NY: Peter Lang.

CHAPTER 18

THE EVIDENCE BEHIND THE UTEACH CAPSTONE COURSE

Does Project-Based Learning Work?

Dawn Woods and Candace Walkington

This systematic literature review examines the evidence behind project-based learning—with traditionally underrepresented students in urban settings—in order to provide teachers, principals, districts, teacher educators, and policy-makers with information to make convincing arguments for the inclusion of project-based learning in secondary STEM classrooms. To this end, we investigate whether and how project-based learning impacts STEM learning outcomes and attitudes towards STEM. Further, we describe six characteristics of project-based learning that may support traditionally underrepresented students in STEM, as well as characterize the kinds of supports that teachers may need in order implement project-based learning in high-needs, urban schools.

Keywords: project-based learning, STEM education, traditionally underrepresented students, urban schools

INTRODUCTION

Project-based learning (PBL; also referred to as project-based instruction) is a hallmark of the UTeach program. PBL is characterized as a learning environment

Preparing STEM Teachers: The UTeach Replication Model, pages 241–258.

where students are engaged in meaningful inquiry as they do similar work to scientists, mathematicians, writers, and historians (Krajcik & Blumenfeld, 2006). Through PBL, students gain knowledge and skills while investigating a driving question related to a real-world problem or phenomena that they find important. They are pressed to make decisions, perform research, and discuss their own and other students' processes and projects. Based on the premise that PBL is a highly desirable addition to a secondary STEM teachers' toolbox, UTeach students, who are science, technology, engineering, and mathematics (STEM) majors, learn the foundational pedagogical and content pedagogical skills needed to implement PBL across the courses and experiences in the UTeach program.

During the program, UTeach students experience PBL as learners through lessons modeled for them in their UTeach courses (e.g., the expert-novice interview project in *Knowing and Learning*). Then, the final education course in the UTeach sequence, *Project-Based Instruction,* challenges UTeach students to formalize and compile this knowledge by designing a project-based unit for their future classrooms. As part of their coursework, they plan inquiry-based instruction using the five key elements of PBL: (1) creating a need to know and then focusing on a driving question related to a real-world problem, (2) engagement in authentic inquiry, (3) collaboration between students, teachers, and community members, (4) project activities scaffolded with a variety of learning technologies, and (5) creation of a final product—a publicly shared artifact—addressing the driving question (Krajcik & Blumenfeld, 2006). They also integrate the big ideas about effective STEM instruction into their PBL lessons that they have been learning about (e.g., scaffolding, technology integration, facilitating group work, teaching through inquiry, communicating STEM content effectively, structuring learning for student engagement and control).

However, the implementation of PBL, particularly in high-needs, urban schools, has been fraught with difficulty and logistical issues, and many UTeach students can graduate from the program and perceive that they are unable to actually implement PBL in their classrooms given practical constraints like time and curriculum restrictions (Marshall, Petrosino, & Martin, 2010; Wasserman & Walkington, 2014). Indeed, around 66% of UTeach graduates are teaching in schools where the majority of students are economically disadvantaged (UTeach Institute, 2015). One step towards a solution is to clarify the evidence supporting PBL—with traditionally underrepresented students in urban settings—such that teachers, principals, districts, teacher educators, and policy-makers can make and communicate convincing arguments for the inclusion of PBL in secondary STEM teaching. Thus, in this literature review, we pose the following research questions: (1) How does PBL impact student STEM learning outcomes? (2) How does PBL impact student attitudes towards STEM? (3) Which characteristics of PBL can support traditionally underrepresented students in STEM? and (4) What kinds of supports do teachers need to implement PBL in high-needs, urban schools?

METHOD

In this systematic literature review we focus on secondary STEM education and define PBL as an innovative learning environment where students learn by doing as they inquire into an open or driving question about a problem or phenomena that is of interest to them (Krajcik & Blumenfeld, 2006; Larmer, Mergendoller, & Boss, 2015). To this end, we conducted an electronic search of eight databases (Academic Search Complete, Education Source, ERIC, MathSciNet, Professional Development Collection, PsycInfo, Science and Technology Collection, and Teacher Reference Center) to locate peer-reviewed studies, from 2007 to 2019, fitting the selection criteria of *project-based learning, STEM education*. This search yielded 261 manuscripts. To narrow the scope of the search, two separate searches were conducted with the additional terms *middle school* and *high school*. Adding the term *middle school* yielded 27 studies, but we eliminated 17 manuscripts because they were repeated citations or did not fit our definition of PBL from Krajcik and Blumenfeld (2006). The term *high school* yielded 56 studies, but we eliminated 45 studies because they were repetitive of (and included in) the middle school search or did not fit our definition of PBL. As we reviewed these manuscripts we noticed that many, if not all, focused on PBL in urban settings and/or with traditionally underrepresented students. Therefore, we performed two additional searches, first using the terms *project-based learning, STEM education*, and *urban schools* which resulted in 11 manuscripts and the terms *project-based learning, STEM education*, and *underrepresented students* which resulted in 15 manuscripts. However, these additional searches only yielded three additional studies that were not included in the previous searches. Therefore, 24 peer-reviewed manuscripts were reviewed for this analysis spanning PBL in sixth through twelfth grade in U.S. schools: ten studies focused on traditionally underrepresented populations, eight studies focused on urban education, three studies focused on both urban and traditionally underserved populations, and three studies did not report a specific focus on these characteristics (see Appendix A).

The decision to use "STEM education" as our search term, rather than "math education" or "science education" narrowed the articles that fell under consideration in this literature review. In particular, using STEM education as the search term returned PBL articles that were more likely to highlight integrated PBL—i.e., studies about projects that involved multiple STEM fields. Recent literature has suggested that STEM integration is an important component of and consideration for PBL (National Academy of Engineering and National Research Council, 2014; Petrosino, Walkington, & Ekberg, in preparation). However, this decision likely resulted in some studies that related to only one STEM field being omitted. Ultimately, the present literature synthesis is aimed towards answering questions about the effectiveness of PBL in STEM when STEM considered in a more integrated fashion.

RESULTS

In this review of literature, we first determine whether and how PBL impacts STEM learning outcomes and attitudes towards STEM disciplines and careers for traditionally underrepresented secondary students. Then, we describe the characteristics of PBL that may promote interest in STEM, as well as summarize the kinds of supports that teachers may need to effectively implement PBL in high-needs, urban schools.

Research Question 1: How Does PBL Impact STEM Learning Outcomes?

Many low income and minority students who live in urban areas are subjected to *pedagogy of poverty* (Haberman, 1991) where opportunities to learn STEM disciplines focus on learning basic skills decontextualized from real-world activities. However, PBL may be an innovation that provides the kind of access to STEM education that can be empowering (National Academy of Engineering and National Research Council, 2014; Young, 2013). Therefore, one way to open up access to students who are typically underrepresented in STEM may be to provide opportunities for meaningful inquiry. This learning environment is established when teachers introduce a "need to know" situation and then guide students' thinking toward an open ended or driving question that gives them purpose and challenge. A few examples of these inquiry-focused innovations are as follows: challenging students to design experiments (with the support of university faculty) that would be replicated on the International Space Station in 2015 (Cwikla, Milroy, Reider, & Skelton, 2014), supporting students to engineer a better world while investigating wind power (Cogger & Miley, 2012), integrating agricultural activities into a subdivision in order to support the life style of its residents (Foutz et al., 2011), or monitoring power consumption in homes and communities using technology (Knezek, Christensen, Tyler-Wood, & Periathiruvadi, 2013). These PBL activities provide students with voice and choice—meaning students are challenged to solve the driving question in a way that is meaningful to them as they collaborate to collect data and create products showcasing their learning. Further, these studies demonstrate that when traditionally underrepresented students participate in PBL their STEM content knowledge increases (Knezek et al., 2013), and they also obtain higher achievement test scores than students in traditional classrooms (Cogger & Miley, 2012; Cwikla et al., 2014; Foutz et al., 2011; Hansen & Gonzalez, 2014; Lynch et al., 2017).

Another example of the potential impact of PBL on student learning outcomes is found in Harris and colleagues' (2015) randomized control trial situated within sixth grade classrooms in an urban district. This study finds that students who participated in a PBL science curriculum—aligned to core science ideas and science practices—outperformed students in the comparison curriculum on outcome measures. Similarly, Erdogan, Navruz, Younes, and Capraro's (2016) study situated within three

urban high schools in the southwest U.S. discovers that when PBL is implemented with fidelity, achievement scores are significantly different (and higher) than students whose schools only partially or minimally implemented PBL. These results also indicate no achievement differences related to ethnicity with full PBL implementation, whereas the schools with partial or minimal PBL implementation show differences in achievement for students of different ethnicities. Further, students in the full PBL implementation school are better situated for university admittance. Relatedly, Han, Capraro, and Capraro (2016) investigate how project-based learning affects high-need students in the southwest U.S. and find that PBL positively influences Hispanic students' achievement in mathematics. Yet, in this same study, students who underperform on state achievement tests, have limited English proficiency, or were in the care of a state agency show no difference in achievement growth rates when compared to the non-PBL group. Thus this study suggests that the benefits of PBL for under-served students may be mixed.

PBL may also have the potential to cultivate 21st century skills—skills such as critical thinking, communication, collaboration, and creativity—during instruction (e.g., Bass, Hu Dahl, & Panahandeh, 2016; Cho & Brown, 2013; Lynch et al., 2017). Researchers across these studies note that attending to these elements of learning encouraged the creative expression and collaboration that is needed to support students in engaging within the PBL learning environment. However, Cho and Brown (2013) note that at times facilitators may not always connect 21st century skill development to projects and often students may become distracted by the collaborative nature of PBL. Therefore, students may need additional supports in order to practice these soft skills so that they could ultimately adopt and infuse them as they collaboratively developed their projects (Lynch et al., 2017).

In sum, PBL has the potential to impact STEM learning outcomes for students traditionally underrepresented in STEM. The innovations reviewed provide students with real-world connections, meaningful inquiry to engage their minds, and a focus on affective elements of learning. Yet, it is important to note that PBL, as a learning environment and curricular innovation, needs to be implemented with fidelity in order to increase learning outcomes.

Research Question 2: How Does PBL Impact Attitudes Towards STEM?

Participation in PBL may have the power to change how traditionally underrepresented students perceive themselves as a STEM learner. Within this review, several studies demonstrate that PBL positively impacts students' confidence and attitudes towards mathematics (Lynch et al., 2017) and increases motivation and interest in engineering as a potential career (Cwikla et al., 2014; Mohr-Schroeder et al., 2014; Mosley, Liu, Hargrove, & Doswell, 2010; Redmond, et al., 2011; Sahin & Top, 2015; Verma, Dickerson, & McKinney, 2011). In these cases, as students have the opportunity to work "hands on" with interesting real-world problems and phenomena, there is a movement toward maturation of attitudes and beliefs about

themselves as a STEM learner. This may be because PBL creates an environment where context, play, and space matter as diverse urban students create solutions to understand real-world problems and/or phenomena (Weisblat & McClellan, 2013). To this end, Wilson, Watson, Kaelin, and Huebner (2018) find that as traditionally underrepresented students engage with real-world problems like a scientist—developing hypotheses, testing them, and evaluating the scientific evidence—that their interest and achievement in scientific fields increases.

Likewise, Beckett and colleagues (2016) discover that low-income urban student engagement as measured by affective (the extent to which students liked, took interest in, and felt positive about the projects) and behavior participation (patterns of engaged or disengaged participation) is significantly enhanced when activities use digital devices during hands-on investigations of real-world projects. In this study, digital devices include (a) Texas Instruments Inspire calculator, (b) Livescribe pen and notebook, (c) laptop, (d) iPod Touch, (e) iPad, (f) digital camera, (g) video camera, and (h) various science probes. Further, in-depth learning using the 5-E Learning Cycle Model, as well as expert adoption of PBL instructional practices support technology integration as students become motivated to learn about their world.

Interestingly, Knezek et al. (2013) quasi-experimental study indicates that when students have the opportunity to participate in real-world activities—like environmental power monitoring activities—they not only improve in their creative tendencies but also their perceptions about STEM subjects and careers become more positive. In this study, the increase in STEM perceptions is more pronounced for underrepresented female middle school students than for male students; however, results also suggest that carefully designed PBL activities may be very effective at the middle school level for all students. Maybe this interest in a STEM career could be because traditionally underrepresented students have opportunities to engage with STEM content and processes that provide hands-on opportunities to connect with real-world activities (e.g., Mohr-Schroeder et al., 2014). Or maybe mentoring by experts in STEM could be another innovation to consider in conjunction with PBL to provide awareness to the possible careers in STEM (Redmond et al., 2011).

In sum, PBL engages students in STEM and this opportunity supports students to see themselves as a STEM learner. In turn, students then begin to consider careers in STEM fields. Yet, future research should consider the other supports and innovations (e.g., mentoring) that may impact students' interest in STEM, as well as pursuing a STEM career.

Research Question 3: Which Characteristics of PBL Support Students in STEM?

We also want to determine which characteristics of PBL support secondary students to engage in STEM subjects. Research literature (Cho & Brown, 2013; Hansen & Gonzalez, 2014; Lynch et al., 2017) provides guidance to which ele-

ments of PBL are essential to STEM learning for traditionally underserved urban students. Particularly salient is the fact that PBL needs to be ubiquitous and embraced by the entire learning community (Lynch et al., 2017). The essential elements of PBL are broad in scope, but include (1) having access to rigorous, authentic PBL opportunities, (2) the purposeful integration STEM disciplines and the use of technological tools, (3) the use of reform instructional strategies—instructional practices informed by research promoting active teaching and learning, (4) community partners to support 21st century skill development, (5) a positive school culture, and (6) dedicated, professional teachers. The following sections describe each of these essential elements in connection to the literature.

Authentic Real-World PBL Projects

Opportunities for authentic, real-world PBL have become a hallmark of STEM-focused schools. PBL immerses students in the process of inquiry and provides many opportunities to develop new knowledge and skills while solving problems (Cho & Brown, 2013; Lynch et al., 2017). Further, well-designed PBL supports the learning of rigorous subject area content (Han et al., 2016). In fact, Edmunds, Arshavsky, Glennie, Charles, and Rice (2016) find that rigorous PBL classrooms are not only marked by opportunities to learn content, but also provide opportunities for active engagement of both students and teachers.

Integrated PBL and Technology

Often, PBL supports deep learning through the integration of STEM disciplines. This purposeful integration of STEM disciplines affords students with opportunities to develop fundamental concepts and content-specific language, while increasing their academic achievement (Han et al., 2016). Similarly, Foutz et al. (2011) discover that when math and science are vertically and horizontally integrated, trends in student performance data (for a five-year period) show an increase on criterion-referenced competency tests in both math and science at the middle school levels, as well as on the end-of-course exams at the high school level. Further, Redmond and colleagues (2011) reveal that when using the engineering design process to enrich science and mathematics content, students' confidence and effort in science and mathematics is positively impacted and students' interest in engineering as a potential career is increased.

Two studies in this review examined technologies that support students in collecting data while in the field (Beckett et al., 2016; Jorgenson, 2018). First, in a mixed methods study, Beckett and colleagues (2016) investigate how the CincySTEM project, which includes the use of wireless weather stations and digital microscopes, support low-income African American students' engagement in STEM education. Findings from this study indicate that the extent to which students liked, took interest in, and felt positive about the projects, as well as the patterns of engaged or disengaged participation, is significantly enhanced when activities utilize digital devices during hands-on investigations of real-world proj-

ect activities. Likewise, Jorgenson (2018) describes sustainability challenges—environmental projects—that provide context for connecting student interests to STEM careers. Here, African American students learn new technical skills by using analytic tools similar to those used by STEM professionals as they develop action plans, use portable technologies (housed in backpacks) that allow students to collect data in the field, and create multi-media presentations to showcase their learning from the project. However, in both of these studies, stability of the PBL projects came into question as grant funding ends (Beckett et al., 2016) and corporate priorities shift (Jorgenson, 2018), thereby loosing important connections and supports needed for the continuation of projects.

In addition, integration of technology—as a tool to support problem solving—can play an important role in PBL beyond the collection of data and creation of products and presentations. For example, using technology (i.e., computers, calculators, or other machines) in math class positively correlates with student gains in math achievement for traditionally underserved students (Hansen & Gonzalez, 2014). Further, as students work on PBL activities that infuse technology applications, integration may become seamless when aligned with school's mission, culture, and learning environment (Lynch et al., 2017). From learning management systems, Google applications, to social networking applications, technology connects students and supports them in communicating, planning projects, and delegating tasks. In some cases, students even learn programming skills in support of the development of the projects that they are creating. Therefore, purposeful integration of technology facilitates student learning while preparing students to excel in an information-based and technology-advanced society (Sahin & Top, 2015).

Reform Instructional Strategies

Reform instructional strategies—teaching practices informed by research promoting active teaching and learning—provide rich opportunities for students to deeply engage in STEM discipline learning. Specifically, these reform instructional practices emphasize active teaching and learning while immersing students in STEM content, processes, habits of mind, and skills (Lynch et al., 2017). To this end, PBL begins with hands-on/minds-on instructional activities (Mohr-Schroeder et al., 2014; Mosley et al., 2010; Sahin & Top, 2015; Verma et al., 2011) where students collaborate (e.g., Cho & Brown, 2013; Lynch et al., 2017) to solve cognitively demanding, real-world problems. Students use a design process to creatively engage in solving a problem (Thompson, Varnado, & Matthew, 2010) while explaining or justifying thinking (Edmunds et al., 2016). As students complete their projects, growth is measured through performance-based assessment practices (Lynch et al., 2017) where students present and share their findings with classmates and visitors (Sahin & Top, 2015).

Community Partners Support the Development of 21st Century Skills

PBL has the potential to cultivate 21st century skills by integrating the workforce needs of local businesses and community organizations and the educational needs of students (Cho & Brown, 2013; Jorgenson, 2018). Students have the opportunity to think like professionals as the school environment reflects the collaborative nature of the workplace. Through active involvement of community partners, projects become authentic and grounded in real-world problems, shifting the role of students from being passive learners to becoming actively involved in projects while developing communication, collaboration, creativity, and critical thinking skills (Cho & Brown, 2013; Cwikla et al., 2014). Further, these partnerships may provide financial support, internship opportunities, mentors, and expert reviewers of projects, all while providing students exposure to their respective STEM fields and possible careers (Lynch et al., 2017).

A Positive School Culture

A positive school culture (Cho & Brown, 2013) is an essential element of PBL that usually results from each member of the learning community's ubiquitous commitment to PBL. This culture provides a sense of place, family, and home displayed by students, their families, faculty and staff; it is a place where students and faculty wanted to be (Lynch et al., 2017). In this culture, faculty and staff are committed to working towards enrolling a diverse student population and preparing them to excel in information-based and technologically advanced society. In turn, students take responsibility for own learning and typically take more STEM coursework of greater depth and breadth than state requirements. Further, the school provides supports (e.g., tutoring, special classes during and outside of school hours) for traditionally underrepresented students to strengthen their STEM content knowledge and skills in preparation for STEM college majors (Lynch et al., 2017). However, it is important to note that district policies sometimes run counter, and may impact school culture (Jorgenson, 2018). Specifically, the number of teachers assigned to a school, the adequacies of the building space, as well as the need to follow strict district pacing guides and accountability measures (state and district tests) may hinder the development of a positive school culture.

Dedicated and Professional Teachers

Highly qualified and dedicated teachers are another important element to the implementation of STEM PBL (Cho & Brown, 2015; Lynch et al., 2017). To this end, teachers possess advanced STEM content knowledge or have practical experience in STEM careers, and see their role in PBL as one that structures activities to motivate students, encourage reflection, and facilitate instruction through scaffolding (Lynch et al., 2017). Although students are charged with taking responsibility for their own learning, teachers are available to help with projects and to give frequent and timely feedback. Further, teachers model continuous improve-

ment for their students thereby creating a culture where assessment is not focused on right answers, but on learning and improved performance.

To this end, several studies show that it may be challenging for schools to recruit teachers who are passionate PBL and are willing to create this learning culture and learning environment for students (e.g., Cho & Brown, 2015; Jorgenson, 2018; Lynch et al., 2017). Further, studies suggest that novice teachers may need to be supported with professional development opportunities in order to take on the role as a facilitator of PBL experiences (e.g., Cho & Brown, 2015; Foutz et al., 2011).

In sum, access to authentic real-world PBL activities, learning that integrated STEM disciplines and technology tools, the use of reform instructional strategies, community partners to support 21st century skill development, a positive school culture, and dedicated, professional teachers are important elements that support students to be successful in STEM PBL.

Research Question 4: What Kinds of Supports do Teachers Need to Implement PBL in High-needs Schools?

As we reviewed the literature, we also wanted to determine the kind of supports teachers need to implement PBL in high-needs, urban schools. To this end, we realize that implementing STEM PBL might have some challenges. Many teachers value PBL, but feel tensions related time constraints, accountability, and personal paradigms (Capraro, et al., 2016; Han, Yalvac, Capraro, & Capraro, 2015). Specifically, teachers report that students' skills impact learning (e.g., difficulty reading impacts the learning of content), more time is needed to successfully dive into the projects, conflicts between students within their small groups, and that PBL is a significant and challenging instructional paradigm shift for students. Further, it is important to note that implementing ambitious curriculum, such as integrated PBL, requires relational trust and vulnerability since by nature, these types of interdisciplinary projects require teachers to reveal the limits of their content knowledge (Jorgenson, 2018). Therefore, teachers may need specialized and ongoing support to fully implement PBL within their STEM classrooms (e.g., Harris et al., 2015). We next discuss three supports that are present in the literature—short-term professional development, sustained professional development, and opportunities for teacher-teacher collaboration.

Short-Term Professional Development

Short-term workshops that typically take place during the summer are a hallmark of professional development (PD). These workshops are sponsored by an agency, school district, or university and are designed to provide teachers with an overview of a curriculum or instructional strategies that they could implement within their classrooms at a later date. For example, Foutz et al. (2011) lead a 5-day workshop to support teachers to develop PBL instructional strategies that integrated math and science using the discipline of agricultural engineering. How-

ever, teachers feel that the 5-day summer workshop is fast paced because they are not familiar with the farm subdivision theme. Further, they feel that they need more support to effectively integrate math and science within the grade level and discipline that they taught. Therefore, teachers may need sustained professional learning opportunities to effectively implement PBL with in their content areas, especially when ideas and content are new to them.

Sustained Professional Development

Studies by Capraro et al. (2016), Han et al. (2016), as well as Han et al. (2015) investigate the impact of sustained PD on classroom enactments of PBL. The intervention program for these studies is a well-structured PD program designed to train high school teachers in preparing PBL lessons and implementing them effectively in classrooms. To this end, teachers participated in 210 hours of PD over the course of three years. Specifically, teachers are required to attend 10 days of PD a year, 7 hours per day, and participate in follow up observations of PD components in their classrooms. The participating school district's curriculum specialists and supervisors are trained to use a PBL observation instrument so that they could provide feedback to teachers when the research team was not in the schools. Also, teachers and content specialists formed professional learning communities within each school to support incorporation of PBL content and pedagogical strategies. Over time, teachers perceive multiple benefits from implementation of PBL such as increased student engagement, real world problems making learning more meaningful, acknowledgement that PBL drew on a larger student skill set than traditional teaching, seeing students take ownership of learning and teachers became facilitators of learning, and improved relationships with students. Yet, teachers' understanding of PBL and positive perceptions of PBL did not guarantee the quality of implementation (Han et al., 2015).

Opportunities for Collaboration

One reason that these later studies (e.g., Capraro et al., 2016; Han et al., 2016) may have had an impact on teaching and learning is that they provide the opportunity for collaboration through professional learning communities. One teacher states, "Improvement in teaching strategies has occurred because of the collaboration between new and old teachers" (Capraro et al., 2016, p. 192). Further, the teachers feel that having time to plan together and communicate with each other also support teachers in adopting and refining the PBL curriculum while giving teachers time to collectively diagnose students' strengths and tailor instruction so that all students could find success as they interact with the various projects. Lynch et al. (2017) find that when professional development is content focused, while providing teachers with the opportunity to work with a highly supportive team of co-workers, that PBL becomes incorporated into the STEM disciplines.

CONCLUSIONS AND RECOMMEDNATIONS

STEM integration is an important consideration for PBL (National Academy of Engineering and National Research Council, 2014; Petrosino, Walkington, & Ekberg, in preparation), yet our decision to use "STEM education" as our search term, rather than "math education" or "science education" may have narrowed the articles that fell under consideration in this literature review. Despite this limitation, the literature reviewed on the effectiveness of PBL in secondary settings offers several insights into the current state of the research base on PBL. First, research has documented how PBL can have positive impacts on student achievement, including students from traditionally under-represented groups. However, one study suggests no achievement benefits for students who were low achieving or English Language Learners within PBL settings. Therefore, additional research is needed to better understand the impact of PBL for different student sub-groups, and to examine the impact of PBL on soft skills like collaboration and communication. Research also suggests PBL had relatively uniform positive effects on students' attitudes towards STEM.

We also examined which elements of PBL are key to supporting students to engage and be successful in PBL environments. We found rigorous, authentic projects, STEM integration and technology-enhanced instruction, reform instructional strategies, community partnerships, a school culture supporting PBL, and qualified, dedicated teachers were all essential elements to PBL's success. These elements may help explain why research on PBL is sometimes mixed, and can provide guidance to stakeholders on how to set teachers and schools up for success.

Finally, we looked at which supports teachers need to be successful in implementing PBL. Findings suggest the importance of sustained professional development and opportunities for collaborative professional learning communities. These findings can help schools and teachers understand the kinds of professional resources and environments that are needed to engage in a complex teaching endeavor like PBL.

Based on these findings, we offer the following recommendations for the UTeach community:

1. Provide UTeachers, as well as partner schools and districts, with information on research literature comparing PBL to other instructional approaches. To the degree evidence exists, make it clear how effects vary for different student groups and different settings and contexts. The articles reviewed in the present chapter represent a starting-off point for this literature base, as they highlight PBL in integrated STEM environments.
2. Acknowledge that like research on any instructional approach, results for PBL are not going to be uniformly positive—there are many contextual variables relating to how PBL is implemented and specific schools, districts, students, and situations, that will cause results to vary (in ad-

dition to methodological and statistical issues). However, the current research base on PBL in integrated secondary STEM environments mirrors broader research in K–12 mathematics and science generally suggesting that PBL can improve student learning, and often has profound effects on students' interests and attitudes.

3. Train UTeachers to understand what supports students need to have to be successful in PBL, as identified here and elsewhere. Discuss both elements that the teacher has more direct control of (content of project, STEM integration, technology, instructional strategies, community involvement) and elements that the teacher may have more limited control of (larger-scale community partnerships, access to highly specialized technology, collaborative atmosphere among teachers in different STEM fields, school culture, teachers' pre-existing qualifications).
4. Help UTeachers to understand how to become a professional teacher leader who works collaboratively with others to influence their school culture, school foci, and school agenda, in the service of enhancing student learning. Whether or not teachers choose to use their leadership roles to advocate for PBL specifically is not as important as them understanding how their schools work and how they can appropriately and realistically shape the environment they are working in. The development of these kinds of skills may not always be appropriate or timely for pre-service training, so continuing UTeach networks and professional development may be a more appropriate space for this work.
5. Communicate to UTeachers that implementing PBL in an effective way will be a lifelong learning pursuit for them as teachers—and encourage them to continue to get professional training on and support for PBL even after they leave the UTeach program. Also model for teachers in UTeach courses, and allow them to observe in the field, the importance of collaborative professional learning communities of teachers that are focused on the implementation of particular instructional strategies.

It has been over 20 years since the UTeach program was founded with PBL as one of its essential components. The expansion of UTeach has afforded the opportunity for teachers trained all over the United States to gain experience with planning and implementing STEM PBL units. Although the educational and research landscape have shifted considerably over the past two decades, PBL remains one of the most promising yet complex instructional strategies in a UTeacher's toolbox.

APPENDIX A: LITERATURE REVIEW SUMMARY

Author(s)	Year	Grade level	Sample size	Subject areas	High needs
Bass, Hu Dahl, & Panahandeh	2016	High school	2 administrators; 4 instructors; 12 students	3D animation and gaming	Low income underrepresented minority students; urban
Beckett, Hemmings, Maltbie, Wright, Sherman, & Sersion	2016	Grades 9 & 10	1097 students	Technology-infused science projects	Low income African-American students; urban
Capraro, Capraro, Scheurich, Jones, Morgan, Huggins, Corlu, Younes, & Han	2016	Grades 9, 10, 11, & 12	3801 students & science, technology, math teachers	Science, technology, or math	Diverse urban schools
Cho & Brown	2013	High school	11 teachers; 12 students; 2 parents; 4 administrators; 2 community partners; director of the Center for Teaching and Learning	All subjects	Urban
Cogger & Miley	2012	Middle school	Sample size not reported	Engineering with science	Urban
Cwikla, Milroy, Reider, & Skelton	2014	High school	2 university professors, 4 teachers, 85 students	Science	Not reported
Edmunds, Arshavsky, Glennie, Charles, & Rice	2016	High school	10 high schools	STEM	Underrepresented students
Erdogan, Navruz, Younes, & Capraro	2016	Grades 8, 10 and 11	3 high schools; 565 students	STEM	Urban
Foutz, Navarro, Hill, Thompson, Miller, & Riddleberger	2011	Middle and high school	Sample size not reported	Agriculture education to integrate science and mathematics	Urban, suburban, & rural students
Han, Capraro, & Capraro	2016	High school	528 students	STEM	Urban
Han, Yalvac, Capraro, & Capraro	2015	High school	5 teachers	STEM	Urban

Hansen & Gonzalez	2014	Middle school	Over 100,000 student-year observations for each year for each sample	Mathematics and science	Underrepresented students
Harris, Penuel, D'Angelo, DeBarger, Gallagher, Kennedy, Cheng, & Krajcik	2015	Grade 6	42 schools	Science	Urban
Jorgenson	2018	High school	7 teachers	STEM	Urban; African American; low income
Knezek, Christensen, Tyler-Wood, & Periathiruvadi	2013	Grades 6, 7, & 8	246 students	STEM	Underrepresented (female students)
Lynch, Spillane, House, Peters-Burton, Behrend, Ross, & Han	2017	High school	8 exemplar ISHSs	STEM	Diverse student populations
Mohr-Schroeder, Jackson, Miller, Walcott, Little, Speler, Schooler, & Schroeder	2014	Middle school	144 students	STEM	Underrepresented students
Mosley, Liu, Hargrove, & Doswell	2010	Grades 11 & 12	70 students	Robotics	Underrepresented students
Redmond, Scott, Thomas, Jordan, High, & Dockers	2011	Grades 6 & 7	1,287 students; 12 teachers	Science & mathematics	Underrepresented (female students)
Sahin & Top	2015	High school	19 students	STEM	Traditionally underserved students
Thompson, Varnado, & Matthews	2010	High school	27 students	Technology & engineering	Not reported
Verma, Dickerson, & McKinney	2011	Grades 8—12	60 teachers	Marine technology	Not reported
Weisblat & McClellan	2013	High school	The school	STEM	Urban
Wilson, Watson, Kaelin, & Huebner	2018	High school	984 students	Biomedical sciences	Underrepresented students

REFERENCES

Bass, K. M., Hu Dahl, I., & Panahandeh, S. (2016). Designing the game: How a project-based media production program approaches STEAM career readiness for under-represented young adults. *Journal of Science Education and Technology*, *25*(6), 1009–1024. https://doi.org/10.1007/s10956-016-9631-7

Beckett, G. H., Hemmings, A., Maltbie, C., Wright, K., Sherman, M., & Sersion, B. (2016). Urban high school student engagement through CincySTEM iTEST projects. *Journal of Science Education and Technology*, *25*(6), 995–1007. http://doi.org/10.1007/s10956-016-9640-6

Capraro, R. M., Capraro, M. M., Scheurich, J. J., Jones, M., Morgan, J., Huggins, K. S., Corlu, M. S., Younes, R., & Han, S. (2016). Impact of sustained professional development in STEM on outcome measures in a diverse urban district. *Journal of Educational Research*, *109*(2), 181–196. http://doi.org/10.1080/00220671.2014.936997

Cho, Y., & Brown, C. (2013). Project-based learning in education: Integrating business needs and student learning. *European Journal of Training and Development*, *37*(8), 744–765. http://doi.org/10.1108/EJTD-01-2013-0006

Cogger, S. D., & Miley, D. H. (2012). Model wind turbine design in a project-based middle school engineering curriculum built on state frameworks. *Advances in Engineering Education*, *3*(2), 1–23. Retrieved from: http://www.scopus.com/inward/record.url?eid=2-s2.0-84872450280&partnerID=40&md5=585e94804872c0f48dc0ab3d35c2d50a

Cwikla, J., Milroy, S., Reider, D., & Skelton, T. (2014). Pioneering Mars. *The American Biology Teacher*, *76*(5), 300–305. http://doi.org/10.1525/abt.2014.76.5.2

Edmunds, J., Arshavsky, N., Glennie, E., Charles, K., & Rice, O. (2016). The relationship between project-based learning and rigor in STEM-focused high schools. *The Interdisciplinary Journal of Problem-Based Learning*, *11*(1), 1–6. http://doi.org/10.7771/1541-5015.1618

Erdogan, N., Navruz, B., Younes, R., & Capraro, R. M. (2016). Viewing how STEM project-based learning influences students' science achievement through the implementation lens: A latent growth modeling. *EURASIA Journal of Mathematics, Science & Technology Education*, *12*(8), 2139–2154. http://doi.org/10.12973/eurasia.2016.1294a

Foutz, T., Navarro, M., Hill, R. B., Thompson, S. A., Miller, K., & Riddleberger, D. (2011). Using the discipline of agricultural engineering to integrate math and science. *Journal of STEM Education: Innovations and Research*, *12*(1&2), 24–32. Retrieved from: http://search.ebscohost.com

Haberman, K. (1991). The pedagogy of poverty versus good teaching. *Phi Delta Kappan, 73*(4), 290–294.

Han, S., Capraro, R. M., & Capraro, M. M. (2016). How science, technology, engineering, and mathematics project based learning affects high-need students in the U.S. *Learning and Individual Differences*, *51*, 157–166. http://doi.org/10.1016/j.lindif.2016.08.045

Han, S., Yalvac, B., Capraro, M. M., & Capraro, R. M. (2015). In-service teachers' implementation and understanding of STEM project based learning. *Eurasia Journal of Mathematics, Science and Technology Education*, *11*(1), 63–76. https://doi.org/10.12973/eurasia.2015.1306a

Hansen, M., & Gonzalez, T. (2014). Investigating the relationship between STEM learning principles and student achievement in math and science. *American Journal of Education*, *120*(2), 139–171. http://doi.org/10.1086/674376

Harris, C. J., Penuel, W. R., D'Angelo, C. M., DeBarger, A. H., Gallagher, L. P., Kennedy, C. A., Cheng, B.H., & Krajcik, J. S. (2015). Impact of project-based curriculum materials on student learning in science: Results of a randomized controlled trial. *Journal of Research in Science Teaching*, *52*(10), 1362–1385.

Jorgenson, S. (2018). Bridges and boundaries to power: How teachers used project-based learning to design a radically inclusive STEM high school. *Critical Education*, *9*(16), 17–39.

Knezek, G., Christensen, R., Tyler-Wood, T., & Periathiruvadi, S. (2013). Impact of environmental power monitoring activities on middle school student perceptions of STEM. *Science Education International*, *24*(1), 98–123.

Krajcik, J. S., & Blumenfeld, P. C. (2006). Project-based learning. In K. Sawyer (Ed.) *Cambridge handbook of the learning sciences*. Cambridge University Press.

Larmer, J., Mergendoller, J. R., & Boss, S. (2015). *Setting the standard for project based learning*. Alexandria, VA: ASCD. 240 pp.

Lynch, S. J., Spillane, N., House, A., Peters-Burton, E., Behrend, T., Ross, K. M., & Han, E. M. (2017). A policy-relevant instrumental case study of an inclusive STEM-focused high school: Manor New Tech High. *International Journal of Education in Mathematics, Science and Technology*, *5*(1), 1–20. http://doi.org/10.18404/ijemst.75656

Marshall, J. A., Petrosino, A. J., & Martin, T. (2010). Preservice teachers' conceptions and enactments of project-based instruction. *Journal of Science Education and Technology*, *19*(4), 370–386.

Mohr-Schroeder, M. J., Jackson, C., Miller, M., Walcott, B., Little, D. L., Speler, L., Schooler, W., & Schroeder, D. C. (2014). Developing middle school students' interests in STEM via summer learning experiences: See blue STEM camp. *School Science and Mathematics*, *114*(6), 291–301. http://doi.org/10.1111/ssm.12079

Mosley, P. H., Liu, Y., Hargrove, S. K., & Doswell, J. T. (2010). A pre-engineering program using robots to attract underrepresented high school and community college students. *Journal of STEM Education: Innovations and Research*, *11*(5), 44–54. Retrieved from: http://eric.ed.gov/?q=robotic+&ff1=dtySince_2006&ff2=subRobotics&pg=11&id=EJ907980

National Academy of Engineering and National Research Council. (2014). *STEM integration in K–12 education: Status, prospects, and an agenda for research*. Washington, DC: The National Academies Press. https://doi.org/10.17226/18612

Petrosino, A., Walkington, C., & Ekberg, D. (in preparation). *Frameworks for integrated project-based instruction in STEM disciplines*. Information Age Publishing, Charlotte, North Carolina.

Redmond, A., Scott, M., Thomas, J., Jordan, P., High, K., & Dockers, J. (2011). Enriching science and math through engineering. *School*, *11*, 399–408. http://doi.org/10.1111/j.1949-8594.2011.00105.x

Sahin, A., & Top, N. (2015). STEM Students on the stage (SOS): Promoting student voice and choice in STEM education through an interdisciplinary, standards-focused, project based learning approach. *Journal of STEM Education*, (September), 24–34. Retrieved from: http://jstem.org/index.php?journal=JSTEM&page=article&op=viewFile&path[]=1911&path[]=1683

Thompson, B., Varnado, T., & Matthews, B. (2010). Design without make: A feasible direction for American technology education. *I-Manager's Journal of Educational Technology*, *7*(1), 19–29.

UTeach Institute. (2015). *Graduates of UTeach STEM teacher preparation programs nationwide.* Retrieved from: https://uteach.utexas.edu/sites/default/files/grad-report-may-2015-final.pdf

Verma, A. K., Dickerson, D., & McKinney, S. (2011). Engaging students in STEM careers with project-based learning—MarineTech Project. *Technology and Engineering Teacher*, *71*(1), 25–31.

Wasserman, N. H., & Walkington, C. (2014). Exploring links between beginning UTeachers' beliefs and observed classroom practices. *Teacher Education and Practice*, *27*(2/3), 376–401.

Weisblat, G. & McClellan, J. (2013). Through the lens of the students: Using narrative inquiry to evaluate an innovative urban high school. *Journal of Urban Learning, Teaching, and Research, 9*, 138–148.

Wilson, R. T., Watson, E., Kaelin, M., & Huebner, W. (2018). Early preparation and inspiration for STEM careers: Preliminary report of the epidemiology challenge randomize intervention, 2014–2015. *Public Health Reports*, *133*(1), 64–74.

Young, M. (2013). Overcoming the crisis in curriculum theory: A knowledge-based approach. *Journal of Curriculum Studies, 45*(2), 101–118.

CHAPTER 19

VCAST LEARNING MODULES

A Functions & Modeling Course Innovation[1]

Laurie O. Cavey, Jason Libberton, Tatia Totorica, Michele Carney, and Patrick R. Lowenthal

The VCAST learning modules are a direct response to calls for teacher preparation programs to improve the mathematical preparation of future secondary teachers and to provide teacher candidates experience with anticipating the different ways that students make sense of mathematics. Using an iterative design-based approach, we developed four video-based learning modules that highlight a range of ways secondary students think about fundamental ideas related to functions and mathematical modeling. While the four modules were developed specifically for use in the UTeach Functions & Modeling course, they can also be integrated into other teacher preparation courses and professional development settings. In this chapter, we describe our design principles for module development and the content focus for each of the modules, and make suggestions for module integration into the Functions & Modeling course.

Over the last twenty years, there has been widespread interest in supporting classroom teachers' ability to attend to students' thinking (Boaler & Staples, 2008; Santagata & Bray, 2016; Star & Strickland, 2008). When teachers attend to students' thinking in ways that progressively develop students' understand-

[1] This research was supported in part by grant #1726543 Preparing Secondary Mathematics Teachers with Video Cases of Students' Functional Reasoning from the National Science Foundation.

Preparing STEM Teachers: The UTeach Replication Model, pages 259–275.

ing of mathematics (Freudenthal, 1973, 1991; Gravemeijer & van Galen, 2003), students' conceptual understanding of mathematics is improved, procedural fluency is maintained (e.g., Fennema et al., 1996), and broader access to mathematical ideas is achieved (Boaler, 2008). As a result, mathematics teacher education programs are encouraged to create opportunities for teacher candidates to learn about student thinking (Association of Mathematics Teacher Educators, 2017; Ball, Thames, & Phelps, 2008). Because developing teacher candidates' knowledge of student thinking through field-based work alone presents limitations (Ko, Milewski, & Herbst, 2017), finding alternative ways to build this knowledge and capacity in candidates is warranted.

One promising approach for engaging teacher candidates in attending to student thinking is the use of video cases (Borko, Jacobs, Eiteljorg, & Pittman, 2008; Friel & Carboni, 2000; Lin, 2005; Llinares & Valls, 2009, 2010). For example, Seago, Mumme, and Branca (2004) developed video cases of high school mathematics teachers, that included lesson materials along with videos and transcripts of classroom interactions, to support teachers' mathematical understanding and pedagogical skills for teaching linear functions. A common finding associated with the use of video-based instructional interventions, in general, is the significant role that organizational frameworks play in supporting teachers' and teacher candidates' abilities to analyze classroom video (Alsawaie & Alghazo, 2010; Santagata, Zannoni, & Stigler, 2007; Star, & Strickland, 2008).

The purpose of the Video Case Analysis of Student Thinking (VCAST) project is to develop an instructional intervention that consists of a series of online video-based modules focused on video cases of individual students solving quantitative reasoning tasks for use in the UTeach Functions & Modeling course. Each VCAST module features secondary students' work on a task and is comprised of both online components and a face-to-face component. Through careful selection and sequencing of video clips and written artifacts, modules are designed to provide candidates opportunities to analyze a range of secondary students' thinking for a task and in turn support secondary mathematics teacher candidates' attentiveness to student mathematical thinking (Carney, Cavey, & Hughes, 2017). In this chapter, we describe the overarching design principles for module development, the primary content focus for each of the four modules, how the modules can be integrated into mathematics courses for secondary mathematics teacher candidates (e.g., the UTeach Functions & Modeling course), and plans for continued design and development.

MODULE DESIGN & DEVELOPMENT

As a design and development project, the modules have been improved through an iterative cycle of data analysis, framework development, and examination of technological design affordances and constraints. The module design principles described here reflect the most recent cycle of development, after the second year of implementation. Our remarks are restricted to a brief description of *attentiveness*, the theoretical lens driving the structural aspects of module design, the related design principles for module development, and other influences on methods

for capturing and representing secondary students' work. For a description of the initial module development process, see Cavey, Totorica, Carney, Lowenthal, and Libberton (2018).

Attentiveness

Attentiveness builds upon ideas from professional noticing (Jacobs, Lamb, & Philipp, 2010), mathematical knowledge for teaching (Ball et al., 2008; Shulman, 1987), and progressive formalization (Freudenthal, 1973; Gravemeijer & van Galen, 2003; Treffers, 1987) and is defined as a teacher's ability to analyze and respond to a particular student's mathematical ideas (Carney, Cavey, & Hughes, 2017). It most closely parallels components from professional noticing, a set of interrelated skills that involve attending, interpreting, and responding to student ideas in the classroom setting (Jacobs et al., 2010), but differs in that its focus is on the analysis, interpretation, and response to the mathematical work of an individual student.

Attending to the mathematical work of an individual student requires three interrelated teacher actions: (1) examine the student's work in relation to the mathematical intent of the task (i.e., the mathematical ideas the task is designed to elicit), (2) situate the student's work within a larger progression of student understanding, and (3) support the student in extending their thinking (e.g., generalize or revise their ideas). In the next section, we describe how each of the three teacher actions have influenced module design.

Module Design Principles

The attentiveness teacher actions described above have different degrees of influence on VCAST module design. The first teacher action (i.e., examine the student's work in relation to the mathematical intent of the task) has the most direct and extensive influence on module design. In each module, video cases are carefully composed to reveal students' approaches and understandings in ways that make it possible for candidates to begin making connections between important features of the task and the elicited student understanding. The second teacher action (i.e., situate the student's work within a larger progression of student understanding for the topic) is indirectly addressed through the selection of video cases that reveal a range of students' approaches and understandings for the task. With these selections, we aim to support candidates' knowledge of the different ways that students might productively work on the task and the different conceptual challenges that students might face. The third teacher action (i.e., support the student in extending their thinking) is indirectly addressed across modules through the selection of video cases that provide different examples of how students might generalize or revise their thinking for different quantitative reasoning tasks.

In general, our goal with this project is to focus module content on the examination of students' mathematical work in relation to a particular task. That is, we purposefully do not address pedagogical practices per se. Our focus on

mathematics is due, in part, to the fact that the modules are intended to be used in mathematics courses. Also, the module content in the first iteration of this project was too broad; we discovered that taking such a broad focus compromised the depth of candidate engagement with the important mathematical ideas. Thus, we focus module content on the analysis of a range of students' ideas, along with examples of how those ideas might be advanced over time, stopping short of engaging candidates in thinking about pedagogical implications. Since most candidates take the Functions & Modeling course after completing Classroom Interactions (a sophomore level UTeach course designed to engage candidates in considering pedagogical principles for inclusive classroom discursive practices), candidates appreciate the value of understanding a range of ways that students might approach and think about a particular mathematical task. Table 19.1 outlines the connections between the three attentiveness teacher actions and the corresponding module design principles.

A number of other factors have also influenced module design, including interest in representing a diverse student population, collecting authentic student work samples from secondary students, and the technological affordances and constraints encountered using different online platforms. For instance, video cases across the modules are intended to represent a diverse student population, providing variance across grade level, gender, race, socioeconomic level, location (urban vs. rural), and formality of mathematical language used. It was important to us that the students featured in the modules not only look like the students that candidates will encounter once they are classroom teachers, but also represent what students say and do when solving quantitative reasoning tasks. Further, our interest in student work samples generated by secondary students makes it possible to convincingly demonstrate how students' informal ways of reasoning about a problem can be very sophisticated, at times more advanced than a student with formal instruction on the topic.

TABLE 19.1. Attentiveness Actions and Module Design Principles

Attentiveness Actions	Module Design Principles
1) Examine student work in relation to the mathematical intent of the task	Within each module, questions focus on identifying student approaches, inferring student reasoning, and thinking about the relationship between task features and the elicited student reasoning.
2) Situate the student's work within a larger progression of student understanding for the topic	Within each module, cases represent a range of student approaches and reasoning with the module task.
3) Support the student in extending their thinking	Across modules, cases provide a range of examples of students generalizing or revising their ideas.

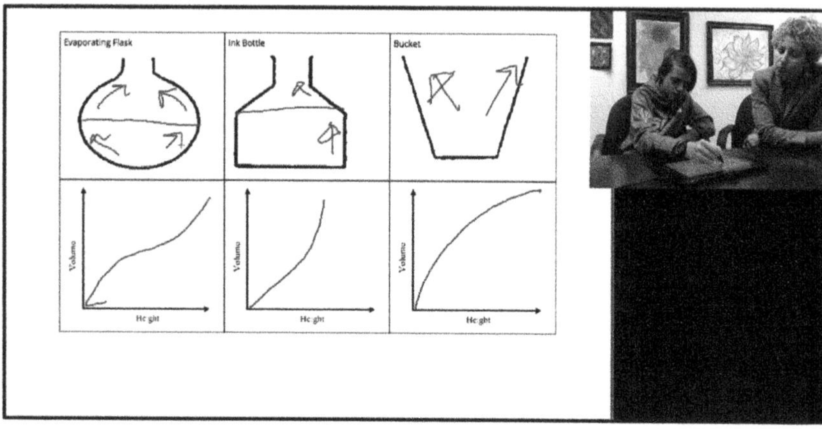

FIGURE 19.1. Screenshot of Candidates' View of Student Work

Student Work Samples

The video-based student work samples for this project were generated through an iterative four-step process. First, we selected mathematical tasks that were both accessible to a full range of learners and had the potential to elicit a wide variety of student approaches. Second, we conducted and recorded one-on-one interviews with secondary students as they worked through the tasks. These interviews took place at four different schools in three districts, and at three different times over an 18 month period. Third, we used the generated data to analyze the range of student strategies and thinking elicited by the task. Last, we used our module design principles to make selections of specific video segments to feature in each module. We also used the first two rounds of interviews to test multiple video recording approaches for capturing students' work. During the third round of interviews, we used a Cintiq tablet with TechSmith Camtasia to capture student work and an external webcam to record video of the teacher and student interactions. Doing so has significantly improved the video production, as the tablet captures what the student writes while the webcam simultaneously captures what the student is saying as they work on the task. Figure 19.1 is a screenshot of what a candidate sees when presented with a video of a middle school student working on the third module's Bottle Filling Task.

MODULE MATHEMATICS & CANDIDATE ENGAGEMENT

Collectively, the modules address a range of key mathematical ideas, including coupling two quantities graphically, working in discrete versus continuous contexts, reasoning with explicit and recursive representations of functions, covariational reasoning, and the role of variance in mathematical modeling. Each module

TABLE 19.2. Summary of Module Mathematics and Candidate Engagement

Module	Mathematical Ideas	Aspects of Candidate Engagement
1	Coupling Two Quantities Graphically	• Complete the Bus Stop Task[1] and describe their own solution and reasoning. • Make and revise predictions about task difficulty. • Observe mathematical challenges students encounter with identifying and comparing quantities represented by points in the Cartesian coordinate plane.
2	Explicit and Recursive Representations of Functions	• Complete the Hexagon Task[2] and describe their own solution and reasoning. • Make and revise predictions about students' approaches. • Encounter a range of student approaches for generalizing relationships associated with figural patterns.
3	Covariational Reasoning	• Complete the Bottle Filling Task[3] and describe their own solution and reasoning. • Make and revise predictions about how students communicate their covariational reasoning graphically. • Compare the ways students communicate their conception of how the bottles' shapes influence how volume and height change together.
4	Variance in Mathematical Modeling	• Complete the Aphids Task and describe their own solution and reasoning. • Anticipate a range of students' approaches to the task before observing video cases. • Compare students' ideas about the meaning of a function model for data.

[1] Adapted from Swan (1985).
[2] Adapted from Hendrickson, Honey, Juehl, Lemon, & Sutorios (2012).
[3] Adapted from Swan (1985).

is designed to focus candidates' attention on specific mathematical ideas and to highlight aspects of student reasoning in relation to the associated mathematical task. The key mathematical idea for each module, along with aspects of candidate engagement are summarized in Table 19.2.

Module 1—The Bus Stop Task: Coupling Two Quantities Graphically

Module 1 engages candidates in observing mathematical challenges students encounter when values of two quantities are represented as discrete points in the Cartesian coordinate plane. The act of coupling two quantities graphically refers to the ability to recognize that a single point simultaneously refers to a specific value for each of the two quantities. This idea is essential to developing a conceptual understanding of interpreting graphical information. In the Bus Stop Task (See Figure 19.2), students match images of people waiting in line at a bus stop

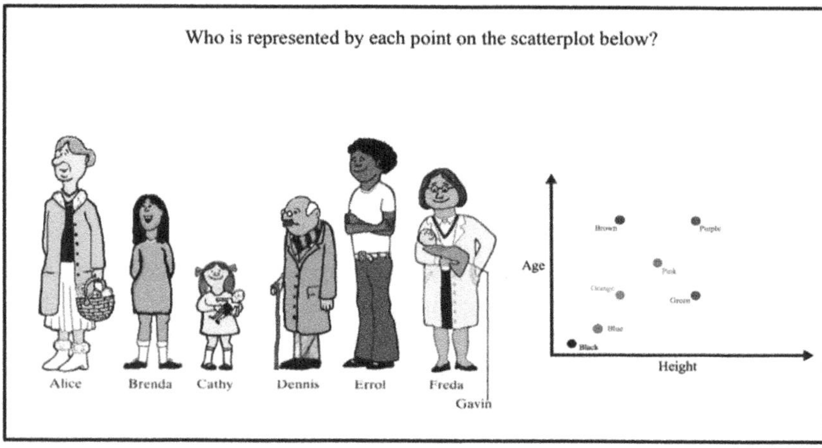

FIGURE 19.2. The Bus Stop Task

with corresponding points in a graph based on the relationally specified quantities height and age, where height is counterintuitively placed along the horizontal axis. In this module, candidates observe conceptual challenges such as interpreting the point labels, attending to both quantities, maintaining axes assignments for height and age, and drawing contextually-based conclusions about points.

Candidates encounter examples of students who approach this task by comparing the people in line only using one quantity, either height or age. Given the visual representation of the people waiting in line, many students naturally notice height and begin to match people's relative heights to points without paying attention to age at all. Contrasting this type of approach with students who account for both quantities (height and age) when matching each point gives candidates an opportunity to think more deeply about what it means to couple two quantities graphically. Video clips also show students who begin matching people to points using only one quantity until they come to two points that share a value for that quantity. As students attempt to use the second quantity to "break the tie," candidates hypothesize how this situation relates to coupling two quantities graphically. In the face-to-face component, candidates are given a set of student quotes to interpret, such as the following dialogue where a student explains to the interviewer the meaning of two vertically-aligned points.

Interviewer: Okay. So if you were going to explain to a couple of your classmates what's going on here with the brown and orange points, what would you say?

Student: Well, there's an age difference (points to age axis), but not a height difference (points to height axis). That's why they're in

the same area (creates vertical segment between the points with fingers). Um, let's see here.

Interviewer: Do you remember who they were?

Student: (Gestures between Brenda and Dennis.) Because she's definitely younger (points to Brenda) ... she doesn't have to use a cane and they are the same height. So agewise, it's not the same but heightwise it's the same.

Candidates work in small groups analyzing, categorizing, and discussing a range of quotes from student work on the Bus Stop Task. By the end of Module 1, we aim for candidates to (1) recognize that students sometimes shift in their approach or understanding as they work through an open-ended task and (2) gain appreciation for how complex the Bus Stop Task is for students learning how to interpret graphed points in context.

Module 2—The Hexagon Task: Explicit and Recursive Representations of Functions

In Module 2, candidates encounter a range of student approaches for generalizing relationships associated with the figural pattern in the Hexagon Task (See Figure 19.3). In this context, explicit formulas (i.e., functions) can arise in several ways. For instance, when trying to determine the perimeter of a figure based upon the figure number, one might generate the formula where p is the perimeter and n represents the figure number. Explicit formulas can also arise when either determining the perimeter based upon the number of hexagons or the number of hexagons based on the figure number. Interestingly, generating an explicit formula can be done either by making connections to the contextual setting (i.e., the figures) or based upon numerical observations devoid of connections to the figures. By making assumptions about how the characteristics of the figures change from one figure to the next figure in the sequence, recursive equations can be written to relate the perimeter of one figure in the sequence to the perimeter of the next figure, or the number of hexagons in one figure of the sequence to the number of hexagons in the next figure.

Early in Module 2, candidates watch partial clips of students explaining their approaches, then make hypotheses about how each student might proceed. As

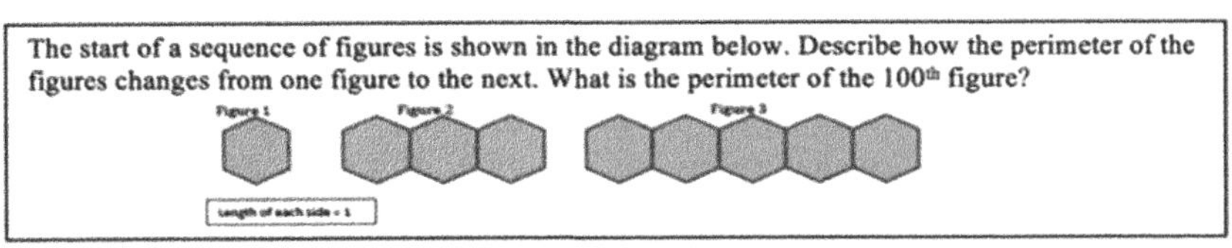

FIGURE 19.3. The Hexagon Task

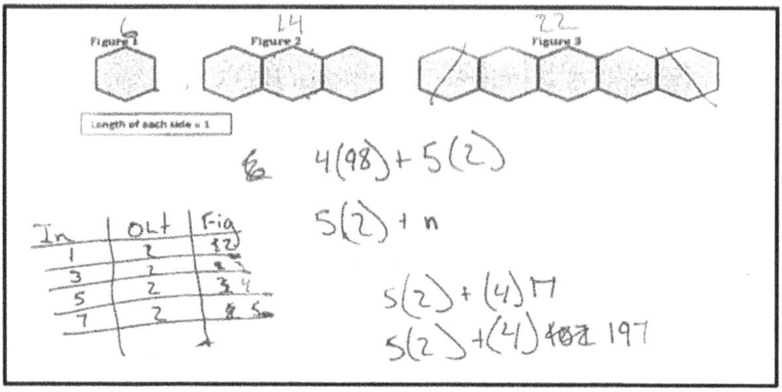

FIGURE 19.4. Sample Student Written Work

they watch the remainder of the clips, candidates explore the depth of reasoning that each student uses in relation to the type of representation (explicit or recursive) being used. They also make comparisons between the two types of representation and the associated reasoning. Later, in the face-to-face component, candidates sort additional examples of student work into categories based on students' approaches.

Candidates recognize, for example, that the student's work in Figure 19.4 illustrates an approach that attends to patterns in the arrangement of the hexagons in each figure to generate a formula to determine the perimeter based upon the number of hexagons. Notice the student appears to have initially assumed that the 100th figure would have 100 hexagons (a common mistake) but then to have later revised this number. By sorting through similar examples of student work, we hope that candidates will recognize that (1) there are a variety of productive approaches that can be used to solve the Hexagon Task that may differ from the candidate's own approach and (2) even students who demonstrate sound reasoning can make errors as they translate their reasoning into "traditional" mathematical expressions.

Module 3—The Bottle Filling Task: Covariational Reasoning

Covariational reasoning is the ability to coordinate the relationship between two quantities that simultaneously vary (Thompson & Carlson, 2017). It provides an alternative perspective on function to the traditionally-taught correspondence relationship between two quantities. Rather than looking at how one quantity can be generated from the other, covariational reasoning entails noticing how quantities are related and, in particular, how changes in one quantity are related to changes in the other quantity. Candidates engage with covariational reasoning, the

key mathematical idea of Module 3, as they view students working on the Bottle Filling Task (Figures 19.1 and 19.5 show student work on the task).

In this task, students are given 2-dimensional drawings of three bottles and asked to create a graph relating the volume of the water in the bottle to the height of the water (with height mapped onto the horizontal axis) as they imagine each bottle being filled with a liquid. One of the common strategies students use to successfully work on tasks similar to the Bottle Filling Task is to conflate one quantity with time (Johnson, McClintock, & Hornbein, 2017). This can be particularly efficient and productive when height is mapped to the vertical axis. To purposefully attempt to elicit covariational reasoning, we asked the students we interviewed to think of height along the horizontal axes.

In Module 3, candidates view video clips demonstrating a variety of student strategies for communicating their covariational reasoning about how the features of each bottle influence the relationship between volume and height of the water. For example, they encounter a student who communicates his ideas with concrete measurements of the quantities, using an unconventional scaling approach to translate his ideas to a graph and middle and high school students who are able to visualize a continuous change in both quantities simultaneously as they imagine one of the bottles filling with a liquid. They also encounter students who communicate their ideas by marking off a bottle into several increments, viewing the changes in quantities as happening in stages. Students using this approach may be thinking of one variable changing at a constant rate (i.e., height) and paying attention to how changes in volume change. During the face-to-face component, candidates engage in a matching activity, where they are given incomplete work for three students (one of the three graphs is missing) along with a set of possible matching graphs. With each proposed match, candidates are asked to provide a sensible explanation for how the given student work supports their selection (See Figure 19.5).

As candidates consider the set of possible graphs shown in Figure 19.5, they may match Card 2 based upon how the curvature of the graph mimics the curvature of the bottle (thinking Student A was using shape-based reasoning), while also considering Card 8 as a potential match (thinking Student A was accounting for all the bottle features based on their success with the constant rate of change sections in the ink bottle), and conclude that Card 6 is not a match (thinking Student A would not use a linear graph when the sides of the bottle are not linear). By doing so, candidates consider how students can have a variety of approaches which result in similar graphs. Towards the end of this activity, candidates learn which graphs the students actually drew. In this case, they learn that Student A drew the graph on Card 2, revealing that the student may have actually been mimicking the shapes of the bottles when sketching their graphs for the bucket and flask. By the end of this module, we aim for candidates to (1) appreciate that students with limited exposure to formal covariational relationships and a lack of

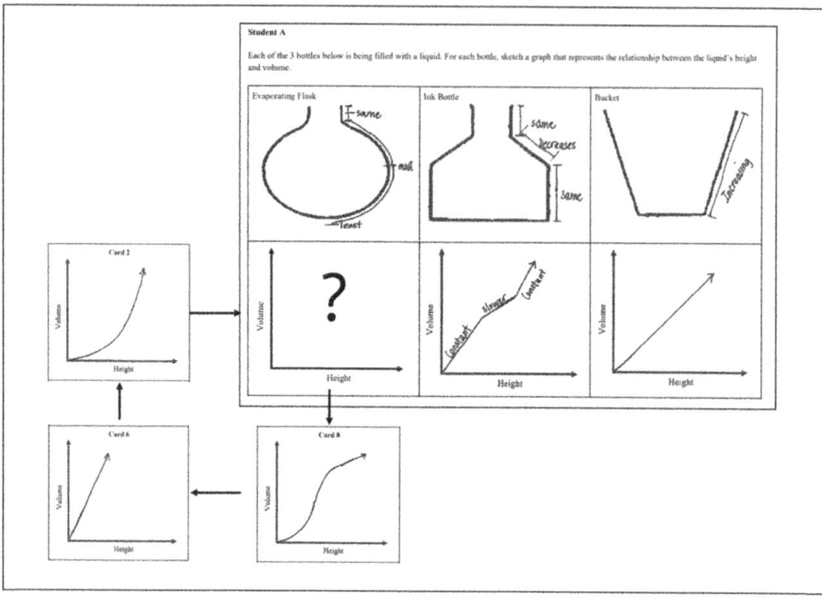

FIGURE 19.5. Example from the Bottle Filling Task MATCHING ACTIVITY

formal language can still reason productively and bring an advanced understanding to the task and (2) develop their own covariational reasoning skills.

Module 4—The Aphids Task: Variance in Mathematical Modeling

Variance is a key mathematical idea in mathematical modeling. In particular, variance plays a role in creating, interpreting, and using mathematical models for real world data. Before creating a mathematical model for a set of bivariate data, it is important to first consider whether there is an overall trend in the data, and then, if a pattern is discovered, to next consider how closely the data are scattered around that pattern. Data that do exhibit a trend are not expected to follow the path of a function model perfectly; rather, a function model is simply one way to capture and perhaps explain the trend in the data. In this case, the function outputs are merely estimates of the associated quantity.

On the other hand, if the data are too scattered and do not follow a trend, there is likely no value in creating a function model. Candidates engage with these ideas as they complete Module 4, which is structured around ways that students complete the Aphids Task (See Figure 19.6). In this task, bivariate data (*days, number of aphids*) are presented in a table and then in a graph. Students are asked to (1) describe what they notice, (2) describe what type of equation they might use to model the data, and (3) extrapolate a prediction just beyond the data given.

Aphids are very small insects that sometimes accumulate on and harm garden plants.

Todd started noticing a problem with aphids on his tomato plants last spring. Todd decided to count the number of aphids on one of his plants each day to see if he could detect a pattern in the number of aphids. Below are the data he collected over 20 days.

1. Take a look at the data collected by Todd. What do you notice about the number of aphids?

Day	1	2	3	4	5	6	7	8	9	10	11	12	13	14	15	16	17	18	19	20
Number of Aphids	15	16	21	20	24	27	28	35	36	41	43	50	57	62	68	79	85	93	100	111

2. If you were going to model the number of aphids as the days go on with an equation, what type of equation would you use? Can you sketch a graph of what it would look like?

3. What can you say about the number of aphids on a tomato plant on day 21?

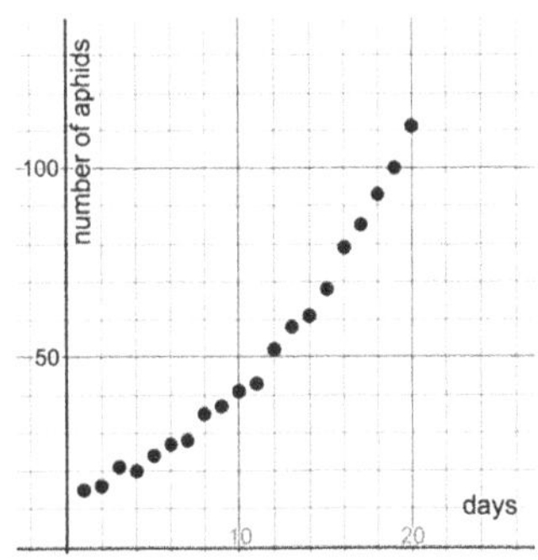

FIGURE 19.6. The Aphids Task

Throughout Module 4, candidates view video clips of students grappling with each type of variation mentioned above. For instance, candidates encounter students who started examining the data by determining the change in the number of aphids between each pair of successive days. This approach led to students making claims about the data being non-linear, possibly exponential, or being random. Candidates then observe what students do when asked to think about an equation that could be used to model the data. One student claims that the aphids data cannot be modeled with an equation because there is no equation that will "go through all of the points." Another student demonstrates a similar expectation for function models by drawing their model to pass through each point, while yet another student demonstrates understanding by stating that they know a correct model will not perfectly match each real-world value. Candidates then observe the different approaches students use when asked to make a prediction for day 21. Some students referred to the table of data to estimate the "jump" from day 20 to day 21, while others judged based on where their drawn model appeared to be intersecting the grid marks on the graph. During the face-to-face component, candidates sort a sample of student responses to the prediction question based on the inferred understanding for each student. Below is a partial justification given by one of the students highlighted in this sort.

Student: I would say 116. About. Just because the numbers aren't really reliable that you can make an actual exact number. So I would say about 116. Because there seems there was a jump from 4 to 7 each time, about on average so I would say 5 is kind of the middle.

In this instance, candidates observe that the student used an informal idea for average rate of change to estimate the increase in aphids from day 20 to day 21. By the end of Module 4, our aim is for candidates to appreciate the three distinct roles described here that variance plays in mathematical modeling.

MODULE INTEGRATION

The modules are designed to be used in a variety of contexts and formats. While we are specifically interested in using them in undergraduate mathematics courses for secondary teacher candidates with the intent that the online component of each module is completed before engaging in the face-to-face component (i.e., in a Flipped Classroom approach), the modules can also be used in other courses and formats, such as in an undergraduate capstone course or in a completely online format. The modules are embedded in the Functions & Modeling course (an upper-division mathematics course for future secondary mathematics teachers) and will be piloted at multiple universities in 2019 and 2020. The modules have also been successfully piloted with classroom teachers participating in a hybrid-format professional development program. In this section, we first address how one might think about integrating the modules in courses for secondary teacher candidates based upon mathematical ideas that are closely related to module content, and then describe logistical aspects of integration.

Module-Related Mathematical Ideas

The two content areas in which the modules most aptly fit in the Functions and Modeling curriculum are *Regressions in Modeling* and *Functions, Rates, and Patterns*. Secondary mathematics teacher education programs that include a capstone course could integrate Modules 1 and 3 with units on functions, including parametric functions, integrate Module 2 with the study of mathematical induction, and integrate Module 4 with a unit on the use of functions as mathematical models. The modules are sequenced based on the difficulty level of module task and the associated key mathematical ideas.

Module 1 is an ideal complement to addressing fundamental ideas about representing and interpreting bivariate data and functions. In our Functions & Modeling course, we use the first two weeks to address a variety of definitions of functions, the use of and connections between different function representations (e.g., verbal, symbolic, graphical, numeric), and characteristics of function families (e.g., linear, quadratic, exponential, sinusoidal). All of this work is done as a

precursor to using characteristics of function families to build function models for bivariate data.

Module 2 is an ideal complement to understanding the connections between recursive and explicit representations of functions, particularly in the context of making generalizations about figural pattern tasks. This module can also be used to enhance candidate understanding of proof by mathematical induction, seeing that the recursive definitions (which includes the starting value of the sequence) provide the information needed to generate a formal proof for the explicit formula. More generally, this module can be used to support candidates in making distinctions between inductive and deductive reasoning.

Module 3 is an ideal complement to the study of parametric functions. In our Functions & Modeling course, after the study of trigonometric models, candidates experiment with writing parametric equations for real world phenomena, including the path of someone riding a Ferris wheel. An important mathematical goal for this part of the course is for candidates to draw connections between situations that yield a linear parametric graph and the rates of change of the situation's two parameterized quantities. In this way, candidates are encouraged to support their ideas about the characteristics of a bottle that would produce a linear graph for volume versus height.

Module 4 is an ideal complement to the more general study of mathematical models. In our Functions & Modeling course, Module 4 is used to address foundational ideas related to variance and mathematical modeling. By doing so, we strive to raise awareness about the statistical nature of mathematical modeling and support candidates in making connections to other statistical ideas in the secondary curriculum. For instance, candidates seem to gain substantial insight by making comparisons between function models as representations of bivariate data and measures of center models for univariate data.

Logistical Aspects of Module Integration

Several decisions must be made when it comes to scheduling implementation of the modules. First, the instructor must decide where the modules fit within the curriculum. We recommend a placement based on the comments above, although arguments could be made for other alignments of the modules to the curriculum. Second, the timing of when candidates complete the online portion of the module in relation to when the candidates experience the face-to-face component must be considered. If too much time passes between the completion of the online and face-to-face components of a module, candidates will not be able to draw on their online experience effectively during the face-to-face activity. On the other hand, if the due date for the online module is too close to the face-to-face component, the instructor will not have time to review candidate responses and plan appropriate adjustments.

A key aspect of this instructional intervention is the VCAST Application (https://vcast.boisestate.edu/), a mobile-friendly, web-based application devel-

oped by the Office of Information Technology at Boise State University specifically to deliver VCAST instructional materials. Access to the modules is free, but to protect the rights of student participants and in accordance with IRB approval, those wishing to use the VCAST modules must register and be approved. The app provides a clean, effective interface for instructors to assign content and monitor candidate (or teacher) responses and offers an intuitive, easily-navigated platform for candidates and teachers to observe, analyze, and make inferences and hypotheses about individual secondary students' mathematics problem-solving approaches and understandings.

FUTURE PLANS

By developing online video modules and face-to-face instructional materials that target the development of attentiveness to student thinking related to functions and modeling, and by offering access and delivery of this content via a flexible, intuitive online platform, the VCAST project team aims to support and improve teachers' mathematical understanding and pedagogical skills for teaching functions and modeling on a large, easily-implemented scale. Through multiple rounds of testing, iterative cycles of revision of both our module design and its online delivery platform, and a working partnership with more than a dozen UTeach program instructors who will assist in our modules' refinement for more widespread use, our goal is to have a final suite of four modules ready for dissemination within two years.

Our hope is that other teacher educators will then use the VCAST modules to supplement their preparation of secondary mathematics teacher candidates or for professional development with classroom teachers, and that engagement with the VCAST modules will foster improved attentiveness to students' thinking and more productive pedagogical practices in mathematics classrooms.

REFERENCES

Alsawaie, O., & Alghazo I. (2010). The effect of video-based approach on prospective teachers' ability to analyze mathematics teaching. *Journal of Mathematics Teacher Education, 13*(3), 223–241.

Association of Mathematics Teacher Educators. (2017). *Standards for Preparing Teachers of Mathematics.* Retrieved from: amte.net/standards.

Ball, D. L., Thames, M. H., & Phelps, G. C. (2008). Content knowledge for teaching: What makes it special? *Journal of Teacher Education, 59*(5), 389–407.

Boaler, J. (2008). *What's math got to do with it? Helping children learn to love their most hated subject—And why it's important for America.* New York, NY: Viking.

Boaler, J., & Staples, M. (2008). Creating mathematical futures through an equitable teaching approach: The case of Railside school. *Teachers College Record, 110*(3), 608–645.

Borko, H., Jacobs J., Eiteljorg, E., & Pittman, M. E. (2008). Video as a tool for fostering productive discourse in mathematics professional development. *Teaching and Teacher Education, 24,* 417–436.

Carney, M., Cavey, L., & Hughes, G. (2017). Assessing teacher attentiveness: Validity claims and evidence. *Elementary School Journal, 118*(2), 281–309.

Cavey, L. O., Totorica, T., Carney, M., Lowenthal, P., & Libberton, J. (2018). Video case analysis of students' mathematical thinking: Initial development process. In L. Liu & D. C. Gibson (Eds.), *Research highlights in technology and teacher education 2018* (pp. 61–70). Waynesville, NC: Association for the Advancement of Computing in Education.

Fennema, E., Carpenter, T., Franke, M., Levi, L., Jacobs, V., & Empson, S. (1996). A longitudinal study of learning to use children's thinking in mathematics instruction. *Journal for Research in Mathematics Education, 27*(4), 403–434.

Freudenthal, H. (1973). *Mathematics as an educational task.* Dordrecht, The Netherlands: Reidel.

Freudenthal, H. (1991). *Revisiting mathematics education.* Dordrecht, The Netherlands: Kluwer.

Friel, S. N., & Carboni, L. W. (2000). Using video-based pedagogy in an elementary mathematics methods course. *School Science and Mathematics, 100*(3), 118–128.

Gravemeijer, K., & van Galen, F. (2003). Facts and algorithms as products of students' own mathematical activity. In J. Kilpatrick, W. G. Martin, & D. Schifter (Eds.), *A research companion to "Principles and Standards for School Mathematics"* (pp. 114–122). Reston, VA: National Council of Teachers of Mathematics.

Jacobs, V. R., Lamb, L. L., & Philipp, R. A. (2010). Professional noticing of children's mathematical thinking. *Journal for Research in Mathematics Education, 41*(2), 169–202.

Johnson, H. L., McClintock, E., & Hornbein, P. (2017). Ferris wheels and filling bottles: A case of a student's transfer of covariational reasoning across tasks with different backgrounds and features. *ZDM Mathematics Education,49*, 851–864.

Ko, I., Milewski, A. E., & Herbst, P. (2017). *How are pre-service teachers' educational experiences related to their mathematical knowledge for teaching geometry?* Paper presented at the annual meeting of the American Educational Research Association Annual Meeting, San Antonio, Texas.

Lin, P. J. (2005). Using research-based video-cases to help pre-service primary teachers conceptualize a contemporary view of mathematics teaching. *International Journal of Science and Mathematics Education, 3,* 351–377.

Llinares, S., & Valls, J. (2009). The building of preservice primary teachers' knowledge of mathematics teaching: Interaction and online video cases. *Instructional Science, 37*(3), 247–271.

Llinares, S., & Valls, J. (2010). Prospective primary mathematics teachers' learning from on-line discussions in a virtual video-based environment. *Journal of Mathematics Teacher Education, 13,* 177–196.

Santagata R., & Bray W. (2016). Professional development processes that promote teacher change: The case of a video-based program focused on leveraging students' mathematical errors. *Professional Development in Education, 42*(4), 547–568.

Santagata, R., Zannoni, C., & Stigler, J. W. (2007). The role of lesson analysis in preservice teacher education: An empirical investigation of teacher learning from a

virtual video-based field experience. *Journal of Mathematics Teacher Education, 10*(2), 123–140.

Seago, N., Mumme, J., & Branca, N. (2004). *Learning and teaching linear functions: Video cases for mathematics teacher professional development, 6–10.* Portsmouth, NH: Heinemann.

Shulman, L. S. (1987). Knowledge and teaching: Foundations of the new reform. *Harvard Educational Review, 57*(1), 1–22.

Star, M. K., & Strickland, S. K. (2008). Learning to observe: Using video to improve preservice mathematics teachers' ability to notice. *Journal of Mathematics Teacher Education, 11,* 107–125.

Swan, M. (1985). *The language of functions and graphs.* Shell Centre & Joint Matriculation Board, Nottingham.

Thompson, P. W., & Carlson, M. P. (2017). Variation, covariation, and functions: Foundational ways of thinking mathematically. In J. Cai (Ed.), *Compendium for research in mathematics education* (pp. 421–456). Reston, VA: National Council of Teachers of Mathematics.

Treffers, A. (1987). *Three dimensions: A model of goal and theory description in mathematics instruction—The Wiskobas Project.* Dordrecht, The Netherlands: Reidel.

CHAPTER 20

BENEFITS OF LIBERAL ARTS CURRICULUM IN STEM TEACHER PREPARATION

Gregory D. Gilson

This chapter provides an overview of the curriculum, outcomes, assessments, and student projects of the UTeach content course *Perspectives on Science and Mathematics (Perspectives)* in order to explain why a liberal arts based course is an important and effective component of the UTeach STEM teacher preparation model. I describe the core course components, including the basic curriculum, outcomes, and assessments, using specific examples of student projects and activities. I point out how Perspectives enhances mathematics and science education with episodes from the history of mathematics and science. I argue that *Perspectives* develops critical reading and writing skills, independent research skills, and the capacity to integrate historical and philosophical content into high school mathematics and science lessons as well as the less tangible benefits of promoting lifelong learning and encouraging STEM students from under-represented populations.

NATURE AND RATIONALE OF *PERSPECTIVES*

Typically in mathematics and science education, scientific theories and mathematical results are presented to students in isolation from their historical, philosophical, cultural, and social context; or more problematically, the history of mathematics and science is presented using fabricated myths and stereotypes that

Preparing STEM Teachers: The UTeach Replication Model, pages 277–290.

misrepresent the nature of science, mathematics, and discovery. Historical presentations in math and science sometimes present the history of science and math as though it follows a path of inevitable progression and can tend to evaluate historical discoveries and theories in light of the current scientific knowledge and contemporary mathematical understanding. While this sort of history can be useful for identifying conceptual predecessors of currently accepted scientific theories and mathematical systems, it also obscures the actual historical events, misrepresents the original conceptual paradigms in which currently accepted theories developed, and entirely ignores scientific theories and mathematical techniques which were widely accepted and highly successful in their time but are now recognized to be incorrect. *Perspectives on Science and Mathematics (Perspectives)* is designed to address this in STEM education by using liberal arts curriculum and methodology to emphasize the conceptual, philosophical, and historical contexts that motivated specific scientific discoveries, theories, and mathematical results. In doing so, the course hopes to give students the critical thinking tools to question and understand more deeply the contexts of their own work and research. The primary purposes of this chapter are to explain why a liberal arts-based course in the history of science and mathematics is an important and effective component of the UTeach STEM teacher preparation model and to assist instructors interested in teaching a course like *Perspectives*.

Perspectives fully engages active learning techniques and the project-based pedagogical environment as it requires students to explore the historical development of science and mathematics from the perspective of its creators, the mathematicians, and scientists themselves. Students confront and try to make sense of historically based problems or puzzles by reading primary sources and engaging in interactive group inquiry, discussion, and class presentation. Students apply this liberal arts content and methodology directly to their pedagogical development when they construct two lesson plans that integrate the history of science and mathematics into STEM curriculum and conduct an extensive independent research project which forms the basis of an expository essay about an episode in the history of science. Explicit peer review of lesson plans and expository essays encourages students to reflect on and critique STEM curriculum and pedagogy.

CORE COURSE COMPONENTS

The classroom activities of *Perspectives* are centered on the investigation of about twenty topics and episodes in the history and philosophy of science and mathematics. The classroom curriculum is assessed though reading comprehension quizzes, in class participation activities, and midterm and final exams. In addition to inquiry-based classroom activities, there are two major student projects. Employing project-based instructional design, students engage in a substantial research project, produce a substantial piece of formal writing, and design, present, and revise two mathematics or science lesson plans in order to incorporate the history or philosophy of mathematics and science directly into STEM curriculum.

Peer, master teacher, and instructor review and provide feedback for revision of both the research paper and lesson plans (UTeach Institute, 2009).

The instructor of *Perspectives* is most often a history or philosophy faculty member, but the course may also be successfully taught by faculty based in a mathematics, science, or education department. One of the ancillary benefits of incorporating a course like *Perspectives* into a teacher preparation program is that it unites faculty and departments in disparate colleges and departments to facilitate collaboration and interdisciplinary approaches to mathematics and science education. The instructor should have expertise in both the history and philosophy of science and mathematics as well as some background in education and pedagogy. While *Perspectives* emphasizes issues in the history and philosophy of mathematics and science, the course content, topics, and student activities should be directly relevant to K–16 science and mathematics education and develop student's pedagogical expertise that enhances their own instruction.

Instructor led modules investigate episodes in the history of science and mathematics that illustrate the relationship between science and mathematics and the various philosophies and analytical frameworks typically used for understanding science, math, and their history. The presentation of these modules should incorporate paradigm STEM instructional practices including instructor modeling of effective direct-teaching and student-centered inquiry and questioning strategies. The primary goal of the instructor led topic discussions and classroom activities is to illustrate the real history of confusion and disagreement among mathematicians and scientists rather than explaining episodes in light of the current scientific knowledge and contemporary mathematical understanding. The course modules also highlight the "non-scientific" factors such as practical, religious, social, and political considerations and motivations that have directly impacted the development of mathematics and science.

There are countless episodes or topics in the history of science and mathematics that might be explored but sample modules might be centered around Pre-Socratic natural philosophy, Platonism and constructivism in mathematics, Aristotle's explanatory system of four causes, Hellenistic mathematical science, medieval Islamic optics, alchemy, the Copernican revolution, the terminological and conceptual development of algebra, theories of the creation and formation of the earth, the Newtonian synthesis, the development and application of non-Euclidian geometry, Dalton's atomic theory, the Darwin-Wallace theory of natural selection, Crick and Watson's discovery of DNA, or the development of quantum mechanics and relativity theory. The topics needn't be explored chronologically and the topics selected probably will not present a comprehensive history of mathematics and science.

I strongly recommend that the course modules integrate mathematics and science topics especially emphasizing the role of mathematics in scientific explanation but there should also be modules dedicated to pure math in addition to applied mathematics. The modules should emphasize and interrogate historical myths

and clichés commonly used in mathematics and science education. The modules should include primary source material written by the historical mathematicians and scientists, secondary sources written by historians and scientists, and employ various digital resources including videos and digital archives maintained by museums, universities, and other academic institutions. The modules should also introduce and perhaps briefly explain, but mostly apply, various analytical and philosophical frameworks that are used to understand science and mathematics. These frameworks might include Platonism and constructivism in mathematics or empiricism, realism, and instrumentalism in science. Finally, modules should examine the historical development of the methods of proof used in science and mathematics including deduction, induction, and abduction. Ideally, most of the module content will be digested by students before class. Class time should be dedicated to interactive group inquiry and discussion.

In addition to classroom activities, *Perspectives* develops independent research and formal writing skills through student research projects that explore student selected episodes in the history of science and mathematics in their historical context, preferably exploring both primary and secondary historical sources. In my *Perspectives* course, students typically engage in a traditional research project and paper based on examining original primary texts written by historical scientists and mathematicians along with journal articles and secondary source books they find using academic databases and search engines. Students develop independent research skills including searching, acquiring, and evaluating the reliability and academic rigor of various sources. Students produce an annotated bibliography to ensure that they reflect upon and evaluate each of their sources. Another research and formal writing project employed by some *Perspectives* instructors is the textbook survey exercise. In this exercise, students survey the way that an important STEM discovery, theory, or innovation is presented in mathematics and science textbooks by examining actual textbooks ideally from different historical periods. The students then conduct a historical investigation of the actual figures, events, and discoveries using primary and secondary historical sources. Comparing the textbook survey and the results of their historical investigation, students write a report that focuses on the differences between the mathematics and science textbook presentations and their understanding of the actual historical events. Students should be encouraged to reflect upon how a mathematics or science instructor might supplement textbook presentations to better inform their students about the history of mathematics, science, and the nature of historical discovery.

The *Perspectives* lesson plan project requires students to summarize, integrate, and apply the knowledge and techniques acquired in both the course modules and their independent research. They design, present, and revise two lesson plans that incorporate the history or philosophy of mathematics and science. These are lesson plans they might actually teach in a mathematics or science classroom and are required to identify and cite the state and national performance standards for mathematics and science education that are promoted in their lesson. The benefits

of focusing on formal education standards are two-fold. First, it illustrates to the students that the history and philosophy of mathematics and science are important aspects of state and national standards, and secondly it affords students the opportunity to practice incorporating outcomes related to liberal arts into mathematics and science lessons.

In the UTeach Program, lesson plans are written using the 5E Instructional Model of Engage, Explore, Explain, Elaborate, and Evaluate. Students will at first likely include history and philosophy only in the beginning Engage section, as is typical in textbook presentations. In *Perspectives*, students are taught to incorporate history and philosophy of mathematics or science into each of the 5Es of the lesson plan. Students usually work in pairs to produce and present their lesson plans to their peers, master teachers, and instructor for feedback and revision. The instructor or master teachers might even observe the lesson presented in the field to actual high school students. An annotated bibliography is required for each lesson plan.

STUDENT OUTCOMES AND ASSESSMENTS

The primary Student Learning Outcomes of *Perspectives* are critical reading and writing and independent research skills. These outcomes are assessed with a formal writing project and the production of two K–12 mathematics and science lessons. Independent research skills and formal writing skills are developed and assessed in two important academic contexts. Students engage formal academic journal style research and writing as well as pedagogical research and professional and vocational writing skills. Just as the instructor led classroom modules and independent student research projects energize and encourage future K–12 teachers to engage in further independent research and lifelong learning, the incorporation of this historical and philosophic content and methodologies into lesson plans makes them better teachers with the capacity to better engage their future students in mathematics and science instruction.

HISTORICAL DISPUTE IN MATHEMATICS AND SCIENCE

In addition to the basic outcomes mentioned above, *Perspectives* develops a host of less tangible critical skills, epistemic capacities, and historical and philosophical appreciations that are particularly important for the academic and professional development of future K–12 mathematics and science teachers. Many of these broad outcomes and benefits are traditionally associated with the liberal arts educational system which has become increasingly isolated from STEM education (Kimball, 1986, 2010). Indeed, STEM students are sometimes surprised to discover the very existence of historical disputes in mathematics and science. STEM students typically understand scientific advances and discoveries through a historical lens which presents discoveries such as helio-centrism and the theory of natural selection as obviously true and without serious scientific objection once

formulated and presented to the scientific community.. Students often believe the false narrative that these theories were resisted only because of closed mindedness or religiously motivated objections. The truth is that there are often serious scientific objections to newly proposed groundbreaking scientific theories. For example, the lack of observed stellar parallax and the lack of an adequate physics of motion for a moving earth were serious scientific problems facing helio-centrism when first proposed by Copernicus, Kepler, and Galileo. Fleeming Jenkin's objection that the variation of individuals would be "swamped" away by blending inheritance of a few generations posed an ultimately incorrect, but understandable and substantial, scientific objection to Darwin's theory of natural selection until the development of population genetics in the 1920s (Morris, 1994). Learning this history provides students with a more accurate understanding of scientific progression and may have the potential to foster the kind of critical thinking that will prompt todays scientists to question their own knowledge and identify gaps or limitations in their own contemporary contexts.

Students in a mathematics class are typically asked to memorize, understand, and apply important mathematical definitions, axioms, and conventions in exclusively mathematical contexts. Students might be taught that "division by zero is undefined," that infinitesimals are "an indefinitely small quantity; a value approaching but not equal to zero," and that complex numbers are "numbers that can be expressed in the form $a + bi$, where a and b are real numbers, and i is a solution of the equation $x^2 = -1$.» While students have memorized, can apply, and in some sense understand these mathematical concepts, they often remain unaware of the long and heated controversies that occurred in the process of establishing these mathematical platitudes. There is rarely space in typical mathematics curriculum for thought about the struggles between mathematicians or the historical criticism and alternatives proposed on the way to establishing accepted axioms and conventions. Asking students to examine the true nature and historical record of these achievements opens space for deeper understanding that has the potential to advance their own research and discoveries.

MYTH IN THE HISTORY OF SCIENCE AND MATHEMATICS

Perspectives encourages students to critically examine the way that history and myth is employed in mathematics and science textbooks and more broadly in mathematics and science education. Instructors in science and mathematics sometimes openly embrace the use of myth or at least are unconcerned about the veracity of historical accounts and examples. It is not uncommon for science instructors to say to their students that whether Galileo *actually* dropped differently weighted balls from the tower of Pisa to determine whether they fell at the same rate doesn't really matter, or whether Darwin was *actually* influenced by differently shaped beaks of the Finches native to the various Galapagos islands is secondary to whether these stories illustrate the scientific or mathematical principles being taught in the physics or biology class. It is not uncommon for mathematics

instructors to recount "Pythagoras' proof" that "the square of the hypotenuse is equal to the sum of the squares of the other two sides," despite the fact that there is no historical evidence that Pythagoras ever engaged in any mathematics and Chinese and Babylonian mathematicians had contemplated "perfect triangles" long before Pythagoras (Martínez, 2011). Myths often obscure the real history of mathematics and science which can be more illustrative of the actual nature of historical development and the nature of science, mathematics, and discovery. Though teaching the myth involving Darwin and the Galapagos finches might seem harmless enough, it obscures Darwin's actual interest in the sea turtles and especially the freshwater amphibians cut off from the mainland by the sea. Emphasis on Galileo's dramatic nonexistent experimental act in Pisa obscures his reliance on thought experiments to develop his theory of free fall motion.

Myth is an important component which motivates mathematicians and scientists themselves and rightfully plays a role in historical explanation. Mathematics students commonly hear that the "golden ratio" or "divine proportion" is a universal mathematical law that can be found in countless natural forms such as the human body, plants, and seashells and works of art such as Leonardo Da Vinci's *Mona Lisa* and *Vitruvian Man*. This folk version of mathematical Platonism is often based on or at least inspired by the discredited work of Aldolf Zeising (Martínez, 2011, Zeising, 1854). Johannes Kepler famously related the golden ratio to the Pythagorean Theorem in Kepler's Triangle.[1] Like Zeising, Kepler also held a deeply Christianized Platonic-Pythagorean and even mystical view about the relation of mathematics to the natural world. An adequate understanding of Kepler's interest in the golden ratio provides the student with invaluable insight into the Christianized Platonic Pythagoreanism that motivated Kepler to discovery his famous Three Laws of Planetary Motion (Kepler, 1981).[2]

CRITICAL THINKING

Perspectives teaches students about critical thinking as a broad methodology across different disciplines. For example, a faculty led classroom module about Thomas Kuhn's analysis of scientific change clearly articulates at least one theory about the way that critical thinking operates in the historical development of

[1] Kepler's Triangle is a right triangle with edge lengths in a geometric progression in which the common ratio is $\sqrt{\varphi}$, where φ is the golden ratio.

[2] Kepler's three laws of planetary motion are commonly represented in astronomy or physics classes as straightforward algebraic equations and generally presented through a Newtonian lens. This modern presentation obfuscates Kepler's religious and metaphysical motivations and methodology. The truth is that Kepler's three laws were the result of his attempts to force all kinds of mathematical conceptions onto nature including his attempt to use the platonic solids to explain the orbits of the planets. Arguably, the central theme to all of Kepler's study of planetary motion was the search for finding harmonic ratios in the cosmos which revealed the Christian God as the Cosmic Geometer. Examination of this episode in the history of science illustrates how natural laws still accepted and taught today have emerged from motivations and historical contexts quite different from those of modern science.

science (Kuhn, 1962). Kuhn contrasts the role that critical thinking plays within the normal science of a paradigm with the role critical thinking plays in the revolutionary science during paradigm shifts. Critical reason operating within a strongly held paradigm is largely puzzle solving, where anomalies and recalcitrant evidence and phenomena are seen as puzzles to be solved, made to fit within the paradigm, and not as a challenge to the paradigm itself. A good example of puzzle solving within a paradigm is the development of the gravitational three-body problem regarding the motions of the moon, earth and sun in the context of Newton's laws of motion, or Tyco Brahe's hybrid model of planetary motion which was designed to save the geocentric model of planetary motion from the recalcitrant evidence provided by the discovery of the phases of Venus.[3] In contrast, critical thinking operates very differently in the context of scientific revolution and paradigm shift. Examples of critical thinking in the context of paradigm shift that students might encounter in *Perspectives* include the fossil record's challenge to the Great Chain of Being, hierarchical system of natural history classification, or how Galileo's quantification of motion challenged the Aristotelian notions of natural and violent motion.

HISTORICAL PERCEPTIONS OF SCIENTIFIC KNOWLEDGE

Perspectives provides students with a valuable historical understanding of how different cultures, at different times, and operating under different epistemological paradigms, understood scientific knowledge and scientists themselves. For example, students who are familiar hearing natural philosophers such as Isaac Newton and Charles Darwin referred to as "paradigm scientists" are often surprised to learn that the very term "scientist" to refer to an individual wasn't coined until the mid-1800s and not widely accepted by natural philosophers themselves until well into the early 1900s.[4] Students are even more surprised to read historically authentic evaluations of even ground breaking scientific theories which downplay the veracity and significance of its scientific claims.[5] It's illuminating for stu-

[3]The Tychonic system maintained that all of the five known planets orbit the sun, but then the sun, moon, and sphere of fixed stars orbit the earth. The model was able to account for the newly observed phases of Venus and make use of the superior mathematical calculations presented by Copernicus but still maintain the philosophic advantages of a stationary earth in the Ptolemaic system.

[4] The term "scientist" is generally thought to have first appeared in print in 1834 in an article by Cambridge University historian and philosopher William Whewell but the terms was not embraced by scientific practitioners until well into the twentieth century as evidenced by Thomas Huxley's 1894 famous public repudiation of the term "scientist." It wasn't until 1924 that the academic journal Nature discontinued the policy of asking authors to not use the term (Ross, 1962).

[5] A good example of a historically authentic text which downplay the veracity and significance of its scientific claims is Andreas Osiander's 1548 preface to Nicholas Copernicus' On the Revolutions of the Celestial Spheres (Copernicus, 1995) which states that "the following book proposes the novel hypotheses ... that the earth moves whereas the sun is at rest in the center of the universe." but then follows with the deflationary announcement that since the astronomer, "cannot in any way attain to the true causes, he will adopt whatever suppositions enable the motions to be computed correctly

dents to discover that the view of scientific knowledge as a paradigm of truth and certainty is comparatively recent and the result of profound epistemological and societal transformation.

RELATIONSHIP BETWEEN MATHEMATICS AND SCIENCE

Perspectives encourages students to critically examine the differences between pure and applied mathematics, the similarities and differences between science and mathematics, and the way that science and mathematics work together to explain the natural world. STEM students are familiar with using deductive methods of proof in mathematics classes and inductive and abductive methods in science classes, but *Perspectives* provides an opportunity to directly compare and contrast these methods of proof as historically developed. Perspectives provides numerous topics which offer this opportunity including the mathematical sciences in antiquity, the philosophy of mathematics, and the use of non-Euclidian geometry in relativity theory. It can be especially fruitful to examine the essential role that mathematic plays in scientific explanation. (Wigner, 1960). *Perspectives* encourages students to critically examine the historical relationship between the STEM disciplines of mathematics and science on the one hand and more technologically based STEM disciplines such as computer science and engineering on the other and encourages students to critically examine the similarities and differences between STEM and non-STEM disciplines such as history, philosophy, education, and art.

Perspectives also helps students to understand the causal role that non-scientific factors such as religion, politics, and economics play in the construction and acceptance of both scientific theories and mathematical systems. Examples of the non-STEM influences on the development of mathematics and science that students might encounter in *Perspectives* include early Hellenistic patronage of academies such as the Museum in Hellenistic Alexandra, the shift of the center of scientific and mathematical activity from Europe to the Islamic West during the early middle ages, how the need for administrators after the fall of the Roman Empire gave rise to the medieval university, how Pope Gregory XIII's motivation for a more accurate calendar for Catholic Easter influenced Copernicus' work, the relative democratization of science that occurred in the *French Academy of Sciences* after the French Revolution as compared to the largely aristocratic and *Royal Society* created by King Charles II in England, and the role that World War Two and the associated emigration of mathematicians and scientists played in the development of university mathematics and science education in the United

from the principles of geometry for the future as well as for the past." and that, "So far as hypotheses are concerned, let no one expect anything certain from astronomy, which cannot furnish it, … ." This historical perspective on science is valuable because in many STEM students' immediate experience, science is the paradigm of literal truth and scientists themselves are respected perhaps above all other professionals.

States. Examining examples such as these enables students to understand the connections between content more commonly associated with liberal arts disciplines and STEM disciplines and helps students to position mathematics and science within the larger historical development of educational systems.

LIFELONG AND INFORMAL LEARNING

History and philosophy also have the potential to increase interest in some STEM students by making their subjects more accessible and can promote lifelong learning and interest in popular presentations of mathematics and science. A commonly used formal writing project in *Perspectives* which directly addresses this issue is to have students construct a review essay of a popularization of the work of an actual scientist or mathematician. For example, in *Relativity: The Special and General Theory*, Albert Einstein wrote ." . . those readers, who from a general scientific and philosophic view, are interested in the theory but are not conversant with the mathematical apparatus of theoretical physics" (Einstein, 1954). Other such books include Stephen Hawking's *A Brief History of Time* (Hawking, 1988), James Watson's *The Double Helix: A Personal Account of the Discovery of the Structure of DNA* (Watson, 1968), Rachel Carson's *Silent Spring* (Carson, 1962), and Richard Dawkins's *The Greatest Show on Earth* (Dawkins, 2009). Academically respectable books written by science journalists and historians promote lifelong learning in mathematics and science and are also illuminating in their own light. Examples of such books include *Turing's Cathedral* (Dyson, 2012), *Infinitesimal: How a Dangerous Mathematical Theory Shaped the Modern World* (Alexander, 2014) or *Burned Alive: Bruno, Galileo, and the Inquisition* (Martínez, 2018).

Perspectives, in addition, provides a good venue for study abroad or study away experiences because of the course's emphasis on the historical and cultural aspects of the development of mathematics and science. Travel to an archival or special collections library will afford students the opportunity to see early editions of scientific and mathematical texts such as Newton's 1687 *Principles of Natural Philosophy* (Newton, 1972), Keplers' 1618 *The Harmony of the World* (Kepler 1977), or Copernicus' 1543 *On the Revolution of Heavenly Spheres* (Copernicus, 1995). Many international destinations provide outstanding city-classrooms for students to immerse themselves in the history of science and mathematics. Many international cities house major historical foundations and institutions that can support academic study abroad.[6] Study Abroad beyond-the-classroom learning is a high impact pedagogical experience that promotes life-long interest and

[6] Museum Galileo https://www.museogalileo.it/en/, The Royal Society of London https://royalsociety.org and the British Museum https://www.britishmuseum.org/research/collection_online/search.aspx are just three examples. These institutions have extensive real and virtual collections of scientific and mathematical artifacts and manuscripts and will often grant special access and privileges to Study Abroad Programs.

learning in students and yet such experiences are rarely aimed at mathematics and science students. Study Abroad programs provide powerful opportunities for students to form long standing professional peer bonds and to help construct an identity and provide valuable branding of STEM Teacher Preparation Programs.

Furthermore, at my own Hispanic Serving institution, I have directly observed how *Perspectives* encourages minority and female students to believe that they might themselves make meaningful and original contributions to mathematics and science. By studying the lives of actual historical figures students realize that many of the greatest scientists and mathematicians come from eccentric, unexpected, underestimated, and under-represented groups. Albert Einstein was working as a patient clerk because he was unable to obtain an academic appointment when he revolutionized physics. Poor, self-educated, fossil hunter Mary Anning was as knowledgeable in anatomy and as influential on the field of paleontology as any of her Victorian male peers including her famous Oxford rival Reverend William Buckland. NASA's human computers Katherine G. Johnson, Dorothy Johnson Vaughan, and their fellow African American computers performed the complex calculations that enabled space flight. Studying the biographies of these sorts of historical features provides role models and inspiration to STEM students from underrepresented populations.

WHY *PERSPECTIVES* IS AN ESSENTIAL COMPONENT OF STEM TEACHER PREPARATION

There are lots of reasons not to include a course like *Perspectives* in the curriculum of your teacher preparation program such as the fact that general education core curriculum requirements of most universities already provide students with exposure to liberal arts content and methodology, STEM degree plans are already tight and continually shrinking in terms of allowable total hours, teacher preparation programs need to devote more hours to pedagogy, classroom management, special learner accommodation, and educational law, and some of the content and methodology discussed above could be integrated into traditional STEM and pedagogy courses. However, I argue that despite these valid objections, incorporating a course like *Perspectives* in your Teacher Preparation program's curriculum is well worth the investment of valuable degree program hours.

It is true that the general education core curriculum requirements of many universities provide students with liberal arts content and methodology, but this content is not focused on STEM and pedagogy whereas *Perspectives* tailors liberal arts content to exactly these contexts. There is substantial empirical evidence that such general core curriculum exposure, while useful, does not offer the full benefits of liberal arts curriculum and methodology (Lewis, 2015). Nevertheless, the university general education core curriculum platform can be usefully engaged to address the problem of reduced allowable total hours in degree plans in the sense that *Perspectives* fits many subcategories of general education core curriculums. For example, *Perspectives* meets the criterial outcomes and objectives of at least

three different sections of the Texas (THECB) general education core curriculum.[7] *Perspectives* might be written into degree programs without adding to the total number of hours in degree plans.

In addition to the purely academic benefits of incorporating wide interdisciplinary faculty, content, and methodology into a STEM teacher preparation program noted above, there are also administrative benefits. STEM teacher education programs are primarily partnerships between colleges of natural sciences and mathematics, education, and sometime engineering. Incorporating a course like *Perspectives* into your curriculum enables tangible buy in and investiture from additional college(s) and multiple departments.

The infusion of arts content within STEM fields is sometimes referred to as STEAM (Science, Technology, Engineering, *Arts*, and Mathematics). I'd like to sketch out why *Perspectives* doesn't exactly fit the STEAM paradigm, not to challenge the value of the STEAM approach, but rather as a way of illustrating the nature and special value of *Perspectives*. STEAM treats the arts as a separate subject matter whose study has cross-over benefit in STEM education. For example, mathematicians and medical students benefit from studying music and makerspaces promote creativity and innovation that are useful in STEM education. In contrast, Perspectives does not treat the arts as a separate discipline but rather illustrates the arts content that is already part of STEM and its history. *Perspectives* is not a supplement to STEM education, but rather *Perspectives* is STEM education in the sense that part of the content of the discipline is surely its history and analytical frameworks. Thus, to not include a course like *Perspectives* in mathematics and science teacher preparation programs is not just to reject the STEAM movement, rather it's to ignore an essential component of mathematics and science itself.

In conclusion, there are many compelling reasons for including a course like *Perspectives* into the curriculum of STEM teacher preparation programs. The core components of the course have broad benefits for both teachers and their students. *Perspectives* can broaden and improve student learning outcomes by providing STEM students with a richer and more nuanced understanding of the history of their subject. Better historical understanding is directly applied to their pedagogy in the development of lesson plans. Understanding the historical and cultural contexts in which math and science has developed may increase student interest and engagement, especially in the case of under-represented groups of students. Deeper historical understanding of the mathematics and science they are learning may improve their ability to think critically about the problems that

---

[7] For example, *Perspectives* easily meets the Language, Philosophy, and Culture subsection of the Texas THECB General Education Core Curriculum requirements. It also meets the Social and Behavior Sciences subsection. Finally, because of its emphasis on formal writing, *Perspectives* meets the criteria for the Communications section especially if the particular university's core requires upper division writing intensive courses. *Perspectives* also meets outcomes for cores that include "signature courses."

math and science raise, and perhaps better equip them with the skills to carry out innovative and cutting-edge work in their fields. As I have argued, *Perspectives* is an essential aspect of STEM education and one that is well worthwhile including in teacher preparation programs.

REFERENCES

Alexander, A. (2014). *Infinitesimal: How a dangerous mathematical theory shaped the modern world.* United States of America: Farrar, Straus and Giroux.

Carson, R. (1962). *Silent spring*: New York: NY. Houghton Mifflin.

Copernicus, N. (1995). *On the revolution of heavenly spheres.* (R. W. Lawson, trans). Amherst, NY: Prometheus Books. (Work originally published as De revolutionibus orbium coelestium in 1543 in Nuremberg, Holy Roman Empire)

Dawkins, R. (2009). *The greatest show on earth.* New York: NY. Simon and Schuster.

Dyson, G. (2012). *Turing's cathedral: The origins of the digital universe.* New York, NY: Pantheon Books.

Einstein, A. (1954). *Relativity: The special and general theory* (R. W. Lawson, trans). New York, NY. Crown Publishers. (Original work published 1916).

Hawking, S. (1988). *A brief history of time.* New York: NY. Bantam Press.

Kepler, J. (1981). *The secret of the universe: Mysterium cosmographicum.* New York, NY: Abaris Books. (Work originally published as Mysterium Cosmographicum in 1597)

Kepler, J. (1997). *The harmony of the world.* Philadelphia, PA: American Philosophical Society. (Work originally published as Harmonice mundi in 1618)

Kimball, B. (1986). *Orators and philosophers: A history of the idea of liberal education.* New York: NY. Teachers College Press.

Kuhn, T S. (1962). *The structure of scientific revolutions.* Chicago: IL. The University of Chicago Press.

Lewis, A. (2015). Full STEAM ahead; putting the "H" in STEAM: paradigms for modern liberal arts education. In Xun Ge, Dirk Ifenthaler, J Michael (Eds.), *Emerging technologies for STEAM education.* Switzerland: Springer International Publishing.

Martínez, A. A. (2011). *Science secrets: The truth about Darwin's finches, Einstein's wife, and other myths.* Pittsburgh: PA. University of Pittsburgh Press.

Martínez, A. A. (2018). *Burned alive: Bruno, Galileo and the inquisition.* London, UK: Reaktion Books.

Morris, S. W. (1994). Fleeming Jenkin and "The Origin of Species": A reassessment, *The British Journal for the History of Science, 27*(3), 313–343.

Newton, I. (1972). *Isaac Newton's philosophiae naturalis principia mathematica: Volume 2: The Third Edition (1726) with Variant Readings.* (Whitman, A., Koyre, A., Cohen, I. B., trans). United Kingdom: Cambridge University Press. (Work originally published 1687).

Ross, S. (1962). Scientist: The story of a word, *Annals of Science, 18*, 65–85.

UTeach Institute. (2009). *Perspectives on science and mathematics: UTeach course curriculum.* Austin, TX. University of Texas at Austin.

Watson, J. (1968). *The double helix: a personal account of the discovery of the structure of DNA.* New York: NY. Simon and Schuster.

Wigner, E. (1960). The unreasonable effectiveness of mathematics in the natural sciences. *Pure and Applied Mathematics*, *13*(1), 1–14.

Zeising, A. (1854). *Neue Lehre von den Proportionen des Menschlichen Körpers*. Leipzig, Germany: Rudolph Weigel.

SECTION III

RESEARCH AND FUTURE DIRECTIONS

CHAPTER 21

DEVELOPING THE UTOP

A Flexible STEM Observation Instrument Based on UTeach Principles

Candace Walkington, Michael Marder, and Elizabeth L. Adams

The UTeach Observation Protocol (UTOP) is a classroom observation instrument developed for STEM (science, technology, engineering, and mathematics) classrooms. The UTOP was originally developed to explore the teaching practices of UTeach graduates in mathematics and science classrooms. The UTOP is intended to be a broad instrument that assesses many different elements of classroom instruction, as teacher preparation is broad-ranging in the skill sets it develops. The purpose of this chapter is to give practical information about using the UTOP to observe teachers in STEM classrooms, to provide an explanation of the UTOP's background and underpinnings, and to present some real-world cases of how the UTOP has been used in practice. This information is aimed at helping the STEM education community make informed decisions about whether the UTOP would be helpful in their context, how to appropriately apply the UTOP, and important considerations to take into account when doing so.

BACKGROUND

The UTeach Observation Protocol (UTOP; Marder et al., 2010) was developed as a classroom observation tool to explore the teaching practices of UTeach gradu-

Preparing STEM Teachers: The UTeach Replication Model, pages 293–310.

ates in mathematics and science classrooms. Although initially developed for secondary (grades 6–12) mathematics and science, the UTOP has been applied in classrooms from 4th grade to college. The UTOP was designed to be aligned to the goals and foci of the UTeach program; thus it is an ideal instrument for assessing the impact of UTeach program sites and for guiding continuous improvement of UTeach teacher preparation and professional development. However, the principles and values captured in the UTOP are broadly applicable to current visions of effective STEM teaching based on the research literature (see Walkington & Marder, 2018 for more details), so the UTOP may be applied in a wide variety of STEM contexts. It is intended to be a broad instrument that assesses many different elements of classroom instruction, as teacher preparation must be broad-ranging in the skill sets teachers are trained to enact. This differentiates the UTOP from other instruments that assess smaller, focused selections of classroom interactions and issues.

The purpose of this chapter is to give practical information about using the UTOP to observe teachers in STEM classrooms and to present some real-world cases of how the UTOP has been used in practice. There are a number of STEM classroom observation protocols in existence, and choosing a protocol for a particular project or program should be a decision based on needs, resources, and alignment to goals. Here we present information aimed at helping the STEM education community make informed decisions about whether the UTOP would be helpful in their context, how to appropriately apply the UTOP, and important considerations and issues to take into account when doing so. We do this through a description of the UTOP, an explanation of its background and underpinnings, and a description of some cases in which it has been used.

Overview of the UTOP

The UTOP has four sections where the observer can describe and rate different elements of classroom instruction—Classroom Environment, Lesson Structure, Implementation, and Mathematics/Science content. Each section has 6–8 different indicators or teaching behaviors that assess one element of a teachers' classroom practice or the classroom dynamics in that area. Example indicators from each section of the UTOP are shown in Table 21.1 below. Each of these indicators is rated on a 1–5 scale (with NA options for some indicators), and observers record supporting evidence justifying their indicator ratings. Each of the four sections of the UTOP concludes with a 1–5 synthesis rating, which is intended to be the observers' overall evaluation of the behaviors associated with the indicators in that section, weighted based on their importance to the current lesson. In addition to rating scales, the UTOP includes observer and teacher demographic forms, a post-observation teacher interview protocol, and sections for summarizing a lesson overview and compiling concluding remarks.

The website https://utop.uteach.utexas.edu/ gives the current version of the UTOP, as well as a variety of background information about how to use it. In

TABLE 21.1. Examples of UTOP Indicators from Each Section

Section of UTOP	Sample Indictor
1. Classroom Environment	The classroom environment established by the teacher reflected attention to issues of access, equity, and diversity for students.
2. Lesson Structure	The structure of the lesson allowed students to engage with and explore important concepts in mathematics or science.
3. Implementation	The instructional strategies and activities used in this lesson clearly connected to students' prior knowledge and experience.
4. Mathematics/Science Content	Appropriate connections were made to other areas of mathematics or science and/or to other disciplines.

particular, the website gives a general rubric for each indicator, describing the kinds of behaviors that are typically associated with 1–5 ratings. The website also gives examples of supporting evidence that an observer might cite for each indicator that would go along with 1–5 ratings for that indicator. The website also provides links to video STEM lessons and full UTOP instruments that have been rated for these lessons, to use for observer training and norming.

The Underpinnings of the UTOP

The UTOP was developed from the Classroom Observation Protocol (COP; Horizons Research Inc., 1999), a classroom observation protocol that was used to measure the impact of K–12 mathematics and science professional development. The indicators on the COP were modified to put a larger emphasis on content knowledge and content knowledge for teaching. Indicators were also added or modified to better value direct instructional strategies that involve active knowledge construction. Some indicators specific to UTeach practices and foci (e.g., connecting mathematics and science to history or current events; lab safety) were also added. Key tenets of the UTOP include classroom indicators that focus on how STEM content is communicated and taken up by students, how a variety of instructional approaches—from inquiry-based learning to direct instruction—are utilized, and how the logistical aspects of teaching (e.g., management, time, organization) make the learning of STEM content possible. Another instrument developers considered as they constructed the UTOP was the Reformed Teaching Observation Protocol (RTOP; Piburn, Sawada, & Turley, 2000). The reason the RTOP was not adopted as the instrument was its high degree of emphasis on reformed teaching and a lesser emphasis on content depth and accuracy.

To ensure the UTOP measured characteristics of "good teaching" valued by the UTeach program, the indicators on the UTOP were compared to the items on previous UTeach instruments that were available at the time (i.e., UTeach Holistic Evaluation Model, the UTeach Portfolio Expectations, the UTeach Focused Proficiency Observation Protocols, the UTeach Apprentice Teacher Final Evaluation,

and the UTeach Teacher Development Rubric). Some significant aspects of the UTeach program's vision that the UTOP assesses include the intellectual engagement of students, the teacher's classroom management and timing, the classroom setup and use of resources, the teacher's attention to issues of equity, the teacher's ability to engage in reflective practice, the teacher's use of investigative or problem-based pedagogical approaches, the teacher's use of questioning strategies to promote understanding and use of formative assessment techniques to monitor student progress, the significance of the content covered and how this is made explicit to students, the fluidity of the teacher's content knowledge, the connections made in the classroom to other disciplines and historical and current events, how the teacher built on students' prior knowledge and experiences, how the teacher adjusted instruction to meet student needs, and how the teacher implemented formal assessments. Further, several constructs of good teaching prominent in the UTeach documentation are measured indirectly through the UTOP—the teacher's use of technology (covered under resource use indicator in Implementation section), the teacher promoting student reflection and student-to-student discussion (covered under Classroom Environment indicators), and the teacher's ability to foster high expectations in a culture of learning (covered under Classroom Environment indicators).

In later work (Walkington & Marder, 2018), we more fully elaborated on the theoretical underpinnings of the UTOP, which reflect the theoretical underpinnings of the UTeach program itself. We describe how the indicators on the UTOP draw from research literature in six key strands: effective communication of STEM content, classroom management, problem-based and discovery learning, lesson structuring and assessment, facilitating intellectual engagement, and equity, diversity, and access.

Technical Information about the UTOP

In the cases later in this chapter, we describe five different case studies that involve the UTOP. In our descriptions, we focus on the process of conducting the observations and the lessons learned. Thus, in this section, we first compile information from across those studies to give an overview of technical information about the UTOP.

In terms of the reliability of the UTOP, our earliest study (Walkington & Marder, 2014) suggested that the four synthesis ratings have an inter-rater reliability (weighted kappa) of 0.63, which corresponds to substantial agreement, while our individual indicators have a kappa of 0.41, which is at the low end of moderate agreement (Landis & Koch, 1977). Reviewers agreed on 58% of the individual indicators, and agreed within a distance of 1 on 97% of the individual indicators. Cronbach's alphas, a measure of internal consistency, ranged from 0.91 to 0.96 for each UTOP section. A larger follow-up study that used a shortened version of the UTOP suggested that the overall UTOP score (an average of the synthesis ratings) had an implied reliability of 0.63 under a schedule where four different observ-

ers observed four lessons of one teacher. This was similar to the reliability level of other widely-used observation protocols to which the UTOP was compared and considerably higher than the reliability of the other mathematics observation instrument in the study. The implied reliability value varied by UTOP section—from 0.65 for Classroom Environment, to 0.57 for Lesson Structure and 0.56 for Implementation, to 0.44 for Mathematics/Science Content.

In terms of the validity of the UTOP, Kane and Staiger (2012) reported the correlation between overall UTOP score and teacher value-added (measured by standardized test score gains) at $r = 0.34$. This value was higher than value-added correlations for other widely-used instruments in the study, such as CLASS (Pianta, La Paro, & Hamre, 2008). This study also used classroom observation instruments to predict scores on the Balanced Assessment of Mathematics (BAM), a standardized test developed to assess conceptual understanding of mathematics. The researchers found that teachers whose overall UTOP score was in the bottom versus the top quartile had a BAM score difference of 0.11, which corresponded to approximately four months of schooling. Follow-up analyses by Walkington and Marder (2014) suggested that UTOP scores on indicators relating to classroom management and student exploration had the highest correlation to gains in student test scores.

We also examined the factorial structure of the indicators on the UTOP (Walkington & Valerius, 2012). In an exploratory analysis, we found one cluster of indicators relating to general teacher pedagogical knowledge—e.g., classroom management and timing. A second cluster included indicators relating to pedagogical and specialized content knowledge teachers demonstrate by engaging students in deep STEM learning opportunities, using inquiry learning, and promoting intellectual engagement. A third cluster included indicators relating to teachers making content connections (e.g., to other disciplines) and using specialized content knowledge. And a fourth cluster included indicators relating to teacher content knowledge—such as their accurate communication of content and use of appropriate abstractions. These clusters may be particularly important to consider in teacher education programs like UTeach. These programs must balance teaching basic pedagogical and management skills, building teachers' STEM content knowledge, facilitating their understanding of the meaning of interconnections between disciplines, and using advanced pedagogical techniques centered around inquiry and exploration. It is notable that these deeper elements of teaching expertise were more salient in the factor loadings than more commonplace distinctions like structuring lessons versus implementing them.

CASE STUDIES

Much of the prior literature on the UTOP has focused on its reliability and validity (e.g., Walkington & Marder, 2014), as well as guiding observers on how to rate specific lessons (Walkington & Marder, 2018). Missing from this literature, however, are accounts of how the UTOP has actually been used in a variety of settings

for a number of different projects with different scales, populations, and goals. A comparison of the applications of the UTOP across a variety of settings can allow readers to glean key insights about STEM teaching observations and the UTeach vision. The UTOP offers unique opportunities for researchers at UTeach replication sites and researchers leading other STEM education projects to examine the impact of what they do on teaching practices.

It is for this purpose that we present case studies that describe how the UTOP has been used in five different contexts—(1) observing UTeach and non-UTeach graduates teaching secondary STEM, (2) observing a large sample of grades 4–8 mathematics teachers mostly using traditional instructional approaches, (3) observing high school teachers at a traditional and a project-based school in the same district, (4) observing middle school science teachers who participated in professional development academies focusing on project-based and maker-based learning, and (5) observing university-level STEM faculty. In each of these contexts, we describe the lessons we learned about using the UTOP to understand STEM teaching, and the ways in which the UTOP was modified and adapted to fit local needs and goals.

Case 1: Using the UTOP to Understand the Impact of the Noyce Scholars Program

The UTOP was originally developed to evaluate the teaching of graduates of the UTeach program who received Noyce scholarships, and compare them both to UTeach graduates who had not been Noyce Scholars, and to non-UTeach teachers from a variety of preparation pathways (see Walkington & Marder, 2014). The requirements for receiving a Noyce scholarship are higher than requirements for entry into the UTeach program, and these students receive extra financial support, so it could be hypothesized that they would be better teachers or teachers who display different teaching practices once in the field. In addition, Noyce Scholars have to commit to teaching in high-needs schools after their graduation.

To set up this study, we selected both UTeach Noyce scholars (7 teachers) and non-Noyce UTeachers (14 teachers) who were teaching at nearby schools. We attempted to "match" each UTeacher with another teacher at their school who had not gone through UTeach, who was teaching the same or similar subjects/grade levels, and who had a similar number of years of teaching experience (15 teachers). Given the limitations in the sizes of schools and the specialization of certain STEM subjects (e.g., a teacher who taught high school classes on medical technology), the matching was not always perfect. This points to a central struggle we faced in this initial endeavor to use the UTOP—using the UTOP for comparative purposes is extremely difficult, given the myriad of background and school factors that would need to be held constant for it to be a fair comparison. We included only teachers with 5 or fewer years of teaching experience.

In addition, multiple observations per teacher each year are needed when making such comparisons, in order to get an accurate picture of what "typical" instruc-

tion for that teacher looks like. These issues were further complicated by the fact that scheduling observations and actually rating with the UTOP was quite time-consuming, meaning that achieving the larger sample size needed for comparisons is problematic without a lot of resources. One insight we gained from looking at the data from this study is that rather than conceptualizing the UTOP as a way to compare teachers, it might be most useful to use the UTOP to understand teachers' growth trajectories. We did see upward trends in UTOP scores for UTeachers who had more experience in the classroom (Walkington & Marder, 2014).

Another issue we confronted in this initial study is that of rater reliability. We had two raters present at almost all observations, one with a background in mathematics and another with a background in science. This was purposeful so there would be a relatively uniform set of raters who together had the expertise to evaluate a variety of STEM lessons. However, given that the raters had different backgrounds, they sometimes picked up on and noticed different aspects of the instruction. After rating independently, the observers would discuss their ratings and supporting evidence and come to a "consensus rating" to be used in analyses. However, their inter-rater reliability when calculated from their initial round of rating was only mediocre. In response to this issue, we began to develop detailed rubrics and examples and made some modifications to the items. As integrated STEM is an idea that is highlighted in UTeach, observer expertise issues of this type may be common when using the UTOP to observe UTeachers.

Finally, when the UTOP is used for comparative purposes, it is important to account for observer bias—in this study, we attempted to blind observers to whether they were observing UTeachers or non-UTeachers. This was not always feasible due to a variety of ways in which teachers could accidentally reveal or suggest their affiliation. Having observers who are truly "external" to the intervention being assessed is important when doing comparative analyses.

Case 2: Using the UTOP in the Measures of Effective Teaching Study

When we conceptualized the Noyce study, measuring the effectiveness of UTeachers through growth in student standardized test scores was not yet feasible, given the limitations of state databases and the lack of testing for many scientific subject areas. However, after the Noyce study was complete, we were given the opportunity to work to further test and validate the UTOP using a video library of grades 4–8 mathematics lessons collected by the Measures of Effective Teaching (MET) Project (Kane & Staiger, 2012). The MET project not only had multiple videoed classroom observations for each of the teachers in the study but also had the teachers' value-added scores (i.e., gains in student standardized test scores) for each of the classes they taught. This study involved rating 1,001 video lessons of 249 teachers with the UTOP, which was a large endeavor involving the work of 99 different observers. We used a shortened version of the UTOP (with 22 indicators) to make the workload more feasible and found that shortening the

instrument did not seem to compromise its use. We also observed that allowing raters to focus on rating fewer indicators did not make them more reliable—contrary to an initial hypothesis.

The general results of the MET study were that the connection between classroom observation scores and student achievement differences are weak (Kane & Staiger, 2012). Although the UTOP had some of the strongest relations of the observation instruments included in the MET study, the correlations were not impressively high. In the case of the UTOP, we felt this relationship (or lack thereof) was partially driven by the lack of the kind of instruction supported by the UTeach program in the videos in the MET sample. There were very few examples of inquiry-based, problem-based, or project-based learning, and the vast majority of the lessons were characterized by relatively typical approaches to direct instruction. This made it difficult for the indicators on the UTOP to distinguish one teacher from another meaningfully and led to conversations about whether the UTOP was setting too high of a bar for teacher practice. The MET study also had the limitation of only looking at mathematics lessons, which may be more slanted towards direct instruction approaches compared to science lessons.

One premise of the MET study was to validate classroom observation instruments by seeing if they related strongly to student achievement gains in those classrooms. We approached this study with the view that this validation went both ways—if teachers observed with the UTOP are seen using strong STEM instructional approaches with students who are intellectually engaged, but teachers' value-added measures are low, this suggests a possible issue with the test or the method of measuring growth. Alternately, it is also problematic if we observe non-interactive classrooms full of disengaged students but see high value-added scores. The UTOP's rating process involved raters formally submitting supporting evidence from the lesson to justify each of their indicator ratings. This led to us having qualitative data about different aspects of instruction in the MET videos and allowed us to take a deeper look at the relationship between teacher value-added scores and UTOP observation scores.

We examined a series of outlier cases where UTOP scores and value-added scores were highly consistent or not consistent at all (Walkington & Marder, 2014). We found that each measure has its limitations—looking at value-added scores alone can mask issues with teachers' respectful treatment of students, a strong focus on isolated test preparation, and teachers making mathematical content errors. Conversely, focusing only on UTOP scores can mask issues with teachers implementing lessons that are relatively monotonous but where the teacher has a strong grasp on the content, or that are well-done in terms of classroom management and surface engagement, but with little intellectual engagement. However, lessons that had both high value-added scores and high UTOP scores were consistently good lessons when examining the qualitative evidence. Thus, having access to value-added data as a way to provide a "check" on UTOP scores might be important for larger-scale uses of the UTOP.

A final issue we looked at with the MET study data was the influence of observer qualifications on UTOP ratings. As the lessons were all mathematics lessons, but many of our raters had a background in science teaching, we were able to look at the influence of rater background characteristics. We found that raters with a science background had lower reliability than raters with a mathematics background when rating mathematics lessons. However, unexpectedly, this did not seem to stem from the science teachers catching on differently to content-related issues and errors. Instead, the science teachers seemed to have higher standards for student engagement and lesson interactivity. When looking at raters with mathematics, science, or mathematics education degrees, and comparing them to raters with education, science education, or other degrees, we did see that raters in the latter categories were less likely to catch content issues. We also generally found that different raters had vastly different reliability levels when rating lessons—with rater weighted kappas ranging from 0.122 to 0.631. This underscored the importance of considering observer qualifications when using the UTOP.

Case 3: Prairie Hill Independent School District

For two years, the UTOP was used to examine two high schools in a socioeconomically disadvantaged suburban area (DeZeeuw, Walker, & Marder, 2015). The local school district is small and has a total of three high schools. We will refer to the district as Prairie Hill Independent School District (PHISD). Two of the schools made for a particularly interesting comparison. One of them was a large high school with a very conventional curriculum and teaching style. The second was a smaller project-based high school, part of a national network of similar schools that admitted students by lottery.

The study began in August 2011. Over the next two years, 28 secondary mathematics and science teachers across two campuses within PHISD consented to participate. Data sources collected included over 130 classroom observations and field notes, surveys from teacher participants and students, teacher chosen areas of focus for peer observations, teacher peer observations, and interviews. The project had several goals. One of them was to develop a process of revealing observation results to teachers in a manner that would allow for growth and development, rather than being construed as punitive. In past studies, UTOP scores had not been shared with teachers and had been used only for evaluation or instrument development purposes. Another purpose was to monitor the extent to which the ostensibly different philosophies of the two different high schools could be detected in the observations of what happened in the classrooms.

Figure 21.1 summarizes the main results of the observational study. For each UTOP indicator, we display a histogram showing the fraction of observation scores at different levels. A score of 5 describes expert behavior, while a score of 1 describes the lowest level of performance on that indicator. Thus, blue shaded bars display a favorable UTOP assessment, and brown shaded bars are unfavorable. Figure 21.1 suggests that differences between instructional practices in the two

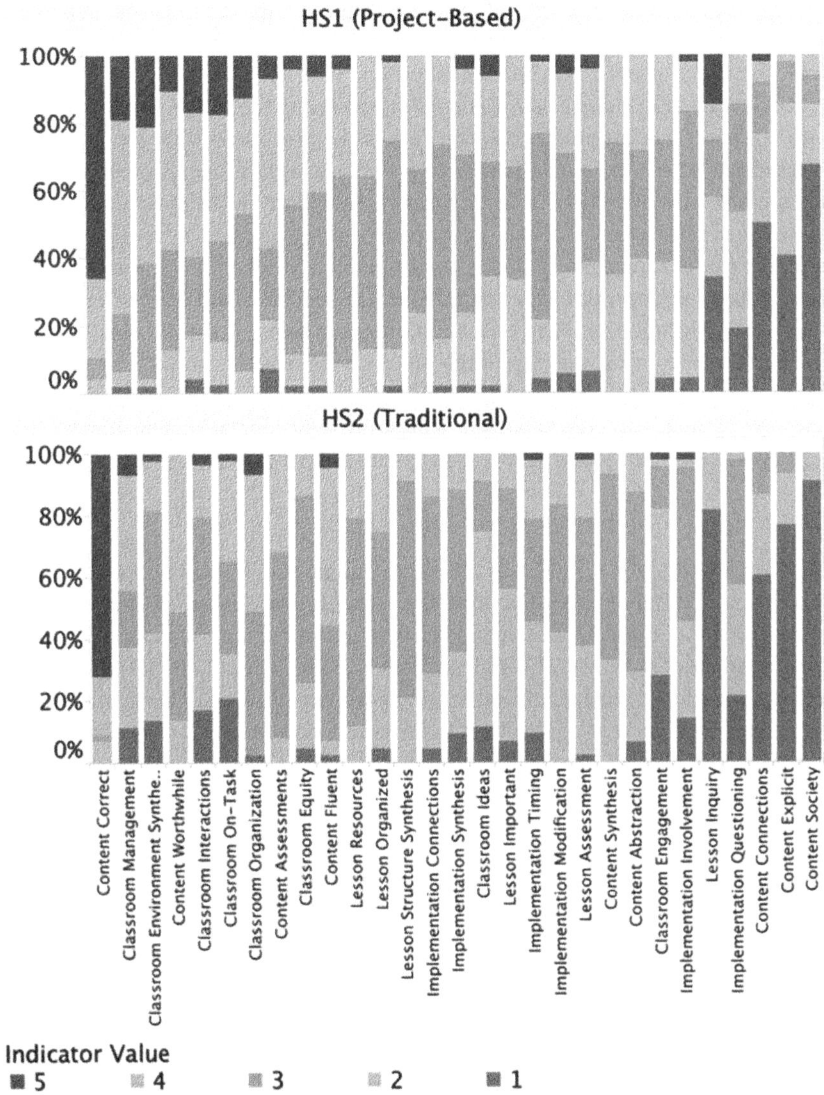

FIGURE 21.1, Comparison of Combined Classroom Observations at HS1 (project-based) and HS2 (traditional)

schools were not as dramatic as might have been anticipated. For both schools, the highest average performance was in Content Correctness, and the lowest was in Content Society. This indicator probes for explicit connections within the lesson to the history of science and mathematics or to current events, and has rarely been observed in all UTOP studies to date. One indicator on which we expected to see

a large difference between the schools was Lesson Inquiry. Indeed, the lessons in the Project-Based school had lessons scoring 3 or better half the time, while the traditional school never did. We noted, however, that even in the Project-Based school, half the lessons did not involve a strong component of student inquiry.

Another indicator that reflects the difference between the schools is Classroom Ideas; in the Project-Based school students generated their own ideas during around 70% of the lessons, while in the conventional school this happened only 20% of the time. There were two indicators where the conventional school may have had better results: Content Correctness and Content Fluency. This may be because project-based instruction can put teachers in situations that are difficult to predict in advance and put a greater strain on content knowledge, compared to what occurs in well-prepared conventional lessons. Thus, the main takeaway from this study was that when making comparisons using the UTOP, individual indicators that are theoretically aligned to important differences between groups may be the most revealing, while other differences may be subtle.

A feature of this project that had not been attempted previously was that the observed teachers were themselves trained on the UTOP and observed each other. They also participated in data reveal sessions; first, overall results for each school were given with all teachers present, followed by sessions with individual teachers where they were given and then interviewed about their individual results. The individual results were held confidential and were used only for formative purposes. Surveys and interviews with teachers, along with statements from the data reveal sessions, revealed a number of trends. First, teachers appreciated receiving UTOP training and found it a demanding but important tool to reflect on their practice. Second, when sharing UTOP scores and related feedback with teachers, the teachers requested additional information—specifically field notes and examples from the UTOP. This suggests that the rich qualitative data written for UTOP observations as supporting evidence was desired by teachers to improve their practice. And third, teachers chose different UTOP indicators to focus on for continued growth, and these areas of interested seemed to be related to teaching experience. This suggests that using the UTOP can allow for differentiated professional development that meets teachers' needs.

Two examples will have to suffice to illustrate the richness of the discussion stimulated by the UTOP data reveal sessions. Concerning the group data reveal, below is an excerpt of a discussion of a case where the group-level UTOP score was unexpected. As can be seen from Figure 21.1, even at the project-based school, the UTOP indicator assessing whether the lesson utilized inquiry or problem-based learning was usually rated a 3 or below.

Researcher: What is the inquiry [score], what does it say about inquiry?

Teacher: Um, it's inquiry oriented or problem based, so investigative, inquiry or project oriented. And you were not alone on that, I

> think the school in general was 3 and teachers' jaws were on the floor.

And from a discussion of how thinking about indicators led to changes in practice, one teacher stated that:

> I just freaked out and changed the entire next project when I was reading one of these, um, content understanding, using mistakes. OK, so, here's one example, there's tons of them I can think of, so 4.2 *Content communicating through direct and indirect instruction is consistent with fluent content knowledge of mathematics and science*, and then it says, *the teachers depth of subject matter knowledge can also be assessed how his understanding of student mistakes, common misconceptions, or alternatives ways of thinking about and solving problems is used to help build student understanding* ..., I can think of multiple examples of misconceptions about the sun, oh it's a big ball of fire, well if so, it would last for 5,000 years... I was like, I could see your note. The whole process of thinking about my teaching is done at a higher and more effective level.

What these comments indicate is that the numerical scores and qualitative comments in the UTOP combined to produce a challenging and productive experience with great potential for professional development.

Case 4: The STEM Academy for Science Teachers and Leaders

Funded by the Texas Instruments Foundation and O'Donnell Foundation, the STEM Academy for Science Teachers and Leaders is a partnership between a large, urban school district and university. During each of up to three years of participation, forty middle grades science teachers engage in summer professional development and on-going coaching during the school year. The primary long-term goal of the STEM Academy is to increase students' interest and achievement in STEM fields. As such, the STEM Academy focuses on increasing the frequency and quality of teachers' instruction emphasizing: (a) active learning strategies including project-based learning (PBL) and maker-based instruction (MBI), (b) the scientific process standards, (c) rigorous content knowledge, and (d) differentiation focused on students' socio-emotional needs (Adams, Hatfield, Cox, & Ketterlin-Geller, 2018; Adams, Hatfield, Cox, Mota, Sparks, & Ketterlin-Geller, 2018; Perry, Reeder, Brattain, Hatfield, & Ketterlin-Geller, 2017). The program implementation followed a cohort model where 15 teachers started in summer 2017 (Cohort 1), and 28 teachers started in summer 2018 (Cohort 2). At the time of this publication, Cohort 1 teachers were in their second year of the STEM Academy and Cohort 2 teachers were in their first year.

To measure teachers' enacted practice across time, our team used the MET-study version of the UTOP, which included 22 of the original 26 indicators. We used the UTOP because of the strong theoretical alignment between the UTOP and the STEM Academy. In our study, eight raters observed the STEM Academy

teachers using the UTOP. Because scoring with the UTOP necessitates adequate pedagogical and content knowledge, we carefully considered raters' backgrounds when hiring. The eight raters had an average of eight years of teaching experience, and most (75%) had a master's degree or higher. During the first year of the STEM Academy, a developer of the UTOP trained four of the eight raters. At that point, calibration statistics indicated that on average 54% of the raters' scores matched the master score exactly, and 76% matched within one. During the second year, one of the previous raters trained four new raters on the UTOP. Following the second training, calibration statistics indicated that on average, 27% of the four new raters' scores matched the master score exactly and 64% matched within one.

At the time of this publication, UTOP raters' scores were available for 15 teachers near the end of their first year of the STEM Academy (Spring 2018) and 40 teachers near the beginning of either their first or second year of the STEM Academy (Fall 2018). Raters observed 12 Cohort 1 teachers at both spring 2018 and Fall 2018. Three teachers did not participate during the second year for various reasons (e.g., exiting the school district). Twenty-eight Cohort 2 teachers were in their first year of the STEM Academy in Fall 2018. Due to the several limitations (e.g., small sample size and absence of baseline data), conclusions based on these data should be made cautiously until further data are collected.

Results suggest that for Cohort 1 teachers, UTOP scores were relatively consistent from spring 2018 to fall 2018. This is not surprising given that teachers engaged with the STEM Academy for almost a full year before they were observed with the UTOP. Similarly, Cohort 2 teachers started participating in summer 2018 and participated in the program for several months before they were observed for the first time in fall 2018. Our team used the UTOP for research observations; whereas, university coaches observed teachers using an observational protocol specifically designed to systematically support practitioners in providing feedback to teachers (Adams, Sparks, & Ketterlin-Geller, 2019). Thus unlike in the previous case, the UTOP was not used directly for teacher professional development. Our coaching observation protocol was, however, informed by several of the indicators on the UTOP, as well as existing PBL and MBI literature. Compared to the UTOP, the coaching observation protocol included fewer, shorter indicators and did not require scorers to include evidence. Coaches were trained to provide systematic and specific teacher feedback based on teachers' scores on the coaching observation protocol. In this way, compared to the UTOP, the coaching observation protocol: (a) required less time to complete, (b) was more tailored to the components of the STEM Academy intervention, and (c) more efficiently informed immediate feedback for teachers.

Based on our experience using the UTOP, we have three recommendations for future users. First, the overall synthesis scores are appealing because they are informed by several indicators and are thus less susceptible to measurement error. However, these scores may not be sensitive enough to capture a change in specific

instructional practices. For the STEM Academy project, we expect to see changes across time on specific indicators that are well-aligned with the training teachers receive (e.g., the lesson included an investigative or problem-based approach to important concepts in mathematics or science).

Second, as with all observational measures, rater bias is an issue. Because observational measures are susceptible to measurement error due to rater perspectives, high-quality training is essential. The rater agreement statistics were higher when raters were trained by a developer of the UTOP rather than a previously-trained rater. Additionally, because raters in our study were typically hired as temporary employees, they were difficult to retain across time. At the same time, they did not have relationships with the teachers or knowledge of teacher cohorts, which minimized bias due to rater perceptions. There is likely no perfect way to address rater bias, but we offer three suggestions: (a) hire raters who are external from the program and available long-term, (b) offer high-quality training before and during data collection, and (c) pair raters as frequently as possible to simultaneously observe live lessons and compare scores.

Third, using observational measures like the UTOP can be time and resource-intensive, but can contribute to in-depth quantitative and qualitative analysis. Researchers recommend collecting multiple observations at each time point to ensure the observed scores capture teachers' practice more broadly (Walkington & Marder, 2018). This is difficult to implement given limitations the time and fiscal resources required for multiple visits to teachers' classrooms. Other research teams may decide to score videos of lessons rather than conduct live observations. In addition, on average, our raters required forty minutes to one hour to score one lesson on the UTOP. Although this is a time commitment, scoring included assigning quantitative scores and providing detailed explanation and justification for their scores, which affords in-depth qualitative analysis. Other research teams may decide to use an abbreviated UTOP protocol or limit score justification, which would require less time to score.

In summary, the UTOP provided an in-depth snapshot of teachers' inquiry-oriented instructional practice that aligned well theoretically with the STEM Academy. Although we are early in collection and analysis of our data, we anticipate that indicator-level analyses and qualitative analysis will contribute to the rich longitudinal examination of teachers' enacted practice within the context of the STEM Academy.

Future Direction (Case 5): Using the UTOP to Observe Undergraduate STEM Instruction

An interesting evolution of the UTOP is taking place in the application to observation of undergraduate STEM teaching. The motivation for this final project was the fact that faculty evaluation has depended largely upon student course surveys. These are known to display a variety of biases and provide little formative assistance to faculty. Peer observation is part of the promotion process but had

largely been carried out by untrained observers without a rubric, without any uniformity over time or between departments, and with a general agreement that any unfavorable statements about teaching were to be avoided. This meant that peer observations were completely ineffective when it came to providing assistance to strong teachers pursuing tenure. For several years discussions with faculty about adopting the UTOP as a solution were fruitless. There were three objections. The first was that the instrument made use of terms and examples that were appropriate in secondary instruction, but not in undergraduate instruction. The second was that the instrument was so long that completing it was burdensome and impractical. The third was that faculty would never consent to have their teaching evaluated with numerical scores.

The first two of these problems were resolved by assembling faculty committees within specific departments and authorizing them to spend several hours to make slight wording changes to UTOP indicators and to pare back the number to indicators that seemed essential. This was done independently by two different units, and in each case, the instrument was reduced to 15 indicators in two sections: Classroom Structure and Content. The committees acquired sufficient familiarity and comfort with the instrument during this process. Consequently, the instrument has been used regularly in observation. The third point concerning numerical scores has been more difficult. Observers and faculty in one of the units have been fairly unanimous that the scores are useful, increase the objectivity of the observations, and assist in the formative discussions that follow observations. The other unit was less persuaded of the value of including numerical scores, although their use has continued. A nearly identical observation instrument to the one developed in these two cases has been employed with several other departments, but so far always without any numerical scores.

Because the UTOP has been modified substantially in this process, its validity and inter-rater reliability are no longer guaranteed by the studies that were carried out previously. Furthermore, the training of faculty observers has so far been less thorough than in earlier uses. Finally, the frequency of observations, typically once per year per faculty member, might be too low to provide reliable results. This is why other instruments such as COPUS have focused on much more objective features of the classroom (Smith, Jones, Gilbert, & Wieman, 2013) technology, engineering, and mathematics (STEM and why the Teaching Practices Inventory (Wieman, 2015) work systematically to improve it, and document that quality. Institutions in turn are unable to incentivize, track, or demonstrate to external stakeholders improvement in the quality of the teaching that they provide. These deficiencies are becoming increasingly conspicuous in light of the calls for greater accountability in higher education and for the adoption of more effective, research-proven teaching practices, particularly in the areas of science, technology, engineering, and mathematics (the STEM disciplines moves away from peer observation altogether. Whether these problems can be overcome by further research and development, or whether an instrument similar to the UTOP

will prove ultimately unsuitable for the improvement of undergraduate teaching remains to be determined.

DISCUSSION AND CONCLUSIONS

The above narrative provides a glimpse at how the UTeach Observation Protocol (UTOP) has been used to observe STEM teachers teaching elementary school to college. In three of the cases (case 1, 2, and 4), the protocol was used strictly for research purposes. We compared groups of teachers (case 1), looked for relationships between teaching behaviors and value-added estimates (case 2), and to tracked teacher change over time as the result of a professional learning intervention (case 4). However, in the other two cases, there have been efforts to use the UTOP as a tool for teacher professional development. In one instance this was in a high school setting where teachers are given both the average school-level UTOP scores and their individual UTOP scores (case 3), while in the other (case 5) was in a university setting where STEM professors are observed and given critical feedback on their teaching through discussions using the UTOP. These cases suggest that the UTOP has the potential to both support research into elements of or changes in teacher classroom behaviors, and to support practitioners in reflecting on their own teaching and receiving feedback from others.

There are some consistent lessons that appear in multiple cases described above. First is that the UTOP is time-intensive due to the number of indicators and amount of supporting evidence observers are expected to cite. However, in our case studies, this rich qualitative evidence allowed for a deeper understanding of teaching when used for research purposes, and for better discussions with teachers when used for professional development. In addition, the cases suggest that while the UTOP has a number of indicators, this is a valuable quality. From a research perspective, growth on a variety of fine-grained behaviors that may be expected to change as the result of an intervention can be tracked, and from a professional development standpoint, teachers have a lot of freedom and choices to determine which indicators to focus their growth on, depending the stage of their teaching career. It is also important to note that some of the cases involved versions of the UTOP that had been shortened to meet local needs. This was particularly true in case 5, examination of undergraduate STEM instruction, where the need to overcome practical barriers to adoption was particularly urgent. It also applied in case 4, where a shorter protocol for coaching was used instead of the UTOP for the purposes of professional development. Thus, while our experience with the UTOP indicates that when used in full it can provide a rich professional development experience, it seems that shortened versions will in practice provide the way forward if we hope for broad adoption.

Another consistent message is that both inter-rater reliability and training are important. The degree to which multiple observations per time point are absolutely required depends in part on the stakes of the observation. The observations might constitute a relatively informal opportunity for teachers to engage

in conversations about their teaching, a high visibility research study seeking to make generalizable claims, or a high-stakes teacher evaluation mechanism tied to future professional opportunities. Training raters on the UTOP may be an excellent professional learning experience in its own right, as we often heard from raters we trained in the MET study (case 2), and as was the case in the PHISD study. Observing cutting edge approaches to STEM instruction, which should be increasingly integrated, inquiry-oriented, and technology-infused, requires careful thought about rater qualifications raters. Three of the cases above (case 1, 3, 4) favored pairing raters, with two of the cases implementing the practice where one observer was an expert in mathematics and the other an expert in science.

The story of the UTOP shows how the vision of the UTeach program was operationalized into a set of classroom-level behaviors and indicators, and how this process let us accomplish goals related to research and teacher professional learning. However, the instrument remains a living document—the vision of any teacher preparation programs necessarily shifts over time as it grows, expands, and learns. A vital element of the UTOP is that it is both broad in its scope and detailed in its particulars—allowing it to be adapted to meet local needs.

ACKNOWLEDGEMENTS

We would like to thank Mary Walker for providing an initial impetus for the project and contributing to the rubrics and examples, Pamela Powell for implementing many versions of the UTOP in student teaching in UTeach and sharing her experiences, Audrey Boklage for conducting many of the early observations, and the Gates Foundation for their financial support through the Measures of Effective Teaching project. We would also like to thank the teachers who participated in our studies.

REFERENCES

Adams, E. L., Hatfield, C., Cox, C. T., Mota, A., Sparks, A., & Ketterlin-Geller, L. R. (2018). *STEM Academy for Teachers and Leaders: 2017–18 Coaching and PLC Evaluation* (Tech. Rep. No. 18-03). Dallas, TX: Southern Methodist University, Research in Mathematics Education.

Adams, E. L., Hatfield, C., Cox, C. T., & Ketterlin-Geller, L. R. (2018). *STEM Academy for Science Teachers and Leaders: 2018 Teacher academy I evaluation* (Tech. Rep. No. 18-02). Dallas, TX: Southern Methodist University, Research in Mathematics Education.

Adams, E. L., Sparks, A., & Ketterlin-Geller, L. (2019, March) *Multi-level exploratory factor analysis: Evidence of internal structure for a measure of STEM instructional practice.* Paper presented at the 7th Annual Texas Universities' Educational Statistics and Psychometrics (TUESAP) Alliance Conference, Dallas, TX.

DeZeeuw, A., Walker, M., & Marder, M. (2015). *A math and science content-specific observation protocol as professional development: Teacher needs, uses, and lessons*

learned. Paper presented at the 2015 Annual Meeting of the American Educational Research Association. Chicago, IL.

Horizons Research Inc. (1999). *Inside the classroom observation and analytic protocol.* Retrieved from: http://www.horizon-research.com/insidetheclassroom/instruments/obs.php

Kane, T. J., & Staiger, D. O. (2012). *Gathering feedback for teaching: Combining high-quality observations with student surveys and achievement gains.* Bill & Melinda Gates Foundation. Retrieved from: http://files.eric.ed.gov/fulltext/ED540960.pdf

Landis, J. R., & Koch, G. G. (1977), The measurement of observer agreement for categorical data. *Biometrics, 33*, 159–174.

Marder, M., Walkington, C., Abraham, L., Allen, K., Arora, P., Daniels, M., Dickinson, G., Ekberg, D., Gordon, J., Ihorn, S. & Walker, M. (2010). *The UTeach observation protocol (UTOP) training guide.* UTeach Natural Sciences, University of Texas Austin.

Perry, L., Reeder, M. J., Brattain, K., Hatfield, C., & Ketterlin-Geller, L. R. (2017). *STEM Academy for Teachers and Leaders: 2017 academy evaluation* (Tech. Rep. No. 17-01). Dallas, TX: Southern Methodist University, Research in Mathematics Education.

Pianta, R. C., La Paro, K. M., & Hamre, B. K. (2008). *Classroom assessment scoring system (CLASS).* Baltimore, MD: Brookes.

Piburn, M., Sawada, D., & Turley, J. (2000). *Reformed teaching observation protocol (RTOP) reference manual.* ACEPT Technical Report No. IN00-3. Tempe, AZ: Arizona State University.

Smith, M. K., Jones, F. H. M., Gilbert, S. L., & Wieman, C. E. (2013). The classroom observation protocol for undergraduate stem (COPUS): A new instrument to characterize university STEM classroom practices. *CBE Life Sciences Education.* https://doi.org/10.1187/cbe.13-08-0154

Walkington, C., & Marder, M. (2014). Exploring excellence in teaching using the UTeach Observation Protocol: Connecting teaching behaviors to teacher value-added on assessments measuring conceptual understanding. In T. Kane, K. Kerr, & R. Pianta (Eds.), *Designing teacher evaluation systems: New guidance from the Measures of Effective Teaching project* (pp. 234–277). San Francisco: Jossey-Bass.

Walkington, C., & Marder, M. (2018). Using the UTeach Observation Protocol (UTOP) to understand the quality of mathematics instruction. *ZDM Mathematics Education, 50*, 507–519. DOI: 10.1007/s11858-018-0923-7.

Walkington, C., & Valerius, M. (April, 2012). *Using classroom observation research to inform debates about teaching effectiveness.* Paper presentation at 2012 Research Pre-session for National Council of Teachers of Mathematics Annual Meeting. Philadelphia, PA.

Wieman, C. (2015). A better way to evaluate undergraduate teaching. *Change: The Magazine of Higher Learning, 47*(1), 6–15

CHAPTER 22

DEVELOPING AN ASSESSMENT OF ATTENTIVENESS FOR PROGRAM EVALUATION

Leslie Atkins Elliott, Michele Carney, Sara Hagenah, and Tatia Totorica

The IDoTeach program at Boise State University is developing an instrument to measure teacher candidates' attentiveness to student thinking situated in key disciplinary-specific contexts; the instrument is designed to inform program improvement and evaluation. In this chapter, we describe the need for such an assessment, its development process, and how the assessment may be used to inform program improvement. This work builds on prior instrument development in mathematics (Carney, Cavey, & Hughes, 2017) and expands to capture the range of disciplines in the IDoTeach program.

Figure 22.1 is an assessment item that IDoTeach, Boise State University's UTeach replication program, has administered to a range of secondary students. We ask our chemistry teaching candidates to analyze the secondary students' answers, and then describe what the student understands and how they would respond, as a teacher, to that student. For example, the candidate might notice whether or not students describe a particulate nature of matter (e.g., describing molecules and their interactions). The candidates might note whether or not the students treat matter as conserved. They could attend to how students justify their

Preparing STEM Teachers: The UTeach Replication Model, pages 311–323.

Enid lights a candle in a small bathroom, and her friend, Amit, warns her that this will leave a waxy residue on everything.

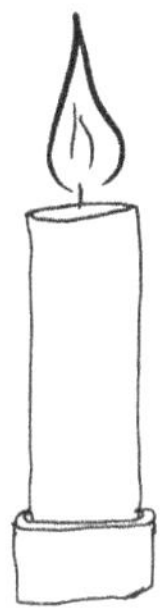

Amit: Oh, don't do that. The wax will evaporate and leave a waxy film on everything and it's hard to clean up.

Enid: No no - the wax gets burned -- it turns into heat. There's no wax leftover after the candle burns.

Do you agree with Amit - the wax evaporates? Or with Enid - the

FIGURE 22.1. IDoTeach Disciplinary Attentiveness Assessment Item

ideas, marshaling evidence from their everyday experiences with flames to suggest that a chemical change is taking place, turning the wax and oxygen into other substances as it burns. In candidates' plans for instruction, they might correct students' misconceptions, build on productive ideas they see in students' responses, or evaluate students' ideas. Our goal is that our teacher candidates in chemistry demonstrate *disciplinary attentiveness for pedagogical practice* (Carney, Cavey & Hughes, 2017): the ability to interpret individual students' ideas in light of disciplinary goals, and respond instructionally to those ideas in ways that connect students' ideas with disciplinary learning targets.

In this chapter, we provide a brief overview of how we developed a program framework that established disciplinary attentiveness as a primary goal, assumptions in the UTeach/IDoTeach structure about how this goal is met, and our steps for constructing these assessments to gauge whether or not this program structure is effective at meeting this goal. Our objective for readers is, first, to provide a model of how we coordinated faculty, teachers, students and stakeholders to develop program goals and assessments, and then to illustrate how the assessment tool we developed can be used to assess and improve our program. For those in UTeach programs, we anticipate that this particular assessment and its development will be of interest to them as they seek assessments that inform program structures. For those not in UTeach, our expectation is that this work provides a model that others might adapt to settings where a range of groups have a stake in student learning outcomes.

PROGRAM FRAMEWORK

As with all teacher preparation programs, IDoTeach is responsible for ensuring that a range of frameworks that outline the knowledge, skills and dispositions for

our teaching candidates (undergraduate preservice teachers) are met, including those required by Council for the Accreditation of Educator Preparation, our state Board of Education, the content knowledge within the state Science and Math standards, and those we set internally, based on empirical research in teacher preparation. These frameworks include: the Charlotte Danielson Framework for Teaching; ISTE Standards; WIDA supports; S-PAT; CAEP Standards; Science/ Math Education Standards (both local and national). Our initial attempts to assign each particular element from the multiple frameworks to a course left us with an overwhelming set of goals to meet in each course, at increasing levels of mastery.

Over several years the core faculty in IDoTeach, including faculty from math, science, philosophy and education departments, along with Master Teachers, engaged in an iterative process of developing, refining, and categorizing a set of goals to develop our own IDoTeach program framework, with four knowledge learning outcomes (KLOs) and four related practice learning outcomes (PLOs) (see Figure 22.2).

From this set of knowledge and practice learning outcomes, we mapped these onto the courses within our program.

Teacher candidates will demonstrate knowledge of and reflective habits of mind related to:
KLO 1: Key ideas in (STEM) discipline(s) and how these ideas emerge from disciplinary practices. **KLO 2:** How and why to engage students in discourse characteristic of STEM disciplines. **KLO 3:** Student thinking related to key disciplinary ideas and practices and how that thinking develops over time. **KLO 4:** Issues of equity and how such issues can promote or inhibit student access to disciplinary ideas.
With professionalism, respect, and enthusiasm, teacher candidates will demonstrate the ability to:
PLO 1: Plan and reflect on instructional sequences that promote progressive development of and connection between disciplinary ideas and practices **PLO 2:** Implement instructional plans at an appropriately high level of cognitive demand with discourse characteristic of STEM disciplines. **PLO 3:** Elicit student reasoning, assess and analyze evidence of student reasoning, and respond in ways that build upon student reasoning. **PLO 4:** Attend to issues of equity in ways that make disciplinary ideas and practices accessible to each and every student.

FIGURE 22.2. IDoTeach Program Learning Outcomes

One clear assumption in the UTeach framework that was highlighted for us in this process is that the disciplinary content knowledge and practices that candidates need for secondary STEM teaching will be met within the courses in their majors. That is, the courses in IDoTeach do not explicitly address KLO 1: "key ideas in STEM disciplines and how these emerge from disciplinary practices." The IDoTeach course Knowing and Learning, for example, introduces students to the sociocultural perspective that students best learn disciplinary content by engaging in practices of the discipline with their peers—but the course does not itself teach this content or these practices. Similarly, the course Classroom Interactions, a field-based course, asks candidates to develop lessons that do this work—that engage students in practices characteristic of geoscience, say, as they learn content standards for geology—but Classroom Interactions does not teach geoscience content or practices but instead assumes this has been learned by students in their geoscience coursework.

The idea that content knowledge as taught in the disciplines may be insufficient for teaching is not a new idea: pedagogical content knowledge (PCK), the "ways of representing and formulating the subject… a veritable armamentarium of alternative forms of representation, some of which derive from research whereas others originate in the wisdom of practice" (Shulman, 1986, p. 9), has long been recognized as critical to classroom teaching. However, explicit incorporation of PCK is not evident in the UTeach coursework. Individual replication programs, of course, can adapt the program to incorporate PCK. For example, IDoTeach has three required courses for secondary mathematics candidates that in large part focus on PCK. However, we do not have parallel offerings for a science candidates, in part due to the smaller number of candidates within each disciplinary area (a course enrollment issue).

Despite explicit instruction in disciplinary content for teaching, anyone who has taught recognizes that, in developing lessons and engaging with students' ideas, we come to understand the content much more deeply. And so, although our courses do not explicitly teach, say, Newton's Laws, we expect that our teacher candidates in physics will, by preparing lessons and assessments of Newton's Laws and listening to students' ideas about this topic, come to understand this content much more deeply and develop a richer PCK in this area. However, a critical question for us is whether or not our teacher candidates have sufficient content and pedagogical knowledge to begin this work.

With this question in mind, and with an eye towards research highlighting the critical role that it plays in student learning, we decided to focus our first program assessment on elements of PLO3: "With professionalism, respect and enthusiasm, teacher candidates will demonstrate the ability to elicit student reasoning, analyze evidence of student reasoning, and respond in ways that build upon student reasoning." In particular, and building on ongoing work in our department with mathematics teacher education, we focus first on candidates' ability to analyze ev-

idence of student reasoning and build on that reasoning. Below, we provide a brief overview of research on this theme before describing our approach to assessment.

DISCIPLINARY ATTENTIVENESS

Attentiveness is a construct that arises from the mathematics education literature, and describes a teacher's ability to analyze and respond to a particular student's disciplinary ideas from a progressive formalization perspective (Freudenthal, 1973; Gravemeijer & van Galen, 2003; Lampert et al., 2013; Pierson, 2008). Or, as Ball (1993) describes it, that a teacher is both *responsive to students* and *responsible to the discipline*. In the science education literature, attentiveness is a core element in *responsive teaching* (e.g., Richards, Atkins, Levin, & Robertson, 2015), an instructional approach in which the "teacher's role is to support [generative] engagement and attend to it—watch and listen to the students' thinking, form a sense of what they are doing, and in this way identify productive beginnings of scientific thinking...build[ing] on what students have begun" (Hammer, Goldberg & Fargason, 2012, p. 55).

To do this well, responsive teaching requires the teacher to recognize the generative scientific ideas within students' informal or incorrect answers, to anticipate connections between those ideas and the targeted scientific content, and to respond with instruction that supports that connection. The construct of attentiveness arises from this necessary skill, and our assessment is an attempt to capture this complex construct that involves both pedagogical and disciplinary content knowledge. In the following section we describe the assessment itself and its development.

Developing an Assessment of Attentiveness

Our multi-phased development of assessment items for teacher candidate attentiveness in the STEM disciplines was modeled after the Mathematical Attentiveness in Pedagogical Practice (MAPP) item-development process (Carney, Cavey, & Hughes, 2017). We expect that the authentic student, candidate, and teacher responses used during this development process will help us and other education programs to efficiently measure teacher candidates' attentiveness.

The four distinct phases of item development, each of which will be described in turn, describe our efforts to engage teacher candidates in the cognitive processes needed for attentiveness: identifying the pedagogical intent and key disciplinary concept(s) of a specific student task; making evidence-based inferences about a student's understanding of the disciplinary concept being targeted; and deciding on an appropriate pedagogical response that builds both on an individual student's thinking and toward the disciplinary intent of the task.

We illustrate our development process by first presenting a final Phase 4 selected-response item to measure teacher candidate attentiveness in the chemistry discipline (see Figure 22.3). *Phase 1*

Phase 1 consists of identifying and then selecting key disciplinary concepts with the potential to elicit a range of secondary student cognition evidence. These

Consider the task given to a task of chemistry students:

Enid lights a candle in a small bathroom, and her friend, Amit, warns her that this will leave a waxy residue on everything.

Amit: *Oh, don't do that. The wax will evaporate and leave a waxy film on everything and it's hard to clean up.*
Enid: *No no - the wax gets burned -- it turns into heat. There's no wax leftover after the candle burns.*

Do you agree with Amit - the wax evaporates? Or with Enid - the wax turns into heat? Or do you have another idea for where the wax goes when a candle burns? Please thoroughly explain your reasoning.

Student A's response is shown below:

"I think that the wax of a candle turns into heat. I am making this inference based on past experiences and the fact that the candle needs a fuel to keep the fire going. I can use oxygen from the air to maintain it but the wax is fuel."

Rank the following statements in terms of your level of agreement with the statement. Each level of agreement can only be selected once.

	High Agreement	Moderate Agreement	Least Agreement
I would help them understand that the wax doesn't disappear, it evaporates. This happens after the candle melts and turns into a liquid.	○	○	○
I would ask them why a fire needs oxygen. If they understand that the oxygen combines with the wax, then I might ask them to model what happens to the elements after the reaction.	○	○	○
I would explain that they are right: wax is a fuel. And then I would explain combustion reactions to them so they would know what the products of that reaction are.	○	○	○

FIGURE 22.3. Selected-Response Chemistry Item for Assessing Attentiveness to Student Ideas

key concepts are considered by both pedagogical and disciplinary experts to be foundational for the discipline and to span the secondary curriculum (e.g., genetics in biology, functions in mathematics, forces in physics, etc.).

Our specific work with this phase entailed recruitment of faculty members from each university STEM department involved in secondary teacher education: Biology, Chemistry, Computer Science, Engineering, Geoscience, Mathematics, and Physics. For each discipline, we assembled teams comprised of at least one IDoTeach instructor and one departmental faculty member from outside the secondary STEM educator preparation program.

To continue with our item used for illustration in Figure 22.1, the chemistry team identified conservation of matter within combustion reactions to be a key disciplinary idea that fit the criteria for inclusion in the assessment development process. Conservation of matter, as a distinct accounting from the conservation of energy, is foundational for chemistry, but also critical across the sciences.

Phase 2

In Phase 2, we develop tasks which target the key disciplinary ideas and have the potential to elicit a broad range of student cognition evidence. After administering these tasks to a range of secondary students in multiple grade levels and school settings, we analyze the resultant evidence to identify common themes in student reasoning and to select exemplar work samples for teacher candidates to consider.

For our specific work with this phase, we engaged in several steps. First, each of the disciplinary teams assembled in Phase 1 created multiple tasks to administer to secondary students that connected to the key pedagogical and disciplinary concepts identified in phase one. To focus their work, we asked that the drafted tasks (a) be accessible to all secondary students, (b) require less than 15 minutes for a typical secondary student to complete, and (c) have the potential to elicit a broad range of student thinking evidence.

Second, we invited in-service secondary teachers from each discipline to provide feedback on and revision suggestions for the tasks. These secondary teacher volunteers (a) verified the key disciplinary idea being targeted by their assigned tasks, (b) assessed the accessibility of each task for a range of secondary students, (c) made informed predictions about the time a typical secondary student would need to complete each task, (d) made recommendations regarding each task's administration feasibility, and (e) suggested revisions to the task in order to improve adherence to and alignment with the established criteria.

Third, we distributed the revised tasks to a wide range of secondary students. Thirty in-service teachers from fifteen schools and four districts administered the tasks to over 1000 students. In an effort to maximize the range of student responses elicited, each task was administered to at least three different secondary grade levels (grades 6 through 12) and in courses ranging from remedial to Advanced Placement.

Last, the IDoTeach instructor and departmental faculty teams sorted and reviewed the collected student work for each secondary task to identify and select exemplar student responses. This selection process focused on the identification of student work samples which were likely to emerge in a typical classroom, and therefore be representative of the future work our candidates would encounter, and to provide evidence of productive, but not necessarily correct, student thinking for candidate analysis.

In the case of our highlighted chemistry item, the original open-ended response task administered to secondary students targeted the key disciplinary idea of con-

servation of matter in combustion reactions (see Figure 22.1), was distributed to students at the middle school, high school, and AP levels, and generated a range of student cognition evidence related to both conservation of matter and combustion reactions.

The resultant secondary student evidence revealed a range of thinking related to conceptions about energy, states of matter, and conservation of matter and also allowed us to select exemplars to represent ways of thinking our teacher candidates are likely to encounter in their future secondary classrooms. For example, one student attends to the reaction of the candle with, noting: "I think that the wax of a candle turns into heat. I am making this inference based on past experiences and the fact that the candle needs a fuel to keep the fire going. I can use oxygen from the air to maintain it but the wax is fuel." Another is less explicit about the chemical reaction taking place, but attends to the conservation of matter, writing, "I agree with Amit sort of. I think that he believes it goes somewhere so I think it evaporates but it goes into the air, it can't just turn into nothing so it has to turn into something."

Phase 3

Phase 3 focuses on the development, administration, and analysis of open-ended response item blocks for teacher candidates. Each item block for each discipline consists of multiple components: the original task presented to secondary students, a prompt asking the candidate to identify the disciplinary intent of the task, two exemplar student responses, and additional prompts asking the candidate to make inferences about the students' understanding of the disciplinary concept being targeted and to describe an appropriate pedagogical response to each student. Following administration, candidate responses are then analyzed and categorized into qualitatively different levels of attentiveness to afford selection of exemplar candidate responses for use in the development of final Phase 4 selected-response items.

Again, to illustrate this phase of development, we feature the open-ended-response item used to elicit a range of chemistry teacher candidate attentiveness evidence (see Figure 22.4).

Disciplinary experts in chemistry analyzed teacher candidate responses to this item block. The purpose of this analysis was to code the qualitatively different categories of teacher candidate responses to the item block prompts as specific, generic, and disparate, and to select exemplars from each of these categories for inclusion in the final selected-response Phase 4 items. In this case, the exemplars selected represent the responses to the prompt 'If you were the teacher of Student A, please describe, in detail, how you would respond.' in the open-ended-response item.

For the chemistry item, the three exemplar candidate responses that emerged from this process represent three different levels of attentiveness evidence. Our rationale for each is described below.

Chemistry: Candle Burning Task

Before responding to the questions below, please work through the following task given to a class of chemistry students:

What important chemistry idea(s) is this task targeting?

Student A's response is shown below:

> *"I think that the wax of a candle turns into heat. I am making this inference based on past experiences and the fact that the candle needs a fuel to keep the fire going. I can use oxygen from the air to maintain it but the wax is fuel."*

Please describe, in detail, what Student A's work indicates to you about their understanding.

If you were the teacher of Student A, please describe, in detail, how you would respond.

Student B's response is shown below:

> *"I agree with Amit sort of. I think that he believes it goes somewhere so I think it evaporates but it goes into the air, it can't just turn into nothing so it has to turn into something."*

Please describe, in detail, what Student B's work indicates to you about their understanding.

If you were the teacher of Student B, please describe, in detail, how you would respond.

Figure 22.4. Open-Ended-Response Item Block Administered to Chemistry Teacher Candidates.

We anticipate that a teacher candidate who indicates a high level of agreement with the statement, "I would help them understand that the wax doesn't disappear, it evaporates. This happens after the candle melts and turns into a liquid" has a low level of attentiveness for this key disciplinary idea because, by describing the

wax as "evaporating," the teacher candidate is not exhibiting a full understanding of combustion reactions, nor does the response build upon the student reasoning.

By contrast, a teacher candidate who indicates a high level of agreement with the statement, "I would explain that they are right: wax is a fuel. And then I would explain combustion reactions to them so they would know what the products of that reaction are" is exhibiting a moderate level of attentiveness because the response addresses the Student A's ideas, but then directs learning, rather than building upon or further eliciting student reasoning related to the task.

Finally, a teacher candidate who indicates a high level of agreement with the specific coded exemplar, "I would ask them why a fire needs oxygen. If they understand that the oxygen combines with the wax, then I might ask them to model what happens to the elements after the reaction to see that they don't disappear" is exhibiting not only their own understanding of conservation of matter in combustion reactions, but they are acknowledging and building upon the student's ideas in their response.

Phase 4

The selection of candidate response exemplars in Phase 3 allows us to develop, administer, and analyze the functioning of the assessment and each selected-response item's functioning as expected and fit the Rasch model. Each Phase 3 item has the potential to generate five final Phase 4 items: one item which asks a teacher candidate to rank agreement with statements regarding the intent of the task, two items which ask the teacher candidate to rank statements of inference about student understanding of the task, and two items which prompt candidates to rank statements about appropriate pedagogical responses. Each item is designed so that respondents can only assign one level of agreement per exemplar. Figure 22.1 provides an example of a chemistry item that targets the first response prompt featured in the Phase 3 item block seen in Figure 22.4.

Due to the need to collect additional data, the functioning of the chemistry item in Figure 22.1 is yet to be analyzed. Previous work in mathematics indicates these items function well in terms of a spread of responses across teacher candidates. In other words, it is not 'obvious' to teacher candidates which responses are ideal (i.e., most and least attentive).

Next Steps

With the goal of developing selected-response Phase 4 items that capture a range of attentiveness in each of the STEM disciplines, we are currently collecting a range of teacher candidate, in-service secondary teacher, and university faculty responses to each phase 3 open-ended-response item. So far, the items were administered to all undergraduate students in each field-based course of our secondary STEM educator preparation program, to Apprentice Teachers in three other UTeach replication programs at different universities, to disciplinary col-

leagues in STEM departments here and at other universities, and to the departmental faculty and in-service secondary teachers who participated in the secondary item development or administration process.

To help widen the net for response collection, we are assembling a grant proposal to help incentivize participation. We are aiming to collect at least 35 responses per item. Following collection, all candidate, secondary teacher, and faculty responses to our current Phase 3 open-ended-response items will be analyzed and categorized into qualitatively different levels of attentiveness. This process will not only enable us to develop more Phase 4 items, but it will also enable us to determine which of our Phase 3 items are functioning as anticipated. We are anticipating that this analysis could perhaps lead to another cycle of development for additional items in specific disciplines.

DISCUSSION

This chapter describes one process for setting an 'evaluation' goal and then developing an assessment to examine progress towards that goal in a teacher education setting. More specifically, we illustrate how an easily administered and scored program assessment, through coordination with a range of program stakeholders and arising from authentic exemplars in the field, can be developed for scalable implementation across programs.

As a teacher education program, we are focused on continually evaluating and improving our program structure, courses, and activities. In addition to our own desire to examine our work to improve practice, we are held accountable to multiple sets of standards across various stakeholder groups. Via both state and national accreditation processes, we must demonstrate we are meeting these various standards and other requirements for the preparation of new teachers. These standards can feel overwhelming in terms of the number of standards we are required to address within limited course offerings, and it can be difficult to provide evidence of having measured and met these standards. Without setting program priorities, this can result in attempting to meet and measure each standard individually. The resultant dilution and diffusion of our efforts can compromise both program implementation and assessment.

To combat this issue, we streamlined the multitude of standards we were responsible for into a framework that also aligned with current research in STEM teacher education, the needs of our community's stakeholders, and the protocols for teacher evaluation currently being enacted in our state. To ensure our program's relevance and to promote collaboration within our local education community, we felt it was important that our evaluation and improvement work was coordinated in conjunction with a variety of stakeholders, such as program faculty, local educators, students, and community members. Even within our framework, we had to narrow to a particular objective to assess. In our case, we focused on the assumption that disciplinary attentiveness was being appropriately developed and cultivated throughout the program because this was something we felt was

particularly critical for our teacher candidates but something we were not sure was being adequately developed due to the design of the program. Once we determined our focus on disciplinary attentiveness, we knew we needed a measure of that construct in order to evaluate if it was increasing as a result of participation in our program.

Assessments in teacher education tend towards two ends of a spectrum: (a) easy to administer and score (e.g., Praxis) but lacking a level of authenticity in relation to classroom practice, or (b) more authentic (e.g., observation of classroom practice) but incredibly time-consuming to achieve reliable and valid score interpretations. In general, the amount of time and money available to teacher education program assessment is limited. Therefore, we needed a measure that would be relatively easy to administer and score while also being explicitly connected to classroom practice. The multi-phased development process we have described in this chapter is extensive but is relatively authentic due to the use of actual responses from students and teacher candidates. In addition, our process takes into account the knowledge of and promotes buy-in across a broad range of stakeholders due to the collaborative nature of the development process. These aspects make the intensive development process worthwhile.

The purpose of our instrument is to examine the trajectory of our teacher candidates' disciplinary attentiveness development through program coursework. We recognize it does not provide a fine-grained analysis of pedagogy or content knowledge. That is not our intent. Instead, it serves as a general indicator of this construct, and a more fine-grained examination would be needed if we find disciplinary attentiveness is not increasing throughout program participation.

CONCLUSION

We see cross-university programs, such as those found in UTeach replications, as particularly situated for larger-scale investigation of how teacher education can be improved. For example, as a first step, other teacher education programs interested in developing attentiveness in teacher candidates could make use of the assessments we are developing to examine candidate progress within their own programs. On a larger scale, teacher education programs could work together to develop common interventions specific to attentiveness/responsiveness and to use the assessments to examine the interventions' impact. Having a larger number of students across different contexts allows for claims of generalizability of the likelihood results would be consistent across multiple site. Engaging in this type of work across teacher education programs has the potential to support a transformation in teacher preparation.

REFERENCES

Ball, D. L. (1993). With an eye on the mathematical horizon: Dilemmas of teaching elementary school mathematics. *The Elementary School Journal*, *93*(4), 373–397.

Carney, M. B., Cavey, L., & Hughes, G. (2017). Assessing teacher attentiveness to student mathematical thinking: validity claims and evidence. *The Elementary School Journal, 118*(2), 281–309.

Freudenthal, H. (1973). *Mathematics as an educational task.* Dordrecht, The Netherlands: Reidel.

Gravemeijer, K., & van Galen, F. (2003). Facts and algorithms as products of students' own mathematical activity. In J. Kilpatrick, W. G. Martin, & D. Schifter (Eds.), *A research companion to Principles and Standards for School Mathematics* (pp. 114–122). Reston, VA: National Council of Teachers of Mathematics.

Hammer, D., Goldberg, F., & Fargason, S. (2012). Responsive teaching and the beginnings of energy in a third grade classroom. *Review of Science, Mathematics and ICT Education, 6*(1), 51–72.

Lampert, M., Franke, M. L., Kazemi, E., Ghousseini, H., Turrou, A. C., Beasley, H., ... & Crowe, K. (2013). Keeping it complex: Using rehearsals to support novice teacher learning of ambitious teaching. *Journal of Teacher Education, 64*(3), 226–243.

Pierson, J. L. (2008). *The relationship between patterns of classroom discourse and mathematics learning.* Ann Arbor, MI: ProQuest.

Robertson, A. D., Atkins, L. J., Levin, D. M., & Richards, J. (2015). What is responsive teaching? In A. D. Robertson, R. Scherr, & D. Hammer (Eds.), *Responsive teaching in science and mathematics* (pp. 19–53). Routledge.

Shulman, L. S. (1986). Those who understand: Knowledge growth in teaching. *Educational researcher, 15*(2), 4–14.

CHAPTER 23

PROMOTING EARLY CAREER TEACHER RESILIENCE THROUGH A CALTEACH PREPARATION PROGRAM

Doron Zinger, Therese Shanahan, and Kris Houston

In this qualitative multiple case study, we examine a UTeach replication program with high graduate retention rate (90% of graduates over the prior 10 years are still teaching). We explore aspects of the preparation program that graduates report helped them persist as teachers. Data were generated from questionnaires and interviews with program graduates. We analyzed aspects of preparation including university coursework and field-based experiences during the program. Findings suggest that the opportunities to design lessons during coursework, then teaching them through field work, and receiving specific feedback from supervisors helped candidates become more resilient. Additionally, the tight coupling of coursework and field experience, along with early career school site support contributed to teacher persistence. Implications for teacher preparation are discussed.

Providing students with equitable learning opportunities is a long-standing goal of the education system in the United States (Next Generation Science Standards Lead States, 2013). Despite this goal and continued reform efforts, primarily students of color, especially those living in poverty, do not experience the same

Preparing STEM Teachers: The UTeach Replication Model, pages 325–338.

learning opportunities that primarily White, more affluent students do (Darling-Hammond, 2010). Teacher preparation programs, specifically, are charged with providing all students with qualified teachers (Darling-Hammond & Berry, 2006). Despite these goals and efforts, and in large part due to teacher turnover and lack of support for beginning teachers, perpetual teacher shortages and low-quality instruction remain pervasive issues in urban schools serving primarily students of color (Ingersoll, 2001; Ronfeldt, Loeb, & Wyckoff, 2013). Teacher turnover is associated with lower student academic and socio-emotional outcomes (Ronfeldt, Loeb, & Wyckoff, 2013). Particularly worrisome for the teacher preparation community is that teacher turnover rates tend to be higher for new teachers in underserved communities, topping 50% in the first three years of teaching in many cases (Oakland Unified School District, 2018). These communities' schools are often labeled as "high-need," reflecting a need for material resources such as curriculum as well as reflecting the perpetual shortage of human resources in the form of qualified teachers.

One approach to address the ongoing teacher shortages and low-quality instruction has been the development of fast track teacher preparation programs (Zeichner, 2016; Zinger & Normore, 2016). These programs aim to quickly prepare teachers and develop their resilience to be successful teachers, especially in hard to staff subjects such as mathematics and science in high-need schools. Successful fast track programs often have components designed to increase teacher resilience through a focus on student learning, development of teacher support networks, and reflective practices (Patterson, Collins, & Abbott, 2004; Zinger & Normore, 2016). But what program designs and features facilitate the development of early career resilience?

To answer this question, we present one program, the University of California Irvine's (UCI) CalTeach program, a UTeach replication program. We focus on this program because since its inception in 2008, 90% of the program's graduates are currently teaching, and 70% of them are teaching in high-need schools. Thus, we set out to explore why graduates from this program remain in the classroom while others often leave. Our goal is to identify program features and experiences which helped graduates persist in high-need schools. In the following sections we provide an overview of the UCI CalTeach program, present data, and analysis about the program's graduates, and identify key program features and experiences that promoted graduates' persistence.

University of California Irvine CalTeach Program Overview

The UCI CalTeach program was designed to meet the urgent need for qualified mathematics and science teachers in Southern California. The program is a collaboration between three schools at UCI: Biological Science, Physical Sciences, and Education. Students in the program are mathematics and science majors and earn both their Bachelor's degrees and preliminary teacher credential in four years. Students take two introductory courses with field work during their first

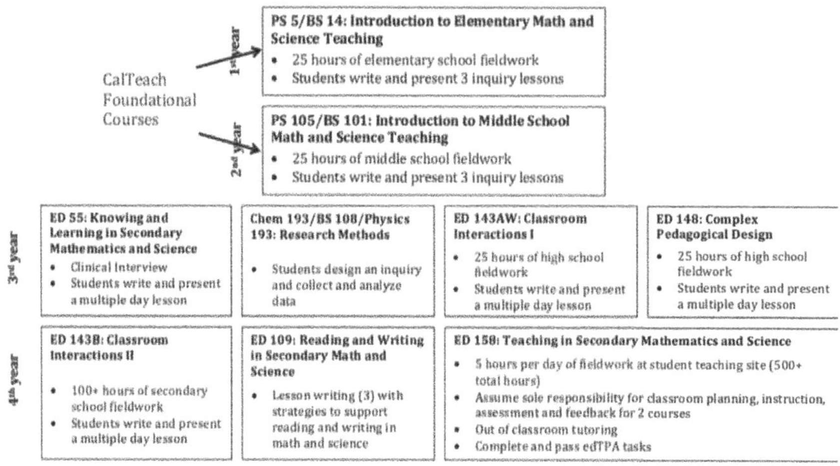

FIGURE 23.1. The Course Sequence for the CalTeach Teacher Credential Program.

two years at UCI and then declare for the CalTeach program as Juniors. While in these introductory courses, the students experience effective, research-based elementary and secondary science and mathematics lessons that are modeled by the instructors who are former and present K–12 classroom teachers. The students experience discovery, hands-on lessons as students so that they can feel what it is like to construct their own conceptual understanding of the content, recognize the benefits of student-to-student talk, and experience English Learner teaching strategies. These model lessons are the foundation for later conversations about teaching and learning. Students also begin early field experience in these courses. In their Junior year, students enroll in methods courses that are tightly coupled with fieldwork and practical teaching experiences. In their Senior year, students complete their field-based student teaching and coursework (See Figure 23.1). By the time they graduate, students complete over 700 hours of field-based observations and teaching in addition to their coursework, almost exclusively in classrooms located in underserved communities.

The program has been running since 2008, with the first graduating class in 2012 and, through 7 cohorts, has graduated a total of 101 students that earned their credential, 91 of whom are still teaching in K–12 schools, 70% of them in Title I schools in underserved communities. The UCI CalTeach program reflects the diversity of the UCI campus and local community, serving a wide body of students (see Table 23.1). Additionally, the program is growing, as it is expected to graduate a total of 70 students over the next two years. The program has also become more diverse, with the current cohorts of Juniors and Seniors comprised of 39% Latinx students and 19% Vietnamese and Filipinx students.

TABLE 23.1. UCI CalTeach Student Demographics 2008–2018

Race/Ethnicity	Percentage of Graduates
Latinx	22.68%
White	17.53%
Chinese-American	16.49%
Vietnamese/Thai	15.46%
Filipinx	8.25%
Korean	8.25%
Japanese	6.19%
East Asian	2.06%
Native American	2.06%
Black	1.03%

Method

We used an anonymous open-ended questionnaire and interviews to contextualize students' experiences in the CalTeach program and their subsequent early careers in an attempt to identify particular program features which supported their persistence in the classroom. The questionnaire contained a total of 20 questions, with four questions about CalTeach early career supports, one question was general and three focused on graduates' university coursework, field experience, and supervision. We sent questionnaires to all program graduates(n=101), with 35 completing the questionnaire, representing all seven graduate cohorts, and half of them working in underserved communities. We conducted interviews with three of the graduates. We selected the interviewees for maximum variation based on their demographics, years teaching, and school characteristics. All graduates taught science, but their backgrounds and teaching contexts varied (see Table 23.2). In the following discussion, we present findings from the questionnaire data, the graduates' interview-based cases, and the key thematic findings of what students identified as the most helpful features in supporting their early career teaching resilience.

Features of CalTeach Program Identified from Questionnaire Based Student Perceptions

The questionnaire was used to identify broad themes of support provided to graduates by the CalTeach program. Questionnaire respondents represented approximately one third of the graduates, with all graduating cohorts being represented. All of the respondents are currently still teaching. Three questions guided our inquiry; they focused on three specific aspects of the CalTeach experience:

TABLE 23.2. Case Study Teacher Demographics and Teaching Context

Pseudonym	Gender	Race/ Ethnicity	Years Teaching	Grade Level Taught	School Setting	Student Demographics
Nathalie	Female	White	6	Middle School	Urban	Latinx 95% Asian 2% White 1% African American 1%
Gloria	Female	Filipina	4	High School	Suburban	Asian 55% White 28% Latinx 8% African American 2%
Nestor	Male	Latino	3	Middle School	Urban	White 50% Latinx 37% African American 4% Asian 2%

coursework, fieldwork, and supervision, asking graduates how these components helped prepare them for their career. In what follows, we provide a description of three main features of the UCI CalTeach program along with the percentage of questionnaire respondents (n = 35) that reported each feature as being beneficial for their learning and teaching.

University Coursework

Four program features related to coursework. Approximately a third (36%) of respondents identified having to design and implement lessons as most helpful from coursework. This included both the practice of designing lessons as well as building a collection of lessons and units over their CalTeach career. One graduate explained this preparation as "so extensive that lesson planning became second nature." The other three features were related to orchestrating classroom discourse (11%) which was grounded in a five practices approach (Smith & Stein, 2011), developing instructional tools (11%), such as tools for classroom routines, and learning theories of effective teaching (8%).

Field Experience

In terms of graduates' field experience, three themes emerged from graduates' responses. Field experience helped candidates develop confidence and provided practical experience (both at 18%). This reflected both the quality of experience and breadth of experiences. One graduate noted that "fieldwork helped me gain the confidence in the classroom to enter feeling competent. I'm not afraid to try new things, even if it goes wrong in my lessons, because it was often pushed in fieldwork to try new things and reflect if it didn't go well." Another graduate shared that "many of my peers [beginning teachers] actually did not have as much hands-on experience as I did going into teaching." The other theme centered on

understanding school site dynamics and negotiating relationships with the other adults at the school site. One graduate explained it as "It gave me insight of how teachers interact with each other, administration, and parents. I was able to see how to collaborate with my colleagues, I also got to see IEP meetings and other student related resources that teachers use." These field experiences reflect the varied skills graduates developed to ease the transition into their own classrooms.

Supervisor Feedback

Through the program, students were observed by as many as four different supervisors. During their senior year, they were observed 10 times by the same supervisor, and worked collaboratively with the supervisor on identified challenges together. Supervisor observations are built on the Teacher Performance Expectations standards, and student teacher growth is evaluated over time. This approach helped provide a lens for observations and improvement, where students would identify areas of improvement and their supervisor focused their observations on those areas. Supervisor's feedback helped students to continuously improve instruction in three main areas. First, supervisors provided specific feedback to Cal-Teach students (40%). One graduate shared that "the supervisors provided timely feedback and good suggestions to bring our lessons and classroom presence to the next level." Second, supervisors helped with emotional support and building confidence (15%). Graduates expressed the importance of having someone who believed in them and built their confidence: "they [supervisors] were very positive and encouraging, they really made me feel like I could do it." Finally, supervisors helped promote reflective practice (9%). Graduates appreciated that "reflective practice became natural and I found myself reflecting during classroom instruction, not just after a class was over." We note that reflective practices were strongly linked around improving teaching through coursework, field experience, and supervisor feedback. Graduates attributed the building of confidence and development of reflective practices to their early career resilience.

Case Study Interviews

We deliberately interviewed a diverse group of graduates to try to capture a potentially wide range of experience in terms of the graduates' backgrounds and teaching contexts. All the names of the teachers in these case studies are pseudonyms.

Nathalie's "Ladder of Growth"

Nathalie is a White female middle school teacher who has been teaching at the same school for six years. She teaches seventh and eighth grade science serving a predominantly Latinx student population. Nathalie grew up in a county adjacent to where she teaches now, approximately 50 miles away. Her mother is a middle school teacher and as a child she spent time in her mother's classroom, getting a sense of what teaching middle schoolers looked like. She enjoyed her entire K–12

experience and liked attending school. She also enjoyed learning new things and was attracted to science as a discipline. Ultimately, she decided to enroll at UCI as a biology major, but "entering UCI, [she] was not thinking about being a teacher at all." She walked around campus and saw signs for CalTeach and recalled her own experiences with learning and being in classrooms. She felt like it was something she should look into: "Maybe I should try the elementary school course... but I was very hesitant." She decided to take the first CalTeach course focusing on elementary students, then the second course focusing on middle school, and subsequently decided to declare for and pursue the program.

In thinking back to the program, Nathalie saw the layering of theories learned through coursework and practice as central to connecting what she should be doing in the classroom and actually doing it. As she stated, "you get all this information in the core courses that you take at night, so you have this big foundation of knowledge of 'this is what I should be doing.'..but what do you do with all that information?" She found the field experience to be where she could bridge what she learned and what she could practice, "this is what I should be doing, now let me try it." Nathalie then connected the experiences in the field back to the classes she was taking. She valued running into challenges in field work and then having a space to bring them up in her courses. Through coursework, she and her peers had to grapple with the challenges and work their way through problems and then go back to their field work site and try it again. She found that engaging in these cycles of learning over the course of four years especially helpful. As she pointed out "I have next quarter to try this same kind of thing. I have these classroom management ideas, I have these check for understanding ideas to assess kids, and I get to try it again, and maybe it goes better this time." Nathalie called it a "ladder of growth" that allowed her to continually refine her skills and develop new tools to improve instruction. Her experiences also highlighted the importance of the duration of the program and numerous cycles of learning.

Another critical component of preparation for Nathalie was the specific and frequent feedback she would get from her field placement supervisor as well as her instructors. She attributed both improvement in her practice as well as developing reflective practices because of the feedback. Having been formally observed 12 times through her UCI CalTeach coursework, she pointed out that "the feedback from the supervisors and the professors in the program was extremely helpful because they made us reflect on it ourselves." Specific to the field observations, she also appreciated the space provided by supervisors as "they let us talk about our feelings about it, before giving us critiques about our lessons."

Finally, through her preparation experiences, Nathalie developed a "large bag of tricks, there is a lot of stuff in that bag" that included curricular resources, classroom management techniques, and a binder of documents. Through her field experience, she "made all these worksheets, my mentor teacher helped me out... and then we had a binder with all the lesson plans... and to this day I still have it in my classrooms...whenever I'm looking for new ideas or trying to do new things,

sometimes I will still go back to that binder." Thus, as she entered the classroom, Nathalie felt extremely well prepared for her own classroom.

Nathalie ultimately decided to teach in the district in which she did her student teaching. Location and her experience student teaching in the district were paramount, "It's not the closest to my family, I guess it's close enough for me, so basically location, and the fact that I did my student teaching here." She selected the specific school based on the school culture. She had multiple job offers, but she reported that the other teachers' vision and instructional approaches matched hers, and that she would fit well with the department. She pointed out that "The majority of our department has similar planning styles and teaching styles, we are all willing to take feedback from other teachers." She also shared that teachers from other departments helped her as she was starting out as well. Nathalie now supports CalTeach in a number of ways including mentoring student teachers.

Gloria's Hard, but "Totally Worth it" Journey

Gloria is a Filipina female whose parents were born in the Philippines and emigrated to the United States. She has been teaching at her current school for four years. She grew up in a county adjacent to UCI, approximately 100 miles away. The K–12 schools she attended were primarily populated by low-income Latinx and Asian students. She points out that her teachers were always supportive and that her family valued academics.

The first class of the program was particularly memorable for Gloria. She thought that starting in elementary school helped her because she got to see students "when they love to learn," before school becomes less fun and engaging. She also felt that beginning to teach at the elementary school level provided a safe space for her to first teach, because "if you mess up in front of these kids, they don't care," suggesting that the children were very forgiving of her mistakes. After the first class, she noted that "working with these kids, it was supposed to be a side gig for me, but it ended up being my life." She then took the middle school class and made a strong connection with her mentor teacher, with whom she is still in contact. She appreciated this early field experience, as it validated her decision to become a teacher. As a junior, she began to take the methods courses and experience teaching in high school, which she preferred to middle school. Nonetheless, her connection with her middle school mentor teacher prompted her to student teach in her middle school mentor's classroom. She appreciated the program's flexibility to select between student teaching in a middle school and a high school setting.

The ability to practice teaching in front of her peers, reflect, and revise before teaching during her methods courses and during student teaching was important for Gloria. Not having to encounter challenges for the first time in a live class by "teaching lessons in front of our peers... it prepared us for when it actually happened in the classroom." On the whole, she saw the collective experiences at CalTeach as building resilience. Part of this related to the heavy workload Cal-

Teach students have as undergraduates, having to balance a load of core classes, their field-based experiences, and personal time. Gloria pointed out that "you just become so good at time management and prioritizing," and "it was hard, it was really really difficult…but I think it's totally worth it, you start teaching right out of undergrad and you don't see that often." Gloria additionally identified a number of tools and strategies which helped her become a teacher. These included approaches to grouping students, cooperative learning strategies, and engaging students in learning by doing. Gloria sees herself as an intentional teacher and the roots of that intentionality come from CalTeach and the lesson design approaches that she learned in the program, including those related to launching instruction with engaging phenomena aligned with the Next Generation Science Standards (NGSS).

As Gloria was completing her student teaching, she was interviewed and subsequently accepted a position at an affluent school in the same county as the university. Although she had opportunities to pursue positions closer to home, she decided to take the job because of the comfort she felt with the school's administrators and the science department's teachers. Gloria pointed out that her first two years in the classroom were challenging, but that the resilience she had built during preparation helped her persist and to become a contributor in the department and school. Gloria has returned to the CalTeach program to speak to candidates and gave the keynote address at a CalTeach graduation dinner. Gloria is currently a mentor for a CalTeach student teacher.

Nestor Finds "Another Window"

Nestor is a Latino male and grew up in the same county as UCI and was a transfer student. When he enrolled at UCI, he found out about the CalTeach program and "liked how you could get your credential and Bachelors at the same time" and thus enrolled in the CalTeach program. While Nestor was taking his core classes for his Biology degree and participating in his CalTeach coursework and field experiences, he was also playing competitive team ice hockey. He noted that he had to learn to manage his time commitments once he started the program.

Nestor's early experiences in the CalTeach program helped him realize that he, indeed, wanted to become a teacher. He shared that his early field experience was important because of "not so much the teaching aspect, but the mentorship, building of relationships, getting to know kids, and helping them out with problems." Spending time in the class and interacting with children confirmed to him that indeed this was something he would enjoy doing as a career. He also appreciated that the first two courses allowed him to see instruction at the elementary and middle school levels, before he transitioned to a high school field-placement in his junior year. As Nestor noted, the methods coursework helped him in several ways. First, he appreciated becoming part of a cohort and building relationships with his peers. Nestor found that in his classes "you see more ways of teaching," and he had opportunities to "realize what works, what doesn't work, and realizing

you need to be very flexible." These experiences helped Nestor develop skills and practices to persist through his first years of teaching.

Overall, Nestor found that having the option to student teach in either a middle school or high school setting was helpful, "seeing each different level, seeing what you like, I initially wanted to do high school, and I probably still could... but when I student taught middle school, I found another window." He further appreciated the way the CalTeach program coordinators worked to find placements that would work best for the students. For example, Nestor shared that the CalTeach coordinator "tried to match you with a teacher that matches your personality." He also felt that CalTeach engaged in more progressive teaching approaches, where "the people and the program try to make it more of a socialized learning." Finally, he reflected that the extensive multi-year preparation "gives you more time to learn what works and what doesn't work" with opportunities to learn and teach.

Nestor had two job offers as he was finishing the program. One was at a diverse high school near the university and the other was in a diverse middle school in an adjacent county. He decided to take the more distant job because he "wanted to try something new." The school serves primarily low income, White and Latinx students. In reflecting on his transition in the classroom, he shared that "I didn't feel over my head, I felt prepared," although he acknowledged that his current year has been difficult due to transitions in administration and staff at the school. Specifically, he has seen some resistance from some veteran teachers to proposed changes. Nonetheless, Nestor has taken an active role in the school's adoption of new NGSS curriculum, by working at the district level and also with teachers at his own school site. He has used what he has learned at CalTeach to help navigate these processes to find curriculum that will engage students with science.

Cross Case Analysis and Critical Moments of Support

Three major themes emerged from the analysis of UCI CalTeach graduate questionnaires and interviews. These themes showcase the program's features that support persistence to graduation and resilience as early career teachers. These features include: a) the duration and rigor of the program; b) the close connection between their in-class learning and field experience, and c) the frequent and specific feedback from instructors and field supervisors.

Duration and Rigor of the Program

All three interviewed graduates highlighted the importance of the duration and course sequence of the CalTeach program. Respondents perceived the first two courses as affirming of their path to teaching. It was during these courses that they realized that teaching was a career that they wanted to pursue. This is reflected by the program retention numbers. Typically, about 50% of students who enroll in the first course (elementary school based) go on to enroll in the second course. Approximately 50% of the students enrolled in the second course (middle school

based) officially declare for the program as juniors and begin taking the methods courses. Once declared, the attrition rate drops significantly over the following two years. The graduates identified the third- and fourth-year courses as being substantive pedagogically, attributing these courses to most of their learning about teaching.

One aspect of the program's rigor was linked to the course load of the UCI CalTeach program. In the cases of Gloria and Nathalie, this rigor was linked to development of resilience. Both graduates pointed to having to develop time management skills and to balance a full science course load, their education-focused CalTeach classes, and their field-based placements and activities. All graduates noted that navigating the program requirements positioned them well for their first year of instruction where they, again, needed to balance the rigor of teaching with their other professional and personal responsibilities. They contrasted their own experiences and those of other first year teachers at their schools with just 12–15 months of preparation and noted that their instruction was more fully developed, and they were better able to cope with the stressors of early career teaching.

Not only do the CalTeach students shadow teachers in each of their field experiences but they are expected to write and teach whole class lessons as well. The UCI CalTeach students have innumerable opportunities for dialogue with CalTeach instructors, supervisors and classmates about teaching and learning over the four years and have time to explore their own teaching persona, growing and refining their own ideas about how they want to teach. Having time to think and practice seems to benefit the students so that they feel prepared. The academic and time demands of simultaneously taking core classes while engaged in field experiences build resilience and a wide array of instructional practices, tools, and skills. Additionally, the program's mantra is: "This is hard. But we can do hard things."

Close Connection between In-Class Learning and Field Experience

The close connection between their in-class learning and field experience helped the CalTeach students coherently link theory and practice. Field experience and being in the classroom was foundational to the graduates' early and late preparation experiences. They attributed the development of their early career practices and survival skills to their later experiences, especially those in the methods courses, and student teaching courses As noted in both the interviews and questionnaires, coursework helped develop theoretical grounding for instruction as well as lesson design approaches. Students then had opportunities to enact what they learned through extensive field-based practice including at least three multi-day instructional sequences prior to beginning student teaching. Two additional components facilitated the close linking between graduates' university coursework and field experience: their ability to practice instruction during class, and their ability to reflect and revise instruction after teaching in the field. This approach reflects the instructional cycles (see Lampert, 2009) that they were able

to engage in which helped bring the course content and practice closer together in coherent ways.

Over their UCI CalTeach coursework, the students intentionally reflected on their field experience in multiple ways. Reflections focused on supporting all learners, lesson design to engage students, making student thinking visible, student-student talk, technology, equitable practices, and differentiating instruction. Many of the courses were taught by the same instructors so there was coherence in the curriculum that built from course to course.

Frequent and Specific Feedback

In CalTeach, instructors provide feedback on lesson plans before instruction and supervisors typically debrief with students immediately after they teach lessons. Conversations begin with asking, "What worked? What did not work?" to promote reflective practices. Frequent and specific feedback from CalTeach instructors and field supervisors provided critical learning opportunities for the students and helped develop reflective practices. During field experience, in addition to being able to enact their instruction, graduates were observed at least 12 times through the CalTeach program. Our cases expanded on questionnaire responses that highlighted the importance of specific feedback from supervisors. From the interviews, two important components of feedback were revealed: receiving productive feedback to improve instruction as well as developing self-reflection practices based on the process and questions raised by supervisors. As Nathalie noted, eventually she began asking herself the questions her supervisor would ask her as a way of reflecting on her instruction. Thus, we see that two aspects of their preparation experience helped graduates become more reflective teachers: their ability to reflect in the courses as well as the modeled reflective practices of their supervisors.

CONCLUSION

Our purpose in this chapter was to identify characteristics of the UCI CalTeach program that likely lead to student retention rates and early career resilience. This program is of importance for study because over 90% of its graduates remain in the teaching profession through seven cohorts. We showcased that the program provided the graduates with a range of experiences. Early coursework and field experience (the first two classes) helped students cement their career path to teaching. Late coursework and field experience (their junior and senior years) provide students with practices, tools, and skills to persist in the classroom. Additionally, we found that the schools at which the graduates were initially hired played an important role in their persistence.

A Broad Range of Preparation Experiences

The CalTeach program exposes students to all grades ranging from K–12, to provide windows into how curriculum and children progress from elementary to middle to high school. In their four years in the program, students have multiple opportunities to teach, be observed, and reflect on their instruction across multiple grade levels. Graduates highlighted the coherence between the CalTeach coursework and the field experiences. Students are expected to incorporate what they experienced and learned in their CalTeach classes into the lessons they teach at their field sites, and they reflect on how what they learn in class is enacted in their field work. Additionally, the feedback from the supervisors reinforces the messages that the CalTeach students receive from their instructors and their classmates. Supervisors are frequently invited to meetings within the program to learn about the curriculum, instructional practices, and assessments, and they often contribute to the vision of the program. Supervisor feedback and the numerous observations also promoted graduates' own reflective practices. Specific tools and skills that graduates took up during preparation that they linked to early career success included lesson design templates and lesson design processes, tools for building community and rapport with students, classroom routine practices, and tools such as the five practices of orchestrating discourse.

Early Career School Site Support

From interviews, we saw that all three candidates became part of a community in their department as they began teaching. In all cases, these communities substantively contributed to the graduates' success and persistence. Nathalie talked about having a shared approach to instruction and planning with her department peers as well as having access to an eighth-grade curriculum to go along with the seventh-grade curriculum she had developed during preparation. Gloria also talked about the collaboration in her department and how she planned with peers and was viewed as a contributor to the team. Nestor initially had some support which was disrupted by some staffing changes, but was developing a new community with some of the newer teachers. Additionally, all three graduates have taken on some leadership roles at their schools, including coaching and either participating in or leading initiatives designed to improve science instruction. Often when they become new faculty members, UCI CalTeach graduates are recognized as experts in the new mathematics and science standards and they are encouraged to share their expertise with their colleagues. Their becoming members of a teaching community seems to contribute to their persistence as well. It could be that the student teachers chose schools and locations where they felt a connection to their departmental colleagues, since they had a strong feeling of belonging in their UCI CalTeach cohort and were comfortable teaching in a collaborative environment that mimicked the one at CalTeach where they worked closely with their peers. In

the future we plan on exploring graduates' early career experiences more closely, including their instruction and interactions with peers at their school site.

REFERENCES

Darling-Hammond, L. (2010). *The flat world and education: How America's commitment to equity will determine our future*. New York, NY: Teachers College Press.

Darling-Hammond, L., & Berry, B. (2006). Highly qualified teachers for all. *Educational Leadership, 64(3),* 14–20.

Ingersoll, R. (2001). Teacher turnover and teacher shortages: An organizational analysis. *American Educational Research Journal, 38*(3), 499–534.

Lampert, M. (2009). Learning teaching in, from, and for practice: What do we mean? *Journal of Teacher Education, 61*(1–2), 21–34.

Next Generation Science Standards Lead States. (2013). *Next generation science standards: For states, by states*. Washington, DC: The National Academies Press.

Oakland Unified School District. (2018). *OUSD retention of teachers at school sites*. Retrieved from: https://dashboards.ousd.org/t/HR/views/RetentionDashboardPublic/Teachersat Sites?%3Aembed=y&%3AshowShareOptions=true&%3Adisplay_count=no&%3AshowVizHome=no

Patterson, J. H., Collins, L., & Abbott, G. (2004). A study of teacher resilience in urban schools. *Journal of Instructional Psychology, 31*(1), 3–11. Retrieved from: http://search.ebscohost.com/login.aspx?direct=true&db=a9h&AN=12670768&site=ehost-live&scope=site

Ronfeldt, M., Loeb, S., & Wyckoff, J. (2013). How teacher turnover harms student achievement. *American Educational Research Journal, 50*(1), 4–36.

Smith, M. S., & Stein, M. K. (2011). *5 Practices for orchestrating productive mathematics discussion*. Reston, VA: National Council of Teacher of Mathematics

Zeichner, K. (2016). *Independent teacher education programs: Apocryphal claims, illusory evidence*. Boulder, CO: National Education Policy Center.

Zinger, D., & Normore, A. (2016). The STEM lab school: Providing support for teacher candidates in alternative certification programs. *Journal of Authentic Leadership in Education*, *5*(2), 2–5.

CHAPTER 24

EXAMINING UTEACH GRADUATES' EMPLOYMENT CHOICES

Outlining a Framework for Future Research

Michael Horvath, Vasilios D. Kosteas, and Joanne E. Goodell

This chapter focuses on the potential effects of the UTeach model on teacher retention. We draw from the literature in education, economics, and psychology to outline a proposed framework of factors that will affect a UTeach graduate's desire to remain in the profession. We first focus on psychological factors related to turnover, including our preliminary work on perceptions of identity as educators or as scientists. Our second focus is on economic factors, discussing how local wages in alternative occupations draw teachers away from teaching, and how these factors interact with psychological variables such as identity and job satisfaction. Our third focus is on experiences during teacher preparation that can increase and sustain desire to remain in the profession. Our early research shows that particular reactions to the student teaching experience are related to the teacher's intent to remain in the profession. We conclude with a number of recommendations for future research, including an analysis of the type of data currently being collected by UTeach and how enhancements to their survey instruments would facilitate a greater understanding of post-graduation outcomes for UTeach students.

Preparing STEM Teachers: The UTeach Replication Model, pages 339–354.

INTRODUCTION

The primary focus of teacher education programs such as UTeach is, and should be, providing students with an optimal background in content knowledge and teaching skills. However, alongside the outcome of teacher quality, it is important to consider other outcomes such as whether and how long graduates apply their skills in the classroom.

The consequences of teacher turnover are significant. For instance, high turnover rates add substantial recruiting and hiring costs to school districts (Barnes, Crowe, & Schaefer, 2007; Milanowski & Odden, 2007). Additionally, if students are consistently taught by new teachers, they do not get to benefit from the knowledge and experience of more seasoned professionals (Kane, Rockoff, & Staiger, 2008; Rivkin, Hanushek, & Kain, 2005). Finally, teacher turnover impacts the teachers themselves, as career doubts and the stress of job or career transitions can exert their toll on life satisfaction and other outcomes.

A number of sources document the general predictors of teacher turnover (e.g., Ingersoll & May, 2012). However, teacher turnover merits particular study in STEM fields such as those covered by the UTeach model. The fact that the US is lagging behind other industrialized nations in producing STEM graduates makes it essential that the next generation is taught by quality educators and not a revolving door of novice teachers. Additionally, we argue that the nature of STEM, and in particular, the model of education proffered by UTeach, may make these teachers uniquely vulnerable to exiting the profession. Contrary to previous studies, Ingersoll and May (2012) did not find a higher turnover rate for STEM teachers. However, their study included STEM teachers in general, not specifically those who—like UTeach graduates—hold a degree in a STEM field and have greater opportunities to pursue STEM careers outside of teaching. Below, we discuss three factors—a psychological sense of identity, local and broader economic influences, and the student teaching experience—that influence turnover. We supplement our theoretical development with descriptions of preliminary research we have conducted, and we conclude with a set of recommendations for UTeach.

IDENTITY

Psychological Definition of Identity.

Most of us have a relatively developed understanding of a teacher. We know what teachers do (e.g., communicating information, reinforcing behavior, assessing performance), but we also have a sense of the type of person who becomes a teacher (e.g., supportive, intelligent). Concepts such as 'teacher,' 'parent,' or 'scientist' are socially constructed schemas that facilitate categorization (see, for instance, Macrae & Bodenhausen, 2000; Turner, Hogg, Oakes, Reicher, & Wetherall, 1987). Although in their extreme form, negative and inaccurate schemas are associated with stereotyping, prejudice, and violence, they also help us organize our lives. For instance, categorizing someone as a 'barista' allows us to predict

certain knowledge and behavior that make interactions more efficient. Similarly, thinking about ourselves as members of particular groups can give us a sense of community, and it can also help us structure our own behavior. Furthermore, we are not limited to a single identity (e. g., Cropanzano, James, & Citera, 1993). We may simultaneously possess identities related to our profession, family, hobbies, or other roles, and the strengths of these identities may vary in response to situational cues. For instance, as we drive home we may take off our 'teacher hats' and put on our 'parent hats.'

Individuals tend to act in accordance with their identities.(e.g., Ashforth & Mael, 1999; Stryker & Burke, 2000). For instance, people who identify as musicians will tend to listen to more music or will play music more frequently. A strong teacher identity will similarly guide behavior. Those who think of themselves as teachers will be more likely to take jobs as teachers, devote their time and effort to teaching, and persist in teaching in the face of obstacles. For instance, using data from the UTeach Institute's end of program surveys as well as follow-up surveys 1–3 years later, we found that the strength of one's identity as a 'teacher-leader' predicted individuals' intent to become teachers, and those with stronger identities were more likely to eventually be employed as teachers (Horvath, Goodell, & Kosteas, 2018). Although these data show the benefits of a teacher identity, they do not address the unique challenges posed by teaching in STEM.

STEM Teacher Identity

As mentioned above, people have more than a single identity. Many identities are compatible or complementary. Some identities, however, can conflict, such as when expectations of working from home during the evening conflict with expectations to spend time with family (i.e., Work-family conflict; Greenhaus, Parasuraman, & Collins, 2001; Kossek & Lee, 2017).

This is not to say that we only have single professional identities. For instance, a manager in a manufacturing organization may identify as both a manager and as an engineer. We believe that STEM teachers are a particularly salient version of this phenomenon. They receive substantial training in both education and a STEM field and thus may develop strong identities in either area, or even in both areas.

Ideally, STEM and teacher identities would reinforce each other. An educator with strong versions of both identities would be engaged as a master teacher, enthusiastic about developing students' STEM knowledge and skills. On the other hand, a strong science identity may pull the individual away from teaching. Compared to other subjects (e.g., art, English), there are plentiful, high-paying alternatives for those with a STEM education in certain fields (see below for a more thorough discussion of this issue). Thus, those with strong science identities (particularly those who network with other STEM professionals outside of teaching) may be more aware of, and more drawn to, jobs outside the teaching profession.

One may argue that it would reduce teacher turnover to select only those teachers whose teaching identity was substantially greater than their STEM identities.

Although we agree that this may have some impact, we are reluctant to abandon the potential positive aspects of a strong science identity. Below, we outline a series of factors that we believe may keep individuals with strong STEM identities within the teaching profession.

First, we examine the interplay between types of identities. As we mentioned above, individuals may simultaneously possess multiple strong identities and may act on the basis of their strongest identity. Nevertheless, we expect that it is much more complicated than that.

To guide our thinking, we draw on the goal-setting literature. Goals represent desired end-states that help to initiate, direct, and sustain behavior (e.g., DeShon & Gillespie, 2005; Kanfer, Chen, & Pritchard, 2009). As such, identities appear to be very similar to goals, if not a type of career goal.

Just as people can possess multiple identities, they can also pursue multiple goals simultaneously, and a subset of goals research has investigated the interplay among competing goals. For instance, Schmidt and DeShon (2007) showed that sometimes people devote more effort to the goals that are farthest from completion. Thus, it is possible that teachers who feel their science identities are 'slipping' or are not being easily satisfied within teaching may decide to find alternate employment in order to satisfy that identity. It may also be that a different goal-related concept, multifinality, may point to times in which this may not happen.

According to Kruglanski et al. (2002), behaviors that can satisfy more than one goal simultaneously may be more likely to be enacted. When applied to teacher and STEM identities, this suggests that individuals with strong forms of both identities may look for behaviors (such as being a STEM teacher) that could fulfill both.

To investigate this idea we examined data from 172 students from 14 different UTeach institutions, measuring their STEM identity, teacher identity, and intent to enter the teaching profession (Horvath, Kosteas, & Goodell, 2018). We found that teacher and STEM identities interacted to predict intentions, such that a strong STEM identity only related to lower teaching intentions when teacher identity was weak.

Thus, programs such as UTeach that emphasize a STEM degree along with teacher preparation may want to find ways to cultivate and strengthen teacher identities. Before they graduate, students may benefit from the explicit shaping of their teacher identities through discussions about the meaning of being a teacher, field placement experiences (discussed later in this chapter), or student memberships in professional societies.

After hire, factors such as professional societies and in-service trainings that emphasize the role of teacher may be beneficial. While such training may strengthen teacher identity, an unfulfilled STEM identity may yet be a detriment to professional engagement. Thus, we recommend that schools also look for ways to satisfy those with strong science identities. Obviously, assigning teachers to STEM courses can complement a STEM identity, but schools may also want to

consider extracurricular assignments. For instance, administrative duties or sports coaching assignments may not be as fulfilling as running a science fair or fielding a team for a Science Olympiad competition (www.scioly.org). Alternatively, schools may encourage STEM teachers to forge partnerships or conduct extracurricular research with industry or local universities—perhaps including students, as appropriate, would serve to reinforce both identities. Finally, we encourage research that explicitly compares the identities of UTeach graduates to those of traditional programs. As UTeach programs have demonstrated an 85% retention rate after four years compared to 60% nationally (UTeach Institute, 2015), it is possible that one reason for this difference is that UTeach may cultivate stronger identities.

In addition to leveraging the strengths of teacher and STEM identities, several other factors may interact with these identities. For instance, as discussed later in this chapter, economic conditions in both the teaching and STEM professions could anchor individuals in their present jobs or pull them irretrievably away. Likewise, expectations of peers or family members may keep some individuals embedded within their jobs, such as the likelihood that teaching positions allow parents to be there when their children come home from parallel school schedules.

We discussed two distinct identities related to one's career above. However, individuals also have other professional identities. Organizational identification (e.g., Ashforth, Harrison, & Corley, 2008) is the extent to which individuals include their employer within their self-definition. This phenomenon may be salient in schools, paralleling school spirit amongst the student body. Schools may be able to leverage this type of identification through language (e.g., talking about 'we' when addressing faculty) or even through tangible items (e.g., clothing or coffee mugs with school logos).

TEACHER TURNOVER AND WAGE DIFFERENTIALS

There is a sizable literature examining the effect of salary differences between teaching and other occupations that teachers might consider as likely alternatives on teacher turnover. Most studies find that while relative salaries do impact teacher turnover, the magnitude of this effect is rather small (see for example Feng, 2009; Gilpin, 2011; Imazeki, 2005; Murnane & Olsen, 1989; Ondrich, Pas, & Yinger, 2008; Podgursky, Monroe, & Watson, 2004), indicating that it would take substantial increases in teacher salaries to significantly reduce turnover This raises questions about whether salary increases provided the most cost-effective method for teacher retention.

Based on these studies, one might reasonably conclude that opportunity wages (the wages teachers could earn in alternative employment) do not play a significant role in teacher exit from the profession. However, other studies indicate that opportunity wages may play a larger role in teacher attrition and that focusing on teachers in general, rather than specific groups, might mask important differences based on area of specialization. Rickman, Wang, and Winters (2017) find that men

are more sensitive to wage differentials, especially those who majored in math, science, or computer education. Stinebrickner (1998) found that teachers with a bachelor's degree in a science field exit teaching more quickly and that a one standard deviation increase in teaching salary increases the probability a teacher will stay in the profession beyond five years by roughly nine percentage points (a substantial difference given that only 40 percent of teachers last six or more years after first entering the profession). These findings suggest that career decisions for UTeach graduates, who major in mathematics or a science discipline, might be more responsive to salary differentials since they are likely to have more high-paying career options compared with students who majored in mathematics or science education. Additionally, there is evidence that higher quality teachers who exit teaching earn higher salaries (Chingos & West, 2010) indicating that even within specializations, with teacher pay schedules that emphasize tenure, higher quality teachers experience a higher wage differential between teaching and alternative occupations. Thus, high quality STEM teachers with degrees in math or science may face the greatest pull of higher earnings outside of teaching.

To get a better sense of the difficulties involved with estimating the effect of opportunity wages on teacher turnover, it helps focus on two dimensions of existing research: 1) how a study identifies the opportunity wage, and 2) the data source and how exit from the profession is measured.

Measuring Opportunity Wages

Ideally, to estimate the effect of outside wages on exits from the teaching profession, we would observe both the teacher's current salary and the salary the teacher would earn if she left the profession and took another full-time job. In reality, we only observe the former, leaving us to define a proxy for the latter. The existing literature has employed two methods to estimate opportunity wages: 1) use average salaries in the region for comparable groups of workers, and 2) predict expected salary outside of teaching for a current teacher by estimating wage equations. Each approach has its flaws.

The second approach proceeds in two steps. First, wage equations are estimated separately for teaching and non-teaching occupations and the parameter estimates are used to construct a predicted wage outside of teaching for each individual currently employed as a teacher. This predicted wage is included in the turnover model as the measure of the opportunity wage. Using this approach, Gilpin (2011) found that the wage differential only matters for teachers with less than six years of experience. Even then, the effect is small relative to the effect of school characteristics. It is important to bear in mind that wage equations are notorious for achieving, at best, modest R^2 values. Thus, as a predictive model, they are likely to produce earnings estimates that are measured with a great deal of error. Including poorly estimated opportunity wage estimates will likely yield coefficients that are biased towards zero, leading to the erroneous conclusion that non- teaching salaries do not matter for teacher turnover.

The vast majority of the literature takes the approach of using estimates of local earnings for workers deemed comparable to those employed as teachers. With this approach, it is not clear how we should determine the relevant comparison group. Existing studies have used either: (1) salaries for college graduates (Ondrich et al., 2008; Rickman et al., 2017); (2) salaries for teachers in occupations or industries selected on an ad hoc basis (Imazeki, 2005); (3) salaries in industries/occupations to which teachers have been observed to move after leaving teaching (Feng, 2009); (4) test score measures to proxy for earnings potential (Murnane & Olsen, 1989); or (5) do not include any measure of opportunity wages and rely on own- salary to determine the effect of salary on turnover (Hendricks, 2014; Stinebrickner, 1998).

Another reason why the existing literature might underestimate the importance of pecuniary factors on teacher turnover is the focus on salary differentials to the exclusion of job benefits. Most research examining the effects of teacher pension systems focus on the timing of (Brown, 2013; Ni & Podgursky, 2016). Further, any discussion of the differences in compensation between teaching and alternative professions on teacher turnover also needs to consider the value of pensions and the potential of generous pensions as an incentive to reduce turnover rates (Gustman & Steinmeier, 1993). In general, the pensions available to public school teachers are far more generous than those provided to workers in the types of occupations teachers might consider as alternatives (so that focusing solely on wages overstates the compensation gap). Additionally, vesting may reduce the incentive to leave for higher wages early in the career. This provides an additional explanation for why teachers with more experience are less likely to leave. Ignoring the relative value of benefits between teaching jobs and alternative occupations likely biases the estimated effect of wage differentials on teacher turnover.

Data Sources

The source of the data being analyzed affects the type of analysis that is performed. We can group the literature into two categories based on type of dataset: 1) administrative data for teachers from a particular state, and 2) national survey data either specifically focused on teachers (i.e., the Schools and Staffing Survey) or yielding a nationally representative sample of households (i.e., the American Community Survey). Each data source has its strengths and weaknesses. Administrative data are generally of high quality and we expect a high degree of accuracy in terms of reported salaries and classroom/school characteristics. Conversely, salary information in survey data is self-reported and subject to significant measurement error (Bound, Brown, & Mathiowetz, 2001) likely causing bias in our estimates. One problem with administrative data is that it only allows the researcher to follow teachers as long as they are employed as teachers in the state from which the data were collected. In this case, teachers who move to another state and continue to teach are incorrectly treated as having exited the profession.

While cross-state moves do not pose the same problem in the Schools and Staffing Survey (SASS), the time dimension is very short and does not allow researchers to use standard approaches for estimating duration of employment spells such as hazard or competing risks models. Administrative data can be collected over several years leading to the creation of longitudinal datasets with a longer time dimension. Nationally representative surveys contain a rich set of information on individual characteristics, but generally do not contain information on career histories, so that we cannot determine how long an individual has been teaching or how many different teaching spells she has had in her career up to that point.

With the exception of nationally representative datasets like the ACS, existing datasets do not allow us to observe whether college students who are enrolled in teacher preparation programs actually enter the teaching profession. The literature assessing the link between opportunity wages and teachers' work decisions almost exclusively focused on exit from the profession and ignored entry into the profession. One notable exception is Podgursky et al. (2004), who used data on college graduates in Missouri, but their study focused on assessing the quality of the candidates who enter the teaching profession and did not test for the effect of relative salaries on the decision to become a teacher.

Relative Wages

Table 24.1 shows mean annual earnings for secondary teachers (excluding special and career/technical education) whose highest degree earned is a Bachelor's and for occupations which are likely to serve as alternative career options for high school STEM teachers with a major in mathematics or a natural science discipline. The data come from the 2017 American Community Survey (United States Census Bureau, 2018). Given that secondary school teacher salaries tend to vary by degree attainment and tenure, but not by discipline, the ratio of teacher salaries to salaries in alternative occupations can vary significantly. Mean annual earnings

TABLE 24.1. Mean Earnings by Occupation

Code	Occupation	Earnings	Observations
1610	Biological Scientist	\$57,396**	414
1700	Astronomers & Physicists	\$90,884**	32
1720	Chemists & Materials Scientists	\$69,693**	442
1210–1240	Mathematical Science Occupations	\$77,704**	874
2320	Secondary School Teachers	\$45,620	3,179

Note. Statistics calculated using data from the American Community Survey. Includes only individuals with a Bachelor's as the highest degree attained.
** Indicates a t-test for the equality of mean earnings for this group compared with secondary teachers is significant at the 1% level.

for biological scientists are only around six thousand dollars higher than average earnings for secondary school teachers. Conversely, astronomers and physicists earn roughly twice as much as secondary school teachers on average. *T*-tests confirm the mean earnings for each occupation is statistically different for mean earnings for Secondary School Teachers. The statistics presented in these tables highlight the importance of developing appropriate measures of opportunity wages for teachers that are based on potential alternative careers and illustrates how UTeach graduates might face higher opportunity wages given they possess science and mathematics degrees.

STUDENT TEACHING SATISFACTION

In this section, we briefly outline the development of current thinking and policies surrounding clinical experiences in educator preparation in the US, and review and critique the extant research directions concerning student teaching and candidate satisfaction.

Since the mid 2000s there has been considerable comment about the value of clinical experiences in teacher preparation programs (Anderson & Stillman, 2013), with a number of high-profile teacher educators calling for significant changes in preparation programs to include more and better clinical experiences (Hammerness, Darling-Hammond, & Bransford, 2005; Zeichner, 2010). In 2010, the National Council for the Accreditation of Teacher Education (NCATE) released its report on transforming teacher education through clinical (NCATE Blue Ribbon Panel on Clinical Preparation and Partnerships for Improved Student Learning, 2010). A few years later, the Council for the Accreditation of Educator Preparation (CAEP) revised its accreditation standards to focus more on clinical components. The Association of Teacher Educators (ATE) release of its Field Experience Standards (Association of Teacher Educators, 2016) was the culmination of work that had begun over twenty years prior.

Given the considerable effort devoted to the development of standards for clinical experiences, it is somewhat surprising that there are not more large-scale studies of candidate experiences in clinical placements. In general, research on teacher education programs focuses mostly on changes in beliefs about teaching resulting from a particular intervention or experience (Anderson & Stillman, 2013), or on the relationships between program features, contexts, courses, or experiences and candidate outcomes (Capraro, Capraro, & Helfeldt, 2010; Cooper & Nesmith, 2013). Few papers report candidate perceptions of specific aspects of their program such as clinical practice or student teaching. One exception is a paper by Darling-Hammond, Chung, and Frelow (2002) who studied the views of almost 3,000 newly-hired teachers in New York City concerning their preparation for teaching, beliefs and practices, and plans to remain in teaching. They found that teachers who were prepared in teacher education programs felt significantly better prepared than those who entered teaching through alternative programs or without preparation. Finally, the extent to which teachers felt well prepared when

they entered teaching was significantly correlated with their sense of teaching efficacy, their sense of responsibility for student learning, and their plans to remain in teaching. The particular experiences that led to their feelings of preparedness were not addressed.

Reflecting on this dearth of literature about candidates' perceptions of their educator preparation programs led us to hypothesize that much of this work is conducted at the program level and does not make it into the teacher education literature. Evaluation data certainly exist, since CAEP accreditation standards require programs to continuously collect and make programmatic changes based on evaluation data. Given the large variations in program type, the huge number of programs across the country, the lack of a shared language around many aspects of teacher education programs and the lack of a common framework for conducting evaluations, collecting or aggregating these data to form a data set worth investigating is impossible. With the advent of the UTeach program replication and the UTeach Institute data collection that follows candidates throughout their program and after graduation, such data sets now exist. As noted earlier, we have analyzed some of those data as well as collected our own data from both UTeach and non-UTeach institutions.

The focus of the two studies we have completed, one published (Horvath, Goodell, et al., 2018) and the other reported at a conference (Horvath, Kosteas, et al., 2018), has been on investigating the variables that are associated with pre-service teachers choosing to enter and stay in the teaching profession. In the published study, we utilized data collected by the UTeach Institute at the end of the candidate's program and from alumni once they enter the workforce, and responses from both surveys were linked. The alumni surveys achieved a response rate of 53 percent. One set of questions asked "Overall, how satisfied were you with each of the following aspects of your student teaching experience(s)?" Participants were asked to rate: the school placement; the learning experiences; support, guidance, and feedback from the cooperating teachers; support, guidance, and feedback from the university facilitator; and support, guidance, and feedback from the university instructors. Another question asked "How likely are you to enter the teaching profession." Analysis showed that the candidates' satisfaction with their student teaching experience predicted both intent and actual entry into the teaching profession. It follows then, that in order to increase the number of candidates who enter teaching after completing their program, teacher preparation programs need to ensure that all teacher candidates have the opportunity to engage in high quality student teaching experiences. In addition, researchers and evaluators need to ensure that measurement instruments are able to adequately capture the candidates' sentiments.

In our second study, we designed a new instrument, based in part on the UTeach Alumni survey but with more robust identity and economic factor items, and administered it to UTeach pre-service teachers in the final year of their teacher preparation program receiving 172 usable responses. To those, we anticipated adding

another dataset of 128 responses from candidates in non-UTeach institutions, but these data were deemed unusable because it was clear that the respondents did not understand our questions about student teaching satisfaction. We attributed this confusion to the wide variation in program experiences and language used to describe those experiences in non-UTeach institutions as compared to the common language and shared experiences of the UTeach candidates. It is this feature of the UTeach replication, and the existence of the repository of data, that holds such promise for future researchers.

IDENTITY, WAGE DIFFERENTIALS, AND STUDENT TEACHING EXPERIENCES

Above we outlined the importance of three, seemingly disparate sets of factors which may significantly impact decisions to exit the teaching profession. How do these concepts relate? Given the findings in Horvath, Kosteas, and Goodell (2019) and the vast salary differentials between teaching and non-teaching professions available to UTeach graduates, one key question is whether possessing a strong teacher identity mitigates the relationship between salary differentials and exit from the teaching profession. Another fundamental question relates to the formation of teacher and science identities and the role played by teacher preparation programs. Specifically, how can the student teaching experience foster a strong teacher identity? Addressing these questions, which bring together insights from psychology, economics, and teacher education will give us a more complete understanding of career decisions for STEM teachers than we can develop by examining the importance of these various factors independently from each other.

RECOMMENDATIONS FOR UTEACH

The UTeach Consortium can make unique contributions to the science on STEM teacher retention and professional entry. It already has the mechanism in place to regularly survey students and alumni, and thus can track the effects of the UTeach model as well as many other factors, such as those discussed above, predicting entry into the profession and short- and long-term turnover. This common language facilitates data collection. We are grateful for the data they have already collected, as it has enabled our preliminary research on these topics. Here, we offer several recommendations for strengthening the process that, if enacted, would allow us and other researchers to extend the science even further.

Recommendations Relating to Identity Measurement

In order to gain a better understanding of identity's role in teacher turnover we need to strengthen UTeach's measurement of identity. Currently, it uses a single item ("to what extent do you identify with being a (future) teacher leader?"). Although we have used this as a proxy for the type of identity discussed in this

chapter, it was not designed to measure this type of identity, and contains other content (e.g., leadership) that is distinct from one's identity as a teacher.

Furthermore, single-item measures may suffer from low reliability and may not capture the full range of a concept. We recommend the development of a short, multiple- item measure of teacher identity. Similarly, as UTeach does not currently measure the strength of STEM identities, we encourage the development of a parallel measure or measures. Ideally, this measure would be used in multiple stages of the surveying process, tracking the development of identity before and after graduation.

Recommendations Relating to Dependent Variable Measurement

Similar to our recommendation for improving the identity measure, we encourage the UTeach Institute to expand its single item assessing intent to enter teaching to a broader multiple-item measure. We also recommend implementing parallel items in the alumni survey, assessing individuals' intent to continue teaching. Such items should capture both the strength of the intent (e.g., "It is important for me to continue teaching") as well as intended duration of teaching (e.g., "I intend to teach for at least 5 / 10 / 15 more years").

Although intent is certainly important, self-reports of actual behavior should also be included. The UTeach alumni survey does ask whether individuals are still employed as teachers, but this may capture one's ultimate decision. More intermediate behaviors (e.g., actually looking for other jobs in teaching or jobs in other professions) should also be assessed.

Although intent to enter teaching and to continue or withdraw from the profession are certainly focal variables, UTeach may also wish to measure related cognitions. For instance, above we talked about the constructs of identification with a specific school-alumni surveys could include this type of commitment as well. Other constructs may include job involvement or work centrality (e.g., Paullay, Alliger, & Stone-Romero, 1994), which tap the extent to which one's position or the notion of work itself are important to the individual.

Recommendations Relating to Economic Variables

We also recommend including questions regarding salary and other forms of compensation on the alumni survey. This information will allow researchers to construct measures of wage differences between the current teaching job and alternative occupations. For individuals who have left teaching, we would like to know how much the decision to switch careers was determined by different aspects of the new job, including salary and other compensation. For those who are currently teaching, questions eliciting perceptions of benefits received by scientists and mathematicians relative to those received by teachers may provide useful insight into the determinants of career intentions. Finally, multi-item job

satisfaction instruments should include questions regarding satisfaction with pay and compensation. The latter can provide especially useful information given the difficulty in obtaining accurate information regarding the level of benefits provided by the employer.

Recommendations Relating to Student Teaching Factors

The questions in the graduate end-of-program survey about experiences during student teaching do provide useful information, although including definitions of key terms could increase instrument reliability. Adding items about other factors that might impact overall satisfaction with the student teaching experience, such as the location (urban, suburban, rural etc.), type (public, private, parochial, magnet, etc.), distance from home, or income level of the students could also provide more insights into how to improve field experiences. Information derived from these questions might also be used by program faculty to help better plan induction program activities.

Future Directions and Other Recommendations

Recent work by Klette and Hammerness (2016) focused on developing a conceptual framework for judging the quality of teacher education programs based on the degree to which a program has a shared vision of good teaching, program coherence and provides opportunities for candidates to put theories into practice in real world settings. We would like to develop a framework to judge the degree to which program activities develop their candidates' identities, and how multiple identities interact to impact entry-level and early career turnover decisions. In order to do that, we need data that has more variation in student teaching experiences and better ways of capturing and measuring other relevant program variations.

CONCLUSION

In this chapter we do not wish to deemphasize or detract from the paramount goal of teacher education: to provide the best possible training for future educators. Indeed, this should be the focus of teacher education programs. Nevertheless, we also realize that the success of a teacher is accumulated over years in the profession, and training a quality teacher is not as impactful if that teacher does not in turn influence many students in the next generation. To that end, we capitalize on our respective strengths in psychological identity, economics, and teacher field experiences by building toward a multidisciplinary framework predicting teacher retention and entry into the field. The UTeach program is a valuable platform from which to test and refine the ideas in this framework, and we hope that our work can similarly strengthen UTeach's ultimate effectiveness.

REFERENCES

Anderson, L. M., & Stillman, J. A. (2013). Student teaching's contribution to preservice teacher development: A review of research focused on the preparation of teachers for urban and high-needs contexts. *Review of Educational Research, 83*(1), 3–69.

Ashforth, B. E., Harrison, S. H., & Corley, K. G. (2008). Identification in organizations: An examination of four fundamental questions. *Journal of Management, 34*(3), 325–374.

Ashforth, B. E., & Mael, F. (1999). Social identity theory and the organization. *Academy of Management Review, 14*(1), 20–39.

Association of Teacher Educators. (2016). *Revised ATE field experience standards*. Retrieved from: https://ate1.org/field-experience-standards. Manassas, VA.

Barnes, G., Crowe, E., & Schaefer, B. (2007). *The cost of teacher turnover in five school districts: A pilot study*. Retrieved from: https://nctaf.org/wp-content/uploads/2012/01/NCTAF-Cost-of-Teacher-Turnover-2007-full-report.pdf. Washington, DC: National Commission on Teaching and America's Future.

Bound, J., Brown, C., & Mathiowetz, N. (2001). Measurement error in survey data. In J. J. Heckman & E. Leamer (Eds.), *Handbook of econometrics* (Vol. 5, pp. 3705–3843). Amsterdam: Elsevier.

Brown, K. M. (2013). The link between pensions and retirement timing: Lessons from California teachers. *Journal of Public Economics, 98*, 1–14. doi:10.1016/j.jpubeco.2012.10.007

Capraro, M. M., Capraro, R. M., & Helfeldt, J. (2010). Do differing types of field experiences make a difference in teacher candidates' perceived level of competence? *Teacher Education Quarterly, 37*(1), 131–154.

Chingos, M. M., & West, M. R. (2010). *Do more effective teachers earn more outside of the classroom?* Retrieved from: http://nrs.harvard.edu/urn-3:HUL.InstRepos:4460855. CESifo Working Paper Series No. 2996, March.

Cooper, S., & Nesmith, S. (2013). Exploring the role of field experience context in preservice teachers' development as mathematics educators. *Action in Teacher Education, 35*(3), 165–185. doi:10.1080/01626620.2013.770376

Cropanzano, R., James, K., & Citera, M. (1993). A goal hierarchy model of personality, motivation, and leadership. In L. L. Cummings & B. M. Staw (Eds.), *Research in organizational behavior* (Vol. 15, pp. 267–322). Greenwich, CT: JAI Press.

Darling-Hammond, L., Chung, R., & Frelow, F. (2002). Variation in teacher preparation: How well do different pathways prepare teachers to teach. *Journal of Teacher Education, 53*(4), 286–302. doi:10.1177/0022487102053004002

DeShon, R. P., & Gillespie, J. Z. (2005). A motivated action theory account of goal orientation. *Journal of Applied Psychology, 90*(6), 1096–1127. doi:10.1037/0021-9010.90.6.1096

Feng, L. (2009). Opportunities, wages, classroom characteristics, and teacher mobility. *Southern Economic Journal, 75*(4), 1165–1190.

Gilpin, G. A. (2011). Reevaluating the effect of non-teaching wages on teacher attrition. *Economics of Education Review, 30*(4), 598–616. doi:10.1016/j.econedurev.2011.03.003

Greenhaus, J. H., Parasuraman, S., & Collins, K. M. (2001). Career involvement and family involvement as moderators of relationships between work–family conflict and

withdrawal from a profession. *Journal of Occupational Health Psychology, 6*(2), 91–100.

Gustman, A. L., & Steinmeier, T. L. (1993). Pension portability and labor mobility: Evidence from the survey of income and program participation. *Journal of Public Economics, 50*(3), 299–323.

Hammerness, K., Darling-Hammond, L., & Bransford, J. D. (2005). How teachers learn and develop. In L. Darling-Hammond & J. D. Bransford (Eds.), *Preparing teachers for a changing world: What teachers should learn and be able to do* (pp. 358–389). San Francisco, CA: Jossey-Bass.

Hendricks, M. D. (2014). Does it pay to pay teachers more? Evidence from Texas. *Journal of Public Economics, 109*, 50–63. doi:10.1016/j.jpubeco.2013.11.001

Horvath, M., Goodell, J. E., & Kosteas, V. D. (2018). Decisions to enter and continue in the teaching profession: Evidence from a sample of US secondary STEM teacher candidates. *Teaching and Teacher Education, 71*, 57–65. doi:https://doi.org/10.1016/j.tate.2017.12.007

Horvath, M., Kosteas, V., & Goodell, J. E. (2018). *Multiple identities and engagement in the teaching profession*. Paper presented at the 33rd Annual Conference of the Society for Industrial and Organizational Psychology, Chicago, IL.

Horvath, M., Kosteas, V. D., & Goodell, J. E. (2019). *Scientists who teach vs. teachers of science: The relationship between multiple identities and intent to enter the teaching profession*. Working paper.

Imazeki, J. (2005). Teacher salaries and teacher attrition. *Economics of Education Review, 24*, 431–449.

Ingersoll, R. M., & May, H. (2012). The magnitude, destinations, and determinants of mathematics and science teacher turnover. *Educational Evaluation and Policy Analysis, 34*(4), 435–464.

Kane, T. J., Rockoff, J. E., & Staiger, D. O. (2008). What does certification tell us about teacher effectiveness? Evidence from New York City. *Economics of Education Review, 27*, 615–631.

Kanfer, R., Chen, G., & Pritchard, R. D. (2009). The three C's of work motivation: Content, context, and change. In R. Kanfer, G. Chen, & R. D. Pritchard (Eds.), *Work motivation: Past, present, and future* (pp. 1–16). New Yor, NY: Routledge.

Klette, K., & Hammerness, K. (2016). Conceptual framework for analyzing qualities in teacher education: Looking at features of teacher education from an international perspective. *Acta Didactica Norge, 10*(2), 26–52.

Kossek, E. E., & Lee, K.-H. (2017). Work-family conflict and work-life conflict. In T. J. Andersen, C. A. Hallin, & R. J. Aldag (Eds.), *Oxford research encyclopedia of business and management* (pp. 1–23). Oxford: Oxford University Press. DOI: 10.1093/acrefore/9780190224851.013.52

Kruglanski, A. W., Shah, J. Y., Fishbach, A., Friedman, R., Chun, W. Y., & Sleeth-Keppler, D. (2002). A theory of goal systems. *Advances in Experimental Social Psychology, 34*, 331–378.

Macrae, C. N., & Bodenhausen, G. V. (2000). Social cognition: Thinking categorically about others. *Annual Review of Psychology, 51*(1), 93–120.

Milanowski, A. T., & Odden, A. R. (2007). *A new approach to the cost of teacher turnover*. Retrieved from: https://www.crpe.org/publications/new-approach-cost-teacher-turnover. University of Washington: Center on Reinventing Public Education.

Murnane, R. T., & Olsen, R. J. (1989). The effect of salaries and opportunity costs on duration in teaching: Evidence from Michigan. *The Review of Economics and Statistics, 71*(2), 347–352.

NCATE Blue Ribbon Panel on Clinical Preparation and Partnerships for Improved Student Learning. (2010). *Transforming teacher education through clinical practice: A national strategy to prepare effective teachers*. Retrieved from: ERIC website: https://eric.ed.gov/?id=ED512807

Ni, S., & Podgursky, M. (2016). How teachers respond to pension system incentives: New estimates and policy applications. *Journal of Labor Economics, 34*(4), 1075–1104.

Ondrich, J., Pas, E., & Yinger, J. (2008). The determinants of teacher attrition in upstate New York. *Public Finance Review, 36*(1), 112–144.

Paullay, I. M., Alliger, G. M., & Stone-Romero, E. F. (1994). Construct validation of two instruments designed to measure job involvement and work centrality. *Journal of Applied Psychology, 79*, 224–228.

Podgursky, M., Monroe, R., & Watson, D. (2004). The academic quality of public school teachers: An analysis of entry and exit behavior. *Economics of Education Review, 23*, 507–518.

Rickman, D. S., Wang, H., & Winters, J. V. (2017). Relative teacher salaries and the decision to teach. *Contemporary Economic Policy, 35*(3), 542–550. doi:10.1111/coep.12195

Rivkin, S. G., Hanushek, E. A., & Kain, J. F. (2005). Teachers, schools and academic achievement. *Econometrica, 73*(2), 417–458.

Schmidt, A. M., & DeShon, R. P. (2007). What to do? The effects of discrepancies, incentives, and time on dynamic goal prioritization. *Journal of Applied Psychology, 92*(4), 928–941.

Stinebrickner, T. R. (1998). An empirical investigation of teacher attrition. *Economics of Education Review, 17*(2), 127–136.

Stryker, S., & Burke, P. J. (2000). The past, present, and future of an identity theory. *Social Psychology Quarterly, 63*(4), 284–297.

Turner, J. C., Hogg, M. A., Oakes, P. J., Reicher, S. D., & Wetherall, M. S. (1987). *Rediscovering the social group. A self-categorization theory*. Oxford: Basil Blckwell Inc.

United States Census Bureau. (2018). *American community survey*. Retrieved from: https://www.census.gov/programs-surveys/acs/data.html

UTeach Institute. (2015). *Graduates of UTeach STEM teacher preparation programs nationwide.* Retrieved from: https://uteach.utexas.edu/sites/default/files/grad-report-may-2015-final.pdf

Zeichner, K. (2010). Rethinking the connections between campus courses and field experiences in college- and university-based teacher education. *Journal of Teacher Education, 61*(1–2), 89–99.

CHAPTER 25

HOW DO INTENTIONS TO TEACH RELATE TO RECRUITMENT AND RETENTION OF UTEACH PROGRAM CANDIDATES?

Melissa Dodson, Whitney Cade, Feng Liu, and Michael Vaden-Kiernan

We analyzed survey and enrollment data from 7 UTeach partners to examine differences in student demographics, enrollment, and perceptions of financial incentives, the program, and teaching as a profession. Two types of students were examined: G1, who enter the program intending to teach, and G2, who enter with no intentions to teach. Approximately 40% of UTeach students enter as G2 students and 30% of G2 students are retained after *Step 1*. A greater proportion of G2 students are Asian and major in the sciences, computer science, and engineering. By the time students have completed *Project-Based Instruction*, there is no significant difference between G1 and G2 students. The majority of students report that financial incentives are important and rate the program positively. UTeach students have concerns about making a career of teaching and these concerns increased over time.

Preparing STEM Teachers: The UTeach Replication Model, pages 355–370.

I. BACKGROUND

According to UTeach administrators, the adoption of the UTeach program has led to a dramatic increase in the number of science, technology, engineering, and mathematics (STEM) majors enrolling in education courses and graduating with teacher certification in STEM disciplines—they estimate that the 46 UTeach programs across the country will graduate over 8,000 teachers by 2023. Furthermore, they report that 90% of UTeach graduates enter the teaching profession and of those, roughly 80% are retained after 5 years (UTeach Institute, 2017). Not only does the program show promise in producing more teachers, but research indicates that UTeach teachers may also be more effective than the average teacher. In a recent study in Texas (Backes, Goldhaber, Cade, Sullivan, & Dodson, 2018), researchers compared UTeach teachers to all other teachers in the state, including those trained through other standard university-based programs and alternative certification programs, and found that students of UTeach teachers outperformed their counterparts in math and science assessments by the equivalent of 2–4 months of "additional" learning. These results are supported by another study in Texas (Marder & Hamrock, 2016) which found that students of UTeach teachers gain around 9 months of schooling in both Algebra I and Biology for gifted students, and 5 months of schooling in Biology for economically disadvantaged and Hispanic students.

There are several plausible explanations for why universities with UTeach programs produce a larger number of STEM teacher candidates compared to other university-based teacher preparation programs, and why these graduates tend to be more effective than the average teacher. Our interest in this study focuses on UTeach's unique recruitment and retention strategies, specifically its ability to recruit and retain STEM majors who otherwise would not consider teaching as a possible profession.

UTeach Recruitment

Undergraduate STEM majors are recruited by Colleges of Science faculty into the UTeach program as early as their freshman year with no selection criteria upon entry. The program offers compact and flexible degree plans that allow students majoring in STEM to complete their degrees and certification in 4 years. In addition, UTeach provides interested undergraduates with two, 1-credit-hour field-based courses ("Step" courses) free of charge, allowing undergraduates to try out teaching before committing to completing the teaching option. In contrast, traditional teacher preparation programs typically require that STEM majors formally apply to teacher preparation programs, usually in the Colleges of Education, after completing the STEM content courses required for their majors. Students in these programs must have a certain standing (juniors or seniors) and meet specific criteria (minimum SAT or ACT scores, minimum GPA, prerequisites, etc.) before formal acceptance.

Role of UTeach Step 1 and Step 2 Field-Based Courses in Program Retention and Completion

Because of the nature of UTeach's two Step courses, retention in the program is often low following the Step courses—an expected, but potentially beneficial outcome because STEM majors can make well-informed decisions regarding their fit in the teaching profession prior to investing substantial time and resources. In theory, students who choose to continue to the next set of UTeach courses do so with increased interest and commitment and a greater chance of completing the program and becoming a teacher. In contrast, teacher candidates in more traditional teacher preparation programs have met highly selective criteria and may have greater retention in the program upon formal entrance, but may graduate fewer candidates overall.

UTeach Teacher Candidates

The key to UTeach's increased production of teachers may be the wide net it casts in recruiting for the Step courses, potentially tapping into a group of students not often recruited. Using the free Step courses, UTeach captures two types of students:

- A group of students who enter the university with the intent to become teachers (G1, equivalent to pre-service teachers in other teacher preparation programs). The Step courses give them an opportunity to test this commitment with a real-world teaching experience.
- A group of students who enter the university with no intention to become teachers (G2). These students may enroll in the courses because of the financial incentives or for other reasons. Like the G1 students, G2 students are exposed to real-world teaching experiences that may either confirm their belief that teaching is not a viable career choice or may alter their view of teaching as a desirable career option. It is this group of students who often remain untapped in traditional teaching programs.

This recruitment strategy may explain the unusually large cohort of STEM majors enrolling in the UTeach program, as well as the unusually large number of students who exit the program. As seen in Figure 25.1, we hypothesize that Step 1 serves as a "screener" to the UTeach teacher preparation program, sorting students into two groups (leavers and stayers). Without this type of screening, it is hypothesized that many teacher candidates will leave the program at later stages of the program, after they have invested substantial time and resources to what, in retrospect, was a poor career decision. As theorized by Wanous (1980), people who enter a career with realistic understanding of it are more likely to be satisfied and retained in that career.

UTeach programs have seen a steady increase in STEM teacher production as a result of the program (UTeach Institute, 2017). Although this appears to be

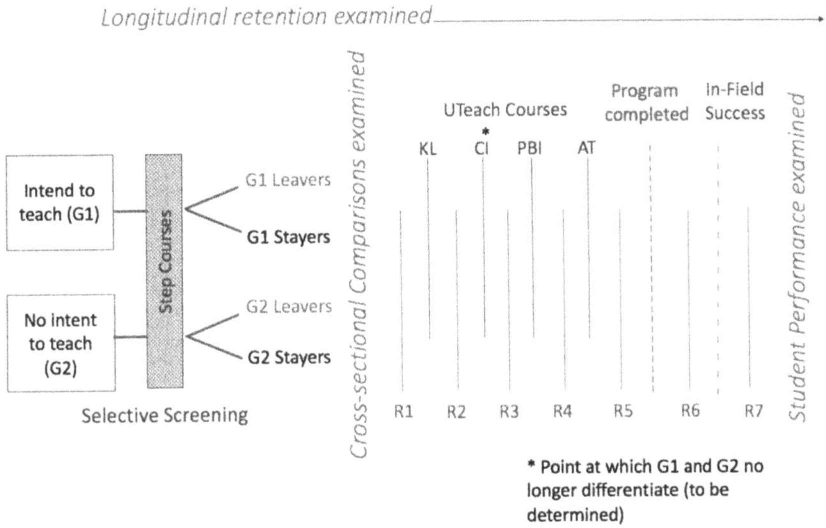

FIGURE 25.1. Theory of Change: Role of UTeach Step Courses in Program Retention and Completion

a cumulative result of recruiting, retaining, and graduating a new population of students (G2), it is still unclear what value the Step courses play. The authors theorize that, for the Step courses to be truly an added value, G2 students must quickly become as committed as the G1 students to teaching, as measured, in part, by their retention and successful completion of the program. If the Step courses let too many students through that ultimately do not complete the program, the Step courses may not be critical in recruiting more students than a standard university teacher preparation program would.

Understanding the Differences Between G1 and G2 Students

Simply examining attrition rates of G1 or G2 students, however, will not help the program understand these students, their perceptions of the program and teaching, and how these perceptions may change as they complete the UTeach program sequence. To better understand these two groups of students, we examine both enrollment data and survey data gathered by UTeach upon completion of the first Step course ("entrance" survey) and at the completion of the UTeach program ("end-of-program" survey). The entrance survey includes an item that defines UTeach teacher candidate as G1 and G2 based upon their response to "Do you plan on teaching in a school?" These surveys also inquire about the student's perceptions of several important factors that may explain their motivations for staying or leaving.

We posit that there are three factors captured by these surveys that may help us to understand the differences between G1 and G2 students and their completion of the program: 1) perceptions of the importance of program financial incentives, 2) perceptions of satisfaction with the program, and 3) perceptions of teaching as a career. How a teacher candidate feels about the teaching program and career may signal their overall satisfaction with the process of becoming a teacher and how they envision a future teaching career. There may be important distinctions between G1 and G2 students along these three factors that help us understand the needs of each group.

II. RESEARCH QUESTIONS

This chapter describes a study of 7 UTeach partners in Texas with the goal of examining potential differences in teacher candidate recruitment and retention in UTeach and perceptions of the program and about teaching. The study investigated the following research questions:

1. What type of students enroll in the Step 1 or Step 1/2 combo course? How do G1 and G2 students compare to each other?
2. What are the trends in retention of G1 and G2 students? At what point do G1 and G2 students no longer differentiate in terms of retention?
3. What are teacher candidates' perceptions of UTeach financial incentives, the program, and teaching as a profession and are there differences in these perceptions between G1 and G2 students? To what degree do these perceptions change between entrance and completion of the program?

III. DATA, SAMPLE, AND METHODS

Data and Samples

The primary data sources for this study are student entrance and end-of-program survey data received from the UTeach Institute at The University of Texas at Austin in February 2019 for the years 2011–2018. Each survey contained encrypted student IDs that allowed the research team to link students across years and to link survey data to course enrollment data.

We used two samples: one to examine group differences in student perceptions in the entrance survey, and one to examine longitudinal group differences in student perceptions in the end-of-program survey as well as differences between entrance and end-of-program surveys.

Sample for Entrance Survey Analysis

The sample of students who took the entrance survey between 2011 and 2018 included 1,623 students from the 7 UTeach partners. Most participants were fe-

male (66%), had a major of Science[1] (41%) or Mathematics (25%), and had previous teaching experience (81%). Thirty-two percent of students were White, 45% were Hispanic, 16% were Asian, and 5% were African-American.

Sample for Combined Entrance and End-of-Program Survey Analysis

To examine the third research question, we explored the differences in the end-of-program survey and differences between entrance and end-of-program surveys for G1 and G2 candidates, which included 328 students from the 7 UTeach partners who completed both entrance and end-of-program surveys. Most participants in this smaller sub-sample were female (71%), had a major of Science[2] (35%) or Mathematics (38%), and came into the program with previous teaching experience (86%). Forty-five percent of students who completed the program were White, 34% were Hispanic, 16% were Asian, and 4% were African American.

Analyses

To address the first research question, we calculated a Chi-square test of independence to compare G1 and G2 to determine whether the two types of students differ with respect to their demographics. To examine the second research question, we explored the differences between G1 and G2 in student retention after the Step 1 or Step 1/2 combo course and program completion using a logistic regression model for the binary outcomes (e.g., whether students remained in the program). We included a group fixed effect, program fixed effects and controlled for student-level characteristics in the model.

To address research question 3, we examined the differences between G1 and G2 students in their perceptions of UTeach program in the entrance survey using a two-level hierarchical linear model (HLM), where students (level 1) were nested within UTeach programs (level 2) to account for the nesting data structure. We employed a cumulative logit model to measure group differences for the outcomes with ordinal distribution when the survey response categories are ordered (e.g., not important, somewhat important, very important). We included a group fixed effect (i.e., G1 or G2) and controlled for student-level characteristics (e.g., gender, race/ethnicity, whether having any teaching experience) and program-level characteristics (e.g., program enrollment, % Female, % White students, % Hispanic students and % students with teaching experience) in the model. We also examined the end-of-program survey results using a one-level cumulative logit model to examine the differences between G1 and G2 students in their perceptions of UTeach program. We included a group fixed effect, program fixed effects and controlled for student-level characteristics in the model. Finally, to further investigate research question three, we analyzed the differences in perceptions between the entrance and end-of-program surveys using a Wilcoxon signed-rank

[1] This includes Astronomy, Biochemistry, Biology, Chemistry, Geosciences and Physics.
[2] This includes Astronomy, Biology, Chemistry, Geosciences and Physics.

test to compare student responses between the surveys. Wilcoxon signed-rank test is a non-parametric statistical hypothesis test alternative to the paired *t*-test used to compare paired data. It is used when the data are not normally distributed such as in student survey responses with ordinal distribution.

IV. RESULTS

Research Question 1: What Type of Students Enroll in the Step 1 or Step 1/2 Combo Course? How do G1 and G2 Students Compare to Each Other?

Of the 1,623 students who had taken the entrance survey after completing Step 1 or the Step 1/2 combo course, approximately 40% indicated that they did not have an intention of teaching as a career, placing them in G2. This confirms claims by UTeach that the program does recruit a sizeable quantity of STEM majors who typically would not consider teaching.

To address research question 1, a Chi-square test of independence was calculated to compare G1 and G2 and determine whether the two types of students differ with respect to their demographics (see Table 25.1 for the results). For the entrance survey, we found that there were no differences in terms of gender nor

TABLE 25.1. Student Demographics in Entrance and End-of-Program (EOP) Surveys

	Entrance				EOP			
Characteristic	**G1**	**G2**	**Chi Square**	**P-Value**	**G1**	**G2**	**Chi Square**	**P-Value**
Female	64.2%	67.5%	1.60	0.206	69.4%	75.3%	0.70	0.403
White	33.8%	29.4%	3.18	0.075	47.0%	38.0%	1.45	0.229
Black/African American	4.9%	5.1%	0.01	0.922	4.8%	0.0%	2.35	0.126
Hispanic	48.0%	40.6%	7.72	0.005	32.5%	38.0%	0.52	0.470
Asian	10.9%	22.9%	39.59	0.000	13.7%	22.5%	2.67	0.103
Have Teacher Experience	84.3%	75.0%	20.83	0.000	85.9%	87.7%	0.04	0.842
Math Major	31.8%	14.8%	58.28	0.000	41.2%	26.0%	4.91	0.027
Science Major	37.7%	46.5%	11.88	0.001	31.4%	47.9%	6.14	0.013
Engineering Major	2.7%	5.5%	7.52	0.006	0.8%	1.4%	0.00	1.000
Education Major	3.7%	0.9%	10.58	0.001	6.3%	1.4%	1.87	0.172
Computer Science Major	0.8%	3.3%	11.84	0.001	0.4%	1.4%	0.01	0.925

in terms of white and black (African American) students. However, G1 students are significantly more likely to be Hispanic (48.0%) than G2 students (40.6%) and less likely to be Asian (10.9%) than G2 students (22.9%). G1 students were more likely to have teaching experience (84.3%) than G2 students (75.0%). G1 students are more likely to major in Math (31.8%) and Education (3.7%) than G2 students (14.8% in Math and 0.9% in Education), while G2 students are more likely to major in Science (46.5%), Engineering (5.5%) and Computer Science (3.3%) than G1 students (37.7% in Science, 2.7% in Engineering, and 0.8% in Computer Science).

By the end of the program, many of these differences disappear, likely due to the attrition of G2 students who significantly differed from G1 students. However, at the end of the program, G2 students are still significantly more likely to be Science Majors (31.4% vs. 47.9%), while G1 students are more likely to be Math majors (41.2% vs. 26.0%).

The differences between these groups indicate that allowing students to test their interest and commitment by "trying out" teaching may be successfully capturing a different group of STEM students and diversifying the initial student-teacher candidate pool, though these differences dwindle as they progress through the program. Although G1 students are more likely to have previous teaching experience, 75% of G2 students have previous experiences suggesting that UTeach may be attractive to STEM students, both G1 and G2, who have taught in some capacity in their past.

Research Question 2: What are the Trends in Retention of G1 and G2 Students? At What Point Do G1 and G2 Students No Longer Differentiate in Terms of Retention?

To better understand the Step courses as a filtering strategy, we examined trends in retention between G1 and G2 students. Table 25.2 shows the proportion of G1 and G2 students who remained (i.e., were retained) in the program immediately after Step 1 or the Step 1/2 combo course (i.e., enrolled in at least one more course after the first course they took). The analysis shows that G1 students were significantly more likely to remain in the program after their first Step course than G2 students. This is expected considering their initial motivation to become a teacher. However, about half of G2 students are still retained at this

TABLE 25.2. Retention after Step 1 or Step 1/2 Combo Course

Group	Retention Rate	N	Odds Ratio (G2/G1)	P-value
G1	79.6%	627	0.19	0.000
G2	49.2%	465		

Notes: Odds ratio is the ratio of the odds of retention between G2 and G1 students.

TABLE 25.3. Retention After Core Courses

Course	Group	Retention Rate	N	G1 VS. G2 Odds Ratio (G2/G1)	G1 VS. G2 P-value
Step 2	G1	77.3%	485	0.32	0.000
	G2	61.0%	228		
Knowing and Learning	G1	82.8%	437	0.57	0.027
	G2	75.9%	162		
Classroom Interactions	G1	86.0%	350	0.50	0.034
	G2	80.2%	116		
Project-Based Instruction	G1	90.6%	234	0.90	0.845
	G2	90.3%	62		

point, indicating that the course was able to convince almost half the students to continue in the program.

We explored whether these groups of students differed in terms of later course taking by examining student retention rates after taking the remaining four "core" courses: *Step 2, Knowing and Learning, Classroom Interactions* and *Project-Based Instruction* (Table 25.3). The data indicate that G1 students were significantly more likely to remain in the program than G2 students after taking *Step 2, Knowing and Learning* and *Classroom Interactions*. No difference was found between G1 and G2 students after *Project-Based Instruction*. The retention rate for both groups increases as the students complete each course in sequence and the differences between G1 and G2 gradually converge at the point of *Project-Based Instruction*. These findings may be helpful in determining the point in the program when G1 and G2 are equally committed, a point comparable to other teacher preparation program's entrance point.

In terms of program completion, Table 25.4 shows the proportion of students who completed the program among those who remained in the program after their first Step course. More than half of G1 students and about one-third of G2 students completed the program. We found that G1 students were significantly more likely to complete the program than G2 students.

TABLE 25.4. Program Completers after Step 1 or Step 1/2 Combo Course

Group	Completion Rate	N	Odds Ratio (G2/G1)	P-value
G1	50.7%	499	0.35	0.000
G2	30.1%	229		

Because UTeach is designed to be "exploratory," this is an expected result but nevertheless suggests that the program allows both groups of students to make well-informed decisions about teaching as a career option and is successful at capturing some part of the additional G2 students.

Research Question 3: What are Teacher Candidates' Perceptions of UTeach Financial Incentives, the Program, and Teaching as a Profession? To What Degree do These Perceptions Change Between Entrance and Completion of the Program?

Another goal of the study is to examine whether there are differences between G1 and G2 students in terms of their perceptions of the program and how they envision teaching as a possible career option; understanding students' perceptions may give us insight into why some G2 students stayed in the program.

Differences in Financial Incentives

To understand the importance of the financial incentives offered by the program, we analyzed the student responses regarding the importance of program-sponsored internships and other program-sponsored financial incentives or support in their decisions to remain in the UTeach program at the time of the entrance survey. As demonstrated in Table 25.5, the majority of students responded that the financial support was somewhat to very important. G2 students were more likely to think *Other Program-Sponsored Financial Incentives or Support* is important than G1 students in their decisions whether to remain in the program, and the difference was marginally significant.

Table 25.6 shows G1 and G2 students' perception of program-sponsored incentives and supports (e.g., internships) on the end-of-program survey. Similar to the entrance survey, the majority of students responded that the incentives and supports were somewhat to very important in their decision to remain in (and presumably, complete) the program, and the groups did not differ in this perception.

TABLE 25.5. Perception of Financial Incentives, Entrance Survey

Stem	Group	Not Important	Somewhat Important	Very Important	N	Odds Ratio (G2/G1)	P-value
Item Stem: How important are the following in your decision whether to remain in UTeach?							
Program-Sponsored Internships	G1	21.1%	40.0%	38.8%	407	1.12	0.510
	G2	20.2%	40.4%	39.3%	183		
Other Program-Sponsored Financial Incentives or Support	G1	19.8%	38.9%	41.3%	409	1.44	0.052
	G2	17.9%	34.2%	47.8%	184		

Notes: Odds ratio is the ratio of the odds of selecting a higher category option in the order of Not Important, Somewhat Important and Very Important between G2 and G1 students.

TABLE 25.6. Perception of Financial Incentives, End-of-Program Survey

Group	Not Important	Somewhat Important	Very Important	N	Odds Ratio (G2/G1)	P-value
Item Stem: How important were program-sponsored financial incentives and supports (e.g., internships) to your decision to remain in the program?						
G1	29.6%	32.7%	37.6%	226	0.79	0.382
G2	38.4%	31.5%	30.1%	73		

These findings indicate that the financial incentives offered by the program may be an important factor in retaining students.

Differences in Program Satisfaction

The study also explored whether students' perceptions of satisfaction with UTeach were a factor in recruitment and retention. Table 25.7 shows the student satisfaction with the UTeach program on the entrance survey. During Step 1 or the Step 1/2 combo course, the majority of students were satisfied with the program. G1 students were significantly more satisfied with the program than G2 students.

By the time students took the end-of-program survey, a majority of G1 students and all G2 students rated the program satisfactorily (see Table 25.8).

Respondents of the end-of-program survey also rated their satisfaction with several aspects of the program, including *program staff support, degree plan flexibility, access to resources and materials,* and *space available for collaborating.* Across the board, a large majority of respondents rated these aspects of the pro-

TABLE 25.7. Satisfaction with the Program, Entrance Survey

Group	Dissatisfied	Neutral	Satisfied	N	Odds Ratio (G2/G1)	P-value
Item Stem: Overall, how satisfied are you with UTeach?						
G1	3.2%	6.1%	90.7%	971	0.38	0.000
G2	4.1%	17.2%	78.7%	639		

Note: Odds ratio is the ratio of the odds of selecting a higher category option in the order of Dissatisfied, Neutral and Satisfied between G2 and G1 students.

TABLE 25.8. Satisfaction with the Program, End-of-Program Survey

Group	Dissatisfied	Neutral	Satisfied	N	Odds Ratio (G2/G1)	P-value
Item Stem: Overall, how satisfied are you with UTeach?						
G1	4.1%	4.1%	91.9%	74	NA	
G2	0.0%	0.0%	100.0%	14		

TABLE 25.9. Satisfaction with the Program Item STEM

	Group	Dissatisfied	Neutral	Satisfied	N	Odds Ratio (G2/G1)	P-value
Item Stem: Please indicate your overall level of satisfaction with the following.							
Program staff support	G1	0.0%	2.2%	97.8%	180	0.65	0.630
	G2	1.7%	1.7%	96.6%	59		
Degree plan flexibility	G1	3.4%	18.0%	78.7%	178	1.65	0.252
	G2	1.7%	13.6%	84.7%	59		
Access to resources and materials needed for teaching	G1	0.6%	2.8%	96.7%	180	2.02	0.521
	G2	0.0%	1.7%	98.3%	59		
Space available for collaborating, practicing lessons, etc.	G1	1.7%	7.8%	90.6%	180	1.00	0.999
	G2	0.0%	11.9%	88.1%	59		

gram satisfactorily (see Table 25.9), but no difference was found between G1 and G2 students.

Difference in Perceptions of Teaching Career

The study also explored whether students' concerns about pursuing a career in teaching were factors in recruitment and completion of the UTeach program. In particular, we were interested in understanding whether G2 students have different concerns about becoming a teacher than G1 students and whether participation in the program changes these concerns over time. UTeach students rated their concerns about the teaching profession on the 4 issues related to pursuing a teaching career (see Table 25.10). On the entrance survey, G2 students were significantly more concerned about the 2 issues, *career path/advancement* and *salary*, than G1 students. G2 students were also more concerned about the *prestige/perception of the teaching profession* than G1 students, and the difference was marginally statistically significant.

Table 25.11 shows student concerns about the 4 issues among students who completed both the entrance and end-of-program surveys. The results show that both G1 and G2 students were significantly more concerned about the 4 issues when they completed the program than at the time of the entrance survey, with the exception of G2 students' concern about *additional time and/or cost to obtain a degree* likely due to the low number of respondents limiting the statistic power to detect the difference. The greatest proportion of students who had concerns (over 70% of students who reported yes on the end-of-program survey) were related to *salary*. Seventy-five percent of G2 students reported concerns on the end-of-program survey for *career path/advancement*.

We saw, for both groups of students, an increase in concerns over time. It may be that, as students become more committed, more invested, and more aware of

TABLE 25.10. Concerns Related to Pursuing a Teaching Career between G1 and G2, Entrance Survey

	Group	No	Yes	N	Odds Ratio (G2/G1)	P-value
Item Stem: Please indicate your level of concern regarding the following issues related to pursuing a teaching career.						
Additional time and/or cost to obtain degree	G1	56.5%	43.5%	976	0.81	0.100
	G2	67.6%	32.4%	642		
Career path/advancement	G1	52.0%	48.0%	976	1.43	0.002
	G2	48.6%	51.4%	642		
Prestige/perception of the teaching profession	G1	64.6%	35.4%	975	1.25	0.077
	G2	65.1%	34.9%	642		
Salary	G1	39.0%	61.0%	976	1.61	0.000
	G2	33.0%	67.0%	640		

Note: This item was asked differently in different years—in one year, it asks for the student's level of concern, and in others, it asks whether the student was concerned about this aspect of teaching. Responses were converted to binaries when the question asked for students' level of concern.

TABLE 25.11. Concerns Related to Pursuing a Teaching Career on Entrance and End-of-Program Surveys

	Group	Survey	No	Yes	N	P-value
Item Stem: Please indicate your level of concern regarding the following issues related to pursuing a teaching career.						
Additional time and/or cost to obtain degree	G1	Entrance	71.7%	28.3%	46	0.004
		EOP	41.3%	58.7%		
	G2	Entrance	80.0%	20.0%	10	0.129
		EOP	40.0%	60.0%		
Career path/ advancement	G1	Entrance	59.6%	40.4%	225	0.000
		EOP	34.2%	65.8%		
	G2	Entrance	56.9%	43.1%	72	0.000
		EOP	25.0%	75.0%		
Prestige/perception of the teaching profession	G1	Entrance	74.8%	25.2%	226	0.000
		EOP	46.9%	53.1%		
	G2	Entrance	72.6%	27.4%	73	0.001
		EOP	43.8%	56.2%		
Salary	G1	Entrance	44.0%	56.0%	225	0.000
		EOP	28.9%	71.1%		
	G2	Entrance	45.2%	54.8%	73	0.020
		EOP	27.4%	72.6%		

the challenges teachers face, concerns about the profession become more real and remarkable.

V. SUMMARY

The goal of this study was to examine UTeach's recruitment and retention strategies, specifically its ability to recruit and retain STEM majors who otherwise would not consider teaching as a possible profession. The results shed light on the value and importance of some of UTeach's unique features that may explain why the UTeach program has been successful at increasing the number of effective STEM teachers.

This study found that:

- UTeach successfully recruits a potentially "untapped" population of STEM majors. Approximately 40% of the UTeach students in the Step 1 or Step 1/2 combo are G2. Approximately 30% of G2 students are retained after their first Step course, a percent add-on to the population that other more traditional programs may produce and a nod to the value of the Step courses.
- UTeach recruitment diversifies the student-teacher population. A greater proportion of G1 students are Hispanic and major in math and education. More G2 students are Asian and major in the sciences, computer science, and engineering. G1 students were more likely to have previous teaching experiences but 75% of G2 students also had previous experiences suggesting that both groups share an initial interest based on practical knowledge of teaching. However, this diversity decreases by the time G2 students finish the program, with the only difference between G1 and G2 students being that G1 students are more likely to be Math majors and G2 students are more likely to be Science majors.
- Although G1 students are more likely to be retained in the program, over time, G2 retention rates gradually become more similar and the two groups converge. By the time students have completed *Project-Based Instruction*, there is no significant difference between the two groups. This may be a point in the program when the UTeach student population can confidently be compared to students in other teacher preparation programs (i.e., GPA, SAT, and other teacher preparation entrance requirements).
- Financial incentives are popular with both groups of students. The majority of students report that program-sponsored financial incentives or support are somewhat to very important upon entrance and completion of the program. This may be helpful knowledge to UTeach administrators who are determining the cost-benefit of student financial support.
- The UTeach program is viewed positively by both groups of students. An overwhelming majority of the students rate the UTeach program as satisfactory; this was true across multiple aspects of the program including

program staff support, degree plan flexibility, access to resources and materials, and *space available for collaborating and practicing lessons.* G1 students tend to be more satisfied after Step 1 or Step 1/2 combo, but there are no differences between the groups upon completion of the program.
- UTeach students have concerns about making a career of teaching, specifically about the *additional time and/or cost to obtain the degree, career path/advancement, prestige/perception of the teaching profession,* and *salary.* Overall, these concerns increased over the course of the program. UTeach administrators may want to consider ways to alleviate these concerns, particularly as students reach completion of the program.

There are a number of limitations that should be considered alongside these findings. First, not every student who enrolls in or completes the UTeach program has taken all of these surveys. We estimate that roughly 1 out of 5 students who have taken one UTeach course have taken the entrance survey. This means that the entrance survey in particular may be prone to response bias. In addition, the entrance survey is completed at the end of the semester and may include students who entered as G1 and G2 but changed their minds as a result of exposure to the Step class and self-reported as being the other category. Additionally, we found less than one-quarter of UTeach students took both the entrance and exit exam (i.e. 328 of 1624 or 20%). This makes interpreting the connections made between the entrance and exit survey inherently limited. This study also takes place in Texas, which has a unique demographic, and includes the flagship university (UT Austin) and other well-established UTeach programs. Findings from these universities may not generalize to other UTeach programs in other states. Furthermore, the survey data do not help us ascertain why certain students persist or not. Finally, this study is descriptive and does not imply causation. This means that some results should be interpreted cautiously.

That being said, UTeach programs can be somewhat confident that the Step courses, financial incentives, and program supports are valuable recruitment and retention strategies. Current and future funders of the program can be assured that their financial support of these strategies are good investments. And policymakers can better understand the UTeach program, specifically when to begin measuring variables of selectivity of program candidates (i.e., SAT, GPA, and prerequisite courses required) for higher education accountability purposes.

REFERENCES

Backes, B., Goldhaber, D., Cade, W., Sullivan, K., & Dodson, M. (2018). Can UTeach? Assessing the relative effectiveness of STEM teachers. *Economics of Education Review, 60,* 284–198. Retrieved from: https://www.sciencedirect.com/science/article/abs/pii/S0272775717306969

Marder, M., & Hamrock, C. (2016). Math and science outcomes for students of teachers from standard and alternative pathways in Texas. Austin, TX: University of

Texas at Austin. Retrieved from: https://pdfs.semanticscholar.org/0693/f28933b-966f68a48813a75894353f954e639.pdf?_ga=2.103896832.1239140174.1510529724-172563159.1510529724

UTeach Institute. (2017). *UTeach and UTeach expansion FAQs*. Retrieved from: https://institute.uteach.utexas.edu/sites/institute.uteach.utexas.edu/files/uteach-expansion-factsheet-faq-may-2017.pdf

Wanous, J. P. (1980). *Recruitment, selection, and socialization of newcomers*. Reading, MA: Addison-Wesley.

CHAPTER 26

COMPUTATIONAL THINKING FOR STEM TEACHER LEADERSHIP TRAINING AT LOUISIANA STATE UNIVERSITY

Fernando Alegre, Juana Moreno, Jerry S. Weltman, David Kirshner, John Arthur Underwood, Frank Neubrander, and James J. Madden

Louisiana State University (LSU) is developing and piloting two new courses: *Computational Thinking for Future Teachers* (CTFT) and *Computer Science Teaching Methods* (CS Methods). These courses build on LSU Gordon A. Cain Center's materials developed for courses at the secondary educator training level. Pedagogical content knowledge gathered from instructor and teacher interactions at multiple high school pilot sites and the summer professional development sessions have contributed to the design of these post-secondary courses. The CTFT course target undergraduates in LSU GeauxTeach educator training program, and will provide pedagogy and content knowledge to prepare them to teach introductory high school CS and CT courses. The CS Methods course targets Computer Science majors who are considering a career in teaching or are interested in CS pedagogy. The Teacher Education STEM graduates who take the envisioned CTFT course will help meet the significant urgent demand for highly qualified CS/CT teachers.

Preparing STEM Teachers: The UTeach Replication Model, pages 371–385.

INTRODUCTION

According to the U.S. Bureaus of Labor Statistics computing occupations are the number one source of all new wages and make up 77% of all projected new jobs in STEM fields. In 2017 Louisiana had 455 computer science graduates, with only 24% of them being female. Only 16% of the public schools in Louisiana offer computer science. In 2017–2018 only 43 high schools offered an AP Computer Science course. Only 536 total exams taken in either of the two AP Computer Science courses in Louisiana for 2018. AP Computer Science examinees represented only 1.5% of the total number of students who took an AP test, which made it the least taken AP exam in Louisiana. Although computer science enrollment at Louisiana State University (LSU) has almost doubled over the last five years, LSU's undergraduate STEM teacher preparation program (GeauxTeach) and LSU 's four graduate level professional degree programs for teachers[1] do not offer introductory computational thinking or programming courses suitable for pre-service teacher training. The two courses we are piloting, *Computational Thinking for Future Teachers* (CTFT) and *Computer Science Teaching Methods (CS Methods),* will enhance the educational experiences offered at LSU.

LSU's move to offer undergraduate computational thinking courses for pre-service teachers complements and builds on the Cain Center's new LSU STEM Certification Pathways Program. The LSU STEM Pathways Program provides high school students an opportunity to enroll in a specialized progression of project-based courses (i.e., pathways) leading to a career-tech diploma or to an enhanced university-prep diploma (e.g., dual-enrolment credit). LSU STEM Pathways Program has computational thinking and programming courses embedded in each of its four branches: (I) Pre-Engineering, (II) Digital Design and Emergent Media, (III) Computing, and (IV) Biomedical Sciences. Yet the push at LSU and across the state to create secondary school pathways involving computer science has not been matched by an increasing supply of qualified computer science teachers. Only a handful of computer science teachers graduated from Louisiana universities in the past few years; of recent note we saw no computer science teachers graduate in 2016.[2] The development of Computational Thinking for Future Teachers (CTFT) and Computer Science Teaching Methods (CS Methods) is a first step to create Computer Science teacher certification programs at LSU.

The LSU STEM Pathways courses, which are aligned with the Next Generation Science Standards[3] and open-sourced, first took shape with field testing occurring at a local high school that collaborated with LSU's faculty. As students take courses in the four "core" subject areas—English language arts, math, science, and social studies—they also engage in electives from the LSU STEM Pathways. These electives provide high school students with the opportunity to

[1] Master of Natural Science, Mathematics for Advanced Secondary Instruction Certificate, Holmes Master of Arts in Teaching and MS in Engineering with a Concentration in Education.
[2] Code.org Advocacy Coalition, 2018, https://code.org/advocacy/state-facts/LA.pdf.
[3] Next Generation Science Standards for States, by States. https://www.nextgenscience.org/.

receive a special STEM endorsement on their high school diploma. The LSU STEM Pathways are fully approved by the Board of Elementary and Secondary Education (BESE) and the state education funding formula provides a school district with additional funds. Districts receive $480 per student per Pathway course completed from the state's Career Development Funds. In order for a school to be able to eligible for this program, the students to receive their STEM endorsements, the schools to receive their accountability bonus points, and to earn the additional career-tech funding BESE requires that all of the course teachers must attend and pass LSU Pathways' teacher training courses. Initial piloting of LSU Pathways' teacher training courses took place in 2017 and continues to date.

The development of the new undergraduate courses is based on materials we previously developed for the LSU STEM Pathways practitioner teacher training. The pedagogical content knowledge gathered from our experience with the piloting year of the high school Pathways courses and the summer professional development was incorporated into the post-secondary pre-service teacher training courses. A brief description of each course follows.

COMPUTATIONAL THINKING FOR FUTURE TEACHERS

Computational Thinking for Future Teachers (CTFT) is oriented to students enrolled in a pre-service teaching program, such as GeauxTeach and to STEM undergraduates considering a teaching career. The course's pedagogical approach and the interdisciplinary connections will also be relevant to undergraduates, in general, because it unifies the often-compartmentalized learning of math and science subjects. Currently only the LSU GeauxTeach Physics students are required to take courses on programming and computational science; the rest of the GeauxTeach participants may complete their degrees without ever being exposed to computing. The CTFT course not only remedies this situation, but it also provides substantial pedagogy and content knowledge to prepare graduates to teach introductory high school CS/CT courses.

The CTFT course assumes that the students have no prior knowledge of computing beyond web browsing. Students learn programming skills in conjunction with associated computational thinking concepts. Codeworld[4], a web-based integrated development environment (IDE) that uses a simplified variant of the Haskell language initially designed for middle school students, is used in many of the activities. Haskell is a high-level programming language that blends well with mathematical notation, which means very little instruction is needed on the language itself. Haskell's syntactic features encourage and promote higher-order thinking in students. Because Haskell is not widely utilized in commercial courses we anticipate that most undergraduate students will not have prior knowledge of it. The language will therefore act as an equalizer between participants regardless of any previously existing programming skills. Although most of the activities require some programming, the focus of the course is geared toward:

[4] Codeworld. https://github.com/google/codeworld, 2018.

- Development of the ability to create and plan computational solutions to problems;
- Understanding of the iterative nature of the software development cycle;
- Creating models and simulations of artifacts following solid design principle;
- Hypothesizing and testing possible solutions to computational problems in a systematic way that follows scientific practice; and
- Formative and summative assessment through project-based learning

Computational Thinking

The course focuses on promoting essential computational thinking skills. Even though computational thinking is emerging as an effective approach to understand many scientific phenomena, its use in instructional settings is limited. Pellegrino and Hilton (2013) list of essential 21st Century Skills includes more than half that are typically considered part of computational thinking: creativity, critical thinking, problem solving, communication, collaboration, information literacy, media literacy and digital citizenship. Computational Thinking is not limited to programming (Angeli & Valanides, 2005; Angeli et al., 2016; Denning, 2017a,b; Grover & Pea, 2013; Grover, Pea, & Cooper, 2015; Valanides & Angeli, 2008; Voogt, Fisser, Good, Mishra, & Yadav, 2015; Weintrop et al., 2016; Wing, 2006, 2008), though it is better understood when using a high-level programming language as the medium to learn it (Buitrago Flórez et al., 2017). Programming is a skill that needs to be learned if it is to be used as a demonstration of computational competencies (Grover & Pea, 2013). Thus, introducing CT effectively in STEM curricula cannot take place without also introducing coding. But launching an approach to integrate CT requires a vision of how this integration would work; the type of coding to be taught must be carefully selected with a broader goal in mind.

To tackle the complexity of contemporary problems in science and engineering, students in all STEM subjects should be exposed to CT concepts and practices as part of their educational experience (Novak & Musonda, 1991; Qualls & Sherrell, 2010). This is especially true for future teachers who have an interest in bringing modern technology to their classrooms in a way that enables not only consumption but also production of new technology (Yadav, Sands, Good, & Lishinki, 2018)

Scope of the Effort

The East Baton Rouge Parish School System administrators requested LSU personnel to develop a secondary school computational thinking curriculum focused on the Algebra 1 and science content ninth graders learn. This request was seated in a desire to improve student performance in STEM. The first course, *Introduction to Computational Thinking (ICT),* uses coding as a new discipline to be learned and to reinforce math concepts within a more applied setting. ICT was designed around hands-on project-based activities in which students write

computer code to run simulations, animations and visualizations. Through the activities, they learn how to use abstraction, decomposition, and pattern recognition to model problems while arriving at an algorithmic solution. ICT uses a functional programming language where all the coding is done by defining mathematical functions. Using similar languages there have been recent interventions which provide promising evidence of transfer between programming and mathematics (Schanzer, Fisler, & Krishnamurthi, 2013; Schanzer, Fisler, Krishnamurthi, & Felleisen, 2015). Our approach built upon this evidence, but focused on a tighter integration with the secondary mathematics and science curricula. All of the activities are programmed in CodeWorld using a limited set of graphical primitives to draw circles, rectangles, and text, and to apply translations, rotations, scalings and colors to them. As previously noted, Haskell follows a syntax very similar to math notation and the semantics of the language follow closely the semantics of algebra. The current ICT topic outline is: (1) Computing and Coding Basics, (2) Transformations and Composition, (3) Symmetry and Regularity, (4) Managing Complexity, (5) Data and Calculations, (6) Functions and Data Structures, and (7) Modeling with Functions.

Practicing practitioner teachers who wish to teach the ICT course must enroll in an intensive summer professional development training. These teachers undergo all parts of the ICT course as students by completing all course assignments and activities. While this training has been effective to make them ready to teach ICT, a preliminary fidelity of implementation study found two main shortcomings of the training that are addressed in the *Computational Thinking for Future Teachers* course. First, future teachers need to cover the high school material in more depth than is required of secondary students. In particular, they should complete similar but more sophisticated activities, learn more about the underlying concepts, and build connections to other science, math and computer science content. Second, the post-secondary course should include a focus on formative and summative assessments implemented in the context of project based learning activities. This component should include a final assignment in which students design and develop a new project suitable to be incorporated into ICT.

In summary, completion of CTFT requires the following modifications to the existing professional development training based on mere exposure to the ICT course: (1) A revised text where explanations are expanded and described in more technical terms, (2) A set of additional activities where the conceptual framework is explored at a deeper level, (3) A new unit covering pedagogical material relevant to the content, and (4) An expanded list of learning objectives along with the corresponding assessment instruments, such as midterm and final tests, and detailed project assignments.

Samples of CTFT Course Activities:

Dots on a line (Figure 26.1): In this activity, students use lines from the point of view of vectors and parametric equations. The goal of this activity is to write

code that places an arbitrary number of dots in an arbitrary line segment within the output window. This activity was deemed too difficult to be included in the secondary curriculum, but variations of it have successfully been used in the professional development sessions. This problem requires students to switch between several representations of a line.

In the Genetics-two factor cross activity, see Figure 26.2, students are asked to implement a program that draws a Punnett square for heterozygous parents with two traits: hair color (blond or dark) and presence of freckles. The main purpose of this activity is to summarize and organize the student knowledge, while at the same time providing a medium for simulating different possibilities, by, for example, extending the code to show more than one generation, or changing the probabilities of each outcome to adjust them to the corresponding proportion of population with a given trait. The code for this activity is mostly symbolic. Here, students need to codify their knowledge as categorical functions instead of numeric ones. For example, the relationship between phenotypes and genotypes for hair color is written in the code as:

```
express :: (GenoType, GenoType) ->
PhenoType express (BlondHairGT,
BlondHairGT) = BlondHair express
(motherOther, fatherOther) = DarkHair
```

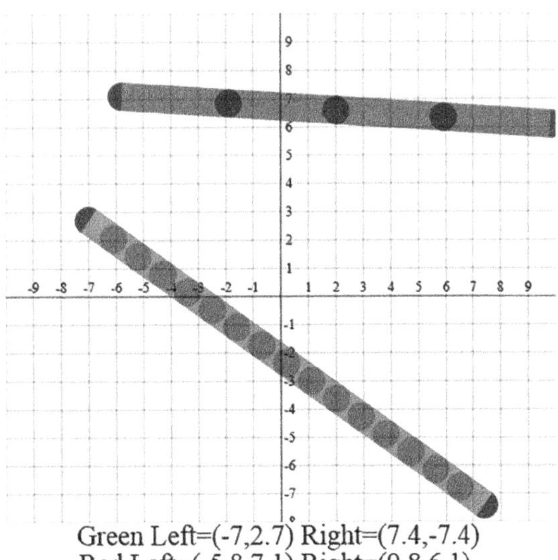

FIGURE 26.1. Dots on a Line Activity. Students Write Code to Place an Arbitrary Number of Dots on a Given Random Line.

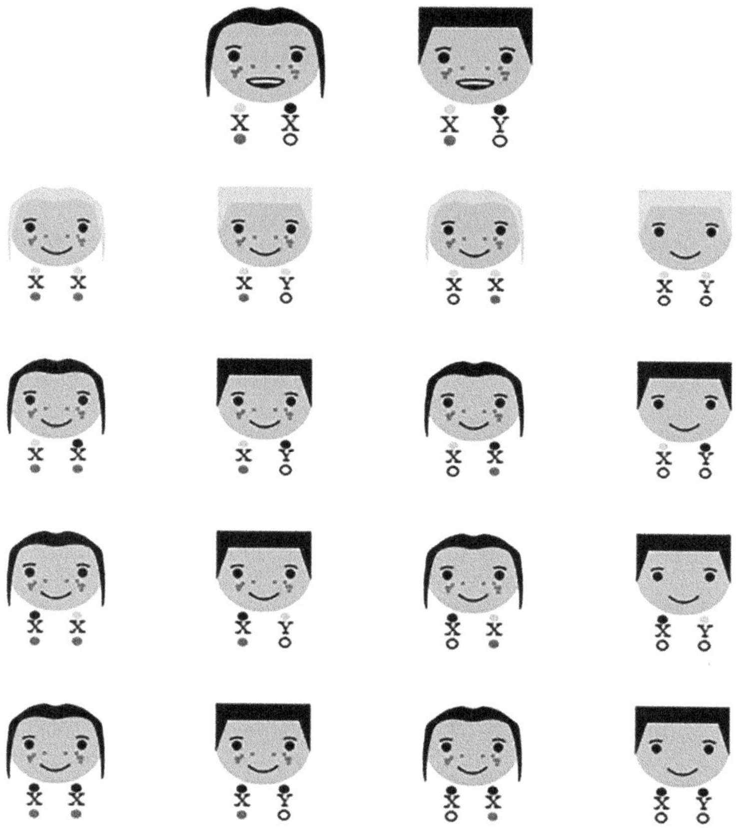

FIGURE 26.2. Genetics Two-Factor Cross Activity. Students Implement a Program to Draw a Punnett Square.

which represents the fact that the BlondHair phenotype is expressed only when both genotypes are BlondHair, and any other combination of alleles will be expressed as DarkHair. And the cross of alleles is given by the following code:

```
cross((f1,f2),(m1,m2))= [(f1,m1),(f1,m2),(f2,m1),(f2,m2)],
```

where f1 and f2 represent the father's alleles, and m1 and m2 represent the mother's alleles.

In the Uniform linear motion activity, see Figure 26.3, students need to reason about the representation of movement as a functional relationship between positions and times. Students are presented with different scenarios that describe in words a particular movement, such as: *A car leaves Baton Rouge towards New*

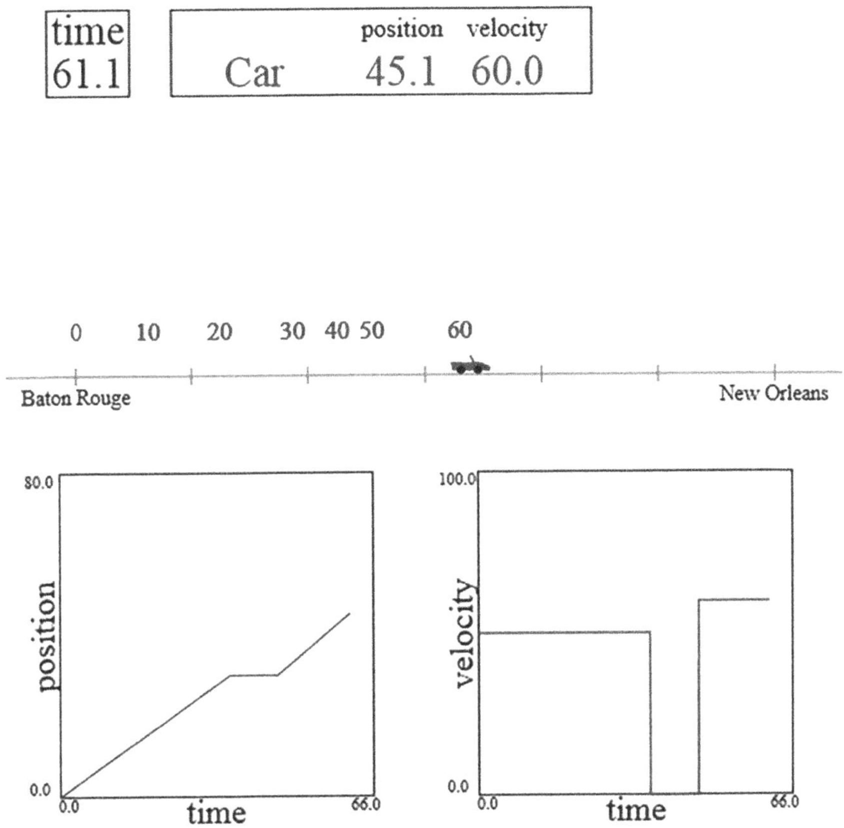

FIGURE 26.3. Uniform Linear Motion Activity. Students Define a Function that Describes the Position or the Velocity as a Function of Time in a Variety of Posed Scenarios.

Orleans, which is located 80 miles Southeast. The car travels at an average speed of 50 miles per hour for the first 30 miles, and then stops for gas for 10 minutes. Then, it resumes its trip, traveling at an average speed of 60 miles per hour. Create a simulation that shows the car as it travels, stops and continues, and find how much time the trip did take.

This exercise illustrates many features of our system. The simulation helps the students by giving them feedback in several ways: a picture of the car moving and stopping, a graph of the distance vs. time, a graph of the velocity vs. time, and a text box showing the state of the car at each time. Students create a simulation by defining a function that describes either the position at each time or the velocity at each time. One of the objectives of this activity is that students understand that

these two ways of describing movement are interlocked, and it is actually enough to define only one of them because the other can be inferred. In some versions of this activity, when a student adds both descriptions, but they do not match, the simulation will show two separate cars, each moving according to either only the position or only the velocity. As an example, a description of the scenario shown above in terms of the velocity would be as follows:

```
time1 = 30/50 * 60
time2 = time1 + 10
time3 = time2 + (80-30)/60 * 60
- velocity is in miles per hour; time is in minutes

car_velocity(time)
  | time < time1 = 50
  | time1 <= time, time < time2 = 0
  | time2 <= time, time < time3 = 60
  | otherwise = 0
```

This code is similar to the standard math notation for defining piecewise functions:

$$v(t) = \begin{cases} 50, \ t < t_1 \\ 0, \ t_1 \leq t < t_2 \\ 60, \ t_2 \leq t < t_3 \\ 0, \ otherwise \end{cases}$$

The variables `time1, time2,` and `time3` represent the times at which each leg of the trip ends. Note how we encourage students to leave calculations unevaluated, because it is clearer to see the reasoning behind them, and the computer will calculate it effortlessly when the simulation is run. While the secondary course contains only elementary scenarios, the post- secondary courses uses more elaborated scenarios.

In the *Multipeg Towers of Hanoi* lesson (Figure 26.4), the *Towers of Hanoi* puzzle is generalized to multiple pegs. In the classical case, there are three pegs, and a number of disks is placed on the left peg. Initially, all disks in the left peg are arranged from smallest on the top to the largest on the bottom, and the puzzle consists of moving all the disks to the rightmost peg according to the following rules:

Rule 1: A disk can be placed only on a larger disk
Rule 2: You can only move one disk at a time
Rule 3: You can only move the topmost disk

Following these rules, a recursive solution can be found, in which moving N disks requires 2^N–1 moves, which makes the problem infeasible for more than a handful of disks, and even when using a fast computer, solutions can be found for just a couple of dozen disks at the most. Surprisingly, adding one extra peg

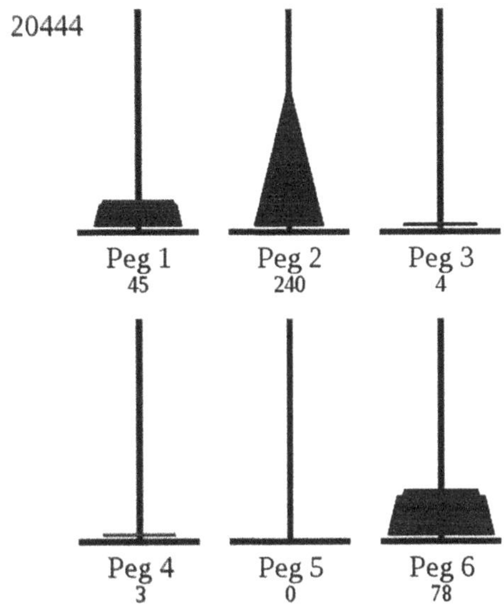

FIGURE 26.4. Multipeg Towers of Hanoi Activity. Students Program a Recursive Solution to a Puzzle Parametrized by Several Variables, Such as the Number of Pegs and the Number of Disks.

makes the problem tractable in many more cases, raising the feasible number of disks to more than a thousand. While the number of moves is still asymptotically exponential, the exact formula does not have a closed form and grows much slower. Even more surprisingly, there is no known proof that the most common given solution, known as the Frame-Stewart algorithm, is actually optimal, and the formula used for solving the 4-peg problem cannot be used for the 5-and-up-peg problem. Furthermore, trying to solve this problem brings up connections with other concepts, such as the Pascal triangle (combinatorial numbers) and the Sierpinski triangle (fractals), and it is itself a very thorough exercise in recursion. Finally, even more surprising is the fact that the algorithm in this puzzle is related to real-world problems, such as the loading and unloading of ship containers.

COMPUTER SCIENCE METHODS COURSE

The CS Methods course is specifically created to recruit CS majors to teach secondary education. However, for the course to be successful and sustain a healthy enrollment, it must appeal to a wider range of students by serving four different groups:

Group 1: CS majors seeking initial teacher certification;
Group 2: STEM majors seeking add-on CS certification;

Group 3: CS majors who want to learn how to design courses for corporate training or online classes;
Group 4: Students seeking basic computing/coding literacy.

The course covers three large units: (1) CS high school pedagogy; (2) self-paced learning and teaching of one secondary school CS curricula; and (3) an application of CS pedagogy. In the first unit on CS high school pedagogy the seven big ideas of the AP Computer Science Principles course, and standards and guidelines for teaching CS are discussed. In addition, specific CS pedagogy, such as how to keep students' attention when they have a screen in front of them, how to manager collaboration in the classroom and make sure students are truly being productive and dealing with homework, is covered.

The second unit is hands-on. Students go over different secondary curricula according to their background. K–12 CS pedagogy is an area of active research with many issues still unknown (Denning, 2017; Guzdial, Kay, Norris & Soloway, 2019). Unlike most high school subjects, CS has not even a glimmer of a standard curriculum. CS majors planning to teach high school (Group 1) follow an abridged ICT curriculum. ICT is currently being taught in 9th grade, but is rigorous enough to challenge 12th graders. Then, Group 1 students are expected to teach a lesson from one of the later units. For the rest of the students the instructor guides them to choose from one of the many online courses. In particular, the CSforAll initiative[5] aims to teach CS as a fundamental discipline to all students, not only those who will become software engineers. Many of these courses are appropriate for students with no CS background, and the course materials are clear and explicit enough that they can be learned with little or no help from the instructor. Regardless of which self-study CS curriculum the student is following, they will create a lesson to teach their fellow students in small groups. Each lesson will be evaluated by the fellow students.

The last unit focuses on an application of the learned pedagogy. First, the students will observe how someone else teaches CS. Students in Group 1 will attend a local high school and observe teaching for two days, all arranged by the instructor. The rest of the students will be allowed to observe a high school class or some other teaching format. For example, they could attend a live corporate seminar, watch a seminar on YouTube, attend meetings of the local .NET User Group, observe an online MOOC course, etc. The course concludes with a teaching project. Students in Group 1 will prepare a lesson from ICT. The instructor will coordinate with a local high school for students to spend three class days teaching the lesson. Depending on the number of Group 1 students, it may be necessary to teach in teams. The rest of the students will prepare a class project in which they teach a lesson to a corporate audience, an online class, or some other appropriate audience. The lesson can be based on the self-paced learning or it can

[5] CSforAll initiative, https://www.csforall.org/.

be based on whatever the student's interest is, but it must involve computational principles. The instructor will help the student set up an appropriate audience for the lesson. For example, the instructor could arrange with LSU's ACM chapter to have a meeting where students give a corporate talk or speak about CS Pedagogical method. Here are some examples:

- A CS major could learn something new about .NET and present it to the local .NET User group.
- A biology student might teach an algorithm from the Human Genome Project.
- A student could do an investigation of the Alice Programming environment and present a report to the local ACM chapter.

ONLINE LEARNING SYSTEM

The high school ICT course follows a hybrid delivery model, where the content is online, and the students submit their assignments through a custom online learning system that is integrated with the content. Teachers log into this system to review the student work and provide feedback to the students. There are several reasons why we chose this model. First, the content and the assessments are standardized across the schools, and it is guaranteed that every school is always using the latest version of the course. Second, as we collect all the data from the course implementations, we are building a corpus of student work that can be used for research in computer science education, especially in relation to teaching programming to novice students, and for review by pre-service and in-service teachers, as explained below. Third, the system also allows us to review the effectiveness of each teacher, based on the feedback they provide to their students and on comparison of student submissions across schools, so that we can provide *just-in-time* professional development to support them. Finally, the system reduces the workload of the teachers by assisting them in automatically grading part of the assignments.

Given the project-based nature of the course and the shallow background on computing that the in-service teachers have, the content delivered online has three purposes: provide conceptual background knowledge about a topic for both teachers and students, provide the basis (such as list of requirements, starting code, sample outcomes and assessment rubrics) that enables students to achieve the learning goals of each project, and provide pedagogical advice to the teachers about each project, including hints to help students with potential roadblocks and commented samples of project deliverables. The design of the course foresees a continuously increasing collection of projects that are made available, and we expect that some of the more experienced teachers become regular contributors of projects. The online system allows us to disseminate each new project to all participants as soon as it is ready.

We intend to leverage this system and use it also for the undergraduate CTFT course. It may seem that some of the reasons cited above would be less critical for a small college-level course with one or two sections each semester. However, this course may be taught not only by computer science faculty but also by faculty from other departments, such as the school of education. Those faculty may be, as far as their computing background is concerned, in a situation similar to the high school teachers, who the first time they teach this course, they may know just slightly more than their students about the subject.

There is also an academic reason for using an online system in the *CT for Future Teachers* course, as it is important for the students to learn about online learning systems not only from the point of view of a student, but also from the point of view of a teacher. Future teachers need to learn how to use an online system effectively to track and assess student work. In the CTFT course, the students will be able to create mock sections, add each other as a student in those sections and grade their peers submissions. At the same time, the instructors are able to seamlessly review student work, run their programs, provide feedback, and present selected works to the class to perform live code reviews. Accumulated work from earlier courses will also be available for students and instructors to review, so that a knowledge base of annotated past work is created on time. In addition, the large corpus of high-school student work will be available for them to review, so that they learn how to review and assess the quality of student code. This feature is particularly important for the courses we are discussing, because the type of programming to be performed has not only technical but also artistic components, and the assignments encourage each student to design and code their own original creations.

CONCLUSION

Education policy reform discussions are increasingly focused on the need for K–12 schools to improve and expand STEM programs, including computer science, as fundamental components of a well-rounded education for all children. With the growing STEM economy and demand for computer-capable citizens, school districts are exploring innovative curricular approaches. One example is the system of four LSU STEM Certification Pathways providing a broad introduction to computational thinking and computer science principles in connection with technology-based occupations (Engineering, Digital Design/Media, Computing, and Biomedical Sciences). Louisiana's critical shortage of teachers and educators with training for such courses severely limits statewide access to the Pathways and similar programs. University-based undergraduate teacher preparation development programs have a critical role to play in reversing this situation by offering computational thinking and programming content and pedagogy courses for all candidates, and setting appropriate participation expectations for program completion. The two new courses proposed here are the first step towards fulfilling the needs of pre-service K–12 teachers to prepare the next generation of com-

puting literate citizens. The two new courses will significantly strengthen LSU's contributions to a Louisiana "CS For All" initiative, and will thereby positively affect LSU relationships with high-tech businesses and industries dependent on the availability of an adequately prepared workforce.

The curricula for the two courses described above will be available in several offline and online formats. The main delivery vehicle of the courses will be a website that integrates lessons and interactive activities, together with an online learning system that will facilitate grading of student work, including automatic grading of most assignments. A printable version of the student-facing content and a printable instructor guide will also be available. The instructional materials will be distributed under an open-source license.

The GeauxTeach Program at LSU dates back to 2002 and was one of the first of the now over 40 replication sites of the UTeach Program. The two courses to be developed for this project will have an obvious positive effect on the variety and quality of curricular offerings and instructional methods within GeauxTeach and other LSU programs for teachers. Because insufficient access to computational thinking and programing courses that fit into the UTeach model is a problem at almost all replication sites, it is our goal that, ultimately, all UTeach replication sites will gain from having access to our open-source courses.

REFERENCES

Angeli, C., & Valanides, N. (2005). Preservice teachers as ICT designers: An instructional design model based on an expanded view of pedagogical content knowledge. *Journal of Computer-Assisted Learning, 21*(4), 292–302.

Angeli, C., Voogt, J., Fluck, A., Webb, M., Cox, M., Malyn-Smith, J., & Zagami, J. (2016). A K–6 computational thinking curriculum framework: Implications for teacher knowledge. *Journal of Educational Technology & Society, 19*(3), 47.

Buitrago Flórez, F., Casallas, R., Hernández, M., Reyes, A., Restrepo, S., & Danies, G. (2017). Changing a generation's way of thinking: Teaching computational thinking through programming. *Review of Educational Research*, *87*(4), 834–860.

Denning, P. J. (2017a). Computational thinking in science. *American Scientist, 105*(1), 13–17.

Denning, P. J. (2017b). Remaining trouble spots with computational thinking. *Communications of the ACM*, *60*(6), 33–39.

Grover, S., & Pea, R. (2013). Computational thinking in K–12: A review of the state of the field. *Educational Researcher, 42*(1), 38–43.

Grover, S., Pea, R., & Cooper, S. (2015). Designing for deeper learning in a blended computer science course for middle school students. *Computer Science Education, 25*(2), 199–237.

Guzdial, M., Kay, A., Norris, C., & Soloway, E. (2019). Computational thinking should just be good thinking. *Communications of the ACM, 62*(11), 28–30.

Novak, J. D., & Musonda, D. (1991). A twelve-year longitudinal study of science concept learning. *American Educational Research Journal, 28*(1), 117–153.

Pellegrino, J. W., & Hilton, M. L. (2013). *Education for life and work: Developing transferable knowledge and skills in the 21st century*. Washington, DC: National Academies Press.

Qualls, J. A., & Sherrell, L. B. (2010). Why computational thinking should be integrated into the curriculum. *Journal of Computing Sciences in Colleges, 25*(5), 66–71.

Schanzer, E., Fisler, K., & Krishnamurthi, S. (2013). Bootstrap: Going beyond programming in after-school comp science. Indianapolis, IN: *SPLASH Education Symposium* October 28, 2013.

Schanzer, E., Fisler, K., Krishnamurthi, S., & Felleisen, M. (2015). Transferring skills at solving word problems from computing to algebra through Bootstrap. In *Proceedings of the 46th ACM Technical Symposium on Computer Science Education* (pp. 616–621). Kansas City, MO: ACM.

Valanides, N., & Angeli, C. (2008). Learning and teaching about scientific models with a computer modeling tool. *Computers in Human Behavior, 24*(2), 220–233.

Voogt, J., Fisser, P., Good, J., Mishra, P., & Yadav, A. (2015). Computational thinking in compulsory education: Towards an agenda for research and practice. *Education and Information Technologies, 20*(4):715–728.

Weintrop, D., Beheshti, E., Horn, M., Orton, K., Jona, K., Trouille, L., & Wilensky, U. (2016). Defining computational thinking for mathematics and science classrooms. *Journal of Science Education and Technology, 25*(1), 127–147.

Wing, J. M. (2006). Computational thinking. *Communications of the ACM, 49*(3), 33–35.

Wing, J. M. (2008). Computational thinking and thinking about computing. *Philosophical Transactions of the Royal Society A: Mathematical, Physical and Engineering Sciences, 366*(1881), 3717–3725.

Yadav, A., Sands, P., Good, J., & Lishinki, A. (2018). Computer science and computational thinking in the curriculum: research and practice. In J. Voogt, G. Knezek, R. Christensen and K.W. Lai (Eds.), *Second handbook of information technology in primary and secondary education* (pp. 89–106). Springer.

CHAPTER 27

INCORPORATING PROJECT-BASED LEARNING INTO THE SECONDARY MATHEMATICS AND SCIENCE CLASSROOM

Is it Pie in the Sky?

David M. Sparks and Debbie Jackson

Recent graduates of two urban universities from the UTeach teacher preparation program were interviewed about their implementation of project-based learning as a teaching pedagogy. The teacher candidates shared the constraints that prevented them from using PBL, the benefits of being trained in the innovative pedagogy, and the circumstances that would allow them to use it more consistently. The mathematics and science teachers also shared the opportunities created by their knowledge and expertise in PBL, including job interview advantages and opportunities for leadership development within their districts.

Colleges of education across the country prepare teachers to utilize particular pedagogies in their classrooms. This chapter reports on the results of interviews and focus groups with recent graduates about whether or not they adopt project-based learning (PBL) as a pedagogy during their first 3–5 years as a mathematics or science teacher. The data revealed that classroom teachers are inclined to

Preparing STEM Teachers: The UTeach Replication Model, pages 387–400.

utilize PBL when they work in a school where PBL is emphasized, feel supported by their principal, and teach in classrooms with high achieving students. Some of their barriers include time constraints and a low number of veteran teachers using PBL. The graduates did report benefits to being trained in PBL strategies in their teacher preparation program, although that has not necessarily translated into using PBL in their *current* classrooms. Much of this is consistent with recent literature about PBL; what this adds is the unique perspective of two urban teacher preparation programs whose curricula were designed to equip the teachers to effectively integrate both inquiry and project-based learning. Even still, there are significant challenges.

LITERATURE REVIEW

What is Project-Based Learning (PBL)?

Project-based learning, also known as project-based instruction, is an instructional strategy that incorporates real-world connections, teamwork, problem-solving, and cooperative learning into student-centered projects focused on solving a complex and provocative scenario. It has been used effectively across a wide range of subject areas. The *project method*, as it is sometimes called, became popular in the early 20th century in the United States. It faded as a preferred teaching method until the 1980s, when its usefulness as an innovative instructional practice was brought back to light (Barron et al., 1998). Over time, it became known as a method of teaching that could enhance the scientific and reflective thinking of students, organize the problem-solving skills as students advanced through their formal education, and stimulate curiosity (Helle, Tynjälä, & Olkinuora, 2006). Although there is some confusion between the terms *project-based learning* and *problem-based learning*, we define both as types of instruction whereby open-ended and authentic projects are assigned to students in a long-term, systematic, and team-led structure, to produce solutions to a complex problem, coordinate a workable and creative solution, and present the process or product for class presentation. When referring to Project-Based Learning (PBL) as an instructional pedagogy taught in UTeach mathematics and science teacher preparation programs, this definition will constitute the teaching methods taught to pre-service teachers during the last year in their training and certification program.

As a teaching method, PBL helps students to bridge knowledge and community service and apply real-world scenarios to explore real-life problems. PBL also makes students prioritize and focus their content knowledge (Baser, Ozden, & Karaarslan, 2017), learn vital 21st century skills such as collaboration, communication, and critical thinking (Buck Institute for Education, 2013), and produce products using skills such as debating, analyzing, synthesizing, and designing (Blumenfeld et al., 1991). For maximum effectiveness, PBL must be structured with a team of students focused on a particular complex problem, whose task it is to come to agreement on the best course of action and present their findings to

the class. This team-based mindset is inspired by the research of Vygotsky (1978), who argued that learning is primarily a social activity, and students can learn and achieve more as a functional team than they can ever learn on their own.

Benefits of PBL for Students

A number of studies have explored the benefits of active-learning, which includes PBL. Freeman et al. (2014) conducted a meta-analysis of over 200 studies in disciplines related to Science, Technology, Engineering, and Mathematics (STEM), and found the benefits of student-centered, active-learning (as opposed to teacher-centered, passive-learning) on exam performance. This study, which included a large effect size, demonstrated increased performance across disciplines, class sizes, and levels of instruction (Beier et al., 2019). The researchers also found that PBL has a positive effect for *all* students, regardless of their race or gender (Beier et al., 2019). Additional studies (Kolodner et al., 2003; Krajcik, Blumenfeld, Marx, & Soloway, 1994) have demonstrated PBLs effectiveness in promoting higher-order thinking, positioning students as authentic researchers, and teaching research skills such as analyzing and interpreting data.

Two common criticisms of PBL are that it is not (a) effective in incorporating and teaching content knowledge, and (b) significantly helpful for lower-ability students. However, research has revealed these characterizations to be inaccurate. A study by Barron et al. (1998) demonstrated that PBL is effective in teaching students to self-assess and take responsibility for their own learning (metacognition), set goals, scaffold knowledge as they solve problems, and use vital social resources to complete a complex task. A study by Boaler (1998) showed that students taught using a type of PBL were just as able as students taught in a more direct manner to answer exam questions, but were even more flexible at adapting their content knowledge toward solving new and unique problems. Yet, content knowledge remains highly important. Students with a deep, conceptual knowledge of the subject matter thrive in PBL classrooms, and are much more able to apply their knowledge to solve complex problems (Dochy, Segers, Vanden Bossche, & Gijbels, 2003).

A study by Holmes and Hwang (2016) showed that the cooperative learning structure of PBL projects can provide vital cognitive and social support for many underrepresented, under-resourced, and marginalized students. While this is encouraging, it important that students who are typically not included in classroom and scientific discourses are given the opportunity to contribute ideas and join in team conversations in substantive and contributory ways (Alozie, Moje, & Krajcik, 2010). Holmes and Hwang (2016) also found that a PBL learning environment in mathematics can decrease the achievement gap among minority students and students from low socio-economic status backgrounds, and was equally effective for students at varied levels of mathematical knowledge (Holmes & Hwang, 2016). Many urban schools have diverse populations of low socio-economic status students, including many ethnic and racial minorities. Because of this reality,

it is crucial that PBL is implemented at urban K–12 schools, but more specifically in urban mathematics and science classrooms (Tal, Krajcik, & Blumenfeld, 2006). Understanding why teachers report not using PBL strategies in their classrooms may relate to concerns and constraints connected to the mathematics and science classroom, as well as school policies and procedures.

Concerns and Constraints of PBL Classroom Implementation

Research by Krajcik et al. (1994) and Ertmer and Simons (2005) point to three levels of challenges for teachers who try to incorporate PBL into their classrooms: 1) teacher (beliefs, previous experiences, pedagogical and content knowledge, commitment to the innovation), 2) classroom (resources, support, class size, class schedule), and 3) school/community (curricular and testing policies, community support and involvement) challenges. Teachers report frustrations with the time required in building and developing projects (Cooper et al., 2015), scaffolding students' work during the PBL process (Brush & Saye, 2001), incorporating standards-driven instruction and assessment (Helm & Beneke, 2003), matching student and task requirements, intra-group dynamics (Helle et al., 2006), and cognitive overload for students in technologically-rich learning environments (Barab et al., 2000).

Lam, Cheng, and Choy (2010) discuss two factors which may predict teachers' implementation of PBL: teacher personal factors and school contextual factors. Both categories relate to the degree to which a teacher will persist in implementing an innovative teaching strategy in their classrooms. *Teacher personal factors* may include how the innovation aligns with their expectations for being successful in implementing the innovation (Abrami, Poulsen, & Chambers, 2004) and the perceived value of the innovation for their own classroom. *School contextual factors* include issues with class size, time constraints, the principal's management style, and the overall school culture (Abrami, Poulsen, & Chambers, 2004). The most important consideration, as it relates to school contextual factors, appears to be school collegiality, which includes an administration that supports a teacher's autonomy and values their expertise (Lam et al., 2010). Teachers who are supported in their competence, autonomy, and collegiality were much more confident to incorporate innovative instructional practices like PBL into their classrooms.

PBL and the Novice Mathematics and/or Science Teacher

Novice mathematics or science teachers, which in this case we define as an in-service teacher in their first 3–5 years of teaching, face a number of challenges that may affect the likelihood of implementing PBL in their classrooms. To be successful in implementing PBL, teachers must make fundamental changes in: (a) the way they facilitate student learning (Krajcik et al., 1994), and (b) their comfort in scaffolding information during PBL projects (Brush & Saye, 2001). For novice teachers to be trained as effective PBL instructors, they must also be

trained in a scaffolded manner (Ertmer & Simons, 2005). This means they should be given ample opportunities to not only develop PBL projects, but also see them in action in other teachers' classrooms, and to develop and teach a PBL project of their own.

Joyce and Weil (1996) found that less than 10% of teachers will effectively adopt a new teaching strategy without support and repeated practice. Scaffolding helps teachers deal with the complexity inherent in learning a new task and gives them opportunities to practice their new knowledge (Reiser, 2004). Ertmer and Simons (2005) believe scaffolding of novice teachers helps prepare them to be more confident in leading their students in collaboration, assessing products, and guiding the students in the reflection of their own learning. Just like K–12 students, novice in-service teachers need scaffolding in their PBL training, which could include being mentored by either expert PBL facilitators or dynamic and confident classroom teachers.

Besides the constraints mentioned earlier, novice teachers may also face an over-abundance of paperwork, inadequate administrative support structures (Brickhouse & Bodner, 1992), ineffective curriculum materials that do not align with inquiry and PBL (Adams & Krockover, 1997), and underdeveloped pedagogical content knowledge (Crawford, 2000). Roehrig and Luft (2004) believe beginning teachers tend to "rely more heavily on one domain of knowledge rather than drawing simultaneously from all domains, as is the case with an expert teacher (p. 5)." The authors also found that novice teachers tend to be teacher-centered, direct-instruction based, concerned about discipline and classroom control, focus on covering required content, and express the need to *survive* their year.

Marshall, Petrosino, and Martin (2010) conducted a study on apprentice math and science teachers (student-teachers), and found that only 49% had implemented PBL projects during their time at the host school. The apprentice teachers mentioned many of the same constraints as practicing teachers, but also said mandates for paperwork and high-stakes testing had decreased their curriculum preparation time. While they had learned to design and implement PBL projects during their teacher preparation program, they never had the opportunity to lead a full PBL in a cooperating teacher's classroom (Marshall et al., 2010). This was due in part because many districts have a *prescribed curriculum* the teachers are required to follow, which does not allow major variations in the classroom teacher's schedule. The authors also found that when teachers implement their first PBL project in their own classroom, they are finally able to see how all the components fit together and apply the principles they were exposed to in their teacher preparation program. In a sense, this is when the PBL *lightbulb* turns on for many novice mathematics and science teachers (Marshall et al., 2010).

METHODS

Recruitment

Both universities are considered research-intensive universities, and both are situated in large metropolitan areas. For the purposes of this study, they will be named Metropolitan University #1 and #2 (MU #1 and MU #2). The participants in this study were recruited from two teacher preparation programs' alumni databases. The advisors for the UTeach programs maintain databases of email addresses for alumni of their programs. The group selected for the surveys included those who had graduated from the two programs between 2012–2017. The program advisors sent emails requesting participation in an online survey, which included the option of later participating in a face-to-face interview and/or focus group. The surveys were distributed in fall 2017 and spring 2018, and 33 out of approximately 130 program alumni from each institution (25%) completed the survey. For purposes of this chapter, we only report on the interview data. The participants in interviews were those who volunteered through the survey.

Participants

This study reports the themes for four graduates of the first university and six graduates of the second university. The participants, and the demographics or their schools, are described in Table 27.1. At the time of data collection, the majority of the participants were teaching in urban or rural high schools with a large population of traditionally underserved students and a high percentage of students qualifying for free and reduced lunch. The focus on participants serving in these types of schools renders this study of particularly importance (see Table 27.1).

Analysis

The study was a qualitative, interpretive case study. Qualitative research is used to guide in understanding a situation (Patton, 2002). In this case, the goal was to understand the impact of the teacher preparation program on the use of PBL as a teaching method. In addition to being qualitative, this research is specifically a case study. A case study is "employed to gain an in-depth understanding of the situation and meaning for those involved" (Merriam, 1998, p. 29). A case study is also defined as a research study that is a single unit or bounded by something (Smith, 1978). This case was defined as "the university's teacher preparation program"; each of university's graduates were treated as individual cases and cross-case analysis (Ayres, Kavanaugh, & Knafl, 2003) was used to discover themes across the cases. Fourteen in-service teachers participated in semi-structured interviews and focus groups. The interview and focus groups were chosen because they allow the researcher to understand the setting more completely. All of the interviews and focus groups were semi-structured; some questions predetermined and others resulted from probing by the researchers (Berg, 2001).

TABLE 27.1. School District Demographics

Pseudonym	School (pseudonym)	School Stats	Overall Stats (combined)[a]
		MU #1 School Districts	
Nathan	GreatTech HS	Black 26.5% Hispanic 37.2% White 31.0%	Black 62.00% Hispanic 19.75% White 13.90%
Anise	ComboNew HS	Black/AA 98.5% Hispanic 1.0% White 1%	
Reagan	SuperLearn HS	Black/AA 96.8% Hispanic 1.0% White 1%	
Noel	SkyBound HS	Black 26.8% Hispanic 40.1% White 22.7%	
		MU #2 School Districts	
Brooke Hayden	CountryWay HS	Black 8% Hispanic 81.9% White 6.4%	Black 20.62% Hispanic 61.32% White 11.72%
Ana	CityLane HS	Black 17.7% Hispanic 47.8% White 22.8%	
Chelsea	MiddleRoad MS	Black 23.1% Hispanic 63.0% White 11.1%	
Lena	LoneAvenue HS	Black 41.3% Hispanic 53.3% White 3.1%	
Mia	EverGreen HS	Black 13.0% Hispanic 60.6% White 15.2%	

[a]Remaining 4–6 % are other/international students and those who chose mixed race/multiracial

Interviews and focus groups were conducted by the authors at both institutions with alumni from the respective universities. At MU #1 the interviews and focus groups took place in the fall 2017 and at MU #2 the interviews and focus groups took place in spring 2018. The decision to conduct a face-to-face interview or focus group was dependent on the availability of the participants. Each interview was transcribed and coded; categories were created, and themes emerged through collaboration by both authors. *In Vivo* coding (Stringer, 1999) was used to analyze the experiences of the teachers in their own words. The researchers conducted member-checking by emailing the students follow-up questions derived from

analysis of the interviews and focus group transcripts (Krueger & Casey, 2000). The authors met frequently to compare codes and build consensus regarding the findings of the study (MacPhail et al., 2015).

Research Questions

In an effort to understand the implementation of PBL in classrooms after graduating from a UTEACH program, we sought to answer these questions by analyzing interviews and case studies from graduates of both mathematics and science teacher preparation programs:

1. How does PBL training and knowledge for pre-service teachers translate from their teacher preparation program to the middle/high school classroom?
2. What are the challenges to teaching using PBL in middle/high school classrooms?
3. What impact does the UTeach curriculum related to PBL instruction have on teachers once they graduate and have their own classrooms?

RESULTS

Research Question 1: How does PBL training and knowledge for pre-service teachers translate from their teacher preparation program to the middle/high school classroom?

Overwhelmingly, the graduates talk about utilizing inquiry in the classroom, but in general, only about half of them are implementing PBL into their classroom curriculum. When they do incorporate PBL, they feel very supported by the school leadership. Reagan said that she supports a balance of inquiry and direct teaching:

> Ideally, I try to balance the two. There's certain things where it's like.. is it worth discovering? Do you know what I mean? There are certain things where I think it's worth knowing what is behind it all, [where] this formula comes from… It's important sometimes, but other times it's like, 'I'm just going to need you to go with me on this one, so that we can start working with it and then later on, maybe we can back off and see what's behind all of this.'

Noel was not a big proponent of inquiry or PBL, for many reasons, but understood why it was the focus of his teaching preparation: "Inquiry; I see it definitely has its parts and certain things where you can really dive into it. It can be really useful. But PBL is, I think… if your whole school is kind of going towards that way…but for one class, it's kind of hard to do I would say." Noel also had a cooperating teacher that pushed direct instruction. Although Noel is a very traditional

teacher, he says that he tried to use some of his training related to inquiry during his apprentice teaching but was discouraged by his host teacher.

Lena also believes in inquiry and says that she contributes a mixture of half inquiry learning and half direct instruction. She reflects:

> It's very split, and I do notice that I tend to be a lot more inquiry-based with my pre-AP kids more than my regular kids. My regular kids have a tendency to be a little more flighty when it comes to behavior. So, I don't like being able to give them something where they have the opportunity to run amuck…I mean it's unfortunate that that happens, but it's a thing.

Chelsea echoed a sentiment expressed by many of the other program alumni, that her curriculum was constrained by its designation as a state-tested subject, which she believed kept her from incorporating PBL into her classroom; however, when asked to do inquiry and 5E lessons, she was way ahead of the curve. Hayden and Brooke are both incorporating PBL. Hayden stated that the district encourages teachers to do both inquiry and PBL, which are both written into their newly revised curriculum.

Nathan works on a full-PBL campus, so he is able to use PBL on a daily basis. His only concern is that students who go to full-PBL schools will not be prepared when they encounter a direct teaching experience in college:

> If that was a point of weakness, I guess I didn't understand how that would happen and even as a teacher now, I am fearful that students that are in a PBL environment, for example a [PBL National Program] school, what do they do when they go to college and they study a more traditional discipline that doesn't do that? Will that hinder them or will they [be] that much more open-minded to trying to figure out how to apply whatever they're learning?

Mia is also able to incorporate PBL and sees the benefits of doing so. Mia found that PBL is effective with English as a Second Language (ESL) students. She explains: "[With] a project, I can see how much they know. I still use PBL a lot and I try to do as much inquiry as I can, but [its] a little harder for me to do so."

Research Question 2: What are the challenges to teaching using PBL in middle/high school classrooms?

The two challenges the graduates discussed during the interviews and focus groups were time and push-back from veteran teachers.

Time

One of the greatest barriers the in-service teachers discussed when talking about the barriers to PBL was time. The talked about time to plan and time to teach the content. Reagan said that she wanted to incorporate PBL units into her classroom, but she was told by administration there was not time to incorporate

it into her regular class time; therefore, she did it as an extra project after school. Ana was unable to use inquiry and PBL in her classroom, mainly because of issues with coverage of required objectives and content. She shares:

> I didn't ever really like PBL, honestly, going through the course. And I know we had to do it and I do see the pros of doing it. Now that I'm in an accelerated program, I could not see putting PBL into place when I have 4 and a half weeks to teach 9 weeks of content.

Lena also included comments about time, specifically planning time: "I think next year with a little bit more planning time, I would like to use a little more inquiry in my lessons. I feel like I'm not doing enough. I think my kids would respond better to it." Her comment relates to teaching using inquiry, not PBL. Time to plan the PBL unit and time to teach the required content are two clear challenges for these teachers.

Push Back from Veteran Teachers

In addition to time, the graduates talked about the lack of support from veteran teachers and other teachers in the building. Anise said she noticed a lot of push-back to try PBL with the older, more veteran teachers:

> There's a lot of push-back from older teachers that [say], "Why are we reinventing the wheel? Why are we changing things?" So I had a lot of ammo [proof] that I could use. I pulled out some of the books from my classes and said, 'Here's the research on this.' I got a couple more teachers to buy into it.

Hayden also said that older teachers who are not knowledgeable about inquiry and PBL are resistant to teach it: "They don't do tons of project-based learning. We teach with so many old-school teachers in the district, that this is really the biggest jump we've had since then. It's kind of like trying to get them there." This is interesting because both Hayden and Anise are in schools were PBL is encouraged by administrators.

Research Question 3: What impact does the UTeach curriculum related to PBL instruction have on teachers once they graduate and have their own classrooms?

Two types of impacts were discussed: (1) participants attributed attaining their first jobs to their experience with PBL in the teacher preparation program and (2) many of the teachers who were using PBL also pursued leadership roles.

Four of the five graduates who are utilizing PBL in their classrooms have pursued leadership opportunities as a result of their expertise. Mia discusses her experience with inquiry and PBL and says it helped prepare her to be a leader in both inquiry and PBL in her department and school district. Many of the teachers said they were better prepared when their principals asked them to teach using inquiry

and the 5E model. They also have shared experiences with other teachers at their schools who have completed their training through a UTeach program. Brooke was able to take a leadership role in her science department and was asked to do training for inquiry-based teaching and PBL within her first few years. In fact, she was asked to rewrite the school's entire science curriculum using the 5E model. She also said she was excited when teachers told her they used some of the materials she wrote. Nathan has also had opportunities to lead teachers as a keynote speaker at a STEM education conference and give presentations at local STEM conferences.

CONCLUSIONS

Graduates from the UTeach teacher preparation programs at MU#1 and MU #2 report similar experiences related to PBL and its incorporation as a classroom teaching pedagogy. Although the majority of the teachers in the survey report they were adequately prepared to develop and teach PBL units, most of the graduates are not doing so, for a variety of reasons. With the constraints of time to plan and the need to adequately teach content and prepare the students for state tests, innovative strategies such as PBL tend to get *left behind*. Although the majority of the teachers reported using inquiry, questioning, and discovery, few discussed consistently using PBL as a learning tool; however, many of the teachers expressed that they felt prepared to use PBL and much more confident than the other new mathematics and science teachers at their schools. Another positive aspect of their PBL training was the ability to use it as a discussion topic during job interviews and to help advance their position as a leader of instruction in their schools and school districts. Research shows that PBL is an effective classroom methodology; therefore, it is the hope of these researchers that the teacher candidates' expertise in PBL will make them effective curriculum designers and position them as leaders of quality instruction in their school districts.

LIMITATIONS

It is possible that a larger number of participants in the interviews and focus groups could have enhanced the analyses of the experiences of new mathematics and science teachers and given a more thorough picture of the particular instances of how and when PBL is used in the novice teachers' classrooms. More examples of teachers using PBL might have presented a clearer delineation of its usage with UTeach recent graduates. Also, as novice teachers, they have much room to grow, and their experience with PBL may be better utilized as they progress from novice to expert teachers in the next few years. Therefore, a follow-up study with these same participants might show how their focus on inquiry and PBL changed as they progressed through their teaching career.

REFERENCES

Abrami, P. C., Poulsen, C., & Chambers, B. (2004). Teacher motivation to implement an educational innovation: factors differentiating users and non-users of cooperative learning. *Educational Psychology, 24*, 201–216.

Adams, P. E., & Krockover, G. H. (1997). Concerns and perceptions of beginning secondary science and mathematics teachers. *Science Education, 81*, 29–50.

Alozie, N., Moje, E., & Krajcik, J. (2010). An analysis of the supports and constraints for scientific discussion in high school project-based science. *Science Education, 94*(3), 395–427. doi:10.1002/sce.20365

Ayres, L., Kavanaugh, K., & Knafl, K. A. (2003). Within-case and across-case approaches to qualitative data analysis. *Qualitative Health Research, 13*(6), 871–883.

Barab, S. A., Hay, K. E., Squire, K., Barnett, M., Schmidt, F., Karragan, K., Yamagata-Lynch, L., & Johnson, C. (2000). Virtual solar system project: Learning through a technology-rich, inquiry-based, participatory learning environment. *Journal of Science Education and Technology, 9*(1), 7–24. doi:10.1023/A:1009416822783

Barron, B. J. S., Schwartz, D. L., Vye, N. J., Moore, A., Petrosino, A., Zech, L., & Bransford, J. D. (1998). Doing with understanding: Lessons from research on problem- and project-based learning. *Journal of the Learning Sciences, 7*(3–4), 271–311. doi: 10.1080/10508406.1998.9672056

Baser, D., Ozden, M., & Karaarslan, H. (2017). Collaborative project-based learning: An integrative science and technological education project. *Research in Science & Technological Education, 35*(2), 131–148. doi:10.1080/02635143.2016.1274723

Beier, M. E., Kim, M. H., Saterbak, A., Leautaud, V., Bishnoi, S., & Gilberto, J. M. (2019). The effect of authentic project-based learning on attitudes and career aspirations in STEM. *Journal of Research in Science Teaching, 56*(1), 3–23. doi:10.1002/tea.21465

Berg, B. (2001). *Qualitative research methods for the social sciences.* Needham Heights, MA: Pearson.

Blumenfeld, P., Soloway, E., Marx, R., Krajcik, J., Guzdial, M., & Palincsar, A. (1991). Motivating project-based learning: sustaining the doing, supporting the learning. *Educational Psychologist, 26*(3–4), 369–398. doi:10.1080/00461520.1991.9653139

Boaler, J. (1998). Open and closed mathematics: Student experiences and understandings. *Journal for Research in Mathematics Education, 29*(1), 41–62. doi:10.2307/749717

Brickhouse, N. W., & Bodner, G. M. (1992). The beginning teacher: Classroom narratives of convictions and constraints. *Journal of Research in Science Teaching, 29*, 471–485.

Brush, T., & Saye, J. (2001). Design, implementation, and evaluation of student-centered learning: A case study. *Educational Technology Research and Development, 48*(3), 79–100. doi:10.1007/BF02319859

Buck Institute for Education. (2013). *What is project-based learning (PBL)?* Retrieved from: http://bie.org/about/what_pbl.

Cooper, M. M., Caballero, M. D., Ebert-May, D., Fata-Hartley, C. L., Jardeleza, S. E., Krajcik, J. S., & Underwood, S. M. (2015). Challenge faculty to transform STEM learning. *Science, 350*(6258), 281–282. doi:10.1126/science.aab0933

Crawford, B. A. (2000). Embracing the essence of inquiry: new roles for science teachers. *Journal of Research in Science Teaching, 37*, 916–937.

Dochy, F., Segers, M., Vanden Bossche, P., & Gijbels, D. (2003). *Effects of problem-based learning: A meta-analysis. Learning and Instruction, 13*, 533–568. doi:10.1016/S0959-4752(02)00025-7

Ertmer, P. A., & Simons, K. D. (2005). Scaffolding teachers' efforts to implement problem-based learning. *International Journal of Learning, 12*(4), 319–328.

Freeman, S., Eddy, S. L., McDonough, M., Smith, M. K., Okoroafor, N., Jordt, H., & Wenderoth, M. P. (2014). Active learning increases student performance in science, engineering, and mathematics. *Proceedings of the National Academy of Sciences of the United States of America, 111*(23), 8410–8415. doi:10.1073/pnas.1319030111

Helle, L., Tynjälä, P., & Olkinuora, E. (2006). Project-based learning in post-secondary education: Theory, practice and rubber sling shots. *Higher Education, 51*(2), 287–314. doi:10.1007/s10734-004-6386-5

Helm, J. H., & Beneke, S. (2003). *The power of projects: Meeting contemporary challenges in early childhood classrooms—Strategies and solutions.* New York, NY: Teachers College Press.

Holmes, V., & Hwang, Y. (2016). Exploring the effects of project-based learning in secondary mathematics education. *The Journal of Educational Research, 109*(5), 449–463. doi:10.1080/00220671.2014.979911

Joyce, B., & Weil, M. (1996). *Models of teaching* (5th ed.). Boston: Allyn and Bacon.

Kolodner, J. L., Camp, P. J., Crismond, D., Fasse, J. G., Holbrook, J., Puntambekar, S., & Ryan, M. (2003). Problem-based learning meets case-based reasoning in the middle school science classroom: Putting learning by design into practice. *Journal of the Learning Sciences, 12*, 495–547. doi:10.1207/S15327809JLS1204_2

Krajcik, J. S., Blumenfeld, P. C., Marx, R. W., & Soloway, E. (1994). A collaborative model for helping middle grade science teachers learn project-based instruction. *The Elementary School Journal, 94*(5), 483–497. doi:10.1086/461779

Krueger, R. A., & Casey, M. A. (2000). *Focus groups: A practical guide for applied researchers* (3rd ed.). Thousand Oaks, CA: Sage.

Lam, S., Cheng, R. W., & Choy, H. C. (2010). School support and teacher motivation to implement project-based learning. *Learning and Instruction, 20*(6), 487–497. doi:10.1016/j.learninstruc.2009.07.003

MacPhail, C., Khoza, N., Abler, L., & Ranganathan, M. (2015). Process guidelines for establishing intercoder reliability in qualitative studies. *Qualitative Research, 16*, 198–212. doi:10.1177/1468794115577012

Marshall, J. A., Petrosino, A. J., & Martin, T. (2010). Preservice teachers' conceptions and enactments of project-based instruction. *Journal of Science Education and Technology, 19*(4), 370–386. doi:10.1007/s10956-010-9206-y

Merriam, S. B. (1998). *Qualitative research and case study applications in education*. San Francisco, CA: Routledge.

Patton, M. (2002). *Qualitative research and evaluation methods*. Thousand Oaks, CA: Sage.

Reiser, B. J. (2004). Scaffolding complex learning: The mechanisms of structuring and problematizing student work. *The Journal of the Learning Sciences, 13*, 273–304. doi:10.1207/s15327809jls1303_2

Roehrig, G., & Luft, U. (2004). Constraints experienced by beginning secondary science teachers in implementing scientific inquiry lessons. *International Journal of Science Education, 26*(1), 3–24. doi:10.1080/0950069022000070261

Smith, L. (1978). An evolving logic of participant observation, educational ethnography and other case studies. In L. Shulman (Ed.), *Review of research in education* (pp. 316–377). Itasca, IL: Peacock.

Stringer, E. T. (1999). *Action research* (2nd ed.). Thousand Oaks, CA: Sage.

Tal, T., Krajcik, J. S., & Blumenfeld, P. C. (2006). Urban schools' teachers enacting project-based science. *Journal of Research in Science Teaching, 43*(7), 722–745. doi:10.1002/tea.20102

Vygotsky, L. S. (1978). *Mind in society: The development of higher psychological processes*. Cambridge, MA: Harvard University Press.

CHAPTER 28

EXPLORING THE EFFECTIVENESS OF FSU-TEACH GRADUATES

An Analysis Through the Lens of Induction

Ellen Granger, Sherry A. Southerland, and Robin R. Smith

The induction period is a pivotal time in the teaching profession. The teaching practices of FSU-Teach graduates were examined through the lens of the classroom visits that are a part of its induction program. Professional trajectories and teaching contexts of graduates were examined. Thematic analysis of induction reports completed during the visits provided insights into areas of strength and weakness in inductees' teaching practice. With respect to the more social classroom challenges that new teachers face, classroom management and discipline, few graduates evidenced areas of weakness and a greater than expected proportion evidenced these practices as strengths. With respect to instructional challenges, an encouraging proportion also evidenced areas of strength in reform-minded instruction—even as novice teachers. Implications and limitations of this examination are discussed.

INTRODUCTION

It is widely recognized that the first years of teaching are particularly challenging; new teachers have obstacles to navigate as they acclimate to the realities of the classroom and their newfound role in that classroom and the broader school

Preparing STEM Teachers: The UTeach Replication Model, pages 401–415.

context (Luft, 2007). The intensity of this struggle is reflected in the attrition rate of teachers; currently, the mode for teacher experience in the US is only one year, and after five years, 40–50% of teachers leave the classroom (Carver & Fieman-Nemser, 2009; Ingersoll & Smith, 2003). This overall rate of losing new teachers is unacceptably high, considering that most novice teachers are not as effective as their more experienced peers. That is, experience does matter, as teachers are the least effective during their first two years of teaching (National Commission on Teaching and America's Future [NCTAF], 2003), with novices attaining maximum effectiveness between years 5–10 (Darling-Hammond, 2010).

Within the overall trends regarding the loss of new teachers, there are disturbing patterns. For example, new teacher attrition is especially high in the areas of special education, mathematics education, and science education, and this attrition contributes significantly to the teacher shortages in these areas. In addition to disciplinary differences, research suggests that school contexts influence attrition, with schools that serve disadvantaged populations experiencing the highest teacher turnover (Ingersoll & May, 2012; Ingersoll & Perda, 2010). Too, new teacher attrition is even more pronounced for secondary science teachers as compared to other teachers (Guarino, Santibanez, & Dailey, 2006; Ingersoll, 2006). While new teacher attrition has been recognized for a number of years, an alarming trend is that the attrition rates for first-year teachers are increasing (Ingersoll & Perda, 2010). Thus, the population of teachers prepared by FSU-Teach and other UTeach programs is particularly at risk as they are science and mathematics teachers, secondary teachers, and frequently work in schools that serve disadvantaged populations.

There are many challenges that teachers face as they enter the profession, the most prominent of which are classroom management and discipline, contextual contradictions, poor administrative support, limited resources, lack of time due to an overloaded schedule, contradictions between theory and practice, and loss of socialization opportunities (Luft & Patterson, 2002; Saka, Southerland, Kittleson, & Hutner, 2013). Such challenges contribute to the high rates of teacher attrition during teachers' early years in the profession, often referred to as the induction period, and have prompted a national call for research. It is hoped that such research can inform the design of effective programs to support beginning teachers (e.g., Darling-Hammond, 2010). Luft (2007) argues that "beginning teachers are different from their preservice and in-service counterparts and deserve some undivided attention." (p. 532). Induction programs that support new teachers have been shown to be critical for increasing the success of teachers as they begin their careers (Ingersoll, 2012). Research indicates that positive impacts of induction include links between beginning teacher induction programs, teacher success, and retention in the profession (Ingersoll, 2012; Smith & Ingersoll, 2004).

Ingersoll and his colleagues (Ingersoll, 2012; Ingersoll, Merrill, Stuckey, & Collins, 2018) note that the positive link between induction programs and teacher retention depends on the types and number of supports that induction programs

provide. The strongest contributors during induction were having a mentor teacher from one's subject area and common planning and collaboration time with the other teachers in one's subject area (Ingersoll, 2012; Ingersol et al., 2018). Furthermore, Ingersoll (2012) reports that being involved in an induction program that consists of multiple components (supports, activities, practices) has a strong positive effect on retention in the profession. Unfortunately, few beginning teachers receive such a comprehensive bundle of induction components.

Novice teachers recognize the importance of teacher preparation programs which include extensive fieldwork and pedagogical coursework (Kee, 2012; Levine, 2006; Ronfeldt, Schwartz, & Jacob, 2014; Stephenson, O'Neill, & Carter, 2012). This is important as "teachers who feel better prepared are more likely to persist" (Preston, Southerland, & Tekkumru-Kisa, 2017). Taken together, this suggests that a novice teacher's participation in a teacher preparation program matters for retention, as novice teachers with little to no preservice preparation leave the profession at much higher rates (NCTAF, 2003).

While extensive teacher preparation programs are important, the support they provide completers is not always straightforward. Indeed, as new teachers navigate the complexities of their induction period they often experience tensions between the reform-minded teaching championed in their preservice development (i.e., the "theory" from above) and their school context which often is not supportive of these research-based pedagogies (i.e., the "practice" from above) (Saka et al., 2013; Wang & Odell, 2007). It is here, where novice teachers struggle to merge the knowledge constructed within their teacher preparation programs with the realities of the classroom and broader work in schools, that induction programs become essential. A review of fifteen empirical studies of induction programs suggests that beginning teachers who participated in some kind of induction performed better at various teaching practices, including "keeping students on task, developing workable lesson plans, using effective student questioning practices, adjusting classroom activities to meet students' interests, maintaining a positive classroom atmosphere, and demonstrating successful classroom management." (Ingersoll, 2012, p. 51; Ingersoll & Strong, 2011). It is our argument that induction programs structured mindful of these particular challenges can support new teachers in implementing the ambitious teaching practices championed by science education research and taught in teacher preparation programs and in so doing, combat attrition at the same time.

In this chapter we will describe the induction program developed at FSU-Teach and share the results of an exploration of the effectiveness of FSU-Teach university teacher preparation as it translates into classroom practice through the lens of the classroom visits that are a part of its induction program.

THE FSU-TEACH INDUCTION PROGRAM

FSU-Teach was one of the original thirteen universities to replicate the UTeach model and graduated its first students in the 2011–12 academic year. Mindful of

the research, the program developed an active induction program for its graduates modeled on that of the parent UTeach program at the University of Texas at Austin. The centerpiece of the FSU-Teach induction program is individual mentoring combined with induction visits for each graduate for their first two years in the classroom. It is important to note that the mentors at the center of this induction program are the master teachers from the FSU-Teach preparation program. This allows for a degree of comfort and familiarity that eases the mentoring process. Other parts of the induction program are a kit of teaching supplies given to each graduate upon graduation including classroom materials for which new teachers often find themselves paying (including everything from paper, markers, and a stapler to books and a USB flash drive of resources). At their request, an introductory membership to the National Science Teachers Association or National Council of Teachers of Mathematics is purchased for them. Monetary support for their attendance at STEM teaching conferences is also available. Finally, professional development for the graduates is offered by FSU-Teach master teachers every other summer along with financial support for them to attend these experiences.

Because induction programs are not tied to credit-bearing classes, funding for these programs must be sought elsewhere. Through a gift to the university from the JP Morgan-Chase Foundation, the program has been able to provide induction visits for our graduates who are teaching in Florida or who are teaching close to FSU in states bordering Florida. This means that the program can visit most, but not all, of its graduates. Most FSU-Teach graduates avail themselves of the mentoring and induction visits, and many attend the summer workshops, but it is interesting to note that only a few take advantage of the national association memberships and state or national conference attendance options.

Classroom Observations and Mentoring Sessions

A cornerstone of the FSU-Teach induction program is the classroom visits and associated mentoring; the observation notes from these visits will provide much of the data herein that speaks to the preparation program's effectiveness. Each graduate who teaches in the state or in a nearby locale is offered the opportunity for on-site mentoring. This includes one to two visits to the classroom by a mentor teacher each year for the first two years post-graduation. This mentoring takes the form of a discussion that revolves around the graduate's views of her/his work and areas for which she/he would like guidance. The master teacher observes a lesson, and following that observation debriefs with the graduate on the lesson to provide guidance and a discussion of how the program can assist through the provision of specific supports. At all times, both before or after the visits, the master teachers provide mentorship and support via phone calls and email and many graduates avail themselves of this option—even those too distant to visit; however, the frequency and content of these contacts are not known as phone and email logs and/or notes of such contacts are not kept.

Following each classroom observation, the master teacher completes an induction report about the visit. The program purposefully decided not to use an observation protocol for the classroom induction visits. While this choice was a difficult one (particularly for the researchers in the program), the master teachers argued that there is a fine line between a beginning teacher feeling supported in their classroom efforts and feeling judged in their classroom efforts. After much discussion, the program as a whole determined that use of an observation protocol would tip this balance toward the latter—increasing the potential that the graduates would view these visits as another effort at evaluating them. Thus, during induction observations master teachers only take notes as they observe the graduates. These notes are used in the follow-up discussion based on the needs articulated by the graduates for assistance with their teaching. During these conversations, the master teachers are very purposeful in their framing of any teaching challenges that they notice so as to engender a feeling of support and not of judgment on the part of the graduate. The format of the form on which notes about the classroom visit are recorded is included at the end of this chapter.

The original purpose of these induction reports was twofold, the first, as described above, was to assist graduates with answers to their teaching questions. The second was program evaluation—to help us continue to enhance our program to better prepare our graduates for the realities of the classroom and school contexts in which they find themselves in their induction years. Thus, this analysis is based on data collected for program evaluation. It is important to note that the individuals completing the induction reports (the master teachers) have worked together in FSU-Teach for almost ten years and are a professional learning community that has developed a shared vision of classroom teaching through multiple avenues. In addition to daily contact and other programmatic activities, these include (1) co-developing and co-teaching FSU-Teach courses with each other and program faculty, (2) co-developing and co-teaching summer professional development workshops for our graduates and local teachers who partner with FSU-Teach, (3) observing and evaluating the teaching of FSU-Teach students together using the FSU-Teach observation protocol (FSU-Teach Math and Science Teaching Evaluation or FMSTE) in all five courses with required classroom teaching events (all students are observed in person and/or recorded for observation in each course), (4) meeting bi-weekly with the entire FSU-Teach team to discuss student issues, solve program challenges as they arise, and ensure continued alignment and articulation in course content, and (5) engaging in a bi-weekly journal club that studies science and mathematics education research and practitioner publications. [Note: The FMSTE protocol was adapted from the Reformed Teaching Observation Protocol (RTOP, Sawada, Piburn, & Alport, 2002), the UTeach Observation Protocol (UTOP, Walkington & Marder, 2014), and Inside the Classroom Observation and Analytic Protocol (Horizon Research, 2000) and satisfactory evaluation with it is required for student completion of the capstone Apprentice Teaching experience.]

FINDINGS

A review of data gleaned from the master teachers' notes taken during induction visits combined with an examination of the retention of FSU-Teach graduates in the field provides some insights into the professional trajectories of the program's graduates and their preparation for teaching through the lens of induction. As the FSU-Teach program gains more longevity, the impact of the induction program itself on retention will be able to be more fully investigated. First, we will examine the professional trajectories of FSU-Teach graduates during their first years of teaching and following that turn the focus to areas of strength and weakness noted for the graduates.

Professional Trajectories

As of spring semester 2019, 87% of FSU-Teach graduates had entered the teaching profession either immediately upon graduation or after some delay (for example, to continue their education or get married) (Table 28.1). Of those choosing not to enter teaching, four immediately enrolled in graduate school, either in a STEM program or in education; four engaged in teaching in informal education programs (science museum or zoo programs); and the remainder pursued non-education career pathways, though many of those report that their teaching skills are important to their success in these careers. Twenty-three who entered secondary teaching have left the profession resulting in a current retention rate of 75%. That said, six of the "leavers" were headed to graduate school with four of their programs being science or mathematics education; these individuals will likely continue education careers in some capacity (and if they do, our retention rate would approach 80%). Thus, the FSU-Teach attrition rate is much lower than that of the national average for STEM secondary teachers; the same is true of the "parent" UTeach program.

An examination of the school contexts in which the graduates are employed during their induction years shows that 56% of FSU-Teach graduates find themselves teaching in a school whose percentage of students from low socioeconomic (SES) backgrounds is higher than the district's percentage of low SES students as measured by the number of students on Free and Reduced Lunch support (Florida DOE, 2018). Using a different SES metric (the National Science Foundation Noyce classification criteria of a high percentage of families below the poverty

TABLE 28.1. FSU-Teach Graduates in Teaching[a]

Percent Entering Teaching	Percent Currently Retained in Teaching	Percent Unaccounted for[b]
87%	75%	1%

[a] 141 total graduates
b No contact post-graduation

line, a high percentage of out-of-field teachers, or a high teacher turnover rate), it was determined that 45% of FSU-Teach graduates are teaching in high-needs schools (Section 201 of the Higher Education Act of 1965: 20 U.S.C. 1021).

Trends in Teaching Practices during Induction

From 2013 through 2019, 165 face-to-face induction visits were completed by FSU-Teach master teachers. The induction reports written by the master teachers for each of these visits served as the primary data source for this analysis. Induction reports include general information on teaching assignments and planning periods, responsibilities outside the classroom, notes on areas of strength, notes on areas of concern (weakness), requests from the graduates, other miscellaneous comments, and school demographics (when available) (see report format provided at the end of the chapter). Given the induction report format, the questions about how FSU-Teach graduates performed in the classroom during their induction period that guided this analysis include:

- What are the strengths of FSU-Teach graduates during their induction years?
- What are the weaknesses of FSU-Teach graduates during their induction years?
- Are FSU-Teach graduates enacting reform-minded, student-centered instruction in the early years of their practice?

For our examination of the induction reports, two rounds of data analysis were used. For the first, a small initial set of themes was identified for analysis based on the research questions above and via a review of the literature on the teaching practices of novice teachers (e.g., Luft & Patterson, 2002; Saka, et al., 2013). These themes were: classroom management, student-centered teaching, and discourse/questioning practices. Induction reports were examined for data which spoke to the graduates' teaching practices around these themes. Examples of master teacher written notes in induction reports related to these themes are: "[Strengths] Good classroom management and rapport with students. Uses time wisely during the class period. Uses knowledge of students and culture to manage students and teach mathematics." "[Strengths] Student-centered teaching: [She] used 'my favorite no' as bell work, which effectively uncovered student thinking about simplifying rational expressions using the laws of exponents by focusing on common student errors rather than 'the right answer.'" and "[Strengths] Promotes classroom discourse with good questioning strategies—students are readily explaining their work in response." These data were recorded in a master file. Other information not initially a part of these three themes was noted in the file and subsequently analyzed for additional themes or sub-themes. Major trends emerging from the "other" data category were identified. Thus, the final set of analytical themes used to examine the induction data represents patterns that consistently

TABLE 28.2. Induction Report Themes and Their Associated Indicators[a]

Theme	Indicators
Rapport with Students	Is there mutual respect established between teacher and students? Has the teacher established a positive relationship with the students? Are students comfortable taking intellectual risks in the classroom?
Classroom Management—Organization	Is the teacher prepared with lesson plans for the week? Are the rules, routines, and procedures well established? Is the classroom an organized and attractive environment? Are there any issues with punctuality or preparedness?
Classroom Management—Discipline	Do the students attend to and follow directions accordingly? Are the students attentive and polite to each other and the teacher? Does the teacher handle off-task behavior promptly and appropriately? Does the teacher fairly and consistently enforce the rules and routines? Can the teacher command the students' attention?
Content Knowledge	Does the teacher use appropriate academic language? Is the teacher able to help students make conceptual connections? Does the teacher display deep understanding of the content (i.e., as evidenced in their talk moves, lesson organization, task selection, and academic language)?
Student-Centered Instruction	Are the students key participants in the sense-making, both in small groups and in whole class discussion? Are the students engaged in developing the concepts with their peers? Is active learning evident?
Discourse/Questioning Skills	Are the teacher's questions an appropriate mix of broad and narrow questions and high and low cognitive demand questions? Are the teacher's questions well-planned to guide productive student thinking? Has the teacher anticipated student responses? Are the teacher's responses to student answers pressing for deeper thinking?

[a]Indicators for which master teachers looked during classroom visits that are associated with the themes found in their induction reports

appeared in the induction reports. These themes were: rapport with students, classroom management—organizational skills, classroom management—discipline, content area knowledge, student-centered instruction, and discourse/questioning skills. Member checking and discussions with the master teachers resulted in indicators for each theme; these indicators are shown in Table 28.2. A second round of analysis was conducted using this final set of themes and their frequency of occurrence was calculated for the entire set of induction reports.

Figure 28.1 shows the results of the thematic analysis of the induction reports in terms of the percentage of graduates falling into each theme as an area of strength or weakness. If neither a strength nor a weakness was noted then the inductee can be considered to fall into a group that could be termed "average" with respect to the theme. That is, the master teachers considered their teaching practice to be acceptable given our program's vision of student-centered teaching which is articulated through the program's FMSTE instrument described above.

With respect to the more social aspects of teaching, this analysis shows that FSU-Teach graduates excel at quickly establishing rapport with their students,

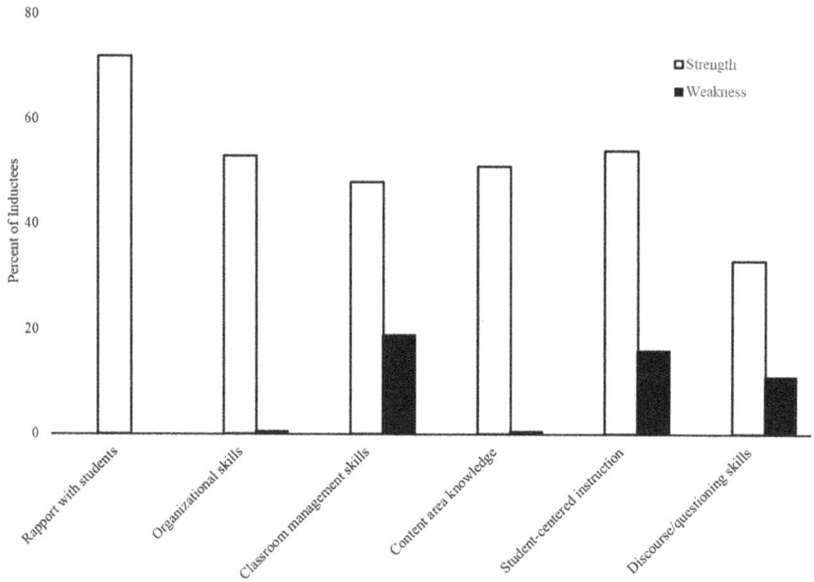

FIGURE 28.1. Percent of Noted as Having a Strength or Weakness on Each Induction Theme as Recorded in Induction Reports Filed by FSU-Teach Master Teachers

with 72% of them having it reported as a strength and 0% as a weakness. Classroom management, an area with which many new teachers struggle (Luft & Patterson, 2002; Saka et al., 2013), naturally split into two themes: classroom organization and classroom discipline. FSU-Teach graduates tended to do well for novice teachers in both of these areas with 53% of them having classroom organizational skills noted as a strength and 48% with classroom discipline noted as a strength. Only 0.6% had weaknesses noted in their classroom organization and 19% in classroom discipline. For almost all of those for whom a classroom management weakness of either type was noted, their observed difficulties improved (though usually did not completely disappear) according to reports from subsequent induction visits. This is important as classroom management, especially "orderly student behavior," is necessary for the success of reform-minded instruction—which is a goal of the FSU-Teach program (Allensworth, Gwynne, Parega, Sebastian, & Stevens, 2014, p. 4).

Turning to the more instructional aspects of teaching, the thematic analysis shows that 51% of FSU-Teach graduates displayed strong content knowledge and only 0.6% showed any deficit in this area. This is not surprising as all FSU-Teach graduates have a double major, a primary major in the content area and a secondary major in science or mathematics teaching. Student-centered instruction was reported as a strength for FSU-Teach graduates in 54% of the reports and as a

weakness in only 16% of the reports. For the remainder, student-centered instruction was not noted as a strength or weakness indicating that the graduate was performing acceptably in this domain of teaching. The data suggest that despite the challenges of the induction period and the high-needs school contexts in which many FSU-Teach graduates find themselves, most were striving to engage in the ambitious teaching practices that were the focus of their university preparation program. Almost every report noting particular strengths in student-centered instruction included a comment by the master teacher that the students in the classroom seemed very comfortable with this method of instruction. Student familiarity with student-centered instruction suggests that this was a common pedagogy in the classroom and not employed for the benefit of the master teacher during the induction visit. Reports from subsequent visits over the two-year induction period indicated that if there was a change in this theme for the graduates it was in a positive direction. Discourse and questioning skills were noted as a strength for 33% or our graduates and as a weakness for 11% of them. Again, for the remainder questioning practices were acceptable indicating that they were neither a strength nor a weakness noted by the master teachers. This is key as discourse and questioning skills are central to supporting reform-minded instruction, and we argue that the theme of discourse is intimately tied to that of student-centered instruction, rapport with students, and to some degree content knowledge, although it would take a larger sample and a quantitative instrument to determine if graduate scores for these themes were statistically related. Likewise, conceptually we can posit connections between student-centered instruction, classroom management (both organization and discipline), and rapport with students.

Master teacher notes around the three areas in which graduates evidenced weakness (i.e., the 19% of graduates in classroom management discipline, 16% in student-centered instruction, and 11% in discourse/questioning skills) were examined for possible insight into the nature of the weakness. Though details about classroom discipline issues were often lacking, when they were present they included the issues of "talking over" the students rather than pressing for attention (noted for 18% of the graduates who displayed this as a weakness) and consistency with discipline (noted for 14% of the graduates who displayed this as a weakness). Less frequently, failure to deal with off-task behavior and lack of follow through on discipline were each noted for 9% of the graduates who displayed discipline as a weakness. Interestingly, for 14% of the graduates for whom discipline was a weakness, it was also noted that there was poor administrative support for teachers in the area of student discipline.

Turning to student-centered teaching weaknesses, there was less additional information noted by the master teachers. Nevertheless, when additional detail was included in the Master Teachers' notes, it was almost always about the graduates who were teaching mathematics. This particular group of graduates tended to be focused on procedures and not on developing conceptual understanding in their students. This mirrors a more traditional approach to mathematics instruc-

tion (e.g., Star, 2014). Finally, for the 11% of graduates who displayed a weakness in Discourse/Questioning Skills, the most frequent note was that their questioning was mostly, if not completely, fact-based without much if any, critical thinking being required of the students. Again, this is the main discourse/questioning default mode found in traditional teaching (e.g., Michaels, Shouse, & Schweingruber, 2008).

For comparison, the classroom management, content, and student-centered instructional abilities displayed in FSU-Teach graduates seem akin to those reported for graduates of the original UTeach program in a study by Walkington & Marder (2014) that used a quantitative instrument, the UTeach Observation Protocol (UTOP). Namely, they showed that the performance of UTeach graduates in their first year in the classroom was no different from that of a comparison group consisting of graduates of non-UTeach programs and alternative certification programs as assessed by the UTOP (Walkington & Marder, 2014). The UTOP is a modification of Horizon's Classroom Observation Protocol (Horizon Research, 2000) to focus more specifically on mathematics and science teaching and includes an examination of classroom management, content knowledge, and instructional skills, among other topics. Despite the similarity they observed in year one between students from the programs that they examined, in their second and third years, the UTeach graduates' scores separated significantly in a positive direction from those of the comparison group (Walkington & Marder, 2014). (Unfortunately, the induction report data we have presented here cannot account for changes in FSU-Teach graduates over years of experience as our data was not taken from a quantitatively scored instrument.)

Other Trends in Graduate's Work

There were other items of interest found in the induction reports that contribute to the picture of how FSU-Teach graduates are assimilating into their school contexts. Even during the induction period, 8% of the graduates reported not having a planning period. This may result from the shortage of mathematics and science teachers and can make a new teacher's assimilation into the profession, particularly when one is attempting to enact reform-minded instruction, even more stressful (Saka et al., 2013). Overall, 72% of FSU-Teach graduates report taking on extracurricular responsibilities in their first two years in the classroom. The most frequently reported responsibilities were athletic team coaching, academic team coaching, club or class sponsorship, yearbook, and department or team leader. With respect to these extra responsibilities, 25% of the graduates took on one, 34% took on two, 8% took on three, and 2% took on four or more responsibilities. While one may look at such responsibilities as "add-ons" that could detract from a novice's focus on her/his instruction, it must be recognized that these roles do allow for novices to more quickly become assimilated into a school culture and can serve as a way to more quickly build rapport with students—something that

can "pay" benefits in terms of student-centered instruction and classroom management (Saka, Southerland, & Brooks, 2009).

Even in their first two years in the classroom, a surprising number of FSU-Teach novice teachers were recognized as leaders in their school contexts. Seven percent of graduates took on leadership roles with all except one becoming either a department chair or a team leader (the other one became the leader of a school technology initiative). Additionally, the master teachers receive regular calls and emails seeking to hire graduates from principals who have already hired one or more FSU-Teach graduates. One principal, who has five FSU-Teach graduates now working in his school, was quoted in a recent article about the program: "[The FSU-Teach graduates] come in and they're ready. Their planning is excellent and that's where teachers often fall short. When I look at resumes, I try to see if anyone has been in FSU-Teach." (Blackburn, 2019, p. 29).

SUMMARY

The data from FSU-Teach graduates' employment and induction reports provide a snapshot of the graduates in the field during their first years of teaching. Implications from this analysis are that the UTeach preservice and induction models as enacted by the FSU-Teach program are successfully preparing graduates for the challenges of the induction period. With respect to the more social classroom challenges that new teachers face, notably classroom management and discipline, few of the FSU-Teach graduates evidence these as areas of weakness and a greater than expected proportion are actually displaying these as areas of strength in their first two years in the classroom. Turning to instructional challenges, a large proportion of FSU-Teach graduates are likewise showing strengths in reform-minded instruction during their induction period. It is heartening to see that this success is not an artifact of placements in less demanding settings, as a significant proportion of FSU-Teach graduates select to teach in high-needs schools.

Ingersoll (2012) and Ingersoll and Strong (2011) noted that beginning teachers who took part in induction programs displayed teaching practices of keeping students on task, developing workable lesson plans, using effective student questioning practices, adjusting classroom activities to meet students' interests, maintaining a positive classroom atmosphere, and demonstrating successful classroom management. The evidence of success in teaching practices demonstrated by the FSU-Teach graduates maps well onto the practices reported in these two studies. Thus, the findings presented here are in line with those of other studies of the influence of teacher induction programs.

There are limitations to this work that could influence its findings. A selection effect is possible in the induction data, as 36.4% of FSU-Teach graduates either did not respond to repeated requests to schedule induction visits or were too far away to visit. Thus, it may be that the graduates who chose to participate in the induction program taught in ways more reflective of their preparation program experiences. Likewise, it is not clear whether or not the 13% of graduates who failed

to enter the profession made that choice because of their abilities in the classroom. Too, the induction data were drawn from subjective notes, and although the master teachers who conduct these observations have worked closely together in the same program for ten years and developed a shared vision of STEM classroom instruction (see Classroom Observations and Mentoring Sessions above), their evaluations of teaching have not undergone interrater reliability training and analysis. Finally, it is impossible to disentangle the influence of the FSU-Teach teacher preparation program from that of the associated induction program.

It would be desirable to visit these teachers in their classrooms after an extended period of time following graduation has elapsed and they are considered to be beyond the induction period (years 5 or 6) to see if they have maintained and/or improved upon their reform-minded practices or if school context and other pressures have changed this. The first graduating class for FSU-Teach was small and they are just now in their sixth year of teaching. Thus, a clear picture of our graduates' classroom practice or their retention rate at the five-year mark cannot yet be accurately determined.

Despite these limitations, the induction reports filed by the master teachers have provided a lens through which it was possible to gain a finer-grained look at

Induction Report Format

(Note: Spaces for notes have been reduced for publication purposes.)

Induction Follow-Up Report Completed by:

Name of Graduate: **Robert E. Noyce Scholar?** Yes/No ?

Date of Induction Visit:

State/School District/School Name:

Teaching Assignment(s)/Grade Level(s):

Does the teacher have a planning period? Yes/No

Does the teacher have any areas of responsibility outside of the classroom, such as coaching, club sponsor, department chair, committees, etc.? Yes/No

If yes, please explain:

Areas of Concern:

Strengths Noted:

Teacher Requests:

Miscellaneous Comments:

Follow Up:

Attach School Demographic Report (if available)

our graduates during their first two years in the classroom than is usually available. The results of this examination are promising and important given the large size of the UTeach replication program nationally. As of September 2017, there were over 3,250 UTeach program graduates nationally (UTeach Institute, 2018)—graduates from over 40 programs whose structure, like that of FSU-Teach, models that of the original UTeach preparation program.

REFERENCES

Allensworth, E. M., Gwynne, J. A., Pareja, A .S., Sebastian, J., & Stevens, W.D. (2014) *Free to fail or on-track to college, setting the stage for academic challenge: classroom control and student support*. Chicago, IL: Research Brief from University of Chicago Consortium on School Research.

Blackburn, D. (2019). From start to STEM: FSU-Teach celebrates 10 years of success. In *Across the spectrum* (pp. 28–29). Tallahassee, FL: Florida State University College of Arts & Sciences.

Carver, C.C., & Fieman-Nemser, S. (2009). Using policy to improve teacher induction: Critical elements and missing pieces. *Educational Policy*, *23*, 295–328

Darling-Hammond, L. (2010). *The flat world and education: How America's commitment to equity will determine our future*. New York, NY: Teachers' College Press.

Florida Department of Education. (2018). *School public accountability reports*. Retrieved 2/10/19 from http://doeweb-prd.doe.state.fl.us/eds/nclbspar/index.cfm.

Guarino, C. M., Santibanez, L., & Dailey, G. A. (2006). Teacher recruitment and retention: A review of the recent empirical literature. *Review of Educational Research, 76*(2), 173–208.

Higher Education Act of 1965. Pub. L. No. 89–329, 79 Stat.1219 (1965)

Horizon Research, Inc. (2000). *Inside the classroom observation and analytic protocol*. Retrieved 2/7/2019 from: http://www.horizon-research.com/inside-the-classroom-observation-and-analytic-protocol

Ingersoll, R. (2006). Understanding supply and demand among mathematics and science teachers. In J. Rhoton & P. Shane (Eds.), *Teaching science in the 21st century* (pp. 197–211). Arlington, VA: NSTA Press.

Ingersoll, R., & May, H. (2012). The magnitude, destinations, and determinants of mathematics and science teacher turnover. *Educational Evaluation and Policy Analysis, 34*(4), 435–464.

Ingersoll, R. M., Merrill, E., Stuckey, D., & Collins, G. (2018). *Seven trends: The transformation of the teaching force*. Philadelphia, PA, Consortium for Policy Research in Education, University of Pennsylvania. RR 2018-2.

Ingersoll, R. M., & Perda, D. (2010). Is the supply of mathematics and science teachers sufficient? *American Educational Research Journal, 47*, 563–594.

Ingersoll, R. M., & Smith, T. M. (2003) The wrong solution to the teacher shortage. *Educational Leadership, 60*(8), 30–33.

Ingersoll R., & Strong M. (2011). The impact of induction and mentoring for beginning teachers: A critical review of the research. *Review of Educational Research, 81*(2), 201–233.

Ingersoll, R. M. (2012). Beginning teacher induction: What data tell us. *Phi Delta Kappan, 93*(8), 47–51.

Kee, A. M. (2012). Feelings of preparedness among alternatively certified teachers: What is the role of program features? *Journal of Teacher Education, 63*(1), 23–38.

Levine, A. (2006) *Educating school teachers*. New York, NY: The Education Schools Project.

Luft, J. A. (2007). Minding the gap: Needed research on beginning/newly qualified science teachers. *Journal of Research in Science Teaching, 44*(4), 532–537.

Luft, J. A., & Patterson, N. C. (2002). Bridging the gap: Supporting beginning science teachers. *Journal of Science Teacher Education, 13*(4), 267–282.

Michaels, S., Shouse, A. W., & Schweingruber, H. A. (2008) *Ready, set, science! Putting research to work in K–8 science classrooms*. Washington, DC: National Research Council. National Academies Press.

National Commission on Teaching and America's Future. (2003). *Launching the next generation of teachers through quality induction*. Retrieved 10/20/2018 from: https://nctaf.org/about-nctaf/

Preston, C., Southerland, S. A., Travis, J., & Tekkumru-Kisa, M. (2017). *Investigating the characteristics of programs that produce persistent STEM teachers with high-quality instruction.* Alexandria, VA: National Science Foundation, Noyce Track 4 #1661400.

Ronfeldt, M., Schwartz, N., & Jacob, B. (2014) Does preservice preparation matter? Examining old questions in new ways. *Teachers College Record, 116*(10), 1–46.

Saka, Y., Southerland, S. A., & Brooks, J. (2009). Becoming a member of a school community while working toward science education reform: Teacher induction through a CHAT perspective. *Science Education, 93*(6), 996–1025.

Saka, Y., Southerland, S. A., Kittleson, J., & Hutner, T. (2013). Understanding the induction of a science teacher: The interaction of identity and context. *Research in Science Education, 43*(3), 1221–1244.

Sawada, D., Piburn, M. D., & Alport, J. M. (2002) Measuring reformed practices in science and mathematics classrooms: The reformed teaching observation protocol. *School Science and Mathematics, 102*, 245–253.

Smith T., & Ingersoll, R. (2004). What are the effects of induction and mentoring on beginning teacher turnover? *American Educational Research Journal, 41*(3), 681–714.

Star, J. (2014) Instrumental and relational understanding in mathematics education. In S. Lerman (Ed.), *Encyclopedia of mathematics education*. Dordrecht, The Netherlands: Springer.

Stephenson, J., O'Neill, S., & Carter, M. (2012) Teaching students with disabilities: A web-based examination of preparation of preservice primary school teachers. *Australian Journal of Teacher Education, 37*(5), 13–23.

UTeach Institute. (2018). *UTeach at 20: Celebrating success and envisioning the future.* Technical Report. Retrieved 1/22/2019 from https://institute.uteach.utexas.edu

Walkington, C., & Marder, M. (2014). Classroom observation and value-added models give complementary information about quality of mathematics teaching. In T. J. Kane, K.A, Kerr, & R. C. Pianta (Eds.), *Designing teacher evaluation systems: New guidance from the measures of effective teaching project* (pp. 234–277). San Francisco, CA: Jossey-Bass.

Wang, J., & Odell, S. (2007). An alternative conception of mentor-novice relationships: Learning to teach in reform-minded ways as a context. *Teaching and Teacher Education, 23*(4), 473–489.

CHAPTER 29

EXAMINING RETENTION OF UTEACH ARLINGTON GRADUATES IN HIGH-NEED CLASSROOMS[1]

Ann M. L. Cavallo, Gregory R. Hale, and Erin Gonzales

Teacher retention has long been a problem for U.S. K–12 schools, with 50% higher attrition in Title I Schools and 70% higher in schools with large minority populations (Carver-Thomas & Darling-Hammond, 2017). The University of Texas at Arlington (UTA) is a diverse, Hispanic-Serving Institution (HSI) in the center of the Dallas-Ft. Worth Metroplex. Most surrounding school districts are economically disadvantaged, high minority population districts. The UTeach model, adopted at UTA in 2010, has been cited as highly effective in science and mathematics teacher recruitment, preparation, induction, and retention (National Academy of Sciences, 2007) and has shown positive impacts on the schools and students where their graduates teach (Marder & Hamrock, 2016). However, little is known about factors within and perhaps outside of the UTeach program that affect the retention of UTeach graduates in classroom teaching. This chapter examines data on UTeach Arlington graduates and their retention in high-need classrooms.

[1] Portions of this material is based upon work supported by the National Science Foundation; Grant Nos. 0833343, 1035483, 1439914, 1758507.

Preparing STEM Teachers: The UTeach Replication Model, pages 417–437.

INTRODUCTION AND BACKGROUND

The nation is acutely aware of the critical need for more, well-prepared science and mathematics teachers and a STEM-educated professional workforce. In the year 2000, the National Commission on Mathematics and Science Teaching for the 21st Century had predicted that over two million qualified teachers would be needed in the coming decade—a prediction that is now a reality (National Commission on Mathematics and Science Teaching [NCMST], 2000). Likewise, the severe shortage of STEM professionals has been of national concern since the landmark publication, *A Nation at Risk* (National Commission on Excellence in Education, 1983), followed by other profound guiding documents, including *Project 2061* (American Association for the Advancement of Science, 1989). On the need for teachers, the National Academy of Sciences (2007) states: "Today there is such a shortage of highly qualified K–12 teachers that many of the nation's 15,000 school districts have hired uncertified or underqualified teachers" (p. 112). In addition, reports indicate the least well-prepared teachers are those teaching in high-need schools, with many teaching out of the field in which they earned their degree and/or teacher certification (National Education Association, 2001). Yet the urgent need for STEM teachers who are well prepared and consequently well positioned to promote a strong, STEM-educated workforce conveyed in these documents is far from being realized, with the U.S. lagging behind most Western countries in student STEM persistence and achievement (e.g., Desilver, 2017).

The shortage of science and mathematics teachers is most severe in high-poverty urban and rural districts (Sutcher, Darling-Hammond, & Carver-Thomas, 2016). A news article published by CNN emphasizes that in the past five years it has been overwhelmingly difficult to find qualified teachers to fill science and mathematics teaching positions, and the situation is only expected to become worse due to the serious compounding problem of teacher attrition (Ostroff, 2017). The U.S. Department of Labor Bureau of Labor Statistics (2014) reports that from 2012 to 2022, a significant number of teachers are expected to reach retirement age leaving significant numbers of vacancies in the high-need subject areas of science and mathematics.

Westervelt (2016) contends that high-poverty schools have the largest numbers of teachers leaving the profession. This same report states that "1 in 5 teachers in high minority, high poverty schools are unprepared for teaching" (p. 1), pointing to a possible explanation for their failure to stay in the profession. Teacher attrition is greatest in certain regions of the U.S. and is 50% higher in Title I schools and 70% higher in schools with a large minority population (Carver-Thomas & Darling-Hammond, 2017). In their report titled *The Opportunity to Excel* the National Education Association (2001) indicates that students in secondary schools with high poverty levels are more likely than students in schools with low poverty levels to be taught core subjects by teachers who have not majored or minored in those subject areas. The Council of Great City Schools urges urban schools to recruit and retain teachers who are well prepared and qualified for teaching in

urban school settings, because these schools are more likely than others to fill vacancies with teachers who have less-than-full credentials (Urban Teacher Collaborative, 2000). Teacher shortages, underprepared teachers, and teacher attrition in our high-need schools widen gaps in student achievement and detrimentally impact their preparation and interest to pursue STEM careers.

Teacher shortages are profound in Texas with students in urban and rural schools having the greatest need for qualified teachers yet being most likely to have the least qualified teachers (Fuller, 2008). Projections on mathematics and science teaching indicate the shortage will continue to increase in the next decades (Stutz, 2012). In 2012, Texas schools needed over 15 thousand new teachers, with the most impact of teacher shortages falling on students in high-need urban schools (Stutz, 2012). In fall 2017, a week after the school year began, Dallas ISD still had 24 posted job openings for science teachers and 19 posted job openings for mathematics teachers. The shortage of science, mathematics, and computer science teachers in Texas is statewide. The Texas Education Agency (TEA) reports that three of the highest teacher shortage areas in the state are computer science/technology applications, mathematics, and science. Another significant shortage area is Bilingual/English as a Second Language (ESL). This is indicative of the challenging classroom environment in the southwest region of the U.S. and Texas.

Population growth exacerbates teacher shortages in the Dallas-Arlington-Fort Worth (DFW) Metropolitan Area. DFW is designated by the U.S. Census Bureau as the 4th largest U.S. metropolitan area with a reported population of 7,399,662 (Hanna, 2018). DFW leads the nation in growth, gaining 146,000 residents in 2017, and the trend is expected to continue (Hanna, 2018). In the next 25 years, DFW is expected to grow 53.5% to 10.9 million residents (Thomas, 2016). The 2017–18 total K–12 enrollment in the three largest DFW was 303,965 students and this number continues to rise (Texas Education Agency, retrieved 2019). Area ISD demographic and standardized test data is summarized in Table 29.1 (Texas Education Agency, 2019).

The University of Texas at Arlington (UTA) is a Carnegie R1 university with nearly 42,000 on-campus students and over 58,000 including online students. UTA is situated between the cities of Dallas and Fort Worth. Most students served by UTA are from the DFW Metroplex. UTA has a large percentage of commuter, non-traditional, and community college transfer students, many of whom work full or part time jobs and face economic hardships in pursuit of a higher education degree. UTA enrolls a majority of undergraduate students from DFW high schools who in large numbers return to their communities to live and work, making the university well-positioned to find a solution to the teacher quality and shortage problems in the DFW area.

Need for UTeach Arlington

The secondary science and mathematics education program prior to UTeach was a general education program, in which STEM students seeking to earn sec-

TABLE 29.1. Standardized Test Scores (SAT and ACT) and Demographic Information of Three Urban School Districts and Three Suburban Comparison Districts in the DFW Metroplex

	Mean SAT and Subject Scores			Mean ACT and Subject Scores			
School District Name	**SAT**	**ELA**	**MATH**	**ACT**	**ELA**	**MATH**	**SCIENCE**
Southlake-Carroll	1261	626	636	27.1	27.2	26.9	26.5
Grapevine-Colleyville	1159	587	572	24.4	24.6	23.8	24.2
Highland Park	1309	652	656	27.6	27.7	27.4	26.9
Suburban Districts: Mean	**1243**	**622**	**621**	**26.4**	**26.5**	**26.0**	**25.9**
Dallas	921	460	461	16.8	15.8	17.6	17.4
Arlington	1026	518	508	20.8	20.5	20.8	21.0
Ft. Worth	916	459	458	19.6	19.1	19.7	19.9
Urban Districts: Mean	**954**	**479**	**476**	**19.1**	**18.5**	**19.4**	**19.4**

	School District Student Population			Percent Underrepresented Groups			
School district name	**Total Students**		**% Economically Disadvantaged**		**African American**	**Hispanic**	**Combined %**
Southlake-Carroll	8,341		1.4		2	9.7	11.7
Grapevine-Colleyville	13,924		22.6		5.7	24	29.7
Highland Park	6,971		0.0		0.8	5.2	6
Suburban Districts: Total	**29,236**	**Mean %**	**8.00**	**Mean %**	**2.8**	**13.0**	**15.8**
Dallas	156,726		86.7		22.5	69.6	92.1
Arlington	61,200		65.4		24.9	46	70.9
Ft. Worth	86,039		77.7		22.7	62.4	85.1
Urban Districts: Total	**303,965**	**Mean %**	**76.6**	**Mean %**	**23.4**	**59.3**	**82.7**

	Percent School District Enrollment By Race/Ethnic Distribution						
School District Name	**African American**	**Hispanic**	**White**	**Native American**	**Asian**	**Pacific Islander**	**Two or More Races**
Southlake-Carroll	2	9.7	66.7	0.2	16.5	0.1	4.7
Grapevine Colleyville	5.7	24	54.8	0.4	9.6	0.1	5.3
Highland Park	0.8	5.2	84.4	0.4	6.7	0.1	2.4
Suburban Districts: Mean %	**2.8**	**13.0**	**68.6**	**0.3**	**10.9**	**0.1**	**4.1**
Dallas	22.5	69.6	5.4	0.3	1.4	0.1	0.7
Arlington	24.9	46	19.7	0.4	6.2	0.2	2.7
Ft. Worth	22.7	62.4	11.3	0.1	1.8	0.1	1.5
Urban Districts: Mean %	**23.4**	**59.3**	**12.1**	**0.3**	**3.1**	**0.1**	**1.6**

ondary teacher certification may have been the lone biology, chemistry, physics, or mathematics students enrolled in teacher preparation courses along with English, social studies, art, music, and PE pre-service teachers. Teacher candidates in student teaching may or may not have been supervised by a science or mathematics education faculty member. In addition, there was no dedicated laboratory/classroom facility for faculty to model or for students to practice laboratory-based, constructivist teaching.

The three UTeach Arlington Co-directors, Cavallo, Hale, and Lopez, began addressing the problem of low STEM teacher production in 2006 by applying for funding support and were awarded their first NSF Robert Noyce Scholarship grant in 2008. The Noyce Scholarship program provides $10K scholarships to STEM majors to complete secondary teacher certification at UTA. Other supports built into the NSF Noyce grant include professional development seminars each year and school-based mentor teachers to provide guidance and support to our teachers in their first years of teaching. This grant helped increase the numbers of students seeking secondary science and mathematics teacher certification by about 100 percent (see Table 29.2). Although a positive result, the overall numbers were still very low, and the teacher preparation program itself had remained unchanged. Students continued to take teacher preparation courses with all other subject areas in education to become certified. The goals to substantially increase our production and change STEM teacher preparation remained the focus, and the UTeach model provided a research-based program fully aligned to the philosophy and practices of teaching and learning of the co-directors/co-founders of UTeach Arlington.

Cavallo, Hale, and Lopez identified the UTeach model as being an ideal means to address the STEM teacher shortage in the region after it was hailed as an exemplary STEM teacher preparation program by multiple bodies. For example, the President's Council of Advisors on Science and Technology (2010) specifically called out UTeach programs, where: "strong pedagogical preparation in STEM, by preparing teachers to convey concepts and lead inquiry in the classroom, and deep content expertise via rigorous majors in STEM disciplines have provided examples of success" (p. 69). The National Academy of Sciences (2007) also cites the UTeach program as highly effective in science and mathematics teacher recruitment, preparation, induction, and retention. Furthermore, UTeach has shown positive impacts on the schools and students where their graduates teach (Backes, Goldhaber, Cade, Sullivan, & Dodson, 2018; Marder & Hamrock, 2016).

UTA was awarded a UTeach replication grant from the National Math + Science Initiative (NMSI) in 2009. The project established a physical location for secondary science and mathematics teachers, including a dedicated classroom/laboratory, advisor, lab materials, and new faculty hires. Once in place, the team's recruitment effort was successful with 142 students enrolling in the Step 1 recruiting/field experience course in the first year of implementation, 2010–11. In 2014, UTeach Arlington graduated its first class of 22 students, and in two years (2014

and 2015) graduated 101 new science and mathematics teachers, a 5-fold increase in teacher production (Table 29.2). Due to budget limitations, the program at UTA has not been able to grow as planned, yet this program has maintained a graduation rate of about 35 finishers per year.

UTeach Arlington has clearly generated much excitement and interest among teachers and administrators in partner high-need school districts with reports of our graduates being the best-prepared first year teachers they have ever had. The principal from one of our Fort Worth partner high schools who recently hired one UTeach graduate stated that he wants only "teachers from our programs prepared just like him from now on." Another urban school principal, who hired 12 of our UTeach Arlington graduates within three years, wrote the following in an email to the program co-directors: "Educators prepared by the UTeach Arlington program have had a profound impact on teaching and learning in math and science at Irving High. Not only is this evident in the continuous increase in state assessment data but it is also reflected in the students' positive engagement in these typically difficult courses. The passion and knowledge of content and pedagogy that these teachers bring with them ensures learning at high levels for ALL students." (Ahna Gomez, Principal, Irving High School).

The UTeach Arlington Certification Program

The UTeach model and our UTeach Arlington implementation are grounded in constructivist, student-centered, inquiry-based teaching and learning. Research-supported models emphasized in our teacher preparation program include the learning cycle/5E model and Project Based Learning (Karplus, Lawson, Wollman, Appel, Bernoff, Howe, Rusch, & Sullivan, 1977; Krajcik, McNeill, & Reiser, 2008; Lawson, Abraham, & Renner, 1989; Marek & Cavallo, 1997). The emphasis of our UTeach program is standards-based, maintaining focus on the dimensions of *practices, crosscutting concepts,* and *disciplinary core ideas* (National Research Council, 2012; Next Generation Science Standards [NGSS], 2013). Studies report that students in science and mathematics experience significant achievement and lower attrition from STEM when taught via the inquiry-based, learning cycle/5E model (Bybee, 2014; Cavallo & Laubach, 2001; Lawson, 2010; Tuna & Kacar, 2013). In UTeach Arlington, teacher candidates *practice* using these inquiry models in early field experiences, with guidance from program faculty and mentor teachers in our partner schools. Within these inquiry models, we integrate other recognized strategies for effective teaching and learning, including meaningful learning strategies and concept mapping (Novak, 2002; Novak & Gowin, 1984), negotiated discourse, creativity, and critical thinking (Zeidler, Lederman, & Taylor, 1992). On campus and in K–12 classrooms, instructors model and teacher candidates engage in practice teaching using active inquiry investigations, interactive technology, classroom research, collaborative group projects, analyses and observations of student learning, experiences with diversity, and self-evaluation/reflective teaching so they gain expertise with the most

TABLE 29.2. Production of Science and Mathematics Teachers at UTA 2005–2017

Science and Mathematics Teacher Production at the University of Texas at Arlington 2005–2017								
Year	Biology	Chemistry	Physics/ Math	Physical Science	Composite Science	Math	Total	Program Explanation/Notes
2005–06	3	0	0	0	1	1	5	1. Non-STEM specific certification program
2006–07	5	0	0	1	0	2	8	2. Non-STEM specific certification program
2007–08	3	0	0	0	0	4	7	3. Non-STEM specific certification program
2008–09	2	0	0	0	0	2	4	4. NSF Noyce I grant award 5. Non-STEM specific certification program
Non-STEM Specific Teacher Certification Program (pre-UTeach) and pre-NSF Robert Noyce scholarship program							**24**	**4-year total STEM teacher production**
2009–10	0	0	0	1	1	0	2	6. Non-STEM specific certification program 7. First year NSF Noyce Scholarships awarded
2010–11	4	1	0	1	1	8	15	8. NSF Noyce II grant award 9. UTeach Arlington program enrolls students 10. Existing non-STEM specific program phasing out, students continue to graduate from program
2011–12	5	3	0	1	1	2	12	11. Existing non-STEM specific program phasing out, remaining students graduate
2012–13	2	1	0	0	2	6	11	12. Existing non-STEM specific certification program phasing out as final students graduate
Non-STEM Specific Teacher Certification Program (pre-UTeach) with NSF Robert Noyce Scholarship Program							**40**	**4-year total STEM teacher production**

2013–14	3	0	0	0	7	12	22	13. UTeach Arlington program produces first cohort of STEM teacher graduates 14. Non-STEM specific teacher undergraduate certification program no longer offered
2014–15	9	0	3	1	9	28	50	5. NSF Noyce III grant award 16. UTeach Arlington program graduates 2 new cohorts of STEM teachers
2015–16	5	0	1	0	10	13	29	17. UTeach Arlington program graduates 2 new cohorts of STEM teachers
2016–17	1	0	1	1	9	24	36	18. UTeach Arlington program graduates 2 new cohorts of STEM teachers
STEM Specific UTeach Arlington Program and NSF Robert Noyce Scholarship Program							**137**	**4-year total STEM teacher production**

effective teaching techniques for *all* students. The guiding philosophy and pedagogical strategies described here best prepare our candidates for teaching science and mathematics, promoting excellence and equity in our urban schools.

EXAMINING IMPACTS OF UTEACH ARLINGTON

Data collected from UTeach Arlington graduates are categorized into three sections as reported here. The first section of these analyses presents data collected on UTeach Arlington graduates maintained over time, and reports where graduates are teaching. This section also presents feedback from school-based mentor teachers assigned to guide a subset of UTeach graduates, Noyce Scholarship awardees within UTeach Arlington, in their first one to two years of teaching. The second section of these analyses presents results of survey data collected on UTeach Arlington graduates relative to retention and attributions to their preparation through UTeach Arlington. The third section examines more long-term retention in teaching among UTeach students who graduated in the first three years of the program.

Demographic and Mentor Teacher Data on UTeach Arlington Graduates

Table 29.3 shows demographic information on the UTeach Arlington program alumni to date.

As shown in Table 29.3, the majority of UTeach Arlington students are females in STEM, with nearly 70% female. UTA reports 61% female enrollment, thus this program recruits and completes a higher percentage of females than in the female population of the university overall (University of Texas at Arlington, Enrollment, 2017). Nearly 55% of students prepared through the UTeach Arlington program become certified in secondary mathematics, whereas 45% are certified in the sciences (including the 8% certified in both physics and mathematics). The race/ethnicity demographics may also be compared with the UTA student population overall (see Table 29.4).

As shown in Table 29.4, a greater percentage of Hispanic students were enrolled in the UTeach Arlington program than enrolled at UTA overall, by over 11%. UTeach Arlington also has greater percentages of the white and Asian students, however a lower percentage (by 10%) of African American students.

Of the 179 graduates from UTeach, we obtained recent data from 114 on the schools and districts where they are teaching. These 114 graduates are teaching in 34 different school districts and 77 different schools, with 106 teaching in the Dallas-Ft. Worth Metroplex. Of the remaining UTeach Arlington graduates from these available data, two are teaching in San Antonio, TX, five are teaching in rural schools in Texas, and one is teaching in Indiana. Ninety-eight percent of the 114 UTeach graduates are teaching in economically disadvantaged, high-need school districts (according to NSF designation). Of the three most economically disadvantaged districts in DFW, six UTeach graduates are teaching in Dal-

TABLE 29.3. Demographic Information Including Certification Area of All UTeach Arlington Alumni Spring 2014–Spring 2019

Gender	n	Percent	Race/Ethnicity	n	Percent	Gender	Certification area	n	Percent
			Hispanic	49	39.2		Mathematics 7–12	68	54.4
			White	48	38.4		Life Science 7–12	22	17.6
Females	125	69.8	Asian	17	13.6	Females	Chemistry 7–12	2	1.6
			African American	7	5.6		Physical Science 6–12	1	0.8
			Native American	2	1.6		Physics/Mathematics 7–12	1	0.8
			Multiple	2	1.6		Science (Composite) 7–12	31	24.8
			Hispanic	15	27.8		Mathematics 7–12	30	55.6
			White	29	53.7		Life Science 7–12	1	1.9
Males	54	30.2	Asian	7	13.0	Males	Chemistry 7–12	0	0.0
			African American	3	5.6		Physical Science 6–12	1	1.9
			Native American	0	0.0		Physics/Mathematics 7–12	7	13.0
			Multiple	0	0.0		Science (Composite) 7–12	15	27.8
			Hispanic	64	35.8		Mathematics 7–12	98	54.7
			White	77	43.0		Life Science 7–12	23	12.8
All	179	100.0	Asian	24	13.4	All	Chemistry 7–12	2	1.1
			African American	10	5.6		Physical Science 6–12	2	1.1
			Native American	2	1.1		Physics/Mathematics 7–12	8	4.5
			Multiple	2	1.1		Science (Composite) 7–12	46	25.7

TABLE 29.4. Comparison of Demographic Information of the UTA Population with UTeach Arlington alumni population (University of Texas at Arlington, Fast Facts, 2018).

UTA	Percent	UTeach Arlington	Percent	Difference UTeach
Hispanic	24.2	Hispanic	35.8	11.6
White	39.5	White	43.0	3.5
Asian	9.6	Asian	13.4	3.8
African American	15.8	African American	5.6	–10.2
International	10.9	Native American	1.1	N/A
		Multiple	1.1	N/A

las ISD (86.7% economically disadvantaged), 14 are teaching in Ft. Worth ISD (77.7%), and 20 are teaching in Arlington ISD (65.4%). Other DFW school districts with high percentages of economically disadvantaged student populations where UTeach graduates are teaching include Irving ISD (16 graduates, 75.5% economically disadvantaged), Grand Prairie ISD (5 graduates, 71.6%), Mesquite ISD (four graduates, 74.9%), Everman ISD (two graduates, 82.9%), Crowley ISD (four graduates, 61%), Birdville ISD (four graduates, 58.1%), and Castleberry ISD (two graduates, 85.2% economically disadvantaged). Of the 179 graduates/UTeach program completers, eleven (6%) chose not to take the state teacher exam to earn certification.

School-based mentor teachers of a subset of UTeach Arlington graduates, those awarded Noyce Scholarships, responded to survey questionnaires on the quality of preparation of our new teacher graduates in their first one to two years of teaching. Toward the end of their first year of teaching or in their second year, mentor teachers of our UTeach Noyce Scholars were asked to rate the new teachers in six areas: content knowledge, classroom management, teaching skills, collegiality, willingness to learn/grow, and ability to connect/build relationships. These six areas were rated on using a Likert scale, where: 1 = well below average; 2 = below average; 3 = average; 4 = above average; and 5 = well above average. Mentor teachers rated UTeach Noyce Scholar graduates with mean scores as follows: Content Knowledge 4.5, Teaching Skills 4.3, Collegiality 4.3, Connect/Build Relationships 4.2, Willingness to Learn/Grow 4.1, and Classroom Management 3.8. The school-based mentor teachers also provided open-ended responses asking them to describe areas of *strength/above average* in teaching and areas they found to be *in need of work/below average* in teaching. Areas of *strength/above average* were reviewed for themes, which emerged as follows: strong content knowledge/understanding, professional disposition, effective with at-risk students, highly organized, works well with others, engages students in learning, personable, and a team player. For areas *in need of work/below average* areas, the common themes

were: classroom management, confidence, and most stated there were no below average areas. With a few exceptions, the results indicated UTeach Noyce graduates were well-prepared for the high-need classroom.

Survey Data from UTeach Graduates on Retention in Teaching

A Qualtrics survey was distributed to all 179 UTeach Arlington graduates to gather data on retention in teaching and possible links to the UTeach Arlington program. A total of 57 graduates responded to the survey (18 male, 39 female). The ethnicity of the respondents was 35.2% Hispanic and 64.8% Non-Hispanic, and the racial backgrounds were 5.5% Black, 80.0% White, 1.8% Native American, 9.1% Asian, and 3.6% Multiple Races. Of the respondents, 47 reported they were currently teaching and ten reported they were not currently teaching. The survey respondent data in which ten out of the 57 are not currently teaching (17.5%) approximate the total UTeach graduate data in which 30 out 179 are not currently teaching (16.8%), though in both groups, a subset of those not teaching is seeking certification and/or is in the process of seeking employment.

The 47 survey respondents currently teaching are employed at 31 different schools, with 40 reporting their school and/or district is 50% or more economically disadvantaged. Table 29.5 shows data on the respondents' graduation data and years teaching.

TABLE 29.5. Data on UTeach Arlington Survey Respondents' Year of Graduation and Years Teaching (n = 57)

Year Graduated	%	**Count**
2013	1.75%	1
2014	19.30%	11
2015	22.81%	13
2016	14.04%	8
2017	15.79%	9
2018	26.32%	15
Have not yet graduated	0.00%	0
Total	100%	57
Years Teaching	%	**Count**
0 to <1	26.32%	15
1	10.53%	6
2	17.54%	10
3	12.28%	7
4	29.82%	17
5	3.51%	2
Total	100%	57

TABLE 29.6. Responses to the Likert Scale Survey Item: To What Extent Were/Are You Prepared to Teach Because of Your Experiences in the UTeach Arlington Program?

		Min	Max	Mean	SD	Variance	Count
	Item Analyses	1	3	1.72	0.59	0.34	57
#	**Answer**	%	**Count**				
1	Extremely well prepared	35.09%	20				
2	Well prepared	57.89%	33				
3	Moderately prepared	7.02%	4				
4	Minimally prepared	0.00%	0				
5	Not at all prepared	0.00%	0				
	Total	**100%**	**57**				

As shown in Table 29.5, 42 UTeach graduates are still teaching beyond their first year, with 19 still teaching 4 and 5 years beyond graduation, the latest year possible, given our first graduates completed the 2013–14 academic year.

An analyses of survey items revealed information on the extent to which graduates were prepared to teach and elements of UTeach that were particularly impactful in their teacher preparation. Survey items also asked graduates to provide reasons they may have chosen not to teach, and reasons why they have chosen to remain in teaching, with possible attributions to the UTeach Arlington program. Results are summarized according to survey item shown in Tables 29.6 through 29.9.

As shown in Table 29.6, 53 of the 57 survey respondents indicated that UTeach Arlington effectively prepared them to teach in secondary schools. Only 4 graduates indicated they were moderately prepared, and no graduates responding to the survey indicated they were minimally or not at all prepared.

TABLE 29.7. Responses to the Survey Item: How Did UTeach Arlington Best Prepare You to Teach in Secondary Schools (choose up to 3)?

Answer	%	Count
Multiple field experiences in addition to student teaching	28.42%	54
Guidance from experienced Master Science and math teachers throughout the program	22.11%	42
Advising for the degree and teacher certification under the UTeach program	18.95%	36
Support of fellow students throughout the program	16.32%	31
Strong content preparation	13.68%	26
Other(s)[a]	0.53%	1
Total	100%	190

Note. [a] Other(s) response was student teaching.

TABLE 29.8. Responses to the Survey Item: Select the Best Reason(s) for Why You Chose Not to Teach in K–12 Schools After Graduating or Started Teaching and Stopped (choose up to 3). (If not applicable because you are teaching, choose N/A)

Answer	%	Count
Went to graduate or professional school	3.33%	2
Accepted a job with a higher salary	0.00%	0
Left for family reasons (ill relative, birth of a child, etc.)	1.67%	1
Decided that teaching was not a good career fit	3.33%	2
Was unable to pass the TExES	3.33%	2
Other(s)[a]	10.00%	6
N/A	78.33%	47
Total	100%	60

Note. a Other(s) responses were, substitute teaching and part time at a museum currently, not going full time yet, graduated and plan to teach, working to complete TExES certification exam, business.

As results in Table 29.7 indicate, the top responses graduates attribute their teaching preparation to are the multiple field experiences in the UTeach program, followed by Master Science and Math teachers who worked with them throughout the program, and dedicated advising specifically for secondary science and mathematics teacher candidates.

As shown in Table 29.8, only 2 respondents found that teaching was not a good fit. At least 3 are working on the TExES (Texas state teacher certification exam), and others left or are not teaching due to being in graduate school or having family or business responsibilities.

Responses to the final item asking UTeach Arlington graduates why they have chosen to remain in teaching were categorized by the research team into positive, neutral, and negative reasons. The data were analyzed for patterns according to positive reasons, neutral reasons, and negative reasons with representative themes shown in Table 29.9. (Note, some responses fell into more than one pattern/theme).

As Table 29.9 indicates, the most frequent reasons UTeach graduates have chosen to stay in teaching are passion or love for teaching their subject, wanting to make a difference, and were inspired by the students. The next most frequent response was passion for the subject, and four graduates directly attributed their teaching preparation to UTeach. The patterns observed in neutral reasons centered on low income, completing certification, teaching as only preparation area, and fulfilling Noyce Scholarship requirement. The negative responses were frustration with students and the school structure/possible lack of support from administration. However, more UTeach graduates provided positive responses on why they chose to remain in teaching (38 positive) compared to neutral (10) and nega-

TABLE 29.9. Patterns in Responses and Number of Responses Within Each Pattern From Open-Ended Responses to the Survey Item: If You Are Still Teaching, List Why You Have Remained in Teaching/Plan to Remain in Teaching. Responses Grouped According to Those Observed as Positive, Neutral, and Negative

#	Positive Response Patterns	# Responses
1	Love teaching, happy with career choice	23
2	Want to make a difference, have an impact on the students, help them learn, they need me	11
3	The students (impressed and inspired by students)	8
4	Passion for the subject (science, mathematics)	5
5	UTeach preparation	4
6	People and/or school where teaching	4
#	**Neutral Response Patterns**	
1	Enjoy teaching but need higher pay or need student loan forgiveness	3
2	Completing state teacher exam, not teaching yet	2
3	Only prepared for teaching as career	2
4	Noyce Scholarship requirement	1
#	**Negative Response Patterns**	
1	Students disrespectful	1
2	Broken school system	1

tive (2), indicating the majority of UTeach graduates are happy with their career choice and committed to teaching.

Long-Term Retention in Teaching among Early Program UTeach Graduates

Data were analyzed on all UTeach students who graduated in the first three years of the program to provide information on more long-term retention in teaching. These results are shown in Table 29.10.

As shown in Table 29.10, 90.9% of the initial UTeach Arlington class became certified secondary science or mathematics teachers. Of the 22 graduates, 86.4%

TABLE 29.10. Certifications, Number of Graduates That Taught, and Number of Graduates Still Teaching in 2018–19 for the First Three UTeach Arlington Graduating Classes

Graduating Class	# Graduates	# Certified Teachers	# Taught	# Teaching 2018–2019
2013–2014	22	20	19	18
2014–2015	50	43	41	38
2015–2016	29	28	28	28

became teachers, and 94.7% of those teachers were still teaching five years after graduating. Out of the 50 graduates in the 2014–2015 graduating class, 86% became certified, and 82% became teachers. Of those that taught, 92.7% of them were still teaching four years after graduating. The 29 graduates from the 2015–2016 class produced 28 teachers (96.6%), and all of them were still teaching three years after graduation.

SUMMARY AND CONCLUSIONS

The UTeach Arlington program at the University of Texas at Arlington has made a major impact on STEM teaching and learning in the Dallas-Ft. Worth Metropolitan area and beyond. The data indicate a five-fold increase in science and mathematics teachers certified through this program who are specifically prepared with the pedagogical content knowledge and skills for teaching their STEM subject in high-need schools. UTeach Arlington now consistently enrolls between 175–200 students in the program—a substantial number compared to our previous program. In addition, the UTeach Arlington co-directors received three additional NSF Noyce grant awards after the first in 2008, with the NSF review panels consistently commenting on the strength of our UTeach Arlington program in making the award.

As findings indicate, most UTeach Arlington graduates receive positive appraisals of their teaching and are highly dedicated to the profession. It is asserted that these results are an outcome of the UTeach program that includes deliberate elements that prepare students to teach in high-need schools, including early field experiences in which students teach in high-need schools in several courses prior to the full semester of student teaching. These field experiences and student teaching place students in the same schools and districts where most graduates will teach, gaining experience and receiving consistent mentoring from school-based mentor teachers and UTA Master Teachers who each have at least 15 years teaching experience in DFW high-need schools and districts. A significant strength of the program are the entry-level Step courses, in which STEM majors can "try out teaching" by enrolling in a 2 credit hour course that places them in elementary and middle school classrooms teaching inquiry lessons to the students. The teaching experiences in these Step courses are highly positive and perhaps enlightening for most students, inspiring many who may not have considered teaching to continue in the UTeach program and earn certification. The Step courses are of no additional cost to students due to flat tuition rates at UTA, and as funding allows, our program provides a small stipend to students who successfully complete each Step course. This strategy is designed to cast a wide net to recruit as many STEM majors as possible into teaching, while also earning their STEM degree. Those who continue on through the UTeach program make the deliberate, informed decision to earn teacher certification and tend to be dedicated to teaching, with the many field experiences solidifying that dedication. The outcome is better prepared, committed teachers who stay in the field at a high rate, as evidenced by

Table 29.10. The very large class of 2014–2015 shows relatively lower teacher retention compared to the smaller class sizes reported, which may be an outcome of having stretched resources in faculty and mentor teacher support during field experiences. Future studies should examine the extent to which frequency and quality of mentoring during early field experiences and student teaching impacts teacher retention.

Available data on our UTeach graduates indicate 112 out of the 114 survey respondents are teaching in economically disadvantaged schools/school districts since our first graduates completed the program in May 2014. With a typical 5 classes per day, and estimating 30 students per class, over 16,800 disadvantaged students are taught by our highly qualified UTeach graduates per year. Remarkably, just these 112 graduates teaching in high-needs districts alone will teach 336,000 students over an estimated 20-year career. This number continues to grow as UTeach Arlington graduates more new, highly qualified STEM teachers each year.

The need for highly qualified STEM teachers is a persistent problem in Texas and across the U.S. We cannot afford to lose the talents of millions of secondary school students who have great potential and passion for STEM fields because they do not have skilled STEM teachers they so desperately need and deserve. UTeach Arlington addressed this issue by establishing this impactful program in the DFW Metroplex where there is significant need in our neighboring school districts. We are already witnessing students of our UTeach graduates enrolling at UTA as STEM majors who report they did so *because* they were educated in STEM and positively influenced by our UTeach teachers. As we continue to graduate new teachers from UTeach, the positive influence of our STEM teachers on secondary school students will continue to cultivate the next generation toward STEM careers.

Further research is needed on the retention of teachers, their attributions to why they remain or leave the profession, and the role of the UTeach program. The research reported here is limited by the number of survey respondents and more work is needed to clarify the reported data and trends. It would be worthwhile to examine, for example, how recruitment, preparation, and retention may compare with other UTeach sites and other teacher preparation programs in the DFW Metroplex. Recruiting physics teachers remains a challenge; however, there were virtually zero physics teachers prepared prior to UTeach, so there has been some improvement with this program. Strategies to recruit more physics majors into teaching remains a challenge, thus more targeted recruitment efforts are needed among students in this field.

Future research will also examine the impact of our UTeach graduates on student achievement and decisions to pursue STEM majors upon graduation. We will collect data on factors such as self-efficacy toward teaching science and mathematics, and the extent to which UTeach graduates continue to implement inquiry-

based/5E and/or Project Based Learning in their teaching over time. The data and trends in findings presented here set the foundation for future research.

REFERENCES

American Association for the Advancement of Science (1989). *Science for all Americans: A Project 2061 report on literacy goals in science, mathematics, and technology*. Washington, DC: Author.

Backes, B., Goldhaber, D., Cade, W., Sullivan, K., & Dodson, M. (2018). *Can UTeach? Assessing the relative effectiveness of STEM teachers*. National Center for Analysis of Longitudinal Data in Education Research. Working Paper 173, Washington, DC: CALDER American Institutes for Research. Retrieved from: https://caldercenter.org/sites/default/files/WP%20173.pdf.

Bybee, R. W. (2014). The BSCS 5E instructional model: Personal reflections and contemporary implications. *Science and Children, 4,* 10–13.

Carver-Thomas, D., & Darling-Hammond, L. (2017). *Teacher turnover: Why it matters and what we can do about it.* Palo Alto, CA: Learning Policy Institute. Retrieved from: https://learningpolicyinstitute.org/product/teacher-turnover-report

Cavallo, A. M. L., & Laubach, T. (2001). Students' science perceptions and enrollment decisions in differing learning cycle classrooms. *Journal of Research in Science Teaching, 38*, 1029–1062.

Desilver, D. (2017, February 15). U.S. students' academic achievement still lags that of their peers in many other countries. *Pew Research Center.* Retrieved from: https://www.pewresearch.org/fact-tank/2017/02/15/u-s-students-internationally-math-science/

Fuller, E. (2008). *Secondary mathematics and science teachers in Texas: Supply, demand, and quality*. Austin, TX: Department of Educational Administration, University of Texas at Austin.

Hanna, B. (2018, March 22). DFW gains a whopping 146,000 residents in 2017 to lead the nation. *Fort Worth Star-Telegram*. Retrieved from: https://www.star-telegram.com/news/local/community/fort-worth/article206357404.html

Karplus, R., Lawson, A., Wollman, W., Appel, M., Bernoff, R., Howe, A., Rusch, J., & Sullivan, F. (1977). *Science teaching and the development of reasoning: A workshop*. Berkeley, CA: University of California.

Krajcik, J., McNeill, K. L., & Reiser, B. J. (2008). Learning-goals-driven design model: Developing curriculum materials that align with national standards and incorporate Project-Based pedagogy. *Science Education, 92*, 1–32.

Lawson, A. E. (2010). *Teaching inquiry science in middle and secondary schools*. Thousand Oaks, CA: Sage.

Lawson, A. E., Abraham, M. R., & Renner, J. W. (1989). *A theory of instruction: Using the learning cycle to teach science concepts and thinking skills* [Monograph, Number One]. Kansas State University, Manhattan, KS: National Association for Research in Science Teaching.

Marder, M., & Hamrock, C. (2016). *Math and science outcomes for students of teachers from standard and alternative pathways in Texas*. UTeach Working Paper. Retrieved from: https://pdfs.semanticscholar.org/0693/f28933b966f68a48813a75894353f954e639.pdf

Marek, E. A., & Cavallo, A. M. L. (1997). *The learning cycle: Elementary school science and beyond* (Rev. ed.). Portsmouth, NH: Heinemann.

National Academy of Sciences. (2007). *Rising above the gathering storm: Energizing and employing America for a brighter economic future.* Committee on Prospering in the Global Economy of the 21st Century Washington, DC: Author.

National Commission on Excellence in Education. (1983). *A nation at risk.* Washington, DC: Author.

National Commission on Mathematics and Science Teaching for the 21st Century. (2000). *Before it's too late.* Washington, DC: Author.

National Education Association. (2001). *The opportunity to excel.* Washington, DC: National Education Association.

National Research Council. (2012). *A framework for K–12 science education: Practices, crosscutting concepts, and core ideas.* Committee on a Conceptual Framework for New K–12 Science Education Standards. Washington, DC: The National Academies Press.

Next Generation Science Standards [NGSS] Lead States. (2013). *Next generation science standards: For states. by states.* Washington, DC: The National Academies Press.

Novak, J. D. (2002). Meaningful learning: The essential factor for conceptual change in limited or inappropriate propositional hierarchies leading to empowerment of learners. *Science Education, 86,* 548–571.

Novak, J., & Gowin, B. (1984). *Learning how to learn.* New York, NY: Cambridge University Press.

Ostroff, C. (2017, August 21). Schools throughout the country are grappling with teacher shortage, data show. *Cable News Network (CNN).* Retrieved from: http://www.cnn.com/2017/08/21/health/teacher-shortage-data-trnd/index.html

President's Council of Advisors on Science and Technology (2010). *Prepare and inspire: K–12 education in science, technology, engineering, and math (STEM) for America's future.* Report to The President. Retrieved from: https://obamawhitehouse.archives.gov/sites/default/files/microsites/ostp/pcast-stem-ed-final.pdf

Stutz, T. (2012, May 28). Texas schools short by 15,000 teachers this year, analysis shows. *Dallas Morning News.* Retrieved from: https://www.dallasnews.com/news/education/2012/05/28/texas-schools-short-by-15000-teachers-this-year-analysis-shows

Sutcher, L., Darling-Hammond, L., & Carver-Thomas, D. (2016). *A coming crisis in teaching? Teacher supply, demand, and shortages in the U.S.* Palo Alto, CA: Learning Policy Institute.

Texas Education Agency (2019). *Texas academic performance reports.* Retrieved from: https://rptsvr1.tea.texas.gov/perfreport/tapr/2018/index.html

Thomas, S. G. (2016, October 11). DFW's population could explode in next 25 years—By more than you'd think. *Dallas Business Journal WFAA.* Retrieved from: http://www.wfaa.com/news/local/dallas-county/dfws-population-could-explode-in-next-25-years-by-more-than-youd-think/334036371

Tuna, A., & Kacar, A. (2013). The effect of 5E learning cycle model in teaching trigonometry on students' academic achievement and the permanence of their knowledge. *International Journal on New Trends in Education and Their Implications, 4,* 7.

United States Department of Labor Bureau of Labor Statistics. (2014). *Occupational outlook handbook.* Retrieved from: http://www.bls.gov/ooh/education-training-and-library/high-school-teachers.htm#tab-6.

University of Texas at Arlington. (2017). *Enrollment.* Retrieved from: https://www.uta.edu/analytics/_downloads/facts/2017-oct-20.pdf

University of Texas at Arlington. (2018). *Fast facts.* Retrieved from: https://www.uta.edu/uta/about/fastfacts/

Urban Teacher Collaborative. (2000). *The urban teacher challenge: Teacher demand in the great city schools.* Washington, D.C.: Council of the Great City Schools.

Westervelt, E. (2016, September 15). Frustration. burnout. attrition. It's time to address the national teacher shortage. *National Public Radio Education (nprED).* Retrieved from: http://www.npr.org/sections/ed/2016/09/15/493808213/frustration-burnout-attrition-its-time-to-address-the-national-teacher-shortage

Zeidler, D. L., Lederman, N. G., & Taylor, S. C. (1992). Fallacies and student discourse: Conceptualizing the role of critical thinking in science education. *Science Education, 76*(4), 437–450.

Printed and bound by CPI Group (UK) Ltd, Croydon, CR0 4YY

15/09/2026

14962336-0007

Acoustics of Fluid Media 1

Series Editor
Frédérique de Fornel

Acoustics of Fluid Media 1

Principles and Applications

Daniel Juvé
Marie-Annick Galland
Vincent Clair

iSTE

WILEY

First published 2024 in Great Britain and the United States by ISTE Ltd and John Wiley & Sons, Inc.

ISTE Ltd
27-37 St George's Road
London SW19 4EU
UK

www.iste.co.uk

John Wiley & Sons, Inc.
111 River Street
Hoboken, NJ 07030
USA

www.wiley.com

Library of Congress Control Number: 2024940982

British Library Cataloguing-in-Publication Data
A CIP record for this book is available from the British Library
ISBN 978-1-78630-932-7

Contents

List of Abbreviations, Acronyms and Symbols

(ρ, φ, z) cylindrical coordinates

(r, θ, φ) spherical coordinates

(x_1, x_2, x_3) Cartesian coordinates

α_m random incidence absorption coefficient

α_r energy reflection coefficient

α_t energy transmission coefficient

β thermal dilation coefficient

$\boldsymbol{I}$ (vector) acoustic intensity

$\boldsymbol{k}$ wave vector

M Mach number

δ Dirac distribution

γ adiabatic index

κ_s adiabatic compressibility coefficient

κ_T isothermal compressibility coefficient

λ wavelength

λ_{th} thermal conductivity coefficient

μ dynamic viscosity coefficient

μ_B volume viscosity coefficient

∇ nabla differential operator

ν kinematic viscosity coefficient

ω angular frequency

ϕ acoustic potential

ρ density

He Helmholtz number

Pr Prandtl number

Re Reynolds number

σ diffusion cross-section

$\zeta = Z/Z_0$ reduced acoustic impedance

c_0 speed of sound

c_f bending wave velocity

c_p constant pressure specific heat

c_v constant volume specific heat

E Young's modulus

E signal energy

e acoustic energy density

f frequency

f_{coin} coincidence frequency

f_{cr} critical frequency

G_0 Green's function of the Helmholtz equation in free space

g_0 Green's function of the free-space wave equation

H Heaviside function

h specific enthalpy

$H_n^{(2)}$ nth-order Hankel function of the second kind

J_n nth-order Bessel function of the first kind

k_0	acoustic wavenumber
k_f	bending wavenumber
L_I	acoustic intensity level (dB)
L_p	sound pressure level (dB)
L_W	sound power level (dB)
$p'(t)$	acoustic pressure
$p(f)$	complex amplitude of sound pressure
p_{ref}	reference sound pressure
p_{rms}	r.m.s. value of sound pressure fluctuations
R	ideal gas constant
r	gas constant
R_{pp}	autocorrelation of pressure fluctuations
s	specific entropy
S_{pp}	power spectral density of pressure fluctuations
T	absolute temperature
t	amplitude transmission coefficient
T_c	temperature in degrees Celsius
T_{60}	reverberation time
$u'(t)$	acoustic velocity fluctuation
$u(f)$	complex amplitude of acoustic velocity
U	flow velocity
v_g	group speed
v_φ	phase speed
Y_n	nth-order Bessel function of the second kind
Z	specific acoustic impedance
Z_0	characteristic impedance
Z_{ray}	radiation impedance
Zs	surface impedance

Preface

This book is based on courses taught by the authors at the École Centrale de Lyon and at the Université de Lyon, both at the undergraduate and graduate levels. It has also benefited from their interactions with the audience of professional training sessions, held in particular at the Collège de Polytechnique.

The book is intended for undergraduate students and engineering students, as well as graduate students and professionals in industry who are increasingly faced with the need to consider acoustic constraints when developing new products. It is limited to acoustics in fluids, with applications to atmospheric and underwater acoustics.

The book is divided into two volumes. The first is devoted to fundamental elements, the knowledge of which allows for a good mastery of acoustics in fluids. The second is an introduction to more advanced aspects, some of which are the subject of active research and whose status is sometimes still evolving (aeroacoustics, propagation in a moving medium, nonlinear acoustics). Some synthesis problems are also presented, focusing on noise control issues.

Volume 1 consists of 10 chapters plus an appendix of fluid mechanics reminders and a second one with some mathematical elements.

The first two chapters establish the equations of acoustics in homogeneous fluids and describe the properties of plane waves and spherical waves, as fundamental elements in the construction of more general solutions.

Chapter 3 is an interlude in the physical analysis offered throughout the book. It is devoted to elements of signal processing useful to the acoustician, to the definition of sound levels and decibel scales, and to notions of human sound perception and the characterization of the associated nuisances.

Chapter 4 describes the phenomena of reflection and transmission of plane and spherical waves at the interface between two fluids or between a fluid and a solid. In particular, the transmission of plane waves through a thin wall subjected to bending vibrations is discussed.

In Chapter 5, volume acoustic sources associated with mass, force or heat contributions within the fluid are introduced. The powerful method of Green's functions is then extensively discussed and used.

Integral methods, which complement the local formulations used so far, are introduced in Chapter 6, as well as their application to radiation from vibrating surfaces and diffraction by obstacles.

Chapter 7 describes these diffraction phenomena in more detail with their application to the characterization of the efficiency of sound barriers. This is followed by a description of wave scattering exerted by rigid obstacles, with emphasis on low-frequency (Rayleigh scattering) and high-frequency (geometric limit) behavior. The effect of fluid inclusions of low contrast relative to the surrounding medium is addressed within the framework of the Born approximation.

Chapters 8 and 9 deal with guided propagation in ducts, first in the general form using the notion of propagation modes, and then in the low-frequency version of one-dimensional networks. This simplified formulation is very useful for defining acoustic filters such as Helmholtz resonators and passive silencers for selective reduction of sound levels.

Chapter 10 is devoted to the acoustics of confined spaces and applications in room acoustics. The concept of diffuse field and the important notion of reverberation time are introduced, as well as elements for characterizing the acoustic quality of rooms from the point of view of human perception.

Each of these chapters is accompanied by a limited number of exercises, ranging from the simple application of definitions and formulas to problems requiring more advanced theoretical analyses or numerical solutions.

Throughout the book, we have striven to illustrate the theoretical results with many figures obtained from measurements and numerical simulations resulting from the evaluation of complex theoretical formulas or the use of a finite element solver. The purpose of these illustrations is to facilitate the physical interpretation of the phenomena involved by making our own Richard Hamming's aphorism, "The purpose of computing is insight, not numbers". They do not, of course, replace the theoretical developments that allow us to highlight the influence of the most influential parameters. However, theoretical formulations are all too often limited to highly simplified or asymptotic situations. Rather than resorting to too often

unintuitive expressions using series of special functions, we have found it preferable, for example, to plot maps of acoustic levels that are much more meaningful.

This aspect distinguishes our work somewhat from the vast existing bibliography. The main works upon which we have relied while writing this book are listed in the references section, which is of course very far from exhaustive. In the text, we sometimes refer to some of these books or articles for additional elements or computational details that we felt it was unnecessary to develop.

We would like to warmly thank all our colleagues at the Acoustic Center of the LMFA at École Centrale de Lyon with whom we have had extensive interactions during our teaching and research activities.

July 2024

1

Equations of Linear Acoustics

In this chapter, we establish the equations that govern the propagation of small-amplitude acoustic waves in fluids under the simplest possible conditions. The fluid is considered as perfect, that is non-viscous and non-heat-conductive. Acoustic disturbances are regarded as small-amplitude perturbations of the ambient state of a time-independent homogeneous fluid at rest. External forces, such as gravity, are not taken into account.

These assumptions may seem very strong, and we will have to relax some of them in later chapters. However, they offer considerable advantages in terms of simplicity, while often remaining unrestrictive in practice. This simplicity makes it possible to emphasize the fundamental properties of acoustic waves, which are generally only slightly modified in more complex situations where, for example, the inhomogeneous nature of the medium will have to be taken into account. They also allow the construction of analytical solutions with very wide applications, which also serve as references for the numerical simulations that become necessary when the geometries of the problems under consideration become complex.

1.1. Validity of the assumptions of linear acoustics and a perfect fluid

It is important to form a first qualitative idea on the legitimacy of the assumptions of a perfect fluid and linearization of fluid dynamics equations by performing order-of-magnitude analyses in a simplified situation. We thus consider a wave of frequency f chosen within the range of audible sounds, for instance, $f = 1$ kHz, and propagating through the air in the x_1 direction only.

To evaluate the validity conditions of the linearization process, we consider the two components of acceleration in the Navier–Stokes equations (these equations are recalled in Appendix 1):

$$\frac{\partial u_1}{\partial t} + u_1 \frac{\partial u_1}{\partial x_1}$$

where u_1 is the component in the x_1 direction of the velocity associated with the propagation of an acoustic wave. The partial derivative of velocity with respect to time $\partial u_1/\partial t$ is linear with respect to fluctuations, while its convective derivative $u_1 \partial u_1/\partial x_1$ exhibits a quadratic nonlinearity. The order of magnitude of the linear term is ωu_1, where $\omega = 2\pi f$ is the angular frequency of the wave. That of the nonlinear term is u_1^2/λ, where λ is the spatial scale of the wave, called the wavelength. The wavelength is related to the angular frequency by $\lambda = 2\pi c_0/\omega$, where c_0 denotes the speed of propagation of acoustic waves in the medium under consideration; in air, the speed of sound is about 340 m.s^{-1}. The ratio of the orders of magnitude of the nonlinear term to the linear term is therefore $u_1/c_0 = \mathrm{M}_a$; this ratio defines the *acoustic Mach number*. We will see in Chapter 3 that a sound wave with a 94 dB level, which is perceived by humans as very intense and painful, is only associated with a root-mean-square (rms) value of pressure fluctuations of only 1 Pa. For this wave, the acoustic Mach number is extremely low, being about 10^{-5}. In the most unfavorable situations, it will hardly exceed 10^{-3}, which fully justifies why terms composed of products of fluctuations in the equations of motion can be neglected, and thus validates the linearization process of the equations.

To evaluate the validity of the perfect fluid assumption, we simply analyze the influence of viscous terms, since the thermal effects are typically of the same order of magnitude as them. The viscous term in the Navier–Stokes equations:

$$\nu \frac{\partial^2 u_1}{\partial x_1^2}$$

has an order of magnitude $\nu u_1/\lambda^2$, where ν is the kinematic viscosity of the fluid. The ratio of the linear acceleration term to the viscous term is therefore of the order $c_0^2/\nu\omega$, or $c_0\lambda/\nu$, some sort of local acoustic Reynolds number. For air, $\nu \approx 15 \times 10^{-6}$ m^2.s^{-1} and for a 1 kHz frequency, this number is about 10^6. This very high value of the Reynolds number ensures that viscous effects are negligible for the usual range of frequencies.

It is important to note, however, that the above reasoning is qualitative and only has a *local* value, that is to say on the scale of a wavelength. Viscous, thermal and nonlinear effects are cumulative, and their consequences can be significant if the distances traveled by waves correspond to a large number of wavelengths. As a case in point, very high frequency acoustic waves (greater than a hundred kilohertz) propagating in the air are very quickly attenuated by visco-thermal effects over distances of only a few meters. It will therefore sometimes be necessary to take into account dissipative effects, often in an approximate way by an a posteriori correction

of a calculation made for a perfect fluid. This is, for example, the case in room acoustics, as we shall see in Chapter 10.

1.2. Linearized equations of fluid dynamics

Since the fluid is now considered perfect and external forces are neglected, the equations of fluid dynamics are reduced to the system of Euler's equations (see Appendix A1.4):

$$\frac{\partial \rho}{\partial t} + \frac{\partial (\rho U_i)}{\partial x_i} = 0 \qquad [1.1]$$

$$\rho \frac{dU_i}{dt} = -\frac{\partial p}{\partial x_i} \qquad [1.2]$$

$$\frac{ds}{dt} = 0 \qquad [1.3]$$

$$p = p(\rho, s) \qquad [1.4]$$

where p, ρ, s and U_i denote fluid pressure, density, entropy and velocity, respectively[1].

The propagation medium is assumed to be time independent, homogeneous and at rest. The variables describing the ambient state (i.e. when there is no acoustic disturbance) are therefore uniform; they will be expressed using a "0" subscript. Since the medium is at rest, the ambient velocity is zero, $\boldsymbol{U}_0 = \mathbf{0}$. Acoustic disturbances, which are time and space dependent, will be indicated by a "prime" symbol. During the propagation of an acoustic wave, the different variables will thus be decomposed according to:

$$p(\boldsymbol{x}, t) = p_0 + p'(\boldsymbol{x}, t) \qquad [1.5]$$

$$\rho(\boldsymbol{x}, t) = \rho_0 + \rho'(\boldsymbol{x}, t) \qquad [1.6]$$

$$U_i(\boldsymbol{x}, t) = 0 + u_i'(\boldsymbol{x}, t) \qquad [1.7]$$

$$s(\boldsymbol{x}, t) = s_0 + s'(\boldsymbol{x}, t) \qquad [1.8]$$

The disturbances are assumed to be small enough so that the products of fluctuations that will appear when the above decompositions are introduced into the equations of fluid dynamics can be neglected. This constitutes the fundamental assumption of linear acoustics.

1 The notations are specified in Appendix A1.1.

We now outline the linearization procedure on the conservation of mass equation (or "continuity equation"):

$$\frac{\partial}{\partial t}(\rho_0 + \rho') + \frac{\partial}{\partial x_i}(\rho_0 + \rho')(0 + u_i') = 0 \qquad [1.9]$$

which reduces to:

$$\frac{\partial \rho'}{\partial t} + \rho_0 \frac{\partial u_i'}{\partial x_i} + \frac{\partial}{\partial x_i}(\rho' u_i') = 0 \qquad [1.10]$$

The last term of the left hand side of [1.10] is of second order as the divergence of a product of fluctuations. It is therefore neglected in the linear approximation, and the linearized continuity equation is finally written as:

$$\frac{\partial \rho'}{\partial t} + \rho_0 \frac{\partial u_i'}{\partial x_i} = 0 \qquad [1.11]$$

To linearize the equations in which the total time-derivative appears, it can be observed that the total derivative of a fluctuation can be reduced to its partial derivative with respect to time, because:

$$\frac{d()'}{dt} = \frac{\partial()'}{\partial t} + u_j' \frac{\partial()'}{\partial x_j} \approx \frac{\partial()'}{\partial t} \qquad [1.12]$$

by neglecting terms of order greater than or equal to 2. To obtain the linearized equations, thanks to the assumptions of homogeneity of the medium and the absence of ambient flow, any total derivative can therefore be replaced by a simple partial derivative with respect to time.

The system of linearized Euler's equations is thus written as:

$$\frac{\partial \rho'}{\partial t} + \rho_0 \frac{\partial u_i'}{\partial x_i} = 0 \qquad [1.13]$$

$$\rho_0 \frac{\partial u_i'}{\partial t} + \frac{\partial p'}{\partial x_i} = 0 \qquad [1.14]$$

$$\frac{\partial s'}{\partial t} = 0 \qquad [1.15]$$

The last equation shows that the entropy fluctuation associated with an acoustic disturbance is identically zero, which leads to an important simplification. The

pressure then depends only on the density and the equation of state takes the simple form $p = p(\rho, s_0) = p(\rho)$.

For the equation of state [1.4], the process of linearization must be carried out differently insofar as there is no explicit form for it, except for the special but important case of ideal gases to which we will return later. The idea is to consider that the presence of acoustic waves only slightly modifies the ambient state; the perturbed variables can then be obtained through a Taylor-series expansion around the ambient state that is limited to the first order. It thus follows that:

$$p(\rho_0+\rho', s_0) = p_0+p' = p(\rho_0, s_0)+\rho' \left(\frac{\partial p}{\partial \rho}\right)_{s,0} + \frac{\rho'^2}{2}\left(\frac{\partial^2 p}{\partial \rho^2}\right)_{s,0} +... \quad [1.16]$$

Partial derivatives are taken at constant entropy (as recalled by the subscript "s") and evaluated at the ambient state. By definition, $p(\rho_0, s_0) = p_0$ and only considering the first order in the linearization assumption, a simple relation of proportionality between pressure and density fluctuations is therefore obtained.

$$p'(\boldsymbol{x}, t) = \left(\frac{\partial p}{\partial \rho}\right)_{s,0} \rho'(\boldsymbol{x}, t) \quad [1.17]$$

It is important to understand that the coefficient $(\partial p/\partial \rho)_{s,0}$ is a thermodynamic quantity characteristic of the medium in which acoustic waves propagate (but that it is independent of them, since it is evaluated at the ambient state). In thermodynamics, it is shown that this coefficient is strictly positive and it will later be seen that it is equal to the square of the propagation speed of acoustic waves (the speed of sound), denoted by c_0:

$$c_0^2 = \left(\frac{\partial p}{\partial \rho}\right)_{s,0} \quad ; \quad p' = c_0^2 \rho' \quad [1.18]$$

1.3. The wave equation

We therefore have a linear system of three first-order equations for three variables, p', ρ' and u_i'. These equations are generally rearranged to eliminate fluctuations in density and velocity to form a second-order equation for the pressure fluctuation alone, called the wave equation. To this end, the density fluctuation is eliminated using the equation of state [1.18], and then the time derivative of [1.13] is subtracted from the divergence of [1.14]. We obtain:

$$\frac{\partial^2 p'}{\partial x_i^2} - \frac{1}{c_0^2}\frac{\partial^2 p'}{\partial t^2} = \nabla^2 p' - \frac{1}{c_0^2}\frac{\partial^2 p'}{\partial t^2} = 0 \quad [1.19]$$

It can be verified that the density fluctuations ρ' and each of the components of the velocity vector fluctuations u_i' satisfy the same equation due to the linearity of the equations and the homogeneity of the medium.

When one variable is known, it is possible to evaluate the other two using the linearized equations [1.13]–[1.15] and [1.18]. The pressure field is often more directly available than other variables, either experimentally or numerically, and the linearized Euler's equation [1.14] can then be used to determine the acoustic velocity.

1.3.1. *The special case of ideal gases*

Many applications focus on acoustic propagation in air. Since air can be considered with an excellent approximation as an ideal gas under the usual pressure and temperature conditions, the following explicit form can then be used for the equation of state:

$$p = \rho r T \qquad [1.20]$$

where $r = R/M$ is the gas constant, ratio of the ideal gas constant $R = 8.314$ S.I. (J.K^{-1}.mol^{-1}) to the molar mass M of the gas under consideration.

Since the behavior is isentropic, as implied by equation [1.15], we also have the classical relation:

$$p\rho^{-\gamma} = \text{constant} = p_0\rho_0^{-\gamma} \qquad [1.21]$$

where $\gamma = c_p/c_v$ is the specific heat ratio, that is, the ratio of the specific heats at constant pressure and volume, respectively. Expanding this last relation to first order and using [1.18] yields:

$$p' = \frac{\gamma p_0}{\rho_0}\rho' = c_0^2\rho' \qquad [1.22]$$

Therefrom the following relations are derived for the speed of sound:

$$c_0 = \sqrt{\gamma\frac{p_0}{\rho_0}} = \sqrt{\gamma r T_0} \qquad [1.23]$$

For air at a temperature of 20°C (293.15 K), with $\gamma = 1.4$ and $M = 29$ g.mol^{-1}, the speed of sound is about 343 m.s^{-1}.

A simple and useful approximate formula, well adapted to the range of common temperatures, makes it possible to express this speed as a function of temperature in degrees Celsius T_c. By writing $T_0 = 273.15 + T_c$ and assuming that $T_c/T_0 \ll 1$, it follows that:

$$c_0 \approx 331 + 0.6T_c \text{ (m.s}^{-1}) \quad [1.24]$$

This formula shows that there is an (approximately) linear relation between the temperature T_c and the speed of sound. This approximation is excellent in the usual temperature range, with the relative error not exceeding 0.3% over the range 0–40°C.

The presence of an acoustic disturbance also causes a temperature fluctuation that can be calculated using the equation of state and the linearized isentropic relation. One obtains:

$$\frac{T'}{T_0} = \frac{\gamma - 1}{\gamma}\frac{p'}{p_0} \quad [1.25]$$

The ratio p'/p_0 rarely exceeds 10^{-5}, so the associated temperature variations will always be extremely small, about a few millikelvin. This formula also shows that there is a linear relationship between temperature fluctuations and pressure fluctuations. It is therefore also possible to write a wave equation similar to [1.19] for the temperature fluctuation.

1.3.2. *The velocity potential*

The linearized Euler's equation [1.14], written as:

$$\frac{\partial u_i'}{\partial t} = -\frac{1}{\rho_0}\frac{\partial p'}{\partial x_i} \quad [1.26]$$

and in vector notation as:

$$\frac{\partial \boldsymbol{u}'}{\partial t} = -\frac{1}{\rho_0}\nabla p' \quad [1.27]$$

shows that the acoustic velocity derives from a potential.

Since the curl of a gradient is zero, taking the curl of [1.27] leads to:

$$\frac{\partial}{\partial t}(\nabla \times \boldsymbol{u}') = 0 \quad [1.28]$$

from which it is deduced that $\nabla \times \boldsymbol{u}' = 0$ if the medium is initially at rest.

Therefore, there exists an *acoustic velocity potential* ϕ defined by:

$$\boldsymbol{u}' = \nabla\phi \qquad [1.29]$$

and then

$$\frac{\partial}{\partial x_i}\left(p' + \rho_0\frac{\partial\phi}{\partial t}\right) = 0 \qquad [1.30]$$

from which one can deduce:

$$p' + \rho_0\frac{\partial\phi}{\partial t} = f(t) \qquad [1.31]$$

and finally

$$p' = -\rho_0\frac{\partial\phi}{\partial t} \qquad [1.32]$$

as it is possible to choose a velocity potential such that $f(t) = 0$ without restricting the generality.

The potential ϕ is a solution to the wave equation, just as all other acoustic variables. Although it has no physical meaning of its own, it is a convenient variable to use in a number of calculations. The physical variables acoustic pressure and velocity can then be determined by a simple time and spatial differentiation, respectively.

1.3.3. *Validity conditions for the linearization of equations*

Broadly speaking, the linearization of the equations of motion assumes that the second-order terms are very small with respect to the first-order terms in each equation. We saw in section 1.1 that an order-of-magnitude analysis of Euler's equations in which the time derivatives are assumed to be equivalent to a multiplication by f and the spatial derivatives to a multiplication by $1/\lambda = f/c_0$ leads to $u'/c_0 = \mathrm{M}_a \ll 1$. By use of the linearized continuity and momentum equations, it is verified that this imposes that $\rho'/\rho_0 = \mathrm{M}_a \ll 1$ and $p'/\rho_0 c_0^2 = \mathrm{M}_a \ll 1$. For an ideal gas, the latter condition is equivalent to $p'/p_0 \ll 1$, but this is not the case for liquids, in which the condition $p'/\rho_0 c_0^2 \ll 1$ is much less restrictive than $p'/p_0 \ll 1$. As an example for water, the product $\rho_0 c_0^2$ is about 2×10^9 Pa. An underwater acoustic wave of amplitude 1 atm (10^5 Pa) is therefore still a "small" signal for which the linearization process remains valid.

1.4. Acoustic energy, acoustic intensity and source power

Since the fluid is assumed to be perfect, there is no mechanism in a volume of fluid that can dissipate energy. Thus, in the absence of energy creation from volume or surface sources (which we will introduce later) and of absorbing materials covering surfaces, the total energy and that associated with fluctuations are conserved.

However, considering only linearized equations to construct an equation of conservation of energy, which involves second-order quantities, raises theoretical problems. Strictly speaking, a general equation for the total energy has to be written and then expanded up to second order (see [PIE 89], section I.1). In the following, we will however use the classic approach, based on a combination of the first-order equations, which yields the correct result.

1.4.1. *Definition of acoustic energy and acoustic intensity*

To construct an acoustic energy conservation equation, we form the product of the momentum equation by the acoustic velocity, that is symbolically the product of a force by a velocity (and therefore a power):

$$\rho_0 u_i' \frac{\partial u_i'}{\partial t} = -u_i' \frac{\partial p'}{\partial x_i} = -\frac{\partial (p' u_i')}{\partial x_i} + p' \frac{\partial u_i'}{\partial x_i} \qquad [1.33]$$

Using the linearized continuity equation and eliminating density fluctuations in favor of pressure fluctuations leads to:

$$\frac{\partial}{\partial t}\left(\frac{1}{2}\rho_0 u_i'^2 + \frac{p'^2}{2\rho_0 c_0^2}\right) + \frac{\partial (p' u_i')}{\partial x_i} = 0 \qquad [1.34]$$

This equation has indeed the general form of a conservation equation that can be written in a condensed manner as:

$$\frac{\partial e}{\partial t} + \nabla \cdot \boldsymbol{I} = 0 \qquad [1.35]$$

with

$$e = \frac{1}{2}\rho_0 u_i'^2 + \frac{p'^2}{2\rho_0 c_0^2} \qquad [1.36]$$

$$\boldsymbol{I} = p' \boldsymbol{u}' \qquad [1.37]$$

The quantity e is called the acoustic energy density. The first term of its expression clearly corresponds to a kinetic energy density; the second can be interpreted as a potential energy density associated with the compression-expansion that the medium experiences when an acoustic wave passes through it. Its form is indeed similar to that of the potential energy of a spring $\frac{1}{2}kx^2$, if the following analogy is introduced: the pressure disturbance p' is identified with the deviation x from the equilibrium position of the spring and $1/\rho_0 c_0^2$ with the stiffness k of the spring. It should also be noted that $1/\rho_0 c_0^2$ is none other than the adiabatic compressibility κ_s of the medium:

$$\kappa_s = -\frac{1}{V^*}\left(\frac{\partial V^*}{\partial p}\right)_{s,0} = \frac{1}{\rho_0}\left(\frac{\partial \rho}{\partial p}\right)_{s,0} = \frac{1}{\rho_0 c_0^2} \qquad [1.38]$$

where V^* denotes the specific volume, the inverse of the density ρ.

$\boldsymbol{I}$ is called the acoustic intensity (vector[2]). This vector is associated with the flux of acoustic energy passing through a surface of unit area, as shown in the integral version of equation [1.35]. If this equation is integrated over a fixed control volume D, bounded by a surface Σ whose external unit normal is denoted $\boldsymbol{n}$ (Figure 1.1(a)), it follows that:

$$\int_D \left(\frac{\partial e}{\partial t} + \nabla \cdot \boldsymbol{I}\right) \mathrm{d}V = 0 \qquad [1.39]$$

and using the divergence theorem (or Gauss' theorem):

$$\frac{\partial}{\partial t}\int_D e \,\mathrm{d}V + \int_\Sigma \boldsymbol{I} \cdot \boldsymbol{n} \,\mathrm{d}\Sigma = 0 \qquad [1.40]$$

This integral form shows that the change over time in the total acoustic energy contained in the volume D is the opposite of the energy flux crossing its boundary Σ.

The acoustic intensity, whose normal component measures the energy transported per unit area and time, is a key quantity for analyzing energy transfers from one spatial region to another and for defining the notion of an acoustic source.

1.4.2. *Acoustic sources*

The acoustic energy conservation equation [1.40] shows that, on average, the flux of acoustic energy through a closed surface is zero.

2 The term *vector* will often be omitted in the following.

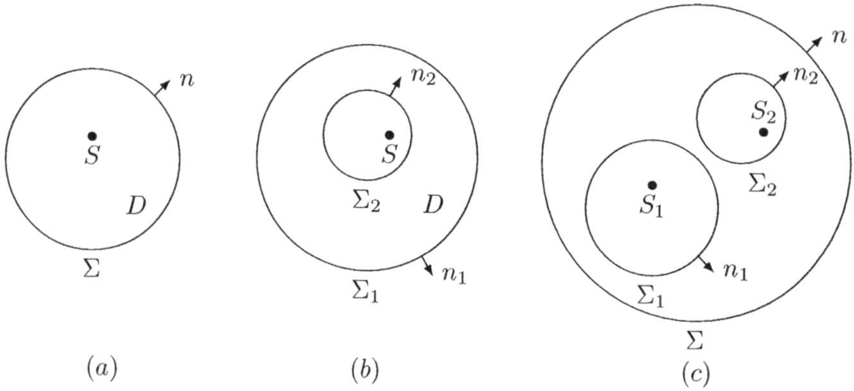

Figure 1.1. *Integration surfaces used to define the power radiated by one or several sources*

We define a time-averaged quantity, for instance, the average intensity denoted $\boldsymbol{I}_{av}$, by:

$$\boldsymbol{I}_{av} = \lim_{T\to\infty} \frac{1}{T} \int_0^T \boldsymbol{I}(t)\,\mathrm{d}t \qquad [1.41]$$

By integrating [1.40] with respect to time, one immediately obtains:

$$\int_\Sigma \boldsymbol{I}_{av} \cdot \boldsymbol{n}\,\mathrm{d}\Sigma = \lim_{T\to\infty} \frac{E(T) - E(0)}{T} = 0 \qquad [1.42]$$

where E designates the total acoustic energy contained in domain D, which is a finite quantity.

The existence of an acoustic source is associated with a breakdown in the conservation properties of mass, momentum or energy, and we will examine several examples later (see Chapter 5). For the time being, and without making any assumption about the nature of the physical phenomena involved, we will say that there exists in D a source of average power W_{av} if:

$$\int_\Sigma \boldsymbol{I}_{av} \cdot \boldsymbol{n}\,\mathrm{d}\Sigma = W_{av} > 0 \qquad [1.43]$$

As long as Σ surrounds the same single source S, the power W_{av} is independent of the position, shape and size of the integration surface, because the divergence of

$\boldsymbol{I}_{av}$ is zero outside of Σ. This very important result is illustrated in Figure 1.1(b). As such:

$$\int_{\Sigma_1} \boldsymbol{I}_{av} \cdot \boldsymbol{n}\, \mathrm{d}\Sigma_1 - \int_{\Sigma_2} \boldsymbol{I}_{av} \cdot \boldsymbol{n}\, \mathrm{d}\Sigma_2 = \int_{\Sigma} \boldsymbol{I}_{av} \cdot \boldsymbol{n}\, \mathrm{d}\Sigma = 0 \qquad [1.44]$$

where $\boldsymbol{n}$ denotes the unit normal to each of the surfaces under consideration, and Σ delimits the volume comprised between Σ_1 and Σ_2, which does not contain any source. This property is often taken advantage of to choose the integration surface that leads to the simplest evaluation of the power of a source; the most common choice is that of a sphere centered on the source and whose radius tends to infinity.

Another important property is that when several sources are involved, the total power is simply the sum of the powers radiated by each of the sources:

$$W_{\text{total}} = \sum_i W_{av_i} = \sum_i \int_{\Sigma_i} \boldsymbol{I}_{av} \cdot \boldsymbol{n}\, \mathrm{d}\Sigma_i \qquad [1.45]$$

where the surface Σ_i is any surface surrounding the source S_i and only that one. The total power can also be evaluated by integrating the normal intensity on a surface surrounding all the sources, as also shown in Figure 1.1(c).

It should be noted, however, that this does not mean that the power radiated by the source S_i is independent of the presence of other sources. What is evaluated by integration on Σ_i is the radiated power from the source S_i *when other sources are present*. In some cases, the change in radiation due to other existing sources will be negligible, but this is not the case in general. As a simple example of the possible changes of the power emitted by a source depending on its environment, the case of a point source placed above a rigid plane will be developed later in detail (section 4.7).

1.5. Harmonic waves

1.5.1. *Definition of harmonic waves*

The time description of waves is the most natural, but it often encounters practical difficulties, especially when introducing absorbing materials. It is therefore often convenient to consider waves whose time dependence is very simple and of sinusoidal form, called harmonic waves. For example, let:

$$p'(\boldsymbol{x}, t) = A(\boldsymbol{x}) \cos(\omega t + \varphi) \qquad [1.46]$$

where ω is the angular frequency, $\omega = 2\pi f$, φ is the phase at $t = 0$ and A is the wave amplitude. The use of the complex representation makes the writing of equations and

their solution much easier. It consists of looking for the pressure fluctuation (as an example) in the form:

$$p'(\boldsymbol{x},t) = \mathrm{Re}(p(\boldsymbol{x})\exp(+\mathrm{i}\omega t)) \qquad [1.47]$$

where Re denotes the real part of a complex number[3]. The quantity $p(\boldsymbol{x})$ is called the *complex amplitude* of the pressure fluctuation. It is related to the real amplitude A and the phase φ by $p = A\exp(\mathrm{i}\varphi)$.

The use of harmonic forms and their complex representation is particularly convenient because of the linear nature of acoustic equations. The superposition principle therefore applies, and if a signal involves several frequencies, the contribution of each one can be independently computed and simply added together to obtain the desired result.

Very generally speaking, the search for harmonic solutions can be compared to that of the Fourier components of a time signal of any form $s'(t)$. The direct transform of $s'(t)$, which corresponds to the decomposition of the signal into elementary components, is denoted $s(\omega)$ or $s(f)$. The inverse transform makes it possible to reconstruct the time signal from its frequency components; this is often referred to as Fourier synthesis.

We define direct and inverse Fourier transforms by the following formulas[4]:

$$s(\omega) = \frac{1}{2\pi}\int_{-\infty}^{\infty} s'(t)\exp(-\mathrm{i}\omega t)\mathrm{d}t\,; \qquad s(f) = \int_{-\infty}^{\infty} s'(t)\exp(-\mathrm{i}2\pi ft)\mathrm{d}t \qquad [1.48]$$

$$s'(t) = \int_{-\infty}^{\infty} s(\omega)\exp(\mathrm{i}\omega t)\mathrm{d}\omega = \int_{-\infty}^{\infty} s(f)\exp(\mathrm{i}2\pi ft)\mathrm{d}f \qquad [1.49]$$

The equations governing harmonic acoustic variables are very easily obtained from the equations previously established in the time domain. We will illustrate the procedure using the linearized continuity equation [1.13]:

$$\frac{\partial\rho'}{\partial t} + \rho_0\frac{\partial u_i'}{\partial x_i} = 0 \qquad [1.50]$$

3 The choice of sign in the complex exponential is arbitrary and differs with authors. If the minus sign is employed instead of the plus sign used in this book, the sign of all the imaginary parts of the complex quantities will have to be changed.

4 Different definitions are sometimes used in which the factor 2π is placed differently.

that gives immediately:

$$\mathrm{Re}\left[\left(\mathrm{i}\omega\rho + \rho_0\frac{\partial u_i}{\partial x_i}\right)\exp(\mathrm{i}\omega t)\right] = 0 \qquad [1.51]$$

where ρ and u_i are the complex density and velocity fluctuation amplitudes, respectively. Since this equation must be satisfied for all values of t, it follows that:

$$\mathrm{i}\omega\rho + \rho_0\frac{\partial u_i}{\partial x_i} = 0 \qquad [1.52]$$

In turn, the linearized Euler's equation [1.14] is transformed into:

$$\mathrm{i}\rho_0\omega u_i + \frac{\partial p}{\partial x_i} = 0 \qquad [1.53]$$

and we also have the relation $p = c_0^2\rho$.

These equations can be rearranged so as to eliminate the velocity and density fluctuations, which leads to the *Helmholtz equation*:

$$\nabla^2 p + \frac{\omega^2}{c_0^2}p = 0 \quad \text{or} \quad \nabla^2 p + k_0^2 p = 0 \qquad [1.54]$$

where we introduced the acoustic wavenumber $k_0 = \omega/c_0$. This last equation could of course have been obtained more directly by taking the Fourier transform of the wave equation.

It should be noted that the solution to these frequency domain equations provides a complex-valued field, from which the real part will have to be taken after multiplying it by $\exp(+\mathrm{i}\omega t)$ in order to reintroduce real variables, as they can be measured.

The use of equations in the frequency domain is very attractive, because of the simplification brought about as a result of eliminating the time derivatives which are replaced by a simple multiplication by $\mathrm{i}\omega$. The Fourier synthesis then yields the solution in the time domain. However, the Helmholtz equation must be solved for a sufficiently large number of frequencies to allow for a good representation of the most energetic part of the signal, which can be very restrictive for "broadband" signals. One should also keep in mind that a harmonic wave is an idealization of the effective behavior of signals due to its time extension from minus infinity to plus infinity; a real signal always has a beginning and an end and then cannot be strictly periodic. The supposed existence of a signal of infinite duration also poses a problem due to the

fact that the physical notion of causality has disappeared (the cause must precede the effect). In some specific situations, it will be necessary to reintroduce this causality condition through an appropriate formulation; this is particularly the case with the Sommerfeld condition, which we will describe later on (section 6.3).

1.5.2. *Average acoustic energy and acoustic intensity*

The expressions of the acoustic energy and the acoustic intensity in the frequency domain can be quite easily inferred from the time-domain formulations. It is nonetheless necessary to specify the calculus rules associated because products of fluctuating quantities have to be addressed. First of all, it should be noted that there is no interest, in general, in observing the changes in these energy quantities over time, and that only their average values are of use. These averages can be calculated equivalently over a signal period, or as a limit over a very long duration (with respect to this period) according to definition [1.41]. It can be easily verified that:

$$e_{av} = \frac{1}{4}\rho_0 u_i u_i^* + \frac{1}{4\rho_0 c_0^2} pp^* \qquad [1.55]$$

where p^* and u_i^*, respectively, denote the complex conjugates of p and u_i.

For the average (or mean) intensity, the calculation is a little more subtle. If we briefly reconsider the real forms:

$$p'(t) = A\cos(\omega t + \varphi) \qquad [1.56]$$

$$u_i'(t) = B_i\cos(\omega t + \varphi') \qquad [1.57]$$

it follows that:

$$\begin{aligned} I_{av_i} &= \lim_{T\to\infty}\frac{1}{T}\int_0^T p'u_i'\mathrm{d}t = \lim_{T\to\infty}\frac{1}{T}\int_0^T AB_i\cos(\omega t + \varphi)\cos(\omega t + \varphi')\mathrm{d}t \\ &= \frac{AB_i}{2}\lim_{T\to\infty}\frac{1}{T}\int_0^T [\cos(2\omega t + \varphi + \varphi') + \cos(\varphi - \varphi')]\mathrm{d}t \\ &= \frac{AB_i}{2}\cos(\varphi - \varphi') \end{aligned}$$

In complex notation, it follows that:

$$I_{av_i} = \frac{1}{2}\mathrm{Re}(pu_i^*) = \frac{1}{4}\mathrm{Re}(pu_i^* + p^* u_i) \qquad [1.58]$$

As previously, the average power of a source is given by:

$$W_{av} = \int_{\Sigma} \boldsymbol{I}_{av} \cdot \boldsymbol{n} \, \mathrm{d}\Sigma \quad [1.59]$$

for any surface Σ surrounding this source and only this source.

1.6. Boundary conditions

The equations that describe acoustic disturbances within a fluid must be complemented by the introduction of boundary conditions and, in time description, of initial conditions.

1.6.1. *Fluid–solid and fluid–fluid interfaces*

For a perfect fluid, the conditions that have to be satisfied when in contact with a rigid solid reflect the impermeability of the solid (Figure 1.2):

$$\boldsymbol{U}_f \cdot \boldsymbol{n} = \boldsymbol{U}_s \cdot \boldsymbol{n} \quad [1.60]$$

where $\boldsymbol{U}_f$ and $\boldsymbol{U}_s$ designate the velocity of a point M of the interface, considered to belong to the fluid and to the solid, respectively, and $\boldsymbol{n}$ is the unit normal vector pointing out of the solid at point M. There are no restrictions on the tangential component of the relative velocity $\boldsymbol{U}_f - \boldsymbol{U}_s$, which will be non-zero in general.

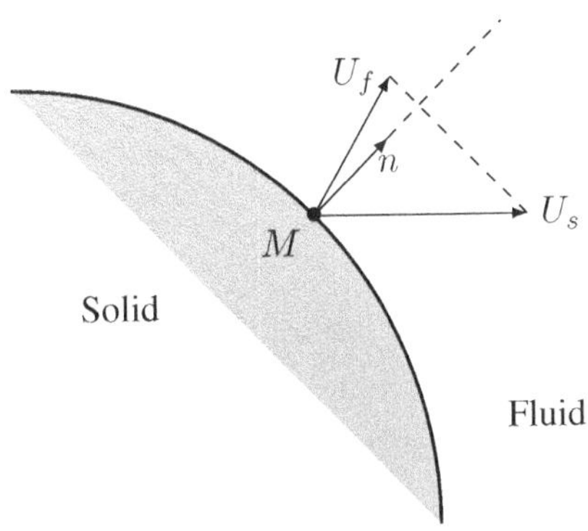

Figure 1.2. *Boundary conditions at the interface between an impermeable solid and a perfect fluid*

If the solid is at rest (in the chosen Galilean frame of reference) and in the approximation of linear acoustics, the condition to be imposed is reduced to a zero

normal acoustic velocity or, equivalently by application of Euler's equation, to a zero normal pressure gradient:

$$\boldsymbol{u}' \cdot \boldsymbol{n} = u'_n = 0 \quad \Leftrightarrow \quad \frac{\partial p'}{\partial n} = 0 \qquad [1.61]$$

If the solid vibrates, the amplitude of the vibrations remaining small, the condition to be imposed is as follows:

$$u'_n = \boldsymbol{u}'_s \cdot \boldsymbol{n} \qquad [1.62]$$

where $\boldsymbol{u}'_s$ is the vibration velocity of the solid.

These continuity conditions have to be applied *on the average position of the interface* between the fluid and the solid. This results from a kind of linearization of the boundary condition over the instantaneous position of the interface that is consistent with the linearization of the equations of motion. The boundary condition is therefore imposed on a (often fixed) time-independent surface and not on the actual position of the interface. This assumption is justified as long as the amplitude of the vibration l remains very small compared to the acoustic wavelength[5]:

$$l \ll \lambda \qquad [1.63]$$

At the interface between two immiscible fluids, the continuity condition of the normal velocity, written on its average position, is also applicable. To this must be added the condition of continuity of the average pressure[6] and of the acoustic pressure fluctuation in each of the fluids:

$$p'_1 = p'_2 \qquad [1.64]$$

For the interface between a perfect fluid and an elastic solid, the continuity of the normal stress must be added to the continuity of the normal velocity. The existence of several types of waves in the solid, compression and shear waves, makes the exercise of writing these conditions relatively complex but without major difficulty. Despite its practical interest, for example, for applications in non-destructive testing by ultrasonic waves or in seismology, we will not address this scenario and refer to the book [ROY 22].

5 See the exercises in Chapter 2.

6 The phenomena of surface tension are very generally negligible in acoustics; the exception are very small bubbles of gas in a liquid, with a radius of less than 10^{-5}m.

1.6.2. *Specific acoustic impedance*

The conditions previously described apply to impermeable surfaces. In acoustics, porous materials are often used, especially to reduce sound levels by partially absorbing acoustic waves. The natural materials encountered during the propagation of sound in the atmosphere, such as vegetated soils, are also permeable to acoustic waves. For these materials, there is a flow of matter crossing the interface and the impermeability condition $u_n = 0$ must be replaced by a new boundary condition.

A commonly used assumption, and often well verified in practice, consists of assuming the existence of a linear relation between the pressure fluctuation p' in the vicinity of the interface and the normal component of the velocity fluctuation, $\boldsymbol{u}' \cdot \boldsymbol{n}_{in} = u'_{in}$, the unit normal $\boldsymbol{n}_{in}$ being conventionally oriented outwards the fluid, that is, inwards the material under consideration. Since the acoustic characteristics of materials often vary greatly with frequency, the description of their properties is usually introduced in the frequency domain.

The specific acoustic impedance of a surface[7] Z is defined by the relation:

$$Z(\omega) = \left(\frac{p}{u_{in}}\right)_\Sigma \qquad [1.65]$$

where the subscript Σ indicates that the quantity is evaluated on the surface Σ. The acoustic impedance Z is a complex number. Its real part is called the acoustic resistance and its imaginary part is called the acoustic reactance. The inverse of Z is the specific acoustic admittance; its real part is called acoustic conductance.

The purpose of introducing impedance as a boundary condition is to avoid a precise description of what happens within a more or less complicated "solid" medium (such as grassy soil, for instance). Such an approach is necessarily reductive, but from a practical point of view it is the one that will be used in the vast majority of applications. To solve a propagation problem above a porous surface, the value of $Z(\omega)$ will of course have to be known, which will be possible either from a model of the propagation of acoustic waves inside the material or from characterization measurements of Z.

We have indicated that the presence of a porous material is often intended to reduce sound levels. To clarify this point, it proves useful to evaluate the average energy flux through a surface Σ characterized by its impedance. This flux is calculated as:

$$\int_\Sigma \boldsymbol{I}_{av} \cdot \boldsymbol{n}_{in} \, \mathrm{d}\Sigma = \int_\Sigma \frac{1}{2} \mathrm{Re}\left(p u_{in}^*\right) \, \mathrm{d}\Sigma = \frac{1}{2} \int_\Sigma \mathrm{Re}(Z) u_{in} u_{in}^* \, \mathrm{d}\Sigma \qquad [1.66]$$

7 In mechanics, an impedance is defined as the ratio of a force to a velocity. Since acoustic impedance is the ratio of a force per unit area to a velocity, the term *specific* is thus added.

This flux corresponds to a reduction in the energy contained in the fluid domain if $\text{Re}(Z) > 0$, and thus to energy transmitted to the porous medium. It can be seen as the energy *absorbed* by the porous solid. Absorption per unit area is expressed as:

$$\frac{1}{2}\text{Re}(Z) u_{in} u_{in}^* \qquad [1.67]$$

It is proportional to the acoustic resistance of the solid $\text{Re}(Z)$. A "passive" interface (which cannot create energy) will therefore be associated with a positive (or possibly zero) acoustic resistance. On the other hand, no constraint is imposed on the imaginary part of the surface impedance.

1.7. Exercises

1) **Isothermal speed of sound**

Newton made the hypothesis (which later turned out to be wrong) that acoustic fluctuations occurred isothermally. Reuse the linearization of the equation of state written this time in the form $p = p(\rho, T)$, and infer the expression for the isothermal speed of sound c_{0_T}. Determine the value of c_{0_T} for propagation in air under standard conditions.

2) **Speed of sound in a gas mixture**

a) Determine the speed of sound in a mixture of helium (25%) and hydrogen (75%) at 300 K and room pressure (see Table A1.1).

b) Dry air consists of approximately 78% nitrogen, 21% oxygen and 1% Argon. From this composition, determine the speed of sound in air at room pressure and at 0°C, and compare with the value given by formula [1.24].

3) **Wave equation in a fluid in uniform motion**

Consider an acoustic wave propagating in a homogeneous fluid in uniform motion (the magnitude of the flow velocity is denoted by U_0). To simplify the notation, the x_1 axis is aligned with the fluid velocity.

a) Write the linearized equations (mass, momentum and entropy) in the fixed coordinate system. Therefrom, deduce that the pressure fluctuation satisfies a *convected* wave equation in which the material derivative $D/Dt = \partial/\partial t + U_0 \partial/\partial x_1$ replaces the partial derivative $\partial/\partial t$.

b) In the Galilean reference frame moving at velocity U_0, it is clear that the acoustic waves satisfy the classical wave equation expressed with the "moving" coordinates,

that is $x_1^* = x_1 - U_0 t^*$, the other coordinates remaining unchanged. Return to the fixed coordinate system to obtain the convected wave equation above.

c) Use the approach of the previous question to obtain the equation of acoustic energy conservation in the fixed frame, which can also be written in the form $\partial e/\partial t + \nabla \cdot (\boldsymbol{I} + e\, U_0 \boldsymbol{x}_1) = 0$. Interpret this result.

2

Plane Waves and Spherical Waves

Both the wave equation and the Helmholtz equation in Chapter 1 are three-dimensional in nature. Before looking for general solutions to these equations, it proves very useful and informative to characterize wave propagation in simpler, one-dimensional situations. The linear nature of the equations makes it possible to build very general solutions from these simple solutions using the superposition principle.

Two types of one-dimensional waves are particularly important, plane waves and spherical waves, which we will describe in detail. Cylindrical waves are a third type of somewhat more limited practical interest, which we will address more briefly.

When outlining these three different types of waves, it is clear that each will be associated with an appropriate orthogonal coordinate system, namely Cartesian, spherical or cylindrical.

We will begin with the study of plane waves in the time domain and then in the frequency domain. They are the simplest to study and their interest will also be apparent when studying spherical or cylindrical waves, and even when considering waves of a more general nature.

2.1. Plane waves

2.1.1. *Plane waves in the time domain*

We are looking for solutions to the wave equation [1.19] written in Cartesian coordinates when the acoustic variables depend on a single spatial variable, x_1, and on time t:

$$p'(\boldsymbol{x},t) = p'(x_1,t) \qquad [2.1]$$

The wave equation then becomes:

$$\frac{\partial^2 p'}{\partial x_1^2} - \frac{1}{c_0^2}\frac{\partial^2 p'}{\partial t^2} = 0 \qquad [2.2]$$

and it can be factored as:

$$\left(\frac{\partial}{\partial x_1} - \frac{1}{c_0}\frac{\partial}{\partial t}\right)\left(\frac{\partial}{\partial x_1} + \frac{1}{c_0}\frac{\partial}{\partial t}\right) p'(x_1, t) = 0 \qquad [2.3]$$

This factorization suggests the introduction of new variables combining spatial and time dependencies:

$$t_r = t - \frac{x_1}{c_0}$$
$$t_a = t + \frac{x_1}{c_0} \qquad [2.4]$$

With these variables, the wave equation can be reduced to:

$$-\frac{4}{c_0^2}\frac{\partial}{\partial t_r}\left(\frac{\partial}{\partial t_a} p'(t_r, t_a)\right) = 0 \qquad [2.5]$$

whose general solution is the sum of two arbitrary functions depending on t_r and t_a, respectively. If we re-introduced the variables x_1 and t, we thus have:

$$p'(x_1, t) = f(t - \frac{x_1}{c_0}) + g(t + \frac{x_1}{c_0}) \qquad [2.6]$$

An analogous form can be obtained for the acoustic velocity whose only non-zero component is carried by the Ox_1 axis. The linearized continuity and Euler's equations can be written as:

$$\frac{\partial p'}{\partial t} + \rho_0 c_0^2 \frac{\partial u_1'}{\partial x_1} = 0$$
$$\rho_0 \frac{\partial u_1'}{\partial t} + \frac{\partial p'}{\partial x_1} = 0 \qquad [2.7]$$

where the linear relation $p' = c_0^2 \rho'$ has been used. These equations can be combined as:

$$\rho_0 c_0 \left(\frac{\partial}{\partial t} \pm c_0 \frac{\partial}{\partial x_1}\right) u_1' = \mp \left(\frac{\partial}{\partial t} \pm c_0 \frac{\partial}{\partial x_1}\right) p' \qquad [2.8]$$

or, using the variables t_r and t_a:

$$\begin{aligned} \frac{\partial}{\partial t_r}(\rho_0 c_0 u_1' - p') = 0 \\ \frac{\partial}{\partial t_a}(\rho_0 c_0 u_1' + p') = 0 \end{aligned} \qquad [2.9]$$

from which it can be deduced that:

$$u_1'(x_1, t) = \frac{1}{\rho_0 c_0}\left(f\left(t - \frac{x_1}{c_0}\right) - g\left(t + \frac{x_1}{c_0}\right)\right) \qquad [2.10]$$

The presence of the combinations $t_r = t - x_1/c_0$ and $t_a = t + x_1/c_0$ is fundamental to the interpretation of these solutions in terms of wave propagation. In the following, the first of these combinations will be called *retarded time* and the second *advanced time*. Consider the signal f associated with the retarded time, observed at two successive times t_1 and t_2. Since time and space only intervene through the $(t - x_1/c_0)$ combination, the shape and amplitude of the wave at time t_2 and at a position $X_2 = X_1 + c_0(t_2 - t_1)$ will be identical to those observed at time t_1 at position X_1:

$$f\left(t_2 - \frac{X_2}{c_0}\right) = f\left(t_1 - \frac{X_1}{c_0}\right) \quad \text{when} \quad X_2 = X_1 + c_0(t_2 - t_1) \qquad [2.11]$$

Figure 2.1 shows the evolution of the wave along the Ox_1 direction when two snapshots are taken at the times t_1 and $t_2 > t_1$. The velocity of the motion of the wave pattern is none other than the speed of sound c_0 since according to [2.11]:

$$\frac{X_2 - X_1}{t_2 - t_1} = c_0 \qquad [2.12]$$

Similarly, the part of the signal associated with the advanced time also corresponds to a wave traveling at speed c_0, but now in direction $-x_1$. The interpretation of the thermodynamic quantity c_0 as being the speed of sound in the medium is thus fully justified.

In many situations, we will consider waves propagating in only one direction, described by the $f(t - x_1/c_0)$ part of the expression [2.6]. In this case, the acoustic velocity fluctuations can be evaluated from the knowledge of the pressure fluctuations alone, and:

$$u_1' = \frac{p'}{\rho_0 c_0} \qquad [2.13]$$

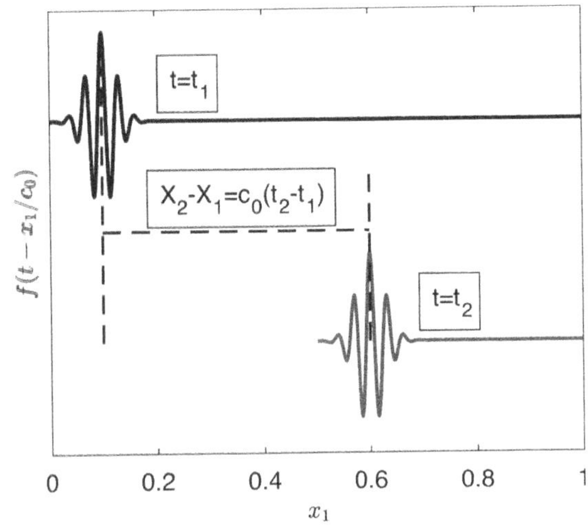

Figure 2.1. *Patterns at two successive instants of a plane wave propagating in the $+x_1$ direction*

The proportionality factor between pressure and velocity fluctuations, $Z_0 = \rho_0 c_0$, is called the *characteristic impedance* of the medium. Two values of this impedance are of particular importance: for air, we have $Z_{\text{air}} \approx 400 \text{ kg} \cdot \text{m}^{-2}\text{s}^{-1}$ or rayls, and for water $Z_{\text{water}} \approx 10^6$ rayls.

When a single plane wave propagates in any direction characterized by a unit normal $\boldsymbol{n}$, the previous expressions take the following form:

$$p'(\boldsymbol{x}, t) = f\left(t - \frac{\boldsymbol{x} \cdot \boldsymbol{n}}{c_0}\right) \qquad [2.14]$$

$$\boldsymbol{u}'(\boldsymbol{x}, t) = \frac{p'}{\rho_0 c_0} \boldsymbol{n} \qquad [2.15]$$

These expressions show that the velocity fluctuation and the direction of propagation of the wave are collinear. Such a wave is said to be longitudinal. It is also a compression wave, since its propagation is associated with density fluctuations by the proportionality relation $\rho' = p'/c_0^2$.

The simplicity of the proportionality relation between pressure and velocity for a single traveling plane wave should not, however, hide the fact that when several of these waves coexist, no simple direct relation can be written between these variables. It is then necessary to use the linearized momentum equation to express the velocity fluctuation from the knowledge of the pressure fluctuations.

The acoustic intensity and acoustic energy can then be determined. When a single plane wave is present, both the acoustic intensity and the energy density are expressed in terms of the pressure fluctuation alone, which is very interesting from a practical point of view, because the acoustic pressure is much easier to measure than the acoustic velocity:

$$e = \frac{p'^2}{\rho_0 c_0^2} \tag{2.16}$$

$$\boldsymbol{I} = p'\boldsymbol{u}' = \frac{p'^2}{\rho_0 c_0}\boldsymbol{n} \tag{2.17}$$

It should be also noted that the kinetic energy and the potential energy are equal at each instant, and that the acoustic intensity and the energy density are related by:

$$\boldsymbol{I} = e c_0 \boldsymbol{n} \tag{2.18}$$

which shows that the acoustic energy is transported at the speed of sound in the direction of the wave propagation.

2.1.2. *Harmonic plane waves*

The great simplicity of plane waves is further enhanced when considering harmonic signals, which satisfy the one-dimensional Helmholtz equation:

$$\frac{\mathrm{d}^2 p}{\mathrm{d}x_1^2} + k_0^2 p = 0\,; \qquad k_0 = \frac{\omega}{c_0} \tag{2.19}$$

whose solution is:

$$p(x_1) = A\exp(-\mathrm{i}k_0 x_1) + B\exp(+\mathrm{i}k_0 x_1) \tag{2.20}$$

where A and B are two complex constants that will be determined when the boundary conditions of the problem are specified. Since the pressure fluctuation is related to the complex amplitude by:

$$p'(x_1, t) = \mathrm{Re}\{p(x_1)\exp(+\mathrm{i}\omega t)\} \tag{2.21}$$

it can be seen that $A\exp(-\mathrm{i}k_0 x_1)$ corresponds to the wave that propagates toward the $+x_1$ direction because:

$$p'(x_1, t) = \mathrm{Re}\{A\exp(+\mathrm{i}(\omega t - k_0 x_1))\} = \mathrm{Re}\{A\exp(+\mathrm{i}\omega(t - x_1/c_0))\} \tag{2.22}$$

and symmetrically $B\exp(+\mathrm{i}\ k_0 x_1)$ represents a wave propagating in the $-x_1$ direction. A single harmonic plane wave propagating in any $\boldsymbol{n}$ direction takes the form:

$$p(x_1) = A\exp(-\mathrm{i}\boldsymbol{k}\cdot\boldsymbol{x}) \qquad [2.23]$$

where the acoustic wave vector $\boldsymbol{k}$ is defined by $\boldsymbol{k} = k_0\boldsymbol{n}$.

Harmonic plane waves are therefore periodic in both time and space. The time period is $T = 2\pi/\omega$, while the spatial period measured in the direction of propagation, and called the wavelength, is given by $\lambda = c_0 T = c_0/f = 2\pi/k_0$. The spatial structure of the wave is shown in Figure 2.2, where the propagation takes place in the (x_1, x_2) plane along the $\boldsymbol{n}$ direction. Of particular note is the presence of planar wavefronts perpendicular to the direction of propagation. In each of these planes, the phase $\varphi = \boldsymbol{k}\cdot\boldsymbol{x}$ and the amplitude A of the wave are constant. It can also be seen that the spatial period along the x_1 direction (for example) is $\lambda/\cos\theta$. The apparent propagation velocity of the wave in this direction is therefore $c_0/\cos\theta$. This *phase velocity* or *trace-velocity* will be used in particular when establishing the conditions for reflection and transmission at the interface between two fluids (Snell's law), which will be discussed in section 4.4).

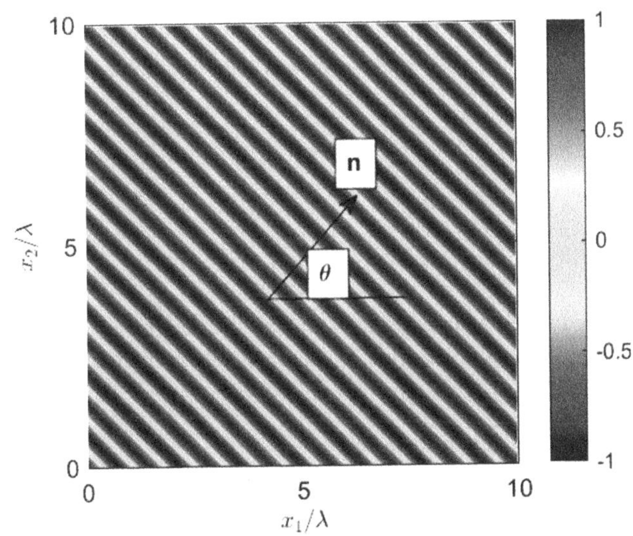

Figure 2.2. *Harmonic plane traveling wave of unit amplitude. The plot of the real part of the complex amplitude* p *highlights the spatial periodicity as well as the surfaces of constant phase, perpendicular to the direction of propagation of the wave. For a color version of this figure, see www.iste.co.uk/juve/acoustics1.zip*

Two propagation velocities associated with wave propagation are often used. The phase velocity in a given direction x_i is defined by:

$$v_{\varphi_i} = \frac{\omega}{k_i} \qquad [2.24]$$

where k_i is the projection of the wave vector in this direction[1]. As shown earlier by an elementary reasoning, $v_{\varphi_1} = c_0/\cos\theta$. Note that the phase velocity does not behave like a vector, unlike its inverse the *wave slowness*, which is often used in seismology.

The group velocity, often associated with the propagation of energy, is defined by:

$$v_{g_i} = \left(\frac{\partial\omega}{\partial k_i}\right)_{\boldsymbol{x}} \qquad [2.25]$$

a formula in which the differentiation is performed for a fixed position $\boldsymbol{x}$, and where the angular frequency has been expressed as a function of the wave vector and, if the medium is not homogeneous, of the position:

$$\omega = \omega(\boldsymbol{k}, \boldsymbol{x}) \qquad [2.26]$$

In the homogeneous situation that we are addressing here, the dispersion relation is simply written as:

$$\omega^2 = c_0^2 k^2 \qquad [2.27]$$

from which can be immediately deduced:

$$\boldsymbol{v}_g = c_0 \boldsymbol{n} \qquad [2.28]$$

This is indeed in accordance with the earlier observation that acoustic energy propagates at the speed of sound in the direction of wave propagation.

Geometrically, the dispersion relation [2.27] shows that the extremity of the wave vector $\boldsymbol{k}$ describes a sphere whose radius $k = k_0 = \omega/c_0$ increases linearly with the frequency; this sphere is called the radiation sphere. This structure is a consequence of the isotropic nature of fluids at rest, where the wave velocity does not depend on the direction of propagation. This property is preserved in inhomogeneous fluids, but is altered in the presence of a non-zero mean flow which breaks the isotropy.

1 When the wave number is complex, the phase velocity is defined by $\omega/\mathrm{Re}(k_i)$.

2.1.3. *Evanescent plane waves*

The dispersion relation of the Helmholtz equation, $\omega^2 = c_0^2 k^2$, is satisfied by any plane wave with the form $A\exp(-\mathrm{i}\boldsymbol{k}\cdot\boldsymbol{x})$, where the components of the wave vector $\boldsymbol{k}$ are real and its magnitude is $k_0 = \omega/c_0$. These waves can be described as traveling homogeneous plane waves.

It is possible to define a second family of plane waves that satisfy the same dispersion relation, but in which one of the components of the wave vector, k_3 for example, is purely imaginary:

$$\boldsymbol{k} = \left| \begin{array}{l} k_1 \\ k_2 \\ k_3 = \pm\sqrt{(k_0^2 - k_1^2 - k_2^2)} = \mathrm{i}\kappa\,; \quad k_1^2 + k_2^2 > \frac{\omega^2}{c_0^2} = k_0^2 \end{array} \right. \qquad [2.29]$$

It then follows that:

$$p(\boldsymbol{x}) = A\exp(\pm\kappa x_3)\exp(-\mathrm{i}(k_1x_1 + k_2x_2))\,; \quad \kappa = \sqrt{(k_1^2 + k_2^2 - k_0^2)} \qquad [2.30]$$

Such a plane wave is said to be *inhomogeneous*. When considering propagation in the half-space $x_3 \geq 0$, only the decreasing exponential $\exp(-\kappa x_3)$ is physically acceptable. We refer to this plane wave as *evanescent* in the x_3 direction. The amplitude and the phase of the wave are constant in any plane orthogonal to x_3, hence the term plane wave; on the other hand, the amplitude of the wave decreases exponentially when x_3 increases, hence the name inhomogeneous wave. Evanescent waves are particularly important when studying the reflection and transmission of waves at a fluid–fluid interface or the radiation of sound from vibrating structures.

2.1.4. *Angular spectrum of plane waves*

A plane wave $\exp(-\mathrm{i}\boldsymbol{k}\cdot\boldsymbol{x})$ can be thought of as a spatial Fourier component of a more complex acoustic field $p(\boldsymbol{x})$. However, a constraint must be respected since the three components of the wave vector are not independent, as they must satisfy the dispersion relation.

In general, the pressure field at any point in space can be written as the following integral:

$$p(x_1, x_2, x_3) = \int_{-\infty}^{+\infty} \mathrm{d}k_1 \int_{-\infty}^{+\infty} P(k_1, k_2)\exp(-\mathrm{i}(k_1x_1 + k_2x_2 + k_3x_3))\mathrm{d}k_2 \qquad [2.31]$$

with $k_3 = \pm\sqrt{k_0^2 - k_1^2 - k_2^2}$; only the + sign is used for propagation in the half-space $x_3 \geq 0$ with sources confined to $x_3 < 0$. The integrations extend from $-\infty$ to $+\infty$, and therefore this decomposition involves both traveling plane waves and evanescent waves (when $k_1^2 + k_2^2 > k_0^2$).

The physical meaning of the quantity $P(k_1, k_2)$ becomes clearer if the pressure field is expressed in the plane $x_3 = 0$, that is:

$$p(x_1, x_2, 0) = \int_{-\infty}^{+\infty} \mathrm{d}k_1 \int_{-\infty}^{+\infty} P(k_1, k_2) \exp(-\mathrm{i}(k_1 x_1 + k_2 x_2)) \mathrm{d}k_2 \quad [2.32]$$

This formula can be interpreted as an inverse two-dimensional Fourier transform in (k_1, k_2). Therefore, $P(k_1, k_2)$ can be expressed as the direct Fourier transform of the physical pressure field in the plane $x_3 = 0$:

$$P(k_1, k_2) = \frac{1}{4\pi^2} \int_{-\infty}^{+\infty} \mathrm{d}x_1 \int_{-\infty}^{+\infty} p(x_1, x_2, 0) \exp(+\mathrm{i}(k_1 x_1 + k_2 x_2)) \mathrm{d}x_2 \quad [2.33]$$

$P(k_1, k_2)$ is called the plane wave angular spectrum of the pressure field. The use of this term becomes clearer when using the spherical coordinates in the spectral domain, (k, θ, φ), to express the components of the wave vector:

$$\boldsymbol{k} = \left| \begin{array}{l} k_1 = k_0 \sin\theta \cos\varphi \\ k_2 = k_0 \sin\theta \sin\varphi \\ k_3 = k_0 \cos\theta \end{array} \right. \quad [2.34]$$

The relations for the components of the wave vector defined by formulas [2.34] can be used for both propagating and evanescent waves, provided that angles with complex values are considered for the latter.

Equations [2.31] and [2.33] show that knowing the pressure field over the entire plane $x_3 = 0$ allows us to calculate the pressure field for any value $x_3 > 0$ using the angular spectrum. This is a remarkable result which is at the basis of the technique known as near-field acoustic holography.

2.1.5. *Near-field acoustic holography*

A basic introduction to this technique is provided is this section. Consider a measurement plane $x_3 = 0$ located at a small distance from sources located in the half-space $x_3 < 0$. From the knowledge of the pressure on the measurement plane, the field at a distance h is to be determined. The problem consists of solving

the Helmholtz equation in the half-space $x_3 > 0$, with the known values of $p(x_1, x_2, x_3 = 0)$ as a boundary condition. The medium is assumed to be infinite, and therefore there are no reflected waves. The problem to solve is then:

$$\nabla^2 p + k_0^2 p = 0 ; \qquad x_3 \geq 0 \qquad [2.35]$$

with the values of the pressure in the plane $x_3 = 0$, $p(x_1, x_2, x_3 = 0)$, as a boundary condition. By performing a Fourier transform with respect to (x_1, x_2), the problem to be solved becomes as follows:

$$\frac{d^2}{dx_3^2} P(k_1, k_2, x_3) + (k_0^2 - k_1^2 - k_2^2) P(k_1, k_2, x_3) = 0 ; \qquad x_3 \geq 0 \qquad [2.36]$$

with the values of $P(k_1, k_2, x_3 = 0)$ in the plane $x_3 = 0$ as a boundary condition.

The solution of the equation is simply given as:

$$P(k_1, k_2, x_3 = h) = P(k_1, k_2, x_3 = 0) \exp\left(-\mathrm{i}\sqrt{(k_0^2 - k_1^2 - k_2^2)}h\right) \qquad [2.37]$$

and an inverse two-dimensional Fourier transform allows $p(x_1, x_2, h)$ to be determined.

As can be seen from equation [2.37], for the wave numbers inside the radiation circle (satisfying $k_1^2 + k_2^2 \leq k_0^2$), P undergoes a simple phase shift and its modulus is not modified. Outside the radiation circle, the waves are evanescent and the phase of P is not modified, while its modulus decreases exponentially when moving away from the measurement plane. These components quickly become negligible when $h >> \lambda$.

The interest of equation [2.37] lies essentially in proceeding the other way around from measurements carried out in the plane $x_3 = h$ and performing a "backpropagation" of the information on the plane $x_3 = 0$, thus getting closer to the region containing the sources. For this purpose, the sign of the complex exponential argument in equation [2.37] just need to be changed, and one obtains:

$$P(k_1, k_2, x_3 = 0) = P(k_1, k_2, x_3 = h) \exp\left(+\mathrm{i}\sqrt{(k_0^2 - k_1^2 - k_2^2)}h\right) \qquad [2.38]$$

Thus, it becomes possible to obtain information in a plane located as close as possible to the sources, which could lead to a better identification of their position and amplitude. However, this requires taking into account part of the evanescent waves during the backpropagation procedure so that the spatial resolution of the

sources is not limited to the wavelength scale, as would be the case if only the wave numbers inside the radiation circle were taken into consideration (and therefore only the propagating waves). However, this method is prone to instabilities related to the exponential amplification of the evanescent components when returning to the source plane. These instabilities must be controlled, for example, by a clever filtering in spectral space. For a more in-depth discussion of this topic, the reader is referred to the book [WIL 99]; an exercise on this aspect is also proposed at the end of this chapter.

One last important remark can be made: In Fourier space, which corresponds to a plane wave decomposition, the acoustic particle velocity $\boldsymbol{V}(k_1, k_2, x_3)$ is obtained directly from the Fourier transform of the pressure according to:

$$\boldsymbol{V}(k_1, k_2, x_3) = \frac{P(k_1, k_2, x_3)}{Z_0} \frac{\boldsymbol{k}}{k_0} \qquad [2.39]$$

After an inverse Fourier transform to obtain the velocity field in the physical space, it is thus possible to determine and plot the sound intensity vector in any plane parallel to the pressure measurement plane, in particular as close as possible to the sources, in order to easily visualize energy flows and transfers.

2.2. Spherical waves

Plane waves have been defined as solutions to the wave equation that depend on only one spatial variable in Cartesian coordinates. This idea can be reconsidered using other coordinate systems, the most useful of which are spherical coordinates and, to a lesser extent, cylindrical coordinates. We thus consider a spherical coordinate system (r, θ, φ) and look for waves that depend only on the radial coordinate r and on time t. It is convenient to use, at least temporarily, the velocity potential $\phi(r, t)$ instead of the pressure fluctuation p'. The wave equation is simply written as:

$$\frac{1}{r}\frac{\partial^2(r\phi)}{\partial r^2} - \frac{1}{c_0^2}\frac{\partial^2 \phi}{\partial t^2} = 0 \qquad [2.40]$$

or

$$\frac{1}{r}\frac{\partial^2(r\phi)}{\partial r^2} - \frac{1}{rc_0^2}\frac{\partial^2(r\phi)}{\partial t^2} = 0 \qquad [2.41]$$

Excluding the origin $r = 0$, we obtain for the variable $\Phi = r\phi$ an equation which is none other than the one satisfied by plane waves, in which the distance r from the origin acts as a Cartesian coordinate:

$$\frac{\partial^2 \Phi}{\partial r^2} - \frac{1}{c_0^2}\frac{\partial^2 \Phi}{\partial t^2} = 0 \qquad [2.42]$$

Therefore, the solution for the velocity potential is:

$$\phi(r,t) = \frac{1}{r}F(t - \frac{r}{c_0}) + \frac{1}{r}G(t + \frac{r}{c_0}) \qquad [2.43]$$

which is interpreted as the combination of a wave propagating away from the origin, or divergent wave, $F(t - r/c_0)$ and a wave propagating toward the origin, or convergent wave $G(t + r/c_0)$. These waves are the analogues of plane waves propagating toward increasing and decreasing values of x_1, respectively.

The physical meaning of the divergent wave is clear. It corresponds to the existence in $r = 0$ of a point source whose radiation is omnidirectional (or isotropic), since the field is independent of the two angles (θ, φ). The existence of a converging wave is less obvious. It can occur, for example, when a spherical shell centered on the origin reflects the waves generated by a point source itself located at the origin, or when this shell is subjected to radial vibrations. Since these situations are very exceptional, we will not consider them in the following, and therefore only the divergent wave will be conserved[2].

We now return to the physical variables, pressure and velocity fluctuations (the latter being necessarily radial). From the definitions [1.29] and [1.32], we get:

$$p'(r,t) = -\rho_0 \frac{\partial \phi}{\partial t} \qquad [2.44]$$

and

$$u'_r(r,t) = \frac{\partial \phi}{\partial r} \qquad [2.45]$$

Therefrom, it can immediately be deduced that:

$$p'(r,t) = \frac{1}{r}f(t - \frac{r}{c_0}) \qquad [2.46]$$

where, to emphasize the formal similarity to plane waves, we have set $f = -\rho_0 \dot{F}$, denoting the time derivative of F by $\dot{F}$. The pressure waveform remains unchanged as it propagates, but its amplitude decreases as the inverse of the distance from the origin.

2 Converging waves also appear when studying the radiation of a source centered on the origin and of finite extent. An example of this will be considered in section 6.5.1.

The velocity fluctuation is written as:

$$u'_r(r,t) = \frac{1}{\rho_0 c_0 r} f(t - \frac{r}{c_0}) - \frac{1}{r^2} F(t - \frac{r}{c_0}) \qquad [2.47]$$

and, unlike plane waves, pressure and velocity are not simply proportional; the relationship between the two depends on the distance from the origin. Note that in the $r \to \infty$ limit, the second term of [2.47] can be neglected; a relation similar to that of plane waves involving the characteristic impedance of the medium, $p' = \rho_0 c_0 u'_r$ is then obtained. Thus, at a large distance from the source (which will be called the *far-field*), the two types of waves, plane and spherical, will exhibit very similar behavior. On the other hand, close to the source ($r \to 0$), it can be seen that the evolution of the pressure and velocity fluctuations will be very different, the velocity varying as $1/r^2$ and the pressure as $1/r$.

The expression of the acoustic velocity also allows us to propose a physical interpretation of the existence of spherical waves. The volume flow through a sphere of radius r centered on the origin is:

$$q_v(r,t) = 4\pi r^2 u'_r = -4\pi F(t - \frac{r}{c_0}) - 4\pi \frac{r}{c_0} \dot{F}(t - \frac{r}{c_0}) \qquad [2.48]$$

If we evaluate this expression for $r = 0$, we get $q_v(t) = -4\pi F(t)$. Thus, a spherical wave is associated with the radiation of an omnidirectional point source of fluctuating volume flow, and the pressure fluctuation is given by:

$$p'(r,t) = \frac{\rho_0}{4\pi r} \frac{\mathrm{d}q_v}{\mathrm{d}t}(t - \frac{r}{c_0}) \qquad [2.49]$$

2.2.1. *Time-averaged intensity and power*

The instantaneous acoustic energy flux of a spherical wave is purely radial and is written as:

$$I_r(t,r) = p' u'_r = \frac{p'^2}{\rho_0 c_0} + \frac{\rho_0}{r^3} F \dot{F} = \frac{p'^2}{\rho_0 c_0} + \frac{\rho_0}{2r^3} \frac{\mathrm{d}F^2}{\mathrm{d}t} \qquad [2.50]$$

and the time-averaged acoustic intensity[3] is simply:

$$\begin{aligned} I_{r_{\mathrm{av}}}(r) &= \lim_{T\to\infty} \frac{1}{T} \int_0^T I_r(r,t)\mathrm{d}t \\ &= \frac{\overline{p'^2}}{\rho_0 c_0} + \frac{\rho_0}{2r^3} \lim_{T\to\infty} \frac{F^2(T) - F^2(0)}{T} = \frac{\overline{p'^2}}{\rho_0 c_0} \end{aligned} \qquad [2.51]$$

3 In the following, we will often omit the term *time-averaged* when there is no possible confusion.

where the notation $\bar{\cdot}$ denotes the time average. The second term in the expression of the intensity has a zero contribution to the mean because $F^2(t)$ is finite. It can therefore be observed that the average (or mean) intensity has the same form as that of a single travelling plane wave and is expressed as a function of the pressure fluctuation alone[4]. The radial evolution of the average intensity as $1/r^2$, which is associated with the variation in $1/r$ in the root-mean-square pressure, is frequently referred to as the spherical spreading law. The average radiated power, which can be calculated from the integral of the acoustic intensity over a sphere of radius r, is equal to:

$$W_{\text{av}}(t) = 4\pi r^2 \frac{\overline{p'^2}}{\rho_0 c_0} = \frac{\rho_0}{4\pi c_0} \overline{\left(\frac{\mathrm{d}q_v}{\mathrm{d}t}\right)^2} \tag{2.52}$$

2.2.2. *Harmonic spherical waves*

From the expression of $p'(r,t)$, it immediately follows that:

$$p(r) = A\frac{\exp(-\mathrm{i}k_0 r)}{r} \tag{2.53}$$

where A is a complex constant. Using the linearized Euler's equation, the acoustic velocity u_r can be calculated as:

$$\mathrm{i}\omega\rho_0 u_r = -\frac{\mathrm{d}p}{\mathrm{d}r} = \left(\frac{1}{r^2} + \frac{\mathrm{i}k_0}{r}\right) A\exp(-\mathrm{i}k_0 r) \tag{2.54}$$

from which:

$$u_r(r) = A\frac{\exp(-\mathrm{i}k_0 r)}{\rho_0 c_0 r}\left(1 - \frac{\mathrm{i}}{k_0 r}\right) = \frac{p(r)}{\rho_0 c_0}\left(1 - \frac{\mathrm{i}}{k_0 r}\right) \tag{2.55}$$

The acoustic intensity is given by:

$$I_{r_{\text{av}}}(r) = \frac{1}{2}\mathrm{Re}(pu_r^*) = \frac{1}{2}\frac{pp^*}{\rho_0 c_0} = \frac{|A|^2}{2\rho_0 c_0 r^2} \tag{2.56}$$

and the total average acoustic energy density, sum of the kinetic and potential energies, is expressed as:

$$e_{\text{av}}(r) = \frac{|A|^2}{4\rho_0 c_0^2 r^2}\left(2 + \frac{1}{(k_0 r)^2}\right) \tag{2.57}$$

4 The mean square value of the pressure fluctuations will be indifferently denoted by $\overline{p'^2}$, p_{av}^2 or p_{rms}^2.

Several comments can be derived from these results. The $1/r^2$ decay of the acoustic intensity and its expression involving only the mean square pressure are only a simple consequence of the expressions obtained in the time domain. On the other hand, the relationship between pressure and velocity can be made more precise for harmonic waves. The far-field (here we can specify *acoustic* far-field) is now defined by the condition $k_0 r \gg 1$ or, equivalently, $r \gg \lambda$. In this region, the term involving $\mathrm{i}/k_0 r$ is negligible. Pressure and velocity are (almost) in phase and the proportionality coefficient is the characteristic impedance of the medium $Z_0 = \rho_0 c_0$. We then have:

$$p(r) = A\frac{\exp(-\mathrm{i}k_0 r)}{r}; \qquad u_r(r) \approx \frac{A}{\rho_0 c_0}\frac{\exp(-\mathrm{i}k_0 r)}{r} \qquad [2.58]$$

In contrast, the acoustic near-field is defined by $k_0 r \ll 1$ or $r \ll \lambda$. In this region, where the term $\mathrm{i}/k_0 r$ dominates, pressure and velocity are (almost) in quadrature:

$$p(r) = A\frac{\exp(-\mathrm{i}k_0 r)}{r}; \qquad u_r(r) \approx -\mathrm{i}\frac{A}{k_0 r}\frac{\exp(-\mathrm{i}k_0 r)}{r} \qquad [2.59]$$

In the far-field, there is an equipartition between kinetic energy and potential energy, and:

$$e_{\mathrm{av}} \approx \frac{|A|^2}{2\rho_0 c_0^2 r^2} = \frac{\overline{p'^2}}{\rho_0 c_0^2}; \qquad I_{r_{\mathrm{av}}} \approx e_{\mathrm{av}} c_0 \qquad [2.60]$$

which are similar to the results obtained for harmonic plane waves. In particular, the relation between intensity and energy density shows that the latter propagates in the radial direction with the speed of sound c_0. In the near-field, the kinetic energy is largely dominant since it varies as $1/r^4$ instead of $1/r^2$ for the potential energy. It is approximately equal to:

$$e_{\mathrm{av}} \approx \frac{|A|^2}{4\rho_0 c_0^2}\frac{1}{k_0^2 r^4} \qquad [2.61]$$

The ratio of intensity to energy density is therefore much smaller than c_0. Near the source ($r \ll \lambda$), strong velocity fluctuations exist, but the associated energy remains essentially confined to the near-field and only a small fraction is radiated as acoustic waves at greater distances from the source.

Let us return for a moment to the very important notion of near-field and far-field. We need to specify *acoustic* near-field here in order to emphasize that this proximity to the source must be judged relative to the wavelength scale; it therefore depends

on the frequency being studied. For example, we can consider that the far-field of a source emitting a 1,000 Hz signal in air is reached when $r > 1$ m for example (the wavelength λ being equal to 0.34 m), but this is still the near-field when the frequency is equal to 100 Hz.

In the acoustic far-field, it can be observed that spherical waves have locally almost identical properties to plane waves (ignoring the $1/r$ decay in the amplitude of the acoustic field). Physically, this is due to the fact that in this region, the curvature effects of the wavefronts become negligible over distances of the order of the wavelength. This very simple nature of the far-field can also be observed for waves of more complex structure, as we will see later.

2.3. Cylindrical waves

It is also helpful to study one-dimensional waves in the cylindrical coordinate system (ρ, φ, z), since the notion of a point source generating spherical waves can be transformed into that of a line source at the origin of cylindrical waves depending only on the radial distance ρ. This concept of a line source is of practical interest in several situations, such as the simulation of the acoustic radiation of a long line of vehicles.

A simple change of variable from p' to rp' allowed us to reduce the study of spherical waves to that of plane waves. By analogy, it makes sense to try changing the variable p' to $p'\sqrt{\rho}$. The acoustic power of a line source aligned with the z axis will actually be uniformly distributed over a cylinder of radius ρ and therefore the acoustic intensity will vary as $1/\rho$ by using the conservation of the radiated power over the surface of cylinders of increasing radius ρ. If we assume from the previous section that the far-field pressure is related to the intensity by $I_{\rho_{\text{av}}} = \overline{p'^2}/\rho_0 c_0$, then $p' \propto 1/\sqrt{\rho}$ in this region. Unfortunately, this idea does not prove to be completely successful, although certain asymptotic behavior can be identified. The structure of cylindrical waves is more complex than that of plane waves or spherical waves.

In cylindrical coordinates, the wave equation is written as:

$$\frac{\partial^2 p'}{\partial z^2} + \frac{1}{\rho^2}\frac{\partial^2 p'}{\partial \varphi^2} + \frac{1}{\rho}\frac{\partial}{\partial \rho}\left(\rho \frac{\partial p'}{\partial \rho}\right) - \frac{1}{c_0^2}\frac{\partial^2 p'}{\partial t^2} = 0 \qquad [2.62]$$

Assuming that the field depends only on ρ and t, that is $p' = p'(\rho, t)$, it simplifies to:

$$\frac{1}{\rho}\frac{\partial}{\partial \rho}\left(\rho \frac{\partial p'}{\partial \rho}\right) - \frac{1}{c_0^2}\frac{\partial^2 p'}{\partial t^2} = 0 \qquad [2.63]$$

and by setting $F = p'\sqrt{\rho}$, the following equation is obtained:

$$\frac{\partial^2 F}{\partial \rho^2} - \frac{1}{c_0^2}\frac{\partial^2 F}{\partial t^2} + \frac{F}{4\rho^2} = 0\,; \qquad \rho \neq 0 \tag{2.64}$$

We recognize the first two terms as the usual components of a one-dimensional wave equation, but the last term is new and prevents from obtaining a simple solution to equation [2.64]. However, the presence of ρ^2 in the denominator of the last term of this equation suggests that it becomes negligible in the far-field ($\rho \to \infty$) and that the pressure field for outgoing waves takes the following simple and intuitive form in the time domain:

$$\lim_{\rho\to\infty} p'(\rho, t) = \frac{1}{\sqrt{\rho}} f\left(t - \frac{\rho}{c_0}\right) \tag{2.65}$$

and

$$\lim_{\rho\to\infty} p(\rho) \approx \frac{A}{\sqrt{\rho}} \exp(-ik_0\rho) \tag{2.66}$$

in the frequency domain.

These two expressions are another illustration of the long-range behavior of sources similar to that of plane waves, with only the amplitude of the wave keeping track of the dimensionality of the problem. In Chapter 5, we will develop exact solutions of the wave equation and the Helmholtz equation in a two-dimensional space that are valid at any distance from the source, and that will justify the heuristic derivation of the far-field proposed above.

2.4. Exercises

1) **Exact boundary condition and validity of its linearization**

A piston oscillating around the origin $x_1 = 0$ is located in an infinite tube. The position of the active surface of the piston is given by $x_p(t) = l \sin \omega t$. The problem is one-dimensional, and the acoustic variables depend only on x_1 and time t.

a) Write the exact boundary condition between the fluid and the solid (use the notation $U_0 = l\omega$).

b) There is only a traveling wave of the form $u'(x_1, t) = f(t - x_1/c_0)$ present in the tube. Find the exact expression for the acoustic velocity at $x_1 = 0$, and then at any

point in the tube. Perform a Taylor series expansion of this expression assuming that the parameter $\epsilon = \omega l/c_0$ is small.

c) Show that up to the first order in ϵ, the solution corresponds to that obtained by "linearizing" the boundary condition (i.e. writing this condition for the mean position of the interface), with an additional constant term and a term that evolves as $\cos 2\omega t$. These last two terms are typical of the effect of a *local* geometric nonlinearity. Verify that the linearized condition is valid when $\lambda \ll l$ or, similarly, when the acoustic Mach number $\mathrm{M}_a = U_0/c_0$ is very small.

2) **Wave attenuation due to viscosity**

We consider the unidirectional propagation of an acoustic wave along the x_1 axis in a viscous but non-heat-conducting fluid. It can then be shown that the acoustic fluctuations remain isentropic, since the viscous dissipation is of the second order in the perturbations. The Navier–Stokes equation is written as (see Appendix 1):

$$\rho \frac{dU_1}{dt} = -\frac{\partial p}{\partial x_1} + \frac{4}{3}\mu \frac{\partial^2 U_1}{\partial x_1^2} \qquad [2.67]$$

a) Write the linearized continuity and Navier–Stokes equations and obtain a wave equation for the pressure fluctuation p' alone.

b) Now consider a harmonic evolution. Write the corresponding second-order differential equation, then find its solution in the form of a damped plane wave $A \exp(-\mathrm{i}Kx_1)$ where the wave number K is complex, with a positive real part and a negative imaginary part for waves propagating in the $+x_1$ direction.

c) It is possible to express analytically and exactly the value of K, but given the small value of the parameter $\nu/\omega c_0^2$ at usual frequencies, a Taylor series expansion limited to the first order is sufficient in practice. Estimate the value of this parameter in the atmosphere for the highest value of the audible frequency range (f = 20 kHz). Give the expressions for the absorption coefficient of the wave $\alpha = -\mathrm{Im}(K)$ and the phase velocity $v_\varphi = \omega/\mathrm{Re}(K)$. Show that, up to the first order, the wave velocity is equal to c_0 and that the absorption increases with the square of the frequency. Estimate the decrease in sound level after propagation over 1 km for a frequency of 10 kHz.

3) **The two-microphone method for the measurement of the acoustic intensity**

Direct measurement of the acoustic particle velocity is difficult. To estimate the acoustic intensity, an approximation of this velocity is usually derived using microphones whose separation is small compared to the wavelength.

To estimate the x_1 component of the acoustic intensity of a harmonic plane traveling wave, two microphones are thus placed at x_1 and $x_1 + l$, where $l \ll \lambda$.

a) Recall the exact expression of the I_1 component of the (average) acoustic intensity, then eliminate the velocity in this expression in favor of the pressure gradient $\partial p/\partial x_1$.

b) Estimate the pressure and the pressure gradient at the midpoint between the two microphones using a Taylor series expansion. Find the correct expression for $I_1(x_1 + l/2)$ up to the second order.

c) To estimate the error introduced by this approximation, consider a plane wave propagating at an angle θ with respect to the x_1 axis. Compare the exact expression of I_1 with its approximation. What is the maximum value of the distance between the two microphones so that the relative error remains less than 10% for frequencies below 1 kHz.

d) Propose a setup to evaluate the two components of the intensity vector $\boldsymbol{I}$ in the plane (x_1, x_2). Estimate the errors in module and direction induced by this setup.

4) **Near-field acoustic holography**

Two harmonic sources of spherical waves are placed in the plane $x_3 = 0$ and separated by a distance d along the Ox_1 axis. We first assume that they are identical (same amplitude and phase). Determine the resulting pressure field in a plane $x_3 = h_2$. The goal is to use this pressure field, spatially sampled as in an experiment, and to "backpropagate" it to obtain more accurate information about the sources.

a) Numerically compute the pressure on a grid of points separated by distances Δx_1 and Δx_2; apply an apodization function (e.g. a Hanning weighting) to gently bring the field to zero at the edges of the domain. Perform a numerical Fourier transform over (x_1, x_2) (e.g. using Matlab®)), then use equation [2.38] and finally apply an inverse Fourier transform to estimate the field in a plane $x_3 = h_1 < h_2$.

b) First use only propagating waves (corresponding to $\sqrt{k_1^2 + k_2^2} < k_0$), then gradually introduce evanescent waves (by limiting $\sqrt{k_1^2 + k_2^2}/k_0$ to 2, 3, ...) and observe the resulting improvement in sources localization, then the appearance of instabilities due to numerical noise. The following values of the parameters can be chosen: $d = \lambda/5$, $h_2 = \lambda/4$, $h_1 = 1$ cm, $\Delta x_1 = \Delta x_2 = 0.1$ m on a grid of 128×128 points.

c) Repeat the analysis of the first question, varying the values of the above parameters, and changing the sources' amplitude and relative phase.

5) **Linear acoustic array**

An array of sensors evenly spaced along a line is used to determine the direction of propagation of an incident wave.

Consider a set of M identical and omnidirectional sensors aligned and separated by a distance d (Figure 2.3; the length of the array is $L = (M-1)d$). A source located

at a large distance from the array produces a harmonic wave that can be modeled as a plane wave arriving at the array with a direction θ in the plane (x_1, x_2).

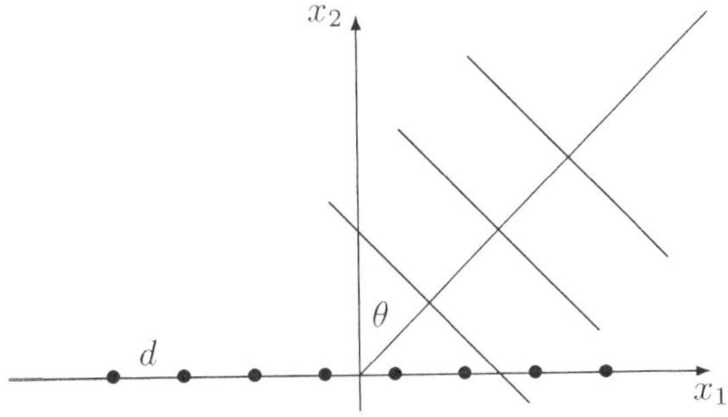

Figure 2.3. *Linear microphone array*

a) Determine the response of the array $R(\theta)$, which is obtained by summing the signals from each of the sensors and dividing it by their number (this quantity can be expressed as the sum of terms in a geometric series). Show that $R(\theta)$ can be written as:

$$R(\theta) = \frac{\sin(\frac{1}{2}k_0 M d \sin\theta)}{M \sin(\frac{1}{2}k_0 d \sin\theta)}$$

and interpret the evolution of this function with θ in terms of the phase shift between the sensors.

b) Plot the array directivity, defined as $G(\theta) = 10\log(|R(\theta)|)$, for an array of 16 sensors separated by 15 cm. Vary the frequency f of the source from 500 Hz to 3 kHz in 500 Hz increments. What is the width of the main lobe (around $\theta = 0$)? What is the level of the first side lobes? What happens when d is greater than the wavelength?

c) In general, the resolving power of the array (i.e. its ability to distinguish between two closely spaced sources) is defined by the width of the main lobe at -3 dB. Show that this corresponds to $\sin\theta \approx 0.5\lambda/L$ when a large number of sensors are used.

d) The in-phase summation of the sensors results in a maximum response, which is obtained for $\theta = 0$. To identify the direction of a source, it is possible to physically rotate the array until the maximum response is obtained; the source direction is then perpendicular to the array. Show that by shifting the phase of the response of each

sensor m by $(m-1)k_0 d \sin\theta_0$, the array is "electronically steered", thus allowing a direct identification of the arrival direction by simply varying the value of θ_0. Show that this phase shift corresponds to a simple constant time delay between the sensors and that this technique can therefore be used for broadband signals.

3

Sound Levels, Spectral Analysis and Notions on Human Sound Perception

This chapter introduces elementary signal processing concepts used in acoustics and their application to the characterization of human sound perception.

The amplitudes of sound pressure signals range over several orders of magnitude and the characterization of noise pollution is usually done using a logarithmic scale that compresses the dynamics, leading to the definition of the "family" of decibel units.

The frequency distribution of sound signals often needs to be analyzed, especially in the frequency range corresponding to human perception. The basic tool for this is the power spectral density (PSD), which we will introduce along with the related notion of correlation functions.

Depending on the application, the frequency description of the acoustic signals can be more or less refined. In environmental acoustics, proportional frequency bands such as octave or one-third octave bands are often used. When a more detailed signal analysis is required, a narrow-band description becomes necessary, for example, to identify the tonal components in the noise generated by fans and other turbomachinery.

Some concepts related to sound perception by the human ear are then introduced. The sensitivity of the ear varies with the frequency of the sound, but also with its level. This effect is accounted for by defining loudness levels (phon scale derived from equal loudness level contours) and their relationship to *subjective loudness* (sone scale) as perceived by humans. The use of frequency-weighted curves allows measured sound levels to be corrected for these perceptual effects in an approximate but simple manner.

While most of the concepts introduced here apply to signals whose average power does not vary with time (stationary signals), the case of non-stationary signals whose

average characteristics evolve with time, in particular transient signals, is briefly discussed at the end of the chapter.

3.1. Energy and average power

Acoustic pressure signals are by definition centered, that is, they have a zero mean value. The quantities of interest to characterize the "strength" of an acoustic signal are quadratic quantities, its energy or its power, depending on the nature of the signal under consideration. The energy[1] associated with the pressure signal $p'(t)$ is defined by:

$$E = \int_{-\infty}^{+\infty} p'(t)^2 \mathrm{d}t \qquad [3.1]$$

The energy is defined only when the integral exists. When this is the case, the signal is said to have finite energy. This is often the case with finite-duration signals, transients or impulses. When this is not the case, for example for infinite duration, random or periodic signals, the appropriate quantity is the average signal power, defined for the pressure signal $p'(t)$ by:

$$P_{\mathrm{av}} = \lim_{D\to\infty} \frac{1}{D} \int_{-D/2}^{D/2} p'(t)^2 \mathrm{d}t = p_{\mathrm{rms}}^2 \qquad [3.2]$$

From this definition, it can be directly seen that the average power of a signal is the square of its root mean square (RMS) value p_{rms}^2, also denoted by $\overline{p'^2}$ or p_{av}^2.

For a periodic signal of period T, the average power can also be defined by:

$$P_{\mathrm{av}} = \frac{1}{T} \int_{-T/2}^{T/2} p'(t)^2 \mathrm{d}t = p_{\mathrm{rms}}^2 \qquad [3.3]$$

When using complex signal notation ($p'(t) = \mathrm{Re}\{p\exp(\mathrm{i}\omega t)\}$), the RMS value can be expressed from the complex amplitude p:

$$p_{\mathrm{rms}}^2 = \frac{1}{2}pp^* = \frac{1}{2}|p|^2 \qquad [3.4]$$

Energy and average power allow us to define the overall "strength" of an acoustic signal, but in many cases, this global value is not sufficient to characterize its impact

1 The signal energy is not to be confused with the acoustic energy.

or harmfulness. The characteristics of the receiver, such as the human ear for instance, must also be taken into account, in particular its overall sensitivity and its variation with frequency.

3.2. Sound levels

As mentioned in the introduction to this chapter, the amplitude of audible pressure fluctuations ranges over several orders of magnitude. In air, the RMS value of the smallest pressure fluctuation detectable by the human ear (the *threshold of hearing* at 1,000 Hz) is estimated to be close to 2×10^{-5} Pa. Conversely, the threshold of pain corresponds to an RMS pressure fluctuation of about 20 Pa; 6 orders of magnitude separate these thresholds. This large range suggests the use of a logarithmic scale to describe sound levels.

The sound pressure level (SPL) L_p in decibels (dB) is defined as follows:

$$L_p = 10 \log \left(\frac{p_{\text{rms}}^2}{p_{\text{ref}}^2} \right) = 20 \log \left(\frac{p_{\text{rms}}}{p_{\text{ref}}} \right) \qquad [3.5]$$

where p_{ref} is a reference value that is used to set the zero of the scale, and where the logarithms used in equation [3.5] and for all decibel scale definitions are decimal logarithms. In air (and also in other gases), the conventional threshold of hearing is chosen as the reference, namely, $p_{\text{ref}} = 2 \times 10^{-5}$ Pa. In liquids, the reference pressure is chosen as $p_{\text{ref}} = 10^{-6}$ Pa. The use of decibels is also well suited to the differential sensitivity of hearing, since a difference in level of 1 dB is approximately equal to the smallest difference that the human ear can detect. A fuller discussion is available in Chapter 6 of [FAH 15].

In addition to sound pressure level, intensity level and power level can also be expressed in dB. These levels are defined from the average acoustic intensity I_{av} and the average acoustic power W_{av}. The intensity level L_I is given by:

$$L_I = 10 \log \left(\frac{I_{\text{av}}}{I_{\text{ref}}} \right) \qquad [3.6]$$

with $I_{\text{ref}} = 10^{-12}$ W/m^2, where this reference is valid for both liquids and gases. This value roughly corresponds to the average intensity of a plane wave propagating in air, whose RMS value of pressure fluctuations is $p_{\text{rms}} = p_{\text{ref}}$, because $I_{\text{ref}} = p_{\text{ref}}^2/\rho_0 c_0$ with $\rho_0 c_0 \approx 400$ kg.m^{-2}.s^{-1}.

The power level L_W is defined by:

$$L_W = 10 \log \left(\frac{W_{\text{av}}}{W_{\text{ref}}} \right) \qquad [3.7]$$

with $W_{\text{ref}} = 10^{-12}$ W, again whatever the fluid under consideration. This reference value corresponds to the acoustic power radiated into air by a plane wave of average intensity I_{ref} over an area of 1 m^2.

It is clear from the above definitions that sound levels in dB do not add up. If a pressure fluctuation signal is defined as the sum of two other signals $p' = p'_1 + p'_2$, then $L_p \neq L_{p1} + L_{p2}$. First, p^2_{rms} should be calculated, and then the level in dB can be derived.

3.3. Energy and power spectral densities

We have already mentioned the need to consider the frequency distribution of the quadratic quantities, energy or average power, in order to correctly estimate the impact of an acoustic signal. For example, the human ear perceives sounds in a frequency range limited to the 20 Hz–20 kHz interval. Therefore, it is essential to locate the frequency content of the pressure signal relative to this range.

For a signal of finite energy, the energy distribution as a function of frequency is determined using Parseval's theorem:

$$E = \int_{-\infty}^{+\infty} p'(t)^2 \mathrm{d}t = \int_{-\infty}^{+\infty} |p(f)|^2 \mathrm{d}f \qquad [3.8]$$

where $p(f)$ is the Fourier transform of the pressure fluctuations $p'(t)$. The quantity $|p(f)|^2$ thus appears as the energy contained in a frequency band of 1 Hz centered around the current value f of the frequency. This is called the energy spectral density of the pressure signal.

The same type of relationship can be obtained for the average power of a signal whose energy is infinite, but in a slightly more complex way. From the signal $p'(t)$, the signal of finite energy $p_D(t)$ is defined by time windowing as:

$$p'_D(t) = p'(t) \quad \text{for} \quad t \in [-D/2, D/2] \quad \text{and} \quad 0 \quad \text{otherwise} \qquad [3.9]$$

The energy of this signal is then expressed according to Parseval's theorem:

$$E_D = \int_{-\infty}^{+\infty} p'_D(t)^2 \mathrm{d}t = \int_{-\infty}^{+\infty} |p_D(f)|^2 \mathrm{d}f \qquad [3.10]$$

where $p_D(f)$ is the Fourier transform of the function $p'_D(t)$. The average power is written as:

$$P_{\text{av}} = \lim_{D\to\infty} \frac{1}{D} \int_{-D/2}^{D/2} p'(t)^2 \mathrm{d}t$$
$$= \lim_{D\to\infty} \frac{1}{D} \int_{-D/2}^{D/2} p'_D(t)^2 \mathrm{d}t = \int_{-\infty}^{+\infty} \lim_{D\to\infty} \frac{1}{D} |p_D(f)|^2 \mathrm{d}f \qquad [3.11]$$

It can then be concluded that the integrand appearing in the second integral of equation [3.11] represents the power spectral density of the pressure signal:

$$S_{pp}(f) = \lim_{D\to\infty} \frac{1}{D} |p_D(f)|^2 \qquad [3.12]$$

3.4. Correlation functions

The correlation function is a tool used for analyzing similarities between two signals. In this chapter, the term "correlation function" is used to refer to the temporal correlation function. Similar definitions can be introduced for the spatial correlation function by replacing the time coordinate with one or more spatial coordinates. Correlation functions are related to the signal energy or PSD introduced in section 3.3 and allow the important notion of uncorrelated signals to be addressed.

The cross-correlation function between two pressure signals $p'_1(t)$ and $p'_2(t)$ of finite average power is defined by:

$$R_{21}(\tau) = \lim_{D\to\infty} \frac{1}{D} \int_{-D/2}^{D/2} p'_2(t) p'_1(t-\tau) \mathrm{d}t \qquad [3.13]$$

An analogous definition can be given for finite energy signals, using the appropriate integration. Two pressure signals $p'_1(t)$ and $p'_2(t)$ are said to be uncorrelated if $R_{21}(\tau) = 0, \forall \tau$.

The autocorrelation function is defined as the correlation of a signal with itself. For the signal $p'_1(t)$, it is expressed as:

$$R_{11}(\tau) = \lim_{D\to\infty} \frac{1}{D} \int_{-D/2}^{D/2} p'_1(t) p'_1(t-\tau) \mathrm{d}t \qquad [3.14]$$

The autocorrelation function is maximum for zero time delay ($\tau = 0$) and is an even function of τ, $R_{11}(-\tau) = R_{11}(\tau)$.

Note that the average power of a signal is equal to the value of the autocorrelation function at zero time delay:

$$P_{1,\mathrm{av}} = p_{1,\mathrm{rms}}^2 = R_{11}(0) \qquad [3.15]$$

It can also be shown that the PSD is the Fourier transform of the autocorrelation function:

$$S_{11}(f) = \int_{-\infty}^{+\infty} R_{11}(\tau)\mathrm{e}^{-\mathrm{i}2\pi f\tau}\mathrm{d}\tau \qquad [3.16]$$

3.5. Random signals

The definitions and concepts introduced earlier in this chapter refer to deterministic signals. In applications, and especially in measurements, some degree of randomness is always present. The purpose here is not to present a comprehensive, rigorous and detailed theory on how to process this type of signals, but to examine how what has been described for deterministic signals can be transposed to random signals. The reader may wish to consult [SHI 08] for more detailed information.

A random signal corresponds to a particular realization of a random process. To study a random process (e.g. measurement noise), the statistical quantities associated with a set of experiments must be considered. Mean values and correlations are thus defined from statistical averages over a set of measurements. This can be tedious and very complex, or even impossible to achieve in practice. Therefore, it is generally assumed that the process is *stationary and ergodic*. Stationarity of a random process means that its statistical properties are independent of time. On the other hand, ergodicity implies that ensemble averages (over a set of independent realizations) can be replaced by time averages over a single realization. More detailed definitions can be found in [BEN 11]. Unless otherwise stated, we will assume that stationarity and ergodicity properties are verified in the wide sense, that is for statistical moments of order less than or equal to 2. It will also be assumed that the signals have a finite average power. Under these assumptions, the previous definitions of average power and correlation can be directly applied to random signals, with time averages replacing statistical averages.

Formally, the PSD $S_{pp}(f)$ of a random process should be computed from a set of realizations as follows:

– The realizations of the random process are windowed for a limited time D.

– Their Fourier transforms are estimated and the squares of their magnitudes are then calculated.

– Their statistical average is computed and the duration D is made to tend to infinity.

$$S_{pp}(f) = \lim_{D\to\infty} \frac{1}{D}\mathbb{E}[p_D(f)p_D^*(f)] \qquad [3.17]$$

where $\mathbb{E}[\cdot]$ is the mathematical expectation computed from the realizations of the random process.

The assumptions of stationarity and ergodicity simplify the procedure and lead to the Wiener–Khintchin theorem, which is the extension of property [3.16]:

$$S_{pp}(f) = \int_{-\infty}^{+\infty} R_{pp}(\tau)\mathrm{e}^{-\mathrm{i}2\pi f\tau}\mathrm{d}\tau \qquad [3.18]$$

The PSD (whose unit is Pa^2/Hz) is therefore also the Fourier transform of the autocorrelation function for stationary and ergodic random signals. By inverse Fourier transform, it follows that:

$$R_{pp}(\tau) = \int_{-\infty}^{+\infty} S_{pp}(f)\mathrm{e}^{\mathrm{i}2\pi f\tau}\mathrm{d}f \qquad [3.19]$$

In particular, since $P_{\mathrm{av}} = p_{\mathrm{rms}}^2 = R_{pp}(0)$, it can be written that:

$$P_{\mathrm{av}} = p_{\mathrm{rms}}^2 = \int_{-\infty}^{+\infty} S_{pp}(f)\mathrm{d}f \qquad [3.20]$$

and thereby it is found again that the total signal power is the integral over all frequencies of the PSD.

The power spectral density level $L_p(f)$, which corresponds to the sound pressure level at frequency f and in a bandwidth of 1 Hz, can be expressed (in dB/Hz) by[2]:

$$L_p(f) = 10\log\left(\frac{S_{pp}(f)}{p_{\mathrm{ref}}^2}\right) \qquad [3.21]$$

In practice, the estimation of a PSD from numerical measurements or simulations can be carried out in different ways. Often we have only a single realization of the process over a long time period T. In this case, a periodogram-based method, such as the Welch method, can be used to compute the PSD. This consists of dividing

2 Similar definitions apply for intensity and power levels.

the signal into a set of shorter time segments of duration T_s, computing the term $p(f)p^*(f)/T_s$ on each segment, and finally averaging over the segments. This average over the segments replaces the mathematical expectation of equation [3.17]. More information can be found in Chapter 4 of [FAH 15] or in [BEN 11].

3.6. Random signals and correlations, some examples

White noise and pink noise are signals often encountered in acoustics. They can be used in calibration procedures of measuring instruments, in experimental protocols (e.g. when measuring reverberation time in a room), or in applications such as acoustic masking. White noise is a centered and random stationary signal whose PSD is constant in frequency, $S_{pp}(f) = C$. The corresponding correlation function is therefore a Dirac delta function at the origin. Pink noise is defined by a PSD that varies as the inverse of the frequency, $S_{pp}(f) = C/f$.

Correlation functions are particularly useful in acoustics when multiple sources contribute to the pressure signal. The presence of a random signal, especially white noise, can mask other sources, and it is often necessary to extract their contribution from the global signal. In many applications, it is also important to determine the degree of interaction between two or more sources: Are they related? Are they independent? Correlation analysis is a powerful tool for answering these questions.

For example, if we assume that the signal $p'(t)$ is the sum of two signals, $p'(t) = p_1'(t) + p_2'(t)$, originating from two sources, the mean square value of $p'(t)$ is given by:

$$p_{\text{av}}^2 = p_{1,\text{av}}^2 + p_{2,\text{av}}^2 + 2R_{21}(0) \qquad [3.22]$$

If the sources are independent, they are uncorrelated and we get:

$$p_{\text{av}}^2 = p_{1,\text{av}}^2 + p_{2,\text{av}}^2 \qquad [3.23]$$

If, in addition, the two signals p_1' and p_2' have the same amplitude, the global sound pressure level is:

$$L_p = 10\log\left(\frac{2p_{1,\text{av}}^2}{p_{\text{ref}}^2}\right) = 10\log\left(\frac{p_{1,\text{av}}^2}{p_{\text{ref}}^2}\right) + 10\log(2) \approx L_{p1} + 3\,\text{dB}$$

In practice, the assumption of uncorrelated sources is very often justified and used when studying the noise emitted by several distinct sources. This is the case, for example, of two machines or two people talking simultaneously.

Examples of autocorrelation and cross-correlation functions are presented below to illustrate the information they can provide about pairs of signals. A first signal p'_1 is constructed by randomly distributing periods of a harmonic signal (with a period $T = 0.1$ s) over time. A second signal p'_2 is defined by $p'_2(t) = Cp'_1(t - \tau_s)$ with $C = 0.3$ and a time delay $\tau_s = 0.4$ s. Finally, a signal p'_3 is generated by adding white noise to the signal p'_2. The amplitude of the noise is so large that the p'_2 signal is completely hidden in the p'_3 signal. The three signals are shown in Figure 3.1.

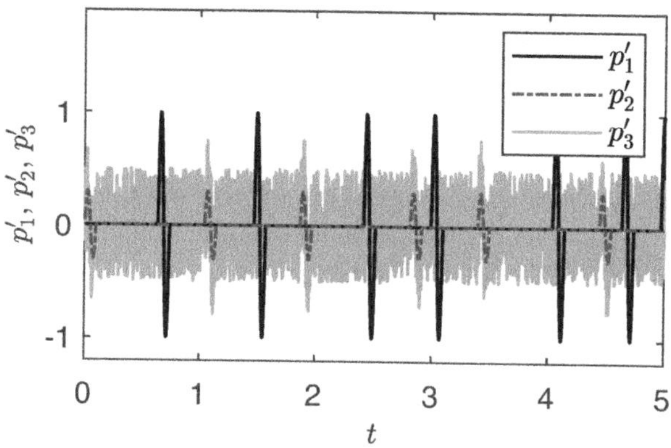

Figure 3.1. *Evolution versus time of the signals p'_1, p'_2, and p'_3 defined in the text*

The autocorrelation functions R_{11} and R_{22} are plotted on the left side of Figure 3.2. Since the signals p'_1 and p'_2 are similar, so are their autocorrelation functions, and a simple proportionality factor C^2 is found between the amplitudes of R_{11} and R_{22}. The cross-correlation functions R_{21} and R_{31} are plotted on the right side of Figure 3.2. The function R_{21} is not zero everywhere, indicating that the signals p'_1 and p'_2 are not completely uncorrelated. The position of the maximum of R_{21} indicates the delay for which the signals p'_1 and p'_2 are most similar. It is located at $\tau = \tau_s = 0.4$ s, as expected from the definition of p'_2. Since the signal p'_2 was constructed from p'_1, the cross-correlation function R_{21} has an identical shape to the autocorrelation functions R_{11} and R_{22}. The function R_{31} is identical to R_{21} because the noise that was added to p'_2 to form p'_3 is uncorrelated with the signal p'_1. Therefore, this noise does not contribute to the cross-correlation function R_{31}, so the part of the signal p'_3 that is correlated with the signal p'_1 is simply p'_2. This illustrates the value of using correlation functions to extract a signal from noise.

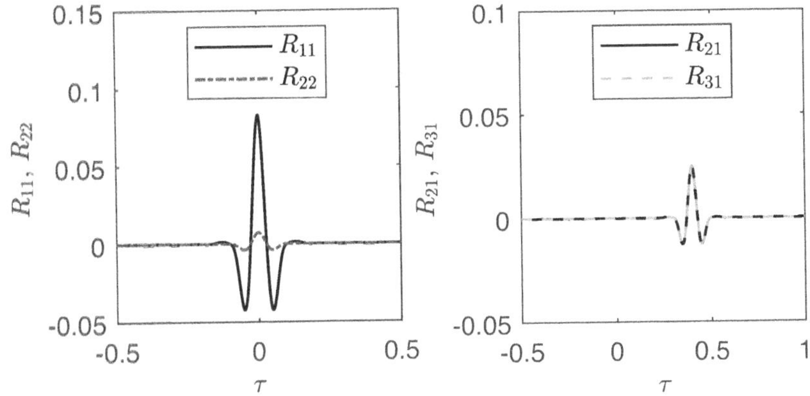

Figure 3.2. *Correlation functions: on the left side, the autocorrelations R_{11} and R_{22}, and on the right side, R_{21} and R_{31}*

3.7. Frequency bands

To characterize acoustic signals in the frequency domain, PSDs have been defined in section 3.3. The total power resulting from the integration of the PSD over the entire frequency range is not sufficient to fully characterize a signal, and in practice the PSD cannot be obtained analytically for each frequency. The frequency domain is then discretized into a series of contiguous bands, whose width is chosen according to the desired precision of the analysis. The frequency spectrum of a signal is then described by a value for each of these frequency bands. This value corresponds to the power of the signal integrated over that band.

More precisely, a set of frequency bands $\{B_b\}$ with $b \in \mathbb{N}$ is defined. Each band has a lower bound $f_1(b)$ and an upper bound $f_2(b)$ such that $B_b =]f_1(b), f_2(b)]$. Since the bands are contiguous, $f_2(b) = f_1(b+1)$.

The signal power in each of the bands b is:

$$p_{b,\mathrm{rms}}^2 = \int_{f_1(b)}^{f_2(b)} S_{pp}(f)\mathrm{d}f \qquad [3.24]$$

and the total signal power is simply the sum of the powers contained in each band:

$$p_{\mathrm{rms}}^2 = \sum_b p_{b,\mathrm{rms}}^2 \qquad [3.25]$$

The sound pressure level in each band b is given by:

$$L_p(b) = 10 \log \left(\frac{p^2_{b,\text{rms}}}{p^2_{\text{ref}}} \right)$$

and the total sound pressure level in dB is obviously obtained by summing the $p^2_{b,\text{rms}}$ before taking the logarithm, and not by summing the $L_p(b)$.

Two types of frequency partitioning are used in acoustics: partitioning by constant bandwidth filters or by proportional bandwidth filters. A set of constant bandwidth filters is defined by $f_2(b) = f_1(b) + \Delta f$. This type of filtering is typically used with narrow bandwidths Δf, which allows for a very detailed description of the power spectrum. For example, it makes it possible to identify tonal components (peaks) in the spectrum and to focus on the physical mechanisms that generate noise.

In many applications, such as environmental noise monitoring, a coarser description of the frequency content of signals is sufficient. This allows one to estimate whether the frequency content of a signal is concentrated in low, mid or high frequencies relative to the audible range (see section 3.8). This description is widely used for setting acoustic standards. It typically uses bands of constant relative width (called proportional bands), such as octave or one-third octave bands. Bands of constant relative width are defined by $f_2(b) = Cf_1(b)$. For octave bands, $C = 2$, and for one-third octave bands, $C = 2^{1/3}$. Thus, each octave band contains three one-third octave bands as shown in Figure 3.3.

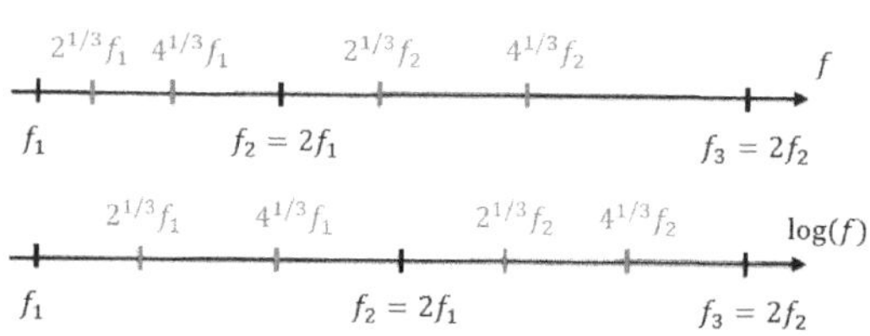

Figure 3.3. *Representation of lower and upper frequencies for octave (black) and one-third octave (gray) bands, in linear and logarithmic scale*

The center frequency f_c is used to denote each frequency band. For constant bandwidth:

$$f_c(b) = \frac{f_1(b) + f_2(b)}{2} = f_1(b) + \frac{\Delta f}{2} \quad [3.26]$$

and for proportional bands:

$$f_c(b) = \sqrt{f_1(b) f_2(b)} = \sqrt{C} f_1(b) \quad [3.27]$$

We thus have $f_c(b) = \sqrt{2} f_1(b)$ for octave bands and $f_c(b) = 2^{1/6} f_1(b)$ for one-third octave bands.

The center, lower and upper frequencies of octave and one-third octave bands are listed in Table 3.1. The center frequencies are defined in a standardized way and one should refer to ISO 266 [ISO 97] or IEC 61260-1 [IEC 14].

Band number	Central frequency (Hz)		Band limits (Hz)	
	Octave	One-third octave	Lower	Upper
14		25	22	28
15	31.5	31.5	28	35
16		40	35	44
17		50	44	57
18	63	63	57	71
19		80	71	88
20		100	88	113
21	125	125	113	141
22		160	141	176
23		200	176	225
24	250	250	225	283
25		315	283	353
26		400	353	440
27	500	500	440	565
28		630	565	707
29		800	707	880
30	1,000	1,000	880	1,130
31		1,250	1,130	1,414
32		1,600	1,414	1,760
33	2,000	2,000	1,760	2,250
34		2,500	2,250	2,825
35		3,150	2,825	3,530
36	4,000	4,000	3,530	4,400
37		5,000	4,400	5,650
38		6,300	5,650	7,070
39	8,000	8,000	7,070	8,800
40		10,000	8,800	11,300
41		12,500	11,300	14,140
42	16,000	16,000	14,140	17,600
43		20,000	17,600	22,500

Table 3.1. *Center, lower and upper frequencies of octave and one-third octave bands*

As an example, consider white noise, which has a frequency-constant PSD. Its narrowband PSD level (in dB) is also constant, while its octave or one-third octave band PSD level increases linearly with $\log(f_c)$ (here 3 dB per octave). Pink noise is

characterized by a PSD that decreases with frequency according to $S_{pp}(f) = C/f$. Its narrowband PSD level decreases at a rate of -3 dB/octave, but it presents identical levels for all octave or one-third octave bands.

Examples of narrowband and octave-band spectra for a white noise and a pink noise are shown in Figure 3.4.

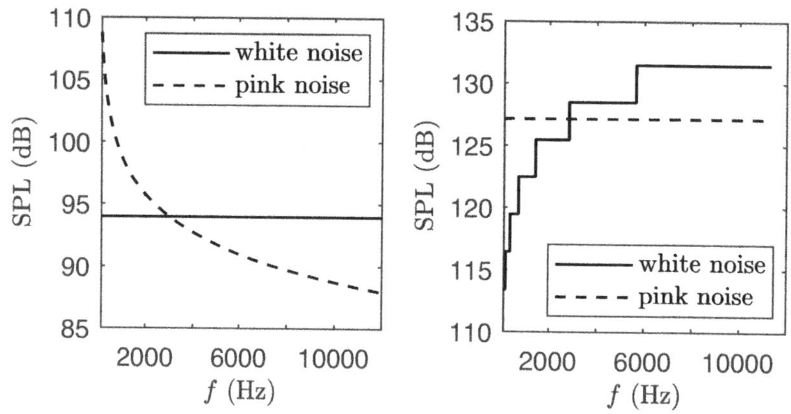

Figure 3.4. *Narrowband (left) and octave-band (right) spectra of white noise and pink noise*

An example of a sound pressure level spectrum calculated in narrowband, one-third octave bands and octave bands for the same signal is shown in Figure 3.5. The signal originates from measurements performed during the study of an air conditioning fan used in aviation. High amplitudes are expected to be observed at frequencies associated with the rotation of the fan and its number of blades (blade passing frequencies). Tonal components (peaks) associated with these frequencies are clearly visible in the narrowband spectrum. On the other hand, they become less visible in the one-third octave band spectrum and they disappear completely in the octave band spectrum.

3.8. Loudness, equal loudness contours and frequency weightings

The human ear's perception of sound level is frequency dependent. Therefore, two sounds with the same amplitude but different frequency content are perceived as having different levels. Equal loudness contours (Figure 3.6) describe the frequency-dependent evolution of the sound pressure level corresponding to a constant perceived level (loudness); the associated unit is the phon. These curves are normalized to the sound pressure level at 1,000 Hz. For example, the 40-phon curve passes through the 40 dB value at 1,000 Hz. These curves are described in the

ISO 226 standard [ISO 03] and are the result of experimental listening studies conducted on population samples.

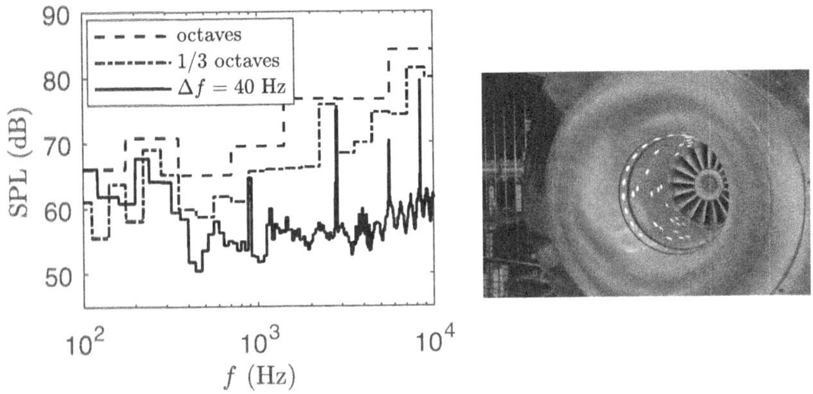

Figure 3.5. *Narrowband, one-third octave and octave band spectra computed from the same sound pressure signal (left) measured on the fan bench shown on the right*

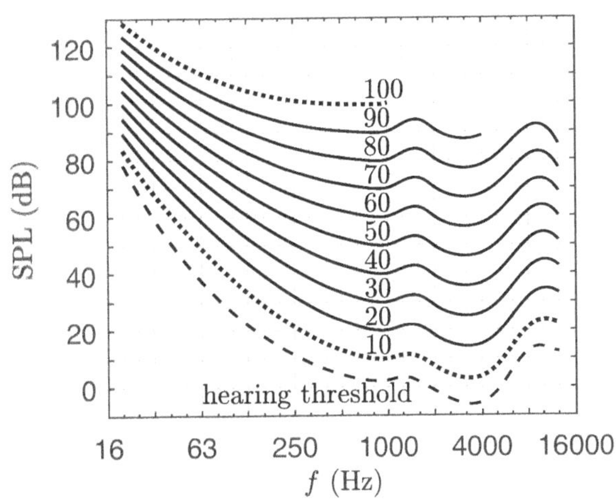

Figure 3.6. *Equal loudness level contours defined in the ISO 226:2003 standard*

These curves show that the human ear is most sensitive between 1,000 Hz and 4,000 Hz. Away from this frequency range, the sensitivity decreases and the sensation of perceived noise is attenuated at high frequencies and even more so at low frequencies. It should also be noted that the frequency evolution of equal

loudness contours is not identical from one curve to another. Therefore, the perceived noise sensation depends not only on the frequency, but also on the sound pressure level. Because equal loudness contours are derived from experimental results on population samples, they may be subject to revision over time. The curves shown in Figure 3.6 thus correspond to the 2003 revision of the ISO 226 standard.

Note that the loudness level in phons corresponds to a logarithmic scale, as it is expressed in dB. This level is not proportional to the perceived sensation of loudness of a sound. A unit called sone has been introduced to characterize this loudness sensation in a linear way (i.e. a sound of 2 sones is perceived as 2 times louder than a sound of 1 sone). The relationship between the loudness S in sones and the loudness level P in phons can be approximated, for loudness levels $P \geq 40$ dB, by:

$$S = 2^{(P-40)/10} \quad \text{thus} \quad \log(S) = \left[\frac{P-40}{10}\right]\log(2) \qquad [3.28]$$

From these definitions, it can be seen that a loudness level of $P = 40$ phons corresponds to a loudness of $S = 1$ sone, a value observed in a quiet environment. An increase of the loudness level of 10 dB ($P' = P + 10$) corresponds to a loudness $S' = 2S$, thus to a doubling of the loudness sensation.

To calculate the loudness of broadband stationary sounds, it is necessary to analyze the sounds with 1/1 or 1/3 octave band filters, and then to convert each band level to loudness on the sone scale. The total loudness of the complex sound is calculated by adding the sone values of each band, taking into account mutual inhibition (masking) between bands (see [FAS 07]).

The interested reader should also consult the website "Journey into the world of hearing" (www.cochlea.eu/en), where some psychoacoustic concepts, such as loudness, frequency discrimination or sound localization can be found, as well as very detailed information on the ear and on pathologies of the auditory system.

When considering human noise exposure, these physiological effects must be taken into account. To do this in a simplified way, frequency-variable weightings are applied to the measured sound spectrum. Different weightings can be used, namely the A, B and C weightings, which are defined from different equal loudness level contours[3]. The A-weighting is defined from the 40-phone contour, the B-weighting from the 70-phone contour and the C-weighting from the 100-phone contour. They are plotted in Figure 3.7. The IEC 61672-1 standard [IEC 13] can be consulted for more information on the definition of these weightings. In principle, therefore, the choice of weighting depends on the sound pressure level, but in practice the

3 The equal loudness level contours used to define these weightings predate the ISO 226:2003 standard.

A-weighting is the most commonly used. When working in octave or one-third octave bands, the values of the weightings to be applied to each band are tabulated (see Table 3.2 for the values per octave band).

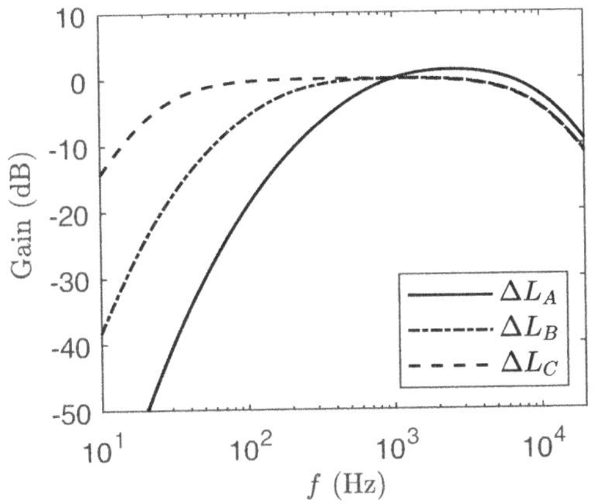

Figure 3.7. *A-, B- and C-weighting curves*

f_c octave (Hz)	31.5	63	125	250	500	1,000	2,000	4,000	8,000	16,000
ΔL_A (dB)	-39.4	-26.2	-16.1	-8.6	-3.2	0	1.2	1.0	-1.1	-6.6

Table 3.2. *A-weighting values for octave bands*

To apply a weighting, the value for each frequency band is added to the spectrum measured in dB in order to obtain a weighted spectrum. From this weighted spectrum, an equivalent value of $p^2_{b,\text{rms}}$ can be calculated for each band. These $p^2_{b,\text{rms}}$ values are summed to obtain a total weighted value p^2_{rms}. This can then be converted in decibels to obtain the total level perceived. Depending on the weighting chosen, the units used are called dB(A), dB(B) or dB(C). An example of calculating the A-weighted pressure level from a spectrum measured in octave bands is given in Table 3.5.

3.9. Characterization of non-stationary acoustic signals

Many acoustic signals do not exhibit the property of stationarity. Some have a long duration, but their statistical properties change over time. This is the case, for example, with noise monitoring measurements near a road, where changes in the amplitude of the signal are observed at different times of the day (rush hour, night, etc.). For such signals, the (weighted or unweighted) sound pressure levels L_p defined in section 3.2 are calculated over a "short" duration (of the order of one second or less). It is thus

possible to study the time evolution of the sound pressure level $L_p(t)$ using indicators such as the statistical levels defined in section 3.9.1 or the equivalent level defined in section 3.9.2[4].

Octave-band spectrum level and overall level							
f_c (Hz)	125	250	500	1,000	2,000	4,000	Overall (dB)
L_p (dB)	66	76	80	79	74	67	84
A-weighting in octave bands:							
f_c (Hz)	125	250	500	1,000	2,000	4,000	
ΔL_A (dB)	-16.1	-8.6	-3.2	0	1.2	1	
A-weighted spectrum and global A-level:							
f_c (Hz)	125	250	500	1,000	2,000	4,000	Overall dB(A)
L'_p (dB)	49.9	67.4	76.8	79	75.2	68	82.4

Table 3.3. *Example of computing the A-weighted sound pressure level from an octave-band spectrum*

Finite energy signals, such as transients, should also be mentioned. Unlike permanent signals, they have a finite energy. These transient signals can be characterized by their sound energy level, which is defined in section 3.9.3, and it is also possible to describe their spectral content from the energy spectral density.

3.9.1. *Statistical levels*

For permanent and non-stationary signals, the distribution of sound levels measured during a recording can be characterized by calculating several statistical levels L_N. The level L_N corresponds to the sound pressure level, expressed in dB, that is exceeded for $N\%$ of the time during the recording. The notation L_N is sometimes used for recordings of sound pressure levels in dB(A); this is indicated by the notation L_{AN}. Typically, the following statistical levels are considered:

– L_{10}, level exceeded for 10% of the time, indicating peak noise level;

– L_{50}, level exceeded for 50% of the time, indicating the median level;

– L_{90}, level exceeded for 90% of the time, indicating background noise.

An example of a time recording of the pressure level is shown in Figure 3.8 and the corresponding levels L_{10}, L_{50} and L_{90} are given.

4 The indicators presented in this section are defined in the ISO 1996-1 standard [ISO 16].

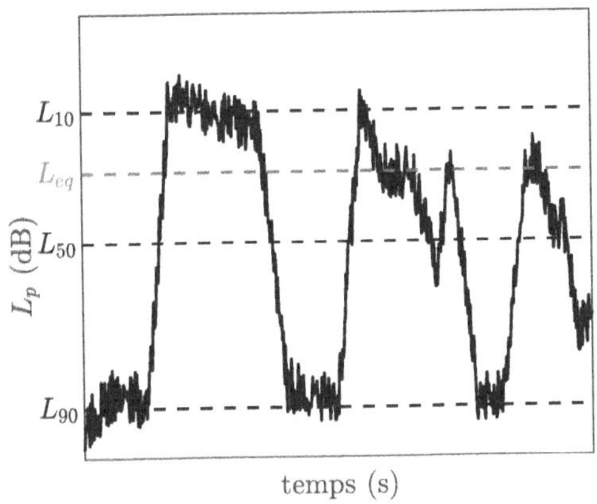

Figure 3.8. *Time recording of sound pressure levels and corresponding statistical levels*

3.9.2. *Equivalent level, "Day ", "Evening" and "Night" levels*

To quantify the average power of a sound pressure level recording such as the one shown in Figure 3.8, the equivalent sound pressure level L_{eq} is calculated. For a recording of duration T, the equivalent level denoted $L_{eq,T}$ is defined by:

$$L_{eq,T} = 10 \log \left[\frac{1}{T} \int_0^T \frac{p_{\text{rms}}^2(t)}{p_{\text{ref}}^2} \text{d}t \right] = 10 \log \left[\frac{1}{T} \int_0^T 10^{L_p(t)/10} \text{d}t \right] \qquad [3.29]$$

where $p_{\text{rms}}(t)$ is the "instantaneous" RMS value (calculated over a short period of time) corresponding to the pressure level $L_p(t)$. The equivalent level can also be calculated from a level recording in dB(A), $L_A(t)$. It is then denoted $L_{Aeq,T}$ and the A-weighted RMS value is denoted by $p_{A,\text{rms}}$.

The equivalent level is used, for example, in occupational health and safety regulations. It is required that workers' exposure to noise does not exceed a certain equivalent level, for example, 80 dB(A), calculated over an 8-h working day. The instantaneous exposure level can therefore be above this threshold at times, but the integrated noise dose over the working day must remain below it. For example, a workstation exposed to a continuous noise level of 83 dB(A) would not allow an operator to work there for more than 4 h per day.

For environmental acoustics applications, such as noise monitoring near major roads, indicators can be built from equivalent levels calculated and weighted over different periods of the day. A 24-h day can be divided into two periods ("day" and

"night") or three periods ("day", "evening" and "night"). Equivalent levels, usually in dB(A), are calculated for each of these periods, for example, L_{day}, L_{evening} and L_{night}. These levels can then be combined, with the "evening" and "night" levels penalized to reflect the increased discomfort during those periods to construct the "day-night" level L_{dn} or the "day-evening-night" level L_{den}. The formulas that give L_{dn} and L_{den} are as follows:

$$L_{\text{dn}} = 10 \log \left[\frac{1}{24} \left(T_{\text{day}} 10^{L_{\text{day}}/10} + T_{\text{night}} 10^{(L_{\text{night}}+10)/10} \right) \right] \quad [3.30]$$

$$L_{\text{den}} = 10 \log \left[\frac{1}{24} \left(T_{\text{day}} 10^{L_{\text{day}}/10} + T_{\text{evening}} 10^{(L_{\text{evening}}+5)/10} + T_{\text{night}} 10^{(L_{\text{night}}+10)/10} \right) \right] \quad [3.31]$$

where T_{day}, T_{evening} and T_{night} are the durations of the day, evening and night periods, respectively. For the L_{dn} level, the day period is typically defined between 7 and 22 h ($T_{\text{day}} = 15$ h) and the night period between 22 and 7 h ($T_{\text{night}} = 9$ h). For the L_{den} level, the day period is typically defined between 7 h and 19 h ($T_{\text{day}} = 12$ h), the evening period between 19 h and 23 h ($T_{\text{evening}} = 4$ h) and the night period between 23 h and 7 h ($T_{\text{night}} = 8$ h). However, the definition of these periods may differ according to the regulations of each country. In the expressions for L_{dn} and L_{den} in equations [3.30] and [3.31], it can be noticed that the noise level in the evening is penalized by 5 dB and the noise level at night is penalized by 10 dB.

3.9.3. *Transient signals: sound exposure level and energy spectral density*

In contrast to permanent signals, transient signals are defined as signals that are non-zero only for a short period of time. They are associated with single events such as an airplane flying overhead, the sound of an explosion, a hammer blow, etc. An example of a transient signal is shown in Figure 3.9.

For such phenomena, the notion of PSD is not appropriate since it requires averaging over a long period of time (in principle tending to infinity) as defined in section 3.1, and this results in a zero power value. On the other hand, the energy contained in the signal, which remains finite and non-zero, is of interest. The energy contained between t_1 and t_2 of a pressure signal $p'(t)$,is defined by:

$$E = \int_{t_1}^{t_2} p'^2(t) \mathrm{d}t \quad [3.32]$$

If the signal is zero for $t < t_1$ and for $t > t_2$, the bounds of the integral can be replaced by $-\infty$ and $+\infty$, and the signal energy remains a finite quantity. Transient signals are sometimes referred to as finite-energy signals, as opposed to permanent signals which have infinite energy. As a reminder, the energy of a pressure fluctuation signal is expressed in Pa^2.s.

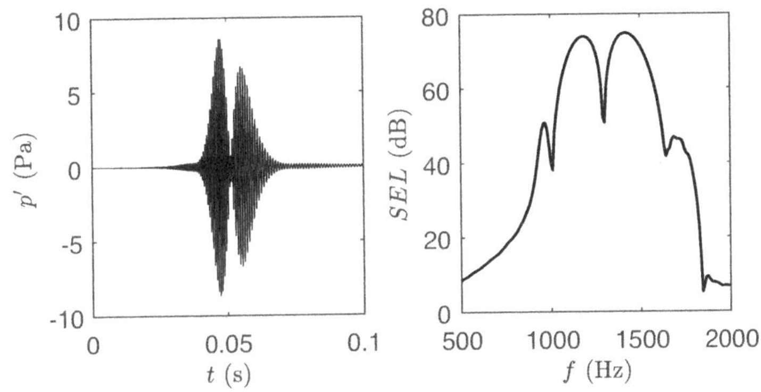

Figure 3.9. *Example of a transient signal (on the left) and of the corresponding energy spectrum, calculated with a constant bandwidth $\Delta f = 10$ Hz (on the right)*

The sound exposure level (SEL) L_E is the level in decibels associated with the signal energy. It is defined by:

$$L_E = 10\log\left(\frac{E}{E_{\text{ref}}}\right) = 10\log\left(\frac{1}{p_{\text{ref}}^2 t_{\text{ref}}}\int_{t_1}^{t_2} p'^2(t)\text{d}t\right) \quad [3.33]$$

The reference energy $E_{\text{ref}} = p_{\text{ref}}^2 t_{\text{ref}}$ is calculated with a reference duration $t_{\text{ref}} = 1$ s. If a recording of the "instantaneous" pressure level $L_p(t)$ (or $L_A(t)$ in dB(A)) corresponding to the instantaneous RMS values $p_{\text{av}}(t)$ (or $p_{A,\text{av}}(t)$ respectively) is provided, the sound exposure level can be written as:

$$L_E = 10\log\left(\frac{1}{p_{\text{ref}}^2 t_{\text{ref}}}\int_{t_1}^{t_2} p_{\text{rms}}^2(t)\text{d}t\right) = 10\log\left(\frac{1}{t_{\text{ref}}}\int_{t_1}^{t_2} 10^{L_p(t)/10}\text{d}t\right) \quad [3.34]$$

The noise exposure level in dB(A) is denoted by L_{AE}.

The definitions of L_E and L_{eq} are very similar, but they should not be confused. L_{eq} is associated with the average power of a permanent signal over a duration T, while L_E is associated with the total sound energy contained in a transient event.

It is also possible to look more closely at the spectral content of a transient signal. The notion of PSD is replaced by that of energy spectral density $E(f)$:

$$E(f) = p(f)p^*(f) \quad [3.35]$$

and the corresponding energy spectral density level (expressed in dB/Hz) is:

$$L_E(f) = 10\log\left(\frac{E(f)}{E_{\text{ref}}}\right) = 10\log\left(\frac{E(f)}{p_{\text{ref}}^2 t_{\text{ref}}}\right) \quad [3.36]$$

The energy spectrum corresponding to the transient signal plotted on the left is shown on the right side of Figure 3.9.

3.10. Exercises

1) **The mother of all speakers**

To study sound propagation over long distances in the atmosphere, the University of Mississippi built a source with an estimated RMS sound power of 20 kW. It is connected to the atmosphere by a horn with an output diameter of 3 m.

a) Assuming that the wave generated in the outer section of the horn is approximately plane, estimate the pressure level L_p in this section, as well as the RMS value of the velocity fluctuations.

b) At a large distance, the source can be considered to generate a spherical wave. What is the maximum distance at which this source can be detected when the background noise level is 30 dB? (it is assumed that a source can be detected when its level is 3 dB higher than that of the background noise). What physical phenomena limit this detection distance in practice?

2) **Fan noise level**

The table below corresponds to the data provided by a fan manufacturer for the pressure level L_p and the power level L_W measured as a function of the frequency (in octave bands) for a centrifugal fan (flow rate $q_v = 0.31$ m^3/s, pressure rise $\Delta p = 3,400$ Pa).

Frequency (Hz)	125	250	500	1,000	2,000	4,000	8,000
L_p (dB)	53	60	69	73	71	67	64
L_W (dB)	66	71	81	85	83	79	76

Table 3.4. *Noise generation by an industrial fan*

The pressure level was measured in free field at a distance of 1.5 m from the fan.

a) What is the overall sound pressure level? What is the overall sound pressure level in dB(A)?

b) Assuming an omnidirectional point source, what should be the difference in dB between the power level and the pressure level (measured at 1.5 m). What do you think of the data in the table?

c) What is the "acoustic efficiency" η of the fan, defined as the ratio between the sound power and its mechanical power, whose expression is $W = 1/2\rho_0 q_v \Delta p$?

3) **Converting noise into electrical energy?**

Since Edison's patent in 1878, the idea of converting sound or noise into usable energy has periodically resurfaced. To evaluate this possibility, an order of magnitude calculation is proposed below.

Consider a noise barrier built along a highway the surface of which has been covered with piezoelectric sensors that convert sound pressure forces into electrical energy. The barrier has a length of 100 m and a height of 4 m; the sound pressure level along the wall is measured to be 84 dB. Assuming incident plane waves, estimate the available sound power in watts) and conclude.

4) **Noise exposure at work**

For the assessment of noise exposure at work, the reference time of the A-weighted equivalent level [3.29] is T= 8h. Two action thresholds are defined from this equivalent level:

– the lower action threshold, of 80 dB(A), which triggers information actions and the provision of personal protective equipment;

– the upper action threshold, of 85 dB(A), which requires the implementation of a noise reduction program.

Where the daily noise exposure includes several periods of exposure to different noise levels, it may useful to use a scoring system of assigning noise exposure points to each period; these points are added up over the day to form an overall noise exposure "score".

The noise exposure $E_{A,T}$, expressed in Pa^2.h, is defined as the integral of the A-weighted square pressure over the time T expressed in hours, $E_{A,T} = \int_0^T p_A^2 dt$. The point system is based on the definition of the value of an exposure point as 1/100th of Pa^2.h.

a) Show that 100 points of exposure are almost equal to the noise exposure corresponding to the upper action threshold.

Phase	1	2	3	Overall results
Equivalent level (dB(A))	98	88	80	
Duration	40 mins	3 h	4 h	
Exhibition points				
Contribution to exposure				

Table 3.5. *Scenario of noise exposure at work*

b) Consider the scenario defined in the table. Fill in the blanks in this table by calculating the exposure points corresponding to each phase, and then the total number of points. Derive the equivalent level for an 8-h period.

Calculate the contribution of each phase in % to the noise exposure and deduce the priorities of the action plan to be implemented.

4

Reflection and Transmission Phenomena

In Chapter 2, we mainly considered the propagation of waves in an infinite medium, although we introduced different boundary conditions. In the following, we develop several applications in which plane or spherical waves are totally or partially reflected at plane interfaces.

In the first part of this chapter, we describe the reflection of plane waves at normal incidence at interfaces separating two fluids, then a fluid and a solid characterized by its surface impedance. The extension to plane wave transmission through any number of fluid layers is then covered using the impedance transfer technique. Finally, the transmission through a rigid wall is addressed; in particular, the well-known *mass law* of partitions for sound insulation between rooms is presented.

In the second part, obliquely incident waves are introduced. At the interface between two fluids, the Snell's relations are established, as well as the possible existence of special regimes associated either with the total reflection of the waves and the existence of evanescent waves in the second medium, or conversely, with the perfect transmission of incident waves therein. For plane waves incident on a solid wall supporting bending waves, the concepts of coincidence frequency and critical frequency are introduced.

Very often, when solving structural acoustic radiation problems, a boundary condition that consists of imposing the velocity of the vibrating surface is used. This is only a simplification, since this velocity can actually be modified by the reaction of the fluid. However, it is sufficient for most cases. To illustrate the consequences of taking into account fluid–structure couplings, we consider the case of a piston subjected to a given force and interacting with the fluid contained in a closed tube.

In a final section, the radiation of a point source located in the vicinity of a rigid wall is addressed by introducing the notion of image sources.

4.1. Reflection and transmission of normally incident plane waves

We first consider the simple case of two fluids with respective properties (ρ_1, c_1) and (ρ_2, c_2) separated by the plane $x_1 = 0$ (Figure 4.1).

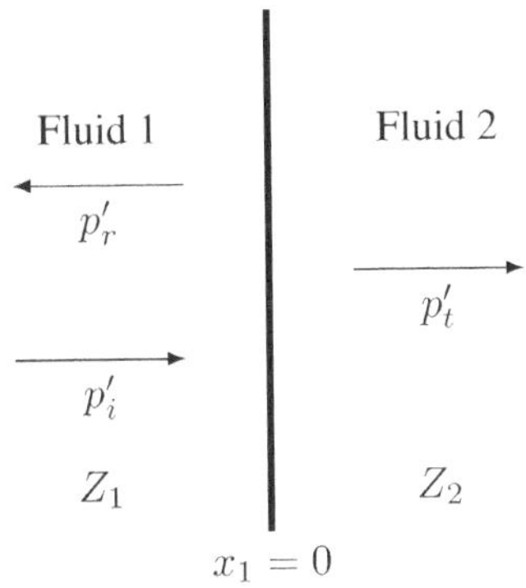

Figure 4.1. *Reflection and transmission of plane waves at normal incidence at the interface between two fluids of characteristic impedance Z_1 and Z_2, respectively*

A plane wave with arbitrary time evolution propagates in fluid 1 in the $+x_1$ direction; it is partially reflected at the interface separating the two fluids, and partially transmitted in the second fluid. The indices i, r and t are associated with each of these waves, respectively. The boundary conditions to be imposed are the continuity of the pressure and normal velocity fluctuations at the interface.

The general form of the incident, reflected and transmitted pressure fluctuations is as follows: $p'_i(t - x_1/c_1), p'_r(t + x_1/c_1), p'_t(t - x_1/c_2)$ and the acoustic velocities are related to the pressures by $p'_i = \rho_1 c_1 u'_i$, $p'_r = -\rho_1 c_1 u'_r$ and $p'_t = \rho_2 c_2 u'_t$.

By defining the pressure-amplitude reflection coefficient by $r = p'_r/p'_i$ and the pressure-amplitude transmission coefficient by $t = p'_t/p'_i$ at $x_1 = 0$, the following formulas are obtained:

$$r = \frac{Z_2 - Z_1}{Z_2 + Z_1} \quad [4.1]$$

and

$$t = \frac{2Z_2}{Z_2 + Z_1} \quad [4.2]$$

where $Z_1 = \rho_1 c_1$ and $Z_2 = \rho_2 c_2$ are the characteristic impedances of the two media. The time evolution of the reflected wave is identical to that of the incident wave, except for the sign if $r < 0$. The transmitted wave has a similar shape to the incident wave, but it is compressed or stretched in time if $c_2 \neq c_1$.

The coefficients r and t are defined from the wave *amplitude* ratios. In many applications, the quantities of interest are the transmission and reflection coefficients of *energy*, α_r and α_t. They are easily determined from the expression for the acoustic intensity of plane waves. It follows that:

$$\alpha_r = \frac{(Z_2 - Z_1)^2}{(Z_2 + Z_1)^2} \qquad [4.3]$$

and

$$\alpha_t = \frac{4Z_1Z_2}{(Z_2 + Z_1)^2} \qquad [4.4]$$

It can be seen that $\alpha_r + \alpha_t = 1$, which simply reflects the conservation of energy at the interface. It can also be noted that these formulas play a symmetrical role with respect to Z_1 and Z_2; energy transmission is the same whether the wave comes from medium 1 or medium 2.

It is important to note that the reflection and transmission coefficients depend only on the impedances of the two media and not on the particular values of the densities and speeds of sound. If the impedances of the two media are very different, the transmission of energy at the interface will be very small. On the other hand, perfect transmission between two different media is possible, provided that their impedances are equal (it is then said that there is *impedance matching*). For example, there are materials whose density and speed of sound are different from those of water, but whose impedance is very close to that of water. The use of such a material (called $Rho-C$ rubber) protects hydrophones without significantly reflecting incident waves, making it possible to accurately measure the local pressure in water.

4.2. Reflection of a harmonic plane wave on an impedance surface

We now consider the important example of a plane wave propagating in fluid 1 and partially reflected by an absorbing material characterized by its surface impedance Z. The analysis is performed for a harmonic wave, since the surface impedance Z usually takes complex values that vary with the frequency. The incident and reflected pressures thus take the form:

$$p_i = I \exp(-ik_1 x_1); \qquad p_r = R \exp(+ik_1 x_1); \qquad k_1 = \frac{\omega}{c_1} \qquad [4.5]$$

where the incident amplitude I can be chosen to be a real quantity without restricting generality. The boundary condition at the surface of the material located at $x_1 = 0$ is written as $p = Zu_1$, where $p = p_i + p_r$ and $u_1 = (p_i - p_r)/Z_1$. We immediately get:

$$r = \frac{R}{I} = \frac{Z - Z_1}{Z + Z_1} \qquad [4.6]$$

This formula can be used to calculate r for a material of known impedance. The reflection coefficient is generally complex, which results in a reduction in amplitude and a phase shift of the reflected wave with respect to the incident wave.

Inverting this formula gives access to the material specific impedance Z as a function of r:

$$Z = Z_1 \frac{1 + r}{1 - r} \qquad [4.7]$$

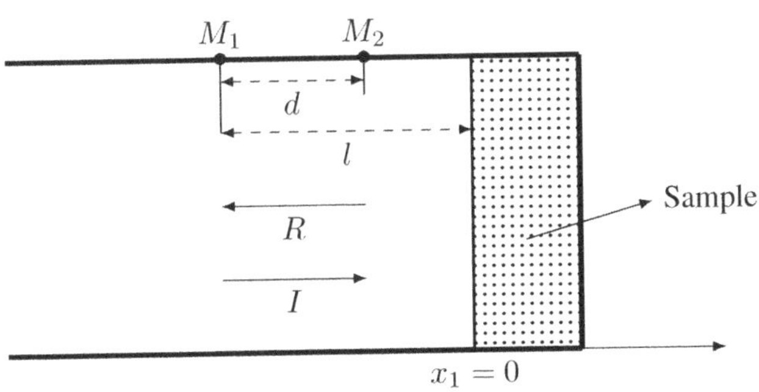

Figure 4.2. *Impedance tube for measuring the surface impedance of a sample using the two-microphone method*

The latter formula is the basis for the use of the impedance tube (or Kundt's tube) to determine the impedance of a material from the measurement of its reflection coefficient (Figure 4.2). Inside a rigid tube, usually cylindrical with a diameter D, a loudspeaker emits a single-frequency wave that is partially reflected by a sample of the material to be characterized; the sample rests on a thick, rigid base. For low frequencies, whose wavelength is greater than $1.71D$ as we will see in Chapter 8, only plane waves can propagate in the tube and the formulas given above can therefore be applied. Wall-mounted microphones measure the pressure fluctuation at two points separated by a distance d (less than half a wavelength). The specific

impedance of the material normalized by the characteristic impedance of the fluid, or *reduced impedance* ζ, is given by:

$$\zeta = \frac{Z}{Z_1} = \mathrm{i}\frac{\sin k_1(l-d) - h_{12}\sin k_1 l}{h_{12}\cos k_1 l - \cos k_1(l-d)} \qquad [4.8]$$

where l is the distance between the sample surface and the microphone the furthest from it. The quantity h_{12} denotes the ratio of the complex pressure amplitudes, $h_{12} = p_2/p_1$. For an emitted broadband signal, h_{12} is the frequency response function of the system consisting of the two microphones.

It is important to note that the measured impedance depends not only on the intrinsic properties of the material, but also on the thickness of the sample and of the substrate on which it is placed. A material behavior model is required to identify the intrinsic properties of the material.

Figure 4.3 shows an example of impedance measurements made using this method on foam samples of different thicknesses and resting on a rigid base. The frequency range studied extends from 200 Hz to 2 kHz.

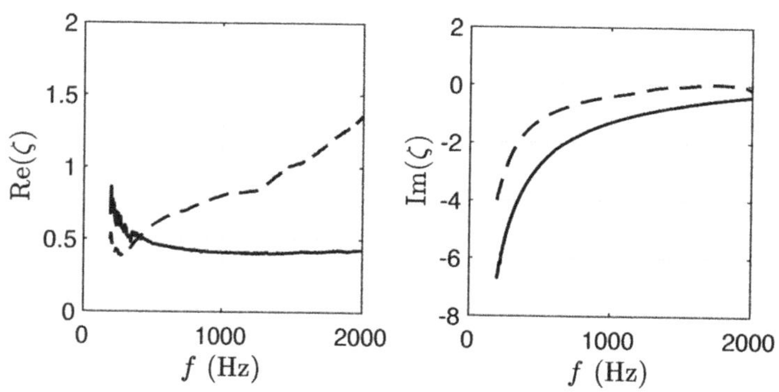

Figure 4.3. *Real (resistance) and imaginary (reactance) parts of the reduced impedance of two samples of the same foam with two different thicknesses, 3 cm (solid line) and 5 cm (dotted lines), and placed on a rigid backing*

The absorption coefficient of a material corresponds to the fraction of the incident acoustic energy that is transmitted to the material. The conservation of the energy flux at the interface leads to:

$$\alpha = 1 - |r|^2 \qquad [4.9]$$

The left side of Figure 4.4 shows the absorption coefficient of these samples over a wider frequency range (200 Hz – 6 kHz). The thicker sample is more efficient,

especially at low frequencies. It can also be seen that the absorption coefficient tends to reach a maximum close to unity before decreasing slightly, exhibiting oscillations probably related to interferences between the direct and reflected waves inside the material. The right side of the figure illustrates the influence of the condition on the back side of the material under test. In the first case, the sample is placed directly on a rigid base, while in the second case, it is separated from the base by an air gap. It can be seen that in this second configuration the absorption coefficient is significantly increased over the low frequency range being studied; on the other hand, the absorption behavior at higher frequencies (not shown in the figure) is oscillatory and degraded. This technique of offsetting an absorbing material from a rigid support is generally used in order to partially compensate for its low efficiency at low frequencies.

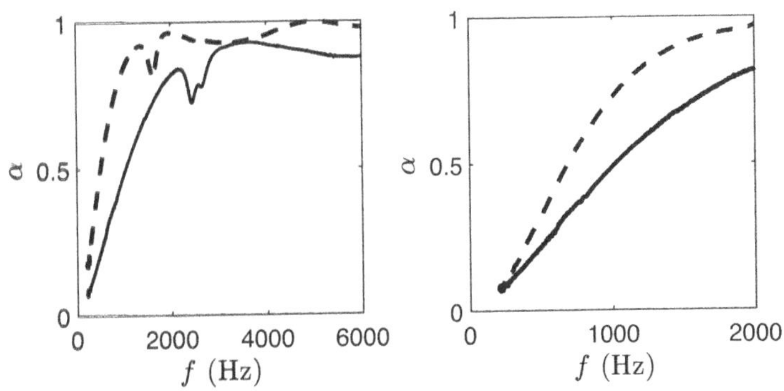

Figure 4.4. *On the left side: Absorption coefficient of two foam samples with a thickness of 3 cm (solid line) and 5 cm (dotted lines). On the right side: Absorption coefficient of the same foam sample with a thickness of 3 cm placed on a rigid base (solid line) or separated from it by an air gap with a thickness of 2 cm (dashed line)*

4.3. Multilayer media

The previous analysis can be extended to the transmission of harmonic plane waves through an arbitrary number of layers of different densities and sound speeds. Determining the reflection and transmission coefficients is not difficult in principle, but expressing the pressure and velocity continuity conditions at the various interfaces quickly becomes a very tedious exercise. An effective technique to avoid this difficulty is to use the concept of impedance transfer.

4.3.1. *Impedance transfer*

The principle of the method consists of expressing the impedance Z_{local} that is seen by an incident wave originating from medium 1 when it reaches the interface

with the second medium. The reflection coefficient can then be calculated using a formula similar to that obtained for two layers:

$$r = \frac{Z_{local} - Z_1}{Z_{local} + Z_1} \quad [4.10]$$

and the fractions of reflected and transmitted energy will be calculated by $\alpha_r = |r|^2$ and $\alpha_t = 1 - \alpha_r$.

Z_{local} is usually determined by starting at the last interface and "bringing" the impedance back to the first interface. In fact, a simple rigidity boundary condition or surface impedance condition can often be introduced at the position of the last interface. If the last medium is assumed to extend to infinity, the pressure-velocity relationship for plane progressive waves can be used. The impedance at $x_1 - e$ can be expressed from its known value at x_1 by the following relation:

$$Z(x_1 - e) = Z\frac{Z(x_1)\cos ke + \mathrm{i}Z\sin ke}{Z\cos ke + \mathrm{i}Z(x_1)\sin ke} \quad [4.11]$$

a formula where $Z = \rho c$ and $k = \omega/c$ are the characteristic impedance and wavenumber in the slice of thickness e comprised between $x_1 - e$ and x_1.

To obtain this formula, the pressure and velocity in the considered slice, where a wave propagating in the $+x_1$ direction and another in the $-x_1$ direction coexist, can be expressed by:

$$p = A\exp(-\mathrm{i}kx_1) + B\exp(+\mathrm{i}kx_1) \quad [4.12]$$

$$Zu_1 = A\exp(-\mathrm{i}kx_1) - B\exp(+\mathrm{i}kx_1) \quad [4.13]$$

where A and B are two arbitrary complex constants. At each interface x_1 and $(x_1 - e)$, the ratio of pressure to velocity is formed, and the ratio $B/A\exp(-2\mathrm{i}kx_1)$ is expressed as a function of the impedance $Z(x_1)$. The value of the impedance of the first interface can thus be obtained back step by step.

4.3.2. *Transmission through three media*

As an application, we consider wave propagation in a system consisting of three media, where the medium 2 has a thickness e. The third medium is assumed to be of infinite extent, so the terminal impedance is $Z(e) = Z_3 = \rho_3 c_3$. The impedance at the first interface (at $x_1 = 0$) is then:

$$Z(0) = Z_2\frac{Z_3\cos k_2e + \mathrm{i}Z_2\sin k_2e}{Z_2\cos k_2e + \mathrm{i}Z_3\sin k_2e} \quad [4.14]$$

and the reflection coefficient is:

$$r = \frac{Z(0) - Z_1}{Z(0) + Z_1} = \frac{(Z_2 Z_3 - Z_1 Z_2)\cos k_2 e + \mathrm{i}\left(Z_2^2 - Z_1 Z_3\right)\sin k_2 e}{(Z_2 Z_3 + Z_1 Z_2)\cos k_2 e + \mathrm{i}\left(Z_2^2 + Z_1 Z_3\right)\sin k_2 e} \quad [4.15]$$

There are several special cases that are interesting to study. First, if the media 1 and 3 are identical (we thus have a layer of material of thickness e that differs from an otherwise homogeneous medium), the reflection coefficient is reduced to:

$$r = \mathrm{i}\frac{\left(Z_2^2 - Z_1^2\right)\sin k_2 e}{2Z_1 Z_2 \cos k_2 e + \mathrm{i}\left(Z_2^2 + Z_1^2\right)\sin k_2 e} \quad [4.16]$$

and with a little algebra, it can be shown that the energy transmission coefficient α_t is equal to:

$$\alpha_t = \frac{1}{1 + \frac{1}{4}\left(\frac{Z_2}{Z_1} - \frac{Z_1}{Z_2}\right)^2 \sin^2 k_2 e} \quad [4.17]$$

which can be used to express the sound level reduction (in dB) caused by the insertion of the layer, or transmission loss factor, $\mathrm{TL} = -10\log\alpha_t$. The reflection coefficient cancels out (and the energy transmission coefficient becomes equal to unity) when the thickness of the intermediate layer is equal to an integer number of half-wavelengths in that medium ($k_2 e = n\pi$). The layer then becomes transparent to acoustic waves for the corresponding frequencies.

A second interesting situation arises when the impedance of medium 3 is very large compared to that of medium 1, which means that the reflection would be quasi-perfect, thus preventing any transmission of acoustic energy. The use of a well-chosen intermediate layer makes it possible, at some specific frequencies, to obtain an *impedance matching* and thus to overcome this difficulty which occurs, for example, when using piezoelectric transducers in liquids or gases. As a matter of fact, formula [4.15] shows that by choosing a layer with a thickness equal to a quarter of the wavelength ($k_2 e = \pi/2$) and an impedance equal to the square root of the product of the impedances of the two media to be coupled, $Z_2 = \sqrt{Z_1 Z_3}$, the reflection coefficient cancels out and the transmission is therefore perfect. The middle layer that allows this matching is often called a *quarter-wave layer* for obvious reasons.

4.3.3. *Transmission of a harmonic plane wave through a thin wall*

We now consider a thin wall that behaves as a pure mass (without stiffness or damping) and on which a plane harmonic wave is incident (Figure 4.5). The noise

reduction provided by this wall is to be calculated; it is given by the transmission loss factor, since the fluid is the same on both sides of the wall. The motion of the wall under the action of the acoustic pressure forces is given by Newton's second law:

$$m_p \frac{du'_p}{dt} = p'_i + p'_r - p'_t \quad \Longrightarrow \quad i\omega m_p u_p = p_i + p_r - p_t = I + R - T \quad [4.18]$$

where m_p denotes the mass per unit area of the wall, that is, the product of its density ρ_p and its thickness h, and u'_p its velocity. If the wall is not porous, the continuity of the velocity at the interfaces implies that:

$$u_p = u_i + u_r = u_t = p_t/Z_1 = T/Z_1 = (I - R)/Z_1 \quad [4.19]$$

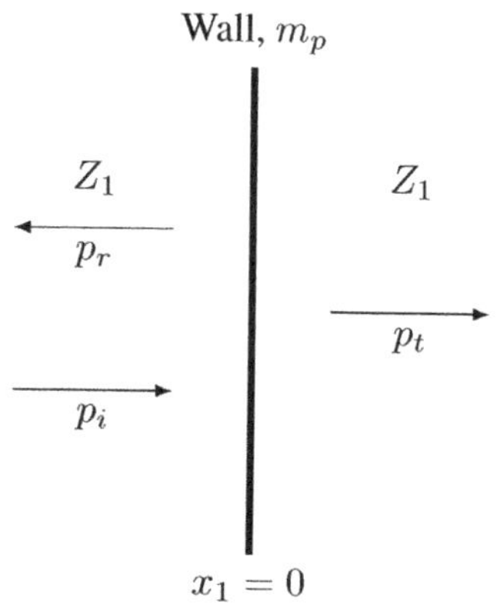

Figure 4.5. *Plane wave reflection at a normal incidence on a thin wall*

When writing equations [4.18] and [4.19], the wall is assumed to be thin enough with respect to the wavelength so that the pressures and the velocities can be evaluated at $x_1 = 0$.

The reflection and transmission coefficients are easily obtained from [4.18] and [4.19]. The transmission coefficient is given by:

$$t = \frac{1}{1 + i\omega m_p / 2Z_1} \quad [4.20]$$

and therefore the transmission loss factor is:

$$\mathrm{TL} = -10\log(|t|^2) = 10\log\left(1 + (\omega m_p/2Z_1)^2\right) \quad [4.21]$$

For a massive wall or for high frequencies when $\omega m_p/2Z_1 \gg 1$, the first term of this expression can be neglected, thus yielding the normal incidence *mass law*: the level reduction increases by 6 dB when the frequency doubles or when the mass per unit area doubles. A generalization to the behavior of walls involving stiffness and damping is suggested in section 4.8.

4.4. Reflection and transmission of plane waves at the interface between two fluids: oblique incidence

This configuration, which is of great practical interest, is often encountered, especially at liquid–gas interfaces, the surface of oceans being a typical case. Consider a plane wave incident on the plane $x_1 = 0$ separating two fluids with characteristics (ρ_1, c_1) and (ρ_2, c_2), respectively. This wave generally gives rise to a reflected plane wave and a transmitted plane wave in the second fluid. The angles of incidence, reflection and transmission, measured with respect to the normal to the interface, are denoted by θ_i, θ_r and θ_t, respectively (Figure 4.6; the directions of propagation are assumed to be in the plane $x_3 = 0$). The complex amplitudes of the incident, reflected and transmitted pressures take the following forms:

$$p_i = I\exp(-\mathrm{i}(k_1^i x_1 + k_2^i x_2)); \quad k_1^i = \omega/c_1 \cos\theta_i; \quad k_2^i = \frac{\omega}{c_1}\sin\theta_i \qquad [4.22]$$

$$p_r = R\exp(-\mathrm{i}(k_1^r x_1 + k_2^r x_2)); \quad k_1^r = -\omega/c_1 \cos\theta_r; \quad k_2^r = \frac{\omega}{c_1}\sin\theta_r \qquad [4.23]$$

$$p_t = T\exp(-\mathrm{i}(k_1^t x_1 + k_2^t x_2)); \quad k_1^t = \omega/c_2 \cos\theta_t; \quad k_2^t = \frac{\omega}{c_2}\sin\theta_t \qquad [4.24]$$

The condition of pressure continuity at the interface $x_1 = 0$ is written as:

$$I\exp(-\mathrm{i}k_2^i x_2) + R\exp(-\mathrm{i}k_2^r x_2) = T\exp(-\mathrm{i}k_2^t x_2) \qquad [4.25]$$

and since this condition has to be satisfied for every x_2, it follows that:

$$I + R = T; \qquad k_2^i = k_2^r = k_2^t \qquad [4.26]$$

These conditions on the components of the wave vectors along the interface can be expressed as a "trace-velocity matching principle", which can be used directly in other contexts: the trace along the interface of the cause (the incident wave) imposes the trace of the effects (the reflected and transmitted waves)[1]. It leads to Snell's relations:

$$\theta_i = \theta_r; \qquad \frac{\sin\theta_i}{c_1} = \frac{\sin\theta_t}{c_2} \qquad [4.27]$$

1 This "principle", which is an exact consequence of the equations of motion, remains valid for any time-dependent wave.

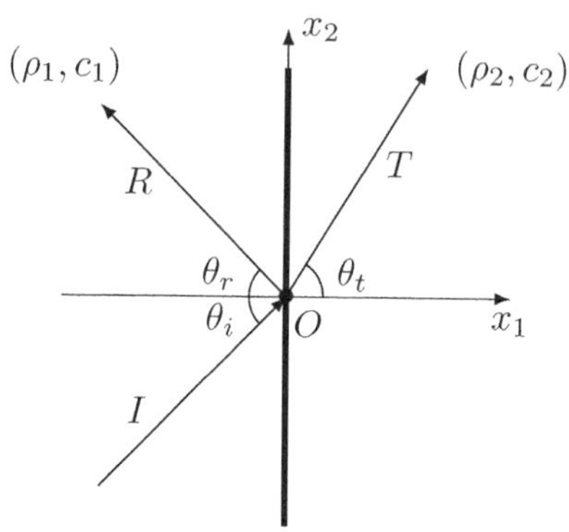

Figure 4.6. *Reflection and transmission for an obliquely incident plane wave on the interface between two fluids with characteristics* (ρ_1, c_1) *and* (ρ_2, c_2)

The continuity of the normal component of the acoustic velocity at the interface provides:

$$\frac{I - R}{\rho_1 c_1} \cos\theta_i = \frac{T}{\rho_2 c_2} \cos\theta_t \qquad [4.28]$$

Introducing the characteristic impedances of the two fluids, Z_1 and Z_2, the reflection and transmission coefficients (for complex amplitudes) are given by:

$$r = \frac{R}{I} = \frac{Z_2 \cos\theta_i - Z_1 \cos\theta_t}{Z_2 \cos\theta_i + Z_1 \cos\theta_t} \qquad [4.29]$$

and

$$t = \frac{T}{I} = \frac{2Z_2 \cos\theta_i}{Z_2 \cos\theta_i + Z_1 \cos\theta_t} \qquad [4.30]$$

Figure 4.7 shows the pressure field associated with the propagation of a plane wave partially reflected at the interface between two fluids of equal density, but with different sound speeds ($c_2/c_1 = 1.2$). The angle of incidence is $45°$, the angle of transmission is about $58°$ and the reflection and transmission coefficients are $r = 0.23$ and $t = 1.23$. In the left part of the figure, the reflected wave has not been plotted to emphasize the refraction of the wavefronts at the interface. The right-hand side shows the total field, in which checkerboards typical of constructive and destructive interference between incident and reflected waves can be seen for $x_1 < 0$.

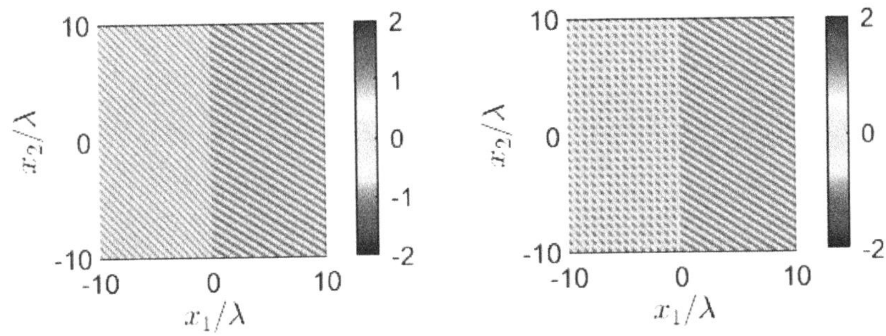

Figure 4.7. *Reflection and transmission of a harmonic plane wave at the interface between two fluids ($\rho_2/\rho_1 = 1$; $c_2/c_1 = 1.2$, $\theta_i = 45°$). On the left, only the incident wave and the transmitted wave are plotted. On the right, the total field is plotted. For a color version of this figure, see www.iste.co.uk/juve/acoustics1.zip*

From these general formulas, several remarkable behaviors can be identified depending on the relative values of the speed of sound and the impedance in the two media. For example, the reflection coefficient cancels out when $Z_2 \cos\theta_i = Z_1 \cos\theta_t$. If this is the case, then there is an angle of incidence at which perfect wave transmission is possible between two fluids with different characteristics. Using Snell's relations, this occurs for a particular value of the angle of incidence called the *angle of intromission* denoted θ_o and defined by:

$$\sin^2\theta_o = \frac{\left(\frac{Z_2}{Z_1}\right)^2 - 1}{\left(\frac{Z_2}{Z_1}\right)^2 - \left(\frac{c_2}{c_1}\right)^2} = \frac{\left(\frac{\rho_2}{\rho_1}\right)^2 - \left(\frac{c_1}{c_2}\right)^2}{\left(\frac{\rho_2}{\rho_1}\right)^2 - 1} \qquad [4.31]$$

For this angle to be real, the following conditions must be met:

$$\frac{\rho_2}{\rho_1} < \frac{c_1}{c_2} < 1 \quad \text{or} \quad \frac{\rho_2}{\rho_1} > \frac{c_1}{c_2} > 1 \qquad [4.32]$$

Thus, the angle of intromission exists only for certain pairs of fluids, and it does not appear for problems involving air–water or water–air interfaces.

A second interesting behavior is that of total reflection, when no propagative wave can exist in the second medium. As a matter of fact, Snell's law states that:

$$\sin\theta_t = \frac{c_2}{c_1}\sin\theta_i \qquad [4.33]$$

The angle of transmission is therefore real only if the angle of incidence is less than the critical angle θ_{cr} given by:

$$\theta_{cr} = \sin^{-1}\frac{c_1}{c_2} \qquad [4.34]$$

The critical angle exists only if the speed of sound in the medium 1 is less than that in medium 2. Beyond the critical angle, the trace-velocity of the incident wave along the interface $c_1/\sin\theta_i$ becomes less than c_2 (the incident wave is said to be subsonic for the medium 2). The transmitted pressure field can be expressed as:

$$p_t(x_1, x_2) = T\exp\left(-\mathrm{i}\frac{\omega}{c_1}\sin\theta_i x_2\right)\exp\left(-a_t\frac{\omega}{c_2}x_1\right) \qquad [4.35]$$

where a_t is defined as:

$$a_t = \sqrt{\left(\frac{c_2}{c_1}\sin\theta_i\right)^2 - 1} \qquad [4.36]$$

This is an evanescent wave that decays exponentially in the direction normal to the interface. The reflection coefficient r is then complex, of unit magnitude and it is written as:

$$r = \frac{Z_2\cos\theta_i + \mathrm{i}a_t Z_1}{Z_2\cos\theta_i - \mathrm{i}a_t Z_1} = \exp(\mathrm{i}2\psi) \qquad [4.37]$$

and the transmission coefficient is:

$$t = 2\cos\psi\exp(\mathrm{i}\psi) \qquad [4.38]$$

with

$$\tan\psi = \frac{Z_1 a_t}{Z_2\cos\theta_i} \qquad [4.39]$$

The reflected wave has the same magnitude as the incident wave, but is phase-shifted with respect to the latter by a value that varies with the angle of incidence but is independent of frequency.

Formula [4.39] shows that the transmission coefficient can be greater than unity (it is exactly 2 at the critical angle), which may seem surprising. However, it should be

remembered that this coefficient is only partially related to the energy transmission, since the impedances of the two media are different. This can be made clearer by evaluating the reflection and energy transmission coefficients α_r and α_t, which are calculated based on the normal sound intensity associated with each incident, reflected and transmitted wave. This normal intensity I_1 is given by $1/2\mathrm{Re}(pu_1^*)$, and we obtain:

$$\begin{cases} \alpha_r = |r|^2 & \text{if} \quad \theta_i < \theta_{cr} \\ \alpha_r = 1 & \text{if} \quad \theta_i \geq \theta_{cr} \end{cases} \qquad [4.40]$$

and

$$\begin{cases} \alpha_t = |t|^2 Z_1 \cos\theta_t / Z_2 \cos\theta_i & \text{if} \quad \theta_i < \theta_{cr} \\ \alpha_t = 0 & \text{if} \quad \theta_i \geq \theta_{cr} \end{cases} \qquad [4.41]$$

where the reflection r and the transmission coefficients t are given by [4.29] and [4.30].

Again, it should be noted that $\alpha_r + \alpha_t = 1$ as a consequence of the conservation of the energy flux at the interface.

Figure 4.8 illustrates the behavior of the magnitudes of the reflection and transmission coefficients in a scenario where both a critical angle and an intromission angle are present.

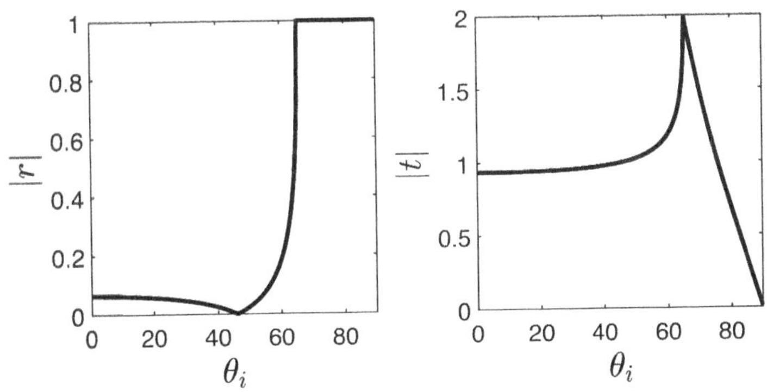

Figure 4.8. *Amplitudes of the reflection and transmission coefficients for a plane wave incident on the interface between two fluids with density and velocity ratios equal to* $\rho_2/\rho_1 = 0.8$ *and* $c_2/c_1 = 1.1$*, respectively. One should note the existence of both an intromission angle* $\theta_o \approx 46^\circ$ *and a critical angle* $\theta_{cr} \approx 65.4^\circ$

Other interesting cases can be mentioned where there is a very strong impedance contrast between the two media. An example of practical interest is that of the

air–water interface. For an incident wave on the air side, the ratio of the velocities is about $c_2/c_1 = 4.4$ and that of the impedances $Z_2/Z_1 = 2,300$. The reflection coefficient is always greater than 0.99, so very close to unity and the pressure transmission coefficient is close to 2. On the other hand, almost no energy is transmitted, because $\alpha_t < 0.002$. The interface thus appears to be infinitely rigid and the zero normal velocity limit condition (Neumann condition) can then be associated with the Helmholtz equation. When the waves originate from the water, the previous relations are reversed and $Z_2/Z_1 = 0.43\ 10^{-3}$. The reflection coefficient, this time, is close to $r = -1$ (and thus the reflected wave is almost in phase opposition relatively to the incident wave), the transmission coefficient is close to zero, $t < 10^{-3}$, and again practically no energy is transmitted. The condition $t \approx 0$ is equivalent to a total fluctuating pressure in the water near the interface close to zero, and thus a simplified boundary condition at the free surface, $p = 0$ (Dirichlet condition), is systematically used for underwater propagation.

4.5. Plane wave transmission through a thin wall: oblique incidence

When a thin wall is exposed to an obliquely incident plane wave, it is subjected to bending vibrations that develop with a trace-velocity imposed by the incident wave. In the plane (x_1, x_2), the displacement ξ along x_1 of the assumed homogeneous and isotropic wall is the solution of the following *thin plate equation* (see, e.g. [FAH 07]):

$$\rho_p h \frac{\mathrm{d}^2\xi}{\mathrm{d}t^2} + D\nabla^4\xi = p'_i + p'_r - p'_t \qquad [4.42]$$

where D is the plate bending stiffness, related to its physical properties by:

$$D = \frac{Eh^3}{12(1-\nu_p^2)} \qquad [4.43]$$

E is the elastic modulus of the material, ν_p is its Poisson's ratio, h is the wall thickness and ρ_p is its density. Since the displacement ξ varies only along x_2, the bilaplacian ∇^4 is reduced to $\mathrm{d}^4\xi/\mathrm{d}x_2^4$.

For a harmonic wave, the dispersion relation of the free plate should first be considered by looking for a solution in the form of a free plane wave propagating along x_2.

The displacement has the form $\xi \propto \mathrm{Re}(\exp(\mathrm{i}\omega t - \mathrm{i}k_f x_2))$, where k_f is called the plate bending wavenumber. For k_f, we get the following expression:

$$k_f = \omega^{1/2}K^{-1/4} = \omega^{1/2}\left(\frac{12\rho_p(1-\nu_p^2)}{Eh^2}\right)^{1/4} \qquad [4.44]$$

where $K = D/\rho_p h = D/m_p$ and m_p is the mass per unit area of the plate. In particular, this formula shows that the propagation of bending waves is dispersive, with the phase velocity varying as the square root of the frequency according to:

$$c_f = \frac{\omega}{k_f} = \omega^{1/2} \left(\frac{Eh^2}{12\rho_p(1-\nu_p^2)} \right)^{1/4} \qquad [4.45]$$

When a harmonic plane wave excites the plate, the trace-velocity matching principle imposes the value of the bending wavenumber, which is therefore equal to $k_{trace} = \omega/c_{trace} = \omega/c_1 \sin\theta$. The equation of motion then becomes:

$$\mathrm{i}\omega m_p(1 - K k_{trace}^4/\omega^2) u_p = I + R - T \qquad [4.46]$$

where $u_p = \mathrm{i}\omega\xi$ is the wall velocity, along x_1.

The continuity of the normal velocities at the interface always gives $I - R = T$ and the wall velocity is related to the transmitted pressure by $u_p = T\cos\theta/Z_1$. From this, the pressure-amplitude transmission coefficient t is derived as:

$$t(\theta) = \frac{1}{1 + (\mathrm{i}\omega m_p \cos\theta/2Z_1)(1 - K\sin^4\theta\,\omega^2/c_1^4)} \qquad [4.47]$$

and the corresponding coefficient of energy transmission α_t is:

$$\alpha_t(\theta) = \frac{1}{1 + (\omega^2 m_p^2 \cos^2\theta/4Z_1^2)(1 - K\sin^4\theta\,\omega^2/c_1^4)^2} \qquad [4.48]$$

This coefficient is equal to one, and therefore the energy transmission is total, for a given value of the frequency, depending on the angle of incidence and the physical characteristics of the wall. This frequency is called *coincidence frequency*, f_{coin} and it is given by the relation:

$$f_{coin} = K^{-1/2} \frac{c_1^2}{2\pi \sin^2\theta} \qquad [4.49]$$

For this frequency, the trace-velocity and the velocity of the free bending wave coincide and the wall appears transparent to acoustic waves. The smallest of the coincidence frequencies is obtained for a grazing incidence ($\theta = \pi/2$). It is called the *critical frequency*, $f_{cr} = K^{-1/2}c_1^2/2\pi$.

The transmission loss factor is expressed as:

$$\begin{aligned} \mathrm{TL}(\theta) &= -10\log(\alpha_t(\theta)) \\ &= 10\log\left(1 + (\omega^2 m_p^2 \cos^2\theta/4Z_1^2)(1 - K\sin^4\theta\,\omega^2/c_1^4)^2\right) \end{aligned} \qquad [4.50]$$

Figure 4.9 shows the transmission loss factor of an 8-mm-thick aluminum plate for different incidences. Below the coincidence frequency, the behavior follows the mass law (the transmission loss doubles when the frequency doubles, see section 4.3.3), but the transmission loss is reduced, especially for high values of the angle of incidence. For frequencies much higher than the coincidence frequency, the evolution is controlled by the stiffness of the plate and shows an increase of 18 dB/octave. The minimum value of the transmission loss that appears around the coincidence frequency is imposed by the damping of the material, which we have not taken into account in this analysis and which can be easily introduced by using a complex elastic modulus, which will be done later.

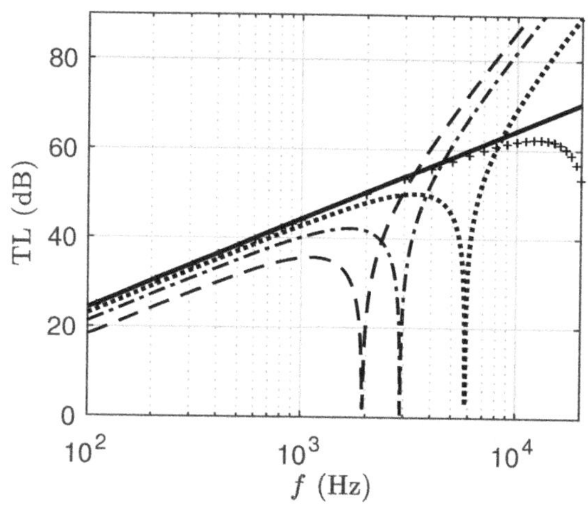

Figure 4.9. *Transmission loss of an aluminum plate with a thickness h =8 mm in air ($\rho_p = 2,700$ kg/m^3, $E = 7.2 \times 10^{10}$ Pa, $\nu_p = 0.34$). The angles of incidence are 15° (+ symbols), 30° (dotted line), 45° (dash-dotted) and 60° (dashed line). The normal incidence mass law is plotted with a solid line*

In practice, incident acoustic waves reach a wall in many directions. This is the case, for example, for a window exposed to road noise. In general, the distribution of incident directions is not known and it is often assumed that there is a homogeneous distribution of these directions, where all directions are equally probable (this is referred to as *diffuse field*). This is particularly used in room acoustics, as we will see in Chapter 10. The diffuse field transmission coefficient τ_r (also known as random incidence transmission coefficient) is obtained by integrating the expression of the transmitted normal intensity divided by the incident normal intensity between 0 and $\pi/2$:

$$\tau_r = \frac{\int_0^{\pi/2} \alpha_t(\theta)\cos\theta\sin\theta \mathrm{d}\theta}{\int_0^{\pi/2} \cos\theta\sin\theta \mathrm{d}\theta} = \int_0^{\pi/2} \alpha_t(\theta)\sin 2\theta \mathrm{d}\theta \quad [4.51]$$

It is not possible to obtain an exact analytical form of this expression over the entire frequency range. However, the mass-controlled portion, below the critical frequency, can be evaluated analytically. The result for the transmission loss factor is as follows (prove it!):

$$\mathrm{TL}_r(\theta) \approx \mathrm{TL}(\theta = 0) - 10\log(0.23\log(\mathrm{TL}(\theta = 0))) \quad [4.52]$$

which shows a reduction in the efficiency of the wall compared to the normal incidence mass law, and this reduction increases slightly with the frequency.

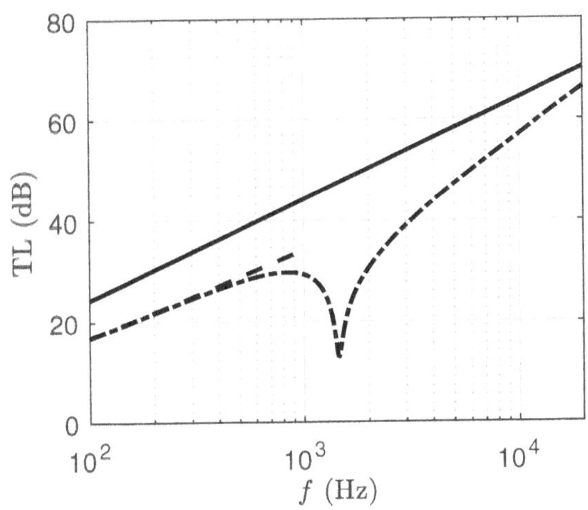

Figure 4.10. *Transmission loss in air of an 8-mm-thick aluminum plate. The dotted dashed line is the random incidence response. The solid black line is the plate response under normal incidence (mass law). The dashed black line is the mass-controlled part under random incidence (formula [4.52])*

Above the critical frequency, for any given value of the angle of incidence within the integration range, a coincidence phenomenon occurs, which reduces the efficiency of the wall. This behavior is illustrated in Figure 4.10 for which a numerical integration has been performed for the aluminum plate considered above, but with the introduction of a small internal damping through the use of a complex elastic modulus $E(1 + \mathrm{i}\eta)$, with $\eta = 0.05$. The reduction of the wall efficiency below the critical frequency varies in this example from 7 to 10 dB and the slope of the

curve above this frequency is about 9 dB/octave. The minimum transmission loss factor appears at the critical frequency, here equal to 1,450 Hz, and its value is controlled by the internal damping of the material.

In practice, these results should be considered as merely indicative. In room acoustics, the experimental measurements of the diffuse sound transmission of a partition separating two rooms do not agree well with [4.52]. In this situation, the finite size of the partitions, the associated boundary conditions and the imperfect diffuse field significantly modify their acoustic behavior. An empirical expression for the observed sound reduction is often used, where the sound reduction of the partition is equal to the transmission loss at normal incidence minus 5 dB. For more details, the reader is referred to [FAH 07] or [BAR 03].

4.6. Piston-tube coupling

Many acoustics problems are presented in the following way: solve the Helmholtz equation (or the wave equation) in a given domain, associated with pressure or normal velocity boundary conditions imposed at its boundaries. This is especially the case when the acoustic radiation of a vibrating structure is to be calculated. While this approach is often sufficient, it should not obscure the fact that the vibration velocity of the structure is rarely imposed in practice. The structure can most often be modeled as a mechanical system *subject to an imposed force* and the resulting structure velocity depends on the "load" associated with the ambient fluid. A *coupled problem* between the motion of the structure and the acoustic behavior of the fluid must then be solved.

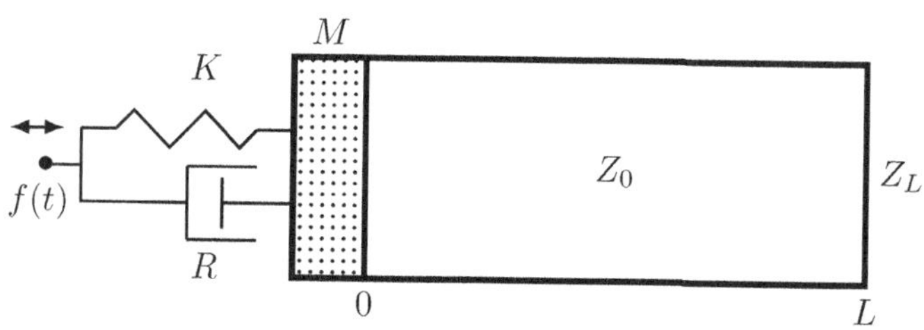

Figure 4.11. *A single degree-of-freedom oscillator coupled to a fluid contained in a tube closed on an impedance Z_L*

In this section, we consider a simple, one-dimensional case of this fluid–structure coupling, which makes it possible to identify a number of typical behaviors [CHA 16]. The system under consideration is depicted in 4.53; it consists of a 1 degree-of-freedom mechanical oscillator that excites the fluid contained in a cavity of length L and cross-section S. The motion of the interface between the structure and the fluid (referred to as the piston) is governed by a second-order differential

equation of the type mass (denoted M) - spring (stiffness K) - damping (denoted by R). The system is subjected to an imposed force $f(t)$ and the reaction of the pressure force exerted by the fluid on the piston surface.

The motion of the piston is the solution of the following equation where ξ is the piston displacement, assumed to be of small amplitude, about its equilibrium position in $x_1 = 0$:

$$M\ddot{\xi} + R\dot{\xi} + K\xi = f(t) - Sp'(0,t) \qquad [4.53]$$

By introducing the resonant angular frequency $\omega_0 = \sqrt{K/M}$ and the reduced damping $r_0 = R/(2M\omega_0)$, we get:

$$M\left(\ddot{\xi} + 2r_0\omega_0\dot{\xi} + \omega_0^2\xi\right) = f(t) - Sp'(0,t) \qquad [4.54]$$

In the tube of length L, the acoustic waves are solutions of the one-dimensional wave equation and the coupling between the two motions takes place via the boundary condition of continuity of the velocity at $x_1 = 0$, $\dot{\xi}(t) = u_1'(0,t)$. In the following, we assume that the applied force is sinusoidal, with a complex amplitude $F(\omega)$, and that the acoustic cavity is closed by an impedance Z_L.

We first consider the typical scenario where this impedance is none other than the characteristic impedance of the medium Z_0, thus representing radiation in an infinite tube. The complex amplitude of the displacement of the piston is given by the following expression:

$$\Xi(\omega) = \frac{F}{M}\frac{1}{(\omega_0^2 - \omega^2) + 2\mathrm{i}\omega_0(r_0 + r_a)\omega} \qquad [4.55]$$

where $r_a = Z_0 S/(2M\omega_0)$. It thus can be seen that the presence of the fluid induces an additional damping r_a, which is related to the energy carried by the acoustic waves propagating to infinity. On the other hand, the resonance frequency of the oscillator, $\omega = \omega_0$, has not changed.

A second very interesting case is when the cavity is closed by a perfectly rigid end. An application example is that of an experiment in which a very high acoustic level is to be obtained in a cavity by taking advantage of its resonant character. A naive approach would be to choose a rigid emitter such as a piezoelectric transducer, resonating at the same frequency as the cavity, but it will be seen that it is not that simple.

The impedance transfer formula (see section 4.3.1) gives the value of the impedance at the piston surface, $-iZ_0 \cot(\omega L/c_0)$, from which the value of the

pressure as a function of the piston displacement is derived; it is then substituted into equation [4.54], and the result is:

$$\Xi(\omega) = \frac{F}{M} \frac{1}{\left(\omega_0^2 - \omega^2 + Z_0 S\omega \cot(\omega L/c_0)/M\right) + 2\mathrm{i}\omega_0 r_0 \omega} \quad [4.56]$$

The cavity acts as an added stiffness that changes the resonances of the coupled system. There is no added damping because no energy is dissipated in the cavity. Resonant frequencies are those that cancel out the real part of the denominator of equation [4.56]. They are solutions of the transcendental equation:

$$\omega_0^2 - \omega^2 + \frac{Z_0 S}{M}\omega \cot(\omega L/c_0) = 0 \quad [4.57]$$

If the acoustic additional stiffness $K_a = \rho_0 S c_0^2/L$ is introduced, it can be shown that the intensity of the coupling depends on the ratio of this stiffness to that of the oscillator, $\epsilon = K_a/K$, and the resonance equation can be rewritten in a dimensionless form for the variable $X = \omega L/c_0$:

$$\tan X = \frac{\epsilon X X_0^2}{X^2 - X_0^2} \quad [4.58]$$

It is helpful to visualize the system response by plotting the left and right sides of this relation separately, which gives the eigenfrequencies of each system individually. These are obtained for $\tan X = 0$, hence $X = m\pi$, and for $X = X_0$, respectively. The eigenfrequencies of the coupled system correspond to the intersections of the two curves associated with each side of equation [4.58]. In Figure 4.12, we plot the response obtained for $\epsilon = 0.02$ and 0.2, when the first natural frequency of the cavity coincides with the resonance frequency of the oscillator ($X_0 = \pi$). Even for the weakest coupling, it is observed that two coupled frequencies appear on either side of $X = X_0 = \pi$, 2.840 and 3.458 in the example. In contrast, the upper natural frequencies of the tube are only slightly modified (6.325 instead of 2π for the second). These effects are obviously very pronounced when the coupling is increased.

Many other examples of fluid-structure interactions in the acoustic context and others can be found in [AXI 06].

4.7. Reflection of spherical waves and image sources

The problem of the reflection and transmission of spherical waves at a plane interface is much more complex than that of plane waves. We will only address situations where reflections are ideal: a source above a rigid plane and (in the

exercise section) a source in the presence of a perfectly resilient surface (also called a "soft" surface).

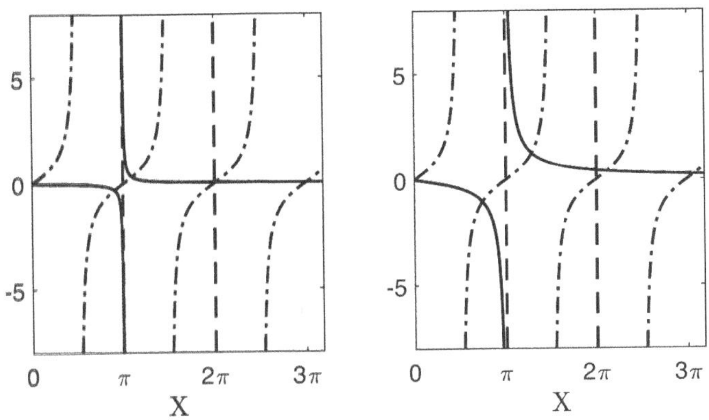

Figure 4.12. *Eigenfrequencies of an oscillator coupled to a tube with a rigid termination. The curves corresponding to the first and second members of equation [4.58] are plotted in dash-dotted and solid lines, respectively. The intersections of these curves define the eigenfrequencies of the coupled system. Two coupling parameter values are considered,* $\epsilon = 0.02$ *on the left side of the figure and* $\epsilon = 0.2$ *on the right side*

We will therefore study the radiation of an omnidirectional point source, emitting a spherical harmonic wave, placed at a distance h above a rigid plane located at $x_3 = 0$ (Figure 4.13). We will use spherical coordinates, where the pressure depends spatially only on (r, θ), given the symmetry of revolution around the x_3 axis where the source is located. The problem consists of solving the Helmholtz equation in the half-space $x_3 > 0$ associated with the boundary condition of rigidity:

$$u_3 = 0 \quad \Longleftrightarrow \quad \frac{\partial p}{\partial x_3} = 0 \quad \text{for} \quad x_3 = 0 \qquad [4.59]$$

We know the shape of a wave propagating from a point source only if the medium is unbounded. We therefore extend the original problem to this situation by introducing an image source, symmetrical to the real one with respect to the plane $x_3 = 0$, and with the same complex amplitude as the real source. A simple symmetry consideration shows that the total velocity, the vector sum of the individual contributions of the sources, has a zero normal component in the plane $x_3 = 0$. The new problem is therefore equivalent to the initial problem in the half-space $x_3 > 0$, since the same Helmholtz equation has to be solved, with a unique and identical singularity in this half-plane (at $x_3 = h$) and with the same boundary conditions.

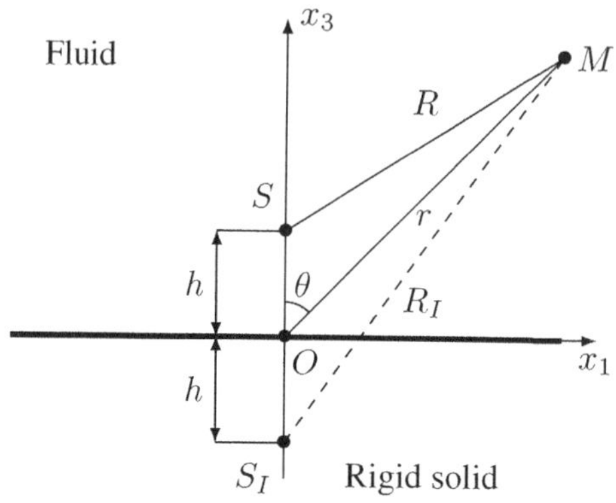

Figure 4.13. *A spherical wave source placed above a perfectly rigid plane. The source S is located at a distance h from the plane, and S_I denotes the image source of the actual source with respect to the plane*

The pressure field at any point M is the simple superposition of the spherical waves propagating from the real source and the image source:

$$p(r,\theta,\varphi) = A\left(\frac{\exp(-\mathrm{i}k_0 R)}{R} + \frac{\exp(-\mathrm{i}k_0 R_I)}{R_I}\right) \qquad [4.60]$$

where R and R_I are the distances from the source and its image to the point M:

$$R = \sqrt{r^2 - 2hr\cos\theta + h^2}; \qquad R_I = \sqrt{r^2 + 2hr\cos\theta + h^2} \qquad [4.61]$$

which formally solves the problem. However, the evaluation of the pressure at larger distances from the sources is very instructive. When $r \gg h$, a Taylor series expansion limited to the first order in h/r for phase terms and to the zeroth order for amplitudes can be performed; this approximation is called the Fraunhofer approximation. It then follows that:

$$p(r,\theta) = \frac{2A\exp(-\mathrm{i}k_0 r)}{r}\cos(k_0 h\cos\theta) \qquad [4.62]$$

This expression shows that at a large distance from the sources ($r \gg h$), which defines the *geometric far-field*, the pressure radiated corresponds to that of a *single but directive spherical wave*. Radial and angular dependence are decoupled and the directivity of the emission can be characterized as the ratio of the radiated pressure to that produced at the same distance by an omnidirectional point source. We define

$D(\theta)$ as the magnitude of this function, whose plot in polar coordinates constitutes the directivity diagram of the source:

$$D(\theta) = 2|\cos(k_0 h \cos\theta)| \qquad [4.63]$$

This function is plotted on the left side of Figure 4.14 for different values of the product $k_0 h$ (i.e. the ratio of source height to wavelength). When this product is very small ($kh \ll 1$), the characteristic size of the source region (here composed of the real source and the image source) is small compared to the wavelength. Then the directivity is nearly uniform, and the source-image pair radiates as a single omnidirectional point source located at the origin (which is the barycenter of the two sources), and its amplitude is the sum of the individual amplitudes. As the frequency increases, and thus as $k_0 h$ increases, the directivity becomes more and more pronounced due to the increasing interference between the fields from each of the sources. We will see later that the behavior illustrated with this relatively simple case is an example of problems involving more complicated distributions of sources, whether discrete or continuous. On the right side of Figure 4.14, the pressure magnitude is plotted for $k_0 h = 10$ in a region covering the near field and the beginning of the far-field. The interference patterns are clearly visible as well as the gradual transition to the far-field directivity lobes.

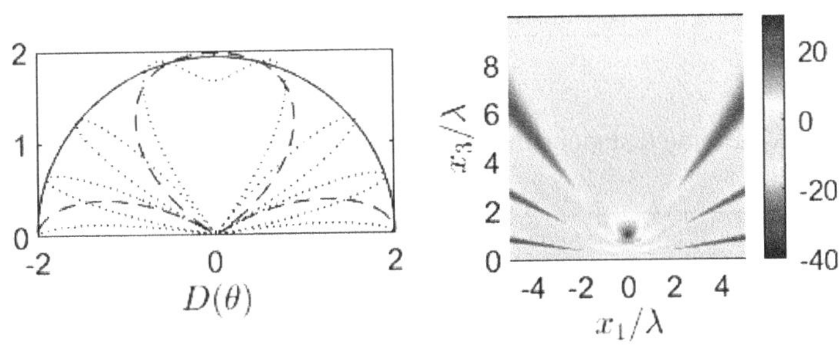

Figure 4.14. *The directivity patterns of a spherical source placed above of a perfectly rigid plane are plotted on the left side of the figure for $k_0 h = 0.25$ (solid line), 3 (dashed line) and 10 (dotted line). The right side of the figure illustrates the transition from the near field to the far-field by plotting the pressure level in dB obtained for $k_0 h = 10$. For a color version of this figure, see www.iste.co.uk/juve/acoustics1.zip*

The calculation of the power radiated by the source as a function of its distance from the reflection plane proves to be very informative. As is often the case, the idea is to integrate the acoustic intensity on a surface located far away from the sources, in this case a hemisphere centered on the origin and with a radius r tending towards infinity. The far-field intensity is purely radial and equal to $pp^*/2Z_0$. Taking the symmetries into account, the radiated power is written as:

$$W = \frac{4\pi|A|^2}{Z_0} \int_0^{\pi/2} \cos^2(k_0 h \cos\theta) \sin\theta d\theta = \frac{2\pi|A|^2}{Z_0} \left(1 + \frac{\sin 2k_0 h}{2k_0 h}\right) \quad [4.64]$$

If this expression is compared to that corresponding to the radiation of the same source in infinite space, $W_0 = 2\pi|A|^2/Z_0$, the following relation is obtained:

$$\frac{W}{W_0} = \left(1 + \frac{\sin 2k_0 h}{2k_0 h}\right) \quad [4.65]$$

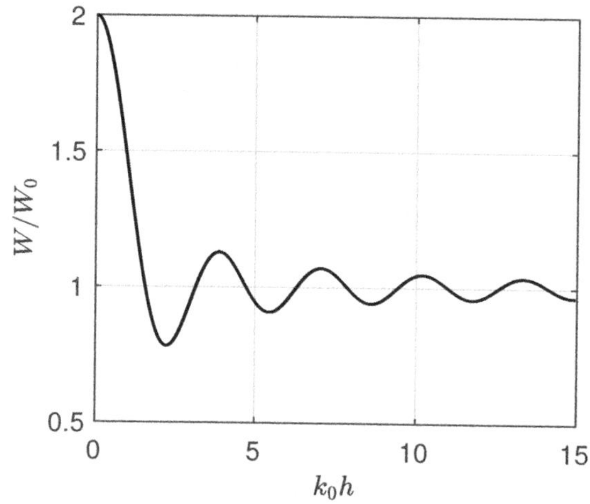

Figure 4.15. *Evolution with $k_0 h$ of the ratio between the power emitted by a source placed above a perfectly rigid plane and that radiated by the same source in unbounded space*

The evolution of this ratio with $k_0 h$ is plotted in Figure 4.15. For low values of this parameter, that is, when the source–image pair is compact, the radiated power is twice that of the real source in an infinite space. As $k_0 h$ increases, the power ratio oscillates around unity, with a minimum around 0.793. It deviates by less than 5% from unity for $k_0 h > 10$, that is, when the distance from the source to the rigid plane is greater than about 1.5λ. It can therefore be seen that the power of the source is significantly modified by its immediate surroundings (on the wavelength scale). Conversely, from the moment the distance to the plane becomes greater than a few wavelengths, the power radiated by the source can be considered to be approximately equal to its value in free space.

4.8. Exercises

1) **Helium layer for noise insulation**

The noise level generated by the Ariane 5 rockets during the launch phase is extremely high and can cause damage to the payload. To reduce this level by a few decibels under the fairing, it was considered to use an intermediate layer of helium. In this gas, the speed of sound and the density are $c_0 = 973$ m/s and $\rho_0 = 0.179$ kg/m^3 at ambient pressure and for $T = 273.15$ K.

A very simple model of the effect of such a device is as follows: consider a layer of helium with a thickness of $e = 0.16$ m separating two air-filled regions.

a) The waves incident on the fairing are assumed to be plane waves arriving at the air-helium interface at normal incidence. Estimate the level reduction that can be obtained. Plot this reduction versus the frequency in the range 0–3 kHz (justify the fact that the influence of a very thin membrane around the gas is negligible).

b) The spectrum of the incident noise is assimilated to white noise in the range $0 - -3,000$ Hz. Express the transmitted acoustic intensity as an integral, and then calculate the overall level reduction that can be obtained. You will need to use the following formula:

$$\int \frac{dx}{1 + a^2 \sin^2 x} = \frac{1}{\sqrt{1 + a^2}} \tan^{-1}\left(\sqrt{1 + a^2}\tan x\right)$$

2) **Sound insulation wall**

The wall in section 4.3.3 has now a stiffness K and a viscous damping R. Determine the transmission coefficient for plane waves at normal incidence by emphasizing the role of the angular resonance frequency of the wall ω_0. Show that the transmission loss factor (TL) is controlled by the stiffness at low frequency, when $\omega \ll \omega_0$ (- 6 dB/octave slope) and by the mass at high frequency (+6 dB/octave slope), the damping imposing the minimum value of TL obtained for $\omega = \omega_0$.

3) **Active control of noise in a duct**

At low frequencies, the radiation from a loudspeaker placed on the wall of a long duct with a cross-section S can be represented schematically (except in the immediate vicinity of the loudspeaker) by a source of plane waves propagating upstream and downstream from the mean position of the source denoted x_s. In this model, a volume flow rate $q'(x_s, t)$ is injected at $x_1 = x_s$.

a) Determine the acoustic particle velocity $u'(x_1, t)$ as a function of the injected flow rate $q'(x_s, t)$ at a point located at $x_1 > x_s$; what is the corresponding expression for $x_1 < x_s$? Express the pressure fluctuation at each point $p'(x_1, t)$.

b) Give the corresponding expressions for the complex pressure and velocity amplitudes for a harmonic excitation of the form $q'(x_s, t) = \text{Re}(q(x_s) \exp(i\omega t))$.

c) A secondary source placed at $x_1 = L$ is used to suppress the acoustic radiation for $x_1 > L$ of the primary source whose position is taken as the origin, $x_s = 0$. What is the value of the secondary flow $q'_s(L, t)$ that cancels out the pressure in $x_1 = L$? Interpret this result physically by considering a primary source that emits at a constant level for a very short time interval. Explain why the pressure is then zero to the right of the secondary source ($x_1 > L$) and give the shape of the pressure field for $x < 0$ and $0 < x_1 < L$. How can a practical application be devised if the primary field is unknown?

d) Give the expressions of the pressure field in each region of the duct for a harmonic primary source. Plot the evolution of the pressure magnitude inside the tube, normalized by its value when only the primary source is operating. Consider different values of the ratio of the distance between the sources L to the wavelength λ. Interpret the results and comment in particular on the upstream radiation of the primary-secondary source assembly.

4) **Interference in underwater acoustics: Lloyd's mirror**

A point harmonic source is located at a depth z_s below the free surface of the sea, which is assumed to be a perfectly resilient boundary ($p(z = 0) = 0$) (Figure 4.16).

a) Introduce a symmetrical image source of the underwater source with amplitude and phase chosen to satisfy the boundary condition at the free surface.

b) Determine the exact pressure field at a point $M(r, \theta)$, then under far-field assumptions $r \gg z_s$.

c) Plot the directivity pattern of the far-field pressure for some values of the ratio z_s/λ, for example, 0.1, 1 and 2.

d) In underwater acoustics, it is often of interest to calculate the evolution of the pressure modulus as a function of the horizontal distance ρ and at constant depth z_r. Show that a pattern of constructive and destructive interferences occurs as ρ increases and that the number of relative minima and maxima is limited.

e) Determine the asymptotic evolution of the pressure magnitude as $\rho \gg z_r$, after the last relative maximum has been reached. Comment on and interpret this evolution in comparison with the radiation from a point source in an unbounded space. Plot the evolution of the pressure magnitude calculated from the exact formulation and from the asymptotic formula for the following parameter values, $z_r = 25$ m, $z_r = 50$ m and $f = 150$ Hz.

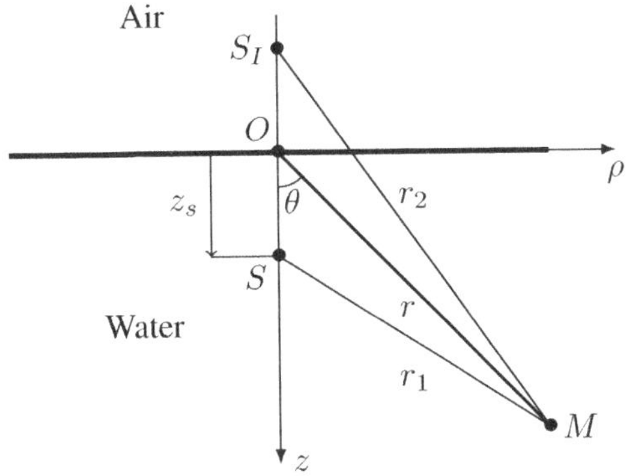

Figure 4.16. *Point source in underwater acoustics: Lloyd's mirror*

5) **Acoustic energy contained in a 1D cavity**

We consider a tube of length L in which harmonic plane waves propagate. They are generated by the motion of a piston located at $x_1 = 0$ with a velocity amplitude U_p.

a) First assume that the termination at $x_1 = L$ is perfectly anechoic (no reflected waves). Give the expressions for the acoustic pressure and velocity at any point in the tube. Deduce therefrom the value of the local acoustic energy density and of the total energy contained in the tube, denoted by E_{ref}.

b) The end of the tube is now assumed to be perfectly reflecting. Find the new expressions for the pressure and velocity inside the tube. Calculate the kinetic and potential energy densities, then their integrated values over the length of the tube, E_c and E_p. Plot the evolution of E_c, E_p and of the total energy E, normalized by E_{ref}, as a function of the product k_0L, for k_0L comprised between 0 and 10. Comment on the curves obtained and their evolution as a function of frequency.

c) For certain values of the frequency (therefore of k_0L), it can be observed that the acoustic energy contained in the rigidly closed tube is lower than that associated with the anechoic termination. Show that this happens for antiresonant conditions. This observation leads to the active control strategy proposed in the following exercise.

6) **Active reduction of acoustic energy contained in a 1D cavity**

The end of the previous tube is replaced by a noise-canceling secondary source, whose velocity amplitude is denoted by U_s.

a) Determine the pressure and velocity fields in the cavity. To perform this calculation, the principle of superposition will be used as follows: the pressure is the sum of the pressure generated by a velocity U_p imposed at $x_1 = 0$ with a rigid end at $x_1 = L$ and the pressure generated by a velocity U_s imposed at $x = L$ with a rigid

end at $x_1 = 0$. The results of question b from the previous exercise can thus be used directly.

b) Calculate the kinetic and potential energy densities, and the total energy contained in the tube.

c) Show that the value $U_s = -U_p \cos k_0 L$ minimizes this energy. Verify that it is then equal to half the energy contained in the tube when its end is anechoic. Show that the "optimal" value of the amplitude of the secondary source U_s makes the tube appear to be working in anti-resonance at all frequencies.

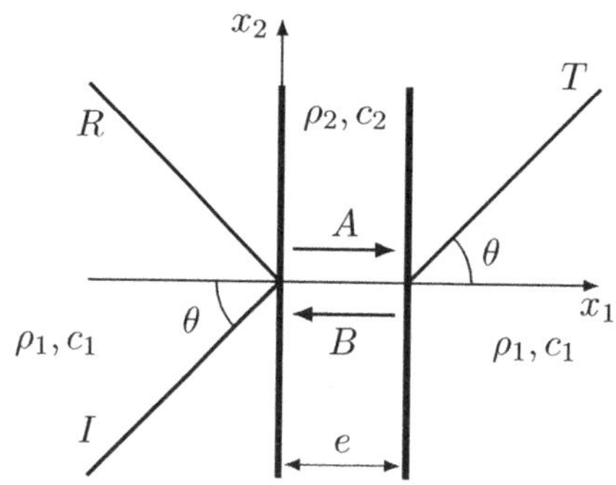

Figure 4.17. *Acoustic tunnel effect*

7) **Acoustic tunnel effect**

We consider a harmonic plane wave at an oblique incidence on a layer of fluid of thickness e and properties (ρ_2, c_2) immersed in a fluid of properties $(\rho_1, c_1 < c_2)$. The angle of incidence, θ, is greater than the critical angle, and therefore only evanescent waves are able to propagate in the intermediate layer. The goal of the exercise is to show that energy can be transmitted through the layer if it is thin enough.

a) Write the continuity conditions of the pressure and the normal component of the velocity at the interface $x_1 = 0$, then at $x_1 = e$.

b) Solve the resulting system of four equations to determine the amplitude of the transmitted wave T (introduce the dimensionless parameter $a_t k_2 e$ where a_t is defined in formula [4.36]). Study the limit behavior for very large and very small values of this parameter.

5

Sound Sources and Green's Functions

In the first chapters, we developed two types of solutions to the wave equation that are fundamental to the understanding of many acoustic phenomena, plane waves and spherical waves. They are solutions of the *homogeneous* wave equation (i.e. with a zero right-hand side), which results from combination of the linearized conservation equations of mass and momentum. Nothing has been said explicitly about how the energy carried by these waves was created. In other words, the *sources of the sound waves* have not been specified, although a connection between spherical waves and point sources of mass has been suggested.

This chapter is devoted to the physical definition of acoustic sources and their associated radiation, first in a three-dimensional space. The formulations applicable to one- and two-dimensions are then given. These sources are related to a violation of the conservation equations at certain points in the fluid (these are called *volume sources*) or to vibrating bodies. These *surface sources* will be discussed in the next chapter.

First, the main types of volume sources are introduced. Then the technique of Green's functions is developed. This allows us to find general solutions of the wave equation and the Helmholtz equation under the action of arbitrary distributions of volume sources.

In many applications, the sources move rapidly; this is especially the case when studying noise from land or air traffic. A final part of the chapter is therefore devoted to the sound of moving sources and the consequences of motion on the evolution of the noise received at a fixed point, both in terms of frequency (Doppler effect) and level (convection amplification).

5.1. Volume sources

The existence of a volume acoustic source is related to a local violation of the conservation equations, caused by the generation of mass, momentum (or forces) or heat within the fluid. In the following, we assume that there is a distribution of fluctuating mass flow rate $q_m(\boldsymbol{x}, t)$ and force $f_i(\boldsymbol{x}, t)$ per unit volume. The linearized continuity and momentum equations are then written as[1]:

$$\frac{\partial \rho'}{\partial t} + \rho_0 \frac{\partial v_i'}{\partial x_i} = q_m(\boldsymbol{x}, t) \quad [5.1]$$

$$\rho_0 \frac{\partial v_i'}{\partial t} + \frac{\partial p'}{\partial x_i} = f_i(\boldsymbol{x}, t) \quad [5.2]$$

By eliminating the velocity fluctuation between these two equations and the density fluctuation using the isentropic relation, we obtain:

$$\nabla^2 p' - \frac{1}{c_0^2} \frac{\partial^2 p'}{\partial t^2} = -\frac{\partial q_m}{\partial t} + \frac{\partial f_i}{\partial x_i} \quad [5.3]$$

In this wave equation, the right-hand side term due to the injected mass flow is isotropic and defines a monopole source. The term due to the presence of a force is in the form of the divergence of a vector and defines a dipole source. These two types of sources are the most frequently encountered. When studying the noise generated by unsteady and turbulent flows in *aeroacoustics*, sources of more complex structure appear, called quadrupoles. In this case, the right-hand side of the wave equation takes the form of the double divergence of a tensor, $\partial^2 q_{ij} / \partial x_i \partial x_j$.

Another type of source is associated with an unsteady local heat addition, which can model, to a first approximation, the noise produced by combustion. The mechanical equations are not modified, but the relationship between pressure and density is altered. The equation of state then actually depends on two thermodynamic variables and not just one. We choose ρ and T as the fundamental variables, and the pressure is then written as $p = p(\rho, T)$. To simplify the notation, we will assume in the following that the fluid is an ideal gas.

By linearizing the equation of state, it follows that:

$$p' = \left(\frac{\partial p}{\partial \rho}\right)_{T,0} \rho' + \left(\frac{\partial p}{\partial T}\right)_{\rho,0} T' = r(T_0 \rho' + \rho_0 T') \quad [5.4]$$

1 The right-hand sides are assumed to be of sufficiently small amplitude to be compatible with the use of linearized equations.

The equation for the evolution of entropy is modified by the presence on the right-hand side of a term related to the unsteady heat source. By specifying the amount of heat supplied per unit mass by $q_c(\boldsymbol{x}, t)$ and using equation [A1.31] for an ideal gas, for which the coefficient of thermal expansion β is equal to $1/T$, the linearized entropy equation is written as:

$$T_0 \frac{\partial s'}{\partial t} = \frac{\partial q_c}{\partial t} = c_p \frac{\partial T'}{\partial t} - \frac{1}{\rho_0} \frac{\partial p'}{\partial t} \qquad [5.5]$$

Eliminating the velocity fluctuation from the mechanical conservation equations provides:

$$\nabla^2 p' - \frac{\partial^2 \rho'}{\partial t^2} = 0 \qquad [5.6]$$

Using equations [5.4] and [5.5] to eliminate the density fluctuation, the following wave equation is obtained:

$$\nabla^2 p' - \frac{1}{c_0^2} \frac{\partial^2 p'}{\partial t^2} = -\frac{\rho_0}{c_0^2} (\gamma - 1) \frac{\partial^2 q_c}{\partial t^2} \qquad [5.7]$$

The unsteady heat addition is therefore at the origin of a monopole source, which is physically the consequence of the isotropic expansion-contraction of the gas induced by the generation of heat.

Now that the propagation equations with sources on the right-hand side have been established, the goal is to solve them using a technique that is general enough to cover all possible cases. This is the purpose of the next section.

5.2. Green's functions for the wave equation

Equation [5.3] is a linear equation with constant coefficients and a non-zero right-hand side (also known as an inhomogeneous equation). Its solution involves calculating the response of the medium at a point $\boldsymbol{x}$ and a time t to an impulse at a time $t_0 < t$ and concentrated at a point $\boldsymbol{y}$. This response is the elementary solution of the wave equation, or *Green's function* g, defined by:

$$\nabla^2 g - \frac{1}{c_0^2} \frac{\partial^2 g}{\partial t^2} = -\delta(t - t_0)\delta(\boldsymbol{x} - \boldsymbol{y}) \qquad [5.8]$$

where δ denotes the Dirac delta function. The dependence of the Green's function on the source $(\boldsymbol{y}, t_0)$ and observer $(\boldsymbol{x}, t)$ pair can be emphasized using the notation $g = g(\boldsymbol{x}, t; \boldsymbol{y}, t_0)$.

There are an infinite number of solutions to equation [5.8]: to any particular solution of the equation with a source term, an arbitrary linear combination of solutions of the homogeneous equation can be added, for example, a combination of plane waves. This degree of freedom in the definition of Green's functions will be used to solve certain problems where the medium is bounded by obstacles of simple shape. The method of image sources introduced in Chapter 4 for the radiation of a spherical wave above a rigid plane is a typical application of this idea.

The simplest Green's function and the one with the most general application corresponds to the radiation of a source in an infinite medium and satisfies the causality condition: the response cannot precede the cause in time. More specifically, the causality condition requires that the Green's function satisfies the following conditions:

$$g(\boldsymbol{x}, t; \boldsymbol{y}, t_0) = 0 \quad \text{and} \quad \frac{\partial g}{\partial t_0}(\boldsymbol{x}, t; \boldsymbol{y}, t_0) = 0 \quad \text{for} \quad t \leq t_0 \qquad [5.9]$$

We will admit that this particular Green's function, called the *free-space Green's function* and denoted by g_0, is given by:

$$g_0(\boldsymbol{x}, t; \boldsymbol{y}, t_0) = \frac{1}{4\pi R}\delta(t - t_0 - R/c_0) \qquad [5.10]$$

where $R = \|\boldsymbol{x} - \boldsymbol{y}\|$ designates the distance between the source and the receiver.

The physical meaning of this expression is very clear: the response in $(\boldsymbol{x}, t)$ reproduces the shape of the source in $(\boldsymbol{y}, t_0)$, with a delay equal to the propagation time between the source and the receiver at the speed of sound, and an amplitude inversely proportional to the source–receiver distance. The similarity to the behavior of a spherical wave is obvious[2].

5.3. General solution of the wave equation in free-space

The general solution of an inhomogeneous equation can be expressed as the convolution product (denoted by $*$) between its elementary solution (the Green's function) and the source term S. Consider:

$$\nabla^2 p'(\boldsymbol{x}, t) - \frac{1}{c_0^2}\frac{\partial^2 p'(\boldsymbol{x}, t)}{\partial t^2} = -S(\boldsymbol{x}, t) \qquad [5.11]$$

2 In the literature, equation [5.8] is often written with a source term without the $-$ sign, or with a factor of 4π multiplying the Dirac delta function. The resulting definition of the Green's function then differs accordingly from [5.10].

The pressure $p'(\boldsymbol{x},t)$ is then given by:

$$p'(\boldsymbol{x},t) = (g_0 * S)_{\boldsymbol{x},t} = \int_\infty \mathrm{d}\boldsymbol{y} \int_\infty \frac{S(\boldsymbol{y},t_0)}{4\pi R}\delta(t - t_0 - R/c_0)\mathrm{d}t_0 \qquad [5.12]$$

The integration with respect to t_0 can be performed by relying on the properties of the Dirac delta function mentioned in Appendix A2.2. This results in:

$$p'(\boldsymbol{x},t) = (g_0 * S)_{\boldsymbol{x},t} = \int_\infty \frac{S(\boldsymbol{y},t - R/c_0)}{4\pi R}\mathrm{d}\boldsymbol{y} \qquad [5.13]$$

This expression is known as the *retarded potential formulation*. Its interpretation is simple: each point in the volume acts as a source of spherical waves. The pressure at the listening position results from the linear superposition of all the contributions, delayed by the propagation time R/c_0 and weighted by the inverse of the distance $R = \|\boldsymbol{x} - \boldsymbol{y}\|$, between each source and the receiver.

5.3.1. *Monopole sources: far-field and compact source region*

When the source term corresponds to a mass-flow fluctuation (monopole source):

$$S(\boldsymbol{y},t_0) = \frac{\partial q_m}{\partial t} \qquad [5.14]$$

the expression [5.13] becomes:

$$p'(\boldsymbol{x},t) = \frac{1}{4\pi}\frac{\partial}{\partial t}\int_\infty \frac{q_m(\boldsymbol{y},t - R/c_0)}{R}\mathrm{d}\boldsymbol{y} \qquad [5.15]$$

Although formally very simple, the use of this formula is often tricky because of the dependence of the integrand on the distance R. The main difficulty in evaluating the integral is to account for time shifts: for a given observation point $\boldsymbol{x}$ at time t, the contribution of each source point must be evaluated at a different emission time $(t - R/c_0)$. Therefore, it is useful to look for approximate formulations that are easier to use. The first idea is to consider an observation point located far away from the source region, which is assumed to be limited in size (the volume integral is therefore restricted to this small region). This is called the "geometric far-field" approximation. If the origin of the reference frame is taken in the source region of characteristic dimension L (ideally at the barycenter of the source distribution, see Figure 5.1),

this is equivalent to simplifying [5.15] under the assumption $x = \|\boldsymbol{x}\| \gg L$. The first-order Taylor series expansion in y/x of the distance R is written as:

$$R^2 = x^2 + y^2 - 2\boldsymbol{x}\cdot\boldsymbol{y} = x^2\left(1 - 2\frac{\boldsymbol{x}\cdot\boldsymbol{y}}{x^2} + \frac{y^2}{x^2}\right) \qquad [5.16]$$

$$R \approx x\left(1 - \frac{\boldsymbol{x}\cdot\boldsymbol{y}}{x^2}\right) \qquad [5.17]$$

In equation [5.15], R is approximated to order 0 for the amplitude (in the denominator) and to order 1 for the retarded time, which appears in the argument of q_m, whose influence is more subtle (this is called the Fraunhofer approximation), and we obtain:

$$p'(\boldsymbol{x},t)_{x\to\infty} = \frac{1}{4\pi x}\frac{\partial}{\partial t}\int_{\infty} q_m\left(\boldsymbol{y}, t - \frac{x}{c_0} + \frac{\boldsymbol{x}\cdot\boldsymbol{y}}{xc_0}\right)\mathrm{d}\boldsymbol{y} \qquad [5.18]$$

It can be seen that the variation of the retarded time involves the projection of the position $\boldsymbol{y}$ of the source points onto the mean direction of observation $\boldsymbol{x}$, a projection that will be denoted y_r in the following (see Figure 5.1).

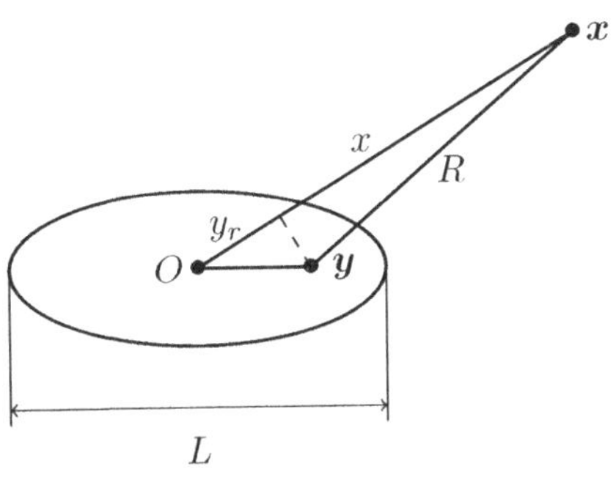

Figure 5.1. *Extended acoustic sources. Geometrical far-field and compact source region*

Expression [5.18] shows a separation between the radial dependence (in $1/x$ as expected) and an angular dependence via the integral of $q_m(\boldsymbol{y}, t - x/c_0 + \boldsymbol{x}\cdot\boldsymbol{y}/xc_0)$ over the source volume. This far-field behavior is characteristic of radiation from extended sources. We already observed such a behavior in the radiation of the source-image pair (see section 4.7), where for the first time a source not reduced to a single point was considered.

Using the spherical coordinates ($x = r, \theta, \varphi$), the far-field pressure can be written as follows:

$$p'(\boldsymbol{x}, t) = p'(x, \theta, \varphi, t) = A(x)D(\theta, \varphi, t) \qquad [5.19]$$

$$A = \frac{1}{4\pi x}; \qquad D(\theta, \varphi, t) = \frac{\partial}{\partial t}\int_{\infty} q_m(\boldsymbol{y}, t - x/c_0 + y_r/c_0)\mathrm{d}\boldsymbol{y} \qquad [5.20]$$

To further simplify expressions [5.18] or [5.20], an additional, and rather restrictive, assumption has to be introduced, which leads to neglecting the variations of the retarded time between the different points of the source volume (corresponding to the variations of y_r/c_0). If this assumption, known as the "source compactness" hypothesis, is satisfied, then we have:

$$p'(\boldsymbol{x}, t) = \frac{1}{4\pi x}\frac{\mathrm{d}Q_m}{\mathrm{d}t}\left(t - \frac{x}{c_0}\right); \qquad Q_m = \int_{\infty} q_m(\boldsymbol{y}, t - x/c_0)\mathrm{d}\boldsymbol{y} \qquad [5.21]$$

where $Q_m(t)$ is the total fluctuating mass flow rate of the source at the current time t. Thus, under the compactness assumption, the acoustic far-field is omnidirectional and equivalent to that of a point source located at the origin and having a mass flow rate $Q_m(t)$.

We will now evaluate the validity of the compact source assumption. To this end, a Taylor series expansion of q_m around $t - x/c_0$ is performed, while maintaining the far-field approximation:

$$q_m\left(t - \frac{x}{c_0} + \frac{y_r}{c_0}\right) = q_m\left(t - \frac{x}{c_0}\right) + \frac{\mathrm{d}q_m}{\mathrm{d}t}\left(t - \frac{x}{c_0}\right)\frac{y_r}{c_0} + O\left(\frac{y_r}{c_0}\right)^2 \qquad [5.22]$$

The value of the time derivative of the mass flow is estimated by introducing a characteristic angular frequency of the fluctuations ω_{char} according to:

$$\frac{\mathrm{d}q_m}{\mathrm{d}t} \approx \omega_{char} q_m = 2\pi f_{char} q_m \qquad [5.23]$$

Variations in the retarded time can be neglected if the second term in the right-hand side of the expansion [5.22] is much smaller than the first one. This condition is satisfied when:

$$\frac{f_{char} L}{c_0} \ll 1 \quad \text{or} \quad \frac{L}{\lambda_{char}} \ll 1 \qquad [5.24]$$

where λ_{char} is the wavelength associated with the characteristic frequency f_{char}.

Retarded time variations are negligible when the size of the source region is small compared to the wavelength of the most energetic part of the pressure signal. The validity of the compactness assumption is therefore limited to a range of low frequencies. By introducing the dimensionless number $\mathrm{He} = f_{char}L/c_0$, called the Helmholtz number, the compactness condition of the source can be expressed as $\mathrm{He} \ll 1$[3].

5.3.2. *Dipole sources*

Thus, subject to a compactness hypothesis, a distribution of mass flow sources is equivalent to a single point source of mass flow $Q_m(t)$ resulting from the integration of distributed local mass flows recorded at the same time. However, it is clear that this approximation is valid only if $Q_m(t) \neq 0$. Indeed, the brutal approximation of the retarded times would lead to the incorrect prediction of a zero radiated pressure in situations where $Q_m(t) = 0$.

When $Q_m(t) = 0$, it is therefore necessary to develop a more accurate approach to calculate the radiated field, even in the low frequency limit. Sources of a dipolar nature associated with a force distribution $f_i(t)$ fall into this category of zero total mass flow sources. This is because the source term appears as a divergence $\partial f_i/\partial x_i$ in equation [5.3]. If this term is integrated, for a fixed value of t, over a volume D bounded by a surface Σ located outside the source domain, along which $f_i = 0$, we indeed obtain:

$$\int_D \frac{\partial f_i}{\partial x_i}(\boldsymbol{y}, t)\mathrm{d}\boldsymbol{y} = \int_\Sigma f_i\, n_i \mathrm{d}\Sigma = 0 \qquad [5.25]$$

Let us now consider the search for the solution of the inhomogeneous wave equation with a dipole source term:

$$\nabla^2 p' - \frac{1}{c_0^2}\frac{\partial^2 p'}{\partial t^2} = \frac{\partial f_i}{\partial x_i} \qquad [5.26]$$

According to the convolution formula [5.13], this solution can be written as:

$$p'(\boldsymbol{x}, t) = -g_0 * \frac{\partial f_i}{\partial x_i} = -\frac{1}{4\pi}\int_\infty \frac{1}{R}\frac{\partial f_i}{\partial y_i}(\boldsymbol{y}, t - R/c_0)\mathrm{d}\boldsymbol{y} \qquad [5.27]$$

This expression, in which the source term appears in the form of a monopole (the integral of a scalar function), is not the best one, because it requires a very

3 When the source region is very elongated, the compactness assumption can be verified in a direction of observation corresponding to the shortest traversal of the source region and not in other directions; the compactness condition is then expressed by $y_r \ll \lambda$.

precise consideration of retarded times in order to correctly evaluate the pressure field. Therefore, it is not suitable for deriving simplified forms, such as for a compact source distribution. To overcome this difficulty, it is necessary to explicitly represent the dipole character of the source. To do this, we rely on the following property of the convolution product (see Appendix A2.2): a differentiation operation can be applied indiscriminately to any of the terms of the convolution product or to the product itself:

$$g_0 * \frac{\partial f_i}{\partial x_i} = \frac{\partial g_0}{\partial x_i} * f_i = \frac{\partial}{\partial x_i}(g_0 * f_i) \qquad [5.28]$$

The last expression is the most interesting here, and it leads to:

$$p'(\boldsymbol{x}, t) = -\frac{1}{4\pi}\frac{\partial}{\partial x_i}\int_\infty \frac{1}{R} f_i(\boldsymbol{y}, t - R/c_0)\mathrm{d}\boldsymbol{y} \qquad [5.29]$$

In the geometric far-field, ($x \gg y$), it is possible to replace R by x in the amplitude factor, and:

$$p'(\boldsymbol{x}, t)_{x \gg y} = -\frac{1}{4\pi}\frac{\partial}{\partial x_i}\left(\frac{1}{x}\int_\infty f_i(\boldsymbol{y}, t - R/c_0)\mathrm{d}\boldsymbol{y}\right) \qquad [5.30]$$

which is decomposed into the sum of two terms according to:

$$p'(\boldsymbol{x}, t)_{x \gg y} = \frac{1}{4\pi}\left(\frac{1}{x}\frac{\partial}{\partial x_i}\int_\infty f_i(\boldsymbol{y}, t - R/c_0)\mathrm{d}\boldsymbol{y} - \frac{x_i}{x^3}\int_\infty f_i(\boldsymbol{y}, t - R/c_0)\mathrm{d}\boldsymbol{y}\right) \qquad [5.31]$$

In the first term on the right-hand side, the spatial derivative can be replaced by a time derivative using the following relation, which is valid for any function of the retarded time $t - R/c_0$:

$$\frac{\partial}{\partial x_i} f_i(t - R/c_0) = -\frac{R_i}{Rc_0}\frac{\partial}{\partial t} f_i(t - R/c_0) \approx -\frac{x_i}{xc_0}\frac{\partial}{\partial t} f_i(t - R/c_0) \qquad [5.32]$$

An order-of-magnitude analysis shows that the second integral in [5.31] is of order λ/x compared to the first one. It is then negligible in the *acoustic* far-field, when $x \gg \lambda$.

By keeping only the first order term in y_r/x in the Taylor series expansion of the retarded time, the pressure takes the following form:

$$p'(\boldsymbol{x}, t)_{x \to \infty} = +\frac{1}{4\pi x c_0}\frac{x_i}{x}\frac{\partial}{\partial t}\int_\infty f_i\left(\boldsymbol{y}, t - \frac{x}{c_0} + \frac{y_r}{c_0}\right)\mathrm{d}\boldsymbol{y} \qquad [5.33]$$

With the additional assumption of compactness of the source region ($L \ll \lambda$), we obtain:

$$p'(\boldsymbol{x},t)_{x\to\infty} \approx +\frac{1}{4\pi x c_0}\frac{x_i}{x}\frac{\partial F_i}{\partial t}\left(t-\frac{x}{c_0}\right) \qquad [5.34]$$

with

$$F_i(t) = \int_{\infty} f_i(\boldsymbol{y},t)\mathrm{d}\boldsymbol{y} \qquad [5.35]$$

The acoustic field is then similar to that produced by a single fluctuating point force, resulting from the integration of the local force distribution taken at the same instant. Therefore, the field has a directivity pattern that varies as $\cos\theta$, where θ is the angle formed by the resultant force $\boldsymbol{F}$ and the direction of observation $\boldsymbol{x}$. This directivity is typical for compact dipoles (Figure 5.2, on the left).

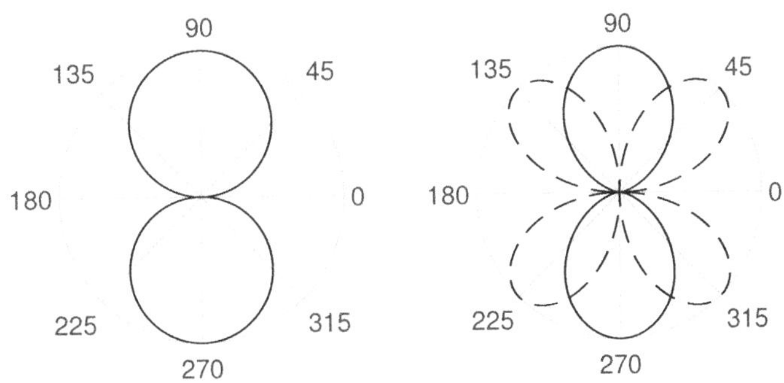

Figure 5.2. *On the left, directivity diagram of a compact dipole with a vertical axis. On the right, directivity diagrams for a lateral quadrupole (dashed line curve) and a longitudinal quadrupole (solid line curve). The diagrams are normalized to the unit and plotted in the plane* (x_2, x_3)

Schematically, a point (or compact) dipole can be represented as the combination of two monopolar mass flow sources of equal magnitude but opposite sign, separated by a very small distance in one direction (the dipole axis); this is shown in Figure 5.3.

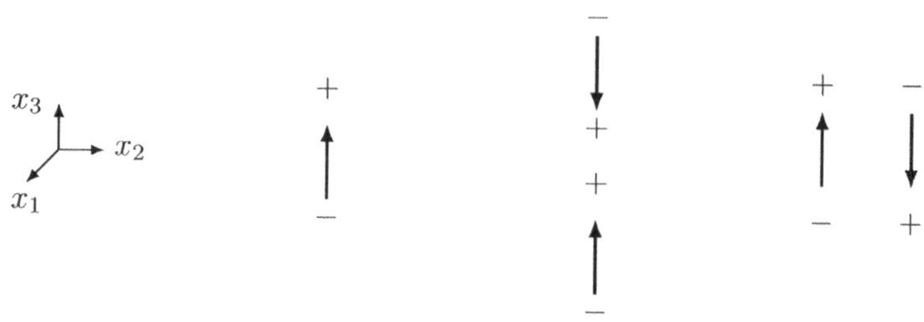

Figure 5.3. *Schematic representation of compact multipole sources as combinations of mass-flow sources: dipole (left), longitudinal quadrupole (center), and lateral quadrupole (right)*

5.3.3. *Quadrupole sources*

The approximation in [5.34] also fails if $F_i(t) = 0$, which occurs when the source term of the wave equation can be written as the double divergence of a tensor q_{ij}:

$$\nabla^2 p' - \frac{1}{c_0^2}\frac{\partial^2 p'}{\partial t^2} = -\frac{\partial^2 q_{ij}}{\partial x_i \partial x_j} \qquad [5.36]$$

This type of source term, known as a quadrupole, does not appear for mass flow or force sources. However, it is a fundamental component in the study of noise generated by turbulent flows. Mass and momentum are conserved within these unsteady flows and only these sources of higher multipolar order are then present. The solution to [5.36] is obtained in a manner similar to that developed for dipole sources. We obtain:

$$p'(\boldsymbol{x}, t) = \frac{1}{4\pi}\frac{\partial^2}{\partial x_i \partial x_j}\int_\infty \frac{1}{R} q_{ij}(\boldsymbol{y}, t - R/c_0)\mathrm{d}\boldsymbol{y} \qquad [5.37]$$

and far-field (geometric and acoustic) and compact source region approximations are given by:

$$p'(\boldsymbol{x}, t)_{x\to\infty} = \frac{1}{4\pi x c_0^2}\frac{x_i x_j}{x^2}\frac{\partial^2}{\partial t^2}\int_\infty q_{ij}\left(\boldsymbol{y}, t - \frac{x}{c_0} + \frac{\boldsymbol{x}\cdot\boldsymbol{y}}{x c_0}\right)\mathrm{d}\boldsymbol{y} \qquad [5.38]$$

and

$$p'(\boldsymbol{x}, t)_{x\to\infty} \approx \frac{1}{4\pi x c_0^2}\frac{x_i x_j}{x^2}\frac{\partial^2 Q_{ij}}{\partial t^2}\left(t - \frac{x}{c_0}\right) \qquad [5.39]$$

with

$$Q_{ij}(t) = \int_{\infty} q_{ij}(\boldsymbol{y}, t)\mathrm{d}\boldsymbol{y} \qquad [5.40]$$

The directivity of a compact quadrupole distribution is complex when all the components of the tensor Q_{ij} are involved. Nevertheless, there are two basic types of quadrupoles: longitudinal quadrupoles and lateral quadrupoles. A longitudinal quadrupole has a single non-zero component located on its diagonal, such as Q_{33}. Using the spherical coordinates (x, θ, φ) centered on the position of the quadrupole, the directivity varies as $\cos^2\theta$. For a lateral quadrupole, only one off-diagonal component is non-zero, for example, Q_{23}. The directivity is then given by $|D(\theta, \varphi)| = |\sin\varphi\cos\theta\sin\theta|$. The directivity of these elementary quadrupoles in the plane (x_2, x_3), that is, for the angle φ of spherical coordinates equal to $\pi/2$, is given in Figure 5.2 and a three-dimensional view of the directivities is shown in Figure 5.4. Compact quadrupoles can be visualized as two dipoles of equal amplitude and opposite direction, that is, as four (monopolar) mass flow sources of equal amplitude and appropriate sign. If all sources are aligned, a longitudinal quadrupole is formed; if the dipoles are separated in a direction perpendicular to their axis, a lateral quadrupole is obtained (Figure 5.3).

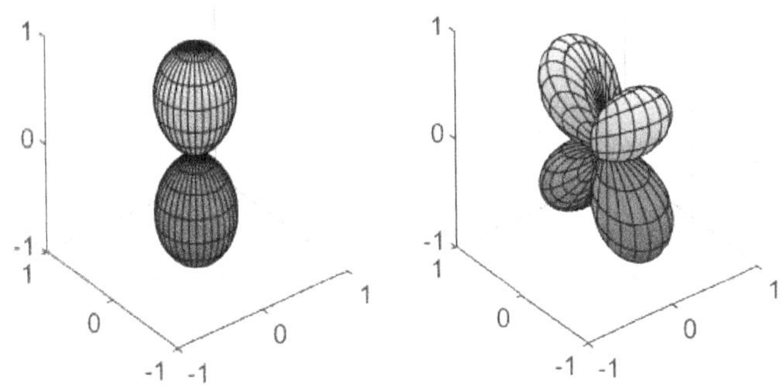

Figure 5.4. *Three-dimensional views of the directivity patterns of a longitudinal quadrupole (left) and a lateral quadrupole (right). The colors are used only for a better understanding of the figure. For a color version of this figure, see www.iste.co.uk/juve/acoustics1.zip*

5.4. Green's functions and general solutions of the Helmholtz equation

The Green's functions for the Helmholtz equation are solutions of the equation:

$$\nabla^2 G + k_0^2 G = -\delta(\boldsymbol{x} - \boldsymbol{y}) \qquad [5.41]$$

They thus correspond to the response of the medium to an isotropic and harmonic point source located at $\boldsymbol{x} = \boldsymbol{y}$. As with the wave equation, the solution to (5.41) is not unique. The "free-space" Green's function G_0 that satisfies the causality condition is obtained as a temporal Fourier transform of the Green's function $g_0(\boldsymbol{x}, t; \boldsymbol{y}, t_0)$:

$$G_0(\boldsymbol{x}, \boldsymbol{y}) = \frac{1}{4\pi R} \exp(-\mathrm{i}k_0 R) ; \qquad R = \|\boldsymbol{x} - \boldsymbol{y}\| \qquad [5.42]$$

The solution of the Helmholtz equation in free space with a right-hand side term:

$$\nabla^2 p + k_0^2 p = -S(\boldsymbol{x} - \boldsymbol{y}) \qquad [5.43]$$

is expressed as the convolution product of the Green's function with the complex-amplitude source term S.

5.4.1. *Monopole sources*

For a fluctuating mass flow source with complex amplitude q_m, the source term is written as $S = \mathrm{i}\omega q_m$, and it follows that:

$$p(\boldsymbol{x}) = \frac{\mathrm{i}\omega}{4\pi} \int_\infty q_m(\boldsymbol{y}) \frac{\exp(-\mathrm{i}k_0 R)}{R} \mathrm{d}\boldsymbol{y} \qquad [5.44]$$

In the far-field, therefore in practice for $R \gg L$ where L characterizes the spatial extent of the source:

$$p(\boldsymbol{x})_{x\to\infty} = \mathrm{i}\omega \frac{\exp(-\mathrm{i}k_0 x)}{4\pi x} \int_\infty q_m(\boldsymbol{y}) \exp\left(\mathrm{i}k_0 \frac{\boldsymbol{x}\cdot\boldsymbol{y}}{x}\right) \mathrm{d}\boldsymbol{y} \qquad [5.45]$$

The separation of the radial dependence in $\exp(-\mathrm{i}k_0 x)/x$ and of the angular dependence $D(\theta, \varphi)$ is thus found again:

$$D(\theta, \varphi) = \int_\infty q_m(\boldsymbol{y}) \exp(\mathrm{i}\boldsymbol{k}_r \cdot \boldsymbol{y}) \mathrm{d}\boldsymbol{y} \qquad [5.46]$$

where $\boldsymbol{k}_r = k_0 \boldsymbol{x}/x$ is the vector of magnitude k_0 pointing in the direction of the observer. Note that the directivity pattern is related to the spatial Fourier transform of the source distribution, evaluated for the particular (and unique) value $\boldsymbol{k}_r = k_0 \boldsymbol{x}/x$.

If the source region is compact ($\lambda \gg L$), then:

$$p(\boldsymbol{x}) \approx \mathrm{i}\omega \frac{\exp(-\mathrm{i}k_0 x)}{4\pi x} Q_m ; \qquad Q_m = \int_\infty q_m(\boldsymbol{y}) \mathrm{d}\boldsymbol{y} \neq 0 \qquad [5.47]$$

5.4.2. *Dipole and quadrupole sources*

With a volume distribution of forces of complex amplitude $f_i(\boldsymbol{y})$, the equation to solve is as follows:

$$\nabla^2 p + k_0^2 p = \frac{\partial f_i}{\partial x_i} \qquad [5.48]$$

whose solution is obtained by differentiation on the convolution product, as seen earlier:

$$p(\boldsymbol{x}) = -\frac{1}{4\pi}\frac{\partial}{\partial x_i}\int_\infty f_i(\boldsymbol{y})\frac{\exp(-\mathrm{i}k_0 R)}{R}\mathrm{d}\boldsymbol{y} \qquad [5.49]$$

and in the far-field,

$$p(\boldsymbol{x})_{x\to\infty} = \frac{\mathrm{i}k_0}{4\pi}\frac{\exp(-\mathrm{i}k_0 x)}{x}\frac{x_i}{x}\int_\infty f_i(\boldsymbol{y})\exp(\mathrm{i}\boldsymbol{k}_r\cdot\boldsymbol{y})\mathrm{d}\boldsymbol{y} \qquad [5.50]$$

which, for a compact source, is reduced to the following form, showing a typical directivity in $\cos\theta = x_i F_i/xF$, where θ is the angle formed by the resultant force $\boldsymbol{F}$ and the direction of observation $\boldsymbol{x}$:

$$p(\boldsymbol{x}) \approx \mathrm{i}k_0\frac{\exp(-\mathrm{i}k_0 x)}{4\pi x}\frac{x_i}{x}F_i\,; \qquad F_i = \int_\infty f_i(\boldsymbol{y})\mathrm{d}\boldsymbol{y} \neq 0 \qquad [5.51]$$

If the monopole Q_m and the dipole F_i contributions are equal to zero, a quadrupole source distribution q_{ij} leads to the following relations:

$$\nabla^2 p + k_0^2 p = -\frac{\partial^2 q_{ij}}{\partial x_i \partial x_j} \qquad [5.52]$$

whose solution is

$$p(\boldsymbol{x}) = \frac{1}{4\pi}\frac{\partial^2}{\partial x_i \partial x_j}\int_\infty q_{ij}(\boldsymbol{y})\frac{\exp(-\mathrm{i}k_0 R)}{R}\mathrm{d}\boldsymbol{y} \qquad [5.53]$$

and in the far-field,

$$p(\boldsymbol{x})_{x\to\infty} = -\frac{k_0^2}{4\pi}\frac{\exp(-\mathrm{i}k_0 x)}{x}\frac{x_i x_j}{x^2}\int_\infty q_{ij}(\boldsymbol{y})\exp(\mathrm{i}\boldsymbol{k}_r\cdot\boldsymbol{y})\mathrm{d}\boldsymbol{y} \qquad [5.54]$$

which, for a compact source, can be reduced to:

$$p(\boldsymbol{x}) \approx -\frac{k_0^2}{4\pi}\frac{\exp(-\mathrm{i}k_0 x)}{x}\frac{x_i x_j}{x^2}Q_{ij}\,; \qquad Q_{ij} = \int_{\infty} q_{ij}(\boldsymbol{y})\mathrm{d}\boldsymbol{y} \neq 0 \qquad [5.55]$$

The directivities of the longitudinal (e.g. $i = j = 3$) and lateral (e.g. $i = 2, j = 3$) quadrupoles, described in section 5.3.3, are also apparent.

5.5. One-dimensional and two-dimensional Green's functions

If the three-dimensional space is the natural space for wave propagation, it proves useful in some situations to use one- or two-dimensional descriptions. The associated Green's functions then correspond to the response of the medium to point impulsive sources in these one- or two-dimensional spaces. More physically, they also correspond to the response in a three-dimensional medium to an impulsive source distributed over an infinite plane or an infinite line, respectively. It is from this idea that it is often easiest to determine the form of these Green's functions in the time or frequency domain.

5.5.1. *Two-dimensional Green's function of the wave equation*

This Green's function corresponds to the field emitted by an impulsive point source in a two-dimensional space or, more physically, by an infinite line of sources emitting an identical impulsive signal in a three-dimensional space. Starting from this idea to calculate the radiation of such a line source, the use of cylindrical coordinates is natural. To simplify the expressions, we assume that the z axis coincides with the source line and that the impulse occurs at time $t_0 = 0$. By simply shifting the time or space origin, a more general expression can be obtained if necessary. Given the axisymmetry of the problem, the Green's function depends only on the normal distance ρ from an observer to the z axis and on time t, $g_{2\mathrm{D}} = g_{2\mathrm{D}}(\rho, t)$. This function results from summing all contributions from source points along the z axis, that is:

$$g_{2\mathrm{D}}(\rho, t) = \frac{1}{4\pi}\int_{-\infty}^{\infty}\frac{\delta\left(t - R/c_0\right)}{R}\mathrm{d}z\,; \qquad R = \sqrt{\rho^2 + z^2} \qquad [5.56]$$

By transforming the integration with respect to z into an integration with respect to R, it follows that:

$$g_{2\mathrm{D}}(\rho, t) = \frac{c_0}{2\pi}\int_{\rho}^{\infty}\frac{\delta(c_0 t - R)}{\sqrt{R^2 - \rho^2}}\mathrm{d}R \qquad [5.57]$$

Using the properties of the Dirac delta function, we obtain:

$$g_{2D}(\rho, t) = 0; \qquad \rho > c_0 t \tag{5.58}$$

$$g_{2D}(\rho, t) = \frac{c_0}{2\pi} \frac{1}{\sqrt{c_0^2 t^2 - \rho^2}}; \qquad \rho < c_0 t \tag{5.59}$$

or in a more condensed way

$$g_{2D}(\rho, t) = H\left(t - \frac{\rho}{c_0}\right) \frac{c_0}{2\pi\sqrt{c_0^2 t^2 - \rho^2}} \tag{5.60}$$

where $H(t)$ is the Heaviside function, which is zero for $t < 0$ and equal to unity for $t > 0$.

This formula calls for several comments. First, note that no signal is received at distance ρ from the source line before time $t = \rho/c_0$: this is a consequence of the causality condition. The response is then discontinuous at $t = \rho/c_0$, but the singularity is much weaker than the one observed in three-dimensions, which involves a Dirac delta function at $t = R/c_0$. Thus, a wavefront at $t = \rho/c_0$ is followed by a "wake", which trails behind it and whose amplitude slowly decreases. The physical interpretation in terms of a line of point sources allows us to understand the existence of this wake: it is associated with the contribution of sources along the z axis that are further and further away from the listening point, and whose amplitude decreases inversely with the distance $\sqrt{\rho^2 + z^2}$. It can also be seen that the shape of the received signal is modified compared to that of the emitted signal (contrary to the three-dimensional behavior), since the signal initially concentrated in time is distributed over an infinite duration for the listener. The temporal behavior of sound waves in a two-dimensional space is therefore very different from that observed in three-dimensions.

It is also difficult to highlight a far-field behavior, since the extent of the line source is infinite. Heuristically, the largest contribution at a given point is expected to occur in the close vicinity of the singularity present in expression [5.60], that is, for $c_0 t = \rho$. By writing $\sqrt{c_0^2 t^2 - \rho^2}$ as the product $(\sqrt{c_0 t - \rho})(\sqrt{c_0 t + \rho})$ and replacing $\sqrt{c_0 t + \rho}$ by $\sqrt{2\rho}$, an approximation of the exact Green's function is obtained, in which the retarded time $t - \rho/c_0$ and a level decrease with the inverse of the square root of the radial distance appear:

$$g_{2D}(\rho, t) \approx \frac{1}{\sqrt{\rho}} f(t - \rho/c_0) \tag{5.61}$$

This heuristic approach will be reinforced when studying the far-field behavior of harmonic waves in a two-dimensional space.

5.5.2. *One-dimensional Green's function of the wave equation*

To obtain the Green's function for a one-dimensional space, one can perform a double spatial integration of the three-dimensional Green's function by considering that it corresponds to the response to impulsive point sources distributed on a plane, in this case the plane $x_1 = 0$. A first spatial integration (over $x_3 = z$) was previously performed to obtain the two-dimensional Green's function as a function of time and distance $\rho = \sqrt{x_1^2 + x_2^2}$. Taking into account the presence of the Heaviside function $H(t - \rho/c_0)$, the integration with respect to x_2 is written as:

$$g_{1\mathrm{D}}(x_1, t) = \frac{c_0}{2\pi} \int_{-\gamma}^{+\gamma} \frac{\mathrm{d}x_2}{\sqrt{c_0^2 t^2 - x_1^2 - x_2^2}}\,; \qquad |x_1| < c_0 t \qquad [5.62]$$

$$g_{1\mathrm{D}}(x_1, t) = 0\,; \qquad |x_1| > c_0 t \qquad [5.63]$$

with $\gamma = \sqrt{c_0^2 t^2 - x_1^2}$. Using the formula $\int \mathrm{d}x_2/\sqrt{\gamma^2 - x_2^2} = \arcsin(x_2/\gamma)$, the integration gives the following result:

$$g_{1\mathrm{D}}(x_1, t) = \frac{c_0}{2} H\left(t - \frac{|x_1|}{c_0}\right) \qquad [5.64]$$

At a distance $|x_1|$ from the source plane, the received signal is therefore zero up to $t = |x_1|/c_0$ (a consequence of causality), at which time a small discontinuity associated with the existence of a wavefront appears, and then the signal remains constant indefinitely. As for the two-dimensional case, the initial impulsive signal is thus distorted by the continuous arrival of contributions from the sources located in the plane $x_1 = 0$.

5.5.3. *Green's functions of the Helmholtz equation in one- and two-dimensions*

To obtain the Green's functions associated with one- or two-dimensional harmonic waves, it is possible to perform a Fourier transform of their time counterpart or, in physical terms, to perform successive spatial integrations of the three-dimensional Green's function G_0. However, it is sometimes easier to solve the Helmholtz equations directly in a one- or two-dimensional space, forced by a point source term.

In two-dimensions, the calculation is not trivial and involves a special function, the Hankel function (see Appendix A2.3). The two-dimensional Green's function can be expressed by an integration along the $z = x_3$ axis of the free-space Green's function G_0 :

$$G_{2\mathrm{D}}(\rho) = \frac{1}{4\pi} \int_{-\infty}^{+\infty} \frac{\exp(-\mathrm{i}k_0 R)}{R} \mathrm{d}z\,; \qquad R = \sqrt{\rho^2 + z^2} \qquad [5.65]$$

It is also a solution of the two-dimensional Helmholtz equation in polar coordinates[4] which, after integration with respect to the variable φ, yields:

$$\frac{1}{\rho}\frac{\partial}{\partial \rho}\left(\rho\frac{\partial G_{2\text{D}}}{\partial \rho}\right) + k_0^2 G_{2\text{D}} = -\frac{\delta(\rho)}{2\pi\rho} \tag{5.66}$$

This is a zero-order Bessel equation, whose solutions are linear combinations of the zero-order Bessel functions of the first and second kind. To satisfy the condition on the behavior at infinity corresponding to a divergent cylindrical wave[5], only the combination leading to the Hankel function of the second kind and of order zero $H_0^{(2)}$ must be retained. Using the following formula, which results from the integration of [5.66] over ρ:

$$\lim_{\rho \to 0} \frac{\partial G_{2\text{D}}}{\partial \rho} = -\frac{1}{2\pi\rho} \tag{5.67}$$

we obtain

$$G_{2\text{D}}(\rho) = -\frac{\mathrm{i}}{4} H_0^{(2)}(k_0\rho) \tag{5.68}$$

The following asymptotic behaviors (see section A2.3) illustrate the nature of the singularity at the origin (the location of the source) on the one hand, and the behavior at infinity of a divergent cylindrical wave on the other hand:

$$\lim_{\rho \to 0} G_{2\text{D}}(\rho) = -\frac{\mathrm{i}}{2\pi} \ln(k_0\rho) \tag{5.69}$$

$$\lim_{\rho \to \infty} G_{2\text{D}}(\rho) = \sqrt{\frac{1}{8\pi k_0 \rho}} \exp^{-\mathrm{i}(k_0\rho + \pi/4)} \tag{5.70}$$

In the far-field, we can observe the expected $1/\sqrt{\rho}$ decay of the amplitude, and apart from this amplitude factor and an additional phase shift of $\pi/4$, the radial evolution is very similar to that of a plane wave with wavelength $\lambda = c_0/f$.

In one-dimension, the Green's function must satisfy the equation:

$$\frac{\mathrm{d}^2 G_{1\text{D}}(x_1)}{\mathrm{d}x_1^2} + k_0^2 G_{1\text{D}}(x_1) = -\delta(x_1) \tag{5.71}$$

4 The expression for the Dirac delta function in polar coordinates is given in Appendix A2.2.

5 See the notion of radiation or Sommerfeld condition in section 6.3.

Given the behavior at infinity and the singularity at the origin, we get:

$$G_{1\mathrm{D}}(x_1) = -\frac{\mathrm{i}}{2k_0}\exp(-\mathrm{i}k_0|x_1|) \tag{5.72}$$

which, as intuitively expected, corresponds to the propagation of a plane wave from the origin, symmetrically to the positive and negative values of x_1.

5.6. Reciprocity of Green's functions

An important property of the Green's functions defined in the previous paragraphs is that of *reciprocity*. For harmonic waves, it can be seen by simple inspection that each Green's function involves only the distance between the source and the receiver $\|\boldsymbol{x} - \boldsymbol{y}\|$. As a result:

$$G_0(\boldsymbol{x}, \boldsymbol{y}) = G_0(\boldsymbol{y}, \boldsymbol{x}) \tag{5.73}$$

and therefore the pressure at $\boldsymbol{x}$ produced by an omnidirectional point source at $\boldsymbol{y}$ is identical to the pressure at $\boldsymbol{y}$ produced by the same source located at $\boldsymbol{x}$. The roles of the source and receiver can therefore be swapped. The Green's function is thus said to have the property of reciprocity.

In the time domain, the generalization of this notion requires reversing the direction of the flow of time in order to respect the condition of causality, and thus:

$$g_0(\boldsymbol{x}, t; \boldsymbol{y}, t_0) = g_0(\boldsymbol{y}, -t_0; \boldsymbol{x}, -t) \tag{5.74}$$

The effect at $\boldsymbol{x}$ at a time t of an impulsive source at $\boldsymbol{y}$ at a time t_0 $(< t)$ is the same as that recorded at $\boldsymbol{y}$ at a time $-t_0$ and due to an impulse at $\boldsymbol{x}$ at a time $-t$ $(< -t_0)$. This is referred to as a *time reversal* principle.

These reciprocity formulas are established by a simple inspection for an infinite, homogeneous medium at rest. It can be shown that they remain valid in the presence of reflecting or diffracting surfaces, even if the properties of the medium are spatially variable. Only the presence of a flow that breaks the isotropy of the medium compromises these properties of reciprocity[6].

In addition to their theoretical interest, reciprocity formulas are often of practical use. In fact, they allow us to choose the most appropriate configuration for placing

6 In mathematics, the reciprocity property is related to the *self-adjoint* nature of the propagation operators [MOR 53].

a source or a microphone, for example, to study acoustic radiation in a complex geometry.

While the reciprocity is quite obvious in an unbounded homogeneous medium, the same cannot be said when, for example, multiple walls are present. The reciprocity property states that the sound pressure produced in the center of a room by a source placed in a corner is equal to the sound pressure produced by the same source in the center of the room and recorded in the same corner; this was far from obvious at first.

5.7. Green's functions for a fluid in uniform subsonic motion

The effect of a uniform flow on the propagation of acoustic waves is a first approximation to the effect of more complex flows. In this situation, it is possible to obtain a Green's function in a relatively simple analytical form, both in the time domain and in the frequency domain. It will thus be possible to develop integral formulations for this type of flow on the same principle as for a medium at rest (see section 6.3). In practice, two applications are of particular interest. The first involves wind tunnel measurements of the radiation from a fixed source using sensors placed in a flow that can reasonably be assumed to be uniform. The second application concerns the noise emission of an aircraft in flight. In the coordinate system attached to the aircraft, the fluid appears to be in uniform motion. In this coordinate system, measurements or computations are naturally described by the wave equation in a medium in uniform motion.

In a medium moving with a uniform velocity U directed along the x_1 axis, the linearized Euler's equations can be manipulated in a similar way as in Chapter 1 to obtain a convected wave equation. The equation defining the Green's function is as follows:

$$\nabla^2 g(\boldsymbol{x}, t; \boldsymbol{y}, t_0) - \frac{1}{c_0^2}\left(\frac{\partial}{\partial t} + U\frac{\partial}{\partial x_1}\right)^2 g(\boldsymbol{x}, t; \boldsymbol{y}, t_0) = -\delta(\boldsymbol{x} - \boldsymbol{y})\delta(t - t_0) \quad [5.75]$$

where the material derivative replaces the partial derivative with respect to time when the fluid is at rest (see the exercises in section 1.7).

To determine this Green's function, the idea is to use a change of variables that reduces the left-hand side of equation [5.75] to that of the wave equation in a medium

at rest. By introducing the factor $\beta = \sqrt{1 - \mathrm{M}^2}$, where $\mathrm{M} = U/c_0$ denotes the Mach number, we define the new spatial and temporal variables by[7]:

$$x_1' = x_1/\beta^2\,; \quad x_2' = x_2/\beta\,; \quad x_3' = x_3/\beta\,; \quad t' = t + \mathrm{M}x_1'/c_0 \qquad [5.76]$$

By transforming the derivatives with respect to $(\boldsymbol{x}, t)$ into derivatives with respect to the new variables $(\boldsymbol{x}', t')$, it follows that:

$$\frac{\partial}{\partial x_1} = \frac{1}{\beta^2}\left(\frac{\partial}{\partial x_1'} + \frac{\mathrm{M}}{c_0}\frac{\partial}{\partial t'}\right)$$

$$\frac{\partial}{\partial x_2} = \frac{1}{\beta}\frac{\partial}{\partial x_2'}$$

$$\frac{\partial}{\partial x_3} = \frac{1}{\beta}\frac{\partial}{\partial x_3'}$$

$$\frac{\partial}{\partial t} = \frac{\partial}{\partial t'}$$

and the material derivative is expressed as:

$$\left(\frac{\partial}{\partial t} + U\frac{\partial}{\partial x_1}\right) = \frac{1}{\beta^2}\left(\frac{\partial}{\partial t'} + U\frac{\partial}{\partial x_1'}\right)$$

The left-hand side of the convected wave equation is therefore transformed into:

$$\frac{1}{\beta^2}\left(\nabla'^2 - \frac{1}{c_0^2}\frac{\partial^2}{\partial t'^2}\right) \qquad [5.77]$$

and the Green's function satisfies:

$$\frac{1}{\beta^2}\left(\nabla'^2 - \frac{1}{c_0^2}\frac{\partial^2}{\partial t'^2}\right) g(\boldsymbol{x}', t'; \boldsymbol{y}', t_0') = -\frac{1}{\beta^4}\delta(\boldsymbol{x}' - \boldsymbol{y}')\delta(t' - t_0') \qquad [5.78]$$

Using the new variables, the Green's function is therefore written as:

$$g(\boldsymbol{x}', t'; \boldsymbol{y}', t_0') = \frac{1}{4\pi\beta^2 R'}\delta\left(t' - t_0' - \frac{R'}{c_0}\right) \qquad [5.79]$$

where $R' = \|\boldsymbol{x}' - \boldsymbol{y}'\|$.

7 This transformation is limited to subsonic flows. It is analogous to the Prandtl–Glauert transformation used in aerodynamics.

Going back to the initial variables, and denoting $R = \|\boldsymbol{x} - \boldsymbol{y}\|$ and $\rho^2 = (x_2 - y_2)^2 + (x_3 - y_3)^2$, the Green's function is finally written as[8]:

$$g(\boldsymbol{x}, t; \boldsymbol{y}, t_0) = \frac{1}{4\pi\sqrt{R^2 - \mathrm{M}^2\rho^2}} \delta\left(t - t_0 - \frac{\sqrt{R^2 - \mathrm{M}^2\rho^2} - \mathrm{M}(x_1 - y_1)}{c_0(1 - \mathrm{M}^2)}\right) \quad [5.80]$$

To interpret this result, a simpler form of [5.80] can be given by introducing the spherical coordinates at the emission point (R, θ, φ), where the axis x_1 corresponds to $\theta = 0$:

$$g(\boldsymbol{x}, t; \boldsymbol{y}, t_0) = \frac{1}{4\pi R\sqrt{1 - \mathrm{M}^2 \sin^2\theta}} \delta\left(t - t_0 - \frac{R}{c_0(1 - \mathrm{M}^2)}\left(\sqrt{1 - \mathrm{M}^2\sin^2\theta} - \mathrm{M}\cos\theta\right)\right) \quad [5.81]$$

As an example, if the observation point is aligned with the flow velocity (with $x_1 > y_1$), this formula can be simplified to give:

$$g(\boldsymbol{x}, t; \boldsymbol{y}, t_0) = \frac{1}{4\pi R}\delta\left(t - t_0 - \frac{R}{c_0(1 + \mathrm{M})}\right) \quad [5.82]$$

where the argument in the delta function simply reflects the reduction in the propagation time between the source and the receiver due to the greater speed of the wave, as the flow velocity is added to the speed of sound.

It should be noted that the Green's function in a uniform flow does not satisfy the usual reciprocity principle, because of the presence of the velocity field, which breaks the isotropy of the medium. However, it is possible to define some kind of reciprocity as follows: the field at $\boldsymbol{y}$ due to a source at $\boldsymbol{x}$ is identical to the one generated at $\boldsymbol{x}$ by a source at $\boldsymbol{y}$, *provided that the direction of the flow (and the direction of time) is reversed*. This property, which can be deduced by a simple inspection of the Green's function [5.81] cannot, however, be generalized to arbitrary flows.

8 It is verified that $\beta^4 R'^2 = R^2 - \mathrm{M}^2\rho^2$.

5.7.1. *Green's function of the convected Helmholtz equation*

In a uniform flow, the Green's function for harmonic waves is written as:

$$\left(\nabla^2 + \left(k_0 - \mathrm{iM}\frac{\partial}{\partial x_1}\right)^2\right) G(\boldsymbol{x};\boldsymbol{y}) = -\delta(\boldsymbol{x}-\boldsymbol{y}) \qquad [5.83]$$

whose solution can be obtained by a Fourier transform of the free space Green's function in the time domain [5.81]:

$$G(\boldsymbol{x};\boldsymbol{y}) = \frac{1}{4\pi R\sqrt{1-\mathrm{M}^2\sin^2\theta}} \exp\left(-\mathrm{i}\frac{k_0 R}{1-\mathrm{M}^2}\left(\sqrt{1-\mathrm{M}^2\sin^2\theta} - \mathrm{M}\cos\theta\right)\right) \qquad [5.84]$$

5.8. Moving sources and the Doppler effect

In noise emission problems, the sources are often in rapid motion. This is especially the case when studying noise from high-speed trains or airplanes. This is also the case for the noise produced by ventilation equipment such as fans. Common experience tells us that the noise perceived by a fixed observer has different frequencies and levels depending on whether the source is moving closer or further away from the listener: this is the classic manifestation of the *Doppler effect*. In this section, we study the pressure field emitted by sources in arbitrary subsonic motion (the speed of the moving object remains less than the speed of sound in the fluid). First, we consider the case of a point mass source, and then the case of a point force. In the last part, the special case of a source in subsonic or supersonic uniform rectilinear motion is presented in detail.

5.8.1. *Point mass source in arbitrary motion*

Let a point mass source move on an arbitrary path defined by $\boldsymbol{x} = \boldsymbol{X}(t)$. The source term of the wave equation is written as:

$$S = -\frac{\partial q_m}{\partial t}; \qquad q_m(\boldsymbol{x},t) = Q(t)\delta(\boldsymbol{x}-\boldsymbol{X}(t)) \qquad [5.85]$$

where $Q(t)$ and $\boldsymbol{X}(t)$ are known functions of time.

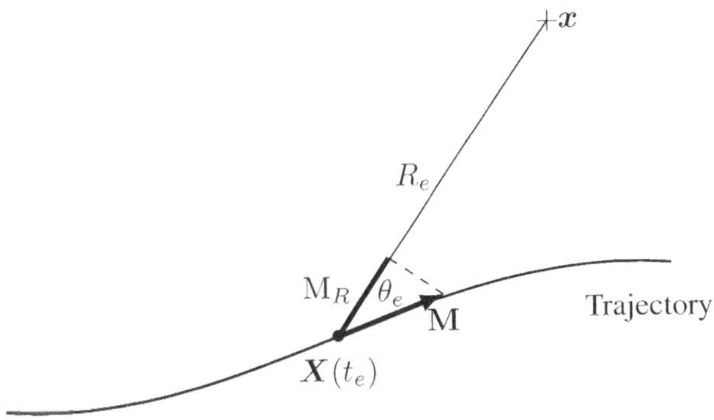

Figure 5.5. *Arbitrarily moving point source and the Doppler effect*

Using the free space Green's function, the pressure is calculated as:

$$p'(\boldsymbol{x},t) = \frac{1}{4\pi}\frac{\partial}{\partial t}\int_{-\infty}^{+\infty} Q(t_e)\frac{\delta\left(t - t_e - R(t_e)/c_0\right)}{R(t_e)}\mathrm{d}t_e \qquad [5.86]$$

$$R_e = R(t_e) = \|\boldsymbol{R}(t_e)\| = \|\boldsymbol{x} - \boldsymbol{X}(t_e)\| \qquad [5.87]$$

The subscript "e" has been used in these expressions to indicate that they are the coordinates at the time of emission of the wave, often referred to as "emission coordinates".

To evaluate the integral appearing in [5.86], the following property of the Dirac delta function is used:

$$\int_{-\infty}^{+\infty} \delta[h(\tau)]f(\tau)\mathrm{d}\tau = \sum_i \frac{f(\tau_i)}{\left|\dot{h}(\tau_i)\right|} \qquad [5.88]$$

where τ_i are the simple zeros of the function h, and $\dot{h}$ designates the derivative of h with respect to the emission time t_e. In our application:

$$h(t_e) = t - t_e - \frac{\|\boldsymbol{x} - \boldsymbol{X}(t_e)\|}{c_0} \qquad [5.89]$$

$$\dot{h}(t_e) = -1 + \frac{\boldsymbol{x} - \boldsymbol{X}(t_e)}{\|\boldsymbol{x} - \boldsymbol{X}(t_e)\|}\cdot\frac{\boldsymbol{U}(t_e)}{c_0}; \qquad \boldsymbol{U} = \frac{\mathrm{d}\boldsymbol{X}}{\mathrm{d}t} \qquad [5.90]$$

whose condensed form can be written as follows:

$$\dot{h}(t_e) = -1 + \mathrm{M}_R = -1 + \mathrm{M}_e\cos\theta_e \qquad [5.91]$$

where the Mach vector is defined by $\mathbf{M} = \boldsymbol{U}/c_0$, and the notation $\mathrm{M}_R = \mathbf{M}_e \cdot \boldsymbol{R}_e/R_e = \mathrm{M}_e \cos\theta_e$ has been introduced for the component of this vector along the source–observer direction (see Figure 5.5). These quantities are to be evaluated at the emission point of the wave, and for the emission time t_e, solution of the equation:

$$c_0(t - t_e) = \|\boldsymbol{x} - \boldsymbol{X}(t_e)\| \qquad [5.92]$$

Equation [5.86] then becomes:

$$p'(\boldsymbol{x},t) = \frac{1}{4\pi}\frac{\partial}{\partial t}\sum_i \frac{Q_e}{R_e|1 - \mathrm{M}_R|} \qquad [5.93]$$

with $Q_e = Q(t_e)$.

For a source in subsonic motion, it can be shown that the equation $h = 0$ has only one zero: for a given $(\boldsymbol{x}, t)$ pair, there is a single emission time and a single associated emission position, and therefore:

$$p'(\boldsymbol{x},t) = \frac{1}{4\pi}\frac{\partial}{\partial t}\frac{Q_e}{R_e(1 - \mathrm{M}_e\cos\theta_e)} \qquad [5.94]$$

This equation seems rather simple, but it should be noted that each of the quantities Q_e, R_e, θ_e is time-dependent. Using the following relations obtained by differentiating equation [5.92]:

$$\frac{\partial t_e}{\partial t} = \frac{1}{1 - \mathrm{M}_e\cos\theta_e} \qquad [5.95]$$

and

$$\frac{\partial R_e}{\partial t} = \frac{-c_0\mathrm{M}_e\cos\theta_e}{1 - \mathrm{M}_e\cos\theta_e} \qquad [5.96]$$

it can be shown that

$$\frac{\partial}{\partial t}(R_e\mathrm{M}_e\cos\theta_e) = \frac{\boldsymbol{R}_e\cdot\dot{\mathbf{M}}_e - c_0\mathrm{M}_e^2}{1 - \mathrm{M}_e\cos\theta_e} \qquad [5.97]$$

and finally

$$p'(\boldsymbol{x},t) = \frac{1}{4\pi}\left(\frac{\dot{Q}_e}{R_e(1 - \mathrm{M}_e\cos\theta_e)^2} + \frac{Q_e\boldsymbol{R}_e\cdot\dot{\mathbf{M}}_e}{R_e^2(1 - \mathrm{M}_e\cos\theta_e)^3} + \frac{Q_e c_0\mathrm{M}_e(\cos\theta_e - \mathrm{M}_e)}{R_e^2(1 - \mathrm{M}_e\cos\theta_e)^3}\right) \qquad [5.98]$$

In this relation, all quantities must be evaluated at the emission time t_e. This formula is said to be expressed in terms of the *emission coordinates* (R_e, θ_e). The last term of [5.98] decays as $1/R_e^2$; it is therefore a near field term that can be overlooked at large distances. Both the first and second terms decays as $1/R_e$ and thus contribute to the far-field, but they are different in nature. The first one is related to the unsteadiness of the source; it is thus similar to what can be observed for a non-moving source. On the other hand, the second term is present even if the source is not time-dependent; it is related to the unsteadiness of the source's motion (in a Galilean reference frame) and more precisely to its acceleration.

Relation [5.95] shows that the time evolution of a quantity s observed in a reference frame attached to the source is different from that observed in the coordinate system of the fixed observer. They are related by:

$$\frac{\partial s}{\partial t} = \frac{\partial s}{\partial t_e}\frac{\partial t_e}{\partial t} = \frac{1}{1 - \mathrm{M}_e \cos\theta_e}\frac{\partial s}{\partial t_e} \qquad [5.99]$$

The observed "instantaneous" angular frequency, heuristically defined by $\partial s/\partial t \propto \omega\, s$, is therefore related to the emission frequency ω_e by:

$$\omega = \frac{\omega_e}{1 - \mathrm{M}_e \cos\theta_e} \qquad [5.100]$$

which is the expression of the Doppler effect: the received frequency is higher as the source approaches the observer ($\theta_e < \pi/2$) and lower as it moves away from the observer. Note that the application of this formula is very general, since it does not assume anything about the motion of the source, which can be arbitrary. In particular, it is absolutely not necessary that the motion be uniform, although this is often how the Doppler effect is introduced. The observed frequency depends only on the value of the projection of the Mach vector in the direction of the receiver at the time of emission.

Another important effect is the amplitude of the received noise, which also depends on the relative motion between the source and the receiver. The amplitude of the pressure increases as the source approaches the observer and decreases as it moves away. This effect depends on a Doppler factor raised at different powers. For a mass flow source moving at constant velocity, the pressure is amplified by a factor of $(1 - \mathrm{M}_e \cos\theta_e)^{-2}$. Figure 5.6 shows the evolution of the frequency and amplitude observed in the far-field when a fluctuating mass flow source moves in a uniform rectilinear motion with a Mach number $\mathrm{M} = 0.25$.

5.8.2. *Arbitrarily moving point forces*

The problem of a point force $\boldsymbol{F}(t)$ in arbitrary motion is probably of greater practical interest than that of the mass flow source. For example, this point force is

a good approximation for estimating the noise generated by a limited portion of an aircraft wing or a compact fan blade. When considering only the far-field, which is the predominant region of interest, the results of taking the motion into account are similar to those obtained in the previous section, except for a directional effect related to the vectorial nature of the source. In the far-field, the relation introduced in section 5.3.2 can be used:

$$\frac{\partial}{\partial x_i} \approx -\frac{x_i}{x c_0}\frac{\partial}{\partial t} \quad [5.101]$$

and using the formula [5.34] we get:

$$p'(\boldsymbol{x},t)_{R_e\to\infty} = \frac{1}{4\pi}\left(\frac{\boldsymbol{R_e}\cdot\dot{\boldsymbol{F}}_e}{c_0R_e^2(1-\mathrm{M}_e\cos\theta_e)^2} + \frac{(\boldsymbol{R_e}\cdot\boldsymbol{F_e})(\boldsymbol{R_e}\cdot\dot{\mathbf{M}}_e}{c_0R_e^3(1-\mathrm{M}_e\cos\theta_e)^3}\right) \quad [5.102]$$

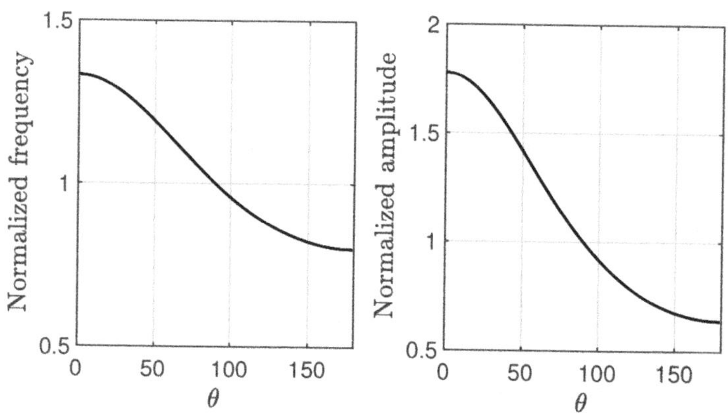

Figure 5.6. *Evolution of the frequency and amplitude measured in the far-field when a fluctuating mass flow source is moving in a uniform rectilinear motion (Mach number* $\mathrm{M} = 0.25$*, corresponding to the operating speed of a high-speed train). Frequency and amplitude are normalized to their values in the absence of source motion;* θ *is the angle formed by the source–observer direction and the source velocity*

This formula once again highlights the existence of two contributions. The first one is due to the unsteadiness of the source and involves the time derivative of the component of the force in the direction of observation. This is called unsteady loading noise. The second is due to the unsteadiness of the motion and involves the derivative of the Mach vector.

A constant force in uniform rectilinear motion, such as that exerted by the steady flow on a lifting surface, produces no far-field noise. On the other hand, if that surface is in accelerated motion, such as a fan blade in uniform rotation, an acoustic radiation is produced; this is called steady loading noise.

5.8.3. *Sources in uniform rectilinear motion*

The special case of a source moving along a straight trajectory at constant (subsonic or supersonic) velocity allows the analytical resolution of the retarded time equation [5.92] and a better visualization of the Doppler effect (see also [DOW 83]).

We consider a point source of mass flow moving along the x_1 axis with a constant velocity U. At a time t, its position is thus $\boldsymbol{X}(t) = (Ut, 0, 0)$. The pressure signal is received at the listening point M with coordinates (x_1, x_2, x_3) and it is convenient to introduce the distance to the axis x_1, $\rho = \sqrt{x_2^2 + x_3^2}$. Equation [5.92] is written as:

$$c_0(t - t_e) = \sqrt{(x_1 - Ut_e)^2 + \rho^2} \quad [5.103]$$

By squaring this expression (which can introduce parasitic solutions), we obtain a quadratic equation in $t - t_e$ which, after some manipulation, can be written as:

$$c_0^2(1-\mathrm{M}^2)(t-t_e)^2 - 2\mathrm{M}c_0(x_1-Ut)(t-t_e) - \left((x_1 - Ut)^2 + \rho^2\right) = 0 \quad [5.104]$$

whose solution is expressed as:

$$\begin{aligned} c_0(t - t_e) &= \frac{\mathrm{M}(x_1 - Ut) \pm \sqrt{(x_1 - Ut)^2 + (1 - \mathrm{M}^2)\rho^2}}{1 - \mathrm{M}^2} \\ &= \frac{\mathrm{M}(x_1 - Ut) \pm \hat{R}}{1 - \mathrm{M}^2} \end{aligned} \quad [5.105]$$

where the notation $\hat{R} = \sqrt{(x_1 - Ut)^2 + (1 - \mathrm{M}^2)\rho^2}$ has been introduced.

A simpler form can be given to this expression by introducing the *reception coordinates* (R_r, θ_r), which involve the position of the source, $X_1 = Ut$, when the acoustic wave arrives at the listening point at time t (Figure 5.7 shows the relations between the different coordinates in the subsonic case). It follows that:

$$c_0(t - t_e) = \frac{R_r}{1 - \mathrm{M}^2}\left(\mathrm{M}\cos\theta_r \pm \sqrt{1 - \mathrm{M}^2 \sin^2\theta_r}\right) \quad [5.106]$$

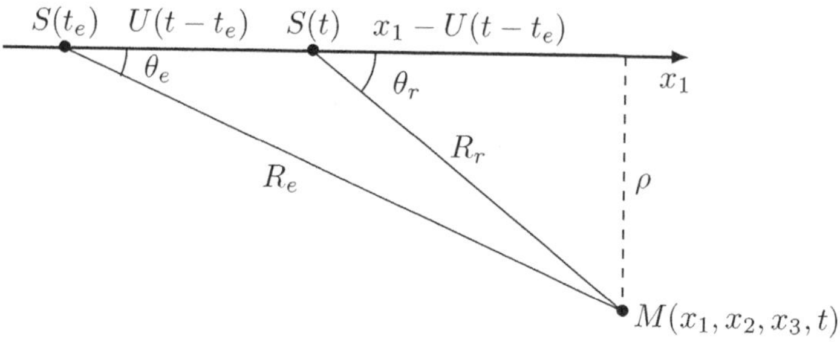

Figure 5.7. *Source in uniform subsonic rectilinear motion. Emission and reception coordinates*

When the motion is subsonic, only the solution associated with the plus sign of the square root is physically acceptable, where $c_0(t - t_e)$ is the source–receiver distance at emission time t_e. The retarded time equation admits a single solution, as indicated above. An illustration of the emission from the moving source is given on the left side of Figure 5.8. The successive positions of the source are indicated by the symbols. The plot of the associated circular wavefronts shows that at any given time, each point in space is reached by a single wavefront, associated with its emission time. It is also very clear that as the source approaches the receiver, successive wavefronts become closer to each other, resulting in a shorter apparent wavelength than if the source were fixed and, correlatively, an increase in the received frequency (Doppler effect).

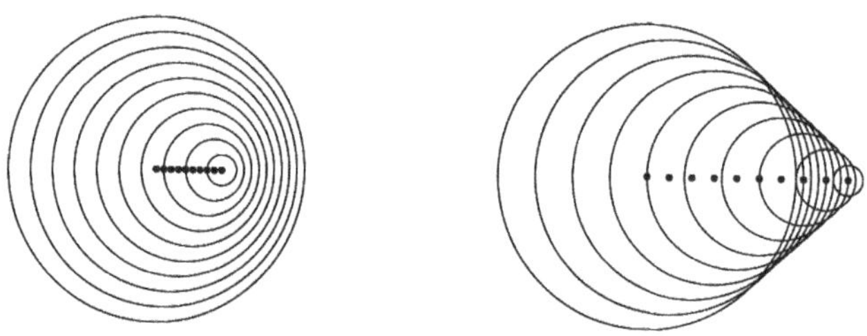

Figure 5.8. *Wavefronts from a source moving from left to right. On the left, the source is in subsonic motion ($\mathrm{M} = 0.5$). On the right, the motion is supersonic $\mathrm{M} = 1.5$ and the formation of the Mach cone (with half angle $\sin^{-1}(1/\mathrm{M})$ is clearly visible, which separates the region not reached by acoustic waves from the region reached by two waves emitted at two different times*

When the source is moving at a supersonic speed, both solutions of the quadratic equation are positive and therefore acceptable, at least as long as the argument under the square root is positive. This is ensured if $\sin\theta_r < 1/\mathrm{M}$. The corresponding limit angle defines the *Mach cone* (with a half-angle $\sin^{-1}(1/M)$, whose apex is attached to the source. An observer located inside the Mach cone receives two contributions from

the source at any given time, produced at two different emission times. One signal is emitted before the source passes the observer's position and the other is emitted after the source has passed through that point. If the emitted signal is harmonic, the received signals will be at different frequencies resulting in level variations over time. A point outside the Mach cone will not initially perceive any pressure fluctuation, and the sound level will therefore change abruptly as the Mach cone boundary reaches the observation point.

It is interesting to note that the reception coordinates are none other than the coordinates in a reference frame that moves with the source. In this coordinate system, the source is emitting in a fluid in uniform motion with velocity $-U$, a situation we have studied in section 5.7. The emission of a source in uniform subsonic motion can therefore be evaluated using the convected Green's function given in that section. For this purpose, it is useful to show the relationship between the emission and reception coordinates by writing the following expressions:

$$\hat{R} = R_e(1 - \mathrm{M}\cos\theta_e) = R_r\sqrt{1 - \mathrm{M}^2\sin^2\theta_r} \qquad [5.107]$$

For example, these relations can be used to express the pressure field created by a mass flow source in uniform rectilinear motion from [5.80]. This calculation is left to the reader as an exercise.

5.9. Exercises

1) **Point dipole and quadrupole**

Calculate the pressure field radiated by a point dipole without using the far-field assumption. Show that there is a near field term that varies as x^{-2}. Perform the same calculations for a point quadrupole (longitudinal or lateral).

2) **Power radiated by combinations of sources**

Calculate the sound power radiated by two monopolar harmonic sources of equal amplitude $|Q|$ and separated by a distance d:

a) When the sources are in phase.

b) When the sources are in phase opposition (so the source is dipolar, why?).

c) Determine the relative efficiency of the sources in phase opposition with respect to the sources in phase by forming the ratio of the powers they radiate when $d \ll \lambda$ (compact dipole).

3) **Green's function of the convected two-dimensional Helmholtz equation**

Using the ideas developed in section 5.7 determine the Green's function for the two-dimensional convected Helmholtz equation in a subsonic flow. Give its form in the far-field.

4) **Point mass source in uniform subsonic motion**

Develop the formula for the far-field pressure produced by a point mass source moving with a constant subsonic speed, using the reception coordinates.

5) **Convection amplification**

Calculate the ratio of the power emitted by a point mass source when it is in uniform rectilinear motion and when it is at rest. Explain the use of the term "convection amplification".

6) **Source in uniform supersonic rectilinear motion**

Determine the values of the two emission times for an observer located inside the Mach cone. A modification of the definition of $\hat{R}$ can be used to account for the fact that $\mathrm{M} > 1$, $\hat{R} = \sqrt{(x_1 - Ut)^2 + (\mathrm{M}^2 - 1)\rho^2}$. Plot the equivalent of Figure 5.7 for the supersonic regime.

Give the far-field expression for the pressure field produced by a point mass flow source by combining the contributions associated with the two emission times.

6

Integral Formulations for Sound Radiation and Diffraction

In the previous chapter, we studied the radiation of acoustic sources in an unbounded medium. This is an idealized situation that is often used because of the simplicity of the expressions obtained, which is reflected in the very clear physical interpretation of the Green's function in free space and the solution of the wave equation for a continuous distribution of sources in the form of retarded potentials. In many applications, however, sources or receivers are located near obstacles that strongly modify the sound field. In some scenarios with simple geometry, it is still possible to use the Green's function in free space by extending the initial problem to infinite space, and relying on the method of images; some examples of this technique have been covered previously. It is also possible, again for simple geometries, to consider the radiation of a vibrating structure by solving the Helmholtz equation, or the wave equation, associated with the continuity of the normal velocity at the interface. We will look at two important examples at the beginning of this chapter.

However, it is important to have a formulation that allows more general geometries and boundary conditions to be considered. For this purpose, a formulation is developed that is complementary to the local approach based on the solution of partial differential equations. This formulation uses an integral expression derived from the local equations in the time domain (starting from the wave equation) or in the frequency domain (starting from the Helmholtz equation) and associated with a set of boundary conditions.

We begin this chapter with the calculation of the acoustic field radiated by a radially pulsating sphere in free space. This allows us to define an important quantity that characterizes the acoustic efficiency of the vibrations, the *radiation impedance*. In a second section, we will describe the acoustic radiation of a planar structure vibrating in a bending mode. Although this example is very simplified, it illustrates the general

behavior of the radiation of structures in "light fluids", that is, in fluids whose presence does not significantly modify the vibration of the structure; this is usually the case for structures immersed in gases.

Moving away from local differential formulations, we then develop the integral formulation associated with the Helmholtz equation, which is the easiest to establish. In the general case, this leads to writing an integral surface equation that requires a numerical solution. This is known as the boundary element method (BEM).

The integral formulation associated with the wave equation is then given. To illustrate its use for an initial-valued problem, we develop the calculation of the acoustic field produced by an overpressure existing in a spherical volume and suddenly released at a given time.

The definition and use of "adapted" Green's functions leads to an explicit integral solution for certain geometries such as wedges, cylindrical or spherical obstacles. A particularly important example concerns the radiation of planar vibrating structures surrounded by an infinite rigid baffle, which leads to the *Rayleigh integral*. The radiation from a circular piston is then discussed in detail. From this, near-field and far-field behaviors are identified that have general application.

6.1. Radially oscillating sphere

We consider the acoustic radiation in free space of a sphere of average radius a vibrating with radial velocity $V_s(t)$. The amplitude of the motion of the surface of the sphere is very small, so that the condition of continuity between the normal vibration velocity and the acoustic velocity can be imposed at the average position $r = a$ at any time. The symmetry of the problem as well as the results of Chapter 2 on spherical waves indicate that the acoustic velocity potential outside the sphere is of the form:

$$\phi(r,t) = \frac{1}{r} F\left(t - \frac{r}{c_0}\right); \qquad r > a \qquad [6.1]$$

Since the physical position of the source is the surface of the sphere $r = a$, it is interesting to replace the function $F\left(t - r/c_0\right)$ by a new function $\psi\left(t - (r-a)/c_0\right)$ to be determined. The acoustic velocity is the radial derivative of the potential, which is composed of a near-field term proportional to ψ and a far-field term associated with its time derivative.

$$u_r'(r,t) = -\frac{1}{rc_0}\frac{\mathrm{d}\psi}{\mathrm{d}t}\left(t - \frac{r-a}{c_0}\right) - \frac{\psi}{r^2}\left(t - \frac{r-a}{c_0}\right) \qquad [6.2]$$

The condition of continuity of normal velocities at $r = a$, $u'_r(a,t) = V_s(t)$, leads to:

$$-\frac{1}{ac_0}\frac{\mathrm{d}\psi}{\mathrm{d}t}(t) - \frac{\psi(t)}{a^2} = V_s(t) \qquad [6.3]$$

This first-order differential equation can be rewritten as follows:

$$\frac{\mathrm{d}}{\mathrm{d}t}(\exp(c_0 t/a)\psi(t)) = -ac_0 \exp(c_0 t/a) V_s(t) \qquad [6.4]$$

which we integrate from $-\infty$ (the vibration being assumed to be initially zero) to the current time t. The final result for the acoustic potential is then:

$$\phi(r,t) = -\frac{ac_0}{r}\exp\left(-\frac{c_0}{a}\left(t - \frac{r-a}{c_0}\right)\right)\int_{-\infty}^{t-(r-a)/c_0} V_s(\tau)\exp(c_0\tau/a)\mathrm{d}\tau \qquad [6.5]$$

Formula [6.5] allows us to determine the transient response of the sound field to the vibration of the sphere. The field at distance r from the origin and at a given time t depends on the values of the vibration velocity at all instants prior to $t - (r-a)/c_0$, which is the observation time corrected for the travel time between the surface of the sphere and the listening point.

6.1.1. *Harmonic vibrations: radiation impedance*

The harmonic solution can be determined from the general solution [6.5] by imposing a vibration velocity such that $V_s(t) = V_0 \exp(\mathrm{i}\omega t)$ for $t > 0$. We get a transient part that tends to zero when $t \to \infty$ and a permanent part, which is what we are interested in. We leave this calculation as an exercise.

It is easier to introduce directly the harmonic form of outgoing spherical waves, $A\exp(-\mathrm{i}k_0 r)/r$, and to determine the unknown constant A using the boundary condition on the average position of the sphere $r = a$. The result for the velocity potential is:

$$\phi(r) = -\frac{a^2 V_0}{1 + \mathrm{i}k_0 a}\frac{\exp(-\mathrm{i}k_0(r-a))}{r} \qquad [6.6]$$

and from this the pressure and velocity fluctuations are derived as:

$$p(r) = \mathrm{i}\rho_0 c_0 \frac{k_0 a^2 V_0}{1 + \mathrm{i}k_0 a}\frac{\exp(-\mathrm{i}k_0(r-a))}{r} \qquad [6.7]$$

$$u_r(r) = \frac{k_0 a^2 V_0}{1 + \mathrm{i}k_0 a}\left(\frac{1}{k_0 r} + \mathrm{i}\right)\frac{\exp(-\mathrm{i}k_0(r-a))}{r} \qquad [6.8]$$

These relations display the typical behavior of spherical waves: proportionality between pressure and velocity in the far-field ($k_0 r \gg 1$) and predominance of velocity fluctuations in the near field in the contribution to the acoustic energy density.

Calculating the radiated sound power in two different ways can give us more information. The first one is to choose a sphere whose radius tends to infinity as the surface for integrating the intensity. Since the intensity is then equal to $pp^*/2\rho_0 c_0$, we obtain:

$$W = \frac{(k_0 a)^2}{1 + (k_0 a)^2}\rho_0 c_0 (4\pi a^2)\frac{1}{2}V_0 V_0^* \qquad [6.9]$$

This formula shows that in the limit of high frequencies ($k_0 a \gg 1$) the power per unit area is equal to the product of the characteristic impedance of the medium $Z_0 = \rho_0 c_0$ by the mean square value of the vibration velocity fluctuations: it corresponds to that which would be radiated by a planar surface with the same vibration velocity. This reflects the fact that the curvature effects become negligible when the distance to the vibrating sphere is much greater than the acoustic wavelength.

The second way to easily calculate the radiated power is to evaluate it at the skin of the vibrating sphere surface. The power is the integral of the intensity at the surface, which is defined by $I = 1/2\mathrm{Re}(p(a)V_0^*)$, since the acoustic velocity is equal to the vibration velocity of the structure. This approach can be used for all structural radiation problems and is often the most effective. The radiation impedance of a vibrating body is defined as:

$$Z_{rad} = \left(\frac{p}{v_{n_{ext}}}\right)_\Sigma \qquad [6.10]$$

where $v_{n_{ext}}$ is the normal component of the vibration velocity (equal to the acoustic normal velocity). By convention, the unit normal points outward from the surface, and thus inward from the fluid. The normal intensity at the surface Σ is therefore $I = 1/2V_0V_0^*\mathrm{Re}(Z_{rad})$.

The radiation efficiency of the structure is obtained by dividing the acoustic intensity by the mean square value of the vibration velocity and by the characteristic impedance of the fluid Z_0. It can be seen that this efficiency, compared to that of a plane wall vibrating in its normal direction, is equal to the real part of the radiation

impedance normalized by Z_0. This impedance is an essential indicator when studying the radiation of vibrating structures. For the pulsating sphere, it follows that:

$$Z_{rad} = Z_0 \frac{(k_0 a)^2 + \mathrm{i} k_0 a}{1 + (k_0 a)^2} \qquad [6.11]$$

The radiation efficiency, which is equal to the real part of the reduced radiation impedance $\zeta_{rad} = Z_{rad}/Z_0$, is therefore:

$$\mathrm{Re}(\zeta_{rad}) = \frac{(k_0 a)^2}{1 + (k_0 a)^2} \qquad [6.12]$$

an expression that could also have been obtained from the formula [6.9] for the radiated power.

Figure 6.1 shows the evolution of the real and imaginary parts of the reduced radiation impedance ζ_{rad} as a function of the reduced wavenumber $k_0 a$. The real part R tends to unity at high frequencies while the imaginary part X tends to zero. At low frequencies, it can be seen that the radiation of the sphere is very inefficient, since it varies as $(k_0 a)^2$.

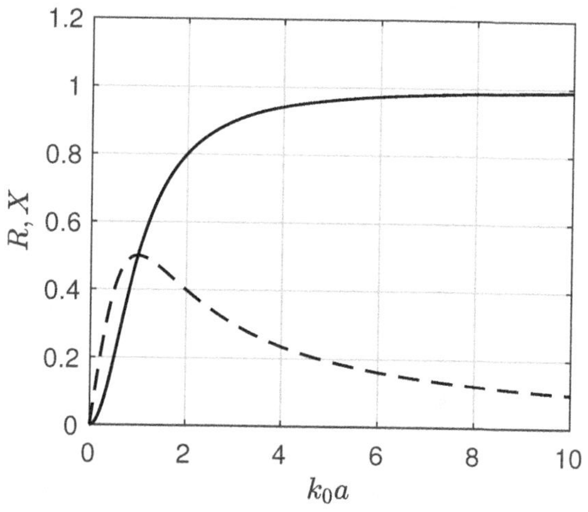

Figure 6.1. *Reduced radiation impedance of a pulsating sphere as a function of the reduced wavenumber. The real part, which reflects the efficiency of the radiation, is plotted in a solid line and the imaginary part is plotted in a dashed line*

6.2. Acoustic radiation from bending vibrations

In the previous section, we saw how a radial vibration can naturally couple to an equally radial sound wave, illustrating a typical compression behavior of acoustic waves in fluids. The same type of phenomenon also occurs when a planar structure vibrates normally (in piston mode), generating waves that propagate normally to the surface in the surrounding fluid. When the vibrations propagate along a plate, the coupling with the adjacent fluid is of a different nature and causes the generation of acoustic waves whose propagation direction is oblique and depends in particular on the frequency.

In this section, we consider the idealized case of an infinite plate in which a single bending (or flexural) wave propagates along x_1. Such a wave is characterized by a displacement $\xi = f(t - x_1/c_f)$ along the x_2 axis and propagates along the x_1 axis with a speed c_f. In the harmonic regime, the corresponding vibration velocity can be written as:

$$V_2(x_1, t) = \mathrm{Re}\{V_0 \exp(-\mathrm{i}k_f x_1) \exp(\mathrm{i}\omega t)\} \quad [6.13]$$

where the bending wavenumber k_f is given by $k_f = \omega/c_f$.

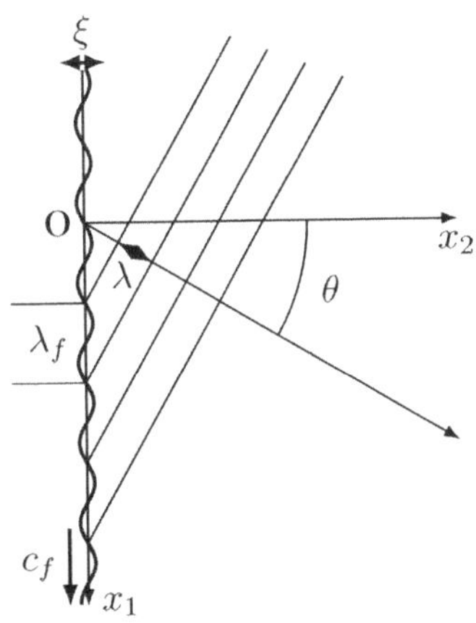

Figure 6.2. *Acoustic radiation from an infinite plate subjected to harmonic bending vibrations. The figure shows the generation of a plane wave propagating at an angle of $30°$ relative to the normal to the plate ($c_f = 2c_0$)*

As we saw in Chapter 4, the velocity of bending waves in a thin plate of thickness h is a function of the mechanical properties of the plate and of the frequency, and is given by:

$$c_f = K^{1/4}\omega^{1/2} \qquad [6.14]$$

$$K = \frac{Eh^2}{12\rho_p(1-\nu_p^2)} \qquad [6.15]$$

where E is the Young's modulus, ρ_p and ν_p are the density and the Poisson's ratio of the material. For example, the numerical values of these parameters are equal to (7.2×10^{10} Pa, 2.7×10^3 kg/m^3 and 0.34) for aluminum and to (2×10^{11} Pa, 7.8×10^3 kg/m^3 and 0.3) for steel.

It is very important to note the dispersive nature of bending waves, whose velocity varies with the square root of the frequency. For a 5-mm thick aluminum plate, it can be calculated that $c_f(1,000 \text{ Hz}) = 223$ m/s and $c_f(4,000 \text{ Hz}) = 446$ m/s, so that the velocity of these waves is first lower and then higher than the velocity of sound waves in air as the frequency increases.

The formulation of the problem (which is clearly two-dimensional) in the plane (x_1, x_2) is simple and involves the Helmholtz equation in the half-space $x_2 > 0$ and the equality between vibration velocity and the normal component of the acoustic velocity at $x_2 = 0$. The following system has thus to be solved:

$$\frac{\partial^2 p}{\partial x_1^2} + \frac{\partial^2 p}{\partial x_2^2} + k_0^2 p = 0 \qquad [6.16]$$

$$u_2(x_1, x_2 = 0) = V_0 \exp(-\mathrm{i}k_f x_1) \qquad [6.17]$$

The geometry of the problem suggests looking for a solution in separate variables of the form $p(x_1, x_2) = X_1(x_1)X_2(x_2)$ and the trace-velocity principle imposes an evolution of the acoustic field along x_1 of the form $X_1(x_1) = A\exp(-\mathrm{i}k_f x_1)$, where the constant A will be determined by the boundary condition. The evolution along the direction normal to the plate is therefore solution of the equation:

$$\frac{\mathrm{d}^2 X_2(x_2)}{\mathrm{d}x_2^2} + (k_0^2 - k_f^2)X_2(x_2) = 0 \qquad [6.18]$$

The nature of the solution depends on the relative value of k_f with respect to k_0 or, in more physical terms, that of c_f with respect to c_0. The previously introduced dispersion relations for air and the aliminum plate, that is, the evolution of the wavenumbers k_0 and k_f with frequency, are plotted in Figure 6.3. The condition

$k_f = k_0$ defines the *critical frequency* f_{cr}, which is about 2,320 Hz. Below this frequency, namely when $c_f < c_0$, the bending waves are subsonic with respect to the surrounding medium; above this frequency they are supersonic ($c_f > c_0$).

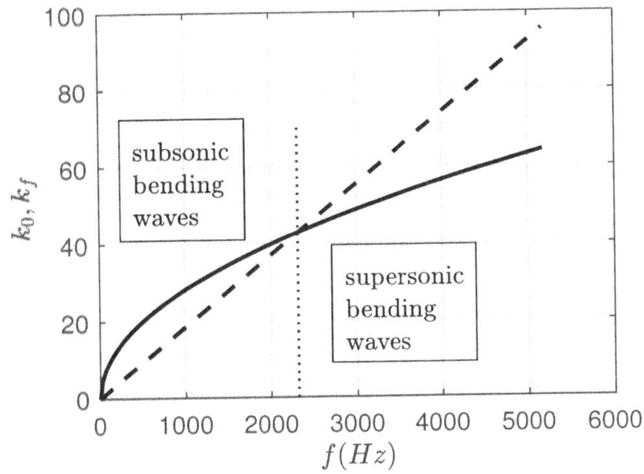

Figure 6.3. *Dispersion relations for air (dashed line) and a 5-mm thick aluminum plate (solid line). The critical frequency f_{cr} is equal to 2,320 Hz*

When bending waves are supersonic, the function $X_2(x_2)$ takes the general form:

$$X_2(x_2) = B\exp(-\mathrm{i}k_2x_2) + C\exp(+\mathrm{i}k_2x_2)\,; \qquad k_2 = +\sqrt{k_0^2 - k_f^2} \quad [6.19]$$

where the constant C must be equal to zero, since the waves propagate to infinity from the source formed by the vibrating plate, without the possibility of the existence of a reflected wave. The real and positive value of k_2 defines the component along x_2 of an acoustic wave vector $\boldsymbol{k}$ of modulus k_0, with $k_1 = k_f$ being its component along x_1. The complete solution is written as:

$$p(x_1, x_2) = AB\exp(-\mathrm{i}k_fx_1)\exp(-\mathrm{i}k_2x_2) \quad [6.20]$$

The multiplicative constant is determined by the boundary condition at $x_2 = 0$, written for the pressure, using the linearized Euler equation projected onto x_2. We obtain:

$$p(x_1, x_2) = V_0Z_0\frac{k_0}{k_2}\exp(-\mathrm{i}\boldsymbol{k}.\boldsymbol{x}) \quad [6.21]$$

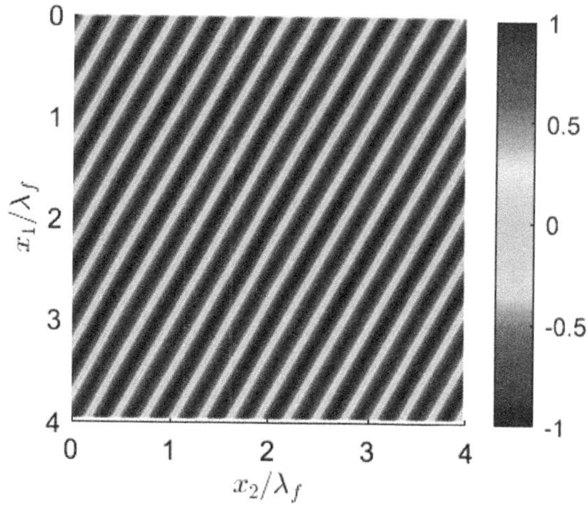

Figure 6.4. *Generation of a propagative plane wave excited by a supersonic bending vibration. The ratio of the bending wave velocity to the speed of sound in the adjacent fluid is equal to 2. The wave propagates at an angle $\theta = 30°$ relative to the normal to the plate. The distances are normalized by the bending wavelength $\lambda_f = c_f/f$. For a color version of this figure, see www.iste.co.uk/juve/acoustics1.zip*

The bending vibration has thus created a plane wave in the adjacent fluid which propagates at an angle θ with respect to the normal; this angle depends on the ratio of the bending velocity in the plate to the speed of sound in the fluid, because $\sin\theta = k_1/k_0 = c_0/c_f$. This angle is real only for supersonic bending waves, when the trace along x_1 of the acoustic wavelength can coincide with the bending wavelength $\lambda_f = c_f/f$. Figure 6.4 depicts the radiation from a plate whose bending wave velocity is equal to twice the speed of sound in the fluid. Finally, it should be noted that the amplitude of the radiated field depends linearly on that of the vibration, which was expected, but also on the ratio of the velocities through:

$$\cos\theta = k_2/k_0 = \sqrt{1 - c_0^2/c_f^2} \qquad [6.22]$$

When the bending waves are subsonic relative to the surrounding fluid ($c_f < c_0$), the nature of the solution changes, because k_f can no longer be the component of an acoustic wavenumber. The pressure in the fluid takes the following form:

$$p(x_1, x_2) = \mathrm{i}\frac{V_0 Z_0}{\sqrt{\left(c_0^2/c_f^2 - 1\right)}} \exp(-\mathrm{i}k_f x_1)\exp(-\kappa x_2)\,; \quad \kappa = +\sqrt{k_f^2 - k_0^2} \qquad [6.23]$$

This is a wave that propagates along x_1 following the shape imposed by the vibration, but whose amplitude is exponentially attenuated in the x_2 direction. This *evanescent wave* decreases all the more rapidly as the velocity of the bending waves deviates from that of the sound waves. The field structure is shown in Figure 6.5 for a ratio c_f/c_0 equal to 0.9.

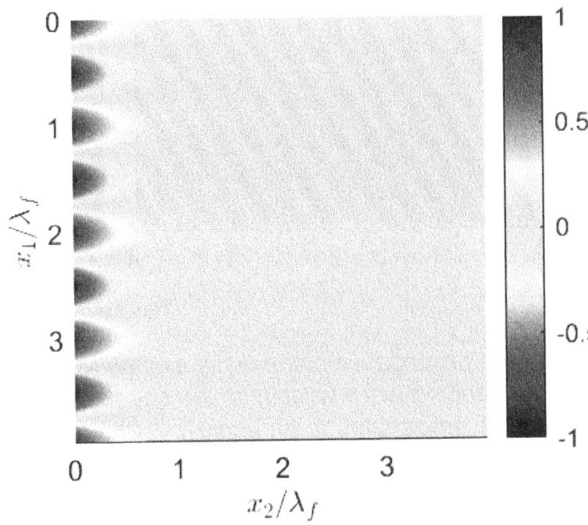

Figure 6.5. *Generation of an evanescent plane wave excited by a subsonic flexural vibration. The ratio of the bending wave velocity to the speed of sound in the adjacent fluid is equal to 0.9. The wave propagates along the interface at speed c_f and is rapidly attenuated in the direction normal to the plate. For a color version of this figure, see www.iste.co.uk/juve/acoustics1.zip*

6.2.1. *Radiated power and radiation impedance*

The energy transfer from the plate vibrations to the acoustic waves and the radiation efficiency can be obtained by calculating the normal sound intensity (which is equal to the radiated power per unit area) at the plate position:

$$I_2 = \frac{1}{2}\mathrm{Re}(p(x_1, x_2 = 0)V_0^* \exp(+\mathrm{i}k_f x_1)) \qquad [6.24]$$

When $c_f > c_0$, the acoustic pressure is given by [6.21] and we get:

$$I_2 = \frac{1}{2}V_0 V_0^* \frac{Z_0}{\cos\theta} \qquad [6.25]$$

This expression shows the influence of the kinetic energy per unit mass of the plate and, more importantly, the dependence on the angle of propagation, which is a function of the ratio of the velocities in the two media.

For subsonic waves, the result is as follows:

$$I_2 = \frac{1}{2} V_0 V_0^* \mathrm{Re}\left(\frac{\mathrm{i}Z_0}{\sqrt{\left(c_0^2/c_f^2 - 1\right)}} \right) = 0 \qquad [6.26]$$

and therefore no energy is transferred to the fluid, as would be expected due to the evanescent nature of the wave created in this case.

The radiation efficiency of the vibration is obtained by calculating the real part of the plate radiation impedance deduced from the previous expressions. The reduced radiation impedance $\zeta_{rad} = p(x_2 = 0)/Z_0 u_2(x_2 = 0)$ is given by:

$$\zeta_{rad} = \frac{1}{\cos\theta}; \qquad c_f > c_0 \qquad [6.27]$$

$$\zeta_{rad} = \frac{\mathrm{i}}{\sqrt{c_0^2/c_f^2 - 1}}; \qquad c_f < c_0 \qquad [6.28]$$

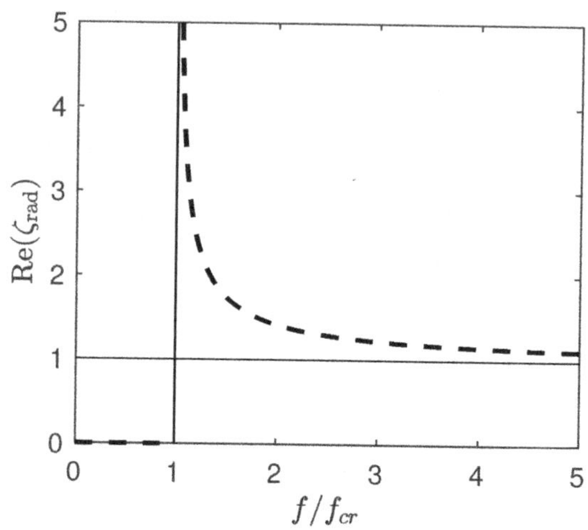

Figure 6.6. *Evolution with the frequency (normalized by the critical frequency) of the real part of the reduced radiation impedance of an infinite plane plate subjected to flexural vibrations*

The change in the behavior of the plate as an acoustic emitter therefore occurs when $c_f = c_0$. Since the plate is a dispersive medium (the bending velocity depends

on the square root of the frequency), this equality occurs when the frequency becomes equal to the critical frequency f_{cr}. At this frequency, the bending wavelength coincides with the acoustic wavelength in the fluid. Below the critical frequency, the plate does not produce any radiation. When $f = f_{cr}$, the efficiency tends to infinity, then it decreases to asymptotically approach unity by higher values. Figure 6.6 illustrates this behavior. It can be observed that the radiation efficiency is not maximal when the propagation is quasi normal to the plate ($f \gg f_{cr}$), which could have been assumed a priori, since the plate would then act as a piston. Conversely, a very intense radiation occurs when almost grazing waves are generated for a very specific value of the frequency ($f \approx f_{cr}$).

6.2.2. *Acoustic radiation from a finite plate*

The extreme simplification of the problem considered in the previous section (infinite plate, vibration with an amplitude independent of the fluid loading) produces unrealistic behavior near the critical frequency. However, the lessons learned from this example are very important for understanding the radiation of vibrating structures and the role of wave propagation velocities. In particular, it is important to note that a strong vibration does not necessarily imply a significant amount of noise radiated: it is necessary to determine how the vibrational energy is distributed along the vibration wavenumbers. In a first analysis, the following important idea can be retained: only bending waves with supersonic wavenumbers relative to the surrounding fluid generate high acoustic levels.

To calculate the radiation of finite structures, it is necessary to use the integral formulations that are described in the following paragraphs (Helmholtz–Kirchhoff or Rayleigh formulation for baffled planar structures). Figure 6.7 gives an example of the radiation efficiency of some of the first vibration modes of a rectangular plate of length a and width b. An infinite rigid baffle surrounds the plate, which is simply supported at its edges. It can be shown that the vibration velocity of the eigenmode (m, n) is given by:

$$u(x_1, x_2) = A \sin\left(\frac{m\pi x_1}{a}\right) \sin\left(\frac{n\pi x_2}{b}\right); \quad 0 \le x_1 \le a; \quad 0 \le x_2 \le b \qquad [6.29]$$

and the associated flexural wavenumber is $k_{mn} = \sqrt{(m\pi/a)^2 + (n\pi/b)^2}$.

In Figure 6.7, the real part of the reduced radiation impedance, that is, the acoustic efficiency, is plotted against the ratio of the acoustic wavenumber to the wavenumber of each of the modes considered. It can be seen that the radiation efficiency is very small at low frequencies ($k_0/k_{mn} \ll 1$); it tends to unity at high frequencies and is maximum in the vicinity of $k_0/k_{mn} = 1$, associated with the critical frequency of the mode under consideration. The maximum takes values of the order of a few units

and depends on the aspect ratio of the plate as well as on the index numbers (m, n) of the modes. This behavior is in qualitative agreement with the results obtained for the infinite plate. A detailed discussion of noise radiation from finite structures can be found in [FAH 07].

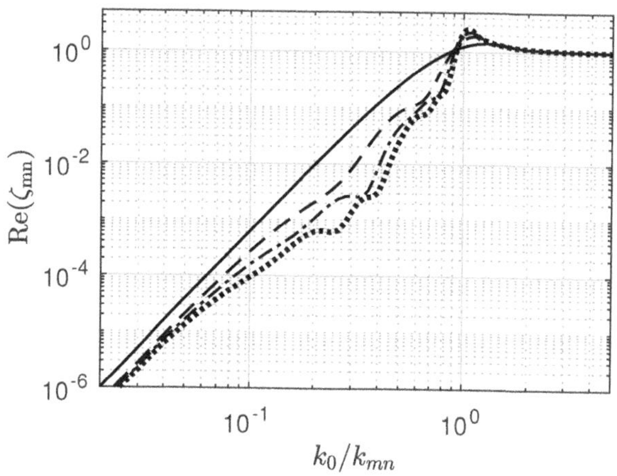

Figure 6.7. *Radiation of the first bending modes of a rectangular baffled plate (the aspect ratio is equal to* 1.5*). The real part of the reduced radiation impedance is plotted against the ratio of the acoustic wavenumber* k_0 *to the flexural wavenumber* k_{mn} *of each mode. Efficiency of the* $(1,2)$ *(solid line),* $(3,4)$ *(dashed line),* $(5,6)$ *(dash-dotted line) and* $(7,8)$ *(dotted line) modes*

6.3. Kirchhoff–Helmholtz integral

We consider a volume distribution of sources $S(\boldsymbol{x}, t)$ radiating in a domain D bounded internally and externally by a surface $\Sigma_{tot} = \Sigma \cup \Sigma_{ext}$ (Figure 6.8). The outer surface Σ_{ext} can be a physical surface or a mathematical boundary closing the domain D. In the first case, we have an *interior* problem, which corresponds, for example, to the computation of the response of a room excited by a loudspeaker. In the second case, we have an *exterior* problem, and Σ_{ext} can be chosen arbitrarily as long as it surrounds all sources; in particular, a surface "at infinity" can be used. It is this latter case that we will consider more particularly in the following.

For harmonic waves, the problem is governed by the Helmholtz equation associated with a set of boundary conditions along Σ_{tot}:

$$\nabla^2 p(\boldsymbol{x}) + k_0^2 p(\boldsymbol{x}) = -S(\boldsymbol{x}) \qquad [6.30]$$

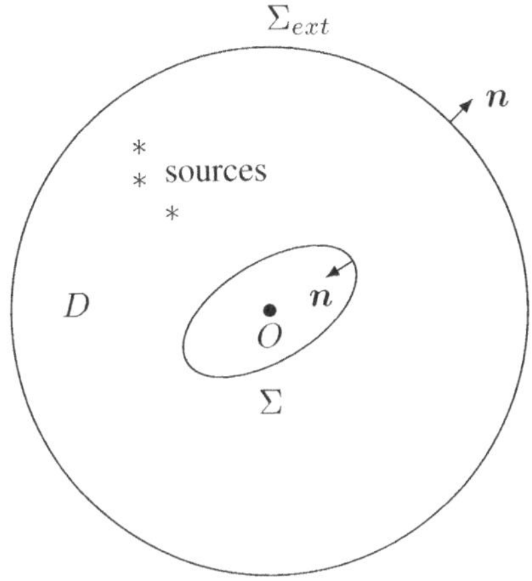

Figure 6.8. *Integral formulation of the Helmholtz equation. Volume sources located in the domain D, bounded by a surface $\Sigma_{tot} = \Sigma \cup \Sigma_{ext}$*

When material surfaces are considered, the boundary conditions are of Neumann, Dirichlet or mixed type; the latter, sometimes called a Robin condition, is associated with the existence of a surface with an impedance Z. These conditions may be homogeneous, or they may have a non-zero right-hand side reflecting, for example, a vibration imposed on a part of the boundary (radiating surface). Therefore, a general form of boundary condition for a material surface can be written as:

$$p(\boldsymbol{x}) - \frac{\mathrm{i}Z}{\rho_0\omega}\frac{\partial p}{\partial \boldsymbol{n}}(\boldsymbol{x}) = f_v(\boldsymbol{x}) \qquad \forall \boldsymbol{x} \in \Sigma \qquad [6.31]$$

where the linearized Euler's equation has been used to replace the normal velocity with the normal pressure gradient, and where $f_v(\boldsymbol{x})$ is a function associated with a possible vibration imposed on the surface.

If the medium is unbounded, the waves can propagate to infinity. It is then necessary to impose a behavior condition at infinity to ensure the uniqueness of the solution. This condition is the translation into the frequency domain of the causal condition in the time domain. From a physical point of view, this condition expresses the fact that, away from sources and obstacles, the field must locally resemble a divergent spherical wave (in three-dimensional space), which can be written as:

$$p(\boldsymbol{x})_{r\to\infty} \propto \frac{\exp(-\mathrm{i}k_0 r)}{r} D(\theta, \varphi) \qquad [6.32]$$

where the spherical coordinates (r, θ, φ) have been used.

This simple and intuitive condition will often be used in practice. More formally, the condition to be imposed is called the *Sommerfeld condition*, or *radiation condition*. It can be shown that it is written as[1]:

$$\lim_{r\to\infty} r\left(\frac{\partial p}{\partial r} + \mathrm{i}k_0 p\right) = 0 \qquad [6.33]$$

As it can be seen, the intuitive form of the behavior at infinity [6.32] does indeed satisfy the Sommerfeld condition.

In order to obtain an integral formulation of the initial problem, we will use the Green's function in free space, as the solution to:

$$\nabla_x^2 G_0(\boldsymbol{x},\boldsymbol{y}) + k_0^2 G_0(\boldsymbol{x},\boldsymbol{y}) = -\delta(\boldsymbol{x}-\boldsymbol{y}) \qquad [6.34]$$

where the subscript x indicates that the derivatives of the Green's function in the Laplacian operator are to be evaluated at the observation point $\boldsymbol{x}$. Let us take the product of [6.30] by $G_0(\boldsymbol{x},\boldsymbol{y})$ and subtract the product of [6.34] with $p(\boldsymbol{x})$. It follows that:

$$G_0(\boldsymbol{x},\boldsymbol{y})\nabla^2 p(\boldsymbol{x}) - p(\boldsymbol{x})\nabla_x^2 G_0(\boldsymbol{x},\boldsymbol{y}) = p(\boldsymbol{x})\delta(\boldsymbol{x}-\boldsymbol{y}) - G_0(\boldsymbol{x},\boldsymbol{y})S(\boldsymbol{x}) \qquad [6.35]$$

Since the Green's function G_0 and the delta function are symmetric, the roles of $\boldsymbol{x}$ and $\boldsymbol{y}$ can be swapped before integrating the relation from [6.35] over the domain D. We then get:

$$\begin{aligned}\int_D (G_0(\boldsymbol{x},\boldsymbol{y})\nabla^2 p(\boldsymbol{y}) - p(\boldsymbol{y})\nabla_y^2 G_0(\boldsymbol{x},\boldsymbol{y}))\mathrm{d}\boldsymbol{y} + \int_D S(\boldsymbol{y})G_0(\boldsymbol{x},\boldsymbol{y})\mathrm{d}\boldsymbol{y} \\ = \int_D p(\boldsymbol{y})\delta(\boldsymbol{x}-\boldsymbol{y})\mathrm{d}\boldsymbol{y}\end{aligned} \qquad [6.36]$$

The right-hand side of [6.36] is equal to $p(\boldsymbol{x})$ if $\boldsymbol{x}$ belongs to the integration domain D; it is otherwise zero (i.e. when $\boldsymbol{x}$ is located inside the surface Σ in Figure 6.8). The volume integral in the right-hand side of [6.36] can be transformed into a surface integral along the boundary Σ_{tot} using the following Green's identity:

$$\int_D (f\nabla^2 g - g\nabla^2 f)\mathrm{d}V = \int_{\Sigma_{tot}} \left(f\frac{\partial g}{\partial n} - g\frac{\partial f}{\partial n}\right)\mathrm{d}\Sigma \qquad [6.37]$$

1 For a two-dimensional problem, the wave must resemble a diverging cylindrical wave $\exp(-\mathrm{i}k_0\rho)/\sqrt{\rho}$ and the Sommerfeld condition is then written as $\lim_{\rho\to\infty}\sqrt{\rho}\,(\partial p/\partial\rho + \mathrm{i}k_0 p) = 0$.

The rest of the discussion focuses on the *exterior problems* of radiation or diffraction. In formula [6.37], the surface integral is then restricted to the single physical boundary Σ because the Sommerfeld condition, which is satisfied by both p and G_0, ensures that the contribution of the surface at infinity Σ_{ext}, which closes the domain D, is zero. Using the Green's identity [6.37] in which the unit normal $\boldsymbol{n}$ points out of the domain D, we thus obtain:

$$p(\boldsymbol{x}) = \int_D S(\boldsymbol{y})G_0(\boldsymbol{x},\boldsymbol{y})\mathrm{d}\boldsymbol{y} + \int_\Sigma \left(G_0(\boldsymbol{x},\boldsymbol{y})\frac{\partial p(\boldsymbol{y})}{\partial n_y} - p(\boldsymbol{y})\frac{\partial G_0(\boldsymbol{x},\boldsymbol{y})}{\partial n_y} \right) \mathrm{d}\Sigma_y \quad [6.38]$$

The physical interpretation of this formula is as follows. The volume integral corresponds to the pressure field that would be generated by the distribution of sources S in free space, that is, in the absence of any obstacle. This contribution is called the *incident field*, denoted by p_i, or the *direct field*. For passive surfaces, the second integral corresponds to the modification of the field induced by the presence of obstacles: spatial redistribution of pressure fluctuations and partial absorption of energy if part of the surface is covered by an absorbing material. This is called the *diffracted field*, denoted by p_d. Sometimes, it is also referred to as *secondary sources* induced by the interaction of the direct, or *primary*, field with the obstacles[2].

This attractive physical interpretation, however, should not hide the fact that the use of formula [6.38] is challenging in practice. The two contributions to the total field are actually very different in nature. The first contribution is a perfectly defined integral from the moment the volume sources are specified. On the other hand, the second contribution is not a simple integral of known functions. The diffracted field is given by:

$$p_d(\boldsymbol{x}) = \int_\Sigma \left(G_0(\boldsymbol{x},\boldsymbol{y})\frac{\partial p(\boldsymbol{y})}{\partial n_y} - p(\boldsymbol{y})\frac{\partial G_0(\boldsymbol{x},\boldsymbol{y})}{\partial n_y} \right) \mathrm{d}\Sigma_y \quad [6.39]$$

and the pressure and its normal derivative that appear in the integral are just as unknown as the value of the pressure at the observation point $\boldsymbol{x}$.

Equation [6.39] is an *integral equation* and not a simple formula. This equation is difficult to use in this form because the observation point $\boldsymbol{x}$ and the integration point $\boldsymbol{y}$ evolve in two different domains, a volume on the one hand and a surface on the other hand. An intermediate computational step is required; it consists of deriving

2 When part of the surface is active, with an imposed vibration velocity, the contribution of the surface integral includes the sound radiation of these vibrations in the presence of diffraction by the rest of the surface.

from [6.39] an integral equation written on the boundary Σ of domain D by making the observation point tend toward this surface. The procedure leading to this equation is a bit delicate, since the Green's function and its normal derivative varies as $1/R$ and $1/R^2$, respectively; thus they become singular during the integration, since the distance $R = \|\boldsymbol{x} - \boldsymbol{y}\|$ cancels out when the observation point and the integration point coincide. We will simply give the result of this calculation by referring to the demonstration and its application to more complex problems (for non-regular surfaces) in [ATA 15]. Assuming that the surface Σ is regular, that is, that it has a tangent plane at each of its points, the following formula is obtained for an observation point on the surface Σ:

$$\frac{1}{2}p(\boldsymbol{x}) = \int_{\Sigma} \left(G_0(\boldsymbol{x}, \boldsymbol{y}) \frac{\partial p(\boldsymbol{y})}{\partial n_y} - p(\boldsymbol{y}) \frac{\partial G_0(\boldsymbol{x}, \boldsymbol{y})}{\partial n_y} \right) \mathrm{d}\Sigma_y + p_i(\boldsymbol{x}) \,; \qquad \boldsymbol{x}, \boldsymbol{y} \in \Sigma \qquad [6.40]$$

The numerical solution of this surface integral equation is the basis of the so-called boundary element method (BEM), which is widely used for solving radiation or diffraction problems in open domains. Again, we refer to the book by Atalla and Sgard[ATA 15] for more details. This method provides the values of $p(\boldsymbol{y})$ and $\partial p/\partial n_y(\boldsymbol{y})$ along Σ. It then becomes possible to use relation [6.38] as a simple integration of known, non-singular functions for $\boldsymbol{x} \in D$.

6.3.1. *Irregular frequencies*

A difficulty of the integral approach is that in this formulation one does not strictly impose that the pressure inside Σ be zero, but only that it satisfies the weak condition:

$$\int_{\Sigma} \left(G_0(\boldsymbol{x}, \boldsymbol{y}) \frac{\partial p(\boldsymbol{y})}{\partial n_y} - p(\boldsymbol{y}) \frac{\partial G_0(\boldsymbol{x}, \boldsymbol{y})}{\partial n_y} \right) \mathrm{d}\Sigma_y = 0 \qquad [6.41]$$

As a result, the solution to the external problem is not unique for specific values of the frequency, called *irregular frequencies*, whereas the solution to the initial Helmholtz problem is on the other hand always unique. For diffraction by a rigid body (Neumann boundary condition imposed on Σ), it is shown that the irregular frequencies correspond to the eigenfrequencies of the internal problem which namely consists of solving the Helmholtz equation (for the same fluid as in the external medium) with Dirichlet boundary conditions [SCH 68]. This property holds regardless of the boundary condition, be it Neumann, Dirichlet, or Robin. On this very technical subject, see [MAR 08].

Many techniques are used in BEM codes to circumvent the issue associated with existing irregular frequencies. The simplest is known as the CHIEF method and was

introduced by Schenck [SCH 68]. It consists of imposing that the pressure be zero at a certain number of control points in the cavity to form an overdetermined system when the surface integral equation has been discretized. This approach can fail if the control points coincide with pressure nodes of the interior problem, and more robust methods have subsequently been proposed which impose either zero pressure on a surface element placed inside the cavity, or absorption by this element, which makes the eigenfrequencies complex. Another technique consists of using a combination of the Kirchhoff–Helmholtz integral equation with its normal derivative. More details on the numerical solution of surface integral equations can be found in [ATA 15].

The approach described above, which involves two successive steps including the delicate exercise of numerically solving a surface integral equation, is necessary when the obstacles are of arbitrary geometry. When their geometry is sufficiently simple (i.e. associated with a coordinate system in which the Laplacian is separable) and the boundary conditions are also simple (typically conditions of rigidity or imposed normal velocity), it is possible to construct an explicit solution through the use of so-called adapted Green's functions (see section 6.4).

6.3.2. *Expressing the surface integral in terms of pressure and velocity*

In the Kirchhoff–Helmholtz equation [6.38], the diffracted part is written as:

$$p_d(\boldsymbol{x}) = \int_\Sigma \left(G_0(\boldsymbol{x},\boldsymbol{y}) \frac{\partial p(\boldsymbol{y})}{\partial n_y} - p(\boldsymbol{y}) \frac{\partial G_0(\boldsymbol{x},\boldsymbol{y})}{\partial n_y} \right) \mathrm{d}\Sigma_y \qquad [6.42]$$

In this expression, the pressure gradient can be replaced by the acoustic velocity using the linearized Euler's equation:

$$\mathrm{i}\omega\rho_0 u_i = -\frac{\partial p}{\partial x_i} \qquad [6.43]$$

On the surface, the normal acoustic velocity is equal to the vibration velocity, $u_n = v_n$, and:

$$p_d(\boldsymbol{x}) = -\int_\Sigma \left(\mathrm{i}\omega\rho_0 v_n G_0(\boldsymbol{x},\boldsymbol{y}) + p(\boldsymbol{y}) \frac{\partial G_0(\boldsymbol{x},\boldsymbol{y})}{\partial n_y} \right) \mathrm{d}\Sigma_y \qquad [6.44]$$

The normal derivative of the Green's function can be developed as:

$$\frac{\partial G_0}{\partial n_y} = \frac{\partial G_0}{\partial R}\frac{\partial R}{\partial n_y} = \frac{1}{4\pi}\left(\frac{\mathrm{i}k_0}{R} + \frac{1}{R^2}\right)(\boldsymbol{e}_r \cdot \boldsymbol{n}_y)\exp(-\mathrm{i}k_0 R) \qquad [6.45]$$

where $\boldsymbol{e_r}$ designates the unit vector pointing in the direction from the current point $\boldsymbol{y}$ to the observer $\boldsymbol{x}$, $\boldsymbol{e}_r = (\boldsymbol{x} - \boldsymbol{y})/R$. We finally get:

$$p_d(\boldsymbol{x}) = -\frac{1}{4\pi} \int_\Sigma \left(\mathrm{i}\omega\rho_0 v_{n_y} + p(\boldsymbol{y}) \left(\mathrm{i}k_0 + \frac{1}{R} \right) (\boldsymbol{e}_r \cdot \boldsymbol{n}_y) \right) \frac{\exp(-\mathrm{i}k_0 R)}{R} \mathrm{d}\Sigma_y \quad [6.46]$$

This expression shows the existence of two contributions to the diffracted pressure: a monopolar contribution associated with the normal component of the vibration velocity, or more precisely with its normal acceleration, and a contribution of dipolar nature, associated with the pressure fluctuation on the obstacle. The dipolar nature becomes more apparent if we bear in mind that the dot product $\boldsymbol{e}_r \cdot \boldsymbol{n}_y$ is none other than the cosine of the angle θ formed by the direction of observation and the normal to the surface for any source point $\boldsymbol{y}$ on the obstacle (Figure 6.9). Note also that the dipole term consists of a far-field term and a near-field term (varying in $1/R^2$), which can be neglected at large distances from the obstacles.

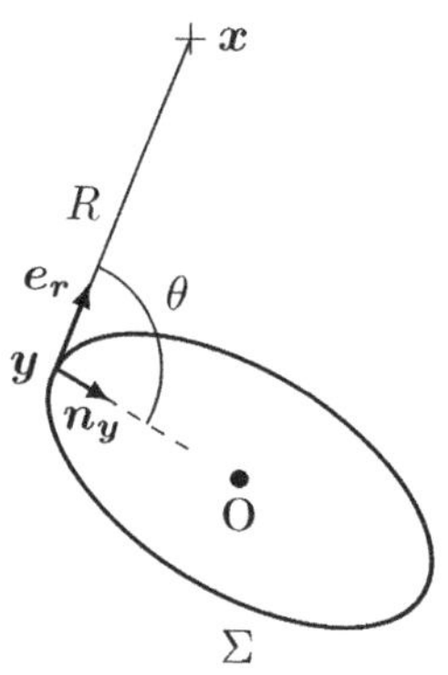

Figure 6.9. *Geometry for the Kirchhoff-Helmholtz formulation*

6.3.3. *Kirchhoff–Helmholtz formula and acoustic field extrapolation*

If it is possible to evaluate the pressure and its gradient on any closed surface surrounding all the primary (volume or surface) and secondary (diffracted) sources, formulas [6.39] or [6.46] can be used to determine the acoustic field at any point outside that surface. This property can be used to combine a solution of the Helmholtz differential equation within a limited domain around the sources with an integral formulation. Solving the Helmholtz equation in a limited region with non-uniform physical properties, for example, using a finite element solver, provides the pressure and pressure gradient along the surface bounding the domain. The values

on this surface can then be propagated (or "extrapolated") over large distances using the Kirchhoff–Helmholtz formula. The coupling surface can be placed in the near field of the sources in order to limit the size of the finite element domain as much as possible, provided only that all direct or indirect sources are surrounded by the coupling surface. This method can be seen as an extension of the extrapolation method using the plane-wave decomposition of the pressure on an infinite plane described in Chapter 2[3].

6.4. Adapted Green's functions

To obtain an integral formulation of the Helmholtz equation, we used the free-space Green's function G_0. However, it is possible to take advantage of the non-uniqueness of the definition of Green's functions by choosing or constructing a function that satisfies the homogeneous version of the boundary conditions imposed on the pressure at the physical boundaries of the domain, as well as the Sommerfeld condition for an exterior problem. Such a Green's function is known as an *adapted* Green's function (meaning adapted to the geometry of the problem).

We will illustrate this process in the particular but common situation where the obstacles are rigid. We thus have $\partial p/\partial n = 0$ along Σ, and the surface integral of the integral formulation is reduced to:

$$p_d(\boldsymbol{x}) = -\int_\Sigma p \frac{\partial G}{\partial n_y} \mathrm{d}\Sigma_y \qquad [6.47]$$

If it is possible to construct a Green's function, denoted G_{adapt}, whose normal derivative is zero at any point of Σ, then this integral is zero and the pressure at the observation point is expressed as a simple integral over the source volume of known quantities:

$$p(\boldsymbol{x}) = \int_D S(\boldsymbol{y}) G_{adapt}(\boldsymbol{x}, \boldsymbol{y}) \mathrm{d}y \qquad [6.48]$$

The effects of the interaction of the incident wave with the obstacles are now fully included in the expression of G_{adapt}. Of course, the difficulty lies in constructing the Green's function adapted to a particular geometry of obstacles, and this is only possible in a limited number of situations. As an example, we can consider sources located inside a wedge with an interior angle equal to π/n, where n is an integer, or near circular cylinders or spheres. Other cases are also possible, but they require the use of sophisticated techniques that fall beyond the scope of this book.

3 This method can also be applied for time-dependent problems.

In fact, we have already implicitly considered an adapted Green's function when we used the method of images to calculate the radiation of a point source located above a rigid plane $y_3 = 0$ (see section 4.7). It can indeed be verified that the following Green's function:

$$G_{adapt}(\boldsymbol{x},\boldsymbol{y}) = \frac{\exp(-ik_0R_s)}{4\pi R_s} + \frac{\exp(-ik_0R_i)}{4\pi R_i} \quad [6.49]$$

$$R_s = \|\boldsymbol{x}-\boldsymbol{y}_s|\,; \qquad R_i = \|\boldsymbol{x}-\boldsymbol{y}_i\| \quad [6.50]$$

where $\boldsymbol{y}_i$ is the symmetric of $\boldsymbol{y}_s$ with respect to the rigid plane, actually satisfies the boundary condition of a rigid wall, $\partial G_{adapt}/\partial y_3 = 0$ for $y_3 = 0$ and has no singularity other than that associated with the real source in the upper half-space. The adapted Green's functions method can then be considered as a generalization of the method of images. Adapted Green's functions for cylinders and spheres can be found in specialized books [MOR 53].

6.5. Integral formulation associated with the wave equation

To obtain an integral formulation in the time domain, we simply have to reproduce the approach mentioned in section 6.3 for harmonic waves. We combine the equations satisfied by a Green's function respecting the causality condition and by a distribution of sources, and we then integrate in space and in time. This calculation is left to the reader, who can refer to Chapter 7 of Morse and Feshbach [MOR 53] for details. For $\boldsymbol{x}$ belonging to the domain D and $t > 0$, we get:

$$\begin{aligned} p'(\boldsymbol{x},t) &= \int_0^t \int_D S(\boldsymbol{y},t_0) g(\boldsymbol{x},t;\boldsymbol{y},t_0) \mathrm{d}\boldsymbol{y}\mathrm{d}t_0 \\ &+ \int_0^t \int_\Sigma \left(g(\boldsymbol{x},t;\boldsymbol{y},t_0) \frac{\partial p(\boldsymbol{y},t_0)}{\partial n_y} - p(\boldsymbol{y},t_0) \frac{\partial g(\boldsymbol{x},t;\boldsymbol{y},t_0)}{\partial n_y} \right) \mathrm{d}\Sigma_y \mathrm{d}t_0 \quad [6.51] \\ &- \frac{1}{c_0^2} \int_D \left[\left(g(\boldsymbol{x},t;\boldsymbol{y},t_0) \frac{\partial p(\boldsymbol{y},t_0)}{\partial t_0} - p(\boldsymbol{y},t_0) \frac{\partial g(\boldsymbol{x},t;\boldsymbol{y},t_0)}{\partial t_0} \right) \right]_0^t \mathrm{d}\boldsymbol{y} \end{aligned}$$

In the last integral, the causality condition [5.9] imposes that the quantity in square brackets is zero for the upper bound, and we finally obtain:

$$\begin{aligned} p'(\boldsymbol{x},t) &= \int_0^t \int_D S(\boldsymbol{y},t_0) g(\boldsymbol{x},t;\boldsymbol{y},t_0) \mathrm{d}\boldsymbol{y}\mathrm{d}t_0 \\ &+ \int_0^t \int_\Sigma \left(g(\boldsymbol{x},t;\boldsymbol{y},t_0) \frac{\partial p(\boldsymbol{y},t_0)}{\partial n_y} - p(\boldsymbol{y},t_0) \frac{\partial g(\boldsymbol{x},t;\boldsymbol{y},t_0)}{\partial n_y} \right) \mathrm{d}\Sigma_y \mathrm{d}t_0 \quad [6.52] \\ &- \frac{1}{c_0^2} \int_D \left(\left(\frac{\partial g}{\partial t_0} \right)_{t_0=0} p(\boldsymbol{y},t_0) - g(t_0=0) \left(\frac{\partial p}{\partial t_0} \right)_{t_0=0} \right) \mathrm{d}\boldsymbol{y} \end{aligned}$$

The last volume integral corresponds to the sound emission caused by non-zero values of the initial pressure or velocity in the volume. The associated "transient" response is very often ignored when considering the long-term behavior of acoustic waves. We will however illustrate the acoustic radiation produced by the existence of an overpressure at $t_0 = 0$ in a confined spherical domain in section 6.5.1.

Finally, it should be pointed out that when the initial pressure and velocity fluctuations are zero, or when their influence is neglected, the integral formulation in the time domain of the acoustic field diffracted by an obstacle can be derived from the formulation in the frequency domain [6.46] using the following substitutions:

$$\mathrm{i}\omega \to \frac{\partial}{\partial t} \quad [6.53]$$

$$\exp(-\mathrm{i}k_0 R) \to t - \frac{R}{c_0} \quad [6.54]$$

Using the free space Green's function, the resulting expression is as follows:

$$\begin{aligned} p'_d(\boldsymbol{x}, t) = \frac{1}{4\pi} \int_\Sigma \bigg(& \frac{1}{R} \frac{\partial p'}{\partial n_y} \left(\boldsymbol{y}, t - \frac{R}{c_0} \right) \\ & + \frac{1}{R} \frac{\partial R}{\partial n_y} \left(\frac{1}{c_0} \frac{\partial}{\partial t} + \frac{1}{R} \right) p' \left(\boldsymbol{y}, t - \frac{R}{c_0} \right) \bigg) \mathrm{d}\Sigma_y \end{aligned} \quad [6.55]$$

or, equivalently, by introducing the normal acceleration to the surface $\gamma'_n = \mathrm{d}v'_n/\mathrm{d}t$ (we recall that the normal is pointing inwards the vibrating surface), using the linearized Euler's equation to relate the time derivative of the velocity to the pressure gradient:

$$\begin{aligned} p'_d(\boldsymbol{x}, t) = & -\frac{\rho_0}{4\pi} \int_\Sigma \frac{1}{R} \gamma'_n \left(\boldsymbol{y}, t - \frac{R}{c_0} \right) \mathrm{d}\Sigma_y \\ & - \frac{1}{4\pi c_0} \int_\Sigma \frac{(\boldsymbol{e}_r \cdot \boldsymbol{n}_y)}{R} \left(\frac{\partial}{\partial t} + \frac{c_0}{R} \right) p' \left(\boldsymbol{y}, t - \frac{R}{c_0} \right) \mathrm{d}\Sigma_y \end{aligned} \quad [6.56]$$

The total field is obtained by adding the incident field to the above expressions, which are sometimes called Kirchhoff formulas.

6.5.1. *Bursting balloon*

As an example of a problem with non-zero initial conditions, consider the following situation: an overpressure is created in a spherical domain centered on the

origin. The pressure is "released" at a given time and the resulting pressure field is to be determined as a function of time and space. A concrete example is that of a balloon filled with gas at a slight overpressure relative to the surrounding fluid and whose thin membrane is destroyed at time $t = 0$ (this is known as the "bursting balloon" problem).

We saw in section 2.2 that the solution of a spherically symmetric problem can be obtained by solving a one-dimensional wave equation for an acoustic variable (potential or pressure) weighted by the distance r from the origin. This radial coordinate then plays a role similar to that of a Cartesian coordinate. We can then use formula [6.52] for the weighted pressure fluctuation rp' and with the one-dimensional Green's function. The pressure at time $t_0 = 0$ is denoted $p_0(r)$[4] and the initial velocity is zero. We obtain:

$$rp'(r,t) = -\frac{1}{c_0^2}\int_0^\infty r_0 p_0(r_0)\left(\frac{\partial g_0}{\partial t_0}\right)_{t_0=0} dr_0 \qquad [6.57]$$

where g_0 is the one-dimensional Green's function [5.64]:

$$g_0(r,t;r_0,t_0) = \frac{c_0}{2} H\left(t - t_0 - \frac{|r - r_0|}{c_0}\right) \qquad [6.58]$$

It can be inferred that:

$$rp'(r,t) = \frac{1}{2}\int_0^\infty r_0 p_0(r_0)\delta\left(c_0 t - |r - r_0|\right) dr_0 \qquad [6.59]$$

and since the argument of the Dirac delta function cancels out for the two values $r_0 = r - c_0 t$ and $r_0 = r + c_0 t$, the final result is as follows:

$$p'(r,t) = \frac{1}{2r}((r - c_0 t)p_0(r - c_0 t) + ((r + c_0 t)p_0(r + c_0 t)) \qquad [6.60]$$

This expression shows that, in general, there exist two contributions depending on a retarded time and an advanced time, respectively. This is the consequence of each point r_0 in the source region emitting waves that propagate to values of r greater than or less than r_0.

4 p_0 here designates the initial pressure difference between the inner and outer regions of the spherical domain.

The example of the gas-filled balloon allows us to be more precise. The initial overpressure in the balloon, denoted now by P_0, is uniform inside the spherical region $r \leq a$ and zero outside. The general formula [6.60] becomes:

$$p'(r,t) = \frac{P_0}{2r}((r - c_0 t)H(a - |r - c_0 t|) + (r + c_0 t)H(a - |r + c_0 t|)) \quad [6.61]$$

For an observation point located outside the sphere of radius a, the second term of this expression is identically zero: only waves propagating toward increasing values of the radial distance exist, provided that $c_0 t > r - a$ and $c_0 t < r + a$. At time $t = (r - a)/c_0$, the pressure undergoes a jump equal to $aP_0/2r$ (it is associated with a wavefront arriving from the nearest source); the signal then decreases linearly with time until $t = (r + a)/c_0$ where a new compression of amplitude $aP_0/2r$ reduces the acoustic pressure to zero (the last received signal originates from the point of the balloon farthest from the observation point). This behavior is shown in Figure 6.10. The evolution observed at a fixed time as a function of the distance is also plotted in this figure.

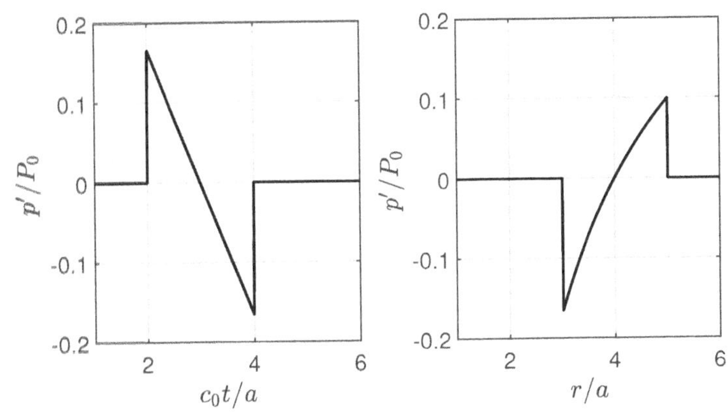

Figure 6.10. *Acoustic response to a uniform overpressure P_0 initially present in a sphere of radius a. On the left, the time evolution of the pressure at a fixed position ($r = 3a$); on the right, the spatial evolution at a fixed time ($t = 4a/c_0$)*

6.6. Radiation from planar structures: Rayleigh integral

A particularly interesting example of the application of integral formulations and adapted Green's functions is provided by the evaluation of the pressure field radiated by a planar vibrating structure surrounded by a rigid and infinite flat rigid baffle (Figure 6.11).

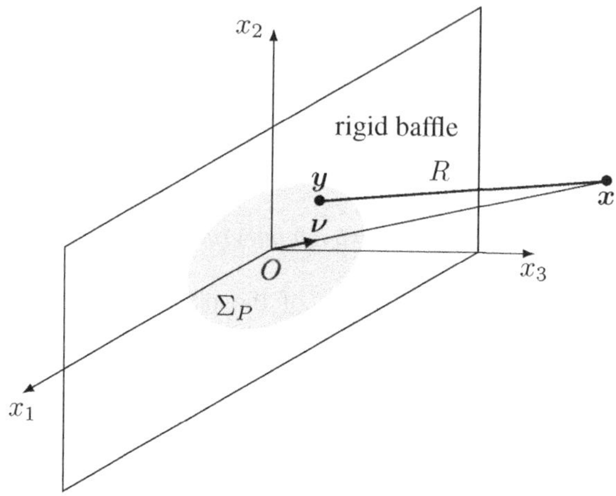

Figure 6.11. *Radiation of a vibrating planar structure surrounded by an infinite rigid baffle*

For harmonic vibrations, formula [6.38] is here reduced to:

$$p(\boldsymbol{x}) = -\int_{\Sigma} \left(G\frac{\partial p}{\partial y_3} - p\frac{\partial G}{\partial y_3} \right) \mathrm{d}\Sigma \qquad [6.62]$$

where G is an arbitrary Green's function satisfying the radiation condition. Note that the x_3 axis in this expression points inside the fluid volume.

Since the velocity (and therefore the pressure gradient) is known on the vibrating part of the structure and is zero on the baffle, it is possible to reduce the integral equation [6.62] to a simple integral of known quantities by choosing a Green's function G_{adapt} having a zero normal derivative in the plane $x_3 = 0$. It then follows that:

$$p(\boldsymbol{x}) = \mathrm{i}\omega\rho_0 \int_{\Sigma_P} G_{adapt} v_3(y_1, y_2) \mathrm{d}y_1 \mathrm{d}y_2 \qquad [6.63]$$

where the integration is restricted to the region of the structure Σ_P vibrating with velocity v_3. The adapted Green's function is easily identified with the one previously obtained by the method of images.

In the plane $x_3 = 0$, the adapted Green's function is equal to twice the free-space Green's function and thus equal to:

$$G_{adapt}(y_1, y_2, 0) = \frac{\exp(-\mathrm{i}k_0 R)}{2\pi R}; \quad R = \sqrt{(x_1 - y_1)^2 + (x_2 - y_2)^2 + x_3^2} \qquad [6.64]$$

With the choice of this adapted Green's function, formula [6.63] is finally written as:

$$p(\boldsymbol{x}) = \frac{\mathrm{i}\omega\rho_0}{2\pi}\int_{\Sigma_P}\frac{\exp(-\mathrm{i}k_0R)}{R}v_3(y_1,y_2)\mathrm{d}y_1\mathrm{d}y_2 \qquad [6.65]$$

This expression is known as the *Rayleigh integral.*

Without further specifying the shape of the planar vibrating structure or the distribution of the vibration velocity, it is possible to obtain a remarkable result if we limit ourselves to the acoustic far-field. When the distance to the origin of the observation point $\boldsymbol{x}$ tends to infinity, [6.65] can be simplified in the Fraunhofer approximation to obtain:

$$\lim_{x\to\infty} p(\boldsymbol{x}) = \frac{\mathrm{i}\omega\rho_0}{2\pi x}\exp(-\mathrm{i}k_0x)\int_{\Sigma_P} v_3(y_1,y_2)\exp\left(\mathrm{i}k_0\frac{\boldsymbol{x}\cdot\boldsymbol{y}}{x}\right)\mathrm{d}y_1\mathrm{d}y_2 \qquad [6.66]$$

or, by denoting $\boldsymbol{k} = k_0\boldsymbol{x}/x$ the wave vector pointing in the direction connecting the origin of the coordinate system to the observation point:

$$\lim_{x\to\infty} p(\boldsymbol{x}) = \frac{\mathrm{i}\omega\rho_0}{2\pi x}\exp(-\mathrm{i}k_0x)\int_{\Sigma_P} v_3(y_1,y_2)\exp(\mathrm{i}\boldsymbol{k}\cdot\boldsymbol{y})\mathrm{d}y_1\mathrm{d}y_2 \qquad [6.67]$$

This expression is characteristic of the radiation from a point source at the origin, affected by a directivity given by the integral over Σ_P. This integral appears as the two-dimensional spatial Fourier transform of the velocity field for the particular value of the wave vector corresponding to the projection $\boldsymbol{k_p}$ of $\boldsymbol{k}$ on the plane structure. Using the spherical coordinates, this vector becomes:

$$\boldsymbol{k_p} = \left|\begin{array}{l} k_0\sin\theta\cos\varphi \\ k_0\sin\theta\sin\varphi \\ 0 \end{array}\right. \qquad [6.68]$$

The far-field value of the acoustic pressure $p(x,\theta,\varphi)$ is thus obtained by "filtering" the vibration velocity in the spectral space for that particular value of the wavevector. For example, if the listening point is located on the axis x_3, it can be seen that the pressure is simply proportional to the integral of the vibration velocity[5].

5 It is important to note that the rapid spatial variations of the vibration, associated with vibration wavenumbers of magnitude greater than k_0, do not contribute to the far-field. As a result, far-field measurements cannot provide any information about source details of size smaller than the acoustic wavelength.

6.6.1. *Radiation from a circular piston*

A particularly important application of the Rayleigh integral concerns the radiation from transducers whose vibration velocity is uniform; they are said to operate in a piston mode. Although the formulation assumes the presence of an infinite rigid baffle around the transducer, it is a very acceptable approximation as soon as the baffle is larger than the wavelength of the emitted sound, which is easily the case for ultrasonic transducers, for example. The most common situation is that of a circular piston of radius a vibrating with a uniform velocity V_0. To estimate the Rayleigh integral, it is natural to use a description in polar coordinates (ρ, φ_s). In the far-field, the following expression is obtained:

$$p(\boldsymbol{x}) = \frac{\mathrm{i}\omega\rho_0 V_0}{2\pi}\frac{\exp(-\mathrm{i}k_0 x)}{x}\int_0^a \rho \mathrm{d}\rho \int_0^{2\pi} \exp(\mathrm{i}k_0\rho\sin\theta\cos(\varphi-\varphi_s))\mathrm{d}\varphi_s \quad [6.69]$$

or, by setting $\phi = \varphi - \varphi_s$:

$$p(\boldsymbol{x}) = \frac{\mathrm{i}\omega\rho_0 V_0}{2\pi}\frac{\exp(-\mathrm{i}k_0 x)}{x}\int_0^a \rho \mathrm{d}\rho \int_0^{2\pi} \exp(\mathrm{i}k_0\rho\sin\theta\cos\phi)\mathrm{d}\phi \quad [6.70]$$

The integral with respect to ϕ is expressed using the Bessel function of the first kind and of order zero J_0 (see Appendix A2.3):

$$\int_0^{2\pi} \exp(-\mathrm{i}k_0\rho\sin\theta\cos\phi)\mathrm{d}\phi = 2\pi J_0(k_0\rho\sin\theta) \quad [6.71]$$

from which

$$p(\boldsymbol{x}) = \mathrm{i}\omega\rho_0 V_0 \frac{\exp(-\mathrm{i}k_0 x)}{x}\int_0^a J_0(k_0\rho\sin\theta)\rho\mathrm{d}\rho \quad [6.72]$$

The radial integral of J_0 can be calculated using the following formula:

$$\int_0^z J_0(x)x\mathrm{d}x = zJ_1(z) \quad [6.73]$$

where J_1 denotes the Bessel function of the first kind and of order 1. We finally get:

$$p(\boldsymbol{x}) = \frac{\mathrm{i}\rho_0 c_0 V_0 k_0 a^2}{2}\frac{2J_1(k_0 a\sin\theta)}{k_0 a\sin\theta}\frac{\exp(-\mathrm{i}k_0 x)}{x} \quad [6.74]$$

The far-field pressure presents a directivity $D(\theta)$ which depends on the reduced wavenumber k_0a such that:

$$D(\theta) = \left| \frac{2J_1(k_0a\sin\theta)}{k_0a\sin\theta} \right| \quad [6.75]$$

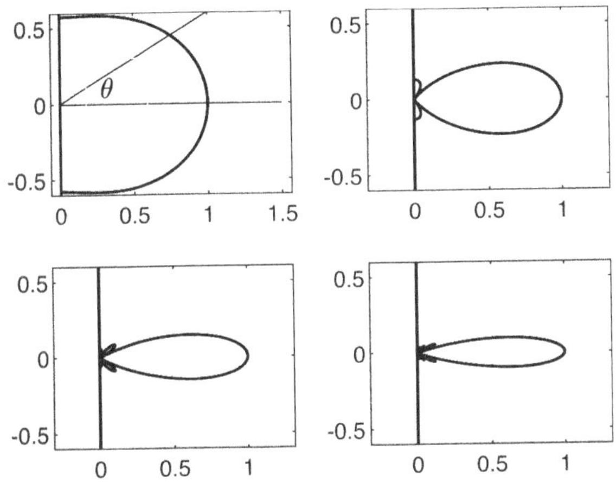

Figure 6.12. *Directivity patterns for the circular piston, plotted for increasing values of the reduced wavenumber, $k_0a = 2$, 5, 8 and 12 (from left to right and top to bottom)*

The directivity pattern of the circular piston is plotted in Figure 6.12 for increasing values of k_0a. For small values of k_0a, the source is compact and therefore its radiation is nearly omnidirectional in the half-space $x_3 > 0$. As k_0a increases, an increasingly narrow main lobe centered on the axis x_3 can be observed. Side lobes of smaller amplitude appear in increasing number when $k_0a > 3.83$, the value of the first zero of J_1.

The circular shape of the piston allows for a calculation of the pressure field on the x_3 axis, valid at any distance. For a listening point on the axis, the Rayleigh integral [6.65] is written as:

$$p(x_3) = \mathrm{i}\omega\rho_0 V_0 \int_0^a \frac{\exp(-\mathrm{i}k_0 R)}{R} \rho \mathrm{d}\rho \quad [6.76]$$

with

$$R = \sqrt{x_3^2 + \rho^2}; \qquad R\mathrm{d}R = \rho\mathrm{d}\rho \quad [6.77]$$

which after integration gives

$$p(x_3) = \rho_0 c_0 V_0 \left(\exp\left(-ik_0\sqrt{x_3^2 + a^2} \right) - \exp(-ik_0 x_3) \right) \qquad [6.78]$$

This expression appears as the difference between the contributions of two "plane waves", one traveling from the edge of the piston ("edge wave") and the other from its center. Rewriting this formula in a different way brings out the existence of zeros for the pressure on the axis under certain conditions. Formula [6.13] can be rewritten as:

$$p(x_3) = 2i\rho_0 c_0 V_0 \exp\left(\frac{-ik_0}{2} \left(\sqrt{x_3^2 + a^2} + x_3 \right) \right) \sin\left(\frac{k_0}{2} \left(\sqrt{x_3^2 + a^2} - x_3 \right) \right) \qquad [6.79]$$

The pressure amplitude, normalized to that of a plane wave with velocity V_0, is then:

$$\frac{|p(x_3)|}{\rho_0 c_0 V_0} = 2 \left| \sin\left(\frac{k_0}{2} \left(\sqrt{x_3^2 + a^2} - x_3 \right) \right) \right| \qquad [6.80]$$

The pressure vanishes when:

$$k_0 x_3 = \frac{(k_0 a)^2 - (2n\pi)^2}{4n\pi} \qquad [6.81]$$

where n is a positive integer smaller than $k_0 a/2\pi$.

For values of the reduced wavenumber $k_0 a$ greater than 2π, a sequence of maxima and zeros is thus obtained, as shown in Figure 6.13 for $k_0 a = 9\pi$. When the position of the most distant zero from the piston (obtained for $n = 1$) has been exceeded, the amplitude reaches a last maximum and then decreases continuously. Asymptotically, for $x_3 \gg k_0 a^2$, the pressure amplitude varies as $k_0 a^2/2x_3$. The transition from the near field to the far-field occurs at a distance $k_0 a^2/2$ called the Rayleigh distance.

For a listening point located on the axis, all the signals coming from the edge of the piston arrive at the same time and, for certain values of the distance from the piston, a completely destructive interference could take place between this edge wave and the plane wave normally emitted by the piston surface. For a piston with a different shape, only a series of minima and maxima will be observed. The study of a rectangular piston is suggested in section 6.8.

Figure 6.14 shows a cross-section of the pressure near field, in a plane passing through the axis and for $k_0 a = 9\pi$. The figure shows in particular the presence of

significant interferences between the contributions of the elements of the piston near its vibrating face, which gradually decrease for axial distances greater than the position of the last zero.

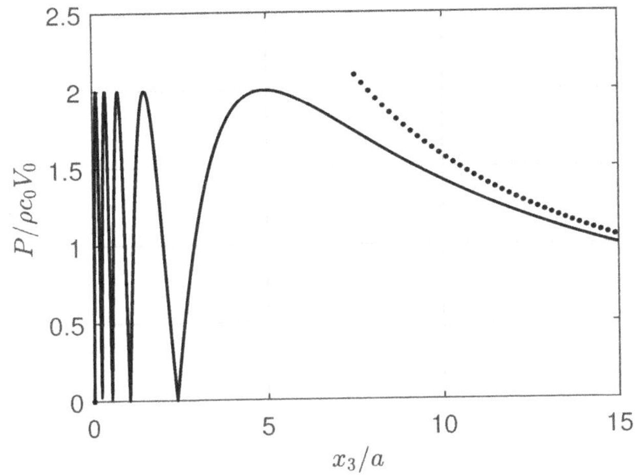

Figure 6.13. *Variation of the acoustic pressure amplitude along the axis of a circular piston for $k_0 a = 9\pi$. Note in particular the presence of four zeros. The dotted line corresponds to the asymptotic evolution of the pressure when x_3 is greater than the Rayleigh distance, equal here to about 14*

The radiation efficiency of the vibrations of the piston can be characterized by evaluating its radiation impedance. To this end, it is necessary to calculate the value of the acoustic pressure at each point on the piston surface and then integrate over the same surface. The result is then divided by the radiating area, namely πa^2. The calculation may look very complicated, since it involves a quadruple integral, but it can be reduced to a double integral by taking into account the symmetries of the problem; for example, Chapter 5 in [PIE 89] should be consulted for details of this calculation. For the reduced impedance $\zeta_{rad} = Z_{rad}/\rho_0 c_0$, the following formula is obtained:

$$\zeta_{rad}(k_0 a) = 1 - \frac{2}{\pi a^2}\int_0^a \left(\int_{-\pi/2}^{\pi/2} \exp(-\mathrm{i}2k_0\rho\cos\varphi)\mathrm{d}\varphi\right)\rho\mathrm{d}\rho \qquad [6.82]$$

We show the result of the numerical integration of formula [6.82] in Figure 6.15[6]. The real part of the impedance, which characterizes the radiation efficiency, is very

6 For an analytical form of ζ_{rad} involving the Bessel and Struve functions, see [PIE 89].

small at low frequencies, since it varies with $(k_0a)^2$ as shown in the approximate formula developed below. It tends to unity at high frequencies, while oscillating on either side of this limit. The imaginary part varies linearly with k_0a at low frequencies and tends slowly toward zero in an oscillating manner as the reduced wavenumber tends to infinity.

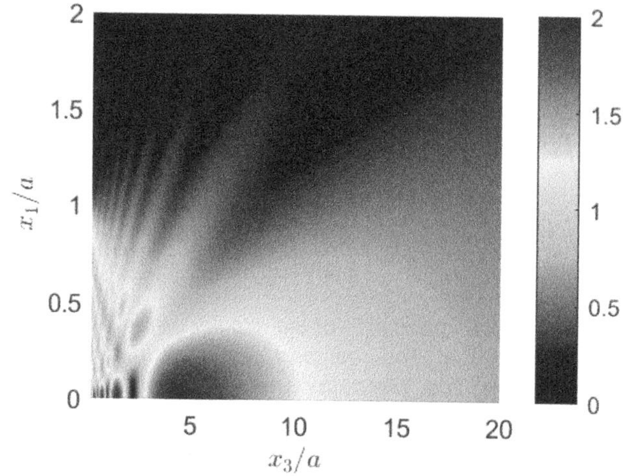

Figure 6.14. *Near-field pressure amplitude generated by a circular piston ($k_0a = 9\pi$). Note the gradual transition from a complicated region where numerous interferences are visible to a more regular field as one moves away from the piston surface. For a color version of this figure, see www.iste.co.uk/juve/acoustics1.zip*

The low frequency behavior of the radiation impedance can be determined analytically from formula [6.82] by expanding the exponential in a Taylor series up to the second order in k_0a. The following expressions are obtained for the real and imaginary parts of the reduced impedance, which will be useful in the following chapters:

$$\zeta_{rad}(k_0a) = R + \mathrm{i}X \approx \frac{(k_0a)^2}{2} + \mathrm{i}\frac{8}{3\pi}k_0a\,; \qquad k_0a \ll 1 \qquad [6.83]$$

This formula shows that the radiation impedance is essentially reactive at low frequencies.

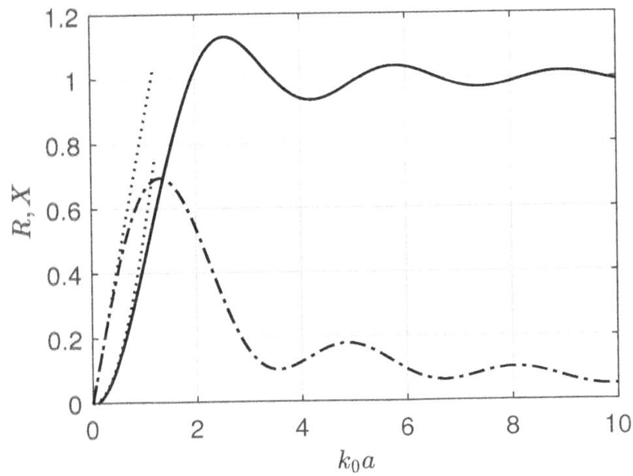

Figure 6.15. *Reduced radiation impedance of the circular piston plotted as a function of the reduced wavenumber k_0a. The real part is plotted in a solid line and the imaginary part in a dash-dotted line. The dotted lines correspond to the low frequency approximation given by formula [6.83]*

6.7. Rayleigh integral in the time domain

The time domain counterpart of the frequency domain Rayleigh's formula [6.65] is given by:

$$p'(\boldsymbol{x},t) = \frac{\rho_0}{2\pi}\int_{\Sigma_P}\frac{1}{R}\gamma_3'\left(y_1,y_2,t-\frac{R}{c_0}\right)\mathrm{d}y_1\mathrm{d}y_2 \tag{6.84}$$

where γ_3' is the acceleration of the vibrating structure. On the axis of a circular piston, this expression is simplified to:

$$p'(x_3,t) = \frac{\rho_0}{2\pi}\int_0^{2\pi}\mathrm{d}\varphi_s\int_0^a\frac{1}{R}\frac{\mathrm{d}v_3'}{\mathrm{d}t}\left(t-\frac{R}{c_0}\right)\rho\mathrm{d}\,\rho \tag{6.85}$$

Integrating this formula leads to:

$$p'(x_3,t) = \rho_0c_0\left(v_3'\left(t-\frac{x_3}{c_0}\right)-v_3'\left(t-\frac{\sqrt{x_3^3+a^2}}{c_0}\right)\right) \tag{6.86}$$

This expression again shows that the pressure on the axis of a circular piston results from the interaction of two plane waves originating from the center and from the edge of the piston surface.

6.8. Exercises

1) **Pulsating sphere**

A sphere of radius a oscillates radially and harmonically. Obtain the frequency domain solution for the radiated pressure, starting from the time domain formula [6.5].

2) **Sphere oscillating at low frequency**

Show that at low frequency the radiated pressure is proportional to the time derivative of the fluctuating mass flow produced by the motion of the surface.

3) **Adapted Green's functions**

a) Use the method of images to construct the adapted Green's function for a point source located inside a rigid wedge of interior angle π/n, where n is an integer. Explicit the far-field pressure generated by a point source when $n = 3$.

b) Construct the adapted Green's function for a perfectly resilient ($p = 0$) infinite plane.

4) **Bursting balloon**

a) Complete the analysis of section 6.5.1 by considering an observation point located *inside* the sphere of radius a. Plot the total field as a function of time and as a function of distance, as well as the incoming (convergent) and outgoing (divergent) contributions.

b) To test the behavior of numerical schemes for solving wave equations, an initial overpressure with a Gaussian shape $P_0(r_0) = \exp(-r_0^2/\sigma^2)$ is frequently used. Plot the time evolution of the pressure for different values of the distance r from the origin. Show that the field at any point in space results from the superposition of an outgoing wave and an incoming wave. Show that the incoming contribution, in which the advanced time appears, decreases rapidly with time when $r > \sigma$.

5) **Radiation impedance of the circular piston**

The power radiated by a vibrating body can be obtained from the integral of its radiation impedance over its surface. It can also be obtained by integrating the acoustic intensity over a surface "at infinity".

a) For the circular piston, write the relation that can be obtained by equating the results of these two methods.

b) Derive the following formula for the real part R of the reduced impedance:

$$R(k_0 a) = \frac{k_0^2 a^2}{4} \int_0^{\pi} \left(\frac{2J_1(k_0 a \sin\theta)}{k_0 a \sin\theta} \right)^2 \sin\theta d\theta$$

c) Obtain the expression of the real part of $R(k_0 a)$ for $k_0 a \rightarrow 0$ using the expansion $J_1(x) \approx x/2 - x^3/16 + O(x^5)$; compare with relation [6.83] and with the exact plot of Figure 6.15.

6) **Rectangular piston**

Consider a rectangular piston with sides a and b. Determine the directivity pattern of the far-field pressure generated by this piston. Particularize to a square piston and compare its directivity to that of a circular piston with the same area.

7) **Near field of a square piston**

a) Plot the amplitude of the pressure near field on the axis of a square piston using a numerical integration. Show the presence of a limited number of successive minima and maxima for a sufficiently large value of $k_0 a$.

b) Plot the near-field pressure amplitude in a symmetry plane of the piston in order to visualize the complex near-field interferences and the progressive transition to a regular far-field.

8) **Impulsive motion of a circular piston**

Determine the time evolution of the pressure field on the axis of a circular piston whose velocity is a step function $v_n'(t) = H(t)$, where H is the Heaviside function.

7

Diffraction and Scattering

Diffraction and scattering are two terms used to describe the interaction of sound waves with obstacles. The former is preferred when the size of the obstacle is large compared to the wavelength, and when the field consists of several relatively well-defined regions that can be partially described in terms of incident wave, reflected wave, interferences and shadow zones. The latter is more likely to be used when the size of the obstacle is small compared to the wavelength, and the term used in this case indicates a less defined, more "diffuse" nature of the interactions. These are nevertheless nuances, since these phenomena are governed by the same equations, in particular the integral formulations developed in the previous chapter.

In this chapter, we give two emblematic examples of these phenomena: the diffraction of plane or spherical waves by acoustic screens, and the scattering by rigid obstacles of simple or generic shape. In particular, the asymptotic behavior at low frequencies (Rayleigh scattering) and at high frequencies is presented in detail.

We then discuss the scattering effects of sound waves that occur when the fluid contains localized non-homogeneous regions. A typical example is the presence of turbulent temperature variations in the atmosphere. The exact volume integral formulation is first presented, and then the Born approximation is introduced. This approximation enables an explicit calculation of the scattered waves when the contrast between the properties of the inhomogeneous zone and those of the surrounding fluid is sufficiently small. Finally, its range of validity is estimated by comparison with the exact solution of the low-frequency scattering by a spherical inclusion.

7.1. Diffraction by a semi-infinite screen

An important area where the modification of the direct field of a source (or distribution of sources) by obstacles has practical applications is that involving the

use of sound barriers to reduce noise pollution, for example in the vicinity of highways or railway tracks. Only numerical simulation[1] is capable of taking into account the geometry and boundary conditions of an often complex barrier placed on an absorbing ground. However, it is instructive to consider the diffraction of a source by a rigid semi-infinite screen. The calculations are complicated and details will not be covered here; one should refer to Wright [WRI 05] for example. An exact solution of the diffraction by a rigid semi-infinite screen can be obtained when the incident wave is plane or spherical, and we describe these two situations successively.

We first consider the diffraction of a harmonic plane wave by a rigid semi-infinite screen. The geometry of the problem is given in Figure 7.1. The screen extends from $-\infty$ to 0 along the x_1 axis, and the incident wave (of unit amplitude) makes an angle θ_i with the x_1 axis in the (x_1, x_2) plane. It therefore takes the form $\exp(+ik_0x_1\cos\theta_i + ik_0x_2\sin\theta_i)$. The position of the observer is defined by its coordinates (r, θ) in the (x_1, x_2) plane.

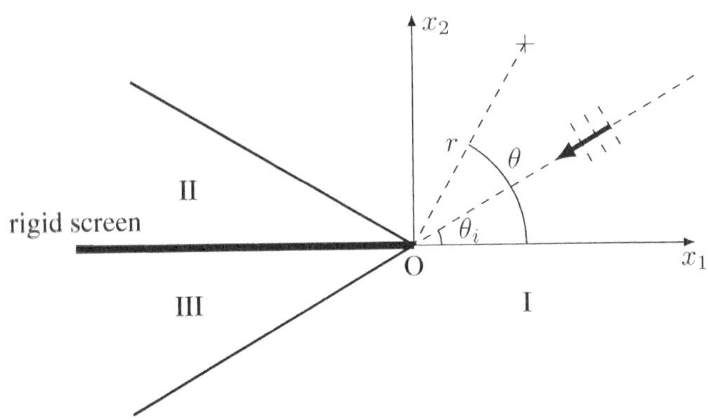

Figure 7.1. *Plane wave diffraction by a semi-infinite rigid thin screen*

By defining the Fresnel function with the following expression:

$$\widetilde{F}(u) = \int_u^{\infty} \exp(\mathrm{i}t^2)\mathrm{d}t\,; \qquad \widetilde{F}(0) = \frac{1+\mathrm{i}}{2}\sqrt{\frac{\pi}{2}} \qquad [7.1]$$

the diffracted field is written exactly as:

1 The boundary element method, based on the integral formulations presented in Chapter 6, is particularly well suited here.

$$
\begin{aligned}
p_d(r,\theta) = &\frac{1-\mathrm{i}}{\sqrt{2\pi}}\left[\exp(+\mathrm{i}k_0 r\cos(\theta_i+\theta))\widetilde{F}\left(\sqrt{2k_0 r}\cos\left(\frac{\theta_i+\theta}{2}\right)\right)\right] \\
&-\frac{1-\mathrm{i}}{\sqrt{2\pi}}\left[\exp(+\mathrm{i}k_0 r\cos(\theta_i-\theta))\widetilde{F}\left(\sqrt{2k_0 r}\cos\left(\frac{\theta_i-\theta}{2}\right)\right)\right]
\end{aligned}
\quad [7.2]
$$

and the total field is the sum of the incident wave and diffracted wave.

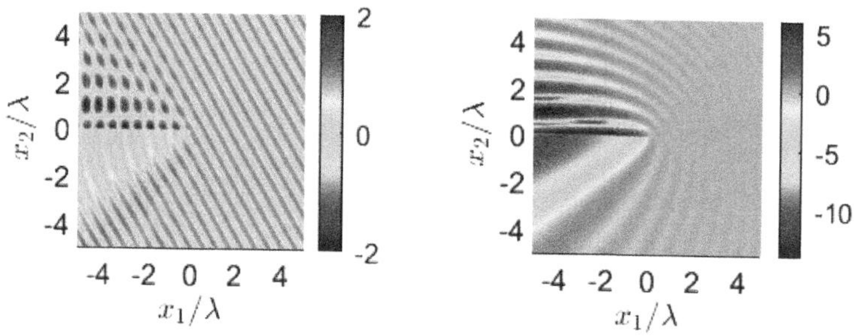

Figure 7.2. *Diffraction of a harmonic plane wave of unit amplitude by a semi-infinite rigid plane. On the left, the real part of the total field; on the right, the sound level in decibels. For a color version of this figure, see www.iste.co.uk/juve/acoustics1.zip*

Figure 7.2 shows the total pressure field obtained for an incident wave making an angle $\theta_i = \pi/6$ with the x_1 axis. The left-hand side of the figure reveals the existence of three regions, defined in Figure 7.1, with significantly different characteristics. In zone I, $(\theta_i - \pi < \theta < \pi - \theta_i)$, the pressure field remains close to that of the incident wave except in the immediate vicinity of the edge. In zone II, $(\pi - \theta_i < \theta < \pi)$, there are pronounced interferences between the incident wave and a wave reflected by the rigid half-plane; the field amplitude varies between 0 and 2. The incident wave does not penetrate zone III, $(\pi < \theta < \pi + \theta_i)$, where only the wave diffracted by the edge is present. In this relative *shadow zone*, the sound level is significantly reduced and the field is regular, as it can be observed on the right-hand side of Figure 7.2. The wave diffracted by the edge, which is present in each region but whose importance is only relevant in the shadow region, can be assimilated to a cylindrical wave originating from the edge and whose amplitude decreases asymptotically as $1/\sqrt{k_0 r}$.

Figure 7.3 shows the evolution of the amplitude of the acoustic field behind the screen, on a line parallel to it and for a normal incident wave $(\theta_i = \pi/2)$. Two values of the distance to the screen normalized by the wavelength have been considered. In the illuminated zone I, the amplitude oscillates around a unit value. These oscillations reflect the interferences between the incident wave and the diffracted wave; they gradually decrease away from the edge. In the shadow zone III, the sound level decreases steadily as the wave penetrates deeper. Directly below the edge (i.e.

for $x_1 = 0$), and for $|x_2| \gg \lambda$, the field amplitude is asymptotically reduced to half the amplitude of the incident wave.

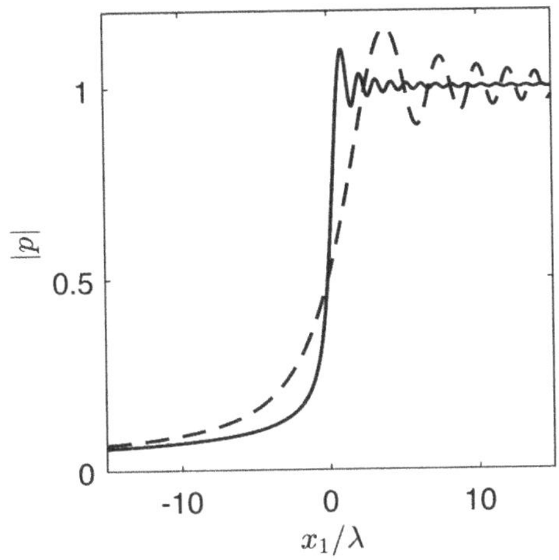

Figure 7.3. *Plane wave diffraction by a semi-infinite rigid screen. The field amplitude at the back of the screen is plotted for two lines parallel to the screen, $x_2/\lambda = -10$ (dashed line) and $x_2/\lambda = -20$ (solid line)*

Although the configuration of the semi-infinite rigid plane is highly idealized, the ideas developed above about the existence of different geometrically defined zones often allow at least a qualitative interpretation of diffraction phenomena due to obstacles of large size relative to the wavelength.

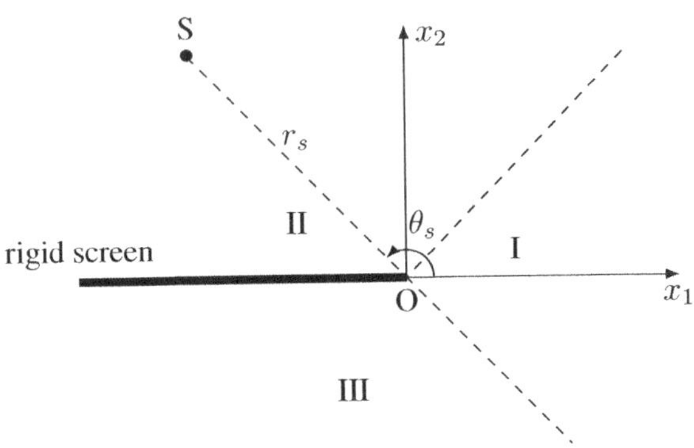

Figure 7.4. *Diffraction of a spherical wave by a semi-infinite rigid thin screen*

When the incident wave is generated by a point source, the exact computation of the resulting field is much more complicated and is expressed in terms of integrals of modified Bessel functions of the second kind [MAC 15]. The geometry of the problem and an example of the pressure field are shown in Figures 7.4 and 7.5, respectively.

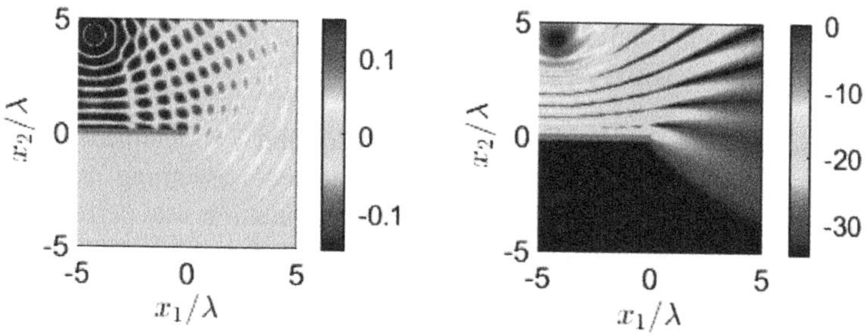

Figure 7.5. *Diffraction of a spherical wave by a semi-infinite rigid screen (source position, $r_s/\lambda = 6, \theta_s = 3\pi/4$). The real part of the total field is plotted in the left side and the sound level in dB is shown in the right side. For a color version of this figure, see www.iste.co.uk/juve/acoustics1.zip*

As for incident plane waves, several regions can be distinguished, depending on the values of the angles formed by the lines connecting the edge of the screen to the source θ_s and to the receiver θ: a geometric shadow zone where only a diffracted wave is present for $\pi < \theta < \pi + \theta_s$, a region where the direct wave is not much modulated by the diffracted wave ($\theta_s - \pi < \theta < \pi - \theta_s$), and finally a region where the direct wave, the wave reflected on the screen, and the diffracted wave interfere ($\pi - \theta_s < \theta < \pi$).

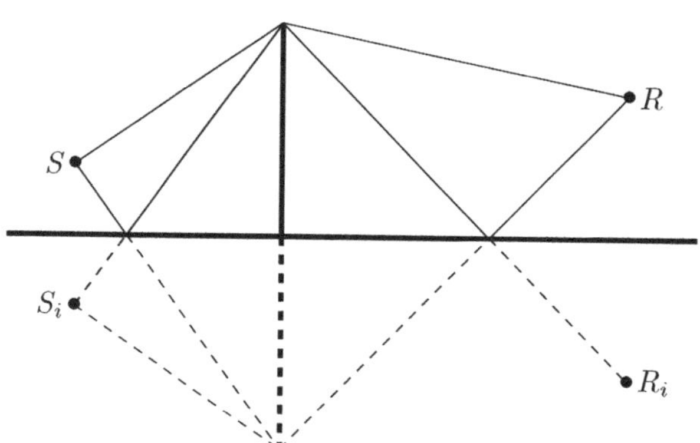

Figure 7.6. *Diffraction by a thin screen. The paths connecting the source and the receiver after diffraction by the edge are plotted as well as their interpretation in terms of images relative to the ground*

This type of interpretation can also be used for the more realistic situation of a thin barrier placed on a reflecting ground, at least when the height of the barrier is much larger than the wavelength and when its thickness is very thin in comparison. A schematic diagram is given in Figure 7.6, which shows the four possible paths connecting the source and the receiver after diffraction at the edge and possible reflection on the ground before or after the barrier, as well as an interpretation in terms of image sources with respect to the ground.

A numerical illustration is given in Figures 7.7 and 7.8 obtained after solving the Helmholtz equation with the finite element method[2], assuming that the screen and the ground are perfectly rigid surfaces. The computation was performed in two dimensions, but the observed behavior is similar to that observed in three-dimensions, in the plane perpendicular to the screen where the source is located. Figure 7.7 shows the real part of the field and allows us to visualize the wavefronts propagating from the source and reflected on the ground and/or the screen. In the shadow zone at the back of the screen, it is very clear that the waves come from the diffracting edge with a cylindrical shape. The reflection from the ground in this region is also visible and is characterized by interference patterns that are slightly inclined to the horizontal. Figure 7.8 quantifies these observations in terms of sound levels; an average reduction of about 30 dB is provided by the screen for the 600 Hz frequency under study.

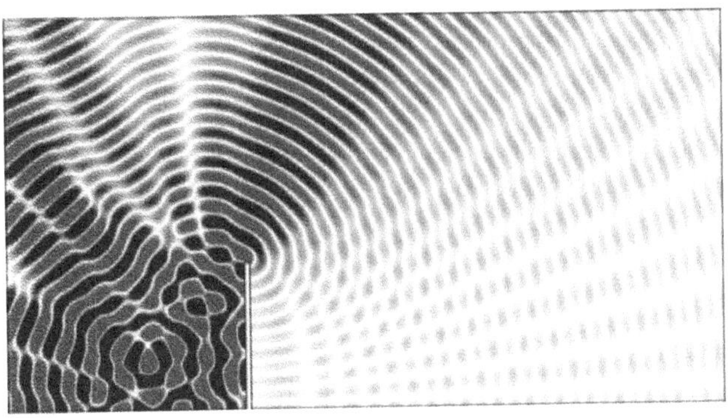

Figure 7.7. *Diffraction of a spherical wave by a thin barrier. The real part of the field highlights the wavefronts propagating from the source and from the edge of the screen. The frequency is 600 Hz, and the height and thickness of the screen are 3 m and 0.1 m, respectively. The source is located at a distance of 2 m from the screen and at a height of 1 m. The ground and the screen are perfectly rigid. For a color version of this figure, see www.iste.co.uk/juve/acoustics1.zip*

2 All the FEM computations presented in this book have been performed using COMSOL Multiphysics®.

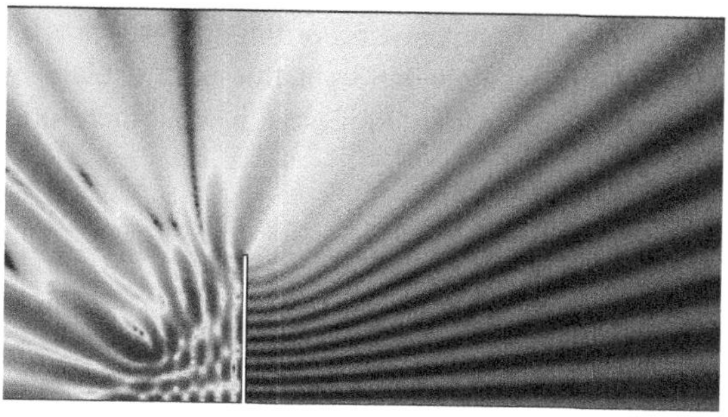

Figure 7.8. *Diffraction of a spherical wave by a thin barrier. The sound levels in dB show especially the noise reduction in the shadow zone. Interference patterns due to the reflection of the diffracted wave propagating from the edge on the ground are also clearly seen. The conditions are identical to those in Figure 7.7 and the plot range is 60 dB (lowest levels in dark). For a color version of this figure, see www.iste.co.uk/juve/acoustics1.zip*

7.2. Scattering by a rigid cylinder

We now consider a two-dimensional problem in which a plane wave of amplitude A is incident on an infinite rigid cylinder of radius a (Figure 7.9). The pressure field is obtained by solving the Helmholtz equation outside the cylinder and imposing a zero normal velocity at $r = a$ and a condition on the pressure behavior when $r \to \infty$. It is often useful to split the total field into the sum of the incident field p_i and of the diffracted, or scattered field p_d. This allows us to better emphasize the role of the obstacle, especially for small values of the reduced wavenumber, and the condition at infinity is also easier to impose.

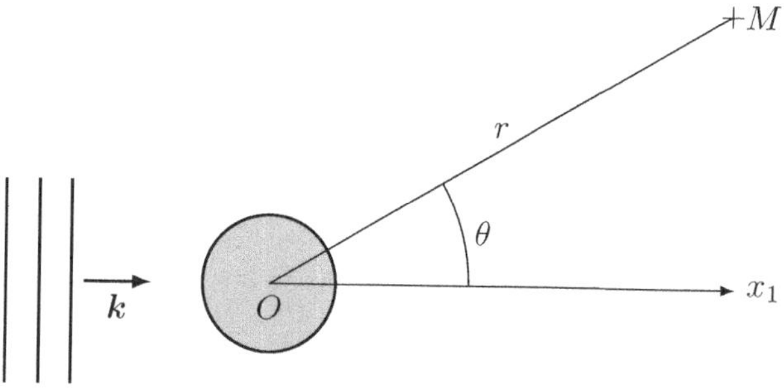

Figure 7.9. *Diffraction of a plane wave by a circular cylinder*

The total pressure is thus decomposed according to $p(r,\theta) = p_i(r,\theta) + p_d(r,\theta)$. Since the total field and the incident field (in this case a plane wave) satisfy the Helmholtz equation, so does the scattered pressure p_d, and therefore:

$$\nabla^2 p_d(r,\theta) + k_0^2 p_d(r,\theta) = 0 \tag{7.3}$$

On the surface of the cylinder, the boundary condition is written as:

$$\frac{\partial p}{\partial r}(a,\theta) = 0 \quad \Longrightarrow \quad \frac{\partial p_d}{\partial r}(a,\theta) = -\frac{\partial p_i}{\partial r}(a,\theta) \tag{7.4}$$

At large distances, the scattered field is seen as propagating from the center of the cylinder and it must satisfy the Sommerfeld radiation condition, which is written in two-dimensions as:

$$\lim_{r\to\infty} r^{1/2}\left(\frac{\partial p_d}{\partial r} + \mathrm{i}k_0 p_d\right) = 0 \tag{7.5}$$

This condition can be replaced by the following behavior at infinity, typical of divergent cylindrical waves:

$$p_d \propto \frac{\exp(-\mathrm{i}k_0 r)}{\sqrt{k_0 r}}; \qquad r \to \infty \tag{7.6}$$

Note that this problem of diffraction by a rigid obstacle is thus formally equivalent to a problem of radiation from a vibrating structure whose velocity would be proportional to $-\partial p_i/\partial r$ for $r = a$. One of the difficulties in the calculation of the scattered field arises precisely from the evaluation of the incident pressure gradient in a way that is adapted to the geometry of the problem. For a circular cylinder, the Helmholtz equation can be split into two separate equations, one for the radial direction and one for the azimuthal direction. By letting $p_d(r,\theta) = R(r)\Theta(\theta)$, the initial equation:

$$\frac{1}{r}\frac{\partial}{\partial r}\left(r\frac{\partial p_d}{\partial r}\right) + \frac{1}{r^2}\frac{\partial^2 p_d}{\partial \theta^2} + k_0^2 p_d = 0 \tag{7.7}$$

is written as

$$\frac{r}{R}\frac{\mathrm{d}}{\mathrm{d}r}\left(r\frac{\mathrm{d}R}{\mathrm{d}r}\right) + \frac{1}{\Theta}\frac{\mathrm{d}^2\Theta}{\mathrm{d}\theta^2} + k_0^2 r^2 = 0 \tag{7.8}$$

that is split according to

$$\frac{\mathrm{d}^2\Theta}{\mathrm{d}\theta^2} + \mu\Theta = 0 \qquad [7.9]$$

and

$$\frac{\mathrm{d}^2R}{\mathrm{d}r^2} + \frac{1}{r}\frac{\mathrm{d}R}{\mathrm{d}r} + \left(k_0^2 - \frac{\mu}{r^2}\right)R = 0 \qquad [7.10]$$

Equation [7.9] yields solutions of the form $\Theta_m(\theta) \propto \mathrm{e}^{\pm\mathrm{i}\mu\theta}$, which must guarantee an angular periodicity of period 2π. Therefore, the separation constant μ is necessarily an integer, denoted by m. On the other hand, the symmetry of the problem with respect to the x_1 axis requires that the solution be even, i.e. of the form $\Theta_m(\theta) \propto \cos(m\theta)$. The radial equation is a Bessel equation that admits two families of independent solutions, the Bessel functions of order m of the first kind $J_m(k_0r)$ and the second kind $Y_m(k_0r)$. The behavior at infinity of these two families is as follows (see Appendix A2.3):

$$J_m(k_0r)_{r\to\infty} = \sqrt{\frac{2}{\pi k_0 r}}\cos\left(k_0r - (2m+1)\frac{\pi}{4}\right) \qquad [7.11]$$

$$Y_m(k_0r)_{r\to\infty} = \sqrt{\frac{2}{\pi k_0 r}}\sin\left(k_0r - (2m+1)\frac{\pi}{4}\right) \qquad [7.12]$$

The general solution of the radial equation is a linear combination of these two families, but the radiation condition shows that only the combination defining the Hankel function of the second kind $H_m^{(2)}(k_0r)$ is acceptable:

$$H_m^{(2)}(k_0r) = J_m(k_0r) - \mathrm{i}Y_m(k_0r) \qquad [7.13]$$

$$H_m^{(2)}(k_0r)_{r\to\infty} = \sqrt{\frac{2}{\pi k_0 r}}\exp(-\mathrm{i}k_0r)\exp\left(\mathrm{i}(2m+1)\frac{\pi}{4}\right) \qquad [7.14]$$

The scattered pressure is thus written as an infinite sum of modes (also called *partial waves*) whose amplitude A_m will be determined by applying the boundary condition at $r = a$:

$$p_d(r,\theta) = \sum_0^\infty A_m\cos(m\theta)H_m^{(2)}(k_0r) \qquad [7.15]$$

This last step is a bit technical and requires first decomposing the incident plane wave into a series of functions adapted to the geometry of the obstacle:

$$p_i = A\exp(-\mathrm{i}k_0x_1) = A\exp(-\mathrm{i}k_0r\cos\theta) = A\left(J_0(k_0r) + 2\sum_1^{\infty}(-\mathrm{i})^m\cos(m\theta)J_m(k_0r)\right) \quad [7.16]$$

Using the formulas for differentiating Bessel functions (see Appendix A2.3), we obtain from [7.16] and [7.15], respectively:

$$\frac{\partial p_i}{\partial r} = -k_0A\left(J_1(k_0r) + \sum_1^{\infty}(-\mathrm{i})^m\cos(m\theta)\left(J_{m+1}(k_0r) - J_{m-1}(k_0r)\right)\right) \quad [7.17]$$

$$\frac{\partial p_d}{\partial r} = -k_0\left(A_0H_1^{(2)}(k_0r) + \sum_1^{\infty}(-\mathrm{i})^mA_m\cos(m\theta)\left(H_{m+1}^{(2)}(k_0r) - H_{m-1}^{(2)}(k_0r)\right)\right) \quad [7.18]$$

and by equating the coefficients of each of the azimuthal modes in the expression of the boundary condition at $r = a$ it follows that:

$$A_0 = -A\frac{J_1(k_0a)}{H_1^{(2)}(k_0a)} \quad [7.19]$$

$$A_m = -2A(-\mathrm{i})^m\frac{J_{m+1} - J_{m-1}}{H_{m+1}^{(2)} - H_{m-1}^{(2)}}(k_0a); \qquad m > 0 \quad [7.20]$$

and the scattered pressure is finally expressed as:

$$p_d(r,\theta) = -A\sum_0^{\infty}\epsilon_m(-\mathrm{i})^m\frac{J_{m+1} - J_{m-1}}{H_{m+1}^{(2)} - H_{m-1}^{(2)}}(k_0a)\cos(m\theta)H_m^{(2)}(k_0r) \quad [7.21]$$

with the following convention: $\epsilon_0 = 1,\ \epsilon_m = 2,\ m > 0$.

This expression is valid at any distance from the obstacle, but most often it is the far-field behavior that we want to know ($k_0r \gg 1$). The asymptotic form of the Hankel

functions [7.14] can then be used to emphasize the directivity pattern of the scattered wave by separating radial and angular dependencies. It follows that:

$$p_d(r,\theta)_{r\to\infty} = A\sqrt{\frac{2}{\pi k_0 r}}\exp(-\mathrm{i}k_0 r + \mathrm{i}\pi/4)\sum_0^\infty \epsilon_m C_m(k_0 a)\cos(m\theta) \qquad [7.22]$$

with

$$C_m(k_0 a) = \frac{J_{m+1} - J_{m-1}}{H^{(2)}_{m+1} - H^{(2)}_{m-1}}(k_0 a) \qquad [7.23]$$

In this formula, a cylindrical waveform propagating from the origin is apparent, with an amplitude decreasing as $1/\sqrt{k_0 r}$. It is modulated in the azimuthal direction by a sum of $\cos(m\theta)$ terms weighted by a function of the reduced wavenumber $k_0 a$.

7.2.1. *Scattering cross-sections*

In order to characterize the directivity of the scattered wave and to specify its evolution with the reduced wave number $k_0 a$, normalized versions of the acoustic intensity scattered by the cylinder in the far-field are introduced. The scattered intensity I_d multiplied by the distance r represents the fraction of the power of the incident wave scattered per unit angle. Normalized by the incident intensity I_i, it defines the *scattering differential cross-section* $\mathrm{d}\sigma/\mathrm{d}\theta$:

$$\frac{\mathrm{d}\sigma}{\mathrm{d}\theta} = \frac{rI_d}{I_i}; \qquad I = \frac{1}{2}\frac{pp^*}{\rho_0 c_0} \qquad [7.24]$$

A dimensionless version of this differential cross-section, which is directly proportional to the usual directivity pattern will be plotted in the following figures; it is defined by:

$$\left(\frac{1}{a}\frac{\mathrm{d}\sigma}{\mathrm{d}\theta}\right)^{\frac{1}{2}} = \sqrt{\frac{2}{\pi k_0 a}}\left|\sum_0^\infty \epsilon_m C_m(k_0 a)\cos(m\theta)\right| \qquad [7.25]$$

The *scattering cross-section* σ is the integral over θ of the differential cross-section. The total scattered power is $W_d = r\int_0^{2\pi} I_d(\theta)d\theta$, and the power incident on the cross-section of the cylinder is $W_i = 2aI_i$. After normalization by the cylinder diameter, the following relation is obtained:

$$\frac{\sigma}{2a} = \frac{W_d}{W_i} \qquad [7.26]$$

This quantity gives the evolution of the fraction of incident power that is scattered throughout the whole space as a function of the reduced wavenumber .

Figure 7.10 depicts the evolution of the scattering directivity, defined by formula [7.25], for increasing values of k_0a. For small values of the reduced wavenumber (less than unity), it can be observed that the wave is mostly backscattered with a maximum obtained for $\theta = \pi$. The graph shows a simple shape, with a single minimum appearing symmetrically with respect to x_1. As k_0a increases, an increasing number of lobes appear and the scattering is preferentially in the forward direction. At high frequency, the main lobe centered around $\theta = 0$ becomes increasingly narrow and intense, while the other contributions are more or less uniformly distributed in the other directions.

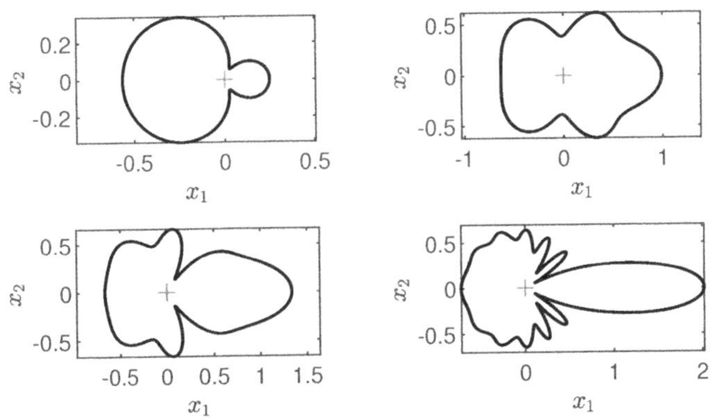

Figure 7.10. *Scattering of a plane wave by a rigid cylinder: directivity patterns obtained for k_0a equal to $0.5, 2.5, 4, 8$ (from left to right and from top to bottom)*

The plot of the scattering cross-section (Figure 7.11) shows that the fraction of scattered energy is very small for $k_0a \ll 1$. It grows rapidly and tends to an asymptotic value of 2 for $k_0a \to \infty$. The asymptotic behavior for $k_0a \to 0$ is called *Rayleigh scattering*. It can be easily evaluated here from the asymptotic forms of the Bessel functions. For $k_0a \ll 1$, that is, when the cylinder is compact ($a \ll \lambda$), only the first two terms C_0 and C_1 of the modal series need to be considered. The higher order terms can be neglected, since C_2 already varies as $(k_0a)^4$. Thus, we have:

$$C_0 = \frac{J_1}{J_1 - \mathrm{i}Y_1} \approx \frac{k_0/2}{k_0a/2 + \mathrm{i}/\pi k_0a} \approx -\mathrm{i}\frac{\pi}{4}(k_0a)^2 \qquad [7.27]$$

$$C_1 = \frac{J_2 - J_0}{(J_2 - J_0) - \mathrm{i}(Y_2 - Y_0)} \approx \frac{J_0}{\mathrm{i}Y_2} \approx \mathrm{i}\frac{\pi}{4}(k_0a)^2 \qquad [7.28]$$

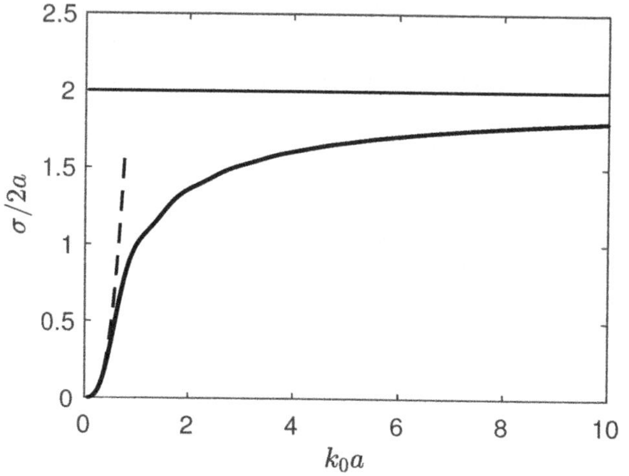

Figure 7.11. *Scattering of a plane wave by a rigid cylinder: evolution of the scattering cross-section with the reduced wavenumber. Comparison of the numerical solution (thick solid line) with the Rayleigh scattering approximation (dashed curve, formula [7.30]) and the geometric limit (thin solid line)*

The dimensionless differential cross-section then takes the limit form:

$$\left(\frac{1}{a}\frac{\mathrm{d}\sigma}{\mathrm{d}\theta}\right)^{\frac{1}{2}} \approx \sqrt{\frac{\pi}{8}(k_0a)^3}|1-2\cos\theta| \qquad [7.29]$$

This formula predicts the existence of a zero for $\theta = \pi/3$, which indeed corresponds to what is observed numerically up to values of k_0a of about 0.2. For the scattering cross-section, we get:

$$\frac{\sigma}{2a} \approx \frac{3}{8}\pi^2(k_0a)^3 \qquad [7.30]$$

which agrees very well with the numerical evaluation up to $k_0a = 0.4$ approximately.

The high frequency limit ($k_0a \to \infty$) is much more challenging to obtain analytically. The limit value of the cross-section is equal to 2 as seen from numerical evaluations. This means that the scattered power is equal to twice the power of the incident wave, calculated on the cylinder cross-section. This seemingly very surprising result can be explained by the partitioning of the total field into incident and scattered fields that we have introduced. At the rear of the obstacle, there is a geometric shadow in which the pressure is asymptotically zero. To cancel out the pressure field in this region, the scattered pressure must be the exact opposite of the

incident pressure; the scattered power is therefore equal to that of the incident wave on the cylinder cross-section. The other fraction of the incident power is reflected specularly by the insonified portion of the obstacle. The reflected power in the whole space is equal to the incident power on the cross-section of the cylinder, which can be verified by integrating the first term of the right-hand side of formula [7.31] over θ.

For large values of k_0a, the differential cross-section is represented by the sum of a frequency-independent term (which can be shown to result from the specular reflection of the incident wave on the cylinder) and a term corresponding to an increasingly narrow beam around $\theta = 0$, which contributes to the formation of the shadow at the back of the cylinder. The following asymptotic formula is given in [MOR 68] as:

$$\frac{1}{a}\frac{\mathrm{d}\sigma}{\mathrm{d}\theta} \approx \frac{1}{2}|\sin(\theta/2)| + \frac{1}{2\pi k_0 a}\cot^2(\theta/2)\sin^2(k_0 a\sin\theta) \qquad [7.31]$$

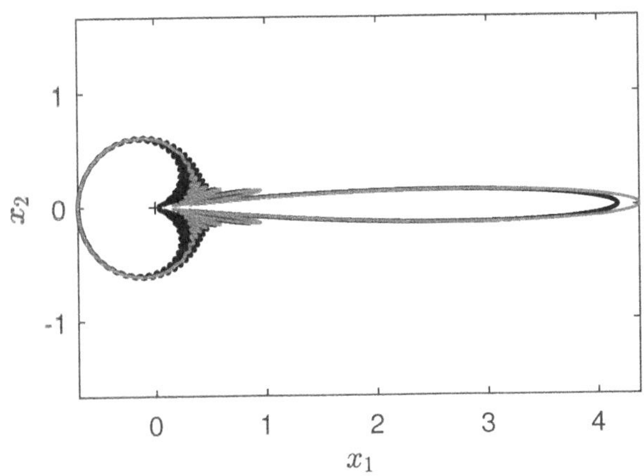

Figure 7.12. *Plane wave scattering by a rigid cylinder: directivity pattern obtained numerically for $k_0a = 30$ (in black) and compared with the asymptotic formula (7.31), valid for very large values of k_0a (in gray)*

The corresponding diagram is compared with the numerical estimation for k_0a =30 in Figure 7.12. The agreement is very good, although there are some differences in the amplitude of the main lobe and the details of the interferential lobes[3].

To complete these asymptotic far-field analyses, we plot in Figure 7.13 the amplitude of the total pressure field in the vicinity of the cylinder for $k_0a = 10$,

3 The expression [7.31] does not take into account terms that evolve very rapidly in the azimuthal direction.

which highlights the formation of the shadow zone (in which the existence of several lobes is seen) and smaller but significant level changes in front of the obstacle.

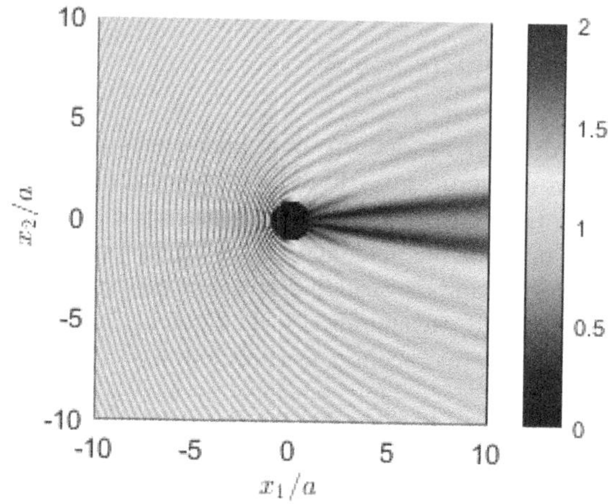

Figure 7.13. *Plane wave scattering by a rigid cylinder. Plot of the amplitude of the total pressure field for $k_0a = 10$. For a color version of this figure, see www.iste.co.uk/juve/acoustics1.zip*

7.3. Rayleigh scattering by a generic obstacle

The results obtained in the previous section are interesting, but one should ask the following question: which properties are general enough to apply to obstacles of any shape? The discussion of the behavior of the scattered field at high frequencies suggests that the limit value of the scattering cross-section as equal to twice the cross-section of the obstacle, must be of general application. But what about the behavior for small values of k_0a?

A first answer is given by studying the scattering of a plane wave by a rigid sphere of radius a, which is calculated similarly to that of the cylinder, where the spherical geometry allows the separation of the variables in the Helmholtz equation.

The differential scattering cross-section for small values of k_0a is given by:

$$\frac{1}{a}\left(\frac{\mathrm{d}\sigma}{\mathrm{d}\Omega}\right)^{\frac{1}{2}} \approx \frac{1}{3}(k_0a)^2\left|1-\frac{3}{2}\cos\theta\right| \qquad [7.32]$$

$$\frac{\sigma}{\pi a^2} \approx \frac{7}{9}(k_0a)^4 \qquad [7.33]$$

where Ω denotes the solid angle.

For high values of k_0a, we have (see [MOR 68]):

$$\frac{\sigma}{\pi a^2} \approx 2 \qquad [7.34]$$

$$\frac{1}{a^2}\frac{\mathrm{d}\sigma}{\mathrm{d}\Omega} \approx \frac{1}{4} + \cot^2(\theta/2) J_1^2(k_0 a \sin\theta) \qquad [7.35]$$

Apart from minor differences due to the three-dimensional nature of the geometry, these behaviors are very similar to those obtained for the rigid cylinder. At low frequency, the first two terms of the modal sum representing the diffracted pressure are sufficient to describe the field, and they are of comparable magnitude. A zero is observed for $\theta \approx \pm 48°$ and the scattered power is very small, as it varies with the fourth power of the reduced wavenumber. It then quickly increases to twice the incident power on the sphere's cross-section, as expected.

For these two obstacles, it can be seen that the two terms defining the directivity of the scattered pressure field for small values of k_0a are of different nature: one is of a monopolar, non-directional, nature, while the second presents a directivity in $\cos\theta$ characteristic of a dipole. We will now show that this behavior is valid for any obstacle.

The main idea to obtain this result is to replace the diffraction problem from a rigid obstacle by a radiation problem with the same obstacle vibrating with a normal velocity opposite to that of the incident wave.

The incident plane wave is written as:

$$p_i = A\exp(-\mathrm{i}\boldsymbol{k}\cdot\boldsymbol{x}) \qquad [7.36]$$

and its gradient in the direction normal to the obstacle's surface is:

$$\frac{\partial p_i}{\partial n} = -\mathrm{i}A\exp(-\mathrm{i}\boldsymbol{k}\cdot\boldsymbol{x})\boldsymbol{k}\cdot\boldsymbol{n} \qquad [7.37]$$

For small values of the reduced wavenumber, the exponential can be approximated by the first two terms of its Taylor series expansion. It can be inferred that:

$$\frac{\partial p_i}{\partial n} \approx -\mathrm{i}A\boldsymbol{k}\cdot\boldsymbol{n}(1 - \mathrm{i}\boldsymbol{k}\cdot\boldsymbol{n}) \qquad [7.38]$$

and the normal velocity to be considered for the radiation of the obstacle is therefore:

$$v_n \approx -\frac{A}{\rho_0 c_0}\boldsymbol{e}_k\cdot\boldsymbol{n}(1 - \mathrm{i}k_0\boldsymbol{e}_k\cdot\boldsymbol{x}) \qquad [7.39]$$

where $\boldsymbol{e}_k = \boldsymbol{k}/k_0$ is the unit vector in the direction of the incident wave.

The first term of this velocity corresponds to a rigid translational motion parallel to $\boldsymbol{e}_k$ which produces a dipole-type radiation. Although the second term is smaller (since it is of order $k_0 a$ with respect to the first), its monopolar character (see [7.40]) induces a higher radiation efficiency, which exactly compensates this smaller magnitude in the far-field. The calculation of these contributions corresponds to that of the acoustic emission of a compact vibrating solid and its development into a series of multipolar terms. It is subtle and uses the method of matched asymptotic expansions. We refer to Pierce ([PIE 89], pp. 425-428) for details.

The monopolar contribution to the scattered field is given by:

$$\begin{aligned} p_{d,monopolar} &= -\frac{Ak_0^2 \exp(-ik_0 r)}{4\pi r} \int_\Sigma (\boldsymbol{e}_k \cdot \boldsymbol{x})(\boldsymbol{e}_k \cdot \boldsymbol{n}) \mathrm{d}\Sigma \\ &= \frac{Ak_0^2 \exp(-ik_0 r)}{4\pi r} V \end{aligned} \qquad [7.40]$$

where V is the volume occupied by the obstacle, because:

$$\int_\Sigma (\boldsymbol{e}_k \cdot \boldsymbol{x})(\boldsymbol{e}_k \cdot \boldsymbol{n}) \mathrm{d}\Sigma = \int_V \nabla \cdot ((\boldsymbol{e}_k \cdot \boldsymbol{x})\boldsymbol{e}_k) \mathrm{d}V = \int_V \mathrm{d}V = V \qquad [7.41]$$

using the identity $\nabla \cdot ((\boldsymbol{e}_k \cdot \boldsymbol{x})\boldsymbol{e}_k) = 1$.

The dipolar contribution corresponds to the unsteady force exerted by the fluid on the obstacle. The computation of that term brings up the added mass tensor of the body. The result is the following:

$$p_{d,dipolar} = \frac{Ak_0^2 \exp(-ik_0 r)}{4\pi r} (\boldsymbol{e}_r \cdot \boldsymbol{M} \cdot \boldsymbol{e}_k) \qquad [7.42]$$

with $M_{ij} = V\delta ij + W_{ij}$.

The term $\rho_0 W_{ij}$ is the added mass tensor, related to the force exerted by the body on the fluid as it accelerates, by $F_i = \rho_0 W_{ij} \dot{v}_j$. The added mass tensor determines the effective mass of the fluid that is driven by the body in its accelerated motion; this mass is "added" to the body's own mass when writing the equations of motion. This tensor can be obtained analytically for bodies of simple shape by solving the equations of motion for an incompressible fluid (see, for example, [BAT 67] or, in the acoustic context, [HOW 03]).

For a fixed, rigid and compact obstacle, the scattering differential cross-section is therefore given by:

$$\frac{\mathrm{d}\sigma}{\mathrm{d}\Omega} = \frac{k_0^4}{16\pi} |V - \boldsymbol{e}_r \cdot \boldsymbol{M} \cdot \boldsymbol{e}_k|^2 \qquad [7.43]$$

and the cross-section, obtained by integration on the solid angle of this expression, is:

$$\sigma = \frac{k_0^4}{4\pi}\left(V^2 + \frac{1}{3}(\boldsymbol{M}\cdot\boldsymbol{e}_k)^2\right) \qquad [7.44]$$

Since the obstacle has a characteristic size a, its volume is of order a^3. We can then observe an evolution in $(k_0 a)^4$ of the scattering cross-section σ divided by the cross-section of the obstacle, whose characteristic dimension is of order a^2. This is the mark of Rayleigh scattering for $k_0 a \ll 1$. It can therefore be seen that for objects of small size compared to the wavelength, only very little acoustic energy is scattered.

7.4. Scattering by non-rigid obstacles and the Born approximation

To complement the previous sections, we now consider objects into which acoustic waves can penetrate[4]. We will be particularly interested in fluid-like inclusions that exhibit low density and compressibility contrasts with respect to the surrounding medium.

7.4.1. *Scattering by inhomogeneities*

A region of space whose properties differ from those of the surrounding medium scatters acoustic waves, as is the case with solid obstacles. A typical example is that of atmospheric turbulence, the properties of which can be derived from acoustic scattering measurements using instruments called SODAR (sonic detection and ranging).

We assume that the inclusion can be modeled by a fluid with spatially variable characteristics $\rho_1(\boldsymbol{x})$ and $c_1(\boldsymbol{x})$. The propagation of acoustic waves is modified by variations in the speed of sound within the fluid, which is intuitively understandable, but also by variations in density. We will assume that the equation for the propagation of harmonic waves in a non-homogeneous medium is as follows ([BER 46]):

$$\rho_1 \nabla\cdot\left(\frac{1}{\rho_1}\nabla p\right) + \frac{\omega^2}{c_1^2}p = 0 \qquad [7.45]$$

or, by introducing the adiabatic compressibility of the fluid $\kappa_1 = 1/\rho_1 c_1^2$:

$$\nabla\cdot\left(\frac{1}{\rho_1}\nabla p\right) + \omega^2\kappa_1 p = 0 \qquad [7.46]$$

4 When the obstacle surface is modeled by an impedance, similar techniques to those employed in section 7.2 can be used. See Chapter 8 of [MOR 68].

In the surrounding fluid, which is assumed to be homogeneous with density ρ_0 and speed of sound c_0, the propagation is governed by the Helmholtz equation:

$$\frac{1}{\rho_0}\nabla^2 p + \omega^2 \kappa_0 p = 0 \qquad [7.47]$$

and at the interface between the inclusion and the surrounding fluid, the conditions of continuity of the acoustic pressure and the normal velocity must be satisfied.

It proves convenient to group these equations, which apply separately inside and outside the obstacle, into a single equation, valid for the whole space, which then leads to an integral formulation of the problem. For this purpose, the following notations are introduced:

$$\gamma_\rho = \frac{\rho_1 - \rho_0}{\rho_1} \quad \text{inside the obstacle, and} \quad 0 \quad \text{outside} \qquad [7.48]$$

$$\gamma_\kappa = \frac{\kappa_1 - \kappa_0}{\kappa_0} \quad \text{inside the obstacle, and} \quad 0 \quad \text{outside} \qquad [7.49]$$

We then use the identities:

$$\nabla \cdot (\gamma_\rho \nabla p) = \nabla \cdot \left(\left(1 - \frac{\rho_0}{\rho_1}\right)\nabla p\right) = \nabla^2 p - \rho_0 \nabla \cdot \left(\frac{1}{\rho_1}\nabla p\right) \qquad [7.50]$$

to convert the wave equation inside the inclusion to the form:

$$\nabla^2 p - \nabla \cdot (\gamma_\rho \nabla p) + \frac{\rho_0}{\rho_1 c_1^2}\omega^2 p = 0 \qquad [7.51]$$

Using the relations:

$$\frac{\rho_0}{\rho_1 c_1^2} = \frac{1}{c_0^2}\frac{\rho_0 c_0^2}{\rho_1 c_1^2} = \frac{1}{c_0^2}\frac{\kappa_1}{\kappa_0} = \frac{1}{c_0^2}(\gamma_\kappa + 1) \qquad [7.52]$$

one finally gets

$$\nabla^2 p + k_0^2 p = -\nabla \cdot (\gamma_\rho \nabla p) + k_0^2 \gamma_\kappa p \qquad [7.53]$$

The right-hand side of this equation cancels out outside the inclusion. It therefore appears as a source term that "excites" the homogeneous surrounding medium. It is

then possible to solve this equation, at least formally, using the free space Green's function $G_0(\boldsymbol{x};\boldsymbol{y})$. This yields:

$$p(\boldsymbol{x}) = p_i(\boldsymbol{x}) + \int_{\infty} (k_0^2\gamma_\kappa(\boldsymbol{y})p(\boldsymbol{y}) - \nabla\cdot(\gamma_\rho(\boldsymbol{y})\nabla p(\boldsymbol{y})))G_0(\boldsymbol{x};\boldsymbol{y})\mathrm{d}\boldsymbol{y} \quad [7.54]$$

where the integration is in practice limited to the volume D of the inhomogeneous region and p_i denotes the incident field, often modeled by a plane wave (see Figure 7.14). The contribution of the integral to the total field corresponds to the pressure scattered by the inclusion, p_d.

This solution remains formal, because the source term is unknown and the pressure that we want to determine, as well as its gradient, appears in the integral. Therefore, we have an *integral equation* and not just a simple formula.

We have seen in section 5.3.2 that the presence of a source term in the form of a divergence leads to a number of difficulties when searching for approximate solutions, for example using a compactness assumption. It is therefore desirable to manipulate the term $\nabla\cdot(\gamma_\rho(\boldsymbol{y})\nabla p(\boldsymbol{y}))G_0(\boldsymbol{x};\boldsymbol{y})$ in such a way that the spatial differentiation is imposed on the Green's function and not on the source term. For this purpose, the following identity is used:

$$\begin{aligned}\nabla\cdot(\gamma_\rho(\boldsymbol{y})\nabla p(\boldsymbol{y}))G_0(\boldsymbol{x};\boldsymbol{y}) &= \nabla\cdot(\gamma_\rho(\boldsymbol{y})\nabla p(\boldsymbol{y})G_0(\boldsymbol{x};\boldsymbol{y})) \\ &\quad - \gamma_\rho(\boldsymbol{y})\nabla p(\boldsymbol{y})\nabla G_0(\boldsymbol{x};\boldsymbol{y})\end{aligned} \quad [7.55]$$

The volume integral of the divergence term that appears on the right-hand side of this formula can be replaced by an integral on a surface Σ located in the homogeneous medium, along which $\gamma_\rho = 0$, and thus:

$$\int_D \nabla\cdot(\gamma_\rho(\boldsymbol{y})\nabla p(\boldsymbol{y})G_0(\boldsymbol{x};\boldsymbol{y}))\mathrm{d}\boldsymbol{y} = \int_\Sigma (\gamma_\rho(\boldsymbol{y})\nabla p(\boldsymbol{y})G_0(\boldsymbol{x};\boldsymbol{y}))\cdot\boldsymbol{n}\mathrm{d}\Sigma = 0 \quad [7.56]$$

and equation [7.54] takes the following form:

$$p(\boldsymbol{x}) = p_{inc}(\boldsymbol{x}) + \int_D (k_0^2\gamma_\kappa(\boldsymbol{y})p(\boldsymbol{y})G_0(\boldsymbol{x};\boldsymbol{y}) + \gamma_\rho(\boldsymbol{y})\nabla p(\boldsymbol{y})\nabla_y G_0(\boldsymbol{x};\boldsymbol{y}))\mathrm{d}\boldsymbol{y} \quad [7.57]$$

where it has been specified that the gradient of the Green's function is taken with respect to the current integration point $\boldsymbol{y}$.

An explicit formula can be constructed from this integral equation when the inclusions are only weakly inhomogeneous. This assumption allows the use of the Born approximation, described in the next section.

7.4.2. *The Born approximation*

In order to transform the integral equation [7.57] into an explicit formula, we assume that the unknown pressure present in the integral can be replaced by the value of the incident pressure at the same point. This assumption seems reasonable if the contrast in properties between the inclusion and the surrounding fluid is small, and therefore the scattered field is small compared to the incident field. This is called the *Born approximation*.

With this assumption, the scattered pressure is written as:

$$p_d(\boldsymbol{x}) = \int_D (k_0^2 \gamma_\kappa p_i(\boldsymbol{y}) G_0(\boldsymbol{x};\boldsymbol{y}) + \gamma_\rho \nabla p_i(\boldsymbol{y}) \nabla_y G_0(\boldsymbol{x};\boldsymbol{y}))\mathrm{d}\boldsymbol{y} \qquad [7.58]$$

Let us consider the scattering of an incident plane wave, $p_i(\boldsymbol{x}) = A\exp(-\mathrm{i}\boldsymbol{k}_i \cdot \boldsymbol{x})$, and place the receiver in the far-field. The Green's function takes the simplified form (see section 5.4.1):

$$G_0(\boldsymbol{x},\boldsymbol{y}) \approx \frac{\exp(-\mathrm{i}k_0 x)}{4\pi x} \exp\left(+\mathrm{i}\boldsymbol{k}_d \cdot \boldsymbol{y}\right) \qquad [7.59]$$

where $\boldsymbol{k}_d = k_0 \boldsymbol{\nu} = k_0 \boldsymbol{x}/x$ is the vector of magnitude k_0 pointing from the origin to the observer, with the origin chosen at the center of the inclusion (see Figure 7.14).

The far-field scattered pressure can be written as:

$$p_d(\boldsymbol{x}) = \frac{A\exp\left(-\mathrm{i}k_0 x\right)}{4\pi x} \int_D (\gamma_\kappa k_0^2 + \gamma_\rho \boldsymbol{k}_i \cdot \boldsymbol{k}_d) \exp\left(\mathrm{i}(\boldsymbol{k}_d - \boldsymbol{k}_i) \cdot \boldsymbol{y}\right)\mathrm{d}\boldsymbol{y} \quad [7.60]$$

or, in a condensed form, by introducing the vector $\boldsymbol{\mu} = \boldsymbol{k_d} - \boldsymbol{k_i}$ (see Figure 7.15):

$$p_d(\boldsymbol{x}) = \frac{A\exp\left(-\mathrm{i}k_0 x\right)}{x} \Phi(\boldsymbol{\mu}) \qquad [7.61]$$

with

$$\Phi(\boldsymbol{\mu}) = \frac{k_0^2}{4\pi} \int_D (\gamma_\kappa + \gamma_\rho \cos\theta) \exp\left(\mathrm{i}\boldsymbol{\mu} \cdot \boldsymbol{y}\right)\mathrm{d}\boldsymbol{y} \qquad [7.62]$$

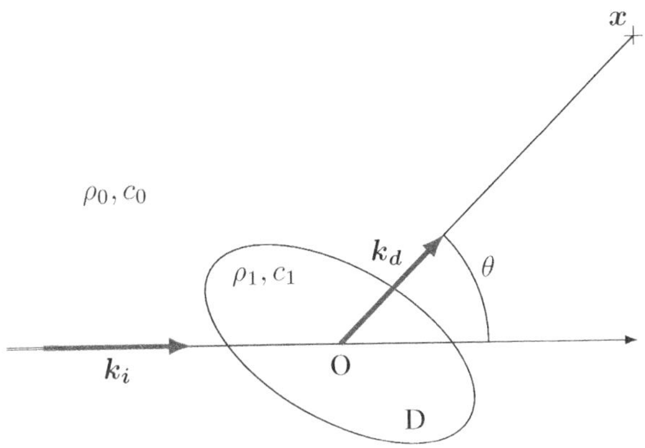

Figure 7.14. *Plane wave scattering by an inhomogeneous fluid inclusion*

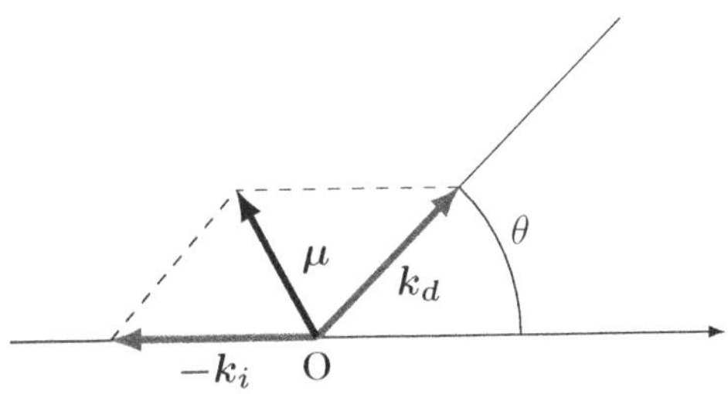

Figure 7.15. *Geometry for the far-field scattering of a plane wave*

The far-field scattered pressure thus corresponds to that from a point source located at the center of the inclusion, associated with a directivity given by relation [7.62]. This directivity is related to the spatial Fourier transform of the density and compressibility contrasts, taken for the particular value $\boldsymbol{\mu}$ of the wave vector, whose magnitude is $\mu = 2k_0 \sin(\theta/2)$.

Note that there are two contributions to the scattered wave: a monopolar contribution associated with the compressibility variation, and a dipolar contribution (directivity in $\cos\theta$) associated with the density variation. These two contributions are of the same order of magnitude and vary with the square of the incident wavenumber. This behavior is very similar to that observed for rigid obstacles in the limit $k_0 a \ll 1$ (Rayleigh scattering), see section 7.3.

Formula [7.62] can also be used to determine the properties of an inhomogeneity from the value of the scattered pressure measured at different frequencies and different angles. This is an example of the use of *inverse methods*. On the general and important topic of diffraction tomography, the book by Devaney [DEV 12] should be consulted, in particular Chapter 8, which is devoted to the use of the Born approximation.

7.4.3. *Validity of the Born approximation*

If the inclusion is a homogeneous sphere with radius a, formula [7.62] can be explicitly integrated using the spherical coordinates system $(r_0, \varphi_0, \theta_0)$, with the z-axis oriented in the direction of propagation of the incident wave $\boldsymbol{k_i}$. This gives:

$$\begin{aligned}\Phi(\boldsymbol{\mu}) &= \frac{k_0^2}{2}(\gamma_\kappa + \gamma_\rho \cos\theta) \int_0^a r_0^2 \mathrm{d}r_0 \int_0^\pi \exp\left(\mathrm{i}\mu r_0 \cos\theta_0\right) \sin\theta_0 \mathrm{d}\theta_0 \\ &= \frac{k_0^2}{2}(\gamma_\kappa + \gamma_\rho \cos\theta) \int_0^a r_0^2 \frac{2\sin\mu r_0}{\mu r_0} \mathrm{d}r_0\end{aligned} \qquad [7.63]$$

If, in addition, the sphere is assumed to be compact, $k_0 a \ll 1$ and therefore $\mu a \ll 1$, then the first-order Taylor series expansion of $\sin \mu r_0$ eventually leads to:

$$\Phi(\theta) \approx \frac{1}{3} k_0^2 a^3 (\gamma_\kappa + \gamma_\rho \cos\theta) = \frac{1}{3} k_0^2 a^3 \left(\gamma_\kappa + \frac{\rho_1 - \rho_0}{\rho_1} \cos\theta\right) \qquad [7.64]$$

The exact expression for the scattering of a plane wave by a compact homogeneous sphere is as follows ([MOR 68]):

$$\Phi^{exact}(\theta) = \frac{1}{3} k_0^2 a^3 \left(\gamma_\kappa + \frac{3\rho_1 - 3\rho_0}{2\rho_1 + \rho_0} \cos\theta\right) \qquad [7.65]$$

Both formulas show the same evolution with the reduced wavenumber and the same influence of the compressibility contrast. They differ in the influence of the density contrast. However, if one introduces into the exact formula the low contrast assumption inherent in the Born approximation, $\gamma_\rho = \epsilon \ll 1$, then $3\left(\rho_1 - \rho_0\right) / \left(2\rho_1 + \rho_0\right) \approx \epsilon$, and the two formulas thus coincide under this condition.

There is no precise criterion for the validity of the Born approximation. However, it is considered that a practical condition for its validity is that the product of the wavenumber by the characteristic dimension of the object L and by its degree of density and compressibility inhomogeneity ϵ must remain very small, that is, $\epsilon k_0 L \ll 1$.

7.5. Exercises

1) **Scattering cross-sections**

Compute the scattering cross-sections for the obstacles whose added mass tensors are given below.

a) Sphere of radius a:

$$M = \frac{3}{2}V \begin{bmatrix} 1 & 0 & 0 \\ 0 & 1 & 0 \\ 0 & 0 & 1 \end{bmatrix}$$

Recover the formulas for the differential scattering cross-section and its integral given in section 7.3

b) Thin disk with radius a and axis x_3:

$$M = \frac{8}{3}a^3 \begin{bmatrix} 0 & 0 & 0 \\ 0 & 0 & 0 \\ 0 & 0 & 1 \end{bmatrix}$$

Physically interpret your result.

2) **Directivity patterns**

Determine numerically the directivity patterns of the scattering of a plane wave by a rigid cylinder. Consider different values of the reduced wavenumber k_0a, ranging from the Rayleigh regime up to the quasi-geometric regime (i.e. up to $k_0a > 30$). Control the convergence of your calculations as a function of the number of azimuthal modes used (typically, this number must be at least twice the value of k_0a).

3) **Scattering from a cylindrical inclusion in the Born approximation**

Consider the scattering of a plane wave of unit amplitude by a cylindrical inclusion of radius a.

a) Based on the developments in section 7.4.2 and using the free field Green's function of the two-dimensional Helmholtz equation, determine the far-field scattered pressure.

b) Now consider that the inhomogeneities depend only on the radial coordinate, denoted by r_0. Show that the pressure scattered in the direction θ takes the form (use formula [A2.45]):

$$p_d(r,\theta) = \frac{\pi}{2}\sqrt{\frac{2}{\pi k_0 r}} \exp\left(-ik_0 r + i\frac{\pi}{4}\right) k_0^2 \int_0^a (\gamma_\kappa(r_0) + \gamma_\rho(r_0)\cos\theta) J_0(\mu r_0) r_0 dr_0$$

c) Show that if the inhomogeneities in the inclusion are uniform and if the cylinder radius is much smaller than the wavelength, the integral above is approximately equal to $a^2/2\,(\gamma_\kappa + \gamma_\rho \cos\theta)$.

d) Compare with the exact solution given by:

$$p_d(r,\theta) = \frac{\pi}{4}\sqrt{\frac{2}{\pi k_0 r}}\,\exp\left(-\mathrm{i}k_0 r + \mathrm{i}\frac{\pi}{4}\right) k_0^2 a^2 \left(\frac{\kappa_1 - \kappa_0}{\kappa_0} + 2\frac{\rho_1 - \rho_0}{\rho_1 + \rho_0}\cos\theta\right)$$

and conclude as to the validity of the Born approximation.

8

Guided Waves

The previous chapters were devoted to the propagation of acoustic waves in free space or in the presence of radiating and diffracting objects. A very different case of practical interest concerns the propagation in ducts, which channel the waves along the axis of the duct. The applications are very diverse, including sound propagation in ventilation ducts, in turbojet nacelles or in some musical instruments. In this chapter, we consider the idealized case of straight, infinite ducts with constant cross-section, acting as acoustic waveguides. The presentation is limited to the propagation of harmonic waves.

In the first part of the chapter, the walls of the duct are assumed to be perfectly rigid and the existence of propagating and evanescent modes is emphasized. The practical case of ducts with a rectangular or a circular cross-section allows us to give explicit formulas for the axial evolution of the pressure field as well as for the associated phase and group velocities.

In the second part, the influence on the sound propagation of walls covered with an absorbing material and modeled by an impedance condition is discussed.

The changes due to the presence of a uniform flow in the duct are discussed in detail in the last part of the chapter.

8.1. Sound propagation in a duct of constant cross-section

When the duct is straight and has a constant cross-section, it makes sense to look for the pressure field as a product of two functions depending separately on the axial variable x_3 and on the variables (x_1, x_2) describing the pressure in the cross-section, that is, $p(\boldsymbol{x}) = \psi(x_1, x_2)X_3(x_3)$. The Helmholtz equation is then split into two equations: one for the behavior of the pressure in the cross-section of the duct and

the second describing its axial evolution. The equations of the problem can then be written as:

$$\nabla_T^2 \psi + \chi^2 \psi = 0 \qquad [8.1]$$

$$\frac{d^2 X_3}{dx_3^2} + \left(k_0^2 - \chi^2\right) X_3 = 0 \qquad [8.2]$$

where χ^2 is a separation constant, and the notation ∇_T^2 corresponds to the transverse Laplacian, the sum of the second-order derivatives with respect to the transverse coordinates.

In the following, we assume that the duct walls are perfectly rigid and the boundary condition to be satisfied along the perimeter of the cross-section is therefore:

$$\nabla_T \psi \cdot \boldsymbol{n}_p = 0 \qquad [8.3]$$

where the unit normal to the wall $\boldsymbol{n}_p$ points outward from the duct. The geometry of the problem is shown in Figure 8.1.

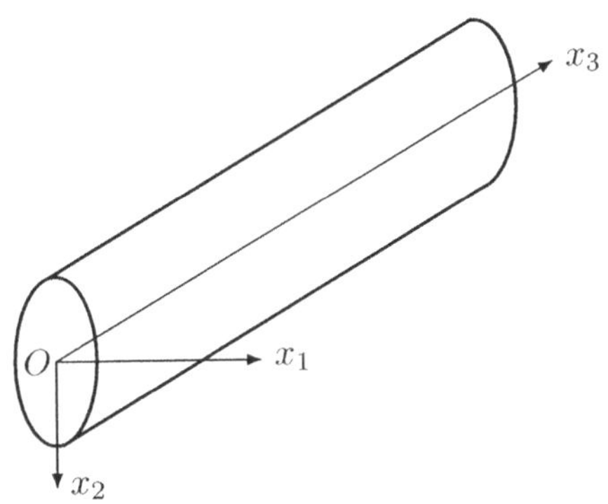

Figure 8.1. *Cylindrical duct with arbitrary cross-section*

It can be shown that equation [8.1] associated with the boundary condition [8.3] admits solutions only for a discrete set of real and non-negative values of the separation constant[1], denoted χ_p^2. They correspond to the eigenvalues of the operator ∇_T^2 with the rigidity condition at the wall; the associated functions ψ_p are the eigenfunctions of the duct cross-section. It can also be shown that the eigenfunctions

1 See the example of the rectangular cross-section pipe in section 8.2

form an orthogonal set and that they can be chosen as real. The following normalization condition is imposed on these functions[2]:

$$\int_S \psi_p \psi_q^* \, \mathrm{d}S = A\delta_{pq} \tag{8.4}$$

where A denotes the area of the duct cross-section and δ_{pq} is the Kronecker symbol. The eigenfunctions form a complete set, and any quantity $f(x_1, x_2)$ in the duct cross-section can be written as a linear combination of these functions.

$$f(x_1, x_2) = \sum_p a_p \psi_p(x_1, x_2) \tag{8.5}$$

$$a_p = \frac{1}{A} \int_S f(x_1, x_2) \psi_p(x_1, x_2) \, \mathrm{d}S \tag{8.6}$$

Another consequence of χ taking only discrete values χ_p is that the solutions of [8.2] are also a set of functions denoted X_{3_p}. The product $\psi_p(x_1, x_2) X_{3_p}(x_3)$ is referred to as the mode of index p of the duct.

8.1.1. *Propagating modes and evanescent modes*

In the following, we will consider that the waves propagate in the positive x_3 direction, after being generated, for example, by a source located at $x_3 = 0$. Since the waveguide is assumed to be infinite, we eliminated the possibility of waves traveling toward decreasing values of x_3, which could only result from a reflection at the end of a duct of finite length.

Equation [8.2] admits two types of solutions, depending on the value of k_0 relative to χ_p. If $\chi_p < k_0$, we set $\kappa_p = +\sqrt{k_0^2 - \chi_p^2}$, and the solution of this equation takes the following form:

$$X_{3_p}(x_3) = X_3(0) \exp(-\mathrm{i}\kappa_p x_3) \tag{8.7}$$

Since κ_p is real and positive, this solution corresponds to a wave propagating toward increasing values of x_3.

If, on the other hand, $\chi_p > k_0$, we set $\kappa_p = -\mathrm{i}\sqrt{\chi_p^2 - k_0^2}$ and the solution of [8.2] is:

$$X_{3_p}(x_3) = X_3(0) \exp(-\left|\kappa_p\right| x_3) \tag{8.8}$$

2 Different normalizations of the eigenfunctions are sometimes used by authors.

It is thus a wave that decays exponentially as it propagates toward increasing values of x_3, that is an evanescent wave.

For a given value of the frequency, and thus of k_0, there are a finite number of values of χ_p for which κ_p is real, and thus a finite number of propagating modes. The mode associated with the value $\chi_p = 0$, which corresponds to $\kappa_p = k_0$, is always propagating. It is called the fundamental mode, or the plane mode because the pressure is then uniform in the cross-section. The cutoff frequency of a mode p is given as $f_p^c = c_0\chi_p/2\pi$. If the frequency of a wave is greater than this value, the mode is said to be cut-on (or propagating), otherwise it is said to be cutoff (or evanescent). These characteristics are very important since, for a source emitting in a given frequency range, they determine the possibility of more or less energy being transmitted down the duct, depending on whether it is cut-on or cutoff modes that are excited. The lowest cutoff frequency of the non-planar modes is called the *duct cutoff frequency*.

When the cross-section of the waveguide is arbitrary, a numerical resolution, by the finite element method for example, must be used to determine the eigenvalues and eigenfunctions. Two cases of practical interest can be solved analytically, where the Helmholtz equation in the cross-section is separable: cylindrical ducts with a rectangular or circular cross-section, which we will study in more detail.

8.2. Duct of rectangular cross-section

We consider a rigid rectangular duct, with sides a and b. We use Cartesian coordinates and choose the origin in one of the corners of the cross-section. We search the eigenfunctions of the problem as the product of a function of x_1 by a function of x_2, $\psi(x_1, x_2) = X_1(x_1)X_2(x_2)$.

The rigid walls conditions are written as:

$$\frac{dX_1}{dx_1} = 0 \quad \text{for} \quad x_1 = 0 \quad \text{and} \quad x_1 = a. \tag{8.9}$$

$$\frac{dX_2}{dx_2} = 0 \quad \text{for} \quad x_2 = 0 \quad \text{and} \quad x_2 = b. \tag{8.10}$$

and the eigenfunctions take the following form:

$$\psi_{mn}(x_1, x_2) = C_{mn} \cos\left(\frac{m\pi x_1}{a}\right) \cos\left(\frac{n\pi x_2}{b}\right) \tag{8.11}$$

where m and n are two integers, positive or equal to zero. The eigenvalues thus form a discrete sequence:

$$\chi_{mn}^2 = \pi^2\left(\frac{m^2}{a^2} + \frac{n^2}{b^2}\right) \tag{8.12}$$

The constant C_{mn} is determined by the normalization condition [8.4]. If both mode indexes are zero, the constant is equal to 1; if one of the integers is zero, it is equal to $2^{1/2}$, and if none of the integers is zero, it is equal to 2.

The cutoff frequency of the mode (m, n) is given by the following formula:

$$f^c_{mn} = \frac{c_0}{2}\sqrt{\frac{m^2}{a^2} + \frac{n^2}{b^2}} \qquad [8.13]$$

With $a > b$, the first non-plane propagating mode is the $(1, 0)$ mode, which propagates without attenuation along the duct if the source frequency is greater than $f^c_{10} = c_0/2a$. This condition is equivalent to saying that this mode is cut-on if the acoustic wavelength is less than twice the largest dimension of the cross-section. Below this cutoff frequency, only the plane mode propagates without attenuation[3]. For low frequencies, it is therefore possible to consider the propagation in the duct as one-dimensional, which greatly simplifies the study and allows variations in the cross-section or the influence of branching to be easily taken into account, as we will see in Chapter 9.

8.2.1. *Phase and group velocities: dispersion of higher modes*

The dispersion relation of the duct modes, which relates the frequency to the components of the wave vector, is as follows:

$$k_0^2 = \omega^2/c_0^2 = \kappa_{mn}^2 + \chi_{mn}^2 = \kappa_{mn}^2 + (\pi m/a)^2 + (\pi n/b)^2 \qquad [8.14]$$

where $(\pi m/a)$ and $(\pi m/b)$ can be considered as the components of the acoustic wave vector $\boldsymbol{k}$ along the x_1 and x_2 axes.

For a propagating mode (m, n) traveling in the $+x_3$ direction, κ_{mn} is real, and defined by $\kappa_{mn} = +\sqrt{k_0^2 - ((\pi m/a)^2 + (\pi n/b)^2)}$. Its axial phase velocity, v_φ, is thus given by:

$$v_\varphi = \frac{\omega}{\kappa_{mn}} = \frac{\omega}{\sqrt{k_0^2 - ((\pi m/a)^2 + (\pi n/b)^2)}} = \frac{c_0}{\sqrt{1 - (f_{mn}/f)^2}} \qquad [8.15]$$

Its axial group velocity is expressed as:

$$v_g = \frac{d\omega}{d\kappa_{mn}} = c_0\sqrt{1 - \left(\frac{f_{mn}}{f}\right)^2} \qquad [8.16]$$

3 Provided that it is excited by the source.

For evanescent modes, κ_{mn}^2 is negative; their phase velocity is infinite and their group velocity is zero.

It can be seen that the plane mode propagates with the speed of sound c_0, but that for higher order modes, the phase and group velocities are frequency dependent; the propagation of these modes is therefore dispersive. Waves of different frequencies propagate at different speeds, resulting in the dispersion (or spreading) of a wave packet consisting of a series of frequencies.

The evolution with the frequency of the axial phase and group velocities of the first five non-planar modes are plotted in Figure 8.2. Note that at high frequencies both velocities tend toward the speed of sound, from below for the group velocity and from above for the phase velocity. Note also that the product $v_\varphi v_g$ is equal to c_0^2, which is a direct consequence of the form of the dispersion relation.

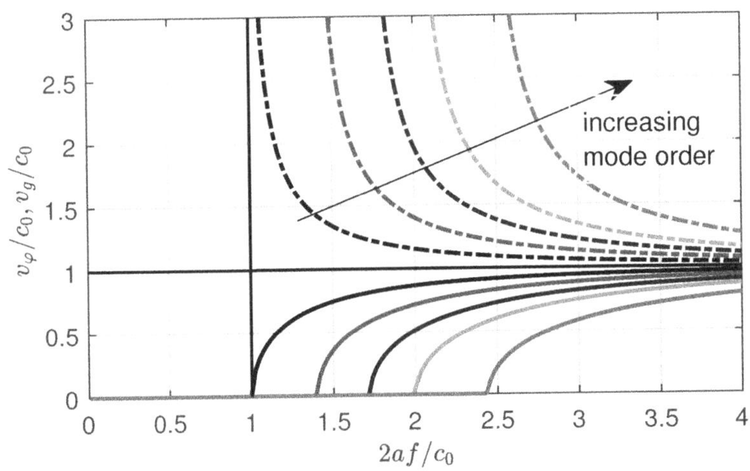

Figure 8.2. *Axial phase velocities (dashed-dotted lines) and group velocities (solid lines) of the first five non-planar modes of a rectangular duct of aspect ratio 1.4. The velocities are normalized by c_0 and plotted as a function of the frequency divided by the duct cutoff frequency*

In Figure 8.3, we illustrate the evolution of the real part of the pressure fluctuation in the cross-section of a rectangular duct for four different modes. Note the existence of nodal lines (along which the pressure is zero) whose number in each direction is equal to the corresponding mode index.

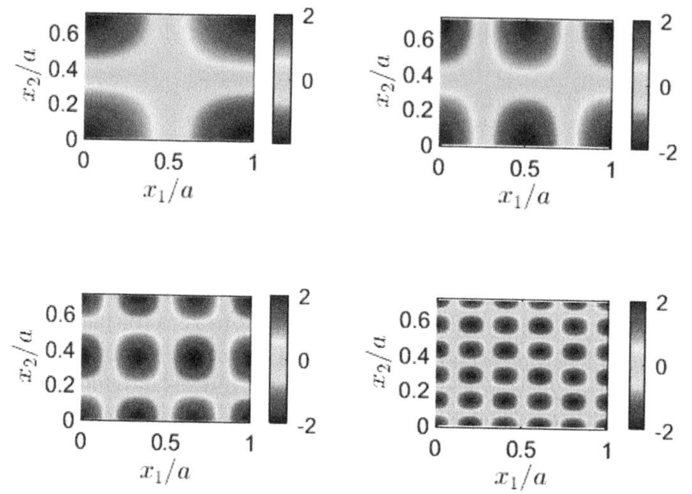

Figure 8.3. *Real part of the pressure fluctuation in the cross-section of a rectangular duct with aspect ratio $a/b = 1.4$ and unit area. From left to right and from top to bottom, modes (1,1), (2,1), (3,2) and (5,5). For a color version of this figure, see www.iste.co.uk/juve/acoustics1.zip*

8.2.2. *Modes and pairs of plane waves*

An interesting interpretation of the notion of modes is provided by the relationship that exists between them and plane waves periodically reflected from the walls. For simplicity, we consider the case of a 2D waveguide consisting of two infinite parallel walls; however, the concept extends to rectangular and circular waveguides. Consider the mode of index m, written as:

$$p(x_1, x_3) \propto \cos\left(\frac{m\pi}{a}x_1\right) \exp\left(-\mathrm{i}\sqrt{k_0^2 - (m\pi/a)^2}x_3\right) \qquad [8.17]$$

If we express the cosine as the sum of two complex exponentials, it follows that:

$$\begin{aligned} p(x_1, x_3) \propto & \exp\left(-\mathrm{i}\left(\frac{m\pi}{a}x_1 + \sqrt{k_0^2 - (m\pi/a)^2}x_3\right)\right) \\ & + \exp\left(-\mathrm{i}\left(-\frac{m\pi}{a}x_1 + \sqrt{k_0^2 - (m\pi/a)^2}x_3\right)\right) \end{aligned} \qquad [8.18]$$

which shows that this mode can be thought of as the sum of two plane waves of wave vector $(\pm k_1, k_3)$, with $k_1 = m\pi/a$ and $k_3 = \sqrt{k_0^2 - (m\pi/a)^2}$. These two waves

propagate symmetrically along the waveguide, forming an angle $\pm\theta_m$ with the axis, and given by:

$$\sin\theta_m = \frac{m\pi}{k_0 a} \quad \text{or} \quad \cos\theta_m = \frac{\sqrt{k_0^2 - (m\pi/a)^2}}{k_0} \qquad [8.19]$$

The angle θ_m is real only for propagating modes. The spatial period of each of the two plane waves along the axis is $\lambda/\cos\theta_m$, which naturally leads to the expression for the phase velocity found above (equation [8.15]). For a given frequency, the angle θ_m increases with the order of the modes; it is equal to zero for the plane mode and increases up to $\pi/2$ before becoming complex. The cutoff frequency of the mode of index m, $f_m^c = mc_0/2a$, corresponds to a wave that reflects indefinitely on the walls in the direction of their normal, thus without any effective propagation along the axis; one can speak of a "resonant" mode.

8.3. Ducts of circular cross-section

The second scenario of practical interest, which is more common than the first one, is that of a circular pipe with radius a. In a plane normal to the x_3 axis, the use of polar coordinates (ρ, φ) is natural, and the transverse Laplacian is written as:

$$\nabla_T^2 = \frac{\partial^2}{\partial\rho^2} + \frac{1}{\rho}\frac{\partial}{\partial\rho} + \frac{1}{\rho^2}\frac{\partial^2}{\partial\varphi^2} \qquad [8.20]$$

The search for a solution with separate variables in the cross-section of the form $\psi_{mn} = \Phi(\varphi)R(\rho)$, immediately gives $\Phi(\varphi) \propto \exp(\pm im\varphi)$, where m must be a positive or zero integer to ensure uniqueness and continuity of the field in the azimuthal direction.

The radial equation is a Bessel equation of order m:

$$\left[\frac{d^2}{d\rho^2} + \frac{1}{\rho}\frac{d}{d\rho} + \left(\chi^2 - \frac{m^2}{\rho^2}\right)\right] R(\rho) = 0 \qquad [8.21]$$

whose solutions are the Bessel functions of order m of the first and second kind (see Appendix A2.3). Since the field must be finite on the axis, only the functions of the first kind J_m must be retained, and:

$$R(\rho) \propto J_m(\chi_{mn}\rho) \qquad [8.22]$$

The rigid wall condition at $\rho = a$ is written as $\chi_{mn} J_m'(\chi_{mn} a) = 0$, where J_m' denotes the derivative of the Bessel function of order m with respect to ρ. The

boundary condition imposes a discrete set of eigenvalues, given by $\chi_{00} = 0$ and the non-trivial successive zeros of the derivative of the Bessel functions of order m. The sequence of values of $\chi_{mn}a$, which correspond to the first zeros of the derivatives of the Bessel functions, is given in Table 8.1.

$\chi_{mn}a$	$m=0$	$m=1$	$m=2$	$m=3$
$n=0$	0	1.841	3.054	4.201
$n=1$	3.832	5.331	6.706	8.016
$n=2$	7.016	8.536	9.969	11.346
$n=3$	10.173	11.706	13.170	14.586

Table 8.1. *Normalized eigenvalues $\chi_{mn}a$ associated with the first modes of a circular duct of radius a*

The eigenfunction of order (m, n) is finally written as[4]:

$$\psi_{mn}(\rho, \varphi) = C_{mn} \exp(\pm im\varphi) J_m(\chi_{mn}\rho) \qquad [8.23]$$

where the multiplicative constant C_{mn} is determined by the normalization of the eigenfunctions, whose integral of the squared magnitude over the cross-section must be equal to πa^2 according to definition [8.4]. We thus obtain using formula [A2.62]:

$$C_{mn} = \left(\frac{1}{2}\left(1 - \frac{m^2}{\chi_{mn}^2}\right) J_m^2(\chi_{mn})\right)^{-1/2} \qquad [8.24]$$

The square of the axial wavenumber is given by $\kappa_{mn}^2 = k_0^2 - \chi_{mn}^2$. As expected, the fundamental mode $(0, 0)$ is plane and always cut-on; it propagates with the free-space speed of sound c_0. The cutoff frequency of a mode (m, n) is given by $f_{mn} = \chi_{mn}c_0/2\pi$, and the first modes to become propagating as the frequency increases are mode pairs $(\pm 1, 0)$ and then $(\pm 2, 0)$ before the first non-planar axisymmetric mode $(0, 1)$. When the frequency is low enough, that is when $f < 1.841c_0/2\pi a$, only the plane mode can propagate along the duct without attenuation. As an example, for a duct with a radius 10 cm and for air under standard conditions, this frequency is about 1,000 Hz. One-dimensional analysis of cylindrical duct propagation is therefore justified for a wide range of applications where only "low frequencies" are excited.

The real part of the pressure field in the cross-section of a circular duct is plotted in Figure 8.4 for four modes of different orders (m, n). There are a total of m nodal

4 In the literature, these modes are often written in a real form, the complex exponential being replaced by $\cos(m\varphi)$ or $\sin(m\varphi)$.

diameters, while n nodal circles can be observed. When a mode is of high azimuthal order m and low radial order n, its energy is strongly concentrated toward the periphery of the duct as can be observed here for mode $(12, 1)$, and this explains why this type of mode is strongly attenuated in the presence of absorbing walls.

Finally, it should be noted that Figure 8.4 gives only a static view of the field at a given time and in a given section. The complete spatial form of the pressure for a propagating mode, $p(\rho, \varphi, x_3) \propto J_m(\chi_{mn}) \exp\left(\mathrm{i}\left(\pm m\varphi - \sqrt{k_0^2 - \chi_{mn}^2}\, x_3\right)\right)$, shows that equiphase surfaces, corresponding to $\pm m\varphi - \sqrt{k_0^2 - \chi_{mn}^2}\, x_3 = \text{cte}$, have a helical shape for $m \neq 0$: These are called spinning modes.

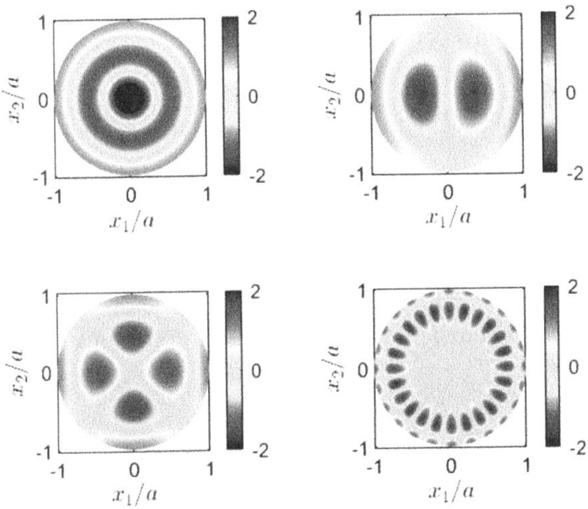

Figure 8.4. *From left to right and top to bottom, plots of the real part of the pressure field in the cross-section of a circular duct of unit area for modes (0,2), (1,1), (2,2) and (12,1). For a color version of this figure, see www.iste.co.uk/juve/acoustics1.zip*

8.4. Point source in a duct and Green's function

In the previous sections, the presence of a source was not considered, and only a pure propagation problem was addressed. In the presence of a sound source in the duct, the previously unknown amplitude of each mode is imposed, and a complete solution for the pressure field can be obtained. In the following, we consider a point source located at position $\boldsymbol{y}$, which is equivalent to finding the Green's function of the duct.

The Green's function we are looking for is the solution to the following equation:

$$\nabla^2 G(\boldsymbol{x}, \boldsymbol{y}) + k_0^2 G(\boldsymbol{x}, \boldsymbol{y}) = -\delta(\boldsymbol{x} - \boldsymbol{y}) \qquad [8.25]$$

associated with the rigid wall condition. It can be expressed on the basis of the eigenfunctions under the following form:

$$G(\boldsymbol{x},\boldsymbol{y}) = \sum_p F_p(x_3, y_i)\psi_p(x_1, x_2) \qquad [8.26]$$

using Cartesian coordinates. This expansion is inserted into the previous differential equation, which is then multiplied by ψ_p[5] and integrated over the duct cross-section. This leads to a set of equations:

$$\left(\frac{d^2}{dx_3^2} + \kappa_p^2\right) F_p = -\frac{\psi_p(y_1, y_2)}{A}\delta(x_3 - y_3); \quad \kappa_p^2 = k_0^2 - \chi_p^2 \qquad [8.27]$$

The solution of each of these equations can be obtained from the expression of the one-dimensional Green's function given by equation [5.72], and we get:

$$F_p = \frac{-\mathrm{i}\psi_p(y_1, y_2)}{2\kappa_p A}\exp(-\mathrm{i}\kappa_p\,|x_3 - y_3|) \qquad [8.28]$$

where, for evanescent modes, κ_p should be taken equal to $-\mathrm{i}\sqrt{\chi_p^2 - k_0^2}$.

The duct Green's function is therefore:

$$G(\boldsymbol{x},\boldsymbol{y}) = \frac{-\mathrm{i}}{2A}\sum_p \frac{\psi_p(x_1, x_2)\psi_p(y_1, y_2)}{\kappa_p}\exp(-\mathrm{i}\kappa_p\,|x_3 - y_3|) \qquad [8.29]$$

This expression can be developed to make the plane mode apparent and separate the contributions of the cut-on modes (for which $n \leq N$, where $\chi_N < k_0 < \chi_{N+1}$) and of the exponentially attenuated cutoff modes, according to:

$$\begin{aligned} G(\boldsymbol{x},\boldsymbol{y}) = {} & \frac{-\mathrm{i}}{2k_0 A}\exp(-\mathrm{i}k_0\,|x_3 - y_3|) \\ & - \frac{\mathrm{i}}{2A}\sum_{p=1}^{N} \frac{\psi_p(x_1, x_2)\psi_p(y_1, y_2)}{\kappa_p}\exp(-\mathrm{i}\kappa_p\,|x_3 - y_3|) \\ & + \frac{1}{2A}\sum_{p=N+1}^{\infty} \frac{\psi_p(x_1, x_2)\psi_p(y_1, y_2)}{\kappa'_p}\exp(-\kappa'_p\,|x_3 - y_3|) \end{aligned} \qquad [8.30]$$

where the notation $\kappa'_p = \sqrt{\chi_p^2 - k_0^2}$ was used.

5 We are considering real eigenfunctions ψ_p ,which is always possible for rigid ducts.

The expression resulting from this series is very different from that of the free-space Green's function and is the consequence of the channeling of the acoustic waves along the x_3 axis, associated with their successive reflections on the walls.

For a point source and using the linearized Euler's equation, the axial acoustic velocity u_p (to the right of the source) of a given mode can be deduced from [8.29] and:

$$u_p = \frac{-\mathrm{i}}{2A} \frac{\psi_p(x_1, x_2)\psi_p(y_1, y_2)}{Z_0 k_0} \exp(-\mathrm{i}\kappa_p |x_3 - y_3|) \tag{8.31}$$

The radiation impedance of a mode p, normalized by the characteristic impedance of the fluid Z_0, can then be obtained as:

$$\zeta = \frac{p_p}{Z_0 u_p} = \frac{k_0}{\kappa_p} \tag{8.32}$$

For the plane mode, this reduced impedance is equal to unity, as expected. It is real for cut-on modes and purely imaginary for cutoff modes. As the frequency increases, the reduced impedance of the point source in the duct is therefore constant and equal to unity up to the value of the first cutoff frequency. It then becomes infinite at this value (corresponding to the resonance of the mode) and each time a cutoff frequency is reached. This behavior is depicted in Figure 8.5 for the rectangular duct studied earlier.

The power radiated in the $+x_3$ direction is the integral over the duct cross-section of the intensity $1/2\mathrm{Re}\,(pu^*)$[6]. The intensity for each mode can be expressed as a function of the real part of the radiation impedance given by [8.32]. Thus, only the cut-on modes contribute, since the impedance of the cutoff modes is purely imaginary. Due to the orthogonality property, all cross products between different modes integrate to zero and the total power is the sum of the powers associated with each cut-on mode taken individually. It yields:

$$W = \frac{1}{8A\omega\rho_0} \sum_{p=0}^{N} \frac{\psi_p^2(y_1, y_2)}{\kappa_p} \tag{8.33}$$

The power radiated therefore exhibits jumps (infinite if there is no dissipative mechanism) each time the frequency reaches one of the cutoff frequencies of the duct.

6 The total power emitted by the source in both directions is twice this value.

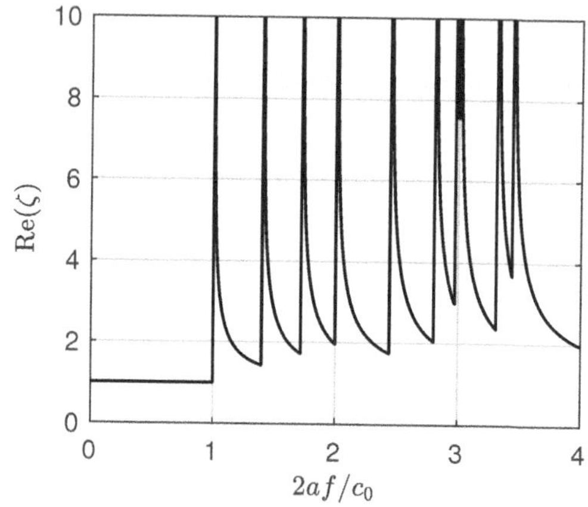

Figure 8.5. *Reduced radiation resistance of a point source in a rectangular duct (a/b=1.4), plotted as a function of the frequency divided by the duct cutoff frequency*

8.5. Propagation in a duct with absorbing walls

The confinement of waves in a long duct makes it possible to reduce the sound emission from sources by covering the walls, at least partially, with absorbing materials. If these materials can be modeled by a uniform impedance, the modal development technique used in the previous sections remains applicable, but with some modifications.

When the real part of the wall impedance is positive, acoustic energy is absorbed by the walls, and the eigenvalues χ_p become complex. The axial wavenumbers also become complex, with a negative imaginary part, causing all modes to attenuate as they propagate toward increasing values of x_3. The modes have complex amplitudes and are not mutually orthogonal in the sense of the Hermitian scalar product as was the case with rigid ducts. However, they satisfy a related condition, known as bi-orthogonality:

$$\int_S \psi_p \psi_q \, \mathrm{d}S = A\delta_{pq} \qquad [8.34]$$

The eigenfunctions are orthogonal to their complex conjugates; when the ψ_p are real, as in the case of rigid walls, the set of eigenfunctions is thus orthogonal.

The axial wavenumber is given by $\kappa_p = \sqrt{k_0^2 - \chi_p^2}$. When the absorption at the wall is strong, the distinction between cut-on and cutoff modes is less clear than for

rigid ducts, since all modes, including the fundamental mode[7], are exponentially attenuated along the duct axis. Nevertheless, from the frequency at which $k_0^2 = \mathrm{Re}\left(\chi_p^2\right)$ on, the axial decay of the mode becomes slower and the mode is said to be cut-on, as in the case of rigid ducts.

To be more specific, we now consider a rectangular duct with two rigid faces and two other faces (in the x_2 direction) characterized by the same value of the wall impedance [MOR 68]. It is somewhat easier for this analysis to take the origin of the coordinates at the center of the duct, and the boundary conditions to be satisfied are then:

$$\frac{dX_1}{dx_1} = 0 \qquad \text{for} \quad x_1 = \mp a/2. \qquad [8.35]$$

$$\frac{dX_2}{dx_2} = \pm \mathrm{i} k_0 \beta X_2 \qquad \text{for} \quad x_2 = \mp b/2. \qquad [8.36]$$

where we have set $p(x_1, x_2, x_3) = X_1(x_1)X_2(x_2)X_3(x_3)$ and where β is the reduced admittance of the wall, that is, the inverse of the reduced impedance of the material.

In this case we will distinguish between even and odd symmetry in each direction; in the following we consider only eigenfunctions that are even in the x_1 direction. They are proportional to $\cos(m\pi x_1/a)$, where m is an even integer. The eigenfunctions of the waveguide are then written as:

$$\psi_{mn}(x_1, x_2) = C_{mn} \cos\left(\frac{m\pi x_1}{a}\right) \cos\left(\frac{q_n \pi x_2}{b}\right) \quad \text{for} \quad n \quad \text{even} \qquad [8.37]$$

$$\psi_{mn}(x_1, x_2) = C_{mn} \cos\left(\frac{m\pi x_1}{a}\right) \sin\left(\frac{q_n \pi x_2}{b}\right) \quad \text{for} \quad n \quad \text{odd} \qquad [8.38]$$

where $q_n = k_2 b/\pi$ is the solution of a transcendental equation that is written for even modes in the x_2 direction (the case of odd modes is left as an exercise) as:

$$q_n \tan \frac{\pi q_n}{2} - \mathrm{i}\frac{k_0 b \beta}{\pi} = 0 \qquad [8.39]$$

This equation can be solved numerically using Newton's method, for example.

For weakly absorbing walls, namely when $\beta \approx 0$, q_n is nearly equal to the integer value n associated with the corresponding mode of the rigid duct. A Taylor series expansion around this value then provides:

$$q_0 = \frac{\sqrt{2\mathrm{i}k_0\beta b}}{\pi}; \quad q_n \approx n + \mathrm{i}\frac{2k_0 b}{n\pi^2}\beta \quad \text{if} \quad n \neq 0 \qquad [8.40]$$

7 For this fundamental mode, the pressure is not uniform in the cross-section; it is therefore no longer "plane".

Figure 8.6 shows the real and imaginary parts of the modal shapes in the direction normal to the absorbing walls. They are plotted for the $n = 2$ mode and for two values of the wall admittance, corresponding to an almost rigid wall and a highly absorbing wall, respectively.

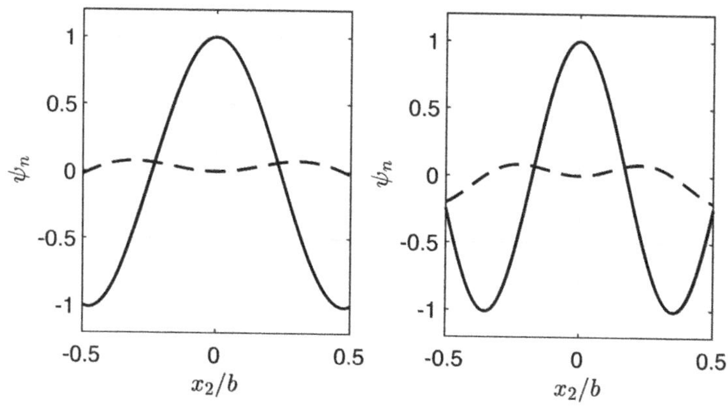

Figure 8.6. *Non-normalized modal shapes (n=2, $k_0 b$ =10) in a lined rectangular duct for two values of the wall admittance, $\beta = 0.1(1 + \mathrm{i})$ on the left and $\beta = 2(1 + \mathrm{i})$ on the right. The real part is displayed by the solid line and the imaginary part by the dashed line. The values of q_2 are equal to $(2.102 + 0.090\mathrm{i})$ and to $(2.847 + 0.130\mathrm{i})$, respectively. For $\beta = 0.1(1 + \mathrm{i})$, the numerical estimate is very close to that obtained from formula [8.40], $(2.101 + 0.101\mathrm{i})$*

If the duct walls are perfectly rigid, a particular mode will have a cutoff frequency that clearly marks the transition from evanescent to propagative behavior. If the walls are absorbing, the transition is less sharp because the wave is exponentially attenuated along the axis at each frequency. However, the notion of a cutoff or cut-on mode can still be used, at least if the absorption at the walls remains moderate. We illustrate this aspect by studying the behavior of mode (0,2) as a function of frequency for the rectangular duct considered above. The imaginary part of the axial wavenumber is plotted in Figure 8.7 for increasing values of the specific admittance of the walls and is compared to its behavior for a rigid duct. The cutoff/cut-on transition remains clearly visible on these plots, with the rate of axial pressure attenuation, measured by the imaginary part of the axial wavenumber. For a rigid duct, this value decreases rapidly with frequency and is zero beyond the cutoff frequency of the duct. For the duct with absorbing walls, the decrease at low frequency remains similar to that of the rigid case, but it transitions to a much slower decrease beyond a "cutoff frequency" whose value increases with that of the admittance.

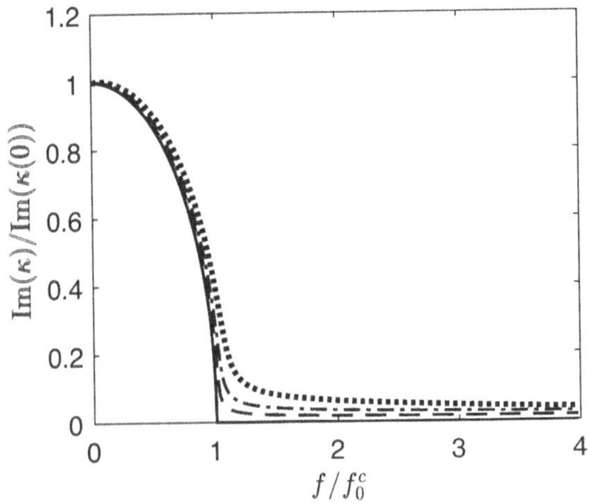

Figure 8.7. *Evolution with the frequency of the imaginary part of the axial wavenumber of mode (0, 2) for a hard-walled duct (solid line), and for increasing values of the wall admittance,* $\beta = 0.05(1 + \mathrm{i})$, *the dashed line,* $0.1(1 + \mathrm{i})$, *the dash-dotted line, and* $0.2(1 + \mathrm{i})$, *the dotted line. The frequency is normalized to the cutoff frequency of the rigid duct mode and the wavenumber is normalized to its zero-frequency value*

8.6. Influence of a uniform flow on modal propagation

In practice, ducts are most often subjected to flow, such as that produced by a fan in an air conditioning system. The presence of flow significantly alters the propagation of acoustic waves, and this alteration increases with the Mach number. In this section, we study the influence of a subsonic uniform flow existing in a hard-walled duct of constant cross-section.

The assumption of a mean flow, uniform in the duct cross-section, is a first approximation to the influence of a real flow. For viscous fluids, the flow velocity at the walls must be zero (no-slip condition). Therefore, there are velocity gradients near the walls whose influence on the propagation of acoustic waves (refraction effects) will be neglected in our analysis. However, the approximation of a uniform flow has the great advantage of allowing a sufficiently simple approach, while highlighting the essence of the observed physical phenomena. It should be noted, however, that the assumption of a uniform mean flow introduces a specific difficulty for ducts with absorbing walls, since the flow velocity at the wall is non-zero. A special formulation of the boundary condition must be used, called the Ingard–Myers condition (see Chapter 3 of [RIE 21]). For an example of the influence of a uniform

mean flow on the propagation of acoustic waves in an annular duct whose walls are absorbing, see Chapter 7 of the same work.

The equation to be solved is the convected Helmholtz equation, which can be obtained by changing the reference frame in order to reduce the problem to that of the acoustic propagation in a medium at rest (see section 5.7). This equation is written as:

$$\nabla^2 p - \left(\mathrm{i}k_0 + \mathrm{M}\frac{\partial}{\partial x_3} \right)^2 p = 0 \qquad [8.41]$$

where M is the Mach number of the flow ($0 < \mathrm{M} < 1$). The acoustic field must satisfy a condition of zero normal pressure gradient at the walls, and a condition of regularity on the axis in the common case of circular ducts.

It is possible to search for a solution of equation [8.41] by separating the variables as in the absence of flow, in the form $p(x_1, x_2, x_3) = \psi(x_1, x_2)X_3(x_3)$, and to construct a solution based on a modal expansion (the modes are orthogonal and form a complete system). The eigenfunctions of the duct cross-section remain unchanged; they are the solutions of:

$$\nabla_T^2 \psi + \chi^2 \psi = 0 \qquad [8.42]$$

and the only change from the no-flow case occurs on the axial evolution of the pressure. Searching for a solution of the form $\exp(-\mathrm{i}\kappa_p x_3)$, where the axial propagation constant κ_p can be real or complex, the following dispersion relation is obtained:

$$\chi_p^2 = (k_0 - \mathrm{M}\kappa_\mathrm{p})^2 - \kappa_p^2 \qquad [8.43]$$

The axial propagation constant is then given by:

$$\kappa_p^{\pm} = \frac{-\mathrm{M}k_0 \pm \sqrt{k_0^2 - \left(1 - \mathrm{M}^2\right)\chi_p^2}}{1 - \mathrm{M}^2} \qquad [8.44]$$

and $\kappa_p^{\pm}$ will be real when $k_0^2 > \left(1 - \mathrm{M}^2\right)\chi_p^2$, and complex otherwise.

The cutoff frequency of a mode is given by $f_p^c = \sqrt{1 - \mathrm{M}^2} f_{p0}^c$, where $f_{p0}^c = c_0\chi_p/2\pi$ is the cutoff frequency of the same mode in the fluid at rest. Thus, a major effect of the flow is to decrease the cutoff mode frequencies by the factor $\sqrt{1 - \mathrm{M}^2}$. As a result, in the presence of flow, more cut-on modes can be excited by a source emitting in a given frequency range.

In Figure 8.8, we plot the evolution with the frequency of the two branches followed by the axial wavenumber κ_p, for a Mach number equal to 0.75. Note that the sign of the axial wavenumber does not always indicate the propagation direction of a mode. When $\sqrt{1-\mathrm{M}^2}\chi_p < k_0 < \chi_p$, the wavenumber associated with the plus sign of the root in equation [8.44], κ_p^+ (which characterizes waves propagating in the $+x_3$ direction in the no-flow case), is negative. The calculation of the phase and group velocities will allow us to explain this somewhat puzzling behavior.

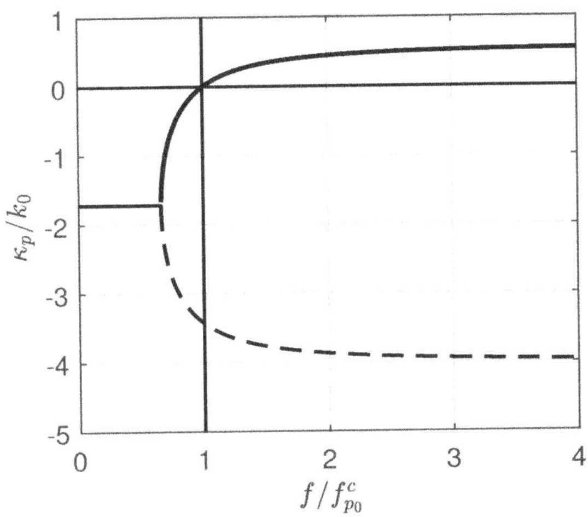

Figure 8.8. *Plots of the two branches of the real part of the axial wavenumber κ_p^+/k_0 (solid line) and κ_p^-/k_0 (dashed line), for cut-on modes in the presence of a $\mathrm{M} = 0.75$ mean flow. The frequency is normalized to the duct cutoff frequency in the absence of flow*

The phase and group velocities of the propagating modes are deduced from [8.44] by $v_\varphi^\pm = \omega/\kappa_p^\pm$ and $v_g^\pm = d\omega/d\kappa_p^\pm = (d\kappa_p^\pm/d\omega)^{-1}$. This yields:

$$v_\varphi^\pm = c_0 \frac{\mathrm{M}k_0^2 \pm k_0\sqrt{k_0^2 - (1-\mathrm{M}^2)\,\chi_p^2}}{k_0^2 - \chi_p^2} \qquad [8.45]$$

and

$$v_g^\pm = \pm c_0 \frac{(1-\mathrm{M}^2)\sqrt{k_0^2 - (1-\mathrm{M}^2)\,\chi_p^2}}{k_0 \mp \mathrm{M}\sqrt{k_0^2 - (1-\mathrm{M}^2)\,\chi_p^2}} \qquad [8.46]$$

Of course, when $\mathrm{M} = 0$, these expressions are identical to those obtained directly from the Helmholtz equation (see the formulas [8.15] and [8.16] obtained for the

rectangular duct). It can also be seen that for the plane mode ($\chi_0 = 0$), the phase and group velocities are equal to $c_0(\mathrm{M} \pm 1)$, corresponding to the convection of a plane wave by the mean flow velocity.

Formulas [8.45] and [8.46] show that for waves associated with the k_p^+ branch and whose frequency falls between the cutoff frequencies without flow, f_{p0}^c, and with flow, $f_p^c = f_{p0}^c\sqrt{1-\mathrm{M}^2}$, the phase and group velocities have opposite signs. The phase velocity is negative, so the wavefronts are moving against the flow, while the group velocity is positive, showing that the energy is actually propagating toward increasing values of x_3.

Figure 8.9 shows the evolution with frequency of the phase and group velocities associated with the two branches of the dispersion relation for a Mach number equal to 0.75. The existence of a region in which the phase velocity and the group velocity have opposite signs for waves propagating toward increasing x_3 can indeed be observed. The high frequency limits of the phase and the group velocities are equal to $c_0(\mathrm{M} \pm 1)$, since in this limit the angle of the modes with respect to the duct axis tends to zero; they thus behave as plane waves convected by the mean flow.

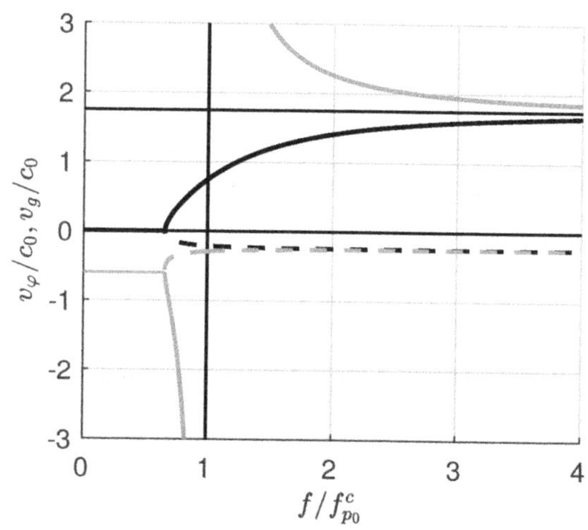

Figure 8.9. *Phase (gray curves) and group (black curves) velocities in a duct with flow ($\mathrm{M} = 0.75$). The branches corresponding to energy propagation toward positive and negative values of x_3 are drawn as solid lines and dashed lines, respectively*

8.7. Exercises

1) **Orthogonality of the eigenfunctions in a hard-walled duct**

Show that the separation constants of equation [8.1] are real and positive and that the eigenfunctions are orthogonal.

2) **Two-dimensional hard-walled duct**

Consider a semi-infinite, two-dimensional rigid duct of height a. This duct is excited by the harmonic vibration of a membrane located at $x_3 = 0$ and whose complex velocity amplitude is given by:

$$V(x_1) = 4V_0\frac{x_1}{a}\left(1 - \frac{x_1}{a}\right)$$

where V_0 is a constant.

a) Give the expression for the amplitudes of the modes propagating in the duct by applying the proper boundary condition at $x_3 = 0$.

b) With $a = 0.08$ m and $f = 1,700$ Hz, show that only one mode is propagating. Express the amplitudes of this mode and that of the first evanescent mode. At what distance from the source does the amplitude of the evanescent mode become negligible (i.e. less than 1% of that of the plane mode)?

c) Calculate and plot the time evolution of the pressure fluctuation along the duct axis, still considering only the contributions of the first two modes.

3) **Odd modes for a duct with absorbing walls**

Write the equation satisfied by the eigenvalues of the odd modes in a rectangular duct with two of its sides covered with an absorbing material (see section 8.5).

4) **Numerical evaluation of eigenvalues for an absorbing duct**

a) Demonstrate formula [8.40], which is valid for nearly rigid walls. Numerically solve the exact transcendental equation [8.39] and estimate the validity of the approximate solution.

b) Compute the eigenvalues and modal shapes for more absorbing walls numerically. Use a variety of values of k_0a corresponding to cut-on or cutoff modes (verify your calculations by referring to Figure 8.6). Calculate the attenuation (in dB) of the wave over a distance equal to the width of the duct for $k_0a = 10$ and $m = 2$.

5) **Modal propagation in underwater acoustics**

Consider a very simplified model of acoustic wave propagation in the ocean: the medium is homogeneous (with uniform sound speed c_0 and density) and has a constant depth H. It is bounded by the free surface, along which $p = 0$, and a perfectly rigid

bottom; we will use the cylindrical coordinates (ρ, φ, z), with the z axis oriented downward. The pressure field is generated by a point source located at depth $z = z_s$. The Helmholtz equation to be solved is as follows:

$$\frac{1}{\rho}\frac{\partial}{\partial \rho}\left(\rho\frac{\partial p}{\partial \rho}\right) + \frac{\partial^2 p}{\partial z^2} + k_0^2 p = -\frac{\delta(\rho)}{\rho}\delta(\varphi)\delta(z - z_s) \qquad [8.47]$$

Given the axisymmetric nature of radiation, integrating this equation over φ leads to:

$$\frac{1}{\rho}\frac{\partial}{\partial \rho}\left(\rho\frac{\partial p}{\partial \rho}\right) + \frac{\partial^2 p}{\partial z^2} + k_0^2 p = -\frac{\delta(\rho)}{2\pi\rho}\delta(z - z_s) \qquad [8.48]$$

We seek a modal separable solution of the unforced equation in terms of radial and vertical coordinates:

$$p(\rho, z) = \sum_n Z_n(z) R_n(\rho) \qquad [8.49]$$

a) The z equation can be written as:

$$\frac{d^2 Z_n}{dz^2} + \chi_n^2 Z_n = 0 \qquad [8.50]$$

where χ_n^2 is a separation constant that is necessarily positive.

Determine the vertical eigenfunctions Z_n (and the corresponding eigenfrequencies) by expressing the boundary conditions at the free surface and at the bottom. Normalize the eigenfunctions by requiring that the integral of Z_n^2 over the depth be unity. Plot the evolution of the first four eigenfunctions with depth.

b) Substitute the development [8.49] in [8.48], then multiply the result by Z_m and integrate over the depth. Use the orthogonality of the vertical eigenfunctions to show that:

$$\frac{1}{\rho}\frac{d}{d\rho}\left(\rho\frac{dR_n}{d\rho}\right) + (k_0^2 - \chi_n^2)R_n = -\frac{\delta(\rho)}{2\pi\rho} \qquad [8.51]$$

c) This is a zero-order Bessel equation, whose solution in an unbounded medium is a Hankel function of the second kind, $R_n = -\mathrm{i}/4H_0^2(\kappa_n)$, with $\kappa_n = \sqrt{k_0^2 - \chi_n^2}$ (see section 5.5.3). Use the asymptotic approximation of the Hankel function to highlight the behavior of the far-field pressure.

d) Show that there exists a finite number of propagating modes (when κ_n is real) at a given frequency. What is the value of the cutoff frequency of the waveguide?

Calculate the phase and group velocities of the modes and plot their evolution as a function of frequency for the first 4 modes.

6) **Phase and group velocities in a duct with flow**

a) Show from equations [8.45] and [8.46] that $(v_g - c_0 M)(v_\varphi - c_0 M) = c_0^2$ and interpret this result.

b) Examine the case of a supersonic flow. Plot the dispersion relations, as well as the evolution with the frequency of the phase and group velocities.

9

One-dimensional Propagation in Ducts

In Chapter 8, we showed that at low frequencies, when the acoustic wavelength is significantly greater than the characteristic dimension of the cross-section of a rigid duct, the plane fundamental mode is the only mode that propagates without being exponentially attenuated along the duct axis. Away from sources or singularities such as abrupt changes in cross-section, the propagation of acoustic waves is then reduced to a one-dimensional problem. In this simplified approach, it is possible to connect different parts of the duct by discrete elements (expansions, bifurcations, resonators and filters) characterized by an impedance matrix. In analogy to the electrical circuit theory, the terms acoustic circuits or acoustic transmission lines are often used. This circuit-based approach makes it easy to define and characterize elements that are widely used in practice to reduce sound levels, such as reactive silencers (where there is no energy dissipation) or selective filters such as Helmholtz resonators.

A second aspect addressed in this chapter concerns ducts whose cross-section varies slowly, and in which the acoustic propagation takes place in the form of waves that can be locally assimilated to plane waves. These ducts or *horns* are used, for example, to adapt the impedance of transducers to that of the surrounding medium in the best possible way, which results in an increased radiation efficiency. They are widely used, especially in electroacoustics.

9.1. Ducts of piecewise constant cross-section: transfer matrices

We consider a straight duct[1] with axis x_1 and constant cross-section S, in which only plane waves propagate in the upstream and downstream directions. Our first goal is to systematically relate the properties of the waves in any cross-section $x_1 = L$ to

---

1 If the duct is slightly curved, the results remain applicable by replacing the x_1 coordinate by the curvilinear coordinate.

their properties at $x_1 = 0$. Instead of considering the usual pressure–velocity pair[2] (p, u), it is interesting to use here the pressure–volume velocity[3] pair $(p, Q = Su)$. It is easy to show that the pairs (p, Q) at $x_1 = 0$ and $x_1 = L$ are related by the following matrix formula:

$$\begin{pmatrix} p(L) \\ Q(L) \end{pmatrix} = \begin{pmatrix} \cos(k_0 L) & -\mathrm{i}Z_{s0} \sin(k_0 L) \\ -\mathrm{i}/Z_{s0} \sin(k_0 L) & \cos(k_0 L) \end{pmatrix} \begin{pmatrix} p(0) \\ Q(0) \end{pmatrix} \quad [9.1]$$

where we have introduced the notation $Z_s = Z/S$ for the acoustic impedance[4] and to avoid the confusion with the previously used *specific* impedance Z. Z_{s0} will denote the characteristic impedance per unit area $\rho_0 c_0 / S$.

The reverse shift from $x_1 = L$ to $x_1 = 0$ is achieved by replacing L with $-L$, thus changing the sign of the off-diagonal terms of the transfer matrix in equation [9.1]. Note that the determinant of this matrix is equal to unity.

9.1.1. *Impedance transfer*

To calculate reflection and transmission coefficients, it is often useful to determine the impedance at $x_1 = 0$ from its value in an arbitrary section $x_1 = L$ (see Chapter 4). A very common example of application corresponds to a duct closed by a rigid termination at $x_1 = L$, where $Q(L) = 0$ and thus $Z_s(L) = \infty$. It follows immediately from [9.1] that $Z_s(0) = -\mathrm{i}Z_{s0} \cot(k_0 L)$. A second common example is that of an open-ended duct; to a first approximation, the pressure fluctuation at the end of the duct can be taken to be zero, and it follows that $Z_s(0) = \mathrm{i}Z_{s0} \tan(k_0 L)$.

9.1.2. *Cross-sectional area discontinuities*

We now consider discrete elements such as rapid changes in the cross-section between two straight ducts. The question that immediately arises is how to link the pressure–volume velocity pairs on either side of the "singularity" under consideration. In the vicinity of an abrupt change in cross-section, it is necessary to take into account the evanescent modes on both sides of the singularity in order to respect the continuity of the velocity and pressure fluctuations. However, if the variation in cross-section is not too large, the contribution of these evanescent modes becomes rapidly negligible at a short distance from the singularity, and only the plane mode can be taken into account.

2 In this chapter, u denotes the only component of the acoustic velocity along the axis of the duct.

3 The volume velocity is nothing other than the flow rate across the duct cross-section.

4 This impedance is similar to a mechanical impedance, the ratio of a force to a velocity.

Consider a duct with cross-section S_1 which connects to a duct with cross-section S_2 at $x_1 = 0$. The second duct extends indefinitely, or is closed by an *anechoic termination* that does not generate any reflected wave (Figure 9.1).

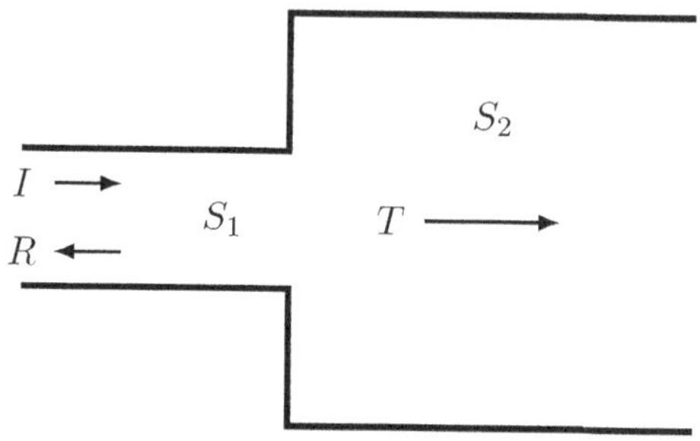

Figure 9.1. *Reflection and transmission of a plane wave at a junction between ducts of different cross-sections. The second duct is assumed to be infinite*

Figure 9.2 shows the pressure field computed by the finite element method for a cross-section variation with a ratio of 4 and for a frequency about half the cutoff frequency of the first non-plane mode of the largest duct. The effects of the evanescent modes are limited to a small region on either side of the junction, and the plane structure of the transmitted wave is obtained at a distance from the singularity equal to a fraction of the largest duct transverse dimension.

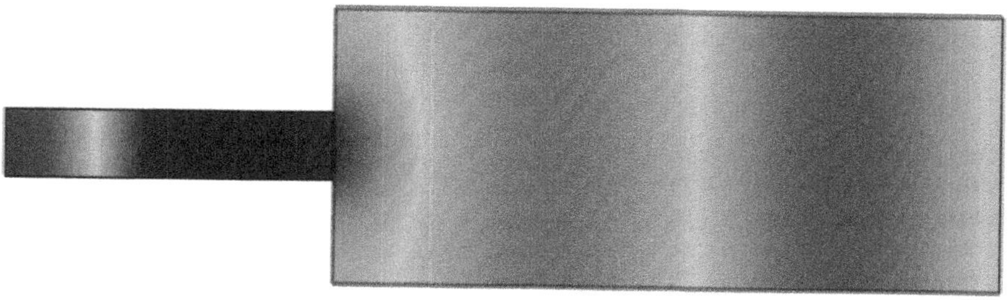

Figure 9.2. *Real part of the pressure field for a sudden change in area in a rectangular duct. The ratio of the cross-section areas is equal to 4, with the dimension of the duct in the 3rd direction remaining unchanged. The frequency is close to half that of the first non-planar mode that can be excited in the largest duct (mode (2,0), considering symmetry). For a color version of this figure, see www.iste.co.uk/juve/acoustics1.zip*

If the region where the plane wave hypothesis is not valid is small compared to the wavelength, the junction is said to be "compact" and the fluid can then be considered

locally incompressible: in this case, the volume flow rate $Q = Su$ is the same on both sides of the junction.

The condition to be imposed on the pressure variation is not so clear. The following reasoning can be carried out. The linearized Euler's equation:

$$\rho_0 \frac{\partial u'}{\partial t} = -\frac{\partial p'}{\partial x_1} \qquad [9.2]$$

is integrated along a streamline between the regions where the acoustic field can be considered as plane waves. It then follows that:

$$p_1' - p_2' = \rho_0 \frac{\partial}{\partial t} \int_1^2 u'(x_1, t)\, \mathrm{d}x_1 \qquad [9.3]$$

For harmonic waves, therefore, the order of magnitude of the variation of the pressure fluctuation across the junction is $\Delta p \approx Z_0 k_0 \Delta x_1 u$, or $\Delta p/p \approx k_0 \Delta x_1$. Since the junction is assumed to be compact, $k_0 \Delta x_1 \ll 1$, the relative change in pressure $\Delta p/p$ is then negligible. At a compact junction, the usual assumption will be that pressure and volume velocity are continuous. This assumption is all the better verified when the frequency is low and the variation in the cross-sectional area is limited.

With these assumptions, the transfer matrix associated with the sudden change in cross-section is simply the identity matrix, and the reflection and transmission coefficients are easy to calculate.

An incident plane mode of amplitude I propagating in the first duct of section S_1 is partially reflected at the junction with an amplitude R and partially transmitted with an amplitude T in the second duct of section S_2. The reflection and transmission coefficients are given by:

$$r = \frac{R}{I} = \frac{S_1 - S_2}{S_1 + S_2} \qquad [9.4]$$

$$t = \frac{T}{I} = \frac{2S_1}{S_1 + S_2} \qquad [9.5]$$

and the transmission loss, defined as the ratio of incident power to transmitted power expressed in decibels, is equal to:

$$TL = -10\log(t^2 S_2/S_1) = -10\log(1 - r^2) = -10\log(4S_1S_2/(S_1 + S_2)^2) \qquad [9.6]$$

This expression shows that the cross-sectional areas of the two pipes play a symmetrical role, indicating that the transmission loss would be the same whether a widening or a narrowing of the cross-sectional area is considered. For a widening by a factor of 4, as shown in Figure 9.2, the theoretical prediction of the transmission loss, given by formula [9.6], TL_{th} = 8 dB, is very close to the value obtained from the numerical simulation: the deviation is almost zero at very low frequencies, and it is less than 0.2 dB under the conditions of this figure.

9.1.3. *Expansion chambers*

A sudden change in cross-sectional area can reduce the transmission of acoustic energy, as we have just seen. However, this method is relatively inefficient because the variations in area are generally limited and because it is not frequency-selective. A more typical method of strongly attenuating the transmission of acoustic waves in certain frequency ranges is to create an expansion chamber: the first duct is enlarged over a length L and its cross-sectional area returns to its initial value. Such a system, which is a simplified version of a commercial muffler, reduces wave transmission in a "reactive" way by using destructive interferences and reflecting part of the acoustic energy back to the source. The total acoustic energy is not changed because no absorbing materials are used.

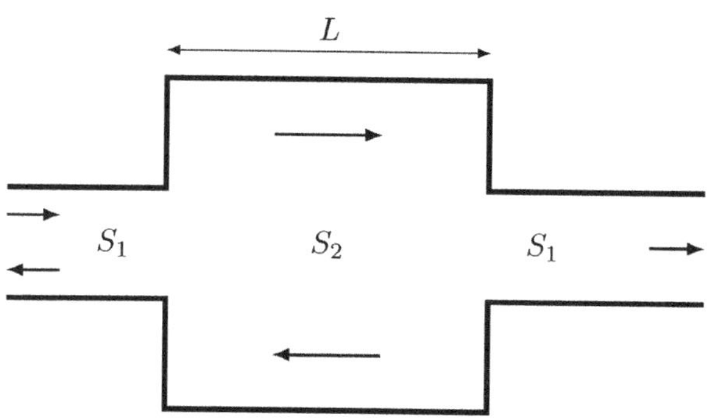

Figure 9.3. *Geometry of an expansion chamber-type reactive silencer. The terminal duct is assumed to be infinite or closed by an anechoic termination*

We consider a silencer consisting of a duct with a section S_1 which widens at $x_1 = 0$ and over a length L into a section S_2 before returning to its initial value; this last section is closed by an anechoic termination (Figure 9.3). In the first two regions, incident and reflected waves coexist, while in the last part of the duct there is only one wave propagating towards increasing values of x_1. Only plane modes are

considered and continuity of the pressure and volume velocity are imposed at the two cross-section changes.

The matrix relationship between the input and output quantities of the expansion chamber is written as:

$$\begin{pmatrix} p(0) \\ Q(0) \end{pmatrix} = \begin{pmatrix} \cos(k_0 L) & \mathrm{i}Z_0/S_2 \sin(k_0 L) \\ \mathrm{i}S_2/Z_0 \sin(k_0 L) & \cos(k_0 L) \end{pmatrix} \begin{pmatrix} p(L) \\ Q(L) = p(L)S_1/Z_0 \end{pmatrix} \quad [9.7]$$

from which the impedance Z_{local} at $x_1 = 0$ can be extracted:

$$\frac{Z_{local}}{Z_0} = \frac{p(0)S_1}{Q(0)} = \frac{\cos(k_0 L) + \mathrm{i}/m \sin(k_0 L)}{\cos(k_0 L) + \mathrm{i}m \sin(k_0 L)} \quad [9.8]$$

where $m = S_2/S_1$.

The calculation of the silencer reflection and transmission coefficients can then be done using the formulas established in Chapter 4 for an incident plane wave on an "equivalent" material with a local impedance equal to Z_e (formula [4.10]). The transmission loss is then given by:

$$TL = -10\log(|t|^2) = 10\log\left(1 + \frac{1}{4}(m - 1/m)^2 \sin^2(k_0 L)\right) \quad [9.9]$$

Formula [9.9] shows that the transmission is perfect when L is a multiple of half the wavelength. This result can be interpreted physically by observing that for the corresponding frequencies, the wave reflected on the $x_1 = L$ side of the silencer and again on the $x_1 = 0$ side is then in phase with the incident wave; it is as if the wave does not see the change in cross-sectional areas. On the other hand, when L is an odd multiple of a quarter of the wavelength, after a double reflection inside the expansion, the incident and the reflected waves are in phase opposition and the transmission loss is maximum; it is then equal to $10\log\left(1 + 1/4(m - 1/m)^2\right)$.

Formula [9.9] shows that the efficiency of the silencer is all the higher with higher values of the ratio of the cross-sectional areas[5]. This efficiency is depicted in Figure 9.4 for four values of the area ratio, m = 4, 8 and 16 and an expanded length L = 0.5 m.

To evaluate the relevance of the one-dimensional approach, we compare in Figure 9.5 the transmission loss estimated from a finite element simulation with the

5 Since in this formula m and $1/m$ play an identical role, a narrowed cross-section presents the same efficiency as a widened cross-section.

value calculated from formula [9.9]. The area ratio is equal to 9 (0.05×0.05 m^2 and 0.15×0.15 m^2 sections) and the length of the expanded area is equal to 0.5 m. The cutoff frequency of the first excitable non-planar mode is here equal to $2,286$ Hz and therefore the plane wave assumption is perfectly justified in the range considered (0–1,500 Hz). The overall shape of the two plots is very similar with the presence of alternating zeros and maxima. The correspondence of the levels for the first two maxima is excellent, but it can be seen that the theoretical formula then slightly underestimates the efficiency of the silencer, and that the deviation gradually increases with the frequency while remaining small and lower than 1 dB. The position of the zeros also differs very slightly from the theoretical value, with the difference increasing approximately linearly with frequency. The explanation for these (small) discrepancies lies in the progressive failure of the assumptions of zero impedance of the junctions and immediate transition to a plane mode in the large section. However, the one-dimensional approach provides a very good first estimate of the efficiency of silencers at low frequencies.

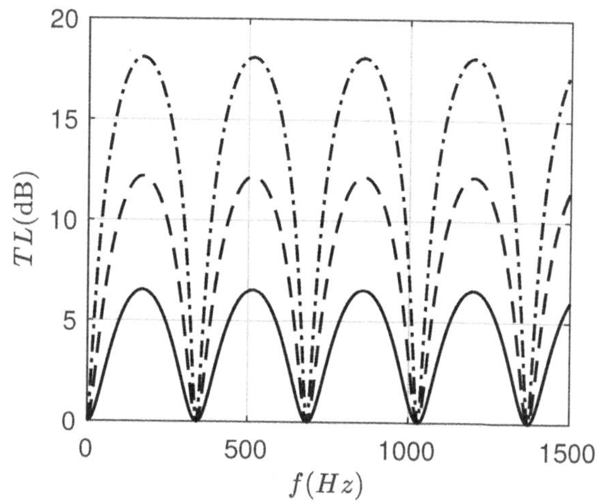

Figure 9.4. *Evolution with the frequency of the transmission loss induced by an expansion chamber of length $L = 0.5$ m and for increasing values of the cross-sectional area ratio: m =4 (solid line), 8 (dashed line) and 16 (dash dotted line). The speed of sound is equal to 343 ms^{-1}*

9.1.4. *Bifurcations and acoustic filters*

The continuity of pressure and volume velocity assumptions can also be used when a duct separates in several tubular elements at a junction, provided that the junction area can be considered compact; the validity is thus limited to a low frequency range (Figure 9.6).

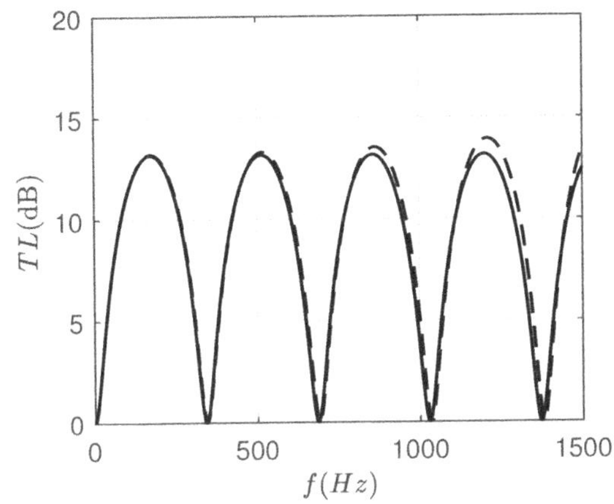

Figure 9.5. *Evolution of the transmission loss due to an expansion chamber ($m = 9$) with the frequency. Comparison between the theoretical estimate plotted in the solid line and the finite element simulation plotted in the dashed line (square ducts with 0.05×0.05 m² and 0.15×0.15 m² sections, $L = 0.5$ m; $c_0 = 343$ ms^{-1})*

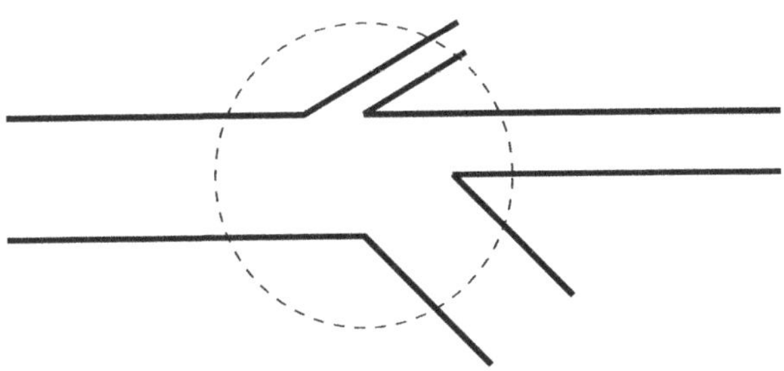

Figure 9.6. *Compact junction between several ducts. The diameter of the dashed circle is small compared to the acoustic wavelength*

The analytical expression of the various continuity conditions leads to a closed system of equations as long as the boundary conditions in the output branches are known, for example, if their terminations are anechoic. A common application example is that of a so-called Y-shaped bifurcation, where one of the branches has an anechoic termination and the other of length l_b is closed by a rigid end (Figure 9.7). A simple qualitative reasoning shows that this system acts as an acoustic filter at certain frequencies. The impedance transfer formula in the closed branch shows that the impedance at the branch inlet, and therefore the pressure is zero for frequencies equal to the odd multiples of $c_0/4l_b$. Due to the continuity of the pressure at the junction, the incident wave will see it as a zero pressure condition, causing a total

reflection (r = -1). For these frequencies, no energy is transmitted into the main duct to the right of the junction.

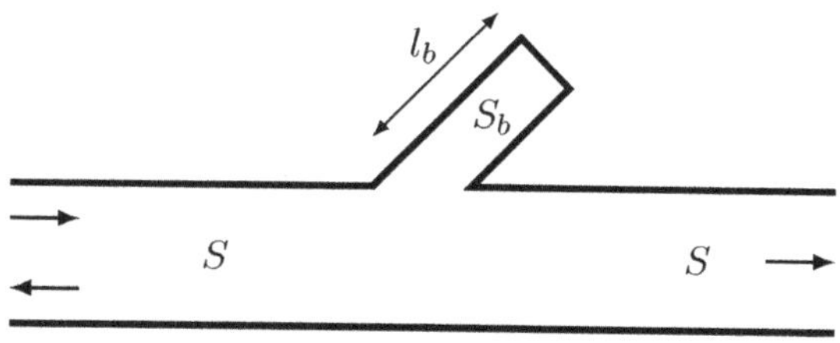

Figure 9.7. *Y-shaped bifurcationy*

The general expression of the transmission loss is given by:

$$TL = 10\log\left(1 + \left(\frac{m}{2}\right)^2 \tan^2(k_0 l_b)\right) \qquad [9.10]$$

where $m = S_b/S$ is the area ratio between the "resonant" bifurcation of section S_b and the main pipe of section S.

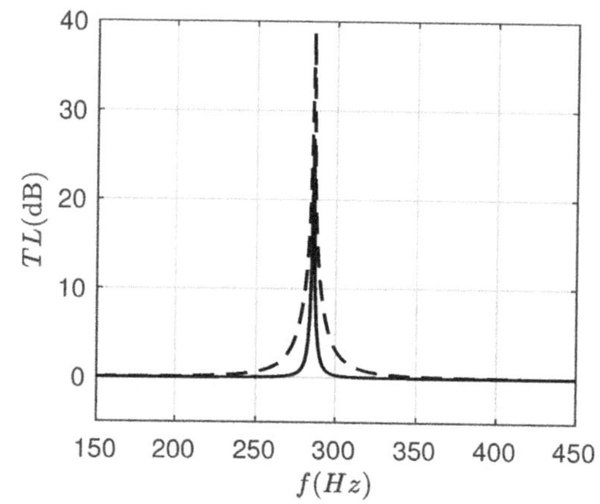

Figure 9.8. *Evolution of the transmission loss due to a resonant bifurcation of length $l_b = 0.3$ m (quarter-wavelength resonator) with the frequency for two values of the area ratio ($m = 0.04$ in the solid line and $m = 0.16$ in the dashed line)*

An example of this type of "quarter-wavelength" filter is shown in Figure 9.8 for two values of the area ratio, $m = 0.04$ and $m = 0.16$. The transmission loss is very

high when the frequency is close to $f = c_0/4l_b = 286$ Hz, since the length of the bifurcation is equal to $l_b = 0.3$ m; note that the "band-rejection" filter is steeper around this frequency for smaller values of the area ratio.

The behavior of a quarter-wavelength resonator at its first resonant frequency is illustrated in Figure 9.9 using a finite element numerical simulation. The bifurcation with length $l_b = 0.3$ m and cross-section 0.05×0.05 m^2 is connected to a rectangular duct with cross-section 0.2×0.3 m^2 (the area ratio is therefore close to 0.04). The deformation of the initial field is very important in the immediate vicinity of the junction, then the wave quickly becomes nearly plane again, but with a greatly reduced amplitude. The storage of energy in the resonator is also clearly seen, as well as the formation of a quasi-standing wave upstream of the bifurcation.

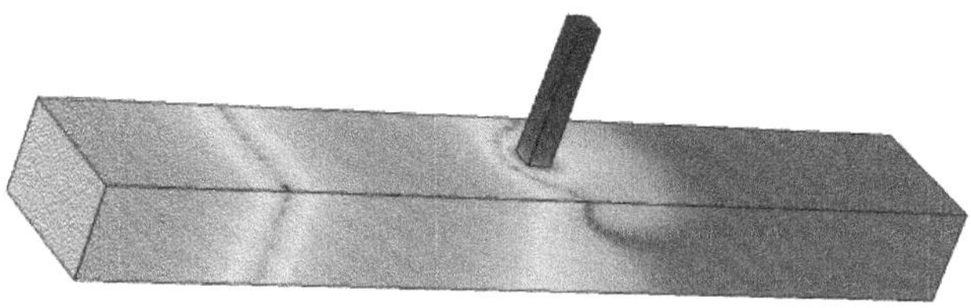

Figure 9.9. *Real part of the pressure field at the resonant frequency of a quarter-wavelength duct of length* $l_b = 0.3$ *m and cross-section* 0.05×0.05 *m*^{2} *opening into a rectangular pipe of cross-section* 0.2×0.3 *m*^{2}. *For a color version of this figure, see www.iste.co.uk/juve/acoustics1.zip*

9.1.5. *Transmission loss and insertion loss*

The transmission loss of a filter is defined when there is no reflected wave in the terminating duct (anechoic termination). In a sense, this value represents the intrinsic characteristic of the filter. In practice, however, what really matters is the reduction in acoustic power which is obtained when a device is *inserted* between a noisy machine and the outside. The *insertion loss* (IL) is defined from sound power levels by:

$$IL = L_{W_1} - L_{W_2} = 10 \log(W_1/W_2) \qquad [9.11]$$

where subscripts 1 and 2 denote the system considered without filter and with filter, respectively. The insertion loss of a filter depends on the internal impedance of the noisy source, and is not easy to calculate, so that the transmission loss is usually preferred in theoretical analyses of the performance of different filters.

For applications of silencers or mufflers to exhaust system design, we refer the reader to Munjal [MUN 14].

9.1.6. *End corrections*

The assumption that a junction between two ducts with very different cross-sections can be represented by a unitary transfer matrix is only a first approximation of the real behavior. Consider the extreme case of a flanged duct opening into an unbounded medium. The simplest boundary condition that can be applied is that of zero pressure fluctuation at the opening. This condition allows us to quickly evaluate resonant frequencies in an open-closed pipe, for example. However, it would lead to a total absence of sound radiation in the exterior region, thus a better formulation is needed. A first method to obtain an acceptable result is as follows: by analogy with the radiation of a vibrating circular piston, it is possible to associate an impedance to the opening, equal to the radiation impedance of a baffled piston. Since we are merely considering a low-frequency regime, it is possible to use only the first terms of the expansion of the radiation impedance of the piston with respect to the reduced wavenumber k_0a, which leads to the following relationship, according to formula [6.83]:

$$Z_{rad}/Z_0 \approx \frac{(k_0a)^2}{2} + \mathrm{i}\frac{8}{3\pi}k_0a \qquad [9.12]$$

For small values of the reduced wavenumber, the impedance is dominated by its reactance $8k_0a/3\pi$. This corresponds to the motion induced by the action of acoustic waves associated with the mass of a fluid element of radius a and length $\Delta l = 8a/3\pi \approx 0.85a$. The duct therefore formally appears as "extended" by this amount (called added length). The resonant frequencies of a closed-open pipe will therefore be lower than predicted by the naive approach, especially for large pipe radii and small lengths.

More rigor can be introduced into the previous reasoning, which can also be extended to unflanged pipes. The following values, obtained in the limit $k_0a \to 0$, give the "end corrections" for a flanged pipe opening, $\Delta l = 0.82a$, or for a thin-walled, unflanged pipe, $\Delta l = 0.61a$ (see, for example, Chapter 7 in [PIE 89])[6].

As an example, the right side in Figure 9.10 shows the evolution of the sound pressure level along the axis of a finite length flanged pipe, excited at its second resonant frequency (247.5 Hz, $k_0a \approx 0.23$). Note the existence of a pronounced minimum inside the pipe due to the interference between the wave emitted by the piston and the one reflected at the opening and the regular decrease in the pressure level outside the duct. The left side of the figure shows the frequency response at a point inside the duct: the first resonance frequency is about 82.5 Hz, which is very close to the theoretical estimate with an effective pipe length $l' = l + 0.82a = 1.041$ m, (f = 82.3 Hz).

6 For a non-circular duct, the radius a is replaced by the square root of the cross-sectional area of the pipe divided by π.

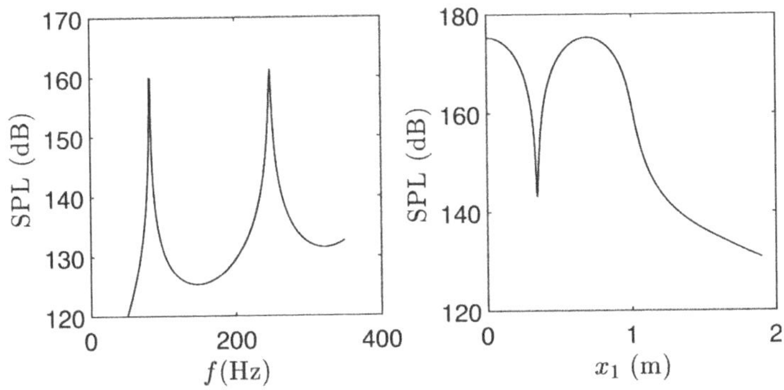

Figure 9.10. *Sound radiation from a flanged duct opening into an otherwise unbounded medium (finite element simulations). On the left, plot of the frequency response at a point located inside the duct. On the right, plot of the sound pressure level along the axis of the pipe for the second resonant frequency ($f = 247.5$ Hz). Circular duct with radius 0.05 m and length 1 m; excitation by a velocity imposed at* $x_1 = 0$

9.1.7. *Helmholtz resonators*

As discussed above, a duct with a rigid end branching into the main duct is an effective means of reducing acoustic propagation at well-defined frequencies. However, the length of such an arrangement is often prohibitive when the frequency to be filtered is low, since the length of the branching must be equal to a quarter of the corresponding wavelength. A much better device for low frequency filtering is the Helmholtz resonator. This device consists of a duct of short length l_c (*the neck*) and cross-section S_c leading into a volume V of arbitrary shape and with rigid walls (Figure 9.11). All dimensions are assumed to be small with respect to the wavelengths of interest.

We consider this system to be excited by an external pressure field. A typical example is a bottle of beer or wine that someone blows over the neck: a relatively pure, low-frequency sound is emitted that exhibits a resonant behavior. The system can be modeled as follows. Assume a nearly uniform pressure throughout the volume of the resonator. The continuity equation is written as:

$$\frac{1}{\rho_0 c_0^2}\frac{\partial p'}{\partial t} = -\nabla \cdot \boldsymbol{u}' \qquad [9.13]$$

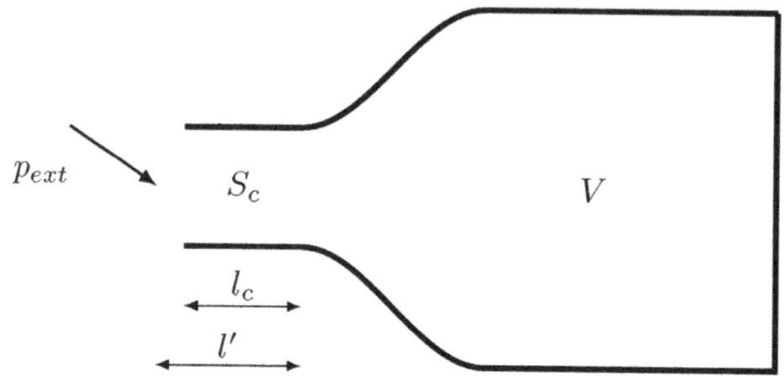

Figure 9.11. *Helmholtz resonator excited by an external pressure field. The main elements are the neck of length l_c and cross-section S_c and the volume V; l' is the effective length of the neck*

When integrated over the volume, the term on the right-hand side is equal to the incoming flow rate $Q'(t) = S_c u'$, and:

$$Q'(t) = \frac{V}{\rho_0 c_0^2} \frac{\partial p'}{\partial t} \qquad [9.14]$$

In the harmonic regime, pressure and flow rate are therefore related by:

$$\frac{p}{Q} = \frac{\rho_0 c_0^2}{\mathrm{i}\omega V} = \frac{1}{\mathrm{i}\omega C_A} \qquad [9.15]$$

where $C_A = V/\rho_0 c_0^2$ is called the acoustic compliance of the volume, the analog of an electrical capacitance.

In the neck and around the openings, the pressure can vary significantly with the position. Since the length of the neck is small relative to the wavelength, the difference in pressure $p_e - p_i$ between the two ends of the neck can be calculated by an approximate integration of Euler's equation along the neck to give:

$$p_e - p_i = \mathrm{i}\omega\rho_0 l_c u = \mathrm{i}\omega\rho_0 l_c \frac{Q}{S_c} = Z_s Q \qquad [9.16]$$

where the impedance Z_s is equal to $\mathrm{i}\omega\rho_0 l_c/S_c$. The neck of the resonator is therefore equivalent to an inertance $\rho_0 l_c/S_c$, the equivalent of an electrical inductance[7]. However, this formula must be adjusted to take into account the end corrections

7 In the electric-circuit analogy, pressure and volume flow rate are analogous to electrical voltage and current, respectively.

associated with the radiation impedance of the resonator opening, by replacing the physical length l_c with an effective length l'.

The whole resonator is therefore equivalent to a circuit whose impedance is the sum of the neck inertance and the volume compliance, and we obtain:

$$Z_{sR} = \mathrm{i}\omega\rho_0 \left(\frac{l'}{S_c} - \frac{c_0^2}{\omega^2 V} \right) \tag{9.17}$$

This impedance cancels out at the resonant frequency given by:

$$\omega_R = c_0 \sqrt{\frac{S_c}{l'V}} \tag{9.18}$$

This formula (which neglects the viscous and thermal losses of the fluid and those due to outward radiation) gives a very good estimate of the resonant frequency once the length correction can be reasonably estimated. For a thin cylindrical neck of radius a, the formula $l' = l + 0.61a$ can be used on the exterior side.

Helmholtz resonators are usually mounted as side branches on long pipes. In the following, we consider a main pipe (with section S) with an anechoic termination to which a resonator is transversely grafted. The continuity conditions of the pressure and the volume velocity at the junction lead to the calculation of the impedance (denoted Z_{sj}) of the resulting filter. This impedance is that of a parallel circuit combining the resonator impedance and the duct impedance. It is calculated as:

$$1/Z_{sj} = 1/Z_{sR} + S/Z_0 \tag{9.19}$$

The reflection and transmission coefficients can then be calculated from formula [4.10]. It gives:

$$r = -\frac{Z_0}{2SZ_{sR} + Z_0} \tag{9.20}$$

or in terms of the specific impedances:

$$r = -\frac{Z_0}{2Z_R S/S_c + Z_0} \tag{9.21}$$

The transmission loss is given by:

$$TL = 10 \log \left(1 - |r|^2 \right) \tag{9.22}$$

Figure 9.12 compares the transmission loss obtained using formula [9.22] with a finite element numerical simulation of a resonator consisting of a cylindrical neck with radius $a = 0.01$ m and length 0.05 m, the volume itself being cylindrical, with height 0.1 m and radius $R = 0.0545$ m. The theoretical resonant frequency is equal to 124.6 Hz.[8]. It can therefore be seen that a low value of the resonant frequency can be obtained with a very compact device compared to a quarter-wavelength resonator whose length would be about 0.7 m. With our somewhat arbitrary choice of the end corrections, the theoretical curve and the one resulting from the simulation practically overlap, except in the extreme vicinity of the resonance. This difference is due to the absence of any damping in the theoretical model, unlike the simulation which takes into account the (low) damping of the resonator due to radiation in the main pipe through the neck (a slight artificial dissipation is also induced by the discretization of the equations in the numerical simulation).

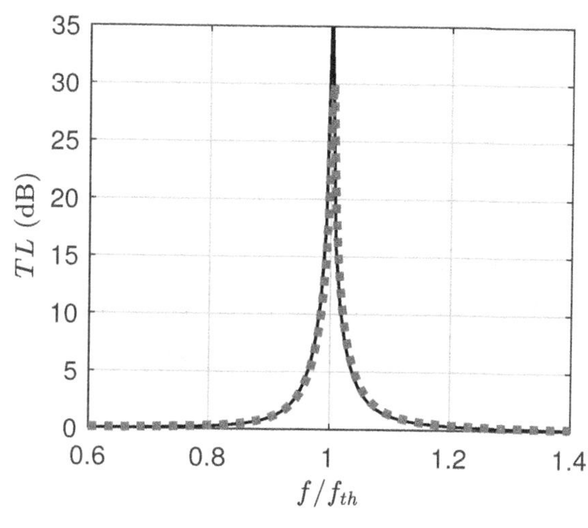

Figure 9.12. *Transmission loss due to a Helmholtz resonator connected to a duct with a square cross-section equal to 0.1×0.1 m^2. Comparison between the theoretical formulation [9.22] in black and the result of a finite element simulation in gray symbols. The frequency is normalized to the theoretical resonance frequency*

Although the agreement obtained here is excellent, it should be noted that larger deviations may occur depending on the geometry of the resonator, since the estimation of the effective neck length is often delicate (progressive flaring, wide neck, etc.). In any case, the theory applies correctly only when the resonator dimensions are much smaller than the wavelength. A rigorous calculation (except for the estimation of

8 The added length of the neck was estimated to be 0.82a on the junction side and 0.82(1-1.24a/R)a on the resonator volume side (see e.g. Chapter 9 of [ATA 15]).

the end corrections) is possible using the transfer matrix method when the resonator geometry is cylindrical; it is suggested as an exercise. In particular, it can be used to estimate the influence of the non-compactness of the resonator volume.

9.2. Webster horn equation

As we have just seen, sudden changes in the cross-section of a pipe induce upstream reflections and only a small fraction of the incident energy is transmitted when the area ratios are important. Therefore, if it is necessary to transmit a high level of sound in the external medium, a gradual evolution of the cross-section between a transducer and the duct opening should be achieved in order to "match" the impedances. It is then desirable to consider the propagation of acoustic waves in ducts of slowly varying cross-section, also called horns. We will see that a good approximation of the propagation in horns can be obtained by solving a second-order differential equation called the Webster horn equation.

9.2.1. *Propagation in ducts with a slowly varying cross-section*

We consider a cylindrical duct whose radius varies slowly and in which low-frequency waves propagate. The acoustic wavelength is assumed to be large with respect to the radius, and then only the fundamental (plane) mode needs to be considered. The geometry of the problem is given in Figure 9.13.

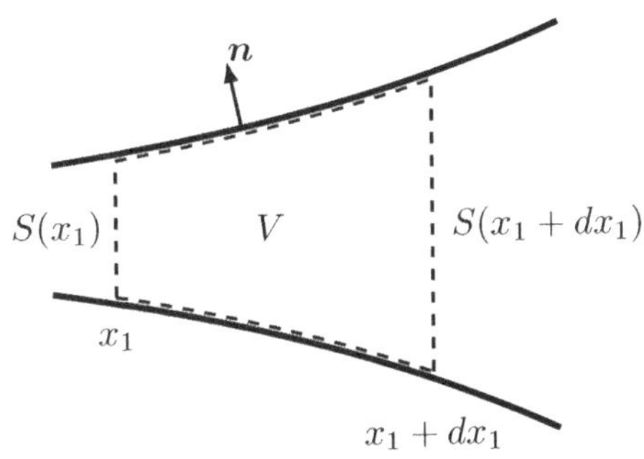

Figure 9.13. *Model for a duct of slowly varying cross-section leading to the development of Webster's equation*

The problem is governed by the Helmholtz equation with the condition of zero normal velocity at the walls. The duct is assumed to be semi-infinite, so that there are

no reflected waves. We integrate the Helmholtz equation over a volume bounded by the duct walls and by two sections S_1 and S_2 separated by a distance dx_1 to obtain:

$$\int_V (\nabla^2 p + k_0^2 p)\,\mathrm{d}V = \int_{S_2} \frac{\partial p}{\partial x_1}\,\mathrm{d}S - \int_{S_1} \frac{\partial p}{\partial x_1}\,\mathrm{d}S + k_0^2 \int_{x_1}^{x_1+dx_1} pS\,\mathrm{d}x_1 = 0 \quad [9.23]$$

where the condition of rigidity at the walls has been taken into account.

Assuming plane wave propagation, the pressure fluctuation depends only on x_1, $p = p(x_1)$, and by making dx_1 tend to zero, the following differential equation is obtained:

$$\frac{1}{S}\frac{d}{dx_1}\left(S\frac{dp}{dx_1}\right) + k_0^2 p = 0 \quad [9.24]$$

which is called the *Webster horn equation*.

Note however that in a pipe of variable cross-sectional area, the acoustic waves cannot be strictly plane. The condition of rigidity requires that the wavefront be normal to the walls and therefore not perpendicular to the axis of the duct. A condition for the validity of the plane wave approximation is that the local opening angle of the pipe be very small, which, for a circular duct with radius $R(x_1)$, can be expressed as $dR/dx_1 \ll 1$.

9.2.2. *Horn families: exponential horns*

For some particular axial variations of the horn circular cross-section, it is possible to obtain simple analytical solutions of Webster's equation [9.24]. It is easy to show that the variable $P = R(x_1)p$ is a solution of:

$$\frac{d^2P}{dx_1^2} + \left(k_0^2 - \frac{1}{R}\frac{d^2R}{dx_1^2}\right)P = 0 \quad [9.25]$$

The simplest condition for obtaining an analytical solution is to impose that the second term of this equation be a constant. Thus, we impose that the duct radius is the solution of the following differential equation:

$$\frac{d^2R}{dx_1^2} = m^2 R \quad [9.26]$$

where m is a constant.

Several families of horns can be obtained according to the values of m and of the slope of the pipe at $x_1 = 0$, $R'_0 = dR/dx_1(0)$.

The value $m = 0$ corresponds to the family of conical horns, $R(x_1) = R_0 + R'_0 x_1$. When $m \neq 0$, we get $R = R_0 \cosh(mx_1) + (R'_0/m)\sinh(mx_1)$. The condition $R'_0 = 0$ defines the catenoidal horns[9], $R(x_1) = R_0 \cosh(mx_1)$. When $R'_0 = m$, we get the family of exponential horns $R(x_1) = R_0 \exp(mx_1)$, which we will study in more detail.

For semi-infinite exponential horns of increasing cross-section area ($m > 0$), the solution of [9.25] leads to:

$$p(x_1) = p_0 \exp(-mx_1)\exp(-ik_0 x_1 \sqrt{1-(m/k_0)^2}) \qquad [9.27]$$

since there are no reflected waves.

The pressure along the duct therefore takes the form of a locally plane wave, whose amplitude decreases as the horn cross-section increases. Two regimes are possible depending on the value of m (called the *flare constant*) relative to the wavenumber k_0. The acoustic waves are propagating when $k_0 > m$, but evanescent at low frequencies when $k_0 < m$.

The transmitted power is zero when $k_0 < m$, and otherwise it is expressed as:

$$W = \frac{|p_0|^2 S_0}{2Z_0}\sqrt{1-(m/k_0)^2};\ k_0 > m \qquad [9.28]$$

where S_0 is the area of the duct at $x_1 = 0$.

An exponential horn thus behaves like a *dispersive* high-pass filter, as shown by the phase and group velocity expressions for $k_0 > m$:

$$v_\varphi = \frac{c_0}{\sqrt{1-(m/k_0)^2}} \qquad [9.29]$$

$$v_g = c_0\sqrt{1-(m/k_0)^2} \qquad [9.30]$$

Below the horn cutoff frequency, $f_c = mc_0/2\pi$, the group velocity is zero, and the phase velocity is infinite. Above the cutoff frequency, the propagation is dispersive, since the phase and group velocities depend on the frequency. The evolution with the

9 Catenoidal horns are often used to give a circular pipe a nearly anechoic termination, as they allow the horn and the pipe to be connected with a continuous slope.

ratio k_0/m of these phase and group velocities, normalized by the speed of sound c_0 is given in Figure 9.14. A broadband signal emitted at $x_1 = 0$ will be distorted as it propagates through the horn; this distortion is particularly pronounced at frequencies near the cutoff frequency, where the phase and group velocities vary rapidly with the frequency.

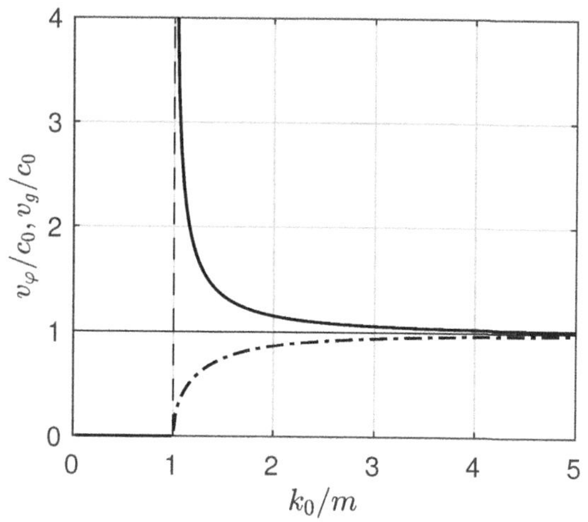

Figure 9.14. *Evolution of the phase (solid line) and group (dotted dashed line) velocities with k_0/m in an exponential horn. Velocities have been normalized by c_0*

The transmission factor of a horn is defined as the ratio of the power it carries to that transmitted in a duct with a constant cross-section S_0 equal to the inlet cross-section of the horn (it is also the real part of the reduced impedance of the source); a unit value corresponds to ideal power transmission. The shape of the horn affects the efficiency of the transmission, which is shown in the following figures comparing exponential and conical horns. Two horns of the same length and the same inlet and outlet cross-sections were considered and their geometries are specified in Figure 9.15. The transmission factors of the two horns are compared in Figure 9.16. It can be seen that, unlike the exponential horn, the conical horn is capable of propagating low frequencies, since it has no cutoff frequency. However, the efficiency of the former is rapidly higher as soon as its cutoff frequency is exceeded.

The results obtained above assume that there are no reflected waves when the horn opens into the surrounding medium. They have the advantage of highlighting the main characteristics associated with each family of horns. However, the finite length of a horn significantly alters its response, in particular due to the existence of resonances

that can be very acute for short lengths and low frequencies, and therefore associated with a high reflection coefficient at the horn outlet.

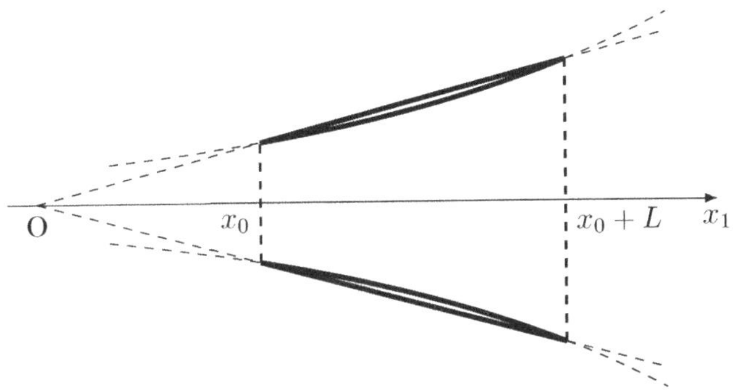

Figure 9.15. *Geometry of the exponential and conical horns used for comparing their efficiency in transmitting acoustic waves*

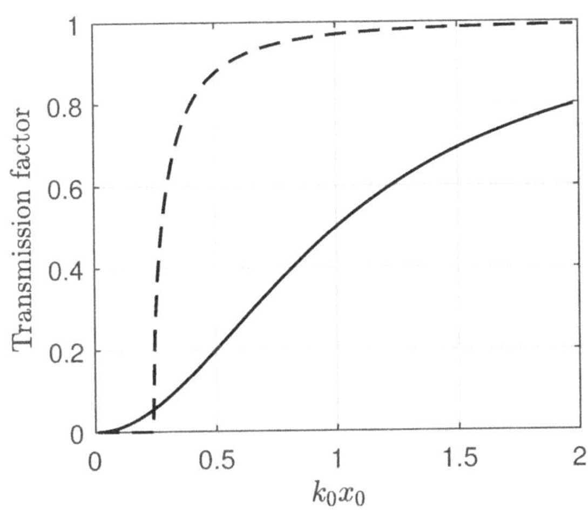

Figure 9.16. *Transmission factors of an exponential horn (dashed curve) and a conical horn (solid line) of the same length and the same inlet and outlet cross-sections. The distance from the first cross-section to the apex of the conical horn is denoted by x_0 and the length L of the horns is equal to 10 x_0*

As an example, Figure 9.17 shows the real part of the radiation impedance of a finite-length, baffled, exponential horn. The output impedance was estimated to be that of a piston vibrating in an infinite rigid baffle and this impedance was transferred to the source via an impedance transfer formula. Note that a weak propagation exists

even below the cutoff frequency of the infinite horn. Note also that the oscillations gradually dampen when the frequency increases, as the impedance of the aperture tends to that of the surrounding medium and the reflection coefficient tends to zero.

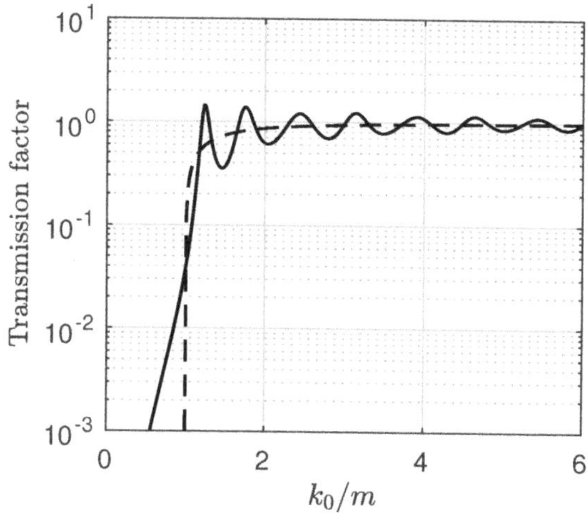

Figure 9.17. *Real part of the reduced radiation impedance of a finite length exponential horn (mL =4, L/r_0 =50, solid line), compared to that of the infinite horn (dashed line)*

9.3. Exercises

1) **Open side-branch**

Consider a Y-bifurcation where the bifurcating pipe of length l_b is open to the outside. Denote the ratio of the cross-sections between the main and the secondary ducts by m and determine the transmission loss factor. Calculate the result using the data in Figure 9.8. Evaluate the impact of the end correction.

2) **Cylindrical Helmholtz resonator**

A more accurate calculation of the resonant frequency of a Helmholtz resonator can be obtained if its geometry is simple. Consider a resonator consisting of two connected circular cylinders. The radius of the neck is denoted by r and its length by l; the radius of the rigid cavity is denoted by R and its length by L.

a) Use the impedance transfer technique to calculate the impedance at the junction between the two cylinders. Determine the value of the total resonator impedance by

considering that the neck is compact; use formula [9.16] taking into account the usual values of the added lengths.

b) Determine the resonance frequencies of the resonator. Perform a Taylor series expansion with respect to k_0L of the cotangent appearing in the exact implicit formula and verify that, up to the first order, one indeed finds the classical formula [9.18]. Show, from the expansion of the cotangent, that the classical value provides a good approximation as long as $L < \lambda/16$ approximately, which corresponds to a very compact resonator. Evaluate the validity of the approximation obtained using the second term of the Taylor series expansion.

c) Canceling out the impedance also causes higher order resonant frequencies to appear; they are associated with "depth modes". Determine the first values of these frequencies and interpret them.

Numerical application : $r = 0.3$, $R = 18$, $l = 0.3$, $L = 18$ (dimensions in cm).

3) **Conical horn**

a) Show that the propagation in a conical horn is not dispersive and that there is no cutoff frequency.

b) Calculate the transmitted power and compare it to that of the exponential horn. Compare their efficiency by considering two horns with the same length and the same inlet and outlet cross-sections (which imposes m when the cone apex angle is given) and excited by a piston with the same vibration velocity.

4) **Finite-length exponential horn**

a) Establish the impedance transfer formula for a finite exponential horn (assuming the terminal impedance is that of a piston vibrating in an infinite baffle).

b) Reproduce the curves in Figure 9.17 and study the influence of the parameters mL and L/r_0.

10

Acoustics of Enclosures: Room Acoustics

In many practical situations, the determination of the acoustic field created in enclosed spaces is of particular importance. An enclosed space is defined as a volume bounded by walls on which well-defined boundary conditions are imposed. The dimensions of these volumes can vary widely, from millimeter cavities or even smaller for microelectromechanical systems (MEMS) to auditoriums or religious buildings of thousands of m^3. The geometry of these spaces can be relatively regular (quasi-rectangular classrooms, for example) or, on the contrary, very complex and cluttered with structures exhibiting absorbing properties (car interiors, airplane cabins and auditoriums, for example). Several theoretical approaches can be used to estimate the sound field in enclosed spaces. They are often chosen according to the size of the volumes to be considered in relation to the wavelengths of the signals of interest, but also according to the nature of the boundary conditions or the purpose of the study: Do we want to estimate only a stationary sound pressure level or rather the time evolution of a signal as in musical applications?

A rigorous theoretical approach is possible in the form of series expansions of modes, which can be extended by finite element simulations when geometries or boundary conditions are too complicated to allow for analytical computation. In practice, however, this approach is limited to small volumes and steady-state analysis for reasons of both computational complexity and physical relevance. For large volumes, and especially for the study of room acoustics, a more global, less rigorous approach is preferred because it is able to account for complex geometries and boundary conditions.

In this chapter, we will successively describe these two approaches, which prove to be very complementary, and we will illustrate these theoretical aspects with some examples of experimental results and numerical simulations. In the last part of the chapter, we propose an introduction to the consideration of subjective effects and

related indicators, which are essential for the characterization of the acoustic quality of rooms.

10.1. Simple-shaped cavities

Consider the rectangular cavity shown in Figure 10.1. An omnidirectional source emits waves that are received by a sensor either directly or after reflection from the walls. Before addressing the problem mathematically, it proves interesting to develop a physical description using the notions of image sources and "paths", or "rays".

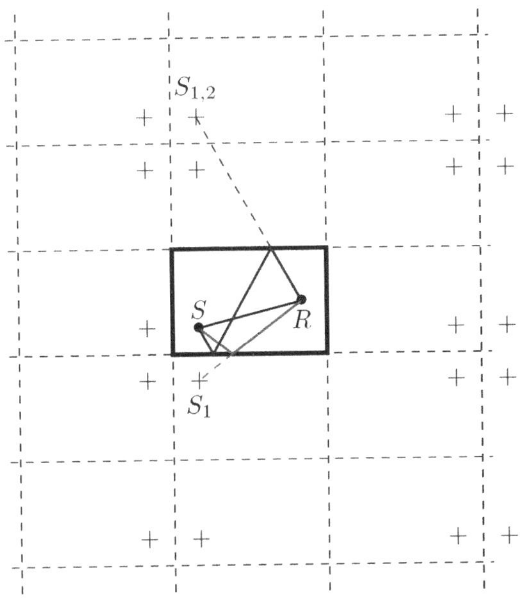

Figure 10.1. *Acoustic field in a rectangular room. Direct path, image sources, and examples of paths that have undergone one reflection or two reflections, and are associated with 1st- and 2nd-order image sources, denoted S_1 and $S_{1,2}$, respectively*

For this very particular geometry, the field associated with the reflected paths connecting the source and the receiver (called the reverberant field) can be considered as originating from an infinite number of image sources symmetrical to the actual source with respect to the walls. If the walls are perfectly reflecting, the contribution of each image decreases simply as the inverse of the distance between the image under consideration and the receiver; if the walls are absorbing, a fraction of energy is lost with each reflection and the contribution of the image sources therefore decreases more rapidly.

This visualization in terms of image sources or paths is very interesting since it allows for an interpretation of the evolution of the received signal as a function of time. When a pulse is emitted by the source, the receiver first detects a direct signal with a travel time related to its distance from the source, then a series of echoes associated with successive reflections on the walls, whose amplitude decreases with time. If the emitted signal is harmonic, the field results from the interference between the direct field and the series of reflected waves. This interference gives rise to a spatial structuring in the form of standing waves or modes, similar to what we saw in the study of the acoustic propagation in ducts in Chapter 8.

10.2. Modal approach

Modal theory is based on the solution of the Helmholtz equation associated with impedance boundary conditions. It is an analytical and rigorous approach whose advantages and limitations can be understood when dealing with simple geometries; for more complicated geometries, a numerical solution, mainly by the finite element method, will be necessary.

In the following, we consider a rectangular cavity with sides L_1, L_2, L_3, whose walls are perfectly rigid (or perfectly "reflecting"). The free oscillations of the cavity (without any sources) are first studied in order to characterize its normal modes. Given the geometry, it makes sense to look for the pressure field as a product of functions depending separately on the coordinates, as $p(x_1, x_2, x_3) = X_1(x_1)X_2(x_2)X_3(x_3)$. The Helmholtz equation can then be written as:

$$\frac{X_1''}{X_1} + \frac{X_2''}{X_2} + \frac{X_3''}{X_3} + k_0^2 = 0 \qquad [10.1]$$

with boundary conditions

$$X_i' = 0 \text{ for } x_i = 0 \text{ and } x_i = L_i \; ; \; i = 1, 2, 3 \qquad [10.2]$$

where X_i' and X_i'' denote the first and second derivatives of the associated functions.

A first separation of variables consists of setting:

$$\frac{X_1''}{X_1} = -\left(\frac{X_2''}{X_2} + \frac{X_3''}{X_3} + k_0^2\right) = -k_1^2 \qquad [10.3]$$

where we anticipated the need for a negative separation constant in order to satisfy the boundary conditions. It immediately follows that $X_1(x_1) = A\exp(-\mathrm{i}k_1x_1) + B\exp(+\mathrm{i}k_1x_1)$. The rigidity condition written at $x_1 = 0$ gives $A = B$. The condition

at $x_1 = L_1$ writes $A\sin(k_1 L_1) = 0$, which imposes a set of discrete values for the separation variable according to $k_1 = m\pi/L_1$ where m is a positive or zero integer; we are therefore faced with an eigenvalue problem. The constant A remains undetermined for this homogeneous problem.

The same separation process can be applied in the other two directions, yielding the pressure field in the form of a (m, n, p) cavity mode:

$$p(x_1, x_2, x_3) = C\cos(m\pi x_1/L_1)\cos(n\pi x_2/L_2)\cos(p\pi x_3/L_3) \qquad [10.4]$$

where C is an arbitrary constant.

The three separation constants cannot be chosen arbitrarily, since they must satisfy the dispersion relation $k_{mnp}^2 = k_1^2 + k_2^2 + k_3^2 = k_0^2$, which is expressed by:

$$k_{mnp}^2 = k_0^2 = \pi^2\left(\left(\frac{m}{L_1}\right)^2 + \left(\frac{n}{L_2}\right)^2 + \left(\frac{p}{L_3}\right)^2\right) \qquad [10.5]$$

The square of the wavenumber is discretized and appears as an eigenvalue of the Helmholtz equation associated with the rigidity condition at the walls. The corresponding frequencies ($f_{mnp} = c_0 k_{mnp}/2\pi$) are the eigenfrequencies (or natural frequencies) of the cavity.

The cavity modes are orthogonal and are defined by the following relationship[1]:

$$\int_V \Psi_{mnp}(\boldsymbol{x})\Psi_{m'n'p'}(\boldsymbol{x})dV = V\delta_{mm'}\delta_{nn'}\delta_{pp'} \qquad [10.6]$$

where V is the volume of the cavity.

The modes form a complete set that allows any regular function to be represented by a linear combination of them. These properties remain valid for a cavity of arbitrary geometry and, with some nuances, with impedance boundary conditions.

In a rectangular room, a (m, n, p) mode is characterized by the existence of m nodal lines in the x_1 direction and n and p, respectively, of these lines in the other directions. As an example, Figure 10.2 shows the pressure field corresponding to mode $(2, 1, 3)$.

1 Ψ_{mnp} is normalized to have a mean square volume average of 1, so that C is now imposed in [10.4].

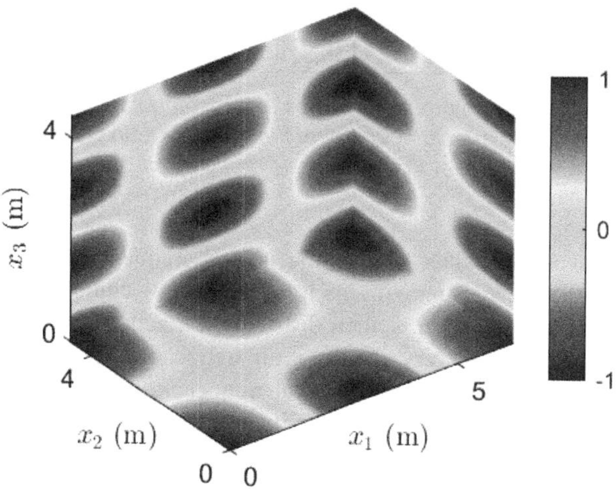

Figure 10.2. *Mode (2,1,3) of a rectangular room which size is close to that of the reverberation chamber of École Centrale de Lyon (L_1 = 6.3 m, L_2 = 4.6 m, L_3 = 4.45 m; V = 129 m^3). For a color version of this figure, see www.iste.co.uk/juve/acoustics1.zip*

The response of a rectangular cavity to a harmonic point source located at $\boldsymbol{x}_0$ is obtained by expanding the Green's function into the series of modes and determining the associated coefficients using formula [10.6]. The following result is obtained:

$$G(\boldsymbol{x}, \boldsymbol{x}_0, \omega) = \frac{1}{V} \sum_{m,n,p} \frac{\Psi_{mnp}(\boldsymbol{x}, \omega)\Psi_{mnp}(\boldsymbol{x}_0, \omega)}{k_{mnp}^2 - k_0^2} \qquad [10.7]$$

The eigenfunctions Ψ_{mnp} are given by formula [10.4] when the walls are perfectly rigid. If each of the walls is characterized by a uniform impedance, formula [10.7] remains valid, but the eigenvalues and eigenfunctions become complex and it is necessary to specify the value of the frequency in expression [10.7] because these quantities depend on it through the value of the impedance of the walls. When the walls are only weakly absorbing, the room response becomes very strong when the source frequency is equal to one of its resonance frequencies (the real part of the natural frequencies).

To illustrate these notions, we consider the reverberation chamber of École Centrale de Lyon in order to evaluate the associated natural frequencies and, in a second step, the acoustic response of the room. This room is nearly rectangular and the walls are very reflecting. As a first approximation, formula [10.4] with the corresponding natural frequencies can be used.

Mode	1,0,0	0,1,0	0,0,1	1,1,0	1,0,1	0,1,1	2,0,0	1,1,1
frequency (Hz)	27.2	37.3	38.5	46.2	47.2	53.6	54.4	60.1

Table 10.1. *First natural frequencies of a rectangular cavity, whose dimensions are close to that of the reverberation chamber of École Centrale de Lyon*

Table 10.1 gives the values of the first natural frequencies of this room, and Figure 10.3 shows the evolution with the frequency of the cumulative number $N(f)$ of these natural frequencies, that is, their distribution function. It can be seen that the number of modes increases very rapidly with frequency; as an example, we can observe the existence of about 1,000 natural frequencies with a value of less than 400 Hz. The stepped structure visible in the inset of the figure (the number of modes increases by one unit each time the frequency value becomes equal to one of the natural frequencies of the room) quickly transforms into an almost continuous shape. This continuous limit is evaluated in the following paragraph.

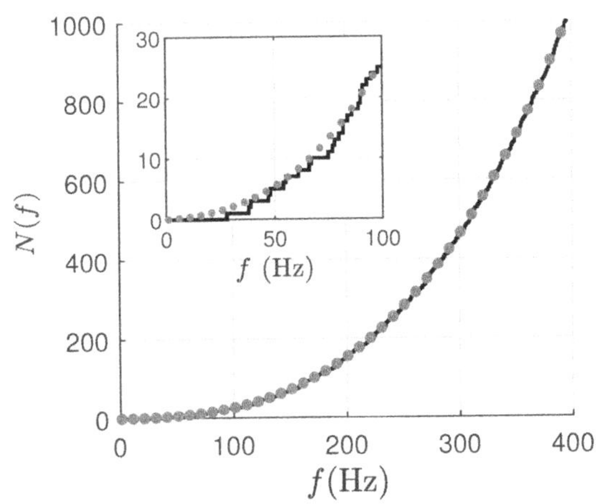

Figure 10.3. *Distribution function of the eigenfunctions in a rectangular room. The staircase form of the plot is the result of the exact formula of [10.5]. The grey symbols correspond to the continuous approximation given by formula [10.9]*

10.2.1. *Distribution of the natural frequencies*

To count the number of natural frequencies below a chosen value of the frequency f, it proves useful to consider the wavenumber space or spectral space (Figure 10.4). In this space, a mode is represented by a point with coordinates (k_1, k_2, k_3).

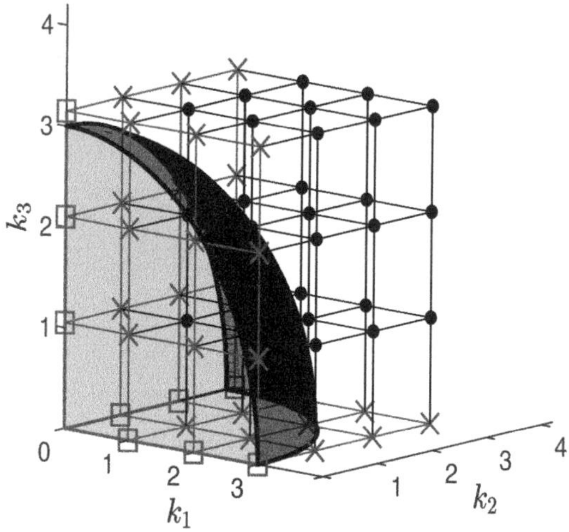

Figure 10.4. *Representation of the modes of a rectangular room in wavenumber, or spectral, space. The different types of modes are depicted in dots (volume modes), crosses (tangential modes) and squares (axial modes). The modes with a frequency lower than $f = \omega/2\pi$ are counted in one octant of a sphere with radius $k_0 = \omega/c_0$. For a color version of this figure, see www.iste.co.uk/juve/acoustics1.zip*

If none of the associated indices is zero, it occupies a "volume" $\Delta k_1.\Delta k_2.\Delta k_3 = \pi^3/(L_1 L_2 L_3) = \pi^3/V$ in the spectral space. Counting the number of natural frequencies is equivalent to counting the number of elementary volumes contained in a sphere of radius k_0, or more precisely in an octant (or eighth) of a sphere. If the frequency is high enough, this number is well approximated by the ratio of the volume of the octant to the volume occupied by a mode, that is:

$$N(f) = \frac{k_0^3 \pi/6}{\pi^3/V} = \frac{4\pi V}{3c_0^3} f^3 \qquad [10.8]$$

The evolution of $N(f)$ with the third power of the frequency is a clear indication of how fast the distribution function grows with frequency.

A finer approximation can be obtained by taking into account the modes with one zero index (tangential modes) or two zero indices (axial modes). We provide hereafter the resulting formula for the number of natural frequencies in which S denotes the total area of the walls and L is the total length of the edges.

$$N_{rect}(f) = \frac{4\pi V}{3c_0^3} f^3 + \frac{\pi S}{4c_0^2} f^2 + \frac{L}{8c_0} f \qquad [10.9]$$

If the geometry of the cavity is arbitrary, an estimate of the number of eigenfrequencies as accurate as formula [10.9] cannot be obtained. However, it can be shown that the approximation [10.8] remains asymptotically valid.

From this expression, the modal density of the room can be computed as the derivative of the distribution function $N(f)$, or its inverse, which is the average distance between two eigenfrequencies $(\Delta f)_m$:

$$\frac{dN}{df} = \frac{4\pi V}{c_0^3} f^2 \qquad [10.10]$$

$$(\Delta f)_m = \frac{c_0^3}{4\pi V f^2} \qquad [10.11]$$

For example, for the room considered above and for a frequency of 400 Hz, the modal density is equal to 6.4 and the average difference between eigenfrequencies is close to $(\Delta f)_m = 0.16$ Hz.

In the following sections, it will be seen that the eigenfrequencies are associated with the presence of resonance peaks in the frequency response of the room. The height and the width of these peaks are influenced by the absorption of the walls of the room. Thus, it is clear that when the difference between the eigenfrequencies $(\Delta f)_m$ becomes less than the width at half height of the resonances, the shape of the room response will appear to be continuous and a discrete modal description, even an accurate one, will not be the most appropriate.

10.2.2. *Room acoustic response to a point source*

The frequency response of the rectangular room representing the reverberation room of École Centrale de Lyon is shown in Figure 10.5. It was calculated using the Green's function given by equation [10.7] and an omnidirectional point source. The frequency range is limited to the 0–100 Hz range in order to emphasize the influence of the first modes of the room. A small damping, proportional to the frequency, was introduced by adding an imaginary part to the squared wavenumbers of the room with rigid walls[2]; this damping is equivalent to that which would be provided by a small value of the random incidence absorption coefficient of the walls α_m (see section 10.5). On the other hand, the modal decomposition of the room with perfectly rigid walls was preserved, which is acceptable when the absorption at the wall is low.

The choice of source and receiver positions affects the frequency response, but it is fairly easy to relate the natural frequencies in Table 10.1 to the first peaks in Figure 10.5.

2 In the modal summation, k_{mnp}^2 was replaced by $k_{mnp}^2 + \mathrm{i}k_0\alpha_m S/4V$; see, for example [PIE 89], Chapter 6.

While the low frequency region clearly shows the first eigenfrequencies, the response at the higher frequencies becomes very dense and difficult to interpret in terms of modes. A response averaged over a bandwidth of a few hertz is clearer and closer to what measurements can provide. The "continuous" shape of the response is the subject of the energy description presented in the next section.

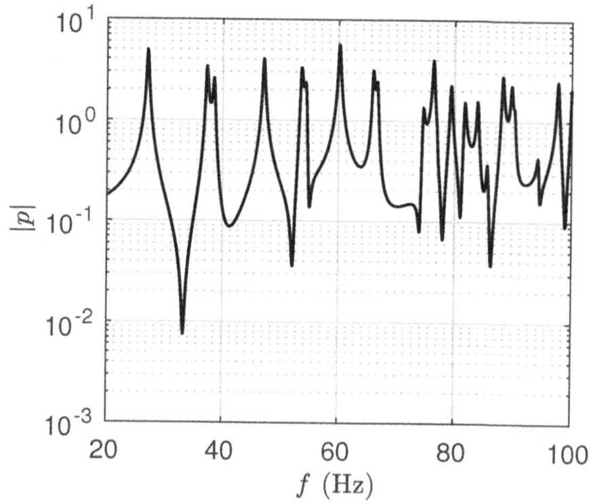

Figure 10.5. *Frequency response of a rectangular room excited by a point source. Source coordinates (0.5 m, 4.2 m, 0.5 m) and receiver coordinates (6 m, 0.5 m, 1 m). For this modal computation, a very small damping proportional to the frequency was introduced; it corresponds to a random incidence absorption coefficient α_m equal to 0.02*

10.3. Energy approach: Sabine's theory

In contrast to the local, modal approach described above, the theory initiated by W. C. Sabine at the beginning of the 20th century consists of evaluating the energy transfer in a room in a global way. The underlying main idea is that if the room is sufficiently "reverberant" (walls only weakly absorbant), the waves fill the volume more or less uniformly. Obviously, this approach is not relevant at low frequencies, where a very limited number of modes are excited, inducing significant level variations between pressure minima and maxima. It is then a "high frequency" approach for which the description in terms of direct and reflected paths or rays is also relevant. An empirical criterion for the validity of the energetic theory, which is often used in practice, is that the wavelength must be less than one-third of the characteristic dimension of the room. For the room previously studied, this corresponds to frequencies greater than 200 Hz. A more precise criterion will be given in section 10.6 (Schröder frequency).

10.3.1. *Global energy balance*

The principle of conservation of energy states that the variation over time of the energy contained in the volume of the room is equal to the difference between the energy supplied by the sources and the energy lost by absorption at the walls. This energy can be associated with the total pressure signal or to the pressure filtered in a frequency band of width Δf. In practice, proportional octave or third-octave filters are often employed. The consideration of fairly large filters actually allows for spatial homogenization of the response, as long as they contain a large number of eigenfrequencies.

The energy balance is written as:

$$\frac{dE}{dt} = \frac{d}{dt}\int_V e_m \, \mathrm{d}V = W_S - W_{abs} \qquad [10.12]$$

where e_m is the average energy density in the considered frequency range, W_S is the power supplied by the sources and W_{abs} is the power dissipated by absorption at the walls[3], here again in the considered frequency range. Formula [10.12] is valid at any time whenever the powers are also evaluated at the same time. It can also be used when considering a moving time average of the different quantities.

The simplicity of formula [10.12] should not, however, hide the fact that the evaluation of energy density and power dissipation is challenging. These quantities are generally a function of both pressure and velocity fluctuations, whereas it is desirable to obtain a formulation in which only the pressure, which is easily measurable, is present.

10.3.2. *Diffuse field*

The energy theory of room acoustics is based on two main assumptions: the already mentioned homogeneity of the pressure field, and its isotropy, that is, the absence of a preferred direction of energy propagation. In room acoustics, this is called the *diffuse field* hypothesis. None of these assumptions can be rigorously verified: in the close vicinity of the source the direct field dominates and the field is therefore neither homogeneous nor isotropic; in the vicinity of absorbing walls, there is an energy flow normal to them, which again does not respect the conditions of validity of the diffuse field assumption. However, if the room is sufficiently reverberant, the diffuse field assumption is acceptable for most of the volume, which we will admit in the following.

---

3 The atmospheric absorption in the room volume must sometimes also be taken into account, as we will see in section 10.4.

If the frequency is high enough so that the number of modes present in the frequency range under consideration is large, the reverberant field can be represented as the superposition of a large number of plane waves (a mode can be associated with a pair of plane waves as we have already pointed out in Chapter 8).

These plane waves propagate in all directions, with variable phases (see Figure 10.6) and the pressure at a point can then be described by the following integral:

$$p(\boldsymbol{x}) = \int_0^{2\pi} \mathrm{d}\varphi \int_0^{\pi} A(\varphi, \theta) \exp(-\mathrm{i}\boldsymbol{k} \cdot \boldsymbol{x}) \sin\theta \, \mathrm{d}\theta \qquad [10.13]$$

where $A(\theta, \varphi)$ denotes the amplitude of each plane wave. The plane waves can be considered as independent and the mean value of the energy density is then the sum of the energy densities associated with each of them. It can be expressed as a function of the mean square value of the pressure fluctuations:

$$e_m = \frac{p_m^2}{\rho_0 c_0^2} = \frac{1}{\rho_0 c_0^2} \int_0^{2\pi} \mathrm{d}\varphi \int_0^{\pi} \frac{|A|^2}{2} \sin\theta \, \mathrm{d}\theta \qquad [10.14]$$

We now calculate the power absorbed at the walls. Per unit area, this power is proportional to the intensity normal to the wall. For a single plane wave, we have $I_m = p_m^2 \cos\theta / \rho_0 c_0$ (see Figure 10.6). The total intensity results from the integration of this expression over all possible angles (θ, φ) and:

$$I_m = \frac{1}{\rho_0 c_0} \int_0^{2\pi} \mathrm{d}\varphi \int_0^{\pi/2} \frac{|A|^2}{2} \cos\theta \sin\theta \, \mathrm{d}\theta = \frac{p_m^2}{4\rho_0 c_0} \qquad [10.15]$$

from which it is deduced that $I_m = e_m c_0 / 4$.

The absorbed intensity is proportional to the incident intensity, and the coefficient of proportionality is called the random incidence absorption coefficient α_m to emphasize that it is an average of the coefficients associated with each plane wave (whose angle of incidence is assumed to be arbitrarily or "randomly" distributed). This coefficient is expressed as the weighted average of the absorption coefficient of each plane wave $\alpha(\theta)$:

$$\alpha_m = \frac{1}{\pi} \int_0^{2\pi} \mathrm{d}\varphi \int_0^{\pi/2} \alpha(\theta) \cos\theta \sin\theta \, \mathrm{d}\theta = 2 \int_0^{\pi/2} \alpha(\theta) \cos\theta \sin\theta \, \mathrm{d}\theta \qquad [10.16]$$

an expression called the Paris formula[4].

4 It is verified that this average coefficient is equal to α if the latter is independent of θ.

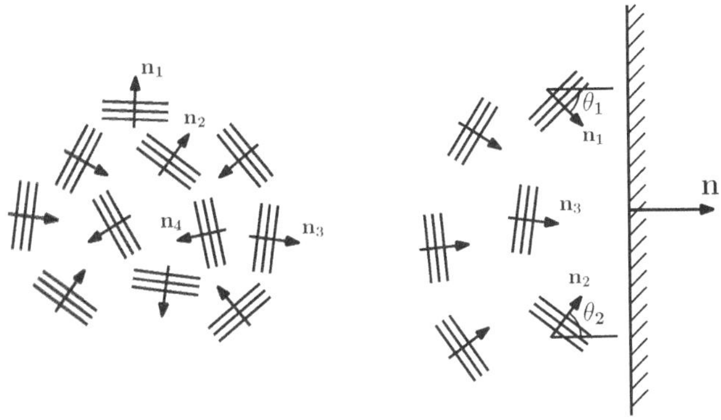

Figure 10.6. *Representation of the reverberant field by superposition of uncorrelated plane waves. On the right, illustration of the waves incident on a wall contributing to the room absorption*

If the absorption is the same on all the walls, then the conservation of energy can then be written as:

$$V\frac{de_m}{dt} = W_S - \frac{e_m c_0}{4}\alpha_m S = W_S - a\frac{e_m c_0}{4} \qquad [10.17]$$

The quantity $a = \alpha_m S$ is called the equivalent absorption area of the room. It corresponds to the area of a perfectly absorbing material that would dissipate the same energy as all the walls of the room combined.

If the walls of the room are made of different materials, the equivalent absorption area is defined by:

$$a = \frac{1}{S}\sum_i \alpha_{m_i} S_i;\ S = \sum_i S_i \qquad [10.18]$$

where the sum includes all the materials covering the walls of the room, with an area S_i and an absorption coefficient α_{m_i}. Equation [10.17] is known as the Sabine equation; it is also valid for the intensity or the mean square pressure, with a simple adjustment of the numerical coefficients.

10.3.3. *Steady-state level: reverberation time*

Equation [10.17] is a first-order differential equation that is integrated according to:

$$Ve(t) = \exp\left(-\frac{ac_0}{4V}t\right)\int_0^t \exp\left(\frac{ac_0}{4V}\tau\right)W_S(\tau)\,\mathrm{d}\tau \qquad [10.19]$$

where it is assumed that the sources are turned on at time $t = 0$. This formula emphasizes a characteristic time $t_c = 4V/ac_0$ which depends on the geometry and absorption properties of the room. The signal received at a given time t essentially depends on the power emitted at times between $(t - t_c)$ and t, due to the presence of the exponential in the integral.

If the emitted power is constant or barely changes on the scale of t_c, the sound level in the room is constant and is given by:

$$e_{stat} = \frac{4W_S}{ac_0} \quad \text{or} \quad p^2_{stat} = \frac{4\rho_0 c_0 W_S}{a} \qquad [10.20]$$

The sound level is therefore proportional to the emitted power and inversely proportional to the absorption introduced by the walls, which makes sense. This formula can be used to determine the sound level caused by a source of known power in a room with known absorption characteristics. Conversely, it can be used to determine the power of an unknown source by measuring the sound level, if the equivalent absorption area of the room is known.

On the other hand, a transient situation can be considered in which the emission from the sources is stopped at time $t = 0$, after reaching the steady-state level. The energy density then decreases exponentially with time and is written as:

$$e_m(t) = e_{stat} \exp(-\frac{ac_0}{4V}t) \qquad [10.21]$$

If we look at the noise level in decibels, it follows that:

$$10 \log(\frac{p^2}{p^2_{stat}}) = -1.09\frac{ac_0}{V}t \qquad [10.22]$$

The reverberation time[5] is defined as the value T_{60} at which the level in the room has decreased by 60 dB, which using the standard value of the speed of sound at 20°C, $c_0 = 343$ ms^{-1}, can be obtained as:

$$T_{60} = \frac{0.161V}{a} \qquad [10.23]$$

Also known as Sabine's formula, this is the most famous and most widely used formula in room acoustics. In practice, it is difficult to achieve a 60 dB decrease in the sound level in a room without the measurement being disturbed by background noise.

5 The reverberation time is sometimes denoted *RT*.

Therefore, the reverberation time is often calculated by observing the sound level decay over a reduced range of 20 or 30 dB, and the result is linearly extrapolated to a range of 60 dB. The corresponding notation, T_{20} or T_{30}, should not be misunderstood.

Although the reverberation time is the simplest and most widely used indicator in room acoustics, we should not limit ourselves to this single global value. For example, the reverberation time does not take into account the evolution of the signal in the first instants of decay, and this evolution is crucial for defining the acoustic quality of a room. Other indicators related to subjective aspects are defined in section 10.11.

10.3.4. *Eyring's formula*

The application of Sabine's formula proves difficult for high values of the absorption at the walls. An extreme case is that of a room that is supposed to be perfectly anechoic ($\alpha_m = 1$ on all walls), for which formula [10.23] predicts a finite reverberation time equal to $T_{60} = 0.161V/S$, when it should be zero. In practice, using this formula to determine the absorption coefficient of materials results in values that are too high, and sometimes even (erroneously) higher than unity.

To overcome this difficulty, the following reasoning can be adopted, considering that the absorption process at the walls is discontinuous and occurs during each reflection. In a diffuse field, it can be shown that the mean time between two successive reflections from the walls is $t_m = 4V/Sc_0$ (see, for example, Chapter 4 of [KUT 00]). During the first reflection the energy density is e_{stat}, but during the second reflection it is only $e_{stat}(1 - \alpha_m)$. After the Nth-order reflection, thereby at time Nt_m, the available energy is $e(Nt_m) = e_{stat}(1 - \alpha_m)^N$. This reasoning implies a stepwise decrease in the sound level. A good approximation is to consider this approach as valid at all times, by writing $e(t) = e_{stat}(1 - \alpha_m)^{t/t_m}$, where t is a continuous variable. The energy density at time t is then given by $e(t) = e_{stat}\exp(-t/T_E)$, where:

$$T_E = \frac{4V}{-c_0 S \ln(1 - \alpha_m)} \qquad [10.24]$$

The reverberation time at -60 dB is then defined by:

$$T_{60} = \frac{0.161V}{-S\ln(1 - \alpha_m)} \qquad [10.25]$$

It can be verified that Eyring's formula [10.25] is equivalent to Sabine's formula when $\alpha_m \ll 1$, because $-\ln(1 - \alpha_m) \approx \alpha_m$. The difference between the two expressions becomes noticeable for $\alpha_m > 0.3$. It can also be seen that Eyring's formula predicts a zero reverberation time for an anechoic room, as it should.

10.4. Influence of the atmospheric absorption

During the propagation of sound in the air, the level of the acoustic waves undergoes a decrease caused by complex mechanisms related to viscosity and thermal conduction, as well as relaxation phenomena of nitrogen and oxygen molecules, that strongly depend on the presence of water vapor. The sound intensity decreases exponentially with the propagation distance x between the source and the receiver, $I = I_0 \exp(-mx)$, or by introducing the propagation time $t = x/c_0$, $I = I_0 \exp(-mc_0 t)$. The parameter m is defined by $m = 2\alpha_{air}$, where α_{air} is the atmospheric absorption coefficient. The value of the atmospheric absorption coefficient as a function of temperature and water vapor concentration is given by international standards based on the work described in [BAS 95].

This atmospheric absorption is added to that of the walls of the room and causes a reduction in the reverberation time compared to that which can be estimated when no volume absorption occurs. To take this effect into account, the Sabine and Eyring formulas should be replaced by the following:

$$T_{60} = \frac{0.161V}{a + 4mV} \qquad [10.26]$$

and

$$T_{60} = \frac{0.161V}{-S\ln(1-\alpha_m) + 4mV} \qquad [10.27]$$

The value of the atmospheric absorption coefficient α_{air} is highly dependent on the temperature and humidity conditions that must be recorded during the reverberation time measurements in order to correctly estimate the equivalent absorption coefficient α_m of the test room walls. Furthermore, normative measurements also require that precise temperature and humidity ranges be considered when characterizing the absorption of acoustic materials.

10.5. Random incidence absorption coefficient

The Paris formula [10.16] can be used to calculate the random incidence absorption coefficient once the plane wave absorption coefficient is known for all angles of incidence. If the absorption of the walls of a room can be modeled by an impedance boundary condition, the absorption coefficient $\alpha(\theta)$ is given by:

$$\alpha(\theta) = \frac{4\mathrm{Re}(\zeta)\cos\theta}{|\zeta|^2\cos^2\theta + 2\mathrm{Re}(\zeta)\cos\theta + 1} = \frac{4\beta_r\cos\theta}{(\beta_r + \cos\theta)^2 + \beta_i^2} \qquad [10.28]$$

where ζ is the reduced impedance and β_r and β_i are the real and imaginary parts of its inverse, the reduced admittance. The Paris formula can then be expressed as:

$$\alpha_m = 8\beta_r \int_0^{\pi/2} \frac{\cos^2\theta \sin\theta}{(\beta_r + \cos\theta)^2 + \beta_i^2} \, d\theta \qquad [10.29]$$

If the walls are only weakly absorbing, formula [10.29] gives approximately $\alpha_m = 8\beta_r$, an expression that was used to calculate the room response shown in Figure 10.5. Formula [10.29] can also be used to determine the optimal wall impedance, the one that maximizes α_m. It can be shown that the optimum is obtained for a value of the reduced impedance that is significantly different from unity[6], $\zeta_r \approx 1.567$ and $\zeta_i = 0$; the optimum absorption coefficient is slightly less than 1, and equals 0.951. Therefore, in a diffuse field, a locally reacting material can never be fully absorbing.

10.6. Schröder frequency

Sabine's energy approach is a "high frequency" approximation, while the modal approach clearly applies to "low frequencies" when the eigenfrequencies are well separated from one another. In order to determine the switchover range between these two descriptions, it is necessary to estimate the frequency range above which room resonances can no longer be reasonably individualized. To clarify this point, the width of the resonances can be estimated and compared to the average distance between two successive resonances, given by formula [10.11]. The Schröder frequency f_{Sch} is defined (somewhat arbitrarily) as the value for which this distance is equal to one-third of the width at mid-height of the resonance. Below f_{Sch}, the modal theory (or the finite element solution) is preferred; above f_{Sch} Sabine's theory is used.

We have previously estimated the average distance between resonances, but the width of these resonances remains to be evaluated. For a weakly absorbing room, the modes are quasi independent and each behaves as a lightly damped one degree-of-freedom system, whose resonance is controlled by a damping related to the imaginary part of the wavenumber; the real part remains practically unaffected by the damping. Near the resonance, the mean square amplitude of the pressure fluctuation is proportional to $\left((k^2 - k_l^2)^2 + k^2 B_l\right)^{-1}$, where $B_l = 2\beta_r S/V$ and $\beta_r \approx \alpha_m/8$ is the real part of the wall admittance[7]. The width at mid-height of the resonance is then given by $(\Delta f)_{res} = ac_0/(8\pi V) = 2.2/T_{60}$. The Schröder frequency, defined by $(\Delta f)_m = (\Delta f)_{res}/3$, is determined by the following relationship:

6 For $\zeta = 1$, we get $\alpha_m \approx 0.91$.

7 See, for example, [PIE 89], pp. 288–293.

$$\frac{c_0^3}{4\pi V f_{Sch}^2} = \frac{1}{3}\frac{2.2}{T_{60}} \qquad [10.30]$$

or, by rounding the numerical coefficient, $f_{Sch} = 2,000\sqrt{T_{60}/V}$. For frequencies above this value, it will be considered that the energy theory can be applied. If we evaluate this formula for the reverberation room studied above and assign an absorption coefficient $\alpha_m = 0.02$, we obtain a value for this "cut-off" frequency close to 450 Hz; it decreases to about 150 Hz for the same room but with an average absorption coefficient $\alpha_m = 0.2$.

10.7. Room critical distance

To determine the steady-state level generated by a source in a room, we used the diffuse field assumptions in the simplest way possible to obtain Formula [10.20]. However, these assumptions cannot be rigorously verified in the entire volume, and especially not in the immediate vicinity of a source. As a matter of fact, near a source, the direct pressure field, which decreases as $1/r$, necessarily dominates the reverberant field. The distance at which the reverberant field becomes dominant is called the critical distance r_c.

This critical distance can be determined as follows. The direct and reverberant contributions are assumed to be independent, and the mean square pressure at a distance r from the source, which is assumed to be omnidirectional, is therefore the sum of the squared pressure associated with the direct field and that due to the reverberant field (formula [10.20]):

$$p_m^2 = 4\rho_0 c_0 W_S \left(\frac{1}{a} + \frac{1}{16\pi r^2}\right) \qquad [10.31]$$

The expression of the critical distance is thus derived as:

$$r_c = \sqrt{\frac{a}{16\pi}} \qquad [10.32]$$

This formula shows that the smaller the equivalent absorption area, the smaller the value of the critical distance; it can also be used to estimate how far a microphone should be placed from a source to essentially measure the reverberant field, with a deviation of less than 1 dB for example.

If the emission of the source is directive, a directivity factor Q_θ is introduced, which is equal to the radial component of the intensity at a distance r from the source,

divided by the value that would be produced at the same point by an omnidirectional source of the same power. The integral of Q_θ over the solid angle is equal to 4π, so that for an omnidirectional source, $Q_\theta = 1$. The critical distance then depends on the source-receiver direction and is given by:

$$r_c(\theta) = \sqrt{\frac{aQ_\theta}{16\pi}} \qquad [10.33]$$

10.8. Coupled rooms: transmission loss of a panel

It is often useful to know the noise level in one room caused by a source in another room by transmission through a dividing wall. The level difference between the two rooms depends on the intrinsic transmission loss factor of the wall, but also on the absorption area of the receiving room. Considering that a diffuse field is dominant in both rooms, the relationships between mean square pressure, mean energy and mean intensity obtained in section 10.3.3 allow us to clarify these links.

In each room, the steady-state mean square pressure is related to the sound power and to the equivalent absorption area by:

$$(p_m^2)_1 = \frac{4\rho_0 c_0 W_1}{a_1} \qquad [10.34]$$

$$(p_m^2)_2 = \frac{4\rho_0 c_0 W_2}{a_2} \qquad [10.35]$$

The power available in the receiving room is the fraction of the power incident on the partition separating the rooms that is transmitted, that is, $\alpha_t I S_p$, where S_p is the area of the partition, $I = (p_m^2)_1/4\rho_0 c_0$ and α_t is the transmission coefficient of the partition. The level difference between the two rooms $L_1 - L_2$ and the transmission loss factor $TL = -10 \log \alpha_t$ are therefore related by:

$$TL = L_1 - L_2 + 10 \log(\frac{S_p}{a_2}) \qquad [10.36]$$

This formula is often used to determine the transmission loss factor of a panel by placing it in an opening between two reverberation rooms, assuming that the rest of the common wall is perfectly rigid. This formula is very simple, but it should be noted that the equivalent absorption area of the receiving room is not the product of the average absorption coefficient of the walls by their total area. As it happens, a fraction of the energy present in the receiving room is actually re-emitted toward the emitting room (it is therefore considered as absorbed when writing the energy balance in the receiving room). The absorption area can be determined experimentally by measuring the reverberation time of the receiving room.

10.9. Measurements in the reverberation room of École Centrale de Lyon

In this section, we present some results of measurements carried out in a reverberation room at École Centrale de Lyon. The room is excited by an omnidirectional source fed by a white noise signal. Figure 10.7 shows the low frequency part (range 0 – 400 Hz) of the power spectral densities measured in the bare room and with an absorbing material of area 4.5 m^2 placed on the floor.

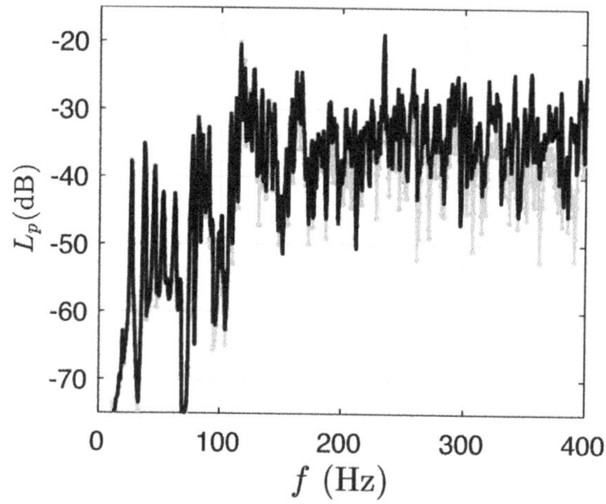

Figure 10.7. *Power spectral density of sound pressure fluctuations measured in the reverberation room. Bare room in black and with* 4.5 *m*^{2} *of absorbing material in gray*

The positions of the first peaks of the spectrum are in excellent agreement with the first natural frequencies calculated by the finite element simulation of the room response, as can be seen in Table 10.2[8]. The discrepancies, which remain modest, increase slightly with frequency, reflecting the necessarily simplified nature of the geometric modeling of the room. It can also be seen that the presence of the absorbing material only slightly modifies the room response at low frequencies. This is due to the very low efficiency of the material in this frequency range, its normal incidence absorption coefficient remaining below 0.1 at 100 Hz. Only a slight reduction in the level of the first peaks, whose frequency has not changed, is noticeable; the level reduction becomes more pronounced from about 350 Hz. At higher frequencies, the efficiency of the absorbing material is better and, for example, the level reduction over the one-third octave band centered at 1 kHz is about 3 dB.

8 The comparison of the measured and calculated peak levels is not possible, because the source frequency response used in the measurements was not taken into account.

f measured (Hz)	27.2	37.7	38.8	46.1	52.8	53.3	62.6
f calculated (Hz)	27.2	37.3	38.5	46.2	53.6	54.4	60.1

Table 10.2. *Comparison between the frequencies of the first peaks of the sound pressure power spectral density measured in a reverberation room and those estimated by a finite element numerical simulation*

A second set of measurements concerns the reduction in reverberation time due to the presence of an absorbing material. The method used is the "historical" interrupted noise method: the one-third octave filtered white noise signal driving the source is interrupted at a given time and the decrease in sound level over time is plotted. Figure 10.8 shows the level decrease in the 1,000 Hz one-third octave band, for the bare room and with the absorbing material. The observed decrease is reasonably linear over a wide range of levels, as predicted by Sabine's theory. From this, a value of the random incidence absorption coefficient α_m of about 0.9 for the 1,000 Hz one-third octave band can be derived using Eyring's formula [10.25]. The reduction in the steady-state level in the presence of the absorbing material as shown in Figure 10.8, holding the source power constant, is consistent with that obtained by applying formula [10.20] with the estimates of the equivalent absorption areas derived from the values of the reverberation times (about -3 dB in the 1 kHz one-third octave band considered above).

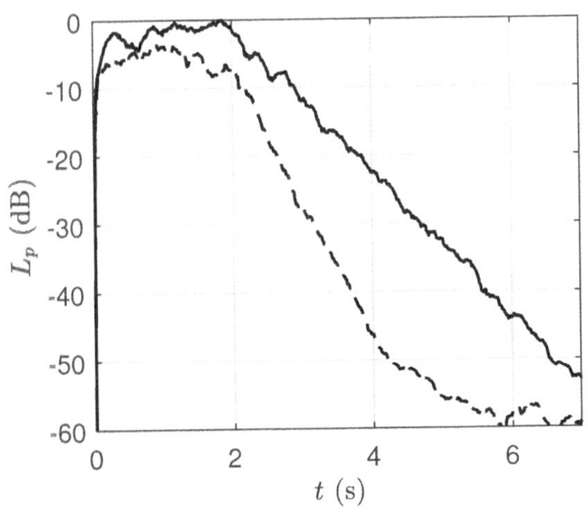

Figure 10.8. *Measurement of the reverberation time using the interrupted noise method. Filtered white noise excitation in the 1 kHz one-third octave band. Bare room (solid line); with absorbing materials (dashed line). The estimated values of the reverberation times are equal to 5.6 s and 3.16 s, respectively*

10.10. Geometric room acoustics

The two approaches studied so far, the modal theory and Sabine's theory, proved to be limited to certain domains:

– The analytical or finite element modal approach is suitable for low frequencies (long wavelengths compared to the dimensions of the room), because the modal density increases very rapidly with frequency and requires a considerable number of modes to be taken into account.

– The energy approach and Sabine's theory are well suited for rooms in which the acoustic field is diffuse, that is at high frequencies (short wavelengths compared to the dimensions of the room), and for "live" rooms with low absorption at the walls, which allows for the required sound field homogeneity and isotropy.

In an ordinary room, and even more so in a very large hall (such as a theater), it is impossible to use the modal approach, except for very low frequencies. For such rooms, the Schröder frequency is often well below 100 Hz. On the other hand, Sabine's theory, which leads to very simple results, assumes that the energy distribution in the room is uniform, which in fact is never verified. For example, the sound level depends on the room architecture and the position of the source and receiver. The two approaches described above also do not take into account the brief transient phenomena that are crucial for assessing certain aspects of human sound perception. However, their consideration is essential when designing a room for a specific use (such as an amphitheater or a concert hall).

A third possible approach is based on the use of geometric acoustics: the propagation of acoustic waves is modeled in terms of rays that carry a certain amount of energy and are incoherent with each other. This last assumption allows us to add directly the mean squared pressures associated with each individual ray, without taking into account any interferences. The sound rays are specularly reflected by the surfaces they encounter and lose part of their energy with each reflection, which is absorbed by these surfaces.

The method of images presented in section 10.1 introduces virtual sources that are symmetrical to the real source with respect to the different walls; it can be used to calculate the travel times along each ray coming from these real and virtual sources. This method is easy to apply to simple rectangular geometries but difficult to implement for complex geometries. In this case, ray tracing methods are more effective. Rays are launched from the source in all directions with an energy that depends on its directivity. Each ray path is then tracked and the position of its reflections on the surfaces detected, as shown in Figure 10.9.

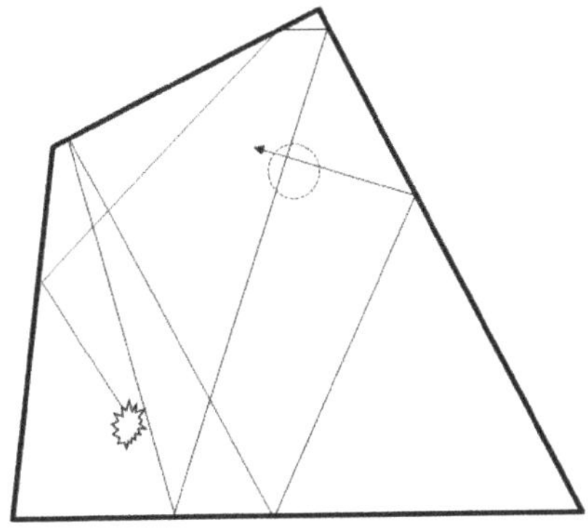

Figure 10.9. *Plot of a ray emitted from a source and arriving at a receiver after multiple reflections on the walls of a room*

An illustration of some paths connecting the source and a receiver in a model of the École Centrale de Lyon reverberation room is given in Figure 10.10[9]. The direct path is plotted in red, and all first-order rays (i.e., those that have undergone a single reflection) and second-order rays (those that have undergone two reflections) are plotted in green and blue, respectively.

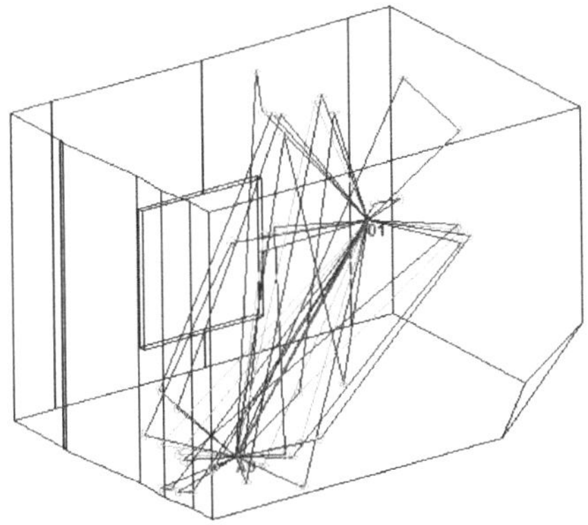

Figure 10.10. *Direct ray and rays joining the source (labeled A0) and a receiver (labeled 01) after undergoing one or two reflections in a model of the École Centrale de Lyon reverberation room. For a color version of this figure, see www.iste.co.uk/juve/acoustics1.zip*

9 The modeling was done with the CATT-Acoustic® software, which was also used to plot Figures 10.10–10.14.

In the geometric approximation, an impulsive source induces a signal at a given point that takes the form of a series of contributions that appear at the travel times associated with each of the rays; the amplitude of each contribution depends on the total distance traveled, the losses associated with each wall reflection via the absorption coefficient of the corresponding wall, and those due to volume absorption.

This *energy impulse response* or *echogram* is illustrated with a model of the movie theater of École Centrale de Lyon. The geometry of the room and the position of the source and receiver chosen for this calculation are shown in Figure 10.11.

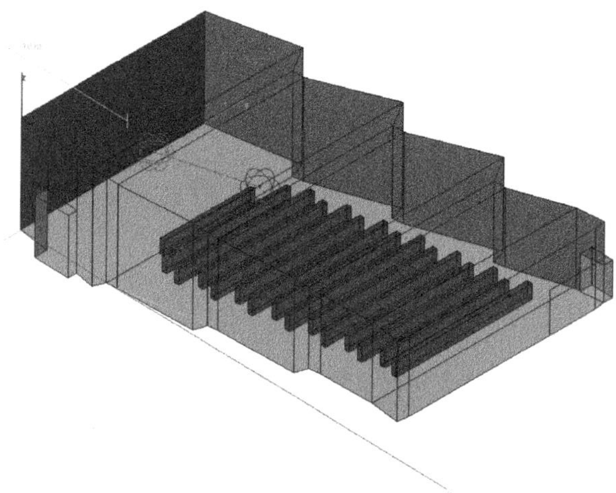

Figure 10.11. *Model of the movie theater at École Centrale de Lyon. The volume of the hall is 640 m^3 with a floor area of 169 m^2. The source is located on the stage and the receiver is on the front row. For a color version of this figure, see www.iste.co.uk/juve/acoustics1.zip*

The first 100 ms of the sound signal are plotted in Figure 10.12, where the colors associated with the different rays correspond to their order of reflection. Note the double time scale used, respectively, corresponding to the travel time from the source and the time difference between the reflected and direct paths.

Some typical characteristics can be clearly seen in this figure:

– The most energetic ray corresponds to the direct path between the source and the receiver and is obviously associated with the shortest arrival time.

– The first reflections produce echoes that are well separated from each other. These reflections are produced by walls close to the source or receiver. They are highly

dependent on the relative positions of the source and receiver in the room and play a fundamental role in acoustic perception.

– Subsequent arrivals originate from rays that have been subjected to an increasing number of reflections. These rays rapidly form a continuum of steadily decreasing level; the precise positions of the source and receiver have no significant effect on this part of the received signal.

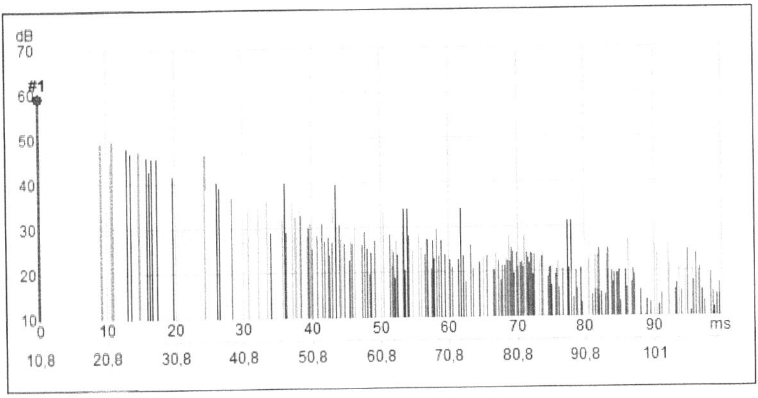

Figure 10.12. *Echogram of the movie theater in the 1-kHz octave. For this plot, only reflections of order less than or equal to seven were considered. The absorption coefficients of the different materials were adjusted to reproduce the experimentally observed reverberation time. For a color version of this figure, see www.iste.co.uk/juve/acoustics1.zip*

To obtain a realistic model of the impulse response of a room, it is necessary to add various contributions to this purely geometric response: diffraction due to the finite size of obstacles, wall roughness and so on. Rays "scattered" by walls or edges are thus taken into account using algorithms specific to each simulation software package [VOR 20].

The impulse response is an essential tool for characterizing a room. As we will see, it is used to calculate certain "subjective" parameters of human perception, that are fundamental to defining the acoustic quality of a room.

10.11. Subjective effects

Some theaters or performance halls "sound" good. Others, on the other hand, sometimes have deplorable acoustics. It should be noted that the definition of the acoustic quality of a room depends first and foremost on its intended use. The expectations are not the same for a library, a theater or a concert hall. In the latter

case, the type of music played should also be taken into consideration. The question then arises as to whether the physical parameters used to calculate the acoustic fields in the room can be related to the "auditory impressions" experienced. How can we quantify the acoustic properties of a room with a set of parameters and thus be able to act in the design phase of the room? The transition from physics to perception is always complex. Here are some elements to understand the main phenomena observed and to define the indicators commonly used in this field.

Some initial criteria are easy to identify:

– The sound pressure level in the room must be high enough and evenly distributed. For example, all audience members must be able to hear properly, regardless of where they sit in the theater.

– The reverberation time must be appropriate for the use. Organ music requires long reverberation times, such as those found in old churches. On the other hand, a spoken message is not always understandable in large spaces (typically in train station halls).

There are tables or figures that can be used to define "optimal" reverberation times, adapted to a given use (see, for example, [HOW 17]). These tables are based on empirical data and show that these recommended values depend on the purpose of the room but also on its volume. The conclusion that can be drawn is that although reverberation time is indeed the first parameter to be taken into account when designing a room, it is far from being sufficient; other indicators related to certain aspects of sound perception must also be considered.

The observation of a typical impulse response such as the one shown in Figure 10.12 illustrates the distinction that can already be made between two types of contributions: those that arrive quickly after the direct contribution and those that arrive later. How are these different contributions perceived by the human ear? Are there reflections that are "useful" and others that would be "detrimental"? How are echoes and reflections perceived?

The answer is not simple. It depends on the characteristics of the echo compared to the direct sound: relative level, origin, time delay, etc. A reflection is not perceived as separate from the direct sound if its delay and pitch do not exceed certain thresholds. It should also be noted that reflections in music are not perceived as unpleasant in the same way as they are in speech. There is a critical delay, 80 ms for music, 50 ms for speech, beyond which an echo is perceived as annoying. Within this critical delay, echoes are generally perceived as beneficial. The only effects are the perceived widening and amplification of the source.

The first reflections make it possible to reach a sufficient level. Therefore, they are very useful and constitute what is called early energy. The reflections that come later

are late energy. They contribute to the reverberation. It is therefore important to be able to evaluate the respective proportions of the different reflections in the impulse response.

Several indicators have been devised from the impulse response g(t) to evaluate the relative contributions of early and late energy. Here are some of the most commonly used ones:

– The definition:

$$D_{50} = \frac{\int_0^{50\text{ ms}} g^2(t)\,\mathrm{d}t}{\int_0^{\infty} g^2(t)\,\mathrm{d}t} \tag{10.37}$$

represents the early energy as a percentage of the total energy.

– The clarity:

$$C_{80} = 10\log\left(\frac{\int_0^{80\text{ ms}} g^2(t)\,\mathrm{d}t}{\int_{80\text{ ms}}^{\infty} g^2(t)\,\mathrm{d}t}\right) \tag{10.38}$$

expressed in decibels, represents the relative proportions of early and late energy. The limit in the above integrals corresponds to the critical delay chosen according to the purpose of the room, the time origin being taken at the arrival of the first ray (the direct ray, if it exists).

– The central time:

$$T_S = \frac{\int_0^{\infty} t g^2(t)\,\mathrm{d}t}{\int_0^{\infty} g^2(t)\,\mathrm{d}t} \tag{10.39}$$

is the barycenter of the energy of the impulse response.

Clarity and definition are obviously related. Indices based on the critical delay of 50 ms can be used to quantify the acoustic properties of the room for speech and are a useful element in assessing intelligibility. A clarity of 0 dB expresses a balance between early and late energy. As a first approach, it is often considered that the value $C_{80} = 0$ dB is appropriate for any type of music, while a value of -3 dB is acceptable. The central time should be compared to the critical delay. It is small when the early reflections are close and intense and the late reflections have a relatively low level in comparison. This situation favors good intelligibility, since the successive syllables of the words are not superimposed as is the case in a highly reverberant room.

Other indicators allow a more precise quantification of intelligibility in a room. They are used to translate the quality of the reproduction of transient speech signals,

taking into account on the one hand the long-term spectral distribution of speech which presents a maximum of energy around 500 Hz (presence of long vowels), and on the other hand the large contribution to the intelligibility of consonants (short sounds) around 2,000 Hz.

The STI (Speech Transmission Index) is defined by examining the "degradation" of a sound consisting of contributions from seven octave bands whose levels follow the energy distribution curve of speech. The sound in each band is modulated by one of 14 low frequencies comprised between 0.63 and 12.5 Hz, representative of the speech rate. In total, there is a matrix of 98 combinations. The degradation of the emitted sound is mainly due to two factors: on the one hand, reverberation, which spreads out over time the short sequences encountered in speech, and, on the other hand, the presence of background noise. It is generally accepted that a signal-to-noise ratio of at least 15 dB(A) is required to ensure a good intelligibility. An STI value close to unity reflects very good intelligibility. A value greater than 0.5 is usually considered to guarantee proper intelligibility.

Other indicators derived from the STI are sometimes mentioned in the literature, such as the RASTI (RApid STI) or the STIPA (STI for Public Address Systems). They are based on reduced modulation octave bands/frequencies matrices (nine elements for RASTI and 14 for STIPA) and have the advantage of considerably reducing the measurement time when a direct measurement is made from the sounds emitted for each of the elements of the matrix. Nowadays, the STI is most often calculated from the knowledge of the impulse response, whether it is measured or calculated. This indirect method of determining the STI is very fast and has made the RASTI obsolete, which is far too approximate because it uses only two octaves. However, the indirect method cannot take into account certain phenomena (such as nonlinearities, for example) because it is based on the assumption of a linear time-invariant system. The advantage of the STIPA is that it is obtained from a direct measurement, in a short time due to the reduction in the size of the matrix. It is a good estimator of the STI and it also allows, by means of an adapted protocol, to take into account the intelligibility of the speech broadcast by a public address system.

Other indicators can be used to refine the characterization of the perception:

– The EDT (Early Decay Time) indicator is related to the perceived "reverberance" associated with the very early reflections. It is estimated by a linear extrapolation over 60 dB of the measured or calculated sound level decrease over the first 10 decibels. Unlike the reverberation time, the EDT is highly dependent on the source and receiver positions.

– The LF (early Lateral energy Fraction) indicator quantifies the feeling of envelopment.

The reader can refer to Vorländer [VOR 20] for a more detailed description. All of these indicators can be calculated as part of a design phase from models using the ray-tracing approach. They can also be measured. They depend on the position of the source and receiver and can be visualized on prediction maps at any point in the room.

As an example, Table 10.7 shows the values of some of the indicators defined above as a function of the frequency for the movie theater modeled with CATT-Acoustic®.

f	125 Hz	250 Hz	500 Hz	1 kHz	2 kHz	4 kHz	8 kHz	16 kHz
D_{50} (%)	80.38	87.89	90.78	93.69	95.44	92.91	95.38	98.46
C_{80} (dB)	15.47	14.17	15.67	18.58	18.48	17.85	19.57	27.11
T_s (ms)	32.88	25.54	20.44	16.26	16.55	16.65	14.24	10.94
T_{60} (s)	0.45	0.35	0.28	0.22	0.23	0.23	0.20	0.15
EDT (s)	0.38	0.36	0.36	0.27	0.25	0.29	0.23	0.20

Table 10.3. *Acoustic indicators computed for the modeled movie theater*

It can be seen that this room is relatively "deaf": the low reverberation time, the high definition and clarity, the low central time as well as the high STI value (0.86) indicate the predominance of early reflections and the limited significance of late contributions. The spatial distribution of the sound level depends strongly on the distance from the source as can be seen in Figure 10.13 where the source is located in the center of the scene. This is not a major drawback if the room is going to be used as a movie theater since an electroacoustic system of loudspeakers placed on the side walls can provide good diffusion throughout the room.

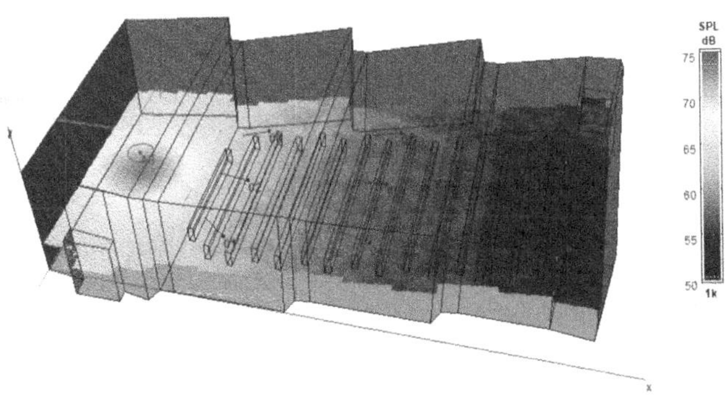

Figure 10.13. *Sound pressure level distribution in the movie theater (1-kHz octave band). For a color version of this figure, see www.iste.co.uk/juve/acoustics1.zip*

However, this hall is also used for acoustic concerts, for which its characteristics are then no longer suitable.

A simulation software can be used to consider different acoustic correction options. From Figure 10.14, it can be observed that the addition of reflective panels on the lateral walls (in gray) leads to a better distribution of the sound pressure levels. Similarly, at 1 kHz, there is a significant decrease in clarity (from 18 dB to 9 dB) and an increase in central time (from 16–36 ms). The improvement is undeniable, although it is not sufficient for the hall to present a good acoustic quality for a concert.

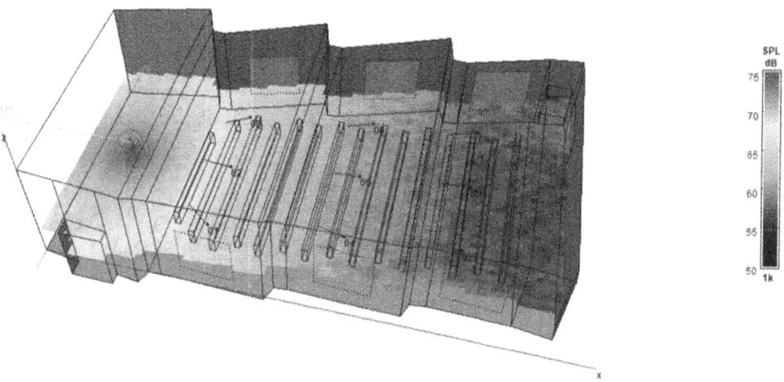

Figure 10.14. *Distribution of the sound pressure level in the movie theater with the addition of reflectors (1 kHz octave band). For a color version of this figure, see www.iste.co.uk/juve/acoustics1.zip*

10.12. Exercises

1) **Sabine's theory – reverberation time**

a) In a room with bare walls 6 m long, 5 m wide and 3 m high, the absorption coefficient of the walls is $\alpha_0 = 0.01$. When the sound sources are switched on, the noise level is 90 dB. To reduce this level, the floor is covered with a carpet with an absorption coefficient α_m. What does the value of α_m need to be to bring the level down to 80 dB?

b) Consider a 20 m$\times$40 m$\times$5 m room. The floor is covered with a carpet with an absorption coefficient of 0.2 and the ceiling and walls have an absorption coefficient of 0.05. Calculate the equivalent absorption area of the room and the reverberation time.

c) Assume that a person speaking represents an acoustic source with a power of 100 μW. Calculate the sound pressure level generated in the large room of question b by a single person and then by 100 people speaking at the same time.

2) The acoustics of the library of École Centrale de Lyon

For a preliminary design of the library, an acoustic analysis based on the estimation of the reverberation times was carried out. The volume of the room is approximately 1,900 m^3; it is assumed that the average reverberation time for a room with a volume greater than 512 m^2 must be less than $0.15V^{1/3}$. The Sabine absorption coefficients for the different constituents of the room are given as a function of in the following table:

f (Hz)	250	500	1,000	2,000
Wooden floor	0.04	0.08	0.12	0.12
Walls	0.03	0.03	0.04	0.05
Glazing	0.03	0.02	0.02	0.02
Ceiling	0.03	0.04	0.08	0.17
Books	0.26	0.22	0.20	0.20

Table 10.4. *Absorption coefficients of materials present in the library*

The corresponding areas are equal to 500, 140, 220, 500 and 300 m^2, respectively.

a) Calculate the reverberation time of the empty room for the different frequencies. Calculate an average reverberation time by defining an average equivalent absorption area, where the areas of each frequency band are weighted by the width of the corresponding octave filter.

b) To improve the room, "acoustic panels" are added, suspended or used as partitions, with a total surface area of 100 m^2 and an absorption coefficient of 0.5, 0.8, 0.9 and 0.8 for the four frequency bands studied. Calculate the new values of the reverberation time and the average time for each frequency band.

c) Estimate the average reverberation time considering a public audience. A value for the equivalent absorption area of a reader will be taken equal to 0.5 m^2 regardless of the frequency, and it can be considered that 25 readers are present simultaneously.

3) Alpha Cabin

The Alpha Cabin is a small, reverberation chamber used in the automotive industry to measure the absorption properties of materials. Its characteristics are as follows: volume 6.44 m^3, total wall area 22.6 m^2, average reverberation time of the bare cabin 1.7 s and surface area of the material to be tested 1.2 m^2.

a) Measurements are usually made between 400 Hz and 10 kHz. Comment on these limits. Give an order of magnitude of the cabin modal density at 400 Hz.

b) Calculate the reverberation time obtained when the material under test has an absorption coefficient of 0.8.

c) Three loudspeakers distributed on the walls of the cabin are used to generate noise inside. Reverberation time measurements are made with an initial pressure level

of 94 dB. Calculate the sound power of each source (it will be assumed that the contributions of the three loudspeakers are uncorrelated, so that the total power is the sum of the individual powers).

4) **Influence of humidity in room acoustics**

The aim is to evaluate the influence of humidity on the measurements made in the reverberation room of École Centrale de Lyon (room volume V = 129 m^3 and wall area S = 155 m^2).

Humidity causes an exponential attenuation of the sound intensity with the distance x traveled by a wave (or the propagation time $t = x/c_0$) according to $I = I_0 \exp(-mx) = I_0 \exp(-mc_0 t)$. It is assumed that the decrease in the intensity in the room is the sum of two independent contributions, one due to the losses on the walls of the room, independent of the climatic conditions and characterized by the equivalent absorption area a, and the other due to the atmospheric absorption.

a) Express the law of decay of the intensity (or energy) with time when considering these two contributions. From this, infer that the reverberation (Sabine) time can be expressed by formula [10.26]:

$$T_{60} = \frac{0.161V}{a + 4mV}$$

b) What is the error made in measuring the absorption coefficient of a material if the atmospheric absorption is not taken into account?

The influence of the relative humidity h of the air at 20°C on the acoustic parameters of the room is studied using the values of the coefficient $4m$ given in the table below. The absorption coefficient of the room walls is 0.025 for the 1-kHz octave and 0.07 for the 8-kHz octave.

Humidity h	**1-kHz octave**	**8-kHz octave**
20%	0.00558	0.18830
50%	0.00388	0.07666
80%	0.00330	0.05323

Table 10.5. *Atmospheric absorption coefficients*

c) Calculate the reverberation time for each octave for the different relative humidity percentages. Comment on the result.

d) Assume that a source of power W_s is emitting in the room. Calculate for each octave the variation of the generated sound pressure level with respect to the reference case ($h = 50\%$) for the different humidities.

e) What change can be expected in an indicator such as definition or clarity with air humidity?

5) **The world record for the longest echo**

In 2012, Trevor Cox (Professor at the University of Salford) measured the reverberation time in an old military fuel oil tank buried in Scotland and used during the Second World War. The total volume of the reservoir is estimated to be 26,000 m^3 and the area of the walls, made of concrete with a thickness of 45 cm, to be 9,000 m^2.

Estimate the absorption coefficient of these walls for different frequency bands using the reverberation time measurements given in the table below.

Given these extreme values, what physical phenomena could have affected the measurements?

f (Hz)	63	125	250	500	1,000	2,000	4,000
T_{60} (s)	163	112	69	43	28	18	7.5

Table 10.6. *Reverberation times as a function of the frequency*

6) **Speech emission and critical distance**

The average sound pressure level produced at 1 m in the open field by a person speaking naturally (without straining their voice) is defined by the table below.

Octave (Hz)	125	250	500	1,000	2,000	4,000
L_p (dB at 1 m)	51.2	57.2	59.8	53.5	48.8	43.8

Table 10.7. *Speech sound pressure levels at 1 m as a function of the frequency*

a) Briefly comment on the spectral distribution of this noise and calculate the overall sound level at 1 m.

b) Calculate the power of the source, first assuming it is omnidirectional.

c) In reality, a person speaking is a source with a pronounced directivity, and it is now assumed that all the pressure is radiated forward (hemispherical radiation). Based on this new assumption, calculate the source power, denoted W_s. This value will be used for the remainder of the exercise.

d) This person is now speaking in a reverberation room, where Sabine's theory is supposed to be verified. The room has a volume of $V = 127$ m^3 and a reverberation time of $T_{60} = 5$ s. It is assumed that the pressure at all points is the result of two uncorrelated contributions: the direct contribution corresponding to the source in free field and the contribution due to the reverberation obtained by Sabine's theory, $p_m^2 = p_{m_d}^2 + p_{m_r}^2$.

Express the mean square pressure at a distance r from the source with respect to the source power W_s and the equivalent absorption area of the room a.

e) Express the critical distance with respect to a.

f) Calculate the sound pressure level at 1 m and plot the evolution of this level with respect to r. Analyze the resulting curve. What observations can be made about the conversation in this room between two people 1 m apart?

7) **Coupled rooms**

Consider a room in which Sabine's theory is supposed to be verified. Its characteristics are as follows: volume V_1, total wall area S_1 and average absorption coefficient α_m.

a) The acoustic energy density in the room is called e_1. Recall the differential equation resulting from the energy balance in the room verified by e_1 in the absence of a source. What is its solution, given an energy density $e_1 = 1$ as initial condition?

b) It is now assumed that this room has an opening to the outside with a small area S_0 compared to S_1. The absorption coefficient associated with this area is denoted α_0. What value can be taken as a first approximation for α_0? Compare the equivalent absorption areas and reverberation times of the closed room and the room with the opening, with the following numerical data: room dimensions 10 m×10 m×4 m, $\alpha_m = 0.03$, $S_0 = 4$ m^2.

c) This room is connected to a second room by the opening described in the previous question. A new energy balance is performed in each room. The notations introduced in question 7.a are kept for room 1 and extended to the second room with index 2.

What are the acoustic powers coming from room 1, incident and then transmitted through the opening? What are the acoustic powers coming from room 2, incident and then transmitted through the opening? (Figure 10.15) Write the energy balance for room 1 and then for room 2. Show that we end up with a system of coupled equations that is written in matrix form as:

$$\frac{d}{dt}\begin{pmatrix} e_1 \\ e_2 \end{pmatrix} + \begin{pmatrix} k_{11} & k_{12} \\ k_{21} & k_{22} \end{pmatrix}\begin{pmatrix} e_1 \\ e_2 \end{pmatrix} = 0$$

where you will make explicit the elements k_{ij}.

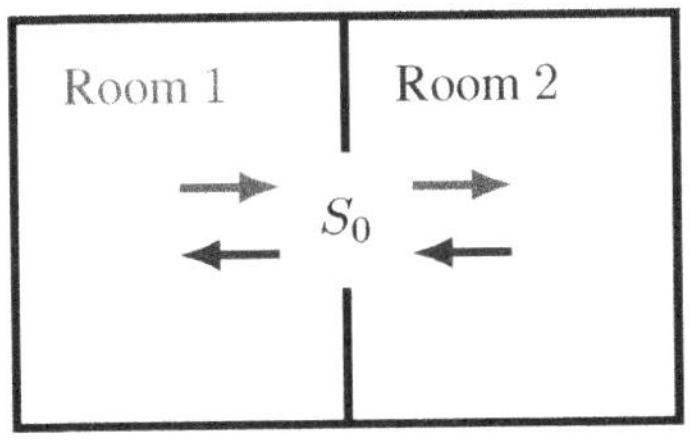

Figure 10.15. *Geometry of coupled rooms*

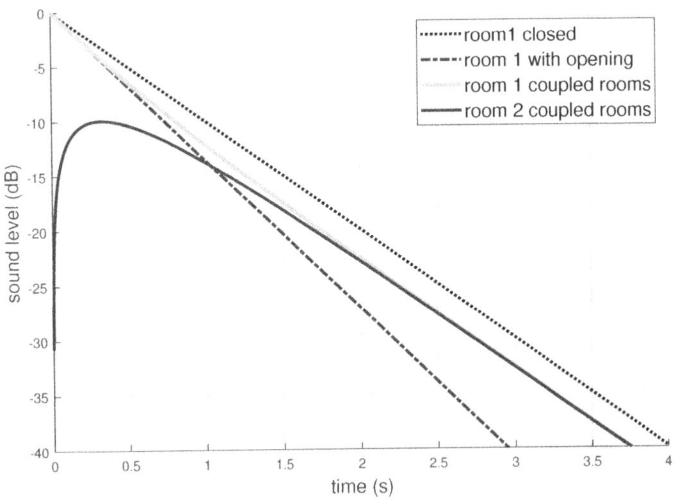

Figure 10.16. *Evolution of energy densities over time*

d) Give the solutions of this differential system when the rooms are identical, namely when they have the same geometry (volume V and surface S) and acoustic characteristics (same average absorption coefficient), and of course a common opening S_0. The initial condition will be an energy density e_1 equal to 1 and e_2 equal to zero.

The results obtained from solving the coupled system are plotted in Figure 10.16, with the numerical values shown above. Comment on this figure, and in particular on the different phases of the time evolution in room 1 coupled with room 2. What conclusions can be drawn from this study regarding measurements in this type of room?

Appendices

Appendix 1

Basic Fluid Mechanics and Thermodynamics

The purpose of this appendix is to recall the equations of fluid mechanics on which the development of the equations governing acoustic disturbances is based. In essence, acoustics is an *unsteady phenomenon* that involves the *compressibility* of the media in which the waves propagate. It is therefore essential to consider the thermodynamic behavior of fluids, as well as the time dependence of the equations of motion[1].

A1.1. The equations of motion

In the following paragraphs, we recall the conservation equations of fluid mechanics and the constitutive law of Newtonian fluids. These equations describe the "ambient" medium on which the acoustic perturbations will develop.

The equations to be considered result from the conservation of mass (continuity equation), momentum (Cauchy equations) and energy. They will be written in Euler variables and given without demonstration. They must be supplemented by experimentally derived relations for the stress tensor (Newtonian fluid) and by an equation of state.

A1.1.1. *Conservation of mass*

The mass-conservation equation, or *continuity equation*, is written as:

$$\frac{\partial \rho}{\partial t} + \frac{\partial (\rho U_j)}{\partial x_j} = 0 \qquad [A1.1]$$

1 The reader is referred to [BAT 67] and [LAN 87] for further reading on fluid dynamics.

where ρ is the density, U_j is the j component of the velocity and the summation convention has been used: when a suffix is repeated in a single term, we sum over all values of that suffix (Einstein summation convention). This equation can be written in a different form by introducing the material derivative:

$$\frac{d}{dt} = \frac{\partial}{\partial t} + U_j \frac{\partial}{\partial x_j} \qquad \text{[A1.2]}$$

from which

$$\frac{d\rho}{dt} + \rho \frac{\partial U_j}{\partial x_j} = 0 \qquad \text{[A1.3]}$$

This relation shows that the divergence of the velocity field $\nabla.\boldsymbol{U}$ is none other than the rate of change in volume of a fluid material element during its motion[2]. The following notation is sometimes used:

$$\Delta = \nabla.\boldsymbol{U} = \rho \frac{d(1/\rho)}{dt}$$

A1.1.2. *Navier–Stokes equations*

In classical continuum mechanics, the momentum conservation equations, or Cauchy equations, are written as:

$$\rho \frac{dU_i}{dt} = \rho f_i + \frac{\partial \sigma_{ij}}{\partial x_j} \qquad \text{[A1.4]}$$

where f_i is the density of the external forces and σ_{ij} is a component of the stress tensor that is symmetric ($\sigma_{ij} = \sigma_{ji}$).

Equivalently, a conservative form of this equation can be used:

$$\frac{\partial(\rho U_i)}{\partial t} + \frac{\partial(\rho U_i U_j)}{\partial x_j} = \rho f_i + \frac{\partial \sigma_{ij}}{\partial x_j} \qquad \text{[A1.5]}$$

For a fluid at rest, we simply have:

$$\sigma_{ij} = -p\delta_{ij} \qquad \text{[A1.6]}$$

where p denotes the pressure and δ_{ij} is the Kronecker symbol.

---

2 See Appendix 2 for the definition of the differential operator nabla ∇.

It is natural to decompose the stress tensor into the sum of an isotropic tensor and a deviatoric (zero trace) tensor d_{ij} according to:

$$\sigma_{ij} = -p_m \delta_{ij} + d_{ij} \quad \text{[A1.7]}$$

Assuming that the fluid is isotropic and that the deviatoric tensor d_{ij} depends linearly on the velocity gradients (Newtonian fluid), it can be shown that [BAT 67]:

$$d_{ij} = \mu \left(\frac{\partial U_i}{\partial x_j} + \frac{\partial U_j}{\partial x_i} \right) + \lambda \Delta \delta_{ij} \quad \text{[A1.8]}$$

where μ and λ are the viscosity coefficients of the fluid (μ is the shear viscosity and λ is the second viscosity). Since the trace of d_{ij} is zero by definition, the two coefficients are related by:

$$3\lambda + 2\mu = 0 \quad \text{[A1.9]}$$

which is a relation often referred to as the Stokes condition. Thereby, the final expression for the stress tensor is:

$$\sigma_{ij} = -p_m \delta_{ij} + \mu \left(\left(\frac{\partial U_i}{\partial x_j} + \frac{\partial U_j}{\partial x_i} \right) - \frac{2}{3} \Delta \delta_{ij} \right) \quad \text{[A1.10]}$$

or by introducing the rate of strain tensor e_{ij}:

$$e_{ij} = \frac{1}{2} \left(\frac{\partial U_i}{\partial x_j} + \frac{\partial U_j}{\partial x_i} \right) \quad \text{[A1.11]}$$

$$\sigma_{ij} = -p_m \delta_{ij} + 2\mu \left(e_{ij} - \frac{1}{3} \Delta \delta_{ij} \right) \quad \text{[A1.12]}$$

A subtlety remains, however, because the pressure that appears in this formula is strictly mechanical in nature: it is defined as the average of the normal stresses σ_{ii}:

$$p_m = -\frac{1}{3} \sigma_{ii} \quad \text{[A1.13]}$$

The difference between this "mechanical" pressure and the thermodynamic equilibrium pressure (obtained using the equation of state, for the local values of density and internal energy, and denoted p in the following) is expressed, assuming a linear dependence on the velocity gradients, by:

$$p_m - p = -\mu_B \Delta \quad \text{[A1.14]}$$

where μ_B is the bulk viscosity of the fluid.

The bulk viscosity represents a thermodynamic imbalance associated with the existence of relaxation phenomena, which we will not discuss in detail here. The bulk viscosity μ_B is zero for monatomic gases; for diatomic gases and air, the generally accepted value is $\mu_B = 0.6\mu$ [PIE 89].

It can be shown that both viscosity coefficients are positive, or zero in the case of perfect fluids. The viscosity coefficients depend on pressure and temperature (Sutherland's formula [A1.53] given below is often used in practice), and as such they can vary spatially in a flow; however, they are assumed to be constant in the following.

The Navier–Stokes equations are obtained by introducing the law of Newtonian fluids into the Cauchy equations. It follows that:

$$\rho\frac{dU_i}{dt} = \rho f_i - \frac{\partial p_m}{\partial x_i} + \frac{\partial}{\partial x_j}\left(2\mu(e_{ij} - \frac{1}{3}\Delta\delta_{ij})\right) \quad [A1.15]$$

and assuming constant viscosity coefficients:

$$\rho\frac{dU_i}{dt} = \rho\left(\frac{\partial U_i}{\partial t} + U_j\frac{\partial U_i}{\partial x_j}\right) = \rho f_i - \frac{\partial p}{\partial x_i} + \mu\frac{\partial^2 U_i}{\partial {x_j}^2} + (\frac{1}{3}\mu + \mu_B)\frac{\partial \Delta}{\partial x_i} \quad [A1.16]$$

This equation can be given a less compact form, which allows us to emphasize the role of the flow vorticity $\boldsymbol{\omega} = \nabla \times \boldsymbol{U}$. We recall that a vector field can be decomposed in a general way into the sum of an irrotational part (derived from a scalar potential ϕ) and a rotational part (with zero divergence, derived from a vector potential $\boldsymbol{\Psi}$):

$$\boldsymbol{U} = \nabla\phi + \nabla \times \boldsymbol{\Psi} \quad [A1.17]$$

The vector notation is more appropriate here than the indicial notation. Using formulas:

$$\frac{D\boldsymbol{U}}{Dt} = \frac{\partial \boldsymbol{U}}{\partial t} + \frac{1}{2}\nabla U^2 - \boldsymbol{U} \times (\nabla \times \boldsymbol{U}) \quad [A1.18]$$

$$\nabla^2 \boldsymbol{U} = \nabla(\nabla.\boldsymbol{U}) - \nabla \times (\nabla \times \boldsymbol{U}) \quad [A1.19]$$

the following expression is obtained for the Navier–Stokes equations:

$$\rho\left[\frac{\partial \boldsymbol{U}}{\partial t} + \frac{1}{2}\nabla U^2 - \boldsymbol{U} \times (\nabla \times \boldsymbol{U})\right] = -\nabla p + \mu'\nabla.(\nabla \boldsymbol{U}) - \mu\nabla \times (\nabla \times \boldsymbol{U}) \quad [A1.20]$$

in which it is defined that:

$$\mu' = \mu\left(\frac{4}{3} + \frac{\mu_B}{\mu}\right) \quad [A1.21]$$

A1.2. The equation of state and thermodynamic relations

We consider simple bivariant fluids in which any thermodynamic quantity can be expressed as a function of two arbitrarily chosen basic variables. These can be density and internal energy (which are defined even in non-equilibrium conditions), density and entropy (the classical choice, adapted to quasi-isentropic flows), or density and temperature. In general, we will choose to write the equation of state as:

$$p = p(\rho, s) \qquad [A1.22]$$

where s is the specific entropy. This formulation is particularly well suited when the fluid can be considered as perfect.

Among the many definitions and formulas of thermodynamics, some are particularly useful for formulating acoustic equations. Some of them are listed below (see for example [TEM 01] for a more detailed description).

A very important quantity is the specific enthalpy given by:

$$h = e + \frac{p}{\rho} \qquad [A1.23]$$

We have the following relations between the entropy s, the temperature T, the internal energy e, and the enthalpy h:

$$de = Tds + \frac{p}{\rho^2}d\rho \qquad [A1.24]$$

$$dh = Tds + \frac{1}{\rho}dp \qquad [A1.25]$$

Pressure and temperature are related to the partial derivatives of the internal energy by:

$$p = \rho^2 \left(\frac{\partial e}{\partial \rho}\right)_s ; \quad T = \left(\frac{\partial e}{\partial s}\right)_\rho \qquad [A1.26]$$

The specific heats (at constant volume and constant pressure, respectively) are given by:

$$c_v = \left(\frac{\partial e}{\partial T}\right)_v ; \quad c_p = \left(\frac{\partial h}{\partial T}\right)_p \qquad [A1.27]$$

The ratio of specific heats, called the adiabatic coefficient or adiabatic index, is defined by $\gamma = c_p/c_v$; it is always greater than or equal to 1.

Other definitions are also useful, such as the coefficient of thermal expansion:

$$\beta = -\frac{1}{\rho}\left(\frac{\partial \rho}{\partial T}\right)_p \qquad \text{[A1.28]}$$

and the adiabatic and isothermal compressibilities:

$$\kappa_s = \frac{1}{\rho}\left(\frac{\partial \rho}{\partial p}\right)_s \; ; \quad \kappa_T = \frac{1}{\rho}\left(\frac{\partial \rho}{\partial p}\right)_T \; ; \quad \kappa_s/\kappa_T = \gamma. \qquad \text{[A1.29]}$$

In order to calculate the evolution of the entropy, it is necessary to be able to evaluate it in terms of measurable quantities such as temperature, pressure or density. This requires the use of relations between partial derivatives of thermodynamic quantities, called Maxwell relations. As an example, we have:

$$\left(\frac{\partial s}{\partial p}\right)_T = \frac{1}{\rho^2}\left(\frac{\partial \rho}{\partial T}\right)_p \qquad \text{[A1.30]}$$

which allows us to write the change in entropy in the form:

$$Tds = c_p dT - \frac{T\beta}{\rho}dp \qquad \text{[A1.31]}$$

It is also useful to recall that for three variables related by a functional expression $c = c(a, b)$, partial derivatives are related by:

$$\left(\frac{\partial a}{\partial b}\right)_c \cdot \left(\frac{\partial b}{\partial c}\right)_a \cdot \left(\frac{\partial c}{\partial a}\right)_b = -1 \qquad \text{[A1.32]}$$

If the equation of state is written in the form $p = p(T, \rho)$, which is useful in the analysis of thermo-viscous effects, then it can be deduced that, for example:

$$\left(\frac{\partial p}{\partial T}\right)_\rho \cdot \left(\frac{\partial T}{\partial \rho}\right)_p \cdot \left(\frac{\partial \rho}{\partial p}\right)_T = -1 \qquad \text{[A1.33]}$$

and it follows that:

$$\left(\frac{\partial p}{\partial T}\right)_\rho = \beta\rho c_0^2/\gamma \qquad \text{[A1.34]}$$

A1.2.1. *Ideal gas*

Gases, and air in particular, can be considered ideal (or "perfect") under the usual pressure and temperature conditions, and it is interesting to rewrite the previous relations in this case. The equation of state takes the explicit form:

$$p = \rho r T \ ; \ r = R/M \qquad [A1.35]$$

where r is the gas constant of the particular gas in question, the ratio of the universal gas constant $R = 8.314$ J.K^{-1}.mol^{-1} to the molar mass of the gas M. For example, the approximate value of the molar mass of air is $M = 29$ g.mol^{-1} and then $r = 287$ J.kg.$^{-1}$K^{-1}.

For an ideal gas, the internal energy and enthalpy are only a function of temperature. In this case, the following relations are obtained:

$$\beta = 1/T \ ; \ \kappa_T = 1/p \ ; \ c_p - c_v = r \qquad [A1.36]$$

from which we derive, for example, $c_v = r/(\gamma - 1)$.

If, in addition, it is assumed that the specific heats are constant over the temperature range under consideration, it yields:

$$e - e_0 = c_v T \ ; \ h - h_0 = c_p T \ ; \ s - s_0 = c_v \ln(p\rho^{-\gamma}) \qquad [A1.37]$$

where e_0, h_0, s_0 are constants. If the behavior is isentropic, it follows that $p\rho^{-\gamma}$ = constant.

A1.2.2. *The equation of energy*

The change in the total energy density of a material element of a fluid is equal to the sum of the powers of the external forces acting on the element, and the density of the surface heat flux supplied.

The heat flux $\boldsymbol{q}$ is assumed to be due solely to thermal conduction and to be related to the temperature gradients in the fluid by Fourier's law:

$$q_j = -\lambda_{th} \frac{\partial T}{\partial x_j} \qquad [A1.38]$$

where the coefficient of thermal conductivity λ_{th} is always positive and generally depends on pressure and temperature; in acoustic applications it can often be considered constant.

The equation for the conservation of the total energy, defined as the sum of internal and kinetic energies, is then written as (volume forces are not considered here):

$$\rho\frac{d}{dt}\left(e+\frac{1}{2}\rho U^2\right)=\frac{\partial}{\partial x_j}(\sigma_{ij}U_i)+\frac{\partial}{\partial x_j}\left(\lambda_{th}\frac{\partial T}{\partial x_j}\right) \quad \text{[A1.39]}$$

Therefrom, an equation for the evolution of the internal energy can be derived as:

$$\rho\frac{de}{dt}=\sigma_{ij}\frac{\partial U_i}{\partial x_j}+\frac{\partial}{\partial x_j}(\lambda_{th}\frac{\partial T}{\partial x_j}) \quad \text{[A1.40]}$$

If we now introduce the Newtonian constitutive relation, we get:

$$\rho\frac{de}{dt}=-p_m\Delta+2\mu(e_{ij}e_{ij}-\frac{1}{3}\Delta^2)+\frac{\partial}{\partial x_j}\left(\lambda_{th}\frac{\partial T}{\partial x_j}\right) \quad \text{[A1.41]}$$

In this expression, the viscous contribution is non-negative, and the rate of dissipation of mechanical energy per unit mass is written as:

$$\Phi=\frac{2\mu}{\rho}(e_{ij}e_{ij}-\frac{1}{3}\Delta^2)=\frac{2\mu}{\rho}(e_{ij}-\frac{1}{3}\Delta\delta ij)(e_{ij}-\frac{1}{3}\Delta\delta_{ij}) \quad \text{[A1.42]}$$

The viscous dissipation is equivalent to an irreversible heat input. The first term on the right-hand side of equation [A1.41], which appears to correspond to a reversible energy of compression, actually contains a dissipative part associated with the bulk viscosity, $\mu_B\Delta^2$.

It is often useful to replace the internal energy evolution equation with an entropy evolution equation. This is done using the thermodynamic relation:

$$Tds=de+pd(1/\rho) \quad \text{[A1.43]}$$

from which

$$T\frac{ds}{dt}=\frac{de}{dt}+p\frac{d(1/\rho)}{dt}=\frac{de}{dt}+\frac{p}{\rho}\Delta \quad \text{[A1.44]}$$

and finally

$$T\frac{ds}{dt}=\Phi+\mu_B\Delta^2+\frac{1}{\rho}\frac{\partial}{\partial x_j}(\lambda_{th}\frac{\partial T}{\partial x_j}) \quad \text{[A1.45]}$$

If the fluid can be considered to be ideal (i.e. non-viscous and non-heat conducting), the entropy of a material element of the fluid remains constant during its motion. Such a flow is said to be isentropic (if the entropy is *uniform*, the flow is said to be *homentropic*).

A1.3. Boundary conditions at a fluid-solid interface

The above equations must be supplemented by the addition of boundary conditions (and also initial conditions), typically at the interface between the fluid under consideration and a rigid solid. For real fluids (viscous and heat-conducting), the no-slip condition must be satisfied at the wall: the relative velocity between the fluid and the solid must be zero at the interface Σ [3]:

$$(\boldsymbol{U}_f)_\Sigma = (\boldsymbol{U}_s)_\Sigma \qquad \text{[A1.46]}$$

A condition of continuity of temperature and heat flow at a fluid-solid interface must also be satisfied:

$$(T_f)_\Sigma = (T_s)_\Sigma\,; \quad \lambda_{th,f}\left(\frac{\partial T_f}{\partial n}\right)_\Sigma = \lambda_{th,s}\left(\frac{\partial T_s}{\partial n}\right)_\Sigma \qquad \text{[A1.47]}$$

Often, the solid has a very high heat capacity and its temperature can be considered constant and uniform; the value of the fluid temperature at the wall is then imposed.

If the fluid is considered as ideal, only the normal component of the velocity at the interface must be continuous (the tangential component is usually discontinuous):

$$(U_{n,f})_\Sigma = (U_{n,s})_\Sigma \qquad \text{[A1.48]}$$

and there are no conditions to be imposed on the temperature.

A1.4. Equations of motion of ideal fluids

In this section, we express the thermo-mechanical equations when fluids can be considered ideal, which is the most common assumption in acoustics. These equations

3 At the interface between two fluids or between a fluid and an elastic solid, the continuity condition of the normal component of the stress must be added.

are often referred to as *Euler's equations* to distinguish them from the Navier–Stokes equations. Euler's equations are written as:

$$\frac{\partial \rho}{\partial t} + \frac{\partial(\rho U_i)}{\partial x_i} = 0 \quad \text{[A1.49]}$$

$$\rho \frac{dU_i}{dt} = -\frac{\partial p}{\partial x_i} \quad \text{[A1.50]}$$

$$\frac{ds}{dt} = 0 \quad \text{[A1.51]}$$

$$p = p(\rho, s) \quad \text{[A1.52]}$$

to which must be added the condition of continuity of the normal component of the velocity at the interface between a fluid and a (non-porous) solid. At the interface between two immiscible fluids, the continuity of pressure should also be imposed, since surface tension effects are usually negligible in acoustic applications.

A1.5. Physical properties of some gases

It is interesting to give the numerical values of some physical properties of common gases. The data in Table A1.1 are mostly taken from [LID 04].

Gas	ρ (kg.m^{-3})	μ (μPa.s)	λ_{th} (mW.m^{-1}.K^{-1})	c_0 (m.s^{-1})
Air	1.17	18.6	26.2	347.6
Nitrogen	1.14	17.9	25.8	353
Oxygen	1.30	20.8	26.3	330
Helium	0.162	20	156.7	1,020
Hydrogen	0.082	9	185	1,310
Argon	1.62	22.9	17.9	323
Carbon dioxide	1.70	15	16.8	271.5

Table A1.1. *Physical properties of some gases at 300 K and atmospheric pressure*

Under normal conditions, gases can be considered ideal. The isentropic speed of sound is calculated with very good accuracy by $c_0 = \sqrt{\gamma r T_0}$, where γ is 5/3 and 7/5 for monatomic and diatomic gases, respectively. At high pressures, the equation of state must be modified (using the virial equation of real gases), but the speed of sound is modified only very slightly. For example, at a pressure of 1 MPa, the speed of sound in helium increases slightly, from 1,020 to 1,024 m.s^{-1}.

If we consider wave propagation in a gas mixture, the speed of sound can be calculated using an equivalent adiabatic index and the molar mass of the mixture according to:

$$\gamma_{eq} = 1 + R(X_1 c_{v1} + X_2 c_{v2} + ...)^{-1}$$
$$M = X_1 M_1 + X_2 M_2 + ...$$

where X_i, M_i and c_{vi} denote the specific molar fractions, molar mass and specific heats of each of the constituents of the mixture.

The Prandtl number Pr= $\mu c_p/\lambda_{th}$ characterizes the relative effect of viscous and thermal diffusion. For gases, these thermal and viscous effects are of the same order, and the Prandtl number is close to 0.7 (it is equal to 0.71 for air and 0.66 for helium at 300 K and atmospheric pressure).

In this book, we assume that the viscosity and heat conductivity are constant. However, if necessary, two semi-empirical formulas can be used to evaluate the variation of these quantities with temperature in the case of air.

For the viscosity, the Sutherland formula is commonly used:

$$\frac{\mu(T)}{\mu_0} = \left(\frac{T}{T_0}\right)^{3/2} \frac{T_0 - T_s}{T_0 + T_s} \qquad [A1.53]$$

where μ_0 is the viscosity at (any) temperature T_0 and $T_s = 110.4$ K. If $T_0 = 300$ K, then $\mu_0 = 1.86 \times 10^{-5}$ kg m^{-1} s^{-1}.

For the thermal conductivity, the following formula can be used ([PIE 89]):

$$\frac{\lambda_{th}(T)}{\lambda_{th_0}} = \left(\frac{T}{T_0}\right)^{3/2} \frac{T_0 + T_A \mathrm{e}^{-T_B/T_0}}{T + T_A \mathrm{e}^{-T_B/T}} \qquad [A1.54]$$

where $\lambda_{th_0} = 2.624 \times 10^{-2}$ W m^{-1}K^{-1}, $T_A = 245.4$ K and $T_B = 27.6$ K.

Appendix 2

Math Refresher

In the following sections, we give a brief introduction to the main mathematical tools and formulas used in this book. For more information and a more precise mathematical perspective, the reader can consult, for example, [STO 09].

Depending on the problem, it is easier to use the vector notation or the indicial notation (using Einstein summation convention). We will therefore use one of these notations interchangeably in this appendix and throughout the book.

A2.1. Differential operators in the main coordinate systems

In this book, we use the nabla (∇) notation[1] to express the gradient, divergence, curl and Laplacian operators. In Cartesian coordinates, the vector operator ∇ is defined by:

$$\nabla = \boldsymbol{e}_1 \frac{\partial}{\partial x_1} + \boldsymbol{e}_2 \frac{\partial}{\partial x_2} + \boldsymbol{e}_3 \frac{\partial}{\partial x_3} \qquad [A2.1]$$

where $\boldsymbol{e}_j$ are the unit vectors along the three directions of space.

The gradient of a scalar p is written as ∇p and its Laplacian is written as$\nabla^2 p$. The divergence of a vector field $\boldsymbol{U}$ is written as $\nabla \cdot \boldsymbol{U}$ and its curl is written as $\nabla \times \boldsymbol{U}$.

1 Bold characters denote vectors.

In Cartesian coordinates, it thus follows that:

$$\nabla p = \frac{\partial p}{\partial x_1}\boldsymbol{e}_1 + \frac{\partial p}{\partial x_2}\boldsymbol{e}_2 + \frac{\partial p}{\partial x_3}\boldsymbol{e}_3 = \frac{\partial p}{\partial x_j}\boldsymbol{e}_j \qquad [A2.2]$$

$$\nabla.\boldsymbol{U} = \frac{\partial U_1}{\partial x_1} + \frac{\partial U_2}{\partial x_2} + \frac{\partial U_3}{\partial x_3} = \frac{\partial U_j}{\partial x_j} \qquad [A2.3]$$

$$\nabla \times \boldsymbol{U} = \left(\frac{\partial U_3}{\partial x_2} - \frac{\partial U_2}{\partial x_3}\right)\boldsymbol{e}_1 + \left(\frac{\partial U_1}{\partial x_3} - \frac{\partial U_3}{\partial x_1}\right)\boldsymbol{e}_2 + \left(\frac{\partial U_2}{\partial x_1} - \frac{\partial U_1}{\partial x_2}\right)\boldsymbol{e}_3 \qquad [A2.4]$$

$$\nabla^2 p = \frac{\partial^2 p}{\partial x_1^2} + \frac{\partial^2 p}{\partial x_2^2} + \frac{\partial^2 p}{\partial x_3^2} = \frac{\partial^2 p}{\partial x_j^2} \qquad [A2.5]$$

While in Cartesian coordinates the expressions of these operators are easy to obtain, in other coordinate systems it becomes a bit more complicated because the basis vectors depend on the position.

A2.1.1. *Cylindrical coordinates*

The position of a point is given by the coordinates (ρ, φ, z), defined in Figure A2.1. The formulas giving the gradient and Laplacian of a scalar and the divergence of a vector field are as follows:

$$\nabla p = \frac{\partial p}{\partial \rho}\boldsymbol{e}_\rho + \frac{1}{\rho}\frac{\partial p}{\partial \varphi}\boldsymbol{e}_\varphi + \frac{\partial p}{\partial z}\boldsymbol{e}_z \qquad [A2.6]$$

$$\nabla.\boldsymbol{U} = \frac{1}{\rho}\frac{\partial(\rho U_\rho)}{\partial \rho} + \frac{1}{\rho}\frac{\partial U_\varphi}{\partial \varphi} + \frac{\partial U_z}{\partial z} \qquad [A2.7]$$

$$\nabla^2 p = \frac{1}{\rho}\frac{\partial}{\partial \rho}\left(\rho\frac{\partial p}{\partial \rho}\right) + \frac{1}{\rho^2}\frac{\partial^2 p}{\partial \varphi^2} + \frac{\partial^2 p}{\partial z^2} \qquad [A2.8]$$

A2.1.2. *Spherical coordinates*

The position of a point is indicated by the coordinates (r, θ, φ), defined in Figure A2.2. The formulas giving the main operators are the following:

$$\nabla p = \frac{\partial p}{\partial r}\boldsymbol{e}_r + \frac{1}{r}\frac{\partial p}{\partial \theta}\boldsymbol{e}_\theta + \frac{1}{r\sin\theta}\frac{\partial p}{\partial \varphi}\boldsymbol{e}_\varphi \qquad [A2.9]$$

$$\nabla.\boldsymbol{U} = \frac{1}{r^2}\frac{\partial\left(r^2 U_r\right)}{\partial r} + \frac{1}{r\sin\theta}\frac{\partial\left(\sin\theta U_\theta\right)}{\partial\theta} + \frac{1}{r\sin\theta}\frac{\partial U_\phi}{\partial\varphi} \qquad [A2.10]$$

$$\nabla^2 p = \frac{1}{r^2}\frac{\partial}{\partial r}\left(r^2\frac{\partial p}{\partial r}\right) + \frac{1}{r^2\sin\theta}\frac{\partial}{\partial\theta}\left(\sin\theta\frac{\partial p}{\partial\theta}\right) + \frac{1}{r^2\sin^2\theta}\frac{\partial^2 p}{\partial\varphi^2} \qquad [A2.11]$$

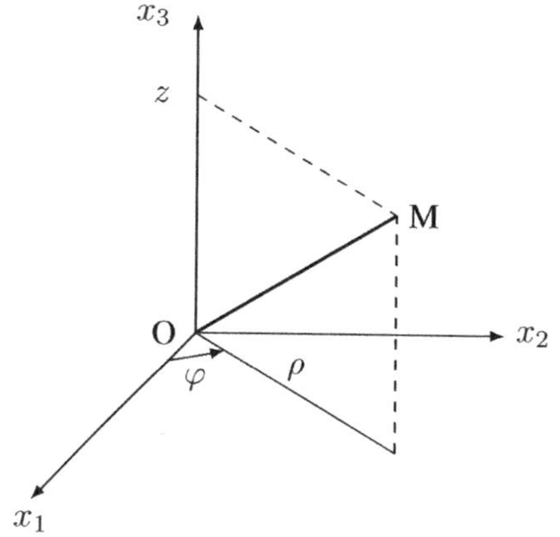

Figure A2.1. *Cylindrical coordinate system*

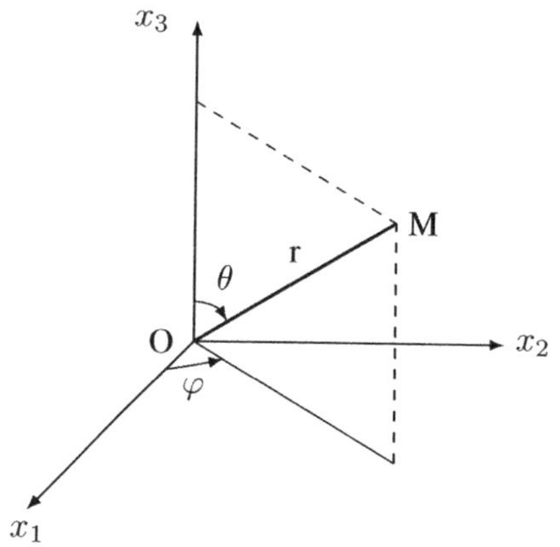

Figure A2.2. *Spherical coordinate system*

The first term of the Laplacian can also be expressed as:

$$\frac{1}{r}\frac{\partial}{\partial r}\left(r\frac{\partial\,(rp)}{\partial r}\right) \qquad [A2.12]$$

A2.2. Dirac delta function, Fourier transforms and convolution product

The Dirac delta function is frequently used in acoustics, especially for the computation of Green's functions. We will use it as a generalized function even though, strictly speaking, no point value can be assigned to it, only its action on functions is defined.

A2.2.1. *The one-dimensional Dirac delta function*

Perhaps the most important property of the Dirac delta function is that it can extract a particular value of a function $g(x)$ using the formula:

$$\int_{-\infty}^{+\infty} g(x)\delta(x-c)\,\mathrm{d}x = g(c) \qquad [A2.13]$$

If the integration is restricted to a finite interval, we have:

$$\int_{a}^{b} g(x)\delta(x-c)\,\mathrm{d}x = g(c) \quad \text{if } c \in [a,b]; \quad 0 \quad \text{if } c \notin [a,b] \qquad [A2.14]$$

Some useful properties are listed below.

Homothety

$$\delta(ax) = \frac{1}{|a|}\delta(x) \qquad [A2.15]$$

Dirac delta function of a function

$$\delta(g(x)) = \sum_i \frac{1}{|g'(x_i)|}\delta(x_i) \qquad [A2.16]$$

where $g'(x)$ is the derivative of $g(x)$ and the sum includes all simple zeros x_i of the function $g(x)$.

Differentiation

The derivative of a Dirac delta function can be defined using the integration by parts formula, which yields:

$$\int_{-\infty}^{+\infty} G'(x)g(x)\,\mathrm{d}x = -\int_{-\infty}^{+\infty} G(x)g'(x)\,\mathrm{d}x \qquad \text{[A2.17]}$$

when $G(x) = 0$ for large values of $|x|$. Applied with $G = \delta$, it follows that:

$$\int_{-\infty}^{+\infty} \delta'(x)g(x)\,\mathrm{d}x = -\int_{-\infty}^{+\infty} \delta(x)g'(x)\,\mathrm{d}x = -g'(0) \qquad \text{[A2.18]}$$

and the derivative of order n of the Dirac delta function is defined by:

$$\int_{-\infty}^{+\infty} \delta^{(n)}(x)g(x)\,\mathrm{d}x = (-1)^n g^{(n)}(0) \qquad \text{[A2.19]}$$

Heaviside function (or unit step function)

The Heaviside function is defined by:

$$H(x) = \left| \begin{array}{l} 0 \text{ if } x < 0 \\ 1 \text{ if } x > 0 \end{array} \right. \qquad \text{[A2.20]}$$

The derivative of the Heaviside function is none other than the Dirac delta function, $H'(x) = \delta(x)$, which can be derived from the relation:

$$\int_{-\infty}^{x} \delta(x)\,\mathrm{d}x = H(x) \qquad \text{[A2.21]}$$

The sign function, $\mathrm{sgn}(x)$:

$$\mathrm{sgn}(x) = \left| \begin{array}{l} -1 \text{ if } x < 0 \\ +1 \text{ if } x > 0 \end{array} \right. \qquad \text{[A2.22]}$$

is related to the Heaviside function by $\mathrm{sgn}(x) = 2H(x) - 1$.

A2.2.2. *Fourier transforms*

We define the direct Fourier transform of a time-dependent function g(t) by:

$$g(f) = \int_{-\infty}^{+\infty} g(t) \exp(-\mathrm{i}2\pi f t) \, \mathrm{d}t \qquad \text{[A2.23]}$$

and its inverse transform by:

$$g(t) = \int_{-\infty}^{+\infty} g(t) \exp(+\mathrm{i}2\pi f t) \, \mathrm{d}t \qquad \text{[A2.24]}$$

The Fourier transform of the Dirac delta function is given by:

$$\delta(f) = \int_{-\infty}^{+\infty} \delta(t) \exp(-\mathrm{i}2\pi f t) \, \mathrm{d}t = 1 \qquad \text{[A2.25]}$$

Therefrom, it can be inferred that the Fourier transform of the constant function 1 is $\delta(f)$.

We also have:

$$\int_{-\infty}^{+\infty} \delta(t - t_0) \exp(-\mathrm{i}2\pi f t) \, \mathrm{d}t = \exp(-\mathrm{i}2\pi f t_0) \qquad \text{[A2.26]}$$

and the Fourier transform of the complex exponential $\exp(-\mathrm{i}2\pi f_0 t)$ is therefore equal to $\delta(f + f_0)$. From this, it can be inferred that the Fourier transforms of $\cos(2\pi f_0 t)$ and $\sin(2\pi f_0 t)$ are expressed as $1/2(\delta(f - f_0) + \delta(f + f_0))$ and $1/2\mathrm{i}(\delta(f - f_0) - \delta(f + f_0))$.

The Fourier transform of the sign function is:

$$\mathrm{sgn}(f) = \frac{1}{\mathrm{i}\pi f} \qquad \text{[A2.27]}$$

and that of the Heaviside function can be derived therefrom as:

$$H(f) = \frac{1}{2\mathrm{i}\pi f} + \frac{1}{2}\delta(f) \qquad \text{[A2.28]}$$

An acoustic source is often modeled by a Gaussian distribution, whether for numerical or analytical studies. The following property is very useful, for example, to switch from the physical to the wavenumber space (or k-space):

$$\mathrm{FT}\left(\exp\left(-\alpha t^2\right)\right) = \sqrt{\frac{\pi}{\alpha}} \exp\left(-\frac{(\pi f)^2}{\alpha}\right) \qquad \text{[A2.29]}$$

where FT denotes the Fourier transform.

If we particularize to $\alpha = \pi$, we can see that the function $\exp(-\pi t^2)$ is its own Fourier transform.

A2.2.3. *The Dirac delta function in higher dimensions*

The properties of the Dirac delta function in one-dimension can easily be extended to higher dimensional space. The delta function $\delta(\boldsymbol{x})$ is defined by its action on a function $g(\boldsymbol{x})$ according to:

$$\int_{-\infty}^{+\infty} g(\boldsymbol{x})\delta(\boldsymbol{x}-\boldsymbol{x_0})\,\mathrm{d}\boldsymbol{x} = g(\boldsymbol{x_0}) \qquad [A2.30]$$

and in particular

$$\int_{-\infty}^{+\infty} \delta(\boldsymbol{x}-\boldsymbol{x_0})\,\mathrm{d}\boldsymbol{x} = 1 \qquad [A2.31]$$

If the meaning of $\delta(\boldsymbol{x})$ in Cartesian coordinates is clear as that of the product $\delta(x_1)\delta(x_2)\delta(x_3)$, its expression in cylindrical or spherical coordinates needs to be specified.

In cylindrical coordinates (ρ, φ, z), we have:

$$\delta(\boldsymbol{\rho}) = \frac{\delta(\rho)}{\rho}\delta(\varphi)\delta(z) \qquad [A2.32]$$

a formula that results from [A2.31] and the expression of the volume element in cylindrical coordinates:

$$\int_{-\infty}^{+\infty} \delta(\boldsymbol{x})\,\mathrm{d}\boldsymbol{x} = 1 = \int_{-\infty}^{+\infty} \delta(z)\,\mathrm{d}z \int_{0}^{2\pi} \delta(\varphi)\,\mathrm{d}\varphi \int_{0}^{\infty} \frac{\delta(\rho)}{\rho}\rho\,\mathrm{d}\rho \qquad [A2.33]$$

Similarly, the expression for the Dirac delta function in spherical coordinates is obtained as:

$$\delta(\boldsymbol{x}) = \frac{\delta(r)}{r^2\sin\theta}\delta(\theta)\delta(\varphi) \qquad [A2.34]$$

It is also useful to define the Heaviside function in three-dimensions, which can be interpreted as an operator for slicing a portion of space. Consider a moving material surface Σ defined by the equation $f(\boldsymbol{x}, t) = 0$. Inside this surface, f is negative, and

outside it is positive. As a simple example, consider the sphere of radius a and center located at $(Ut, 0, 0)$; by defining $f(x_1, x_2, x_3, t) = (x_1 - Ut)^2 + x_2^2 + x_3^2 - a^2$, the surface of the sphere is $f = 0$, and $f > 0$ outside the sphere and $f < 0$ inside the moving sphere. The associated Heaviside function $H(f(\boldsymbol{x}, t))$ takes the value $+1$ outside Σ and 0 inside.

By using H, the integration of a function g defined only in the volume D internally bounded by Σ can be extended to all space, since:

$$\int_D g(\boldsymbol{x})\,\mathrm{d}\boldsymbol{x} = \int_\infty H(f(\boldsymbol{x}, t))g(\boldsymbol{x})\,\mathrm{d}\boldsymbol{x} \qquad \text{[A2.35]}$$

When evaluating the gradient of H along Σ, the Dirac delta function naturally appears, and:

$$\nabla(H(f(\boldsymbol{x}, t))) = \nabla f(\boldsymbol{x}, t).\delta(f(\boldsymbol{x}, t)) = |\nabla f|\delta(f)\boldsymbol{n} \qquad \text{[A2.36]}$$

where $\boldsymbol{n}$ is the unit normal pointing away from Σ.

When Σ moves at a velocity $\boldsymbol{V}$, then since the material derivative of $H(f(\boldsymbol{x}, t))$ is zero, the time and space derivatives of H are related by:

$$\frac{\partial H(f(\boldsymbol{x}, t))}{\partial t} + V_i \frac{\partial H(f(\boldsymbol{x}, t))}{\partial x_i} = 0 \qquad \text{[A2.37]}$$

A2.2.4. *The convolution product*

The convolution product of two functions is defined by[2]:

$$(g * h)(t) = \int_{-\infty}^{+\infty} g(\tau)h(t - \tau)\,\mathrm{d}\tau = (h * g)(t) \qquad \text{[A2.38]}$$

The identity element of the convolution product is the Dirac delta function, since:

$$(g * \delta)(t) = \int_{-\infty}^{+\infty} g(\tau)\delta(t - \tau)\,\mathrm{d}\tau = g(t) \qquad \text{[A2.39]}$$

2 This definition extends very naturally to functions that depend on several space or space–time variables.

The derivative of a convolution product can be evaluated by deriving any of its elements according to:

$$(g * h)' = g' * h = g * h' \qquad [A2.40]$$

The Fourier transform allows us to replace the convolution product in one space by a simple product in the conjugate space, according to:

$$\mathrm{FT}(g * h)(t) = g(f).h(f) \qquad [A2.41]$$

Conversely, the Fourier transform of a product is equal to the convolution product of Fourier transforms:

$$\mathrm{FT}(g(t).h(t)) = (g * h)(f) \qquad [A2.42]$$

A2.3. Bessel functions

When the geometry of a problem involves cylindrical coordinates (or polar coordinates in planar geometry), a decomposition of the acoustic field into a series of special functions called Bessel functions is often used. In a sense, they are an “equivalent” of the plane wave decomposition in Cartesian geometry.

Bessel functions of order μ are the solutions of Bessel's equation:

$$\frac{d^2y}{dx^2} + \frac{1}{x}\frac{dy}{dx} + \left(1 - \frac{\mu^2}{x^2}\right) y = 0 \qquad [A2.43]$$

The order μ can be arbitrary, real or complex, but here we will only consider integers, $\mu = m$.

Since Bessel's equation is of the second order, it admits two families of independent solutions, the Bessel functions of the first and second kind, denoted, respectively, by $J_m(x)$ and $Y_m(x)$[3]. The general solution of equation [A2.43] is a linear combination of the functions J_m and Y_m.

The functions of the first kind are regular at the origin, while those of the second kind are singular at the origin. Both decrease as $1/\sqrt{x}$ to infinity.

Figure A2.3 shows the evolution of the Bessel functions of the first and second kind from zeroth to third-order with x.

3 The function Y is sometimes denoted with N and called the Neumann function.

The functions of the first kind admit the following integral representations:

$$J_m(x) = \frac{\mathrm{i}^{-m}}{2\pi} \int_0^{2\pi} \exp(\mathrm{i}x\cos\theta)\cos(m\theta)\,\mathrm{d}\theta \qquad [A2.44]$$

and in particular

$$J_0(x) = \frac{1}{2\pi} \int_0^{2\pi} \exp(\mathrm{i}x\cos\theta)\,\mathrm{d}\theta \qquad [A2.45]$$

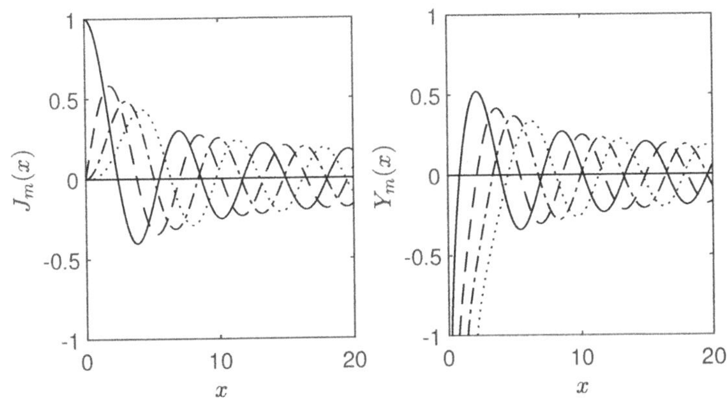

Figure A2.3. *Illustration of the mth-order Bessel functions of the first kind $J_m(x)$ and the second kind $Y_m(x)$ for m=0 (solid line), m=1 (dashed line), m=2 (dash-dotted line) and m=3 (dotted line)*

The asymptotic behavior of Bessel functions for small or large values of the argument is as follows:

$$\lim_{x\to 0} J_m(x) = \frac{1}{m!}\left(\frac{x}{2}\right)^m \qquad [A2.46]$$

$$\lim_{x\to 0} Y_m(x) = \frac{-(m-1)!}{\pi}\left(\frac{2}{x}\right)^m \; ; \; m \neq 0 \qquad [A2.47]$$

$$\lim_{x\to 0} Y_0(x) = \frac{2}{\pi}\left(\ln(\frac{x}{2}) + 0.5772...\right) \qquad [A2.48]$$

$$\lim_{x\to \infty} J_m(x) = \left(\frac{2}{\pi x}\right)^{1/2} \cos\left(x - \frac{\pi}{4} - \frac{m\pi}{2}\right) \qquad [A2.49]$$

$$\lim_{x\to \infty} Y_m(x) = \left(\frac{2}{\pi x}\right)^{1/2} \sin\left(x - \frac{\pi}{4} - \frac{m\pi}{2}\right) \qquad [A2.50]$$

In particular, these expressions highlight the similarity between Bessel functions and decreasing sine functions for the large values of the argument.

A certain linear combination of the functions J_m and Y_m defines the Hankel functions of the first and second kind, $H_m^{(1)}$ and $H_m^{(2)}$, according to:

$$H_m^{(1)} = J_m(x) + \mathrm{i}Y_m(x) \qquad \text{[A2.51]}$$

$$H_m^{(2)} = J_m(x) - \mathrm{i}Y_m(x) \qquad \text{[A2.52]}$$

Their asymptotic behavior for large values of the argument is then:

$$\lim_{x \to \infty} H_m^{(1)}(x) = \left(\frac{2}{\pi x}\right)^{1/2} \exp\left(\mathrm{i}\left(x - \frac{\pi}{4} - \frac{m\pi}{2}\right)\right) \qquad \text{[A2.53]}$$

$$\lim_{x \to \infty} H_m^{(2)}(x) = \left(\frac{2}{\pi x}\right)^{1/2} \exp\left(-\mathrm{i}\left(x - \frac{\pi}{4} - \frac{m\pi}{2}\right)\right) \qquad \text{[A2.54]}$$

The Hankel function of the second kind is the one that, together with the convention for representing harmonic waves as $\exp(+\mathrm{i}\omega t)$, allows us to describe the behavior of cylindrical waves that satisfy the radiation condition at infinity (Sommerfeld condition).

The following properties are also often used:

$$J_{-m}(x) = (-1)^m J_m(x) \qquad \text{[A2.55]}$$

$$Y_{-m}(x) = (-1)^m Y_m(x) \qquad \text{[A2.56]}$$

If we denote by B any function J, Y, or H, we have the following recurrence relations between Bessel functions and their derivatives:

$$B_{m-1}(x) + B_{m+1}(x) = \frac{2m}{x} B_m(x) \qquad \text{[A2.57]}$$

$$B_{m-1}(x) - B_{m+1}(x) = 2\frac{d}{dx} B_m(x) \qquad \text{[A2.58]}$$

Bessel functions satisfy various orthogonality conditions involving the zeros of these functions or their derivatives. If we denote by λ_i and λ_j two zeros of the Bessel function $J_m(x)$, then we have:

$$\int_0^1 x J_m(\lambda_i x) J_m(\lambda_j x)\,\mathrm{d}x = 0 \quad \text{if} \quad i \neq j \qquad \text{[A2.59]}$$

$$\int_0^1 x J_m(\lambda_i x) J_m(\lambda_j x)\,\mathrm{d}x = \frac{1}{2} J_m'^2(\lambda_i) = \frac{1}{2} J_{m+1}^2(\lambda_i) \quad \text{if} \quad i = j \qquad \text{[A2.60]}$$

and if λ_i and λ_j denote two zeros of $J'_m(x)$:

$$\int_0^1 xJ_m(\lambda_i x)J_m(\lambda_j x)\,\mathrm{d}x = 0 \quad \text{if} \quad i \neq j \qquad \text{[A2.61]}$$

$$\int_0^1 xJ_m(\lambda_i x)J_m(\lambda_j x)\,\mathrm{d}x = \frac{1}{2}\left(1 - \frac{m^2}{\lambda_i^2}\right) J_m^2(\lambda_i) \quad \text{if} \quad i = j \qquad \text{[A2.62]}$$

References

[ATA 15] ATALLA N., SGARD F., *Finite Element and Boundary Methods in Structural Acoustics and Vibration*, CRC Press, Boca Raton, FL, 2015.

[AXI 06] AXISA F., ANTUNES J., *Modelling of Mechanical Systems: Fluid-Structure Interaction*, Butterworth-Heinemann, Oxford, 2006.

[BAR 03] BARRON R.F., *Industrial Noise Control and Acoustics*, Marcel Dekker Inc., New York, 2003.

[BAS 95] BASS H.E., SUTHERLAND L.C., ZUCKERWAR A.J., et al., "Atmospheric absorption in air: Further developments", *J. Acoust. Soc.*, vol. 99, pp. 680–683, 1995.

[BAT 67] BATCHELOR G.K., *An Introduction to Fluid Dynamics*, Cambridge University Press, Cambridge, 1967.

[BEN 11] BENDAT J.S., PIERSOL A.G., *Random Data, Analysis and Measurement Procedures*, Wiley, Hoboken, NJ, 2011.

[BER 46] BERGMANN P.G., "The wave equation in a medium with a variable index of refraction", *J. Acoust. Soc.*, vol. 17, pp. 329–333, 1946.

[CHA 16] CHAIGNE A., KERGOMARD J., *Acoustics of Musical Instruments*, Springer, New York, 2016.

[DEV 12] DEVANEY A.J., *Mathematical Foundations of Imaging, Tomography and Wavefield Inversion*, Cambridge University Press, Cambridge, 2012.

[DOW 83] DOWLING A.P., FFOWCS-WILLIAMS J.E., *Sound and Sources of Sound*, E. Horwood, Chichester, 1983.

[FAH 07] FAHY J., GARDONIO P., *Sound and Structural Vibration*, Academic Press, Amsterdam, 2007.

[FAH 15] FAHY F., THOMPSON D., *Fundamentals of Sound and Vibration*, 2nd edition, CRC Press, Boca Raton, FL, 2015.

[FAS 07] FASTL H., ZWICKER E., *Psychoacoustics: Facts and Models*, Springer, Berlin, 2007.

[HOW 03] HOWE M.S., *Theory of Vortex Sound*, Cambridge University Press, Cambridge, 2003.

[HOW 17] HOWARD D.M., ANGUS J.A.S., *Acoustics and Psychoacoustics*, 5th edition, Routledge, New York, 2017.

[IEC 13] IEC 61672-1:2013, Standard, Electroacoustics – sound level meters – part 1: specifications, International Electrotechnical Commission, 2013.

[IEC 14] IEC 61620-1:2014, Standard, Electroacoustics – octave-band and fractional-octave-band filters – part 1: specifications, International Electrotechnical Commission, 2014.

[ISO 97] ISO 266:1997, Standard, Acoustics – preferred frequencies, International Organization for Standardization, 1997.

[ISO 03] ISO 226:2003, Standard, Acoustics – normal equal-loudness-level contours, International Organization for Standardization, 2003.

[ISO 16] ISO 1996-1:2016, Standard, Acoustics – description, measurement and assessment of environmental noise – part 1: basic quantities and assessment procedures, International Organization for Standardization, 2016.

[KUT 00] KUTTRUFF H., *Room Acoustics*, 4th edition, Spon Press, London, 2000.

[LAN 87] LANDAU L.D., LIFSHITZ E.M., *Fluid Mechanics*, 2nd edition, Elsevier, Cambridge, 1987.

[LID 04] LIDE D.R., *Handbook of Chemistry and Physics*, 84th edition, CRC Press, Boca Raton, FL, 2004.

[MAC 15] MACDONALD H.M., "A class of diffraction problems", *Proc. London Math. Soc.*, vol. s2–14, pp. 410–425, 1915.

[MAR 08] MARBURG S., WU T.W., "Treating the phenomenon of irregular frequencies", in MARBURG S., NOLTE B. (eds). *Computational Acoustics of Noise Propagation in Fluids-Finite and Boundary Element Methods,* Springer, Berlin, 2008.

[MOR 53] MORSE P.M., FESHBACH H., *Methods of Theoretical Physics*, McGraw-Hill, New York, 1953.

[MOR 68] MORSE P.M., INGARD U.K., *Theoretical Acoustics*, Princeton University Press, Princeton, NJ, 1968.

[MUN 14] MUNJAL L.M., *Acoustics of Ducts and Mufflers*, 2nd edition, Wiley, Hoboken, NJ, 2014.

[PIE 89] PIERCE A., *Acoustics: An Introduction to its Physical Principles and Applications*, Acoustical Society of America, Woodbury, NY, 1989.

[RIE 21] RIENSTRA S., HIRSCHBERG A., *An Introduction to Acoustics*, Eindhoven University of Technology, Eindhoven, 2021 [Online]. Available at: www.win.tue.nl/~sjoerdr/papers/boek.pdf.

[ROY 22] ROYER D., VALIER-BRAZIER T., *Elastic Waves in Solids 1: Propagation*, ISTE Ltd, London and John Wiley & Sons, New York, 2022.

[SCH 68] SCHENCK H.A., "Improved integral formulation for acoustic radiation problems", *J. Acoust. Soc.*, vol. 44, pp. 41–58, 1968.

[SHI 08] SHIN K., HAMMOND J., *Fundamentals of Signal Processing for Sound and Vibration Engineers*, John Wiley & Sons, Chichester, 2008.

[STO 09] STONE M., GOLDBART P., *Mathematics for Physics: A Guided Tour for Graduate Students*, Cambridge University Press, Cambridge, 2009.

[TEM 01] TEMKIN S., *Elements of Acoustics*, Acoustical Society of America, Woodbury, NY, 2001.

[VOR 20] VORLÄNDER M., *Auralization: Fundamentals of Acoustics, Modelling, Simulation, Algorithms and Acoustic Virtual Reality*, Springer, Cham, 2020.

[WIL 99] WILLIAMS E.G., *Fourier Acoustics*, Academic Press, San Diego, CA, 1999.

[WRI 05] WRIGHT M.C.M., *Lectures Notes on the Mathematics of Acoustics*, Imperial College Press, London, 2005.

Index

V, W

Other titles from

iSTE

in

Waves

2024

RÉVEILLAC Jean-Michel
Synthesizers and Subtractive Synthesis 1: Theory and Overview
Synthesizers and Subtractive Synthesis 2: Application and Practice

SAKHO Ibrahima
Nuclear Physics 2: Radiochronometers and Radiopharmaceuticals
Nuclear Physics 3: Radiopharmaceuticals used in Nuclear Medicine

2022

GONTRAND Christian
Electromagnetism: Links to Special Relativity

LACAZE Pierre Camille, LACROIX Jean-Christophe
Nanotechnology and Nanomaterials for Energy

RÉVEILLAC Jean-Michel
Recording and Voice Processing 1: History and Generalities
Recording and Voice Processing 2: Working in the Studio

ROYER Daniel, VALIER-BRASIER Tony
Elastic Waves in Solids 1: Propagation

SAKHO Ibrahima
Nuclear Physics 1: Nuclear Deexcitations, Spontaneous Nuclear Reactions

VALIER-BRASIER Tony, ROYER Daniel
Elastic Waves in Solids 2: Radiation, Scattering, Generation

2021

DAHOO Pierre-Richard, LAKHLIFI Azzedine
Infrared Spectroscopy of Symmetric and Spherical Top Molecules for Space Observation 1
(Infrared Spectroscopy Set – Volume 3)
Infrared Spectroscopy of Symmetric and Spherical Top Molecules for Space Observation 2
(Infrared Spectroscopy Set – Volume 4)

2020

DANIELE Vito G., LOMBARDI Guido
Scattering and Diffraction by Wedges 1: The Wiener-Hopf Solution - Advanced Applications
(Waves and Scattering Set – Volume 1)
Scattering and Diffraction by Wedges 2: The Wiener-Hopf Solution - Advanced Applications
(Waves and Scattering Set – Volume 2)

SAKHO Ibrahima
Introduction to Quantum Mechanics 2: Wave-Corpuscle, Quantization & Schrödinger's Equation

2019

BERTRAND Pierre, DEL SARTO Daniele, GHIZZO Alain
The Vlasov Equation 1: History and General Properties

DAHOO Pierre-Richard, LAKHLIFI Azzedine
Infrared Spectroscopy of Triatomics for Space Observation
(Infrared Spectroscopy Set – Volume 2)

RÉVEILLAC Jean-Michel
Electronic Music Machines: The New Musical Instruments

ROMERO-GARCIA Vicente, HLADKY-HENNION Anne-Christine
Fundamentals and Applications of Acoustic Metamaterials: From Seismic to Radio Frequency
(Metamaterials Applied to Waves Set – Volume 1)

SAKHO Ibrahima
Introduction to Quantum Mechanics 1: Thermal Radiation and Experimental Facts Regarding the Quantization of Matter

2018

SAKHO Ibrahima
Screening Constant by Unit Nuclear Charge Method: Description and Application to the Photoionization of Atomic Systems

2017

DAHOO Pierre-Richard, LAKHLIFI Azzedine
Infrared Spectroscopy of Diatomics for Space Observation
(Infrared Spectroscopy Set – Volume 1)

PARET Dominique, HUON Jean-Paul
Secure Connected Objects

PARET Dominque, SIBONY Serge
Musical Techniques: Frequencies and Harmony

RÉVEILLAC Jean-Michel
Analog and Digital Sound Processing

STAEBLER Patrick
Human Exposure to Electromagnetic Fields

2016

ANSELMET Fabien, MATTEI Pierre-Olivier
Acoustics, Aeroacoustics and Vibrations

BAUDRAND Henri, TITAOUINE Mohammed, RAVEU Nathalie
The Wave Concept in Electromagnetism and Circuits: Theory and Applications

PARET Dominique
Antennas Designs for NFC Devices

PARET Dominique
Design Constraints for NFC Devices

WIART Joe
Radio-Frequency Human Exposure Assessment

2015

PICART Pascal
New Techniques in Digital Holography

2014

APPRIOU Alain
Uncertainty Theories and Multisensor Data Fusion

JARRY Pierre, BENEAT Jacques N.
RF and Microwave Electromagnetism

LAHEURTE Jean-Marc
UHF RFID Technologies for Identification and Traceability

SAVAUX Vincent, LOUËT Yves
MMSE-based Algorithm for Joint Signal Detection, Channel and Noise Variance Estimation for OFDM Systems

THOMAS Jean-Hugh, YAAKOUBI Nourdin
New Sensors and Processing Chain

TING Michael
Molecular Imaging in Nano MRI

VALIÈRE Jean-Christophe
Acoustic Particle Velocity Measurements using Laser: Principles, Signal Processing and Applications

VANBÉSIEN Olivier, CENTENO Emmanuel
Dispersion Engineering for Integrated Nanophotonics

2013

BENMAMMAR Badr, AMRAOUI Asma
Radio Resource Allocation and Dynamic Spectrum Access

BOURLIER Christophe, PINEL Nicolas, KUBICKÉ Gildas
Method of Moments for 2D Scattering Problems: Basic Concepts and Applications

GOURE Jean-Pierre
Optics in Instruments: Applications in Biology and Medicine

LAZAROV Andon, KOSTADINOV Todor Pavlov
Bistatic SAR/GISAR/FISAR Theory Algorithms and Program Implementation

LHEURETTE Eric
Metamaterials and Wave Control

PINEL Nicolas, BOURLIER Christophe
Electromagnetic Wave Scattering from Random Rough Surfaces: Asymptotic Models

SHINOHARA Naoki
Wireless Power Transfer via Radiowaves

TERRE Michel, PISCHELLA Mylène, VIVIER Emmanuelle
Wireless Telecommunication Systems

2012

LALAUZE René
Chemical Sensors and Biosensors

LE MENN Marc
Instrumentation and Metrology in Oceanography

LI Jun-chang, PICART Pascal
Digital Holography

2011

BECHERRAWY Tamer
Mechanical and Electromagnetic Vibrations and Waves

BESNIER Philippe, DÉMOULIN Bernard
Electromagnetic Reverberation Chambers

GOURE Jean-Pierre
Optics in Instruments

GROUS Ammar
Applied Metrology for Manufacturing Engineering

LE CHEVALIER François, LESSELIER Dominique, STARAJ Robert
Non-standard Antennas

2010

BEGAUD Xavier
Ultra Wide Band Antennas

MARAGE Jean-Paul, MORI Yvon
Sonar and Underwater Acoustics

2009

BOUDRIOUA Azzedine
Photonic Waveguides

BRUNEAU Michel, POTEL Catherine
Materials and Acoustics Handbook

DE FORNEL Frédérique, FAVENNEC Pierre-Noël
Measurements using Optic and RF Waves

FRENCH COLLEGE OF METROLOGY
Transverse Disciplines in Metrology

2008

FILIPPI Paul J.T.
Vibrations and Acoustic Radiation of Thin Structures

LALAUZE René
Physical Chemistry of Solid-Gas Interfaces

2007

KUNDU Tribikram
Advanced Ultrasonic Methods for Material and Structure Inspection

PLACKO Dominique
Fundamentals of Instrumentation and Measurement

RIPKA Pavel, TIPEK Alois
Modern Sensors Handbook

2006

BALAGEAS Daniel *et al.*
Structural Health Monitoring

BOUCHET Olivier *et al.*
Free-Space Optics

BRUNEAU Michel, SCELO Thomas
Fundamentals of Acoustics

FRENCH COLLEGE OF METROLOGY
Metrology in Industry

GUILLAUME Philippe
Music and Acoustics

GUYADER Jean-Louis
Vibration in Continuous Media

Printed and bound by CPI Group (UK) Ltd, Croydon, CR0 4YY

16/09/2026

14963082-0008